Interplay of Connexins and Pannexins in Tissue Function and Disease

Interplay of Connexins and Pannexins in Tissue Function and Disease

Special Issue Editors

Brenda R. Kwak
Patricia Martin

MDPI • Basel • Beijing • Wuhan • Barcelona • Belgrade

MDPI

International Gap Junction Conference

Special Issue Editors
Brenda R. Kwak
University of Geneva
Switzerland

Patricia Martin
Glasgow Caledonian University
UK

Editorial Office
MDPI
St. Alban-Anlage 66
4052 Basel, Switzerland

This is a reprint of articles from the Special Issue published online in the open access journal *International Journal of Molecular Sciences* (ISSN 1422-0067) in 2018 (available at: https://www.mdpi.com/journal/ijms/special_issues/IGJC2017).

For citation purposes, cite each article independently as indicated on the article page online and as indicated below:

LastName, A.A.; LastName, B.B.; LastName, C.C. Article Title. *Journal Name* **Year**, *Article Number, Page Range.*

ISBN 978-3-03897-392-8 (Pbk)
ISBN 978-3-03897-393-5 (PDF)

Contents

About the Special Issue Editors

Brenda R. Kwak is full professor in the Department of Pathology and Immunology at the University of Geneva Medical School. After her medical studies at the University of Amsterdam in the Netherlands, she has joined the connexin research field as a PhD student in 1988 and has not left it since. In 2003, she established her research group in Switzerland supported by a Swiss National Science Foundation professorship. Her research principally focuses on the role of connexins in cardiovascular diseases with a strong immuno-inflammatory component, such as atherosclerosis, restenosis, cardiac reperfusion injury, and thrombosis. More recently, she has extended her research interests towards pannexins, as well as to translational research investigating the role of biomechanical forces in vascular disease.

Patricia Martin was awarded her PhD in Molecular Virology at the Department of Biological Sciences, University of Warwick before taking up a Royal Society European Fellowship at the Institute of Virology, Würzburg, Germany. She then spent a further 18 months as a visiting Research Associate in the Department of Entomology, Wageningen Agricultural University, The Netherlands, where she worked on parasitic nematodes. In 1994, she returned to the UK to Cardiff University, where she joined MRC supported research teams studying the assembly and function of gap junction intercellular communication channels. In January 2004, she joined the Department of Biological and Biomedical Sciences and is Team Leader of the Connexin Research Group with linking roles in the Diabetes Research Group and a strong interest in molecular mechanisms underlying disease. Research focus: to develop integrative approaches to study the role of Connexin-mediated communication and intercellular signalling in diverse tissue networks.

International Journal of
Molecular Sciences

MDPI

Editorial

An Overview of the Focus of the International Gap Junction Conference 2017 and Future Perspectives

Patricia E. Martin [1],* and Brenda R. Kwak [2],*

[1]   Department of Biological and Biomedical Sciences, School of Health and Life Sciences,
     Glasgow Caledonian University, Glasgow G4 0BA, UK
[2]   Department of Pathology and Immunology, University of Geneva, CH-1211 Geneva, Switzerland
*   Correspondence: Patricia.Martin@gcu.ac.uk (P.E.M.); Brenda.KwakChanson@unige.ch (B.R.K.);
     Tel.: +44-141-331-3726 (P.E.M.); +41-22-379-5737 (B.R.K.)

Received: 13 September 2018; Accepted: 17 September 2018; Published: 18 September 2018

This Special Issue relates to the 18th biannual International Gap Junction Conference (IGJC2017), held at the Crowne Plaza Hotel, Glasgow, U.K., from the 29 July–2 August 2017. The special issue, entitled: Interplay of Connexins and Pannexins in Tissue Function and Disease focused on six key state of the art reviews written by chairs of the sessions highlighting the assembly and functional interactions of connexins and pannexins in diverse tissues and disease states, with translational outputs emerging. A further 14 articles detail specific contributions that were presented as oral communications (Table 1). The meeting was attended by over 200 delegates from 24 different countries and celebrated 50 years of Gap Junction Research.

Table 1. Publication contents of this special issue.

Topic	Reviews	Participant Contributions
Trafficking, Assembly, Gating and Protein-Protein Interactions	[1,2]	[3–7]
Cardiovascular System	[8]	[9–13]
Connexins and Tumorigenesis	[14]	[15,16]
Epithelial Tissue, Wound Healing	[17]	[18,19]
Connexin Therapy Translates to Clinic	[20]	

Connexins and pannexins are tetramembrane spanning channel proteins with a shared topology and related functional properties. They oligomerise to form channels in the plasma membrane with two extracellular loops that project into the extracellular space and an intracellular carboxyl tail is subject to post-translational modification. Within the connexin family (21 members in man), these extracellular loops interdigitate and dock with loops from neighbouring cells to form dodecameric intercellular gap junction channels (GJCs) that link the cytoplasm's of neighbouring cells. These GJCs enable the regulated exchange of over 300,000 metabolites of less than 1000 Da in size, in so doing co-ordinating specific cell and tissue homeostasis. Connexin compatibility is highlighted by phylogenetic classification into three–four specific subgroups, with only specific heteromeric channel combinations possible and in so doing facilitating the segregation of tissue compartments.

Over the last 15 years, it has emerged that connexin hemichannels can be triggered to open under conditions of cell stress releasing signalling molecules such as ATP, glutamate and Nicotinamide adenine dinucleotide (NAD) into the extracellular environment and subsequently elicit localised extracellular signalling cascades via purinergic receptors. Pannexins, proteins forming hemichannel-like structures, identified about 15 years ago, share a common topology but no sequence homology with connexins, are thought to be evolutionary related to the innexins, gap junction forming proteins in invertebrates. Unlike connexins (and innexins) pannexins are highly glycosylated proteins and act to release ATP, engaging with downstream signalling pathways. In particular, pannexin signalling has been closely linked with inflammatory mediated events, where caspase1 cleavage of the carboxyl tail renders these channels constitutively open and triggers cell death 'find me' signals.

Due to the diversity of connexin and pannexin expression in tissue networks, their importance cannot be underestimated and they are now firmly established as key proteins in diverse tissue networks including the cardiovascular system, the skin, the nervous system, the liver, the ocular, respiratory and immune system. Further, changes in connexin expression and function occur in disease states associated with all of these tissues: Tumorigenesis, diabetic retinopathy, cardiac arrhythmia, atherosclerosis, stroke, Alzheimer's disease, chronic non–healing skin wounds, and inflammation of epithelial tissue, to name a few. A range of mutations in connexins are associated with clinical disease, where mutations in Cx26 are among the most common in recessively inherited hearing impairment, Cx32 with the demyelinating disorder Charcot Marie Tooth-Linked disease, Cx43 with oculodentodigital dysplasia, and Cx46 and Cx50 with familial cataract formation. All these diseases impact on the quality of life, healthcare resources, ageing populations and many with conditions that can be managed with current therapies, but not cured.

Thus, connexins and pannexins have emerged as prime therapeutic targets for a diverse range of disease states. These channels are amenable to targeting therapeutically and a range of antisense and peptidomimetic strategies have emerged as key regulators of channel function. Such regulators were first identified over 25 years ago when Evans and Warner synthesised the first mimetic peptides and antibodies that mimicked amino acid sequences on the extracellular loops. These tools are now widely used by the research community to define the role of connexins and pannexins in tissue function and are proving successful in translational research where the success of clinical trials of a connexin targeted therapy, from bench to bedside, for improving wound healing events is reported in this special issue.

The Editors thank all contributors to this special issue, delegates and in the conference support team for the successful running of IGJC2017. The follow-up meeting will be held in Victoria, Vancouver Island, July 2019 (#IGJC2019).

1. Contents of the Special Issue

1.1. Trafficking, Assembly, Gating and Protein-Protein Interactions

Reviews

- Aasen, T.; Johnstone, S.; Vidal-Brime, L.; Lynn, K.S.; Koval, M. Connexins: Synthesis, Post-Translational Modifications, and Trafficking in Health and Disease. *Int. J. Mol. Sci.* **2018**, 19(5). [1]
- Sorgen, P.L.; Trease, A.J.; Spagnol, G.; Delmar, M.; Nielsen, M.S. Protein–Protein Interactions with Connexin 43: Regulation and Function. *Int. J. Mol. Sci.* **2018**, 19(5). [2]

Participant Contributions

- Ek-Vitorin, J.F.; Pontifex, T.K.; Burt, J.M. Cx43 Channel Gating and Permeation: Multiple Phosphorylation-Dependent Roles of the Carboxyl Terminus. *Int. J. Mol. Sci.* **2018**, 19(6). [3]
- Spagnol, G.; Trease, A.J.; Zheng, L.; Gutierrez, M.; Bazu, I.; Sarmiento, C.; Moore, G.; Cervantes, M.; Sorgen, P.L. Connexin43 Carboxyl-Terminal Domain Directly Interacts with β-Catenin. *Int. J. Mol. Sci.* **2018**, 19(6). [4]
- Sanchez-Pupo, R.E.; Johnston, D.; Penuela, S. N-Glycosylation Regulates Pannexin 2 Localization but Is Not Required for Interacting with Pannexin 1. *Int. J. Mol. Sci.* **2018**, 19(7). [5]
- Batissoco, A.C.; Salazar-Silva, R.; Oiticica, J.; Bento, R.F.; Mingroni-Netto, R.C.; Haddad, L.A. A Cell Junctional Protein Network Associated with Connexin-26. *Int. J. Mol. Sci.* **2018**, 19(9). [6]
- Schadzek, P.; Helmes, D.; Stahl, Y.; Dilger, N.; Ngezahayo, A. Concatenation of human connexin26 (hCx26) and human connexin46 (hCx46) for the analysis of heteromeric gap junction hemichannels and heterotypic gap junction channels. **2018**, 19(9). [7]

1.2. Cardiovascular System

Review

- Molica, F.; Figueroa, X.F.; Kwak, B.R.; Isakson, B.E.; Gibbins, J.M. Connexins and Pannexins in Vascular Function and Disease. *Int. J. Mol. Sci.* **2018**, 19(6). [8]

Participant Contributions

- Carballo, S.; Pfenniger, A.; Carballo, D.; Garin, N.; James, R.W.; Mach, F.; Shah, D.; Kwak, B.R. Differential Association of Cx37 and Cx40 Genetic Variants in Atrial Fibrillation with and without Underlying Structural Heart Disease. *Int. J. Mol. Sci.* **2018**, 19(1). [9]
- Noureldin, M.; Chen H, Bai D. Functional Characterization of Novel Atrial Fibrillation-Linked GJA5 (Cx40) Mutants. *Int. J. Mol. Sci.* **2018**, 19(4). [10]
- Viczenczova, C.; Kura, B.; Egan Benova, T.; Yin, C.; Kukreja, R.C.; Slezak, J.; Tribulova, N.; Szeiffova Bacova, B. Irradiation-Induced Cardiac Connexin-43 and miR-21 Responses Are Hampered by Treatment with Atorvastatin and Aspirin. *Int. J. Mol. Sci.* **2018**, 19(4). [11]
- Boucher, J.; Simonneau, C.; Denet, G.; Clarhaut, J.; Balandre, A.C.; Mesnil, M.; Cronier, L; Monvoisin, A. Pannexin-1 in Human Lymphatic Endothelial Cells Regulates Lymphangiogenesis. *Int. J. Mol. Sci.* **2018**, 19(6) [12]
- Htet M, Nally, J.E.; Shaw, A.; Foote, B.E.; Martin, P.E.; Dempsie, Y. Connexin 43 Plays a Role in Pulmonary Vascular Reactivity in Mice. *Int. J. Mol. Sci.* **2018**, 19(7). [13]

1.3. Connexins and Tumorigenesis

Review

- Graham, S.V.; Jiang, J.X.; Mesnil, M. Connexins and Pannexins: Important Players in Tumorigenesis, Metastasis and Potential Therapeutics. *Int. J. Mol. Sci.* **2018**, 19(6). [14]

Participant Contributions

- Busby, M.; Hallett, M.T.; Plante, I. The Complex Subtype-Dependent Role of Connexin 43 (GJA1) in Breast Cancer. *Int. J. Mol. Sci.* **2018**, 19(3). [15]
- Iikawa, N.; Yamamoto, Y.; Kawasaki, Y.; Nishijima-Matsunobu, A.; Suzuki, M.; Yamada, T.; Omori, Y. Intrinsic Oncogenic Function of Intracellular Connexin26 Protein in Head and Neck Squamous Cell Carcinoma Cells. *Int. J. Mol. Sci.* **2018**, 19(7). [16]

1.4. Epithelial Tissue, Wound Healing

Reviews

- Chanson, M.; Watanabe, M.; O'Shaughnessy, E.M.; Zoso, A.; Martin, P.E. Connexin Communication Compartments and Wound Repair in Epithelial Tissue. *Int. J. Mol. Sci.* **2018**, 19(5). [17]

Participant Contributions

- Faniku, C.; O'Shaughnessy, E.; Lorraine, C.; Johnstone, S.R.; Graham, A.; Greenhough, S.; Martin, P.E. The Connexin Mimetic Peptide Gap27 and Cx43-Knockdown Reveal Differential Roles for Connexin43 in Wound Closure Events in Skin Model Systems. *Int. J. Mol. Sci.* **2018**, 19(2). [18]
- Chen, J.; Liang, C.; Zong, L.; Zhu, Y.; Zhao, H.B. Knockout of Pannexin-1 Induces Hearing Loss. *Int. J. Mol. Sci.* **2018**, 19(5). [19]

1.5. Connexin Therapy Translates to Clinic

Review

- Montgomery, J.; Ghatnekar, G.S.; Grek, C.L.; Moyer, K.E.; Gourdie, R.G. Connexin 43-Based Therapeutics for Dermal Wound Healing. *Int. J. Mol. Sci.* **2018**, 19(6). [20]

Conflicts of Interest: The authors declare no conflict of interest.

References

1. Aasen, T.; Johnstone, S.; Vidal-Brime, L.; Lynn, K.S.; Koval, M. Connexins: Synthesis, Post-Translational Modifications, and Trafficking in Health and Disease. *Int. J. Mol. Sci.* **2018**, *19*, 1296. [CrossRef] [PubMed]
2. Sorgen, P.L.; Trease, A.J.; Spagnol, G.; Delmar, M.; Nielsen, M.S. Protein–Protein Interactions with Connexin 43: Regulation and Function. *Int. J. Mol. Sci.* **2018**, *19*, 1428. [CrossRef] [PubMed]
3. Ek-Vitorín, J.F.; Pontifex, T.K.; Burt, J.M. Cx43 Channel Gating and Permeation: Multiple Phosphorylation-Dependent Roles of the Carboxyl Terminus. *Int. J. Mol. Sci.* **2018**, *19*, 1659. [CrossRef] [PubMed]
4. Spagnol, G.; Trease, A.J.; Zheng, L.; Gutierrez, M.; Basu, I.; Sarmiento, C.; Moore, G.; Cervantes, M.; Sorgen, P.L. Connexin43 Carboxyl-Terminal Domain Directly Interacts with β-Catenin. *Int. J. Mol. Sci.* **2018**, *19*, 1562. [CrossRef] [PubMed]
5. Sanchez-Pupo, R.E.; Johnston, D.; Penuela, S. N-Glycosylation Regulates Pannexin 2 Localization but Is Not Required for Interacting with Pannexin 1. *Int. J. Mol. Sci.* **2018**, *19*, 1837. [CrossRef] [PubMed]
6. Batissoco, A.C.; Salazar-Silva, R.; Oiticica, J.; Bento, R.F.; Mingroni-Netto, R.C.; Haddad, L.A. A Cell Junctional Protein Network Associated with Connexin-26. *Int. J. Mol. Sci.* **2018**, *19*, 2535. [CrossRef] [PubMed]
7. Schadzek, P.; Hermes, D.; Stahl, Y.; Dilger, N.; Ngezahayo, A. Concatenation of Human Connexin26 (hCx26) and Human Connexin46 (hCx46) for the Analysis of Heteromeric Gap Junction Hemichannels and Heterotypic Gap Junction Channels. *Int. J. Mol. Sci.* **2018**, *19*, 2742. [CrossRef] [PubMed]
8. Molica, F.; Figueroa, X.F.; Kwak, B.R.; Isakson, B.E.; Gibbins, J.M. Connexins and Pannexins in Vascular Function and Disease. *Int. J. Mol. Sci.* **2018**, *19*, 1663. [CrossRef] [PubMed]
9. Carballo, S.; Pfenniger, A.; Carballo, D.; Garin, N.; James, R.W.; Mach, F.; Shah, D.; Kwak, B.R. Differential Association of Cx37 and Cx40 Genetic Variants in Atrial Fibrillation with and without Underlying Structural Heart Disease. *Int. J. Mol. Sci.* **2018**, *19*, 295. [CrossRef] [PubMed]
10. Noureldin, M.; Chen, H.; Bai, D. Functional Characterization of Novel Atrial Fibrillation-Linked GJA5 (Cx40) Mutants. *Int. J. Mol. Sci.* **2018**, *19*, 977. [CrossRef] [PubMed]
11. Viczenczova, C.; Kura, B.; Egan Benova, T.; Yin, C.; Kukreja, R.C.; Slezak, J.; Tribulova, N.; Szeiffova Bacova, B. Irradiation-Induced Cardiac Connexin-43 and miR-21 Responses Are Hampered by Treatment with Atorvastatin and Aspirin. *Int. J. Mol. Sci.* **2018**, *19*, 1128. [CrossRef] [PubMed]
12. Boucher, J.; Simonneau, C.; Denet, G.; Clarhaut, J.; Balandre, A.C.; Mesnil, M.; Cronier, L.; Monvoisin, A. Pannexin-1 in Human Lymphatic Endothelial Cells Regulates Lymphangiogenesis. *Int. J. Mol. Sci.* **2018**, *19*, 1558. [CrossRef] [PubMed]
13. Htet, M.; Nally, J.E.; Shaw, A.; Foote, B.E.; Martin, P.E.; Dempsie, Y. Connexin 43 Plays a Role in Pulmonary Vascular Reactivity in Mice. *Int. J. Mol. Sci.* **2018**, *19*, 1891. [CrossRef] [PubMed]
14. Graham, S.V.; Jiang, J.X.; Mesnil, M. Connexins and Pannexins: Important Players in Tumorigenesis, Metastasis and Potential Therapeutics. *Int. J. Mol. Sci.* **2018**, *19*, 1645. [CrossRef] [PubMed]
15. Busby, M.; Hallett, M.T.; Plante, I. The Complex Subtype-Dependent Role of Connexin 43 (GJA1) in Breast Cancer. *Int. J. Mol. Sci.* **2018**, *19*, 693. [CrossRef] [PubMed]
16. Iikawa, N.; Yamamoto, Y.; Kawasaki, Y.; Nishijima-Matsunobu, A.; Suzuki, M.; Yamada, T.; Omori, Y. Intrinsic Oncogenic Function of Intracellular Connexin26 Protein in Head and Neck Squamous Cell Carcinoma Cells. *Int. J. Mol. Sci.* **2018**, *19*, 2134. [CrossRef] [PubMed]
17. Chanson, M.; Watanabe, M.; O'Shaughnessy, E.M.; Zoso, A.; Martin, P.E. Connexin Communication Compartments and Wound Repair in Epithelial Tissue. *Int. J. Mol. Sci.* **2018**, *19*, 1354. [CrossRef] [PubMed]
18. Faniku, C.; O'Shaughnessy, E.; Lorraine, C.; Johnstone, S.R.; Graham, A.; Greenhough, S.; Martin, P.E. The Connexin Mimetic Peptide Gap27 and Cx43-Knockdown Reveal Differential Roles for Connexin43 in Wound Closure Events in Skin Model Systems. *Int. J. Mol. Sci.* **2018**, *19*, 604. [CrossRef] [PubMed]

19. Chen, J.; Liang, C.; Zong, L.; Zhu, Y.; Zhao, H.B. Knockout of Pannexin-1 Induces Hearing Loss. *Int. J. Mol. Sci.* **2018**, *19*, 1332. [CrossRef] [PubMed]
20. Montgomery, J.; Ghatnekar, G.S.; Grek, C.L.; Moyer, K.E.; Gourdie, R.G. Connexin 43-Based Therapeutics for Dermal Wound Healing. *Int. J. Mol. Sci.* **2018**, *19*, 1778. [CrossRef] [PubMed]

International Journal of
Molecular Sciences

MDPI

Review

Connexins: Synthesis, Post-Translational Modifications, and Trafficking in Health and Disease

Trond Aasen [1,*], Scott Johnstone [2,3,*], Laia Vidal-Brime [1], K. Sabrina Lynn [4] and Michael Koval [4,5,*]

1   Translational Molecular Pathology, Vall d'Hebron Institute of Research (VHIR),
    Autonomous University of Barcelona, CIBERONC, 08035 Barcelona, Spain; laia.vidal.b@gmail.com
2   Robert M. Berne Cardiovascular Research Center, University of Virginia School of Medicine, P.O. Box 801394,
    Charlottesville, VI 22908, USA
3   Institute of Cardiovascular and Medical Sciences, College of Medical, Veterinary and Life Sciences,
    University of Glasgow, Glasgow G12 8TT, UK
4   Division of Pulmonary, Allergy, Critical Care and Sleep Medicine, Department of Medicine,
    Emory University School of Medicine, Atlanta, GA 30322, USA; k.s.lynn@emory.edu
5   Department of Cell Biology, Emory University School of Medicine, Atlanta, GA 30322, USA
*   Correspondence: trond.aasen@vhir.org (T.A.); srj6n@virginia.edu (S.J.); mhkoval@emory.edu (M.K.);
    Tel.: +34-93489-4168 (T.A.)

Received: 19 March 2018; Accepted: 21 April 2018; Published: 26 April 2018

Abstract: Connexins are tetraspan transmembrane proteins that form gap junctions and facilitate direct intercellular communication, a critical feature for the development, function, and homeostasis of tissues and organs. In addition, a growing number of gap junction-independent functions are being ascribed to these proteins. The connexin gene family is under extensive regulation at the transcriptional and post-transcriptional level, and undergoes numerous modifications at the protein level, including phosphorylation, which ultimately affects their trafficking, stability, and function. Here, we summarize these key regulatory events, with emphasis on how these affect connexin multifunctionality in health and disease.

Keywords: connexins; gap junctions; transcription; translation; post-translational modifications; trafficking

---

1. Introduction

Since the cloning of the first connexins in the 1980s, steady progress towards elucidating their regulation and function as signaling hubs and mediators of direct intercellular communication has been made [1–3]. All connexins share a conserved four-transmembrane domain structure that assembles into hexameric pores known as connexons that can integrate into the cell membrane (Figure 1). Hundreds to thousands of these connexons typically dock with opposing connexons in an adjacent cell, creating intercellular channels forming a clustered gap junction plaque that permits direct flux of ions and small cytosolic signaling molecules between cells, commonly referred to as gap junctional intercellular communication (GJIC) (Figure 1). More recently, connexons have been shown to act as "hemichannels" to facilitate direct exchange of molecules between the cell cytosol and the extracellular milieu under specific conditions [4]. Additionally, numerous noncanonical channel-independent functions have been described, in particular for connexin 43 (Cx43), which are mediated through direct protein interactions and modulation of signaling pathways [5]. The complexity and isoform-specificity of the connexin gene family is reflected by their links to numerous human diseases, many of which are rare syndromes with unique genotype–phenotype associations [6,7]. This latter phenomenon is underscored by the observation that mutations in different connexins can cause the same disease, whereas varying mutations in one connexin gene can result in vastly divergent diseases and phenotypes. Dysregulation of connexins is also increasingly linked to many common and often morbid medical conditions—such as stroke, heart attack, and cancer—which have been linked to the discovery of an expanding number of new functional attributes through both gap junction-dependent and -independent mechanisms [2,3,6–9]. As such, exploring the clinical and therapeutic potential of connexins as drug targets is pertinent and ongoing [10–12]. Towards this, a deeper understanding of how these genes and proteins are regulated and function is essential. This review aims to summarize and underscore important and unique mechanisms that regulate connexin function in healthy and diseased states, which ultimately shed light on clinical observations and future therapeutic opportunities.

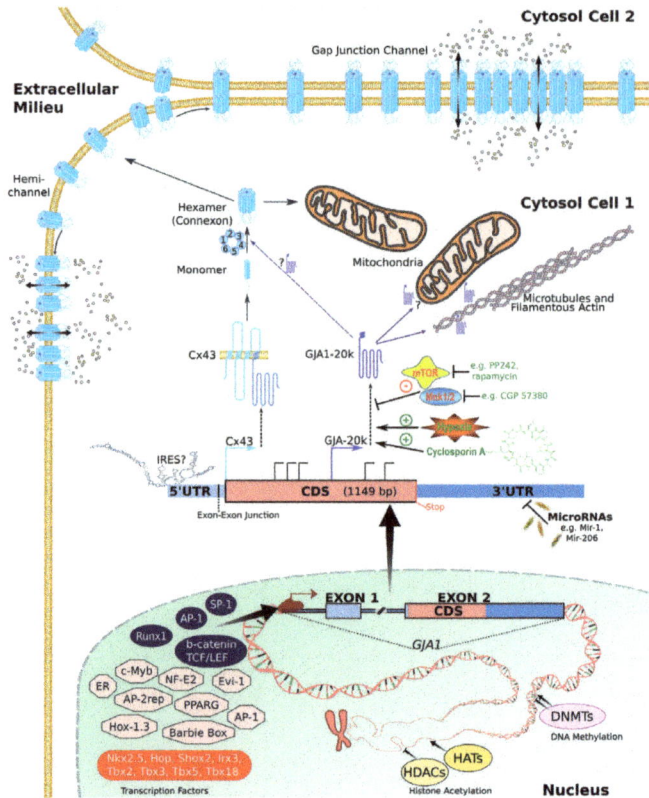

Figure 1. Connexins form hexameric connexons permeable to small molecules acting either as hemichannels or as intercellular channels. The human *GJA1* gene, encoding for connexin 43 (Cx43), contains two exons spanning a genomic region of 14,168 bp. Exon 1 contains 256 bp of the 5′ UTR (untranslated region), whereas exon 2 encompasses 16 bp of the 5′ UTR, the entire coding region (1149 bp), and the entire 3′ UTR region (1748 bp). Transcription of mRNA (3169 bp) is under regulation by numerous transcription factors as indicated in this figure and in the main text. Notably, Sp-1 and AP-1 are key regulators of Cx43 mRNA expression (grouped in blue). Multiple tissue-specific promoters are active, which has been well described in the heart (grouped in red). Additional transcription factors (grouped in light red) are derived from promoter analysis using the online Lasagna-Search tool (using a very strict cut-off of $p < 0.0001$ and Transfac transcription factor binding sites) [13]. Epigenetics regulate transcription, including through promoter hypermethylation by DNA methyltransferase enzymes (DNMTs). Acetylation by histone acetyltransferase enzymes (HATs) promote transcription, and the reverse reaction is mediated by histone deacetylases (HDACs). The transcript is also regulated by numerous microRNAs (see main text for details). In addition to full-length Cx43 (43 kilodalton (kDa)), the same mRNA can produce multiple truncated forms via internal translation initiation (indicated by arrows within the CDS (coding DNA sequence) of the mRNA, most notably the 20 kDa form named GJA1-20k). Truncated forms are also under translational regulation by a number of pathways such as mechanistic target of rapamycin (mTOR) and mitogen-activated protein kinase (MAPK)-interacting serine/threonine-protein kinase 1 (MNK1) and 2 (MNK2), and can be induced by inhibitors of these pathways as well as by other specific drugs such as cyclosporin A (the positive regulators are depicted in green). GJA1-20k is also induced by pathological states such as hypoxia. The function of GJA1-20k may include interaction with mitochondria and regulation of the actin cytoskeleton as well as regulation of Cx43 oligomerization and trafficking to the membrane. See main text for further details related to the figure.

2. Connexins: From Gene to Protein

2.1. Gene Structure and Splicing

Twenty-one human genes and 20 mouse genes encoding for connexin proteins have been identified, of which 19 are considered orthologous pairs [14,15]. The genes tend to have distinct chromosomal locations, although there are some regions of the genome containing clusters of connexin genes [14]. Most connexin genes share a common structure consisting of two exons separated by an intron of variable size. The majority of the 5′ UTR (untranslated region) is localized on exon 1, whereas the entire coding region and the 3′ UTR are found in exon 2. Some connexin genes contain more than two exons (for the 5′ UTR of the transcript), such as human *GJA5* (Cx40) [16], which contains three exons producing two distinct and tissue-specific transcripts, and *GJB6* (Cx30), described to contain six exons that allows for tissue-specific splicing [17]. Mouse connexin genes with three or more exons include *Gjb1* (Cx32) [18], *Gja1* (Cx43) [19], and *Gjc1* (Cx45) [20]. In a few cases, the coding region is also distributed over more than one exon [21–24]. A basal promoter (P1) is typically found within 300 bp upstream of the transcription initiation site of exon 1 [25]. However, splice isoforms have been reported due to alternate promoter usage, yielding different transcripts with the coding region being unaltered. As such, a deeper understanding of connexin gene structure, promoter usage, and splicing pattern is required for a full understanding of their impact in connexin-related diseases. For example, the human *GJB1* gene encoding Cx32 contains at least three exons (E1, E1B, and the coding exon E2) and produces two different alternatively spliced transcripts by using two tissue-specific promoters (P1 and P2) [26]. It is thus pertinent to include this region in mutational screening of dominant X-linked Charcot-Marie-Tooth (CMTX1) disease, a type of neuropathy that can be caused by mutations in Cx32 leading to defects in Schwann cell function, at least in cases where no mutations are found in the Cx32 coding region. Indeed, recent studies have identified mutations affecting *GJB1* splicing [27], and even deletion of the *GJB1* P2 promoter [28], as underlying causes of CMTX1. Others have shown that splicing mutations in *GJC2* encoding Cx47 can cause a severe form of Pelizaeus-Merzbacher-like disease [29]. Another splice-site mutation, in *GJB2* encoding Cx26, has been suggested to cause a mild postlingual onset form of hearing loss [30].

In addition to these more well-described biological phenomena, a few connexin pseudogenes (genes thought to originate from decay of genes that stems from duplication through evolution) have been identified in the human genome. The *GJA1* pseudogene (GJA1P) is located on human chromosome 5, whereas the regular *GJA1* gene encoding for Cx43 is located on chromosome 6. Although most pseudogenes are thought to be nonfunctional, GJA1P appears to be transcribed, possibly even translated, and may regulate tumor growth [31,32]. Mutations in GJA1P have also been associated with nonsyndromic deafness [33]. Functionally, GJA1P may influence *GJA1* expression levels by acting as a microRNA sponge [34]. In contrast, GJA6P seems to be a nonfunctional pseudogene, originated from the mouse *Gja6* connexin gene encoding Cx33, which has no human counterpart (Gene ID: 100126825). Another potential pseudogene has been inferred for *GJA4* (Gene ID: 100421028) encoding Cx37. The role of pseudogenes in disease is an emerging field, particularly among genes causing multiple different diseases or syndromic diseases, such as connexins.

2.2. Transcription Factors and Epigenetics

Connexins are expressed distinctively in almost all vertebrate cell types (excluding erythrocytes, mature sperm cells, and differentiated skeletal muscle cells) [35]. Some connexins (notably Cx43) are expressed in numerous cell types, whereas others show a more restricted expression profile (e.g., Cx50 that is mainly found in lens cells). Most tissues express multiple connexins. The epidermis of the skin, for example, is thought to express at least 10 different connexins whose expression partially overlap during keratinocyte stratification and differentiation [36–38]. Five of these connexins underlie 11 clinically different cutaneous disorders [37,39]. This spatiotemporal expression pattern is in large part controlled by transcription factors and epigenetic mechanisms. Several transcription factors acting as regulators of basal (ubiquitous) or cell-specific gene activity, and their upstream signal transduction pathways, have been implicated in the control of connexin expression (Figure 1). Notably, specificity protein 1 (Sp1), an important basal transcription factor that binds to GC box sequences in promoter regions, has been reported to favor transcriptional initiation of several connexin genes, including Cx26 [40], Cx32 [41,42], Cx40 [16,43–47], and Cx43 [45,48–53]. Examples of other important regulators that control connexin gene expression include: (i) Activator protein 1 (AP1) transcription factor, composed of proteins belonging to the c-Fos, c-Jun, activating transcription factor (ATF), and J domain containing protein(JDP) families that typically promote positive regulation. AP-1 sites have mainly been described in Cx43 [48,54,55], whereas putative sites have been identified in the Cx45 promoter [56]. (ii) The Wnt pathway: activation of this pathway leads to the formation of nuclear β catenin/TCF (T-cell factor) complexes that act as transcription factors by binding to specific TCF/LEF (lymphoid enhancer-binding factor) motifs present in the promoter of human *GJA1* and mouse *Gja1* encoding Cx43 [57]. From a physiological and disease point of view, this may also be relevant. For example, one study showed Wnt signaling could modulate Cx43-dependent GJIC in the heart, which ultimately may contribute to altered impulse propagation and arrhythmia in the myopathic heart [58]. The importance of GJIC in the heart is well documented and several cell-specific transcription factors have been shown to either activate or repress connexin gene expression in this setting (Figure 1 (red box), reviewed in [25,59]). These studies have revealed a role of: (i) homeobox proteins, transcription factors with a unique DNA-binding domain that target gene promoter sequences by self-complementarity (e.g., Nkx2.5, Hop, Shox2, Irx3); (ii) T-box (Tbx) proteins, transcription factors that possess a domain that recognizes a DNA binding element (e.g., Tbx2, Tbx3, Tbx5, Tbx18); and (iii) GATA proteins, important regulators of specific gene expression in different tissue (e.g., GATA-4) [25,59].

Besides the well-described transcriptional regulation of the cardiac connexins, other cases of tissue-specific regulation have been reported (for an overview, see [25]). Cx32 transcription has been found to be positively regulated by hepatocyte nuclear factor-1 (HNF-1) via Sp1 in liver cells [60], by the transcription factor Mist1 in secretory pancreatic acinar cells [61], and by the Sox10 in synergy with the early growth response-2 gene (Egr2) in Schwann cells [62]. This exemplifies how different transcription factors act in a tissue-dependent fashion. Complex transcriptional control thus allows for tissue-specific regulation of connexin expression. It also facilitates rapid response to environmental changes, for example, progesterone and estrogen act as positive and negative regulators, respectively, of Cx43 transcription in the myometrium during pregnancy and labor [63]. Transcription factors are also important during pathological states, such as in ischemia, where multiple connexins are emerging as important injury response mediators. Their roles in complex disease, such as cancer, are also being unraveled. In breast cancer, for example, Cx43 has been proposed to play a biphasic role acting both as a tumor promotor and a tumor suppressor depending on context such as cancer subtype and stage [3]. In this setting, the aforementioned role of progesterone and estrogen as regulators of Cx43 expression may be of importance. Recent evidence also suggests that the transcription factor FOXP3 directly binds to and inhibits RUNX1 in mammary epithelial cells, whereas in the absence of FOXP3 in breast tumors, RUNX1 downregulates Cx43 expression [64]. Understanding the role of transcription factors will provide further insight into loss and overexpression of connexins during tumor progression and other pathological states.

Connexin expression is also under significant epigenetic regulation (for recent extensive reviews see [25,65]). Two major epigenetic mechanisms have been described to regulate transcriptional control: DNA methylation and histone acetylation. Connexin gene inactivation due to hypermethylation of CpG islands in the promoter region has been described in various human carcinomas, including Cx26 in lung [66] and breast [67], Cx32 in a renal cell carcinoma cell line [68], and Cx43 in breast cancer [69]. In addition, a gradual decrease in Cx32 and Cx43 mRNA expression levels is associated with promoter hypermethylation in *Helicobacter pylori*-associated gastric tumorigenesis [70]. Transcriptional silencing via promoter hypermethylation is mediated by the enzyme DNA methyltransferase (DNMT). The use of demethylating drugs (DNMTs inhibitors), such as 5-aza-2-deoxycytidine and 5-azacytidine, has been proposed as a potential therapeutic solution in cancer, as an increase connexin expression and/or GJIC has been demonstrated in specific cases [68,71,72]. However, the correlation between hypermethylation and gene expression is not always direct and differs between connexin isoforms [72].

Histone acetylation and deacetylation—causing chromatin decondensation and condensation, respectively—constitute other important mechanisms of epigenetic regulation of connexin transcription [25,65]. While acetylation is catalyzed by histone acetyltransferase (HAT) enzymes and promotes transcription, the reverse reaction is mediated by histone deacetylase (HDAC) enzymes. Histone acetylation also affects connexin expression, and inhibitors of HDAC enzymes (HDACi)—such as trichostatin A, sodium butyrate, and 4-phenylbutarate—have been shown to enhance connexin and GJIC in a variety of cell populations, including in cancer cells [73], in which therapeutic and preventive roles for specific HDACi have been proposed. Histone deacetylase inhibition has also been shown to reduce Cx43 expression and gap junction communication in cardiac cells [74], which has implications with regards to potential side effects such as slow ventricular conduction or arrhythmias. Therefore, the action of HDACi seems to be connexin- and cell type-dependent. Curiously, Cx43 has been shown to influence histone acetylation of other genes; in a human pulmonary giant cell carcinoma cell line, the follistatin-like 1 (FSTL1) promoter was shown to be associated with acetylated histones H3 and H4 upon Cx43 transfection. Cx43 was proposed to act as a "histone deacetylase inhibitor" that modulates gene expression and inhibits tumor invasion [75].

The potential therapeutic role of epigenetic regulations has broad interest, particularly in complex diseases such as cancer, as exemplified above. However, the nonspecific nature of this gene regulatory mode complicates more direct and specific therapeutic targeting. Moreover, research is needed to determine if connexin levels are mainly mediated via HDACi histone modification [73,76], via non-histone protein modification of transcription factors, or via direct or indirect connexin protein modification such as Cx43 acetylation or phosphorylation [77–79].

2.3. RNA Stability and MicroRNAs

MicroRNAs (miRNAs) are short single-stranded noncoding RNAs that can regulate expression at a post-transcriptional level by base pairing to mRNA sequences (usually located at the 3′ UTR region), reducing protein expression levels via mRNA degradation, translational inhibition, or transient mRNA sequestration. Numerous microRNAs have been predicted to downregulate the expression of different connexin genes (for recent reviews see [65,80]). Cx43 is by far the best studied connexin, and a number of functional microRNAs targeting this gene have been identified, including miR-1, miR-23a, miR-186, miR-200a, miR-206, and miR-381 in human breast cancer [81], miR-20a in human prostate cancer [82], miR-221/222 in glioblastoma multiforme [83], and miR-206, miR-1, and miR-133 in cardiac myocytes and during skeletal myoblast differentiation [84–86].

Regulation of connexin expression by miRNAs has been described to be active in various disease states (for example, in cancer) by affecting hallmarks such as proliferation and invasion [82,83]. In therapeutic settings, options include targeting miRNAs that regulate connexins in order to reverse the malignant phenotype. This has been shown in several studies, including in human glioblastoma cells, where inhibition of miR-221/222 activity with antisense oligonucleotides led to the upregulation of Cx43 and restoration of GJIC [83].

As mentioned above, miR-1 acts in cardiac muscle and downregulates Cx43 expression. This has been related to several cardiopathologies in humans, including the regulation of cardiac arrhythmogenic potential [86]. In contrast, loss of miR-1, and thus increased Cx43 expression, has been linked to myotonic dystrophy [87]. Interestingly, a severe congenital heart defect, tetralogy of Fallot, is associated with downregulation of miR-1 and miR-206, which is thought to lead to an increase in Cx43 protein levels [88]. miR-1 downregulation of Cx43 in the bladder musculature has also been reported to have a role in overactive bladder syndrome [89].

Connexins are implicated in joint and bone disease [90]. Cx43 has an important role in osteoblast growth and differentiation, and various miRNAs (including miR-23a [91] and miR144-3p [92]) have been shown to target Cx43 in this setting. Cx43 can also influence the expression of miRNAs themselves, notably miR-21 in osteocytes, a pathway linked to osteocyte apoptosis and osteoclast formation/recruitment [93]. Moreover, direct transfer of miRNAs—through gap junctions—has been described, and is thought to play a role in bone development [94] as well as in aspects of tumor growth and tumor dormancy [3].

In addition to miRNAs, connexin transcript stability can be regulated by RNA-binding proteins (RBPs), such as human antigen R (HuR) that stabilizes the Cx43 mRNA by binding adenylate/uridine-rich elements (AREs) present in the Cx43 3' UTR [95]. Other examples include S1516-binding protein elements, which may regulate Cx43 expression, particularly in Ras-transformed cancers [96]. For further insight into the epigenetic regulation of connexins, including by miRNAs, we refer to other more exhaustive recent reviews [65,80].

2.4. Translational Regulation

2.4.1. Internal Ribosome Entry Site (IRES)

Due to the key role of connexins in sustaining many cellular functions and tissue physiology, it has been suggested that connexin expression needs to be maintained at all times, even under conditions where the classical cap-dependent mRNA translation pathway is suppressed, such as during mitosis, apoptosis, differentiation, senescence, or cell stress [97,98]. Several internal ribosome entry site (IRESs) elements have been reported in the mRNA of connexins, notably in Cx43 [99] (Figure 1), Cx32 [100], and Cx26 [101]. An IRES is a nucleotide sequence, usually located within the 5' UTR of the mRNA, which—in contrast with the canonical translation mechanism—allows for cap-independent translation initiation, a process regulated by specific RBPs also known as IRES trans-acting factors (ITAFs) [102,103]. However, numerous other translation initiation mechanisms are thought to exist [104] and whether true IRES-mediated translation occurs in the aforementioned connexins and other family members, is subject to caution, and additional specific molecular assays are warranted [105]. Additional work is also needed towards elucidating their functional relevance. One study suggests that IRES-mediated translation of Cx26 and Cx43 occurs in density-inhibited cancer cells (where cap-dependent translation is reduced), thus leading to the induction of GJIC and potentially reduced tumor growth [101]. Some data also points towards an important role of IRES-translation of connexins in human physiology. Notably, a specific mutation in the 5' UTR IRES sequence of Cx32 is linked to neurodegenerative Charcot–Marie–Tooth disease [100].

2.4.2. Alternative Translation of Truncated Connexin Forms

Most IRES sequences are located in the 5' UTR, yet a few examples exist (notably Notch2 [106]) where an IRES sequence is located within the coding region, allowing translation of truncated protein forms. A similar mechanism has been proposed for Cx55.5 in zebrafish, in which an 11 kilodalton (kDa) truncated C-terminal form is produced and localizes to the nucleus of outer retina cells [107,108].

In mammalian cells, the presence of truncated forms of Cx43 is often observed in immunoblots. In particular, a 20 kDa form (named GJA1-20k) is highly prevalent in cultured cells, which was described to arise from the Cx43 coding sequence and correspond to the C-terminal tail [109]. More recently, Smyth and Shaw described that GJA1-20k and several other less prevalent truncated forms can occur in normal tissue, and is due to internal translation initiation events [110]. Multiple groups have now confirmed this observation and further delineated key regulatory pathways, such as the mechanistic target of rapamycin (mTOR) [110,111] and the mitogen-activated protein kinase (MAPK)-interacting serine/threonine-protein kinase 1 (MNK1) and 2 (MNK2) [111] signaling cascades. Additionally, regulation occurs in response to important physiological conditions such as hypoxia [112] (Figure 1). Although an internal IRES element has been suggested [112], evidence suggests a highly unusual cap-dependent mechanism is critical for the efficient synthesis of these truncated forms [80,111].

The C-terminus of Cx43 has been extensively studied and is implicated in the regulation of a variety of biological events such as cell migration and proliferation, neuronal differentiation, and cytoskeletal changes (for a recent review see [5]). However, functional roles for specific internally truncated forms of Cx43 are currently being elucidated. Thus far, GJA1-20k has been shown to act as a potential chaperone for Cx43 [110,113] that facilitates microtubule-based mitochondrial transport and mitochondrial network integrity [114] (for details, see Section 4.3). Additionally, loss of GJA1-20k (but not full-length Cx43) has been reported in early-stage human breast cancers, followed by its re-expression in cell lines regulated by p53 activation via miR-125b [115]. Roles for these truncated forms of connexins in complex genetic disease is of future interest considering recent advancements in the potential for pharmacologic modulation of internal translation [80].

3. Post-Translational Regulation of Connexins

Post-translational modification of connexin proteins regulates many important aspects of their life-cycle, including synthesis, trafficking, channel gating, and protein–protein interactions. While highly conserved, variations occur throughout the connexin family in protein sequence, size of intracellular N-/C-terminus, and loop regions. Connexin extracellular loop regions contain disulfide bridges that form between cysteines to maintain membrane topology and facilitate docking with opposing connexons, allowing the formation of gap junctions [116]. Unlike many other membrane-bound proteins, connexins are not glycosylated, with membrane trafficking and protein folding being regulated through alternative pathways (for details, see Section 4). The relatively unstructured nature of intracellular connexin domains makes for an ideal environment for post-translational modification to induce conformational changes that regulate protein–protein interactions. The majority of connexins contain multiple consensus sites for modifications through phosphorylation, *S*-nitrosylation, SUMOylation, and others. There have been several recent and comprehensive reviews on connexin post-translational modifications [5,117–119]. Instead of recapitulating these articles, we will highlight some of the main aspects of post-translational modifications of connexins and discuss their relevance in human disease.

3.1. Phosphorylation

Phosphorylation is a key regulator of connexin proteins, hemichannels, and gap junction channels [120–122]. The addition of phosphate groups to specific amino acids—including serine (Ser, S), threonine (Thr, T), or tyrosine (Tyr, Y)—leads to changes in charge, hydrophobicity, and potential alterations in protein structure resulting from formation of hydrogen bond networks [123]. These can alter the way the connexin protein interacts with itself (e.g., channel regulation) or with other proteins (e.g., trafficking and protein–protein interactions).

Phosphorylation has been reported in a large number of connexins, e.g., Cx31 [124], Cx32 [125–127], Cx37, Cx40 and Cx45 [128,129], Cx43 [130,131], Cx46 and Cx50 [132–134], and Cx47 [135]. The majority of phosphorylation events are reported within the connexin C-terminus, with the exception of Cx26, which is not phosphorylated in its short 11 a.a. C-terminus [136,137]. However, mass spectrometry has demonstrated multiple potential Cx26 phosphorylation sites in the N-terminus, which are differentially regulated by hydroxylation, and further putative sites in the cytoplasmic loop, although the functions of these Cx26 phosphorylation sites are unknown [138,139]. There are some reports of intracellular loop phosphorylation—such as Cx56 [140] and Cx35 [141]—although this does not appear to be the case for Cx43 or other connexins [5,142,143]. There are no reports of N-terminus phosphorylation in other connexins, although Cx43–Ser5 is a potential candidate site [144]. The C-terminus of connexins are intrinsically disordered protein (IDP) regions with a high Ser/Thr/Tyr content, as described for Cx32, Cx40, Cx43, Cx45, and Cx50 [145–149]. Stable α-helical regions have been identified by nuclear magnetic resonance (NMR) and circular dichroism (CD) in the C-terminus of Cx43 [146,150] and other connexins, for example, Cx37, Cx45, and Cx50 [151–154]. However, stable alpha-helices are not a common feature of the connexin C-terminus. For instance, Cx40 only forms dynamic alpha-helices between Cys267–Gly285 [155]. Several lines of evidence—such as electrophoretic shifts on SDS-PAGE gels and NMR analysis—suggest that phosphorylation by enzymes, such as MAPK and Protein Kinase C (PKC), result in differential, transient increases in connexin C-terminal alpha-helical content [128,149,156–158].

The significance of the formation of alpha-helical domains is the potential for higher order secondary structures that regulate channel gating and protein partner binding. In Cx43, it has been demonstrated that the C-terminus interacts with the intercellular loop to regulate channel functions in a "ball-and-chain" type mechanism [159,160], although other factors relating to phosphorylation (e.g., charge and hydrophobicity) may also influence channel gating. Multisite phosphorylation of proteins is known to alter protein half-life, docking, and intracellular localization, which may also influence gap junction signaling [148,149,161]. Connexin 43, the most widely studied of the connexin family, has 30 putative phosphorylation sites which have been extensively demonstrated to be post-translationally modified, leading to alterations in gap junction signaling. For detailed reviews of these phosphorylation sites and their effects on channel regulation see [5,143,144,162–164]. The effects of post-translational modifications on connexins are also shown in Table 1.

Table 1. Connexin post-translational modifications (PTMs) and functional effects.

Connexin/Residue	PTM	GJIC	Expression	Refs.
Cx26 [m.s./a]:				
M1/K15/K102/ K103/K105/ K108/K112/K116	Acetylation	ND	ND	[138,139]
N14/N113/ N170/N176	Hydroxylation	ND	ND	
E42/E47/E114	carboxylation	ND	ND	
K61/R75/ K221/K223	Methylation	ND	ND	
T123/T177/S183/ T186/(Y233 or Y235 or Y240)	Phosphorylation	ND	ND	
Cx31(m):				
263 [b]	CK1	No change	No change	[124]
266 [b]	CK1	No change	No change	[124]
Cx32:				
S229	PKC	Increase/Decrease	Increase/Decrease	[165]
S233	PKA/PKC	Increase/Decrease	Increase/Decrease	[120,138,165,166]
S240	ND	ND	ND	[138]
Y7/Y243	EGFR tyrosine kinase	ND	ND	[167]
Cx35/Cx36:				
S110	PKA/PKG	No change	Decrease	[141,168–170]
S276/293	PKA/PKG	No change	Decrease	[141,168–171]
S289	PKG (NO mediated)	ND	Decrease	[170]
Cx37:				
S275/S285/ S302/S319/S321/ S325/S329	PKC	Increased	Decrease	[128]

Table 1. *Cont.*

Connexin/Residue	PTM	GJIC	Expression	Refs.
Cx43:				
S5 [m.s.]	ND	ND	ND	[144]
K144	SUMO	Increase	Increase	[172]
K237	SUMO	Increase	Increase	[172]
S244 [m.s.]	CAMKII	ND	ND	[173]
Y247 [c]	Src	Decrease	Decrease [c]	[120,146,174–178]
S255 [m.s.]	CAMKII	ND	ND	[173]
	P34cdc2	Decrease	Decrease	[179,180]
	MAPK	No change/Decrease	No change	[120,131,148,181]
S257 [m.s.]	PKG/CAMKII	ND	ND	[173]
S262 [d]	P34cdc2	Decrease	Decrease	[179,180]
	MAPK	Decrease	Decrease/no change	[120,131,148,182,183]
	PKCε [a]	Decrease	Decrease	[131,181,182]
Y265 [c]	Src	Decrease	Decrease [c]	[120,146,174–178]
C271	Nitrosylation	Increase	No change	[184]
S279 [e]	MAPK	Decrease	Decrease/no change	[131,148,174]
	CDK5		Decrease	[185]
S282 [e]	MAPK	Decrease	Decrease/no change	[131,148,174]
	CDK5	Decrease	Decrease	[185]
S296 [m.s.]	CAMKII	ND	No change	[173,186]
S297 [m.s.]	CAMKII/PKCε	ND	No change	[173,186]
S306 [m.s.]	CAMKII	Decrease	Decrease associated with De-Phosph.	[173,186,187]
S314 [m.s.]	CAMKII	ND	ND	[173]
S325 [m.s.]	CAMKII	ND	ND	[173]
	CK1	Increase	Increase	[188]
S328 [m.s.]	CAMKII	ND	ND	[173]
	CK1	Increase	Increase	[188]
S330 [m.s.]	CAMKII	ND	ND	[173]
	CK1	Increase	Increase	[188]
S364 [m.s.]	CAMKII	ND	ND	[173]
	PKA	Increase	Increase	[120,189–191]
S365 [m.s.]	CAMKII	ND	ND	[173]
	PKA	Increase	Increase	
	PKC	Decrease	Decrease	[120,192–194]
S368 [f]	PKCα	Increase/Preserved/	Increase	[192–196]
	PKCε	Decrease [g]	Decrease	[120,193–199]
S369 [m.s.]	CAMKII	ND	ND	
	PKA	Increase	No change	[173]
	PKC	Increase	Increase	[5,120,192–194]
S372 [m.s.]	CAMKII	ND	ND	[173]
	PKC	Decrease	Decrease	[5,148,192,193,200,201]
	Akt	Increase	Increase [e]	[200,201]
S373 [g,m.s.]	CAMKII	ND	ND	[173]
	PKC	Decrease	Decrease	
	PKA	Increase	Increase	[120,148,192–194]

Table 1. *Cont.*

Connexin/Residue	PTM	GJIC	Expression	Refs.
Cx45:				
S326/Y337/ S381/S382/S384/ S385/S387/S393 [m.s.]	CAMKII	ND	ND	[129]
S326/S382/S384/ S387/S393 [m.s.]	CK1	ND	ND	[129]
Cx46 (Cx56 Chick homologue):				
S118	PKCε	ND	Decrease	[140,202]
Cx50:				
S363	CK1	Increase	Increase	[120,193]

Notes: [a] mass spec identified a number of potential phosphorylation sites in Cx26 but did not test functions, although mutations at many of these sites are associated with disease pathology [139]. [b] Direct phosphorylation not shown, S263 and S266 on Cx31 contain consensus sequence for Ck1 which, when deleted, alters functions. [c] Src may not alter function of formed gap junctions. [d] Currently debated as to whether Cx43-S262 is a CDK1/CDC2/PKC/MAPK site, and several lines of evidence indicate that this is most likely an ERK-regulated site [148]. [e] Functions of S279/S282 typically shown by single phosphorylation antibodies or multiple site directed mutagenesis including both residues. Decrease GJIC as a result of reduced open probability. [f] Phosphorylation of S368 by phorbyl esters, e.g., TPA, are associated with PKCε phosphorylation and reduced communication. In ischemia, treatment by peptides, e.g., rotagaptide, increase S368 phosphorylation by PKCα, leading to increases in GJIC. [g] While initial phosphorylation at S373 is associated with a temporal increase in GJ size, it is thought to be the start in the process that leads to internalization. Abbreviations: ND, not demonstrated; (m), mouse; m.s., mass spectrometry-based identification approach. GJIC, gap junction intercellular communication; EGFR, epidermal growth factor receptor; PKA/PKC, Protein kinase A/ Protein kinase C; NO, nitric oxide; SUMO, small ubiquitin-like modifier.

Phosphorylation is a key regulator of physiological states in tissues, and changes in the phosphorylation status has been observed in several disease states. Within the vasculature, heterocellular endothelial cell–smooth muscle cell contacts, called the myoendothelial junctions (MEJs), express Cx37, Cx40, and Cx43, allowing for the direct exchange of intercellular signaling ions and molecules such as Ca^{2+} and IP_3 [203,204]. At the MEJ, Cx43 and Cx37 are regulated by post-translational modifications, including phosphorylation and *S*-nitrosylation (for details, see Section 3.2). In vitro and ex vivo data demonstrate that gap junctions at MEJs allow for the movement of Ca^{2+} and IP_3 between endothelial and vascular smooth muscle cells, which is in part regulated via Cx43-Ser368 [205].

In the healthy heart, Cx43 is primarily localized to the intercalated disc region of cardiomyocytes. Opening of Cx43-containing channels and signal conduction is facilitated by phosphorylation at residues including Ser365, 325, 328, and 330 [206–208]. Phosphorylation acts as a molecular switch, regulating gap junction opening. In ventricular arrhythmias following myocardial infarction, raised intracellular Ca^{2+} concentration leads to de-phosphorylation of Cx43-Ser365, which acts as the gatekeeper to phosphorylation of Cx43-Ser368. This resulting increase in Cx43-Ser368 reduces GJIC and promotes a redistribution of Cx43 to lateral regions of the cardiac myocytes, disrupting signaling in the heart [208–210].

Formation of large cardiac gap junction plaques at the intercalated disc is modulated through Cx43 interactions with zonula occludens 1 (ZO-1) [211–214]. In turn, this protein–protein interaction is regulated by PKC phosphorylation of Cx43 at Ser368, which inhibits ZO-1-mediated disassembly of gap junctions [215]. In ischemic heart disease, Cx43 is lost at the intercalated disc, but Cx43-Ser368 phosphorylation can act to indirectly stabilize the protein [196]. Multiple studies have investigated the effects of targeting the C-terminus of Cx43 in ischemia/reperfusion injuries, reducing infarct size, and other diseases [11,216]. A peptide that mimics the terminal region of the Cx43 known as ACT1 can disrupt Cx43/ZO-1 interaction [214,217]. This peptide promotes phosphorylation of Cx43-Ser368 via upregulation of PKCε activity, inhibits Cx43-ZO-1 binding, and improves cardiac function following ischemic insult in mice [215]. Similar results have been found for other connexin mimetic peptides targeting the Cx43 C-terminus (e.g., antiarrhythmic peptide 10 (AAP10) and rotigaptide (ZP123)), causing increases in Cx43-Ser368 phosphorylation through PKCα (reported to stabilize protein expression and increase GJIC) associated with improved cardiac functions in experimental animal models and early tests demonstrating no adverse effects in humans [192,218–222]. However, it should be noted that a similar peptide, danegaptide (a stabilized form of rotigaptide), failed to change clinical outcomes in ischemic reperfusion injuries in human Phase II testing (NCT01977755, completed 2016) [223].

In vascular disease, phosphorylation-mediated connexin–protein interactions and GJIC have been found to regulate disease state. Oxidized phospholipids found within atherosclerotic plaques increase MAPK and PKC phosphorylation of Cx43 and are associated with increased inflammation and cellular proliferation [130,224–227]. In response to the release of growth factors in disease, Cx43 is phosphorylated at MAPK residues (Cx43-Ser255, -Ser262, -Ser279, -Ser282), promoting direct interactions with the cyclin E/CDK2 complex and enhancing smooth muscle cell proliferation [131]. Conversely, PKC phosphorylation of Cx37 alters GJIC, which is linked with growth suppressive effects (e.g., reducing vasculogenesis and angiogenesis) [228–233]. Mutation of all seven Cx37 Ser > Ala essentially closes Cx37 GJs and hemichannels and inhibits both proliferation and cell death, whereas mutation of only three (Cx37-Ser275, -Ser302 and -Ser328) partially inhibits channel opening and decreases cellular death in rat insulinoma cells [128].

Phosphorylation also plays an important role in altered localization and function of connexins in cancer [3]. Several oncogenes and proto-oncogenes robustly inhibit GJIC, including HRAS [234], c-Src [235], and v-Src [178]. Curiously, the tyrosine-protein kinase c-Src has a reciprocal relationship with Cx43 that regulates its activity, where Cx43 is shown to bind with phosphatases (e.g., PTEN and Csk) reducing c-Src activity [236]. Conversely, Src phosphorylation of tyrosine residues on Cx43 (Cx43-Tyr243/-Tyr265) mediates interactions with endosomal machinery, leading to internalization of Cx43 and reduced expression [118,237]. Numerous tumor promoters, such as phorbol esters, also rapidly inhibit Cx43-mediated GJIC [238–240] through PKC- and ERK-mediated phosphorylation events [181,241]. On the contrary, loss of phosphorylation can also negatively affect GJIC. One recent study showed that the levels of total Cx43 protein and Cx43 phosphorylated at Ser368 and Ser279/282 were high in normal tissue but low to absent in malignant pancreatic tissue [79]. Altered Cx43-phosphorylation can be indicative of prognosis in some tumors, such as gliomas [242]. Phosphorylation of other connexins can also affect GJIC and the cancer phenotype, notably PKC-mediated phosphorylation of Cx37 [128]. Targeting dysregulated phosphorylation events of connexins in cancer may be one therapeutic angle towards restoring connexin function or GJIC. Indeed, the chemotherapeutic drug gefitinib has been suggested to upregulate GJIC by inhibiting Src and PKC-modulated Cx43 phosphorylation [243]. However, resistance to cisplatin-based chemotherapy has been suggested to be due to Src-induced Cx43 phosphorylation and loss of GJIC [244].

During wound healing, phosphorylation may also play a role in coordinating GJIC and connexin redistribution [245–247]. Initial responses to wounding include a generalized loss in Cx43, which may be modulated by increases in cyclic adenosine monophosphate (cAMP). In wound models, 8-bromo-cAMP-treated embryonic stem cells promote enhanced wound repair associated with reduced membrane bound Cx43, disruption in Cx43-ZO-1 interactions, and reduced GJIC [248]. However, the mechanisms regulating this are unclear, since cAMP-associated kinases have been previously described to increase PKA-mediated Cx43 synthesis, phosphorylation (Cx43-Ser364), GJ assembly, and GJIC in other model systems [189,249]. Phosphorylation is extremely dynamic within the wound and appears to be coordinated with the stage of repair. Initial increases in Cx43-Ser373, driven by AKT, can be seen between 1 and 30 min after wounding occurs, disrupting interactions with ZO-1, initially stabilizing Cx43 at the membrane, but is followed by rapid internalization of Cx43 [200]. Following wounding, transient increases (24–72 h) in PKC-mediated Cx43-Ser368 phosphorylation in regions proximal to the injured sited are associated with a loss of GJIC [250,251]. These data and others suggest that a combination of phosphorylation events sequentially regulate connexin signaling during wound repair [252].

In disease states such as diabetes, nonhealing wounds lead to complications, including ulcerations in skin tissues. In streptozotocin-induced diabetic mice, Cx43 dynamics are different from normal skin tissues, with increased expression of dermal Cx43 associated with reduction in keratinocyte migration [253]. Similar observations have been made in human diabetic ulcers, with Cx43 found to remain at elevated levels as compared to normal skin wounds [254]. In vitro and ex vivo evidence suggests that peptides aimed at disrupting gap junction and hemichannel communication (e.g., Gap27) can improve wound healing, which is associated with increased Cx43-Ser368 phosphorylation [251]. Recent studies have also shown that increases in Cx43-Ser368 phosphorylation following topical application of the ACT1 peptide is associated with clinically significant improvements in scar reduction and wound closure rates [255].

3.2. S-Nitrosylation

S-Nitrosylation occurs through covalent binding of nitric oxide (NO) to reactive cysteine(s) and can result in structural alterations of proteins leading to functional changes [256]. Protein *S*-nitrosylation is highly dependent on the cysteine oxidation state and surrounding amino acids, meaning that not all cysteines in a protein can be *S*-nitrosylated. While there are cysteine residues on the extracellular loops of all connexins, these have not been demonstrated to be *S*-nitrosylation targets [257]. Within the C-terminus of Cx43, there are three cysteines (Cx43-Cys260, -Cys271, and -Cys298), but only Cx43-Cys271 has been demonstrated to be *S*-nitrosylated, leading to an increase in GJ permeability in endothelial cells and at the MEJ [184]. Direct *S*-nitrosylation of other connexins has not been demonstrated, although there are multiple lines of evidence demonstrating that nitric oxide (NO) activation leads to regulation of gap junction and hemichannel signaling [258]. Within the vasculature, NO plays an important role in vasodilation. Figueroa et al. found that vascular connexins channels formed by Cx37, Cx40, and Cx43 are activated by and directly permeable to NO, and have suggested that this is an alternative method to NO transfer across plasma membranes [259]. Cx37 is enriched at the MEJ of resistance arteries and is reported to be important in the regulation of NO-mediated Ca^{2+} regulation via reducing Cx37-mediated gap junctional coupling between endothelial cells and smooth muscle cells [260]. However, unlike Cx43, the effects of NO on Cx37 gap junction channels are thought to be indirect, with no known cysteine modification occurring. Rather, the phosphorylated tyrosine residue (Cx37-Tyr332) is protected from de-phosphorylation by Src homology region 2 (Shp2) phosphatase, which is inhibited in the presence of NO, reducing MEJ transfer of Ca^{2+} signaling through Cx37 GJ [261]. Thus, *S*-nitrosylation appears to have diverse effects, depending on GJ composition particularly at the MEJ [261].

3.3. Other Post-Translational Modifications: SUMOylation, Ubiquitination, and Acetylation

A number of post-translational modifications are associated with regulation of connexin protein turnover, for example, ubiquitination, SUMOylation, and acetylation. Small ubiquitin-like modifier proteins (e.g., SUMO-1/-2/-3) interact with lysine residues on proteins, altering protein targeting and turnover [262]. So far, there is only evidence for direct Cx43 SUMOylation at lysine residues (Cx43-Lys144, -Lys237) within its intracellular loop and C-terminus [172]. Overexpression of all three SUMO proteins in HeLa cells increases Cx43 expression, promotes gap junction formation, and increases signaling. However, the exact mechanism by which SUMOylation regulates protein expression is not known. The amino acids sequences surrounding Cx43-Lys144 and Cx43-Lys237 are not common motifs associated with SUMOylation, although the same motifs of a conserved Lys144 followed by an upstream large hydrophobic amino acid (valine) are found in at least six other connexins, suggesting a common regulatory pathway [172].

Once at the plasma membrane, the majority of connexins are rapidly turned over with half-lives estimated between 1.5 and 5 h for Cx43 and Cx26 and up to 24 h for other isoforms such as Cx46 [124,263–266]. While connexins use a multitude of pathways for internalization and degradation, the process typically involves formation of an endosome (termed connexosome [267]), where older gap junctions are internalized to be targeted to the lysosome for degradation, although there is also evidence for endosomal recycling back to the membrane [214,268,269]. Endosomal formation is driven by multiple proteins in complex, including interactions with ZO-1, tubulin, and others. In the case of Cx43, this interaction (with ZO-1) is regulated via Cx43-Ser373 and Cx43-Ser368 phosphorylation [200,214,270,271]. Monoubiquitinylation typically acts as a signal for internalization of proteins via endosomes to lysosomes, leading to degradation [272,273]. Multiple covalently linked ubiquitin molecules bind lysine residues within the target protein, which are then recognized by receptors and targeted for degradation by the 26S proteasome [274–276] and by autophagy [277–279]. Recent evidence has demonstrated a complementary role for Cx43 in regulating autophagy, in that Cx43 at the plasma membrane interacts with several pre-autophagosomal proteins, including Atg16, but not other autophagosome proteins, such as LC3 [280]. When the cells are under stress, such as nutrient depletion, Cx43 becomes ubiquitinylated and internalized, causing recruitment of other factors (Atg5, Atg12, and LC3) to form fully functional autophagosomes. While regulated autophagy can have a protective effect in stressed cells, there is also evidence linking aberrant autophagy and Cx43 degradation from intercalated discs to heart failure [281], suggesting the potential for a novel pharmacologic approach to treat cardiac failure.

Proteasomal–ubiquitin pathways have been proposed to indirectly regulate Cx43 through interaction with the ZO-1 protein, thus disrupting part of the process that is critical for Cx43 membrane organization [214,282]. Multiple studies suggest that other connexin proteins (e.g., Cx50, Cx43, and Cx31.1) are regulated by ubiquitination [283]. Several studies show that ubiquitin regulates internalization of Cx43 via clathrin-mediated endocytosis, by both tyrosine (Y)-dependent sorting signal (YXXΦ, where X is any amino acid and Φ is an amino acid with a bulky hydrophobic side chain) and tyrosine-independent, EPS15-dependent pathways [284,285]. However, the route through which ubiquitin regulates the connexins has not been fully delineated, with studies in Cx43 demonstrating that the C-terminal lysines are dispensable for protein turnover [286]. Despite this, there is increasing evidence that Cx43 is modified in response to ubiquitin, and corresponding ligases are controlled in part by phosphorylation events, such as MAPK and PKC phosphorylation [287,288]. A number of ubiquitin-binding proteins (e.g., EPS15, p62, Hrs, and TSG101) are rec ruited to Cx43 to facilitate its internalization and sorting to the lysosome [289,290]. In addition, TSG101 has been found to interact with Cx30.2, Cx31, Cx36, and Cx45 [290]. While classic lysine-based motifs may not be responsible for direct ubiquitin binding, more recent studies have shown that proline-rich regions of the Cx43 C-terminus (xPPxY) bind to ubiquitin ligase. A number of ubiquitin ligases have also been associated with direct binding, internalization, and degradation of Cx43 (e.g., Trim21 [291], WWP1 [292], SMURF2 [293], and NEDD4 [287–289,294]). NEDD4 also has been directly associated with loss of Cx43 at the plasma membrane in experimental models [287].

The process of degradation may be further regulated by connexin N-terminal acetylation, which can act to regulate protein stability in the membrane. In mouse cardiac myocytes, N-terminal acetylation through binding of P300/CBP-associated factor with Cx43 leads to a loss of Cx43 at the intercalated disc, a lateral reorganization of the protein, reduced gap junction formation in cardiac myocytes, and internalization in NIH-3T3 (mouse embryo) fibroblasts [77]. These patterns of disorganization of Cx43 are similar to those seen in mouse models of Duchenne cardiomyopathies, where NO and oxidative stress lead to an imbalance in acetylation/deacetylation and alterations in cardiac conduction. Similarly in dogs, cardiac pacing leads to increased Cx43 acetylation, suggesting that this mechanism is important in regulating signaling in physiology and pathology of the cardiac system [77,295,296].

4. Connexin Trafficking

Formation of gap junctions by connexins is regulated by the delivery of newly synthesized channels to the plasma membrane and is balanced by the removal of channels via endocytosis [263,297,298]. As mentioned above, since connexin turnover is generally quite rapid and influenced by post-translational modifications, the dynamic regulation of connexins by secretion and turnover provides a means to control gap junction formation, composition, and, thus, GJIC.

4.1. Control of Oligomerization

Secretion of newly synthesized connexins from the endoplasmic reticulum (ER) through the Golgi apparatus is coordinately regulated with oligomerization into hexameric hemichannels [299]. Based on structural homology, connexins can be separated into two distinct oligomerization groups. *GJB1–GJB7* (so-called β connexins, including Cx26 and Cx32) follow a more traditional pathway, where full oligomerization into hexamers is required prior to transport from the ER to the cis-Golgi apparatus [300–302]. By contrast, other connexins are stabilized by a connexin-specific quality control apparatus as monomers that are subsequently transported to the trans-Golgi network (TGN) where they then have the capacity to oligomerize [301,303]. The best-studied connexin known to oligomerize in the TGN is Cx43, although there is also experimental evidence for Cx40 and Cx46 oligomerization late in the secretory pathway [304,305]. By homology, it is likely that most non-beta connexins will also follow the late oligomerization pathway that has been demonstrated for Cx43 [299].

Several lines of evidence suggest that the transition from monomeric to hexameric Cx43 requires a conformational change, largely centered on the third transmembrane domain (TM3) where it is stabilized in a monomeric conformation by motifs containing charged amino acids on both ends of the TM domain (Figure 2) [300,305]. At the cytoplasmic interface of the Cx43 TM3 domain is an LR motif containing a highly charged arginine residue, and at the extracellular interface is a glutamine-containing motif with a QYFLYGF amino acid consensus sequence. The extracellular loop domain of Cx43 also interacts with a chaperone protein, ERp29, that is required to stabilize monomeric Cx43 [300].

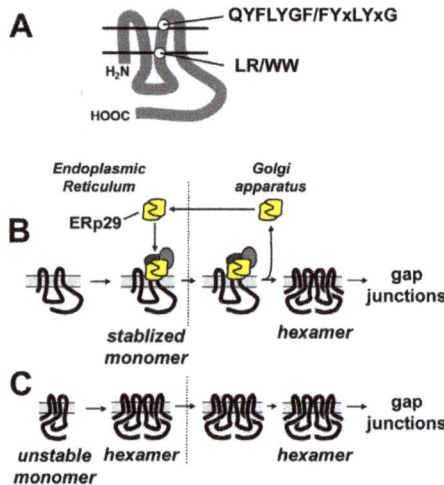

Figure 2. Differential connexin oligomerization. (**A**) Line diagram showing two key connexin motifs adjacent to the third transmembrane domain. Connexins (such as Cx43), which oligomerize in the Golgi apparatus (**B**), have a cytosolic LR and extracellular QYFLYGF motif that interacts with ERp29 (yellow) and other putative chaperones (grey ovals) that stabilize monomeric connexins until they transition from the endoplasmic reticulum (ER) to the Golgi apparatus (delineated by the dashed lines). In the Golgi apparatus, ERp29 dissociates from monomeric connexins and then recycles back to the ER, enabling connexins to oligomerize into hexameric hemichannels. By contrast, connexins (such as Cx32) that have a WW and a FYxLYxG motif cannot interact with ERp29—they are inserted into the ER membrane as unstable monomers and thus immediately oligomerize (**C**).

Int. J. Mol. Sci. **2018**, *19*, 1296

By contrast, beta connexins lack charged residues adjacent to the TM3 domain. They instead have a di-tryptophan (WW) motif that is less stringently localized to the membrane/cytosol interface, and they lack the ability to interact with ERp29. Thus, beta connexins are not stable as monomers and instead oligomerize in the ER (Figure 2) [300–302]. Since motifs associated with the TM3 domain also have been implicated in regulating connexin hetero-oligomerization [299,306], this implicates a role for spatial separation of connexin oligomerization in regulating the extent and stoichiometry of heteromeric channel formation.

4.2. Connexin Quality Control

The differences in quality control for Cx26 and Cx43 were directly observed for native connexins in human airway epithelial cells derived from a cystic fibrosis (CF) patient expressing the CF transmembrane conductance regulator (CFTR) protein harboring the Fdel508 mutation [307]. In these cells, Cx43 trafficking and function is impaired, yet Cx26 transport and assembly into gap junction channels is normal. Interestingly, CFTR also interacts with ERp29 [308], and Cx43-mediated GJIC by Fdel508-CFTR-expressing cells is restored by treatment with 4-phenylbutyrate, a drug that upregulates ERp29 expression [307,308]. In addition, 4-phenylbutyrate has been shown to upregulate GJIC in several other contexts [309–314], further underscoring a role for ERp29 and other 4-phenylbutyrate-sensitive factors in connexin quality control.

Aberrant accumulation of connexins in the ER clearly decreases the pool of connexins available to produce gap junction channels at the cell surface. However, ER accumulation of connexins has also been found to induce an unfolded protein response (UPR) that, in turn, has the capacity to impair cell function and lead to human disease. UPR induced by mutant connexins has been directly demonstrated for Cx50 mutations associated with cataract [315–317] and Cx31 mutations that cause the skin disease erythrokeratoderma variabilis (EKV) [318] or hearing impairment [319]. The association of UPR with human diseases related to misfolded connexins suggests the possibility that treatments alleviating ER stress, such as 4-phenylbutyrate, may have therapeutic value by promoting proper protein folding and trafficking as well as increasing GJIC. Also, as mentioned above, the ability of 4-phenylbutyrate to enhance GJIC also may contribute to its potential as an anticancer therapeutic, and may be related to increased ERp29 activity [320].

4.3. Connexin Cytoplasmic Domains and the Cytoskeleton

In addition to motifs adjacent to the TM3 domain, there are several lines of evidence in support of connexin C-terminal domains in regulating connexin trafficking. As described above, in addition to containing several motifs that can be post-translationally modified, the semi-structured nature of the C-terminus [153,155,321] enables it to be conformationally labile and to interact with several different classes of cytosolic scaffold proteins and the cytoskeleton that can influence connexin targeting (reviewed in [5] for Cx43). For instance, several truncated connexins lack the ability to be efficiently trafficked to the plasma membrane or be endocytosed [322,323]. The connexin C-terminal domains also have the capacity to homo- and hetero-dimerize [153,155,159,324] as well as interact with other connexin domains, including the cytoplasmic loop [155,325,326] that can influence connexin targeting, oligomerization, and function.

Interestingly, it was determined that there is reciprocal regulation of Cx43 and Cx46 in the lens, where conditions such as activation of PKC caused an increase in Cx46 transcription and expression that was associated with a concomitant decrease in Cx43, via ubiquitination and proteasomal degradation [327]. In fact, transfecting cells with Cx46 was sufficient to induce Cx43 degradation and this effect required the C-terminus of Cx46, since a Cx46 tail truncation mutant had no effect on Cx43 expression. Increased Cx50 also had no effect on Cx43. However, transfecting cells with a soluble Cx46 tail construct had the ability to decrease Cx43 expression. Since the decrease in Cx43 was induced by an intracellular pool of Cx46, this raises the possibility that crosstalk between Cx46 and Cx43 may be related to differential oligomerization [304]. However, this remains to be determined.

As another instance where the C-terminus plays a key role in regulating Cx43 trafficking, it has been shown that amino N-terminal truncated forms of Cx43 are also expressed by cells, through alternative internal translation via one of six different AUG initiation sites (see Section 2.4.2) [328]. The most prominent of these is GJA1-20k, which consists of a portion of the TM4 domain as well as the entire C-terminus [110] (Figure 1). GJA1-20k expression promotes formation of Cx43 gap junction channels, resulting in an increase in intercellular communication [110,113]. As discussed in Section 2.4.2, alternative translation of Cx43, including production of GJA1-20k, is inhibited by mTOR [110,111] and Mnk1/2 kinases [111], suggesting that metabolic stress regulates gap junctional coupling through mTOR- and Mnk1/2-mediated pathways as a means to protect cells both by enabling scarce metabolites to be distributed via intercellular communication as well as limiting damage by restricting generation of reactive oxygen species [329].

How GJA1-20k regulates channel formation by Cx43 is still under investigation. One intriguing possibility is that GJA1-20k acts as a chaperone protein that promotes Cx43 oligomerization, as was recently demonstrated to regulate the decrease in gap junction formation and function that can occur in the epithelial to mesenchyme transition [330] (Figure 1).

Another likely role for GJA1-20k relates to cytoskeletal control of Cx43 trafficking, since it has been shown that the C-terminus of Cx43 and, therefore, GJA-20k as well, interacts with both microtubules and filamentous actin [331–333]. Microtubules and actin perform complementary functions in regulating connexin trafficking, where microtubules help facilitate rapid transport of Cx43-containing vesicles to sites of junction formation [332], whereas actin has a more subtle role in regulating connexin trafficking, since quantitative live cell-imaging shows that transport of Cx43-containing vesicles temporarily pauses when they interact with actin filaments, perhaps as a means to enhance sorting or to remodel vesicle composition [331]. Also, transfecting HeLa cells with GJA1-20k nucleates the formation of actin filaments [113], suggesting a role for GJA1-20k in altering the itinerary of Cx43 trafficking in the cell. Reverse regulation is also suggested by studies where gap junction inhibitors resulted in misalignment of actin filaments across the monolayer and reduced calcium signaling in rat astrocytes [334]. Furthermore, treatment of astrocytes with an actin polymerization inhibitor cytochalasin D or anti-actin antibodies reduced GJIC, as visualized by a reduction in the spread of microinjected neurobiotin between cells [335].

4.4. Regulation of Gap Junction Plaque Morphology

Actin has also been implicated in regulating gap junction plaque morphology. Double knockout of the actin capping protein tropomodulin 1 and intermediate filament protein CP49 in lens fiber cells led to a significant decrease in Cx46 plaque volume and increase in plaque number, affecting gap junction coupling and function in the lens tissue [336]. Regulation of plaque size by actin is likely to be coordinated by interactions involving the C-terminus of connexins and zonula occludens 1 (ZO-1). For example, enhanced green fluorescent protein (EGFP)-tagged Cx43 incapable of interacting with ZO-1 produces plaques that are not size regulated [337]. By contrast, the perimeter of gap junction plaques (the perinexus) is ringed by Cx43/ZO-1 complexes, whereas the center of plaques is largely devoid of ZO-1 [338]. Inhibition of Cx43/ZO-1 interactions cause an increase in gap junction plaque size. Consistent with this possibility, Cx43 phosphorylation inhibits ZO-1 binding and facilitates connexin channel endocytosis [339]. Additional roles for ZO-1, connexin phosphorylation, and ubiquitinylation in regulating connexin endocytosis and degradation are described in Sections 3.2 and 3.3, above.

Although the precise mechanism whereby ZO-1 limits plaque formation is still under investigation, it seems plausible that it may be analogous to the role of ZO-1 in regulating tight junctions, where claudin/ZO-1/actin interactions have a junction-stabilizing influence on the apical junctional complex, whereas, in the absence of ZO-1, there is increased access of myosin that increases tight junction dynamics and tension [340,341]. Consistent with this possibility, myosin VI has also been found to have a specific role in increasing gap junction plaque size, analogous to treatments inhibiting Cx43/ZO-1 interactions [342].

Whether regulation of plaque assembly strictly follows the perinexus model has recently been challenged by observations of Cx36 plaque formation [343]. Pulse-chase experiments with Cx36 indicated addition of Cx36 to both the ends and the middle of preexisting gap junction plaques, with diffusion of Cx36 throughout the plaque. When the experiments were repeated with Cx43, there appeared to be less diffusion of newly added Cx43 in preexisting plaques [343]. Targeted delivery of connexins has only recently been observed. Through interactions with plus-end binding protein EB1 and the dynein/dynactin complex, microtubule plus-ends are tethered to adherens junctions at the plasma membrane, leading to the targeted deposition of connexin hemichannels and gap junction plaque formation [332]. These two models begin to bring to light the vast complexity of connexin trafficking and gap junction formation, suggesting a network of cytoskeleton and protein-binding partners tailored to specific connexins that was previously unrealized.

A less understood but intriguing role for the cytoskeleton in gap junction biology is the creation of unique junctional subregions involved in gap junction dynamics. Using an EGFP-tagged Cx32 construct, particularly dynamic regions at the edges of gap junction plaques were observed as invaginated tubular structures, where plaque fragments pinched off into the cytoplasm [344]. These tubulovesicular extensions of gap junction plaques were recently observed with Cx36 and termed filadendrites [345]. Filadendrites at the edges of gap junction plaques appeared to be the same thickness as the plaque, suggesting that the filadendrites were a continuation of the gap junction plaques themselves. Filadendrites were also observed in interior regions of the gap junction plaques, but appeared to be much thinner than the gap junction plaques. From pulse-chase labeling of Cx36, it was observed that filadendrites exhibited some of the same dynamic properties as the earlier observed Cx32 invaginations, constantly pinching off and fusing with the gap junction plaque. Labeling of actin filaments showed colocalization with Cx36 filadendrites, suggesting that the actin cytoskeleton could be one of the drivers behind the formation of these dynamic structures. Treatment with the actin polymerization inhibitor Latrunculin A or actin depolymerization inducer cytochalasin D reduced the presence of filadendrites, indicating that the driving force behind the dynamic gap junction plaques requires actin polymerization [345].

Similar structures have been noted at other junctions. Primary human keratinocytes treated with pemphigus vulgaris (PV) IgG containing antibodies targeted to the adherens junction protein desmoglein 3 (Dsg3) exhibited reorganized Dsg3 at the membrane into projections perpendicular to the membrane plane. These projections, termed linear arrays, are similar to the filadendrites in that they are sites of disassembly of junction components and active endocytosis at the junctions. Linear arrays also colocalized with actin filaments oriented perpendicular to the plasma membrane, similar to those observed in filadendrites. Furthermore, linear arrays were associated with decreased cell adhesion, suggesting a functional effect of these junctional subregions [346,347]. A comparable structure formed by tight junction proteins, termed tight junction spikes, have been observed to correlate with treatments that enhance junction disassembly and paracellular leak, including oxidative stress induced by chronic alcohol exposure, transforming growth factor (TGF)-β1 treatment, and inhibition of NF-κB [348–350]. In alveolar epithelial cells, actin filaments colocalized with the tight junction protein claudin-18 in tight junction spikes. Spikes were also found to be sites of budding and fusion of vesicles carrying tight junction proteins, both indicators of active tight junction remodeling. Treatment of lung alveolar epithelial cells with granulocyte-macrophage colony-stimulating factor (GM-CSF) reduced actin filament colocalization with claudin-18 containing tight junction spikes, whereas keratinocyte growth factor treatment inhibited spike formation and instead promoted formation of cortical actin as opposed to actin stress fibers [350,351]. Taken together, these findings indicate that these similar junctional subregions observed universally across several different classes of intercellular junctions, including gap junctions, could represent a common mechanism of junction protein turnover, where the junctions partition themselves into unique filamentous structures. Whether these structures serve to restrict turnover of junction proteins to specific subdomains or whether they nucleate the formation of signaling complexes that recruit specialized subsets of cytosolic binding partners remains to be determined.

5. Conclusions and Future Perspectives

In order to fully understand the complex role of connexins in health and disease, it is essential to elucidate their regulation at all steps, from gene transcription, protein synthesis, post-translational modifications, and trafficking to their regulation at the cell membrane. This review is intended to highlight some of the progress made in these areas in relation to health and disease, giving examples of how this knowledge is pertinent for future therapeutic applications. Going forward, understanding how modulation of connexins occurs at any of these stages will require additional work and insight, which over time may lead to more fruitful and safer strategies to alleviate patient suffering. For example, the danegaptide trials that were based on strong preclinical data suggested that alterations to the trafficking and increased Cx43 signaling in the heart would have a profound effect in reducing ischemic reperfusion injury and reduce cardiac tissue damage. However, Phase II clinical trials in humans failed to show an effect, highlighting the complex nature of targeting gap junctions as a treatment modality including differences in how connexins are regulated in model systems as opposed to human disease. Additional caution is also needed for therapeutic approaches in cancer, where it is now clear that connexins have distinct roles that both promote and inhibit cell growth and metastasis.

Despite substantial progress, it is important to acknowledge the complexity of gap junctions that serve as a conduit that enables cells to share thousands of different signaling molecules. Additionally, the complex connexin protein interactome underscores the non-junctional functions of connexins, including their ability to act as a signaling platform. In particular, it is critical to identify connexin-specific functions that are unique and targetable. This is best approached by understanding how connexins are regulated at multiple levels by multiple mechanisms, from gene transcription and translation to post-translational modification, and as a specifically localized multiprotein complex.

Acknowledgments: Trond Aasen acknowledges support from Instituto de Salud Carlos III grants PI13/00763, PI16/00772 and CPII16/00042, co-financed by the European Regional Development Fund (ERDF). Supported by NIH R01-AA025854 and R01-HL137112 (Michael Koval) and F31-HL139109 (K. Sabrina Lynn).

Conflicts of Interest: The authors declare no conflict of interest. The funding sponsors had no role in the design of the study; in the collection, analyses, or interpretation of data; in the writing of the manuscript, and in the decision to publish the results.

References

1. Goodenough, D.A.; Paul, D.L. Gap junctions. *Cold Spring Harb. Perspect. Biol.* **2009**, *1*, a002576. [CrossRef] [PubMed]
2. Esseltine, J.L.; Laird, D.W. Next-generation connexin and pannexin cell biology. *Trends Cell Biol.* **2016**, *26*, 944–955. [CrossRef] [PubMed]
3. Aasen, T.; Mesnil, M.; Naus, C.C.; Lampe, P.D.; Laird, D.W. Gap junctions and cancer: Communicating for 50 years. *Nat. Rev. Cancer* **2016**, *16*, 775–788. [CrossRef] [PubMed]
4. Saez, J.C.; Leybaert, L. Hunting for connexin hemichannels. *FEBS Lett.* **2014**, *588*, 1205–1211. [CrossRef] [PubMed]
5. Leithe, E.; Mesnil, M.; Aasen, T. The connexin 43 C-terminus: A tail of many tales. *Biochim. Biophys. Acta* **2018**, *1860*, 48–64. [CrossRef] [PubMed]
6. Delmar, M.; Laird, D.W.; Naus, C.C.; Nielsen, M.S.; Verselis, V.K.; White, T.W. Connexins and disease. *Cold Spring Harb. Perspect. Biol.* **2017**. [CrossRef] [PubMed]
7. Srinivas, M.; Verselis, V.K.; White, T.W. Human diseases associated with connexin mutations. *Biochim. Biophys. Acta* **2018**, *1860*, 192–201. [CrossRef] [PubMed]
8. Kelly, J.J.; Simek, J.; Laird, D.W. Mechanisms linking connexin mutations to human diseases. *Cell Tissue Res.* **2015**, *360*, 701–721. [CrossRef] [PubMed]
9. Aasen, T. Connexins: Junctional and non-junctional modulators of proliferation. *Cell Tissue Res.* **2015**, *360*, 685–699. [CrossRef] [PubMed]

10. Leybaert, L.; Lampe, P.D.; Dhein, S.; Kwak, B.R.; Ferdinandy, P.; Beyer, E.C.; Laird, D.W.; Naus, C.C.; Green, C.R.; Schulz, R. Connexins in cardiovascular and neurovascular health and disease: Pharmacological implications. *Pharmacol. Rev.* **2017**, *69*, 396–478. [CrossRef] [PubMed]

11. Becker, D.L.; Phillips, A.R.; Duft, B.J.; Kim, Y.; Green, C.R. Translating connexin biology into therapeutics. *Semin. Cell Dev. Biol.* **2016**, *50*, 49–58. [CrossRef] [PubMed]

12. Grek, C.L.; Rhett, J.M.; Ghatnekar, G.S. Cardiac to cancer: Connecting connexins to clinical opportunity. *FEBS Lett.* **2014**, *588*, 1349–1364. [CrossRef] [PubMed]

13. Lee, C.; Huang, C.H. Lasagna-search: An integrated web tool for transcription factor binding site search and visualization. *BioTechniques* **2013**, *54*, 141–153. [CrossRef] [PubMed]

14. Söhl, G.; Willecke, K. An update on connexin genes and their nomenclature in mouse and man. *Cell Commun. Adhes.* **2003**, *10*, 173–180. [CrossRef] [PubMed]

15. Beyer, E.C.; Berthoud, V.M. Gap junction gene and protein families: Connexins, innexins, and pannexins. *Biochim. Biophys. Acta* **2018**, *1860*, 5–8. [CrossRef] [PubMed]

16. Dupays, L.; Mazurais, D.; Rücker-Martin, C.; Calmels, T.; Bernot, D.; Cronier, L.; Malassiné, A.; Gros, D.; Théveniau-Ruissy, M. Genomic organization and alternative transcripts of the human connexin40 gene. *Gene* **2003**, *305*, 79–90. [CrossRef]

17. Essenfelder, G.M.; Larderet, G.; Waksman, G.; Lamartine, J. Gene structure and promoter analysis of the human gjb6 gene encoding connexin30. *Gene* **2005**, *350*, 33–40. [CrossRef] [PubMed]

18. Sohl, G.; Theis, M.; Hallas, G.; Brambach, S.; Dahl, E.; Kidder, G.; Willecke, K. A new alternatively spliced transcript of the mouse connexin32 gene is expressed in embryonic stem cells, oocytes, and liver. *Exp. Cell Res.* **2001**, *266*, 177–186. [CrossRef] [PubMed]

19. Pfeifer, I.; Anderson, C.; Werner, R.; Oltra, E. Redefining the structure of the mouse connexin43 gene: Selective promoter usage and alternative splicing mechanisms yield transcripts with different translational efficiencies. *Nucleic Acids Res.* **2004**, *32*, 4550–4562. [CrossRef] [PubMed]

20. Anderson, C.L.; Zundel, M.A.; Werner, R. Variable promoter usage and alternative splicing in five mouse connexin genes. *Genomics* **2005**, *85*, 238–244. [CrossRef] [PubMed]

21. Cicirata, F.; Parenti, R.; Spinella, F.; Giglio, S.; Tuorto, F.; Zuffardi, O.; Gulisano, M. Genomic organization and chromosomal localization of the mouse connexin36 (*mCx36*) gene. *Gene* **2000**, *251*, 123–130. [CrossRef]

22. Von Maltzahn, J.; Euwens, C.; Willecke, K.; Söhl, G. The novel mouse connexin39 gene is expressed in developing striated muscle fibers. *J. Cell Sci.* **2004**, *117*, 5381–5392. [CrossRef] [PubMed]

23. Hombach, S.; Janssen-Bienhold, U.; Sohl, G.; Schubert, T.; Bussow, H.; Ott, T.; Weiler, R.; Willecke, K. Functional expression of connexin57 in horizontal cells of the mouse retina. *Eur. J. Neurosci.* **2004**, *19*, 2633–2640. [CrossRef] [PubMed]

24. Söhl, G.; Joussen, A.; Kociok, N.; Willecke, K. Expression of connexin genes in the human retina. *BMC Ophthalmol.* **2010**, *10*, 27. [CrossRef] [PubMed]

25. Oyamada, M.; Takebe, K.; Oyamada, Y. Regulation of connexin expression by transcription factors and epigenetic mechanisms. *Biochim. Biophys. Acta Biomembr.* **2013**, *1828*, 118–133. [CrossRef] [PubMed]

26. Neuhaus, I.M.; Bone, L.; Wang, S.; Ionasescu, V.; Werner, R. The human connexin32 gene is transcribed from two tissue-specific promoters. *Biosci. Rep.* **1996**, *16*, 239–248. [CrossRef] [PubMed]

27. Murphy, S.M.; Polke, J.; Manji, H.; Blake, J.; Reiniger, L.; Sweeney, M.; Houlden, H.; Brandner, S.; Reilly, M.M. A novel mutation in the nerve-specific 5′ UTR of the GJB1 gene causes X-linked charcot-marie-tooth disease. *J. Peripher. Nerv. Syst. JPNS* **2011**, *16*, 65–70. [CrossRef] [PubMed]

28. Kulshrestha, R.; Burton-Jones, S.; Antoniadi, T.; Rogers, M.; Jaunmuktane, Z.; Brandner, S.; Kiely, N.; Manuel, R.; Willis, T. Deletion of p2 promoter of GJB1 gene a cause of charcot-marie-tooth disease. *Neuromuscul. Disord. NMD* **2017**, *27*, 766–770. [CrossRef] [PubMed]

29. Al-Yahyaee, S.A.; Al-Kindi, M.; Jonghe, P.D.; Al-Asmi, A.; Al-Futaisi, A.; Vriendt, E.D.; Deconinck, T.; Chand, P. Pelizaeus-merzbacher-like disease in a family with variable phenotype and a novel splicing GJC2 mutation. *J. Child Neurol.* **2013**, *28*, 1467–1473. [CrossRef] [PubMed]

30. Gandia, M.; Del Castillo, F.J.; Rodriguez-Alvarez, F.J.; Garrido, G.; Villamar, M.; Calderon, M.; Moreno-Pelayo, M.A.; Moreno, F.; del Castillo, I. A novel splice-site mutation in the GJB2 gene causing mild postlingual hearing impairment. *PLoS ONE* **2013**, *8*, e73566. [CrossRef] [PubMed]

31. Kandouz, M.; Bier, A.; Carystinos, G.D.; Alaoui-Jamali, M.A.; Batist, G. Connexin43 pseudogene is expressed in tumor cells and inhibits growth. *Oncogene* **2004**, *23*, 4763–4770. [CrossRef] [PubMed]

32. Bier, A.; Oviedo-Landaverde, I.; Zhao, J.; Mamane, Y.; Kandouz, M.; Batist, G. Connexin43 pseudogene in breast cancer cells offers a novel therapeutic target. *Mol. Cancer Ther.* **2009**, *8*, 786–793. [CrossRef] [PubMed]

33. Hong, H.M.; Yang, J.J.; Shieh, J.C.; Lin, M.L.; Li, S.Y. Novel mutations in the connexin43 (*GJA1*) and *GJA1* pseudogene may contribute to nonsyndromic hearing loss. *Hum. Genet.* **2010**, *127*, 545–551. [CrossRef] [PubMed]

34. Poliseno, L.; Salmena, L.; Zhang, J.; Carver, B.; Haveman, W.J.; Pandolfi, P.P. A coding-independent function of gene and pseudogene mRNAs regulates tumour biology. *Nature* **2010**, *465*, 1033–1038. [CrossRef] [PubMed]

35. Rackauskas, M.; Neverauskas, V.; Skeberdis, V.A. Diversity and properties of connexin gap junction channels (review). *Medicina (Kaunas Lithuania)* **2010**, *46*, 1–12. [PubMed]

36. Di, W.L.; Rugg, E.L.; Leigh, I.M.; Kelsell, D.P. Multiple epidermal connexins are expressed in different keratinocyte subpopulations including connexin 31. *J. Investig. Dermatol.* **2001**, *117*, 958–964. [CrossRef] [PubMed]

37. Scott, C.A.; Tattersall, D.; O'Toole, E.A.; Kelsell, D.P. Connexins in epidermal homeostasis and skin disease. *Biochim. Biophys. Acta* **2012**, *1818*, 1952–1961. [CrossRef] [PubMed]

38. Faniku, C.; Wright, C.S.; Martin, P.E. Connexins and pannexins in the integumentary system: The skin and appendages. *Cell. Mol. Life Sci. CMLS* **2015**, *72*, 2937–2947. [CrossRef] [PubMed]

39. Lilly, E.; Sellitto, C.; Milstone, L.M.; White, T.W. Connexin channels in congenital skin disorders. *Semin. Cell Dev. Biol.* **2016**, *50*, 4–12. [CrossRef] [PubMed]

40. Tu, Z.J.; Kiang, D.T. Mapping and characterization of the basal promoter of the human connexin26 gene. *Biochim. Biophys. Acta* **1998**, *1443*, 169–181. [CrossRef]

41. Bai, S.; Spray, D.C.; Burk, R.D. Identification of proximal and distal regulatory elements of the rat connexin32 gene. *BBA Gene Struct. Expr.* **1993**, *1216*, 197–204. [CrossRef]

42. Field, J.M.L.; Tate, L.A.; Chipman, J.K.; Minchin, S.D. Identification of functional regulatory regions of the connexin32 gene promoter. *Biochim. Biophys. Acta Gene Struct. Expr.* **2003**, *1628*, 22–29. [CrossRef]

43. Seul, K.H.; Tadros, P.N.; Beyer, E.C. Mouse connexin40: Gene structure and promoter analysis. *Genomics* **1997**, *46*, 120–126. [CrossRef] [PubMed]

44. Bierhuizen, M.F.A.; Van Amersfoorth, S.C.M.; Groenewegen, W.A.; Vliex, S.; Jongsma, H.J. Characterization of the rat connexin40 promoter: Two Sp1/Sp3 binding sites contribute to transcriptional activation. *Cardiovasc. Res.* **2000**, *46*, 511–522. [CrossRef]

45. Echetebu, C.O.; Ali, M.; Izban, M.G.; MacKay, L.; Garfield, R.E. Localization of regulatory protein binding sites in the proximal region of human myometrial connexin 43 gene. *Mol. Hum. Reprod.* **1999**, *5*, 757–766. [CrossRef] [PubMed]

46. Teunissen, B.E.J.; van Amersfoorth, S.C.M.; Opthof, T.; Jongsma, H.J.; Bierhuizen, M.F.A. Sp1 and Sp3 activate the rat connexin40 proximal promoter. *Biochem. Biophys. Res. Commun.* **2002**, *292*, 71–78. [CrossRef] [PubMed]

47. Linhares, V.L.F.; Almeida, N.A.S.; Menezes, D.C.; Elliott, D.A.; Lai, D.; Beyer, E.C.; Campos De Carvalho, A.C.; Costa, M.W. Transcriptional regulation of the murine connexin40 promoter by cardiac factors nkx2-5, gata4 and tbx5. *Cardiovasc. Res.* **2004**, *64*, 402–411. [CrossRef] [PubMed]

48. Geimonen, E.; Boylston, E.; Royek, A.; Andersen, J. Elevated connexin-43 expression in term human myometrium correlates with elevated C-jun expression and is independent of myometrial estrogen receptors. *J. Clin. Endocrinol. Metab.* **1998**, *83*, 1177–1185. [CrossRef] [PubMed]

49. Fernandez-Cobo, M.; Stewart, D.; Drujan, D.; De Maio, A. Promoter activity of the rat connexin 43 gene in nrk cells. *J. Cell. Biochem.* **2001**, *81*, 514–522. [CrossRef]

50. Teunissen, B.E.J.; Jansen, A.T.; Van Amersfoorth, S.C.M.; O'Brien, T.X.; Jongsma, H.J.; Bierhuizen, M.F.A. Analysis of the rat connexin 43 proximal promoter in neonatal cardiomyocytes. *Gene* **2003**, *322*, 123–136. [CrossRef] [PubMed]

51. Vine, A.L.; Leung, Y.M.; Bertram, J.S. Transcriptional regulation of connexin 43 expression by retinoids and carotenoids: Similarities and differences. *Mol. Carcinog.* **2005**, *43*, 75–85. [CrossRef] [PubMed]

52. Hernandez, M.; Shao, Q.; Yang, X.J.; Luh, S.P.; Kandouz, M.; Batist, G.; Laird, D.W.; Alaoui-Jamali, M.A. A histone deacetylation-dependent mechanism for transcriptional repression of the gap junction gene *cx43* in prostate cancer cells. *Prostate* **2006**, *66*, 1151–1161. [CrossRef] [PubMed]

53. Villares, G.J.; Dobroff, A.S.; Wang, H.; Zigler, M.; Melnikova, V.O.; Huang, L.; Bar-Eli, M. Overexpression of protease-activated receptor-1 contributes to melanoma metastasis via regulation of connexin 43. *Cancer Res.* **2009**, *69*, 6730–6737. [CrossRef] [PubMed]

54. Geimonen, E.; Jiang, W.; Ali, M.; Fishman, G.I.; Garfield, R.E.; Andersen, J. Activation of protein kinase c in human uterine smooth muscle induces connexin-43 gene transcription through an AP-1 site in the promoter sequence. *J. Biol. Chem.* **1996**, *271*, 23667–23674. [CrossRef] [PubMed]

55. Ghouili, F.; Martin, L.J. Cooperative regulation of GJA1 expression by members of the AP-1 family cjun and cfos in TM3 leydig and TM4 sertoli cells. *Gene* **2017**, *635*, 24–32. [CrossRef] [PubMed]

56. Baldridge, D.; Lecanda, F.; Shin, C.S.; Stains, J.; Civitelli, R. Sequence and structure of the mouse connexin45 gene. *Biosci. Rep.* **2001**, *21*, 683–689. [CrossRef] [PubMed]

57. Van der Heyden, M.A.; Rook, M.B.; Hermans, M.M.; Rijksen, G.; Boonstra, J.; Defize, L.H.; Destree, O.H. Identification of connexin43 as a functional target for Wnt signalling. *J. Cell Sci.* **1998**, *111 Pt 12*, 1741–1749. [PubMed]

58. Ai, Z.; Fischer, A.; Spray, D.C.; Brown, A.M.; Fishman, G.I. Wnt-1 regulation of connexin43 in cardiac myocytes. *J. Clin. Investig.* **2000**, *105*, 161–171. [CrossRef] [PubMed]

59. Teunissen, B.E.; Bierhuizen, M.F. Transcriptional control of myocardial connexins. *Cardiovasc. Res.* **2004**, *62*, 246–255. [CrossRef] [PubMed]

60. Koffler, L.D.; Fernstrom, M.J.; Akiyama, T.E.; Gonzalez, F.J.; Ruch, R.J. Positive regulation of connexin32 transcription by hepatocyte nuclear factor-1α. *Arch. Biochem. Biophys.* **2002**, *407*, 160–167. [CrossRef]

61. Rukstalis, J.M.; Kowalik, A.; Zhu, L.; Lidington, D.; Pin, C.L.; Konieczny, S.F. Exocrine specific expression of connexin32 is dependent on the basic helix-loop-helix transcription factor mist1. *J. Cell Sci.* **2003**, *116*, 3315–3325. [CrossRef] [PubMed]

62. Bondurand, N.; Girard, M.; Pingault, V.; Lemort, N.; Dubourg, O.; Goossens, M. Human connexin 32, a gap junction protein altered in the X-linked form of charcot-marie-tooth disease, is directly regulated by the transcription factor SOX10. *Hum. Mol. Genet.* **2001**, *10*, 2783–2795. [CrossRef] [PubMed]

63. Petrocelli, T.; Lye, S.J. Regulation of transcripts encoding the myometrial gap junction protein, connexin-43, by estrogen and progesterone. *Endocrinology* **1993**, *133*, 284–290. [CrossRef] [PubMed]

64. Recouvreux, M.S.; Grasso, E.N.; Echeverria, P.C.; Rocha-Viegas, L.; Castilla, L.H.; Schere-Levy, C.; Tocci, J.M.; Kordon, E.C.; Rubinstein, N. RUNX1 and FOXP3 interplay regulates expression of breast cancer related genes. *Oncotarget* **2016**, *7*, 6552–6565. [CrossRef] [PubMed]

65. Vinken, M. Regulation of connexin signaling by the epigenetic machinery. *Biochim. Biophys. Acta* **2015**, *1859*, 262–268. [CrossRef] [PubMed]

66. Chen, Y.; Huhn, D.; Knosel, T.; Pacyna-Gengelbach, M.; Deutschmann, N.; Petersen, I. Downregulation of connexin 26 in human lung cancer is related to promoter methylation. *Int. J. Cancer* **2005**, *113*, 14–21. [CrossRef] [PubMed]

67. Tan, L.W.; Bianco, T.; Dobrovic, A. Variable promoter region cpg island methylation of the putative tumor suppressor gene connexin 26 in breast cancer. *Carcinogenesis* **2002**, *23*, 231–236. [CrossRef] [PubMed]

68. Hirai, A.; Yano, T.; Nishikawa, K.; Suzuki, K.; Asano, R.; Satoh, H.; Hagiwara, K.; Yamasaki, H. Down-regulation of connexin 32 gene expression through DNA methylation in a human renal cell carcinoma cell. *Am. J. Nephrol.* **2003**, *23*, 172–177. [CrossRef] [PubMed]

69. Chen, J.T.; Cheng, Y.W.; Chou, M.C.; Sen-Lin, T.; Lai, W.W.; Ho, W.L.; Lee, H. The correlation between aberrant connexin 43 mRNA expression induced by promoter methylation and nodal micrometastasis in non-small cell lung cancer. *Clin. Cancer Res.* **2003**, *9*, 4200–4204. [PubMed]

70. Wang, Y.; Huang, L.H.; Xu, C.X.; Xiao, J.; Zhou, L.; Cao, D.; Liu, X.M.; Qi, Y. Connexin 32 and 43 promoter methylation in helicobacter pylori-associated gastric tumorigenesis. *World J. Gastroenterol. WJG* **2014**, *20*, 11770–11779. [CrossRef] [PubMed]

71. Yi, Z.C.; Wang, H.; Zhang, G.Y.; Xia, B. Downregulation of connexin 43 in nasopharyngeal carcinoma cells is related to promoter methylation. *Oral Oncol.* **2007**, *43*, 898–904. [CrossRef] [PubMed]

72. Sirnes, S.; Honne, H.; Ahmed, D.; Danielsen, S.A.; Rognum, T.O.; Meling, G.I.; Leithe, E.; Rivedal, E.; Lothe, R.A.; Lind, G.E. DNA methylation analyses of the connexin gene family reveal silencing of *GJC1* (connexin45) by promoter hypermethylation in colorectal cancer. *Epigenetics* **2011**, *6*, 602–609. [CrossRef] [PubMed]

73. Ogawa, T.; Hayashi, T.; Tokunou, M.; Nakachi, K.; Trosko, J.E.; Chang, C.C.; Yorioka, N. Suberoylanilide hydroxamic acid enhances gap junctional intercellular communication via acetylation of histone containing connexin 43 gene locus. *Cancer Res.* **2005**, *65*, 9771–9778. [CrossRef] [PubMed]

74. Xu, Q.; Lin, X.; Andrews, L.; Patel, D.; Lampe, P.D.; Veenstra, R.D. Histone deacetylase inhibition reduces cardiac connexin43 expression and gap junction communication. *Front. Pharmacol.* **2013**, *4*, 44. [CrossRef] [PubMed]

75. Zhao, W.; Han, H.B.; Zhang, Z.Q. Suppression of lung cancer cell invasion and metastasis by connexin43 involves the secretion of follistatin-like 1 mediated via histone acetylation. *Int. J. Biochem. Cell Biol.* **2011**, *43*, 1459–1468. [CrossRef] [PubMed]

76. Hohl, M.; Thiel, G. Cell type-specific regulation of RE-1 silencing transcription factor (REST) target genes. *Eur. J. Neurosci.* **2005**, *22*, 2216–2230. [CrossRef] [PubMed]

77. Colussi, C.; Rosati, J.; Straino, S.; Spallotta, F.; Berni, R.; Stilli, D.; Rossi, S.; Musso, E.; Macchi, E.; Mai, A.; et al. Nepsilon-lysine acetylation determines dissociation from gap junctions and lateralization of connexin 43 in normal and dystrophic heart. *Proc. Natl. Acad. Sci. USA* **2011**, *108*, 2795–2800. [CrossRef] [PubMed]

78. Colussi, C.; Berni, R.; Rosati, J.; Straino, S.; Vitale, S.; Spallotta, F.; Baruffi, S.; Bocchi, L.; Delucchi, F.; Rossi, S.; et al. The histone deacetylase inhibitor suberoylanilide hydroxamic acid reduces cardiac arrhythmias in dystrophic mice. *Cardiovasc. Res.* **2010**, *87*, 73–82. [CrossRef] [PubMed]

79. Forster, T.; Rausch, V.; Zhang, Y.; Isayev, O.; Heilmann, K.; Schoensiegel, F.; Liu, L.; Nessling, M.; Richter, K.; Labsch, S.; et al. Sulforaphane counteracts aggressiveness of pancreatic cancer driven by dysregulated CX43-mediated gap junctional intercellular communication. *Oncotarget* **2014**, *5*, 1621–1634. [CrossRef] [PubMed]

80. Salat-Canela, C.; Munoz, M.J.; Sese, M.; Ramon y Cajal, S.; Aasen, T. Post-transcriptional regulation of connexins. *Biochem. Soc. Trans.* **2015**, *43*, 465–470. [CrossRef] [PubMed]

81. Ming, J.; Zhou, Y.; Du, J.; Fan, S.; Pan, B.; Wang, Y.; Fan, L.; Jiang, J. Identification of miR-200a as a novel suppressor of connexin 43 in breast cancer cells. *Biosci. Rep.* **2015**, *35*. [CrossRef] [PubMed]

82. Li, X.; Pan, J.H.; Song, B.; Xiong, E.Q.; Chen, Z.W.; Zhou, Z.S.; Su, Y.P. Suppression of CX43 expression by miR-20a in the progression of human prostate cancer. *Cancer Biol. Ther.* **2012**, *13*, 890–898. [CrossRef] [PubMed]

83. Hao, J.; Zhang, C.; Zhang, A.; Wang, K.; Jia, Z.; Wang, G.; Han, L.; Kang, C.; Pu, P. MiR-221/222 is the regulator of CX43 expression in human glioblastoma cells. *Oncol. Rep.* **2012**, *27*, 1504–1510. [CrossRef] [PubMed]

84. Anderson, C.; Catoe, H.; Werner, R. MiR-206 regulates connexin43 expression during skeletal muscle development. *Nucleic Acids Res.* **2006**, *34*, 5863–5871. [CrossRef] [PubMed]

85. Hak, K.K.; Yong, S.L.; Sivaprasad, U.; Malhotra, A.; Dutta, A. Muscle-specific microRNA miR-206 promotes muscle differentiation. *J. Cell Biol.* **2006**, *174*, 677–687. [CrossRef]

86. Yang, B.; Lin, H.; Xiao, J.; Lu, Y.; Luo, X.; Li, B.; Zhang, Y.; Xu, C.; Bai, Y.; Wang, H.; et al. The muscle-specific microRNA miR-1 regulates cardiac arrhythmogenic potential by targeting *GJA1* and *KCNJ2. Nat. Med.* **2007**, *13*, 486–491. [CrossRef] [PubMed]

87. Rau, F.; Freyermuth, F.; Fugier, C.; Villemin, J.P.; Fischer, M.C.; Jost, B.; Dembele, D.; Gourdon, G.; Nicole, A.; Duboc, D.; et al. Misregulation of miR-1 processing is associated with heart defects in myotonic dystrophy. *Nat. Struct. Mol. Biol.* **2011**, *18*, 840–845. [CrossRef] [PubMed]

88. Wu, Y.; Ma, X.J.; Wang, H.J.; Li, W.C.; Chen, L.; Ma, D.; Huang, G.Y. Expression of CX43-related microRNAs in patients with tetralogy of fallot. *World J. Pediatr. WJP* **2014**, *10*, 138–144. [CrossRef] [PubMed]

89. Imamura, M.; Sugino, Y.; Long, X.; Slivano, O.J.; Nishikawa, N.; Yoshimura, N.; Miano, J.M. Myocardin and microRNA-1 modulate bladder activity through connexin 43 expression during post-natal development. *J. Cell. Physiol.* **2013**, *228*, 1819–1826. [CrossRef] [PubMed]

90. Donahue, H.J.; Qu, R.W.; Genetos, D.C. Joint diseases: From connexins to gap junctions. *Nat. Rev. Rheumatol.* **2017**, *14*, 42–51. [CrossRef] [PubMed]

91. Gindin, Y.; Jiang, Y.; Francis, P.; Walker, R.L.; Abaan, O.D.; Zhu, Y.J.; Meltzer, P.S. MiR-23a impairs bone differentiation in osteosarcoma via down-regulation of *GJA1. Front. Genet.* **2015**, *6*, 233. [CrossRef] [PubMed]

92. Sun, Y.X.; Zhang, J.F.; Xu, J.; Xu, L.L.; Wu, T.Y.; Wang, B.; Pan, X.H.; Li, G. MicroRNA-144-3p inhibits bone formation in distraction osteogenesis through targeting connexin 43. *Oncotarget* **2017**, *8*, 89913–89922. [CrossRef] [PubMed]

93. Davis, H.M.; Pacheco-Costa, R.; Atkinson, E.G.; Brun, L.R.; Gortazar, A.R.; Harris, J.; Hiasa, M.; Bolarinwa, S.A.; Yoneda, T.; Ivan, M.; et al. Disruption of the CX43/miR21 pathway leads to osteocyte apoptosis and increased osteoclastogenesis with aging. *Aging Cell* **2017**, *16*, 551–563. [CrossRef] [PubMed]

94. Plotkin, L.I.; Pacheco-Costa, R.; Davis, H.M. MicroRNAs and connexins in bone: Interaction and mechanisms of delivery. *Curr. Mol. Biol. Rep.* **2017**, *3*, 63–70. [CrossRef] [PubMed]

95. Ale-Agha, N.; Galban, S.; Sobieroy, C.; Abdelmohsen, K.; Gorospe, M.; Sies, H.; Klotz, L.O. Hur regulates gap junctional intercellular communication by controlling β-catenin levels and adherens junction integrity. *Hepatology* **2009**, *50*, 1567–1576. [CrossRef] [PubMed]

96. Lee, K.W.; Chun, K.S.; Lee, J.S.; Kang, K.S.; Surh, Y.J.; Lee, H.J. Inhibition of cyclooxygenase-2 expression and restoration of gap junction intercellular communication in h-ras-transformed rat liver epithelial cells by caffeic acid phenethyl ester. *Ann. N. Y. Acad. Sci.* **2004**, *1030*, 501–507. [CrossRef] [PubMed]

97. Ul-Hussain, M.; Dermietzel, R.; Zoidl, G. Connexins and cap-independent translation: Role of internal ribosome entry sites. *Brain Res.* **2012**, *1487*, 99–106. [CrossRef] [PubMed]

98. Werner, R. IRES elements in connexin genes: A hypothesis explaining the need for connexins to be regulated at the translational level. *IUBMB Life* **2000**, *50*, 173–176. [CrossRef] [PubMed]

99. Schiavi, A.; Hudder, A.; Werner, R. Connexin43 mRNA contains a functional internal ribosome entry site. *FEBS Lett.* **1999**, *464*, 118–122. [CrossRef]

100. Hudder, A.; Werner, R. Analysis of a charcot-marie-tooth disease mutation reveals an essential internal ribosome entry site element in the connexin-32 gene. *J. Biol. Chem.* **2000**, *275*, 34586–34591. [CrossRef] [PubMed]

101. Lahlou, H.; Fanjul, M.; Pradayrol, L.; Susini, C.; Pyronnet, S. Restoration of functional gap junctions through internal ribosome entry site-dependent synthesis of endogenous connexins in density-inhibited cancer cells. *Mol. Cell. Biol.* **2005**, *25*, 4034–4045. [CrossRef] [PubMed]

102. Martinez-Salas, E.; Lozano, G.; Fernandez-Chamorro, J.; Francisco-Velilla, R.; Galan, A.; Diaz, R. RNA-binding proteins impacting on internal initiation of translation. *Int. J. Mol. Sci.* **2013**, *14*, 21705–21726. [CrossRef] [PubMed]

103. Faye, M.D.; Holcik, M. The role of IRES trans-acting factors in carcinogenesis. *Biochim. Biophys. Acta* **2015**, *1849*, 887–897. [CrossRef] [PubMed]

104. Komar, A.A.; Mazumder, B.; Merrick, W.C. A new framework for understanding IRES-mediated translation. *Gene* **2012**, *502*, 75–86. [CrossRef] [PubMed]

105. Thompson, S.R. So you want to know if your message has an IRES? *Wiley Interdiscip. Rev. RNA* **2012**, *3*, 697–705. [CrossRef] [PubMed]

106. Lauring, A.S.; Overbaugh, J. Evidence that an IRES within the Notch2 coding region can direct expression of a nuclear form of the protein. *Mol. Cell* **2000**, *6*, 939–945. [CrossRef]

107. Ul-Hussain, M.; Zoidl, G.; Klooster, J.; Kamermans, M.; Dermietzel, R. IRES-mediated translation of the carboxy-terminal domain of the horizontal cell specific connexin CX55.5 in vivo and in vitro. *BMC Mol. Biol.* **2008**, *9*, 52. [CrossRef] [PubMed]

108. Ul-Hussain, M.; Dermietzel, R.; Zoidl, G. Characterization of the internal IRES element of the zebrafish connexin55.5 reveals functional implication of the polypyrimidine tract binding protein. *BMC Mol. Biol.* **2008**, *9*, 92. [CrossRef] [PubMed]

109. Joshi-Mukherjee, R.; Coombs, W.; Burrer, C.; de Mora, I.A.; Delmar, M.; Taffet, S.M. Evidence for the presence of a free C-terminal fragment of cx43 in cultured cells. *Cell Commun. Adhes.* **2007**, *14*, 75–84. [CrossRef] [PubMed]

110. Smyth, J.W.; Shaw, R.M. Autoregulation of connexin43 gap junction formation by internally translated isoforms. *Cell Rep.* **2013**, *5*, 611–618. [CrossRef] [PubMed]

111. Salat-Canela, C.; Sese, M.; Peula, C.; Ramon y Cajal, S.; Aasen, T. Internal translation of the connexin 43 transcript. *Cell Commun. Signal.* **2014**, *12*, 31. [CrossRef] [PubMed]

112. Ul-Hussain, M.; Olk, S.; Schoenebeck, B.; Wasielewski, B.; Meier, C.; Prochnow, N.; May, C.; Galozzi, S.; Marcus, K.; Zoidl, G.; et al. Internal ribosomal entry site (IRES) activity generates endogenous carboxyl-terminal domains of CX43 and is responsive to hypoxic conditions. *J. Biol. Chem.* **2014**, *289*, 20979–20990. [CrossRef] [PubMed]

113. Basheer, W.A.; Xiao, S.; Epifantseva, I.; Fu, Y.; Kleber, A.G.; Hong, T.; Shaw, R.M. GJA1-20k arranges actin to guide CX43 delivery to cardiac intercalated discs. *Circ. Res.* **2017**, *121*, 1069–1080. [CrossRef] [PubMed]

114. Fu, Y.; Zhang, S.S.; Xiao, S.; Basheer, W.A.; Baum, R.; Epifantseva, I.; Hong, T.; Shaw, R.M. Cx43 isoform GJA1-20k promotes microtubule dependent mitochondrial transport. *Front. Physiol.* **2017**, *8*, 905. [CrossRef] [PubMed]

115. Maqbool, R.; Rashid, R.; Ismail, R.; Niaz, S.; Chowdri, N.A.; Hussain, M.U. The carboxy-terminal domain of connexin 43 (CT-CX43) modulates the expression of p53 by altering miR-125b expression in low-grade human breast cancers. *Cell. Oncol.* **2015**, *38*, 443–451. [CrossRef] [PubMed]

116. Foote, C.I.; Zhou, L.; Zhu, X.; Nicholson, B.J. The pattern of disulfide linkages in the extracellular loop regions of connexin 32 suggests a model for the docking interface of gap junctions. *J. Cell Biol.* **1998**, *140*, 1187–1197. [CrossRef] [PubMed]

117. Johnstone, S.R.; Billaud, M.; Lohman, A.W.; Taddeo, E.P.; Isakson, B.E. Posttranslational modifications in connexins and pannexins. *J. Membr. Biol.* **2012**, *245*, 319–332. [CrossRef] [PubMed]

118. Solan, J.L.; Lampe, P.D. Specific Cx43 phosphorylation events regulate gap junction turnover in vivo. *FEBS Lett.* **2014**, *588*, 1423–1429. [CrossRef] [PubMed]

119. Pogoda, K.; Kameritsch, P.; Retamal, M.A.; Vega, J.L. Regulation of gap junction channels and hemichannels by phosphorylation and redox changes: A revision. *BMC Cell Biol.* **2016**, *17* (Suppl. 1), 11. [CrossRef] [PubMed]

120. Lampe, P.D.; Lau, A.F. The effects of connexin phosphorylation on gap junctional communication. *Int. J. Biochem. Cell Biol.* **2004**, *36*, 1171–1186. [CrossRef]

121. Solan, J.L.; Lampe, P.D. Connexin phosphorylation as a regulatory event linked to gap junction channel assembly. *Biochim. Biophys. Acta* **2005**, *1711*, 154–163. [CrossRef] [PubMed]

122. Saez, J.C.; Martinez, A.D.; Branes, M.C.; Gonzalez, H.E. Regulation of gap junctions by protein phosphorylation. *Braz. J. Med. Biol. Res.* **1998**, *31*, 593–600. [CrossRef] [PubMed]

123. Stultz, C.M.; Levin, A.D.; Edelman, E.R. Phosphorylation-induced conformational changes in a mitogen-activated protein kinase substrate. Implications for tyrosine hydroxylase activation. *J. Biol. Chem.* **2002**, *277*, 47653–47661. [CrossRef] [PubMed]

124. Diestel, S.; Eckert, R.; Hülser, D.; Traub, O. Exchange of serine residues 263 and 266 reduces the function of mouse gap junction protein connexin31 and exhibits a dominant-negative effect on the wild-type protein in hela cells. *Exp. Cell Res.* **2004**, *294*, 446–457. [CrossRef] [PubMed]

125. Qin, J.; Chang, M.; Wang, S.; Liu, Z.; Zhu, W.; Wang, Y.; Yan, F.; Li, J.; Zhang, B.; Dou, G.; et al. Connexin 32-mediated cell-cell communication is essential for hepatic differentiation from human embryonic stem cells. *Sci. Rep.* **2016**, *6*, 37388. [CrossRef] [PubMed]

126. Ghosh, P.; Ghosh, S.; Das, S. Self-regulation of rat liver gap junction by phosphorylation. *Biochim. Biophys. Acta* **2002**, *1564*, 500–504. [CrossRef]

127. Ghosh, P. Self-phosphorylation modulates the gating of rat liver gap junction channels: A nonstationary noise analysis. *Biophys. Chem.* **2007**, *127*, 97–102. [CrossRef] [PubMed]

128. Jacobsen, N.L.; Pontifex, T.K.; Li, H.; Solan, J.L.; Lampe, P.D.; Sorgen, P.L.; Burt, J.M. Regulation of cx37 channel and growth-suppressive properties by phosphorylation. *J. Cell Sci.* **2017**, *130*, 3308–3321. [CrossRef] [PubMed]

129. Bao, M.; Kanter, E.M.; Huang, R.Y.; Maxeiner, S.; Frank, M.; Zhang, Y.; Schuessler, R.B.; Smith, T.W.; Townsend, R.R.; Rohrs, H.W.; et al. Residual Cx45 and its relationship to Cx43 in murine ventricular myocardium. *Channels (Austin)* **2011**, *5*, 489–499. [CrossRef] [PubMed]

130. Johnstone, S.R.; Ross, J.; Rizzo, M.J.; Straub, A.C.; Lampe, P.D.; Leitinger, N.; Isakson, B.E. Oxidized phospholipid species promote in vivo differential Cx43 phosphorylation and vascular smooth muscle cell proliferation. *Am. J. Pathol.* **2009**, *175*, 916–924. [CrossRef] [PubMed]

131. Johnstone, S.R.; Kroncke, B.M.; Straub, A.C.; Best, A.K.; Dunn, C.A.; Mitchell, L.A.; Peskova, Y.; Nakamoto, R.K.; Koval, M.; Lo, C.W.; et al. Mapk phosphorylation of connexin 43 promotes binding of cyclin e and smooth muscle cell proliferation. *Circ. Res.* **2012**, *111*, 201–211. [CrossRef] [PubMed]

132. Pelletier, R.M.; Akpovi, C.D.; Chen, L.; Kumar, N.M.; Vitale, M.L. Complementary expression and phosphorylation of Cx46 and Cx50 during development and following gene deletion in mouse and in normal and orchitic mink testes. *Am. J. Physiol. Regul. Integr. Comp. Physiol.* **2015**, *309*, R255–R276. [CrossRef] [PubMed]

133. Walter, W.J.; Zeilinger, C.; Bintig, W.; Kolb, H.A.; Ngezahayo, A. Phosphorylation in the C-terminus of the rat connexin46 (rCx46) and regulation of the conducting activity of the formed connexons. *J. Bioenergy Biomembr.* **2008**, *40*, 397–405. [CrossRef] [PubMed]

134. Liu, J.; Ek Vitorin, J.F.; Weintraub, S.T.; Gu, S.; Shi, Q.; Burt, J.M.; Jiang, J.X. Phosphorylation of connexin 50 by protein kinase a enhances gap junction and hemichannel function. *J. Biol. Chem.* **2011**, *286*, 16914–16928. [CrossRef] [PubMed]

135. May, D.; Tress, O.; Seifert, G.; Willecke, K. Connexin47 protein phosphorylation and stability in oligodendrocytes depend on expression of connexin43 protein in astrocytes. *J. Neurosci.* **2013**, *33*, 7985–7996. [CrossRef] [PubMed]

136. Traub, O.; Look, J.; Dermietzel, R.; Brummer, F.; Hulser, D.; Willecke, K. Comparative characterization of the 21-kD and 26-kD gap junction proteins in murine liver and cultured hepatocytes. *J. Cell Biol.* **1989**, *108*, 1039–1051. [CrossRef] [PubMed]

137. Elvira, M.; Díez, J.A.; Wang, K.K.; Villalobo, A. Phosphorylation of connexin-32 by protein kinase C prevents its proteolysis by mu-calpain and m-calpain. *J. Biol. Chem.* **1993**, *268*, 14294–14300. [PubMed]

138. Locke, D.; Koreen, I.V.; Harris, A.L. Isoelectric points and post-translational modifications of connexin26 and connexin32. *FASEB J.* **2006**, *20*, 1221–1223. [CrossRef] [PubMed]

139. Locke, D.; Bian, S.; Li, H.; Harris, A.L. Post-translational modifications of connexin26 revealed by mass spectrometry. *Biochem. J.* **2009**, *424*, 385–398. [CrossRef] [PubMed]

140. Berthoud, V.M.; Beyer, E.C.; Kurata, W.E.; Lau, A.F.; Lampe, P.D. The gap-junction protein connexin 56 is phosphorylated in the intracellular loop and the carboxy-terminal region. *Eur. J. Biochem.* **1997**, *244*, 89–97. [CrossRef] [PubMed]

141. Ouyang, X.; Winbow, V.M.; Patel, L.S.; Burr, G.S.; Mitchell, C.K.; O'Brien, J. Protein kinase a mediates regulation of gap junctions containing connexin35 through a complex pathway. *Brain Res. Mol. Brain Res.* **2005**, *135*, 1–11. [CrossRef] [PubMed]

142. Johnstone, S.; Isakson, B.; Locke, D. Biological and biophysical properties of vascular connexin channels. *Int. Rev. Cell Mol. Biol.* **2009**, *278*, 69–118. [CrossRef] [PubMed]

143. Solan, J.L.; Lampe, P.D. Connexin43 phosphorylation: Structural changes and biological effects. *Biochem. J.* **2009**, *419*, 261–272. [CrossRef] [PubMed]

144. Chen, V.C.; Gouw, J.W.; Naus, C.C.; Foster, L.J. Connexin multi-site phosphorylation: Mass spectrometry-based proteomics fills the gap. *Biochim. Biophys. Acta* **2013**, *1828*, 23–34. [CrossRef] [PubMed]

145. Sorgen, P.L.; Duffy, H.S.; Spray, D.C.; Delmar, M. Ph-dependent dimerization of the carboxyl terminal domain of cx43. *Biophys. J.* **2004**, *87*, 574–581. [CrossRef] [PubMed]

146. Sorgen, P.L.; Duffy, H.S.; Sahoo, P.; Coombs, W.; Delmar, M.; Spray, D.C. Structural changes in the carboxyl terminus of the gap junction protein connexin43 indicates signaling between binding domains for c-Src and zonula occludens-1. *J. Biol. Chem.* **2004**, *279*, 54695–54701. [CrossRef] [PubMed]

147. Bouvier, D.; Kieken, F.; Kellezi, A.; Sorgen, P.L. Structural changes in the carboxyl terminus of the gap junction protein connexin 40 caused by the interaction with c-Src and zonula occludens-1. *Cell Commun. Adhes.* **2008**, *15*, 107–118. [CrossRef] [PubMed]

148. Solan, J.L.; Lampe, P.D. Spatio-temporal regulation of connexin43 phosphorylation and gap junction dynamics. *Biochim. Biophys. Acta* **2018**, *1860*, 83–90. [CrossRef] [PubMed]

149. Grosely, R.; Kopanic, J.L.; Nabors, S.; Kieken, F.; Spagnol, G.; Al-Mugotir, M.; Zach, S.; Sorgen, P.L. Effects of phosphorylation on the structure and backbone dynamics of the intrinsically disordered connexin43 C-terminal domain. *J. Biol. Chem.* **2013**, *288*, 24857–24870. [CrossRef] [PubMed]

150. Sorgen, P.L.; Duffy, H.S.; Cahill, S.M.; Coombs, W.; Spray, D.C.; Delmar, M.; Girvin, M.E. Sequence-specific resonance assignment of the carboxyl terminal domain of connexin43. *J. Biomol. NMR* **2002**, *23*, 245–246. [CrossRef] [PubMed]

151. Shi, Q.; Banks, E.A.; Yu, X.S.; Gu, S.; Lauer, J.; Fields, G.B.; Jiang, J.X. Amino acid residue val362 plays a critical role in maintaining the structure of C terminus of connexin 50 and in lens epithelial-fiber differentiation. *J. Biol. Chem.* **2010**, *285*, 18415–18422. [CrossRef] [PubMed]

152. Kopanic, J.L.; Sorgen, P.L. Chemical shift assignments of the connexin45 carboxyl terminal domain: Monomer and dimer conformations. *Biomol. NMR Assign.* **2013**, *7*, 293–297. [CrossRef] [PubMed]

153. Kopanic, J.L.; Al-mugotir, M.H.; Kieken, F.; Zach, S.; Trease, A.J.; Sorgen, P.L. Characterization of the connexin45 carboxyl-terminal domain structure and interactions with molecular partners. *Biophys. J.* **2014**, *106*, 2184–2195. [CrossRef] [PubMed]

154. Kyle, J.W.; Berthoud, V.M.; Kurutz, J.; Minogue, P.J.; Greenspan, M.; Hanck, D.A.; Beyer, E.C. The n terminus of connexin37 contains an α-helix that is required for channel function. *J. Biol. Chem.* **2009**, *284*, 20418–20427. [CrossRef] [PubMed]

155. Bouvier, D.; Spagnol, G.; Chenavas, S.; Kieken, F.; Vitrac, H.; Brownell, S.; Kellezi, A.; Forge, V.; Sorgen, P.L. Characterization of the structure and intermolecular interactions between the connexin40 and connexin43 carboxyl-terminal and cytoplasmic loop domains. *J. Biol. Chem.* **2009**, *284*, 34257–34271. [CrossRef] [PubMed]

156. Grosely, R.; Kieken, F.; Sorgen, P.L. ^{1}h, ^{13}c, and ^{15}n backbone resonance assignments of the connexin43 carboxyl terminal domain attached to the 4th transmembrane domain in detergent micelles. *Biomol. NMR Assign.* **2013**, *7*, 299–303. [CrossRef] [PubMed]

157. Sosinsky, G.E.; Solan, J.L.; Gaietta, G.M.; Ngan, L.; Lee, G.J.; Mackey, M.R.; Lampe, P.D. The C-terminus of connexin43 adopts different conformations in the golgi and gap junction as detected with structure-specific antibodies. *Biochem. J.* **2007**, *408*, 375–385. [CrossRef] [PubMed]

158. Grosely, R.; Kieken, F.; Sorgen, P.L. Optimizing the solution conditions to solve the structure of the connexin43 carboxyl terminus attached to the 4(th) transmembrane domain in detergent micelles. *Cell Commun. Adhes.* **2010**, *17*, 23–33. [CrossRef] [PubMed]

159. Hirst-Jensen, B.J.; Sahoo, P.; Kieken, F.; Delmar, M.; Sorgen, P.L. Characterization of the PH-dependent interaction between the gap junction protein connexin43 carboxyl terminus and cytoplasmic loop domains. *J. Biol. Chem.* **2007**, *282*, 5801–5813. [CrossRef] [PubMed]

160. Duffy, H.S.; Sorgen, P.L.; Girvin, M.E.; O'Donnell, P.; Coombs, W.; Taffet, S.M.; Delmar, M.; Spray, D.C. Ph-dependent intramolecular binding and structure involving Cx43 cytoplasmic domains. *J. Biol. Chem.* **2002**, *277*, 36706–36714. [CrossRef] [PubMed]

161. Cohen, P. The regulation of protein function by multisite phosphorylation—A 25 year update. *Trends Biochem. Sci.* **2000**, *25*, 596–601. [CrossRef]

162. Axelsen, L.N.; Calloe, K.; Holstein-Rathlou, N.H.; Nielsen, M.S. Managing the complexity of communication: Regulation of gap junctions by post-translational modification. *Front. Pharmacol.* **2013**, *4*, 130. [CrossRef] [PubMed]

163. Ek-Vitorin, J.F.; Burt, J.M. Structural basis for the selective permeability of channels made of communicating junction proteins. *Biochim. Biophys. Acta* **2013**, *1828*, 51–68. [CrossRef] [PubMed]

164. Moreno, A.P. Connexin phosphorylation as a regulatory event linked to channel gating. *Biochim. Biophys. Acta* **2005**, *1711*, 164–171. [CrossRef] [PubMed]

165. Saez, J.C.; Nairn, A.C.; Czernik, A.J.; Spray, D.C.; Hertzberg, E.L.; Greengard, P.; Bennett, M.V. Phosphorylation of connexin 32, a hepatocyte gap-junction protein, by cAMP-dependent protein kinase, protein kinase C and Ca^{2+}/calmodulin-dependent protein kinase II. *Eur. J. Biochem.* **1990**, *192*, 263–273. [CrossRef] [PubMed]

166. Takeda, A.; Saheki, S.; Shimazu, T.; Takeuchi, N. Phosphorylation of the 27-kDa gap junction protein by protein kinase c in vitro and in rat hepatocytes. *J. Biochem.* **1989**, *106*, 723–727. [CrossRef] [PubMed]

167. Diez, J.A.; Elvira, M.; Villalobo, A. The epidermal growth factor receptor tyrosine kinase phosphorylates connexin32. *Mol. Cell. Biochem.* **1998**, *187*, 201–210. [CrossRef] [PubMed]

168. Kothmann, W.W.; Li, X.; Burr, G.S.; O'Brien, J. Connexin 35/36 is phosphorylated at regulatory sites in the retina. *Vis. Neurosci.* **2007**, *24*, 363–375. [CrossRef] [PubMed]

169. Li, H.; Chuang, A.Z.; O'Brien, J. Photoreceptor coupling is controlled by connexin 35 phosphorylation in zebrafish retina. *J. Neurosci.* **2009**, *29*, 15178–15186. [CrossRef] [PubMed]

170. Patel, L.S.; Mitchell, C.K.; Dubinsky, W.P.; O'Brien, J. Regulation of gap junction coupling through the neuronal connexin Cx35 by nitric oxide and cGMP. *Cell Commun. Adhes.* **2006**, *13*, 41–54. [CrossRef] [PubMed]

171. Li, H.; Chuang, A.Z.; O'Brien, J. Regulation of photoreceptor gap junction phosphorylation by adenosine in zebrafish retina. *Vis. Neurosci.* **2014**, *31*, 237–243. [CrossRef] [PubMed]

172. Kjenseth, A.; Fykerud, T.A.; Sirnes, S.; Bruun, J.; Yohannes, Z.; Kolberg, M.; Omori, Y.; Rivedal, E.; Leithe, E. The gap junction channel protein connexin 43 is covalently modified and regulated by sumoylation. *J. Biol. Chem.* **2012**, *287*, 15851–15861. [CrossRef] [PubMed]

173. Huang, R.Y.; Laing, J.G.; Kanter, E.M.; Berthoud, V.M.; Bao, M.; Rohrs, H.W.; Townsend, R.R.; Yamada, K.A. Identification of CaMKII phosphorylation sites in connexin43 by high-resolution mass spectrometry. *J. Proteome Res.* **2011**, *10*, 1098–1109. [CrossRef] [PubMed]

174. Cottrell, G.T.; Lin, R.; Warn-Cramer, B.J.; Lau, A.F.; Burt, J.M. Mechanism of v-Src- and mitogen-activated protein kinase-induced reduction of gap junction communication. *Am. J. Physiol. Cell Physiol.* **2003**, *284*, C511–C520. [CrossRef] [PubMed]

175. Lin, R.; Warn-Cramer, B.J.; Kurata, W.E.; Lau, A.F. v-Src phosphorylation of connexin 43 on Tyr247 and Tyr265 disrupts gap junctional communication. *J. Cell Biol.* **2001**, *154*, 815–827. [CrossRef] [PubMed]

176. Giepmans, B.N.; Hengeveld, T.; Postma, F.R.; Moolenaar, W.H. Interaction of c-Src with gap junction protein connexin-43. Role in the regulation of cell-cell communication. *J. Biol. Chem.* **2001**, *276*, 8544–8549. [CrossRef] [PubMed]

177. Zhou, L.; Kasperek, E.M.; Nicholson, B.J. Dissection of the molecular basis of pp60(v-Src) induced gating of connexin 43 gap junction channels. *J. Cell Biol.* **1999**, *144*, 1033–1045. [CrossRef] [PubMed]

178. Swenson, K.I.; Piwnica-Worms, H.; McNamee, H.; Paul, D.L. Tyrosine phosphorylation of the gap junction protein connexin43 is required for the pp60v-Src-induced inhibition of communication. *Cell Regul.* **1990**, *1*, 989–1002. [CrossRef] [PubMed]

179. Lampe, P.D.; Kurata, W.E.; Warn-Cramer, B.J.; Lau, A.F. Formation of a distinct connexin43 phosphoisoform in mitotic cells is dependent upon p34cdc2 kinase. *J. Cell Sci.* **1998**, *111 Pt 6*, 833–841. [PubMed]

180. Kanemitsu, M.Y.; Jiang, W.; Eckhart, W. Cdc2-mediated phosphorylation of the gap junction protein, connexin43, during mitosis. *Cell Growth Differ.* **1998**, *9*, 13–21. [PubMed]

181. Sirnes, S.; Kjenseth, A.; Leithe, E.; Rivedal, E. Interplay between pkc and the map kinase pathway in connexin43 phosphorylation and inhibition of gap junction intercellular communication. *Biochem. Biophys. Res. Commun.* **2009**, *382*, 41–45. [CrossRef] [PubMed]

182. Doble, B.W.; Dang, X.; Ping, P.; Fandrich, R.R.; Nickel, B.E.; Jin, Y.; Cattini, P.A.; Kardami, E. Phosphorylation of serine 262 in the gap junction protein connexin-43 regulates DNA synthesis in cell-cell contact forming cardiomyocytes. *J. Cell Sci.* **2004**, *117*, 507–514. [CrossRef] [PubMed]

183. Srisakuldee, W.; Jeyaraman, M.M.; Nickel, B.E.; Tanguy, S.; Jiang, Z.S.; Kardami, E. Phosphorylation of connexin-43 at serine 262 promotes a cardiac injury-resistant state. *Cardiovasc. Res.* **2009**, *83*, 672–681. [CrossRef] [PubMed]

184. Straub, A.C.; Billaud, M.; Johnstone, S.R.; Best, A.K.; Yemen, S.; Dwyer, S.T.; Looft-Wilson, R.; Lysiak, J.J.; Gaston, B.; Palmer, L.; et al. Compartmentalized connexin 43 S-nitrosylation/denitrosylation regulates heterocellular communication in the vessel wall. *Arterioscler. Thromb. Vasc. Biol.* **2011**, *31*, 399–407. [CrossRef] [PubMed]

185. Qi, G.J.; Chen, Q.; Chen, L.J.; Shu, Y.; Bu, L.L.; Shao, X.Y.; Zhang, P.; Jiao, F.J.; Shi, J.; Tian, B. Phosphorylation of connexin 43 by cdk5 modulates neuronal migration during embryonic brain development. *Mol. Neurobiol.* **2016**, *53*, 2969–2982. [CrossRef] [PubMed]

186. Procida, K.; Jorgensen, L.; Schmitt, N.; Delmar, M.; Taffet, S.M.; Holstein-Rathlou, N.H.; Nielsen, M.S.; Braunstein, T.H. Phosphorylation of connexin43 on serine 306 regulates electrical coupling. *Heart Rhythm* **2009**, *6*, 1632–1638. [CrossRef] [PubMed]

187. Hund, T.J.; Decker, K.F.; Kanter, E.; Mohler, P.J.; Boyden, P.A.; Schuessler, R.B.; Yamada, K.A.; Rudy, Y. Role of activated CaMKII in abnormal calcium homeostasis and i(na) remodeling after myocardial infarction: Insights from mathematical modeling. *J. Mol. Cell. Cardiol.* **2008**, *45*, 420–428. [CrossRef] [PubMed]

188. Cooper, C.D.; Lampe, P.D. Casein kinase 1 regulates connexin-43 gap junction assembly. *J. Biol. Chem.* **2002**, *277*, 44962–44968. [CrossRef] [PubMed]

189. Paulson, A.F.; Lampe, P.D.; Meyer, R.A.; TenBroek, E.; Atkinson, M.M.; Walseth, T.F.; Johnson, R.G. Cyclic AMP and LDL trigger a rapid enhancement in gap junction assembly through a stimulation of connexin trafficking. *J. Cell Sci.* **2000**, *113 Pt 17*, 3037–3049. [PubMed]

190. Darrow, B.J.; Fast, V.G.; Kleber, A.G.; Beyer, E.C.; Saffitz, J.E. Functional and structural assessment of intercellular communication. Increased conduction velocity and enhanced connexin expression in dibutyryl camp-treated cultured cardiac myocytes. *Circ. Res.* **1996**, *79*, 174–183. [CrossRef] [PubMed]

191. TenBroek, E.M.; Lampe, P.D.; Solan, J.L.; Reynhout, J.K.; Johnson, R.G. Ser364 of connexin43 and the upregulation of gap junction assembly by camp. *J. Cell Biol.* **2001**, *155*, 1307–1318. [CrossRef] [PubMed]

192. Axelsen, L.N.; Stahlhut, M.; Mohammed, S.; Larsen, B.D.; Nielsen, M.S.; Holstein-Rathlou, N.H.; Andersen, S.; Jensen, O.N.; Hennan, J.K.; Kjolbye, A.L. Identification of ischemia-regulated phosphorylation sites in connexin43: A possible target for the antiarrhythmic peptide analogue rotigaptide (ZP123). *J. Mol. Cell. Cardiol.* **2006**, *40*, 790–798. [CrossRef] [PubMed]

193. Shah, M.M.; Martinez, A.M.; Fletcher, W.H. The connexin43 gap junction protein is phosphorylated by protein kinase a and protein kinase c: In vivo and in vitro studies. *Mol. Cell. Biochem.* **2002**, *238*, 57–68. [CrossRef] [PubMed]

194. Yogo, K.; Ogawa, T.; Akiyama, M.; Ishida, N.; Takeya, T. Identification and functional analysis of novel phosphorylation sites in Cx43 in rat primary granulosa cells. *FEBS Lett.* **2002**, *531*, 132–136. [CrossRef]

195. Zou, J.; Yue, X.Y.; Zheng, S.C.; Zhang, G.; Chang, H.; Liao, Y.C.; Zhang, Y.; Xue, M.Q.; Qi, Z. Cholesterol modulates function of connexin 43 gap junction channel via PKC pathway in H9c2 cells. *Biochim. Biophys. Acta* **2014**, *1838*, 2019–2025. [CrossRef] [PubMed]

196. Ek-Vitorin, J.F.; King, T.J.; Heyman, N.S.; Lampe, P.D.; Burt, J.M. Selectivity of connexin 43 channels is regulated through protein kinase C-dependent phosphorylation. *Circ. Res.* **2006**, *98*, 1498–1505. [CrossRef] [PubMed]

197. Xie, Y.; Liu, S.; Hu, S.; Wei, Y. Cardiomyopathy-associated gene 1-sensitive PKC-dependent connexin 43 expression and phosphorylation in left ventricular noncompaction cardiomyopathy. *Cell. Physiol. Biochem.* **2017**, *44*, 828–842. [CrossRef] [PubMed]

198. Bao, X.; Altenberg, G.A.; Reuss, L. Mechanism of regulation of the gap junction protein connexin 43 by protein kinase C-mediated phosphorylation. *Am. J. Physiol. Cell Physiol.* **2004**, *286*, C647–C654. [CrossRef] [PubMed]

199. Liao, C.K.; Cheng, H.H.; Wang, S.D.; Yeih, D.F.; Wang, S.M. Pkcvarepsilon mediates serine phosphorylation of connexin43 induced by lysophosphatidylcholine in neonatal rat cardiomyocytes. *Toxicology* **2013**, *314*, 11–21. [CrossRef] [PubMed]

200. Dunn, C.A.; Lampe, P.D. Injury-triggered akt phosphorylation of Cx43: A ZO-1-driven molecular switch that regulates gap junction size. *J. Cell Sci.* **2014**, *127*, 455–464. [CrossRef] [PubMed]

201. Park, D.J.; Wallick, C.J.; Martyn, K.D.; Lau, A.F.; Jin, C.; Warn-Cramer, B.J. Akt phosphorylates connexin43 on ser373, a "mode-1" binding site for 14-3-3. *Cell Commun. Adhes.* **2007**, *14*, 211–226. [CrossRef] [PubMed]

202. Berthoud, V.M.; Westphale, E.M.; Grigoryeva, A.; Beyer, E.C. Pkc isoenzymes in the chicken lens and TPA-induced effects on intercellular communication. *Investig. Ophthalmol. Vis. Sci.* **2000**, *41*, 850–858.

203. Isakson, B.E. Localized expression of an Ins(1,4,5)P3 receptor at the myoendothelial junction selectively regulates heterocellular Ca2+ communication. *J. Cell Sci.* **2008**, *121*, 3664–3673. [CrossRef] [PubMed]

204. Isakson, B.E.; Ramos, S.I.; Duling, B.R. Ca2+ and inositol 1,4,5-trisphosphate-mediated signaling across the myoendothelial junction. *Circ. Res.* **2007**, *100*, 246–254. [CrossRef] [PubMed]

205. Straub, A.C.; Johnstone, S.R.; Heberlein, K.R.; Rizzo, M.J.; Best, A.K.; Boitano, S.; Isakson, B.E. Site-specific connexin phosphorylation is associated with reduced heterocellular communication between smooth muscle and endothelium. *J. Vasc. Res.* **2010**, *47*, 277–286. [CrossRef] [PubMed]

206. Revel, J.P.; Karnovsky, M.J. Hexagonal array of subunits in intercellular junctions of the mouse heart and liver. *J. Cell Biol.* **1967**, *33*, C7–C12. [CrossRef] [PubMed]

207. Solan, J.L.; Marquez-Rosado, L.; Sorgen, P.L.; Thornton, P.J.; Gafken, P.R.; Lampe, P.D. Phosphorylation at s365 is a gatekeeper event that changes the structure of Cx43 and prevents down-regulation by PKC. *J. Cell Biol.* **2007**, *179*, 1301–1309. [CrossRef] [PubMed]

208. Lampe, P.D.; Cooper, C.D.; King, T.J.; Burt, J.M. Analysis of connexin43 phosphorylated at s325, s328 and s330 in normoxic and ischemic heart. *J. Cell Sci.* **2006**, *119*, 3435–3442. [CrossRef] [PubMed]

209. Jabr, R.I.; Hatch, F.S.; Salvage, S.C.; Orlowski, A.; Lampe, P.D.; Fry, C.H. Regulation of gap junction conductance by calcineurin through Cx43 phosphorylation: Implications for action potential conduction. *Pflugers Arch.* **2016**, *468*, 1945–1955. [CrossRef] [PubMed]

210. Schulz, R.; Gres, P.; Skyschally, A.; Duschin, A.; Belosjorow, S.; Konietzka, I.; Heusch, G. Ischemic preconditioning preserves connexin 43 phosphorylation during sustained ischemia in pig hearts in vivo. *FASEB J.* **2003**, *17*, 1355–1357. [CrossRef] [PubMed]

211. Rhett, J.M.; Gourdie, R.G. The perinexus: A new feature of Cx43 gap junction organization. *Heart Rhythm* **2012**, *9*, 619–623. [CrossRef] [PubMed]

212. Palatinus, J.A.; O'Quinn, M.P.; Barker, R.J.; Harris, B.S.; Jourdan, J.; Gourdie, R.G. ZO-1 determines adherens and gap junction localization at intercalated disks. *Am. J. Physiol. Heart Circ. Physiol.* **2011**, *300*, H583–H594. [CrossRef] [PubMed]

213. Zhu, C.; Barker, R.J.; Hunter, A.W.; Zhang, Y.; Jourdan, J.; Gourdie, R.G. Quantitative analysis of ZO-1 colocalization with Cx43 gap junction plaques in cultures of rat neonatal cardiomyocytes. *Microsc. Microanal.* **2005**, *11*, 244–248. [CrossRef] [PubMed]

214. Hunter, A.W.; Barker, R.J.; Zhu, C.; Gourdie, R.G. Zonula occludens-1 alters connexin43 gap junction size and organization by influencing channel accretion. *Mol. Biol. Cell* **2005**, *16*, 5686–5698. [CrossRef] [PubMed]

215. O'Quinn, M.P.; Palatinus, J.A.; Harris, B.S.; Hewett, K.W.; Gourdie, R.G. A peptide mimetic of the connexin43 carboxyl terminus reduces gap junction remodeling and induced arrhythmia following ventricular injury. *Circ. Res.* **2011**, *108*, 704–715. [CrossRef] [PubMed]

216. Schulz, R.; Görge, P.M.; Görbe, A.; Ferdinandy, P.; Lampe, P.D.; Leybaert, L. Connexin 43 is an emerging therapeutic target in ischemia/reperfusion injury, cardioprotection and neuroprotection. *Pharmacol. Ther.* **2015**, *153*, 90–106. [CrossRef] [PubMed]

217. Soder, B.L.; Propst, J.T.; Brooks, T.M.; Goodwin, R.L.; Friedman, H.I.; Yost, M.J.; Gourdie, R.G. The connexin43 carboxyl-terminal peptide act1 modulates the biological response to silicone implants. *Plast. Reconstr. Surg.* **2009**, *123*, 1440–1451. [CrossRef] [PubMed]

218. Su, G.Y.; Wang, J.; Xu, Z.X.; Qiao, X.J.; Zhong, J.Q.; Zhang, Y. Effects of rotigaptide (ZP123) on connexin43 remodeling in canine ventricular fibrillation. *Mol. Med. Rep.* **2015**, *12*, 5746–5752. [CrossRef] [PubMed]

219. Stahlhut, M.; Petersen, J.S.; Hennan, J.K.; Ramirez, M.T. The antiarrhythmic peptide rotigaptide (ZP123) increases connexin 43 protein expression in neonatal rat ventricular cardiomyocytes. *Cell Commun. Adhes.* **2006**, *13*, 21–27. [CrossRef] [PubMed]

220. Dhein, S.; Larsen, B.D.; Petersen, J.S.; Mohr, F.W. Effects of the new antiarrhythmic peptide ZP123 on epicardial activation and repolarization pattern. *Cell Commun. Adhes.* **2003**, *10*, 371–378. [CrossRef] [PubMed]

221. Xing, D.; Kjolbye, A.L.; Nielsen, M.S.; Petersen, J.S.; Harlow, K.W.; Holstein-Rathlou, N.H.; Martins, J.B. ZP123 increases gap junctional conductance and prevents reentrant ventricular tachycardia during myocardial ischemia in open chest dogs. *J. Cardiovasc. Electrophysiol.* **2003**, *14*, 510–520. [CrossRef] [PubMed]

222. Kjølbye, A.L.; Haugan, K.; Hennan, J.K.; Petersen, J.S. Pharmacological modulation of gap junction function with the novel compound rotigaptide: A promising new principle for prevention of arrhythmias. *Basic Clin. Pharmacol. Toxicol.* **2007**, *101*, 215–230. [CrossRef] [PubMed]

223. Skyschally, A.; Walter, B.; Schultz Hansen, R.; Heusch, G. The antiarrhythmic dipeptide ZP1609 (danegaptide) when given at reperfusion reduces myocardial infarct size in pigs. *Naunyn-Schmiedeberg's Arch. Pharmacol.* **2013**, *386*, 383–391. [CrossRef] [PubMed]

224. Cherepanova, O.A.; Pidkovka, N.A.; Sarmento, O.F.; Yoshida, T.; Gan, Q.; Adiguzel, E.; Bendeck, M.P.; Berliner, J.; Leitinger, N.; Owens, G.K. Oxidized phospholipids induce type viii collagen expression and vascular smooth muscle cell migration. *Circ. Res.* **2009**, *104*, 609–618. [CrossRef] [PubMed]

225. Kadl, A.; Meher, A.K.; Sharma, P.R.; Lee, M.Y.; Doran, A.C.; Johnstone, S.R.; Elliott, M.R.; Gruber, F.; Han, J.; Chen, W.S.; et al. Identification of a novel macrophage phenotype that develops in response to atherogenic phospholipids via nrf2. *Circ. Res.* **2010**, *107*, 737–746. [CrossRef] [PubMed]

226. Leitinger, N. Oxidized phospholipids as triggers of inflammation in atherosclerosis. *Mol. Nutr. Food Res.* **2005**, *49*, 1063–1071. [CrossRef] [PubMed]

227. Chatterjee, S.; Berliner, J.A.; Subbanagounder, G.G.; Bhunia, A.K.; Koh, S. Identification of a biologically active component in minimally oxidized low density lipoprotein (MM-LDL) responsible for aortic smooth muscle cell proliferation. *Glycoconj. J.* **2004**, *20*, 331–338. [CrossRef] [PubMed]

228. Good, M.E.; Nelson, T.K.; Simon, A.M.; Burt, J.M. A functional channel is necessary for growth suppression by Cx37. *J. Cell Sci.* **2011**, *124*, 2448–2456. [CrossRef] [PubMed]

229. Good, M.E.; Ek-Vitorín, J.F.; Burt, J.M. Extracellular loop cysteine mutant of Cx37 fails to suppress proliferation of rat insulinoma cells. *J. Membr. Biol.* **2012**, *245*, 369–380. [CrossRef] [PubMed]

230. Good, M.E.; Ek-Vitorin, J.F.; Burt, J.M. Structural determinants and proliferative consequences of connexin 37 hemichannel function in insulinoma cells. *J. Biol. Chem.* **2014**, *289*, 30379–30386. [CrossRef] [PubMed]

231. Traub, O.; Hertlein, B.; Kasper, M.; Eckert, R.; Krisciukaitis, A.; Hülser, D.; Willecke, K. Characterization of the gap junction protein connexin37 in murine endothelium, respiratory epithelium, and after transfection in human hela cells. *Eur. J. Cell Biol.* **1998**, *77*, 313–322. [CrossRef]

232. Morel, S.; Burnier, L.; Roatti, A.; Chassot, A.; Roth, I.; Sutter, E.; Galan, K.; Pfenniger, A.; Chanson, M.; Kwak, B.R. Unexpected role for the human Cx37 C1019T polymorphism in tumour cell proliferation. *Carcinogenesis* **2010**, *31*, 1922–1931. [CrossRef] [PubMed]

233. Larson, D.M.; Seul, K.H.; Berthoud, V.M.; Lau, A.F.; Sagar, G.D.; Beyer, E.C. Functional expression and biochemical characterization of an epitope-tagged connexin37. *Mol. Cell. Biol. Res. Commun.* **2000**, *3*, 115–121. [CrossRef] [PubMed]

234. Vanhamme, L.; Rolin, S.; Szpirer, C. Inhibition of gap-junctional intercellular communication between epithelial cells transformed by the activated h-ras-1 oncogene. *Exp. Cell Res.* **1989**, *180*, 297–301. [CrossRef]

235. Azarnia, R.; Reddy, S.; Kmiecik, T.E.; Shalloway, D.; Loewenstein, W.R. The cellular *SRC* gene product regulates junctional cell-to-cell communication. *Science* **1988**, *239*, 398–401. [CrossRef] [PubMed]

236. Gonzalez-Sanchez, A.; Jaraiz-Rodriguez, M.; Dominguez-Prieto, M.; Herrero-Gonzalez, S.; Medina, J.M.; Tabernero, A. Connexin43 recruits PTEN and Csk to inhibit c-Src activity in glioma cells and astrocytes. *Oncotarget* **2016**, *7*, 49819–49833. [CrossRef] [PubMed]

237. Johnson, K.E.; Mitra, S.; Katoch, P.; Kelsey, L.S.; Johnson, K.R.; Mehta, P.P. Phosphorylation on ser-279 and ser-282 of connexin43 regulates endocytosis and gap junction assembly in pancreatic cancer cells. *Mol. Biol. Cell* **2013**, *24*, 715–733. [CrossRef] [PubMed]

238. Brissette, J.L.; Kumar, N.M.; Gilula, N.B.; Dotto, G.P. The tumor promoter 12-O-tetradecanoylphorbol-13-acetate and the ras oncogene modulate expression and phosphorylation of gap junction proteins. *Mol. Cell. Biol.* **1991**, *11*, 5364–5371. [CrossRef] [PubMed]

239. Oh, S.Y.; Grupen, C.G.; Murray, A.W. Phorbol ester induces phosphorylation and down-regulation of connexin 43 in wb cells. *Biochim. Biophys. Acta* **1991**, *1094*, 243–245. [CrossRef]

240. Asamoto, M.; Oyamada, M.; el Aoumari, A.; Gros, D.; Yamasaki, H. Molecular mechanisms of TPA-mediated inhibition of gap-junctional intercellular communication: Evidence for action on the assembly or function but not the expression of connexin 43 in rat liver epithelial cells. *Mol. Carcinog.* **1991**, *4*, 322–327. [CrossRef] [PubMed]

241. Ruch, R.J.; Trosko, J.E.; Madhukar, B.V. Inhibition of connexin43 gap junctional intercellular communication by tpa requires erk activation. *J. Cell. Biochem.* **2001**, *83*, 163–169. [CrossRef] [PubMed]

242. Ye, X.Y.; Jiang, Q.H.; Hong, T.; Zhang, Z.Y.; Yang, R.J.; Huang, J.Q.; Hu, K.; Peng, Y.P. Altered expression of connexin43 and phosphorylation connexin43 in glioma tumors. *Int. J. Clin. Exp. Pathol.* **2015**, *8*, 4296–4306. [PubMed]

243. Wu, J.F.; Ji, J.; Dong, S.Y.; Li, B.B.; Yu, M.L.; Wu, D.D.; Tao, L.; Tong, X.H. Gefitinib enhances oxaliplatin-induced apoptosis mediated by Src and PKC-modulated gap junction function. *Oncol. Rep.* **2016**, *36*, 3251–3258. [CrossRef] [PubMed]

244. Peterson-Roth, E.; Brdlik, C.M.; Glazer, P.M. Src-induced cisplatin resistance mediated by cell-to-cell communication. *Cancer Res.* **2009**, *69*, 3619–3624. [CrossRef] [PubMed]

245. Wong, P.; Tan, T.; Chan, C.; Laxton, V.; Chan, Y.W.; Liu, T.; Wong, W.T.; Tse, G. The role of connexins in wound healing and repair: Novel therapeutic approaches. *Front. Physiol.* **2016**, *7*, 596. [CrossRef] [PubMed]

246. Becker, D.L.; Thrasivoulou, C.; Phillips, A.R. Connexins in wound healing; perspectives in diabetic patients. *Biochim. Biophys. Acta* **2012**, *1818*, 2068–2075. [CrossRef] [PubMed]

247. Cogliati, B.; Vinken, M.; Silva, T.C.; Araújo, C.M.M.; Aloia, T.P.A.; Chaible, L.M.; Mori, C.M.C.; Dagli, M.L.Z. Connexin 43 deficiency accelerates skin wound healing and extracellular matrix remodeling in mice. *J. Dermatol. Sci.* **2015**, *79*, 50–56. [CrossRef] [PubMed]

248. Kim, M.O.; Ryu, J.M.; Suh, H.N.; Park, S.H.; Oh, Y.M.; Lee, S.H.; Han, H.J. Camp promotes cell migration through cell junctional complex dynamics and actin cytoskeleton remodeling: Implications in skin wound healing. *Stem Cells Dev.* **2015**, *24*, 2513–2524. [CrossRef] [PubMed]

249. Mehta, P.P.; Yamamoto, M.; Rose, B. Transcription of the gene for the gap junctional protein connexin43 and expression of functional cell-to-cell channels are regulated by camp. *Mol. Biol. Cell* **1992**, *3*, 839–850. [CrossRef] [PubMed]

250. Richards, T.S.; Dunn, C.A.; Carter, W.G.; Usui, M.L.; Olerud, J.E.; Lampe, P.D. Protein kinase C spatially and temporally regulates gap junctional communication during human wound repair via phosphorylation of connexin43 on serine368. *J. Cell Biol.* **2004**, *167*, 555–562. [CrossRef] [PubMed]

251. Pollok, S.; Pfeiffer, A.C.; Lobmann, R.; Wright, C.S.; Moll, I.; Martin, P.E.; Brandner, J.M. Connexin 43 mimetic peptide gap27 reveals potential differences in the role of Cx43 in wound repair between diabetic and non-diabetic cells. *J. Cell. Mol. Med.* **2011**, *15*, 861–873. [CrossRef] [PubMed]

252. Solan, J.L.; Lampe, P.D. Kinase programs spatiotemporally regulate gap junction assembly and disassembly: Effects on wound repair. *Semin. Cell Dev. Biol.* **2016**, *50*, 40–48. [CrossRef] [PubMed]

253. Wang, C.M.; Lincoln, J.; Cook, J.E.; Becker, D.L. Abnormal connexin expression underlies delayed wound healing in diabetic skin. *Diabetes* **2007**, *56*, 2809–2817. [CrossRef] [PubMed]

254. Brandner, J.M.; Houdek, P.; Hüsing, B.; Kaiser, C.; Moll, I. Connexins 26, 30, and 43: Differences among spontaneous, chronic, and accelerated human wound healing. *J. Investig. Dermatol.* **2004**, *122*, 1310–1320. [CrossRef] [PubMed]

255. Grek, C.L.; Montgomery, J.; Sharma, M.; Ravi, A.; Rajkumar, J.S.; Moyer, K.E.; Gourdie, R.G.; Ghatnekar, G.S. A multicenter randomized controlled trial evaluating a Cx43-mimetic peptide in cutaneous scarring. *J. Investig. Dermatol.* **2017**, *137*, 620–630. [CrossRef] [PubMed]

256. Stamler, J.S.; Simon, D.I.; Osborne, J.A.; Mullins, M.E.; Jaraki, O.; Michel, T.; Singel, D.J.; Loscalzo, J. S-nitrosylation of proteins with nitric oxide: Synthesis and characterization of biologically active compounds. *Proc. Natl. Acad. Sci. USA* **1992**, *89*, 444–448. [CrossRef] [PubMed]

257. Retamal, M.A.; García, I.E.; Pinto, B.I.; Pupo, A.; Báez, D.; Stehberg, J.; Del Rio, R.; González, C. Extracellular cysteine in connexins: Role as redox sensors. *Front. Physiol.* **2016**, *7*, 1. [CrossRef] [PubMed]

258. Contreras, J.E.; Sánchez, H.A.; Eugenin, E.A.; Speidel, D.; Theis, M.; Willecke, K.; Bukauskas, F.F.; Bennett, M.V.; Sáez, J.C. Metabolic inhibition induces opening of unapposed connexin 43 gap junction hemichannels and reduces gap junctional communication in cortical astrocytes in culture. *Proc. Natl. Acad. Sci. USA* **2002**, *99*, 495–500. [CrossRef] [PubMed]

259. Figueroa, X.F.; Lillo, M.A.; Gaete, P.S.; Riquelme, M.A.; Sáez, J.C. Diffusion of nitric oxide across cell membranes of the vascular wall requires specific connexin-based channels. *Neuropharmacology* **2013**, *75*, 471–478. [CrossRef] [PubMed]

260. Pogoda, K.; Füller, M.; Pohl, U.; Kameritsch, P. No, via its target Cx37, modulates calcium signal propagation selectively at myoendothelial gap junctions. *Cell Commun. Signal. CCS* **2014**, *12*, 33. [CrossRef] [PubMed]

261. Pogoda, K.; Mannell, H.; Blodow, S.; Schneider, H.; Schubert, K.M.; Qiu, J.; Schmidt, A.; Imhof, A.; Beck, H.; Tanase, L.I.; et al. No augments endothelial reactivity by reducing myoendothelial calcium signal spreading: A novel role for Cx37 (connexin 37) and the protein tyrosine phosphatase SHP-2. *Arterioscler. Thromb. Vasc. Biol.* **2017**, *37*, 2280–2290. [CrossRef] [PubMed]

262. Gareau, J.R.; Lima, C.D. The sumo pathway: Emerging mechanisms that shape specificity, conjugation and recognition. *Nat. Rev. Mol. Cell Biol.* **2010**, *11*, 861–871. [CrossRef] [PubMed]

263. Laird, D.W. Life cycle of connexins in health and disease. *Biochem. J.* **2006**, *394*, 527–543. [CrossRef] [PubMed]

264. Jiang, J.X.; Paul, D.L.; Goodenough, D.A. Posttranslational phosphorylation of lens fiber connexin46: A slow occurrence. *Investig. Ophthalmol. Vis. Sci.* **1993**, *34*, 3558–3565.

265. Beardslee, M.A.; Laing, J.G.; Beyer, E.C.; Saffitz, J.E. Rapid turnover of connexin43 in the adult rat heart. *Circ. Res.* **1998**, *83*, 629–635. [CrossRef] [PubMed]

266. Fallon, R.F.; Goodenough, D.A. Five-hour half-life of mouse liver gap-junction protein. *J. Cell Biol.* **1981**, *90*, 521–526. [CrossRef] [PubMed]

267. Norris, R.P.; Baena, V.; Terasaki, M. Localization of phosphorylated connexin 43 using serial section immunogold electron microscopy. *J. Cell Sci.* **2017**, *130*, 1333–1340. [CrossRef] [PubMed]

268. Boassa, D.; Solan, J.L.; Papas, A.; Thornton, P.; Lampe, P.D.; Sosinsky, G.E. Trafficking and recycling of the connexin43 gap junction protein during mitosis. *Traffic* **2010**, *11*, 1471–1486. [CrossRef] [PubMed]

269. Piehl, M.; Lehmann, C.; Gumpert, A.; Denizot, J.P.; Segretain, D.; Falk, M.M. Internalization of large double-membrane intercellular vesicles by a clathrin-dependent endocytic process. *Mol. Biol. Cell* **2007**, *18*, 337–347. [CrossRef] [PubMed]

270. Hunter, A.W.; Gourdie, R.G. The second pdz domain of zonula occludens-1 is dispensable for targeting to connexin 43 gap junctions. *Cell Commun. Adhes.* **2008**, *15*, 55–63. [CrossRef] [PubMed]

271. Chen, C.H.; Mayo, J.N.; Gourdie, R.G.; Johnstone, S.R.; Isakson, B.E.; Bearden, S.E. The connexin 43/ZO-1 complex regulates cerebral endothelial f-actin architecture and migration. *Am. J. Physiol. Cell Physiol.* **2015**, *309*, C600–C607. [CrossRef] [PubMed]

272. Leithe, E.; Sirnes, S.; Fykerud, T.; Kjenseth, A.; Rivedal, E. Endocytosis and post-endocytic sorting of connexins. *Biochim. Biophys. Acta* **2012**, *1818*, 1870–1879. [CrossRef] [PubMed]

273. Falk, M.M.; Fong, J.T.; Kells, R.M.; O'Laughlin, M.C.; Kowal, T.J.; Thevenin, A.F. Degradation of endocytosed gap junctions by autophagosomal and endo-/lysosomal pathways: A perspective. *J. Membr. Biol.* **2012**, *245*, 465–476. [CrossRef] [PubMed]

274. Leithe, E.; Rivedal, E. Ubiquitination of gap junction proteins. *J. Membr. Biol.* **2007**, *217*, 43–51. [CrossRef] [PubMed]

275. Willis, M.S.; Townley-Tilson, W.H.; Kang, E.Y.; Homeister, J.W.; Patterson, C. Sent to destroy: The ubiquitin proteasome system regulates cell signaling and protein quality control in cardiovascular development and disease. *Circ. Res.* **2010**, *106*, 463–478. [CrossRef] [PubMed]

276. Voges, D.; Zwickl, P.; Baumeister, W. The 26s proteasome: A molecular machine designed for controlled proteolysis. *Annu. Rev. Biochem.* **1999**, *68*, 1015–1068. [CrossRef] [PubMed]

277. Bejarano, E.; Girao, H.; Yuste, A.; Patel, B.; Marques, C.; Spray, D.C.; Pereira, P.; Cuervo, A.M. Autophagy modulates dynamics of connexins at the plasma membrane in a ubiquitin-dependent manner. *Mol. Biol. Cell* **2012**, *23*, 2156–2169. [CrossRef] [PubMed]

278. Fong, J.T.; Kells, R.M.; Gumpert, A.M.; Marzillier, J.Y.; Davidson, M.W.; Falk, M.M. Internalized gap junctions are degraded by autophagy. *Autophagy* **2012**, *8*, 794–811. [CrossRef] [PubMed]

279. Lichtenstein, A.; Minogue, P.J.; Beyer, E.C.; Berthoud, V.M. Autophagy: A pathway that contributes to connexin degradation. *J. Cell Sci.* **2011**, *124*, 910–920. [CrossRef] [PubMed]

280. Bejarano, E.; Yuste, A.; Patel, B.; Stout, R.F., Jr.; Spray, D.C.; Cuervo, A.M. Connexins modulate autophagosome biogenesis. *Nat. Cell Biol.* **2014**, *16*, 401–414. [CrossRef] [PubMed]

281. Hesketh, G.G.; Shah, M.H.; Halperin, V.L.; Cooke, C.A.; Akar, F.G.; Yen, T.E.; Kass, D.A.; Machamer, C.E.; Van Eyk, J.E.; Tomaselli, G.F. Ultrastructure and regulation of lateralized connexin43 in the failing heart. *Circ. Res.* **2010**, *106*, 1153–1163. [CrossRef] [PubMed]

282. Leithe, E.; Rivedal, E. Ubiquitination and down-regulation of gap junction protein connexin-43 in response to 12-O-tetradecanoylphorbol 13-acetate treatment. *J. Biol. Chem.* **2004**, *279*, 50089–50096. [CrossRef] [PubMed]

283. Zhu, X.; Ruan, Z.; Yang, X.; Chu, K.; Wu, H.; Li, Y.; Huang, Y. Connexin 31.1 degradation requires the clathrin-mediated autophagy in NSCLC cell h1299. *J. Cell. Mol. Med.* **2015**, *19*, 257–264. [CrossRef] [PubMed]

284. Catarino, S.; Ramalho, J.S.; Marques, C.; Pereira, P.; Girao, H. Ubiquitin-mediated internalization of connexin43 is independent of the canonical endocytic tyrosine-sorting signal. *Biochem. J.* **2011**, *437*, 255–267. [CrossRef] [PubMed]

285. Girao, H.; Pereira, P. The proteasome regulates the interaction between Cx43 and ZO-1. *J. Cell. Biochem.* **2007**, *102*, 719–728. [CrossRef] [PubMed]

286. Dunn, C.A.; Su, V.; Lau, A.F.; Lampe, P.D. Activation of AKT, not connexin 43 protein ubiquitination, regulates gap junction stability. *J. Biol. Chem.* **2012**, *287*, 2600–2607. [CrossRef] [PubMed]

287. Totland, M.Z.; Bergsland, C.H.; Fykerud, T.A.; Knudsen, L.M.; Rasmussen, N.L.; Eide, P.W.; Yohannes, Z.; Sorensen, V.; Brech, A.; Lothe, R.A.; et al. The E3 ubiquitin ligase nedd4 induces endocytosis and lysosomal sorting of connexin 43 to promote loss of gap junctions. *J. Cell Sci.* **2017**, *130*, 2867–2882. [CrossRef] [PubMed]

288. Spagnol, G.; Kieken, F.; Kopanic, J.L.; Li, H.; Zach, S.; Stauch, K.L.; Grosely, R.; Sorgen, P.L. Structural studies of the NEDD4 WW domains and their selectivity for the connexin43 (Cx43) carboxyl terminus. *J. Biol. Chem.* **2016**, *291*, 7637–7650. [CrossRef] [PubMed]

289. Girao, H.; Catarino, S.; Pereira, P. Eps15 interacts with ubiquitinated Cx43 and mediates its internalization. *Exp. Cell Res.* **2009**, *315*, 3587–3597. [CrossRef] [PubMed]

290. Auth, T.; Schluter, S.; Urschel, S.; Kussmann, P.; Sonntag, S.; Hoher, T.; Kreuzberg, M.M.; Dobrowolski, R.; Willecke, K. The TSG101 protein binds to connexins and is involved in connexin degradation. *Exp. Cell Res.* **2009**, *315*, 1053–1062. [CrossRef] [PubMed]

291. Chen, V.C.; Kristensen, A.R.; Foster, L.J.; Naus, C.C. Association of connexin43 with E3 ubiquitin ligase TRIM21 reveals a mechanism for gap junction phosphodegron control. *J. Proteome Res.* **2012**, *11*, 6134–6146. [CrossRef] [PubMed]

292. Basheer, W.A.; Harris, B.S.; Mentrup, H.L.; Abreha, M.; Thames, E.L.; Lea, J.B.; Swing, D.A.; Copeland, N.G.; Jenkins, N.A.; Price, R.L.; et al. Cardiomyocyte-specific overexpression of the ubiquitin ligase WWP1 contributes to reduction in connexin 43 and arrhythmogenesis. *J. Mol. Cell. Cardiol.* **2015**, *88*, 1–13. [CrossRef] [PubMed]

293. Fykerud, T.A.; Kjenseth, A.; Schink, K.O.; Sirnes, S.; Bruun, J.; Omori, Y.; Brech, A.; Rivedal, E.; Leithe, E. Smad ubiquitination regulatory factor-2 controls gap junction intercellular communication by modulating endocytosis and degradation of connexin43. *J. Cell Sci.* **2012**, *125*, 3966–3976. [CrossRef] [PubMed]

294. Fang, W.L.; Lai, S.Y.; Lai, W.A.; Lee, M.T.; Liao, C.F.; Ke, F.C.; Hwang, J.J. CRTC2 and NEDD4 ligase involvement in FSH and TGFβ1 upregulation of connexin43 gap junction. *J. Mol. Endocrinol.* **2015**, *55*, 263–275. [CrossRef] [PubMed]

295. Colussi, C.; Gurtner, A.; Rosati, J.; Illi, B.; Ragone, G.; Piaggio, G.; Moggio, M.; Lamperti, C.; D'Angelo, G.; Clementi, E.; et al. Nitric oxide deficiency determines global chromatin changes in duchenne muscular dystrophy. *FASEB J.* **2009**, *23*, 2131–2141. [CrossRef] [PubMed]

296. Meraviglia, V.; Azzimato, V.; Colussi, C.; Florio, M.C.; Binda, A.; Panariti, A.; Qanud, K.; Suffredini, S.; Gennaccaro, L.; Miragoli, M.; et al. Acetylation mediates Cx43 reduction caused by electrical stimulation. *J. Mol. Cell. Cardiol.* **2015**, *87*, 54–64. [CrossRef] [PubMed]

297. Carette, D.; Gilleron, J.; Denizot, J.P.; Grant, K.; Pointis, G.; Segretain, D. New cellular mechanisms of gap junction degradation and recycling. *Biol. Cell Auspices Eur. Cell Biol. Org.* **2015**, *107*, 218–231. [CrossRef] [PubMed]

298. Koval, M. Pathways and control of connexin oligomerization. *Trends Cell Biol.* **2006**, *16*, 159–166. [CrossRef] [PubMed]

299. Koval, M.; Molina, S.A.; Burt, J.M. Mix and match: Investigating heteromeric and heterotypic gap junction channels in model systems and native tissues. *FEBS Lett.* **2014**, *588*, 1193–1204. [CrossRef] [PubMed]

300. Das, S.; Smith, T.D.; Sarma, J.D.; Ritzenthaler, J.D.; Maza, J.; Kaplan, B.E.; Cunningham, L.A.; Suaud, L.; Hubbard, M.J.; Rubenstein, R.C.; et al. ERP29 restricts connexin43 oligomerization in the endoplasmic reticulum. *Mol. Biol. Cell* **2009**, *20*, 2593–2604. [CrossRef] [PubMed]

301. Maza, J.; Das Sarma, J.; Koval, M. Defining a minimal motif required to prevent connexin oligomerization in the endoplasmic reticulum. *J. Biol. Chem.* **2005**, *280*, 21115–21121. [CrossRef] [PubMed]

302. Jara, O.; Acuna, R.; Garcia, I.E.; Maripillan, J.; Figueroa, V.; Saez, J.C.; Araya-Secchi, R.; Lagos, C.F.; Perez-Acle, T.; Berthoud, V.M.; et al. Critical role of the first transmembrane domain of Cx26 in regulating oligomerization and function. *Mol. Biol. Cell* **2012**, *23*, 3299–3311. [CrossRef] [PubMed]

303. Musil, L.S.; Goodenough, D.A. Multisubunit assembly of an integral plasma membrane channel protein, gap junction connexin43, occurs after exit from the ER. *Cell* **1993**, *74*, 1065–1077. [CrossRef]

304. Koval, M.; Harley, J.E.; Hick, E.; Steinberg, T.H. Connexin46 is retained as monomers in a trans-golgi compartment of osteoblastic cells. *J. Cell Biol.* **1997**, *137*, 847–857. [CrossRef] [PubMed]

305. Smith, T.D.; Mohankumar, A.; Minogue, P.J.; Beyer, E.C.; Berthoud, V.M.; Koval, M. Cytoplasmic amino acids within the membrane interface region influence connexin oligomerization. *J. Membr. Biol.* **2012**, *245*, 221–230. [CrossRef] [PubMed]

306. Lagree, V.; Brunschwig, K.; Lopez, P.; Gilula, N.B.; Richard, G.; Falk, M.M. Specific amino-acid residues in the N-terminus and TM3 implicated in channel function and oligomerization compatibility of connexin43. *J. Cell Sci.* **2003**, *116*, 3189–3201. [CrossRef] [PubMed]

307. Molina, S.A.; Stauffer, B.; Moriarty, H.K.; Kim, A.H.; McCarty, N.A.; Koval, M. Junctional abnormalities in human airway epithelial cells expressing F508DEL CFTR. *Am. J. Physiol. Lung Cell. Mol. Physiol.* **2015**, *309*, L475–L487. [CrossRef] [PubMed]

308. Suaud, L.; Miller, K.; Alvey, L.; Yan, W.; Robay, A.; Kebler, C.; Kreindler, J.L.; Guttentag, S.; Hubbard, M.J.; Rubenstein, R.C. Erp29 regulates deltaf508 and wild-type cystic fibrosis transmembrane conductance regulator (CFTR) trafficking to the plasma membrane in cystic fibrosis (CF) and non-CF epithelial cells. *J. Biol. Chem.* **2011**, *286*, 21239–21253. [CrossRef] [PubMed]

309. Das Sarma, J.; Kaplan, B.E.; Willemsen, D.; Koval, M. Identification of Rab20 as a potential regulator of connexin43 trafficking. *Cell Commun. Adhes.* **2008**, *15*, 65–74. [CrossRef] [PubMed]

310. Asklund, T.; Appelskog, I.B.; Ammerpohl, O.; Ekstrom, T.J.; Almqvist, P.M. Histone deacetylase inhibitor 4-phenylbutyrate modulates glial fibrillary acidic protein and connexin 43 expression, and enhances gap-junction communication, in human glioblastoma cells. *Eur. J Cancer* **2004**, *40*, 1073–1081. [CrossRef] [PubMed]

311. Hattori, Y.; Fukushima, M.; Maitani, Y. Non-viral delivery of the connexin 43 gene with histone deacetylase inhibitor to human nasopharyngeal tumor cells enhances gene expression and inhibits in vivo tumor growth. *Int. J. Oncol.* **2007**, *30*, 1427–1439. [CrossRef] [PubMed]

312. Khan, Z.; Akhtar, M.; Asklund, T.; Juliusson, B.; Almqvist, P.M.; Ekstrom, T.J. Hdac inhibition amplifies gap junction communication in neural progenitors: Potential for cell-mediated enzyme prodrug therapy. *Exp. Cell Res.* **2007**, *313*, 2958–2967. [CrossRef] [PubMed]

313. Berthoud, V.M.; Minogue, P.J.; Guo, J.; Williamson, E.K.; Xu, X.; Ebihara, L.; Beyer, E.C. Loss of function and impaired degradation of a cataract-associated mutant connexin50. *Eur. J. Cell Biol.* **2003**, *82*, 209–221. [CrossRef] [PubMed]

314. Kaufman, J.; Gordon, C.; Bergamaschi, R.; Wang, H.Z.; Cohen, I.S.; Valiunas, V.; Brink, P.R. The effects of the histone deacetylase inhibitor 4-phenylbutyrate on gap junction conductance and permeability. *Front. Pharmacol.* **2013**, *4*, 111. [CrossRef] [PubMed]

315. Berthoud, V.M.; Minogue, P.J.; Lambert, P.A.; Snabb, J.I.; Beyer, E.C. The cataract-linked mutant connexin50D47A causes endoplasmic reticulum stress in mouse lenses. *J. Biol. Chem.* **2016**, *291*, 17569–17578. [CrossRef] [PubMed]

316. Lichtenstein, A.; Gaietta, G.M.; Deerinck, T.J.; Crum, J.; Sosinsky, G.E.; Beyer, E.C.; Berthoud, V.M. The cytoplasmic accumulations of the cataract-associated mutant, connexin50P88S, are long-lived and form in the endoplasmic reticulum. *Exp. Eye Res.* **2009**, *88*, 600–609. [CrossRef] [PubMed]

317. Alapure, B.V.; Stull, J.K.; Firtina, Z.; Duncan, M.K. The unfolded protein response is activated in connexin 50 mutant mouse lenses. *Exp. Eye Res.* **2012**, *102*, 28–37. [CrossRef] [PubMed]

318. Tattersall, D.; Scott, C.A.; Gray, C.; Zicha, D.; Kelsell, D.P. Ekv mutant connexin 31 associated cell death is mediated by er stress. *Hum. Mol. Genet.* **2009**, *18*, 4734–4745. [CrossRef] [PubMed]

319. Xia, K.; Ma, H.; Xiong, H.; Pan, Q.; Huang, L.; Wang, D.; Zhang, Z. Trafficking abnormality and ER stress underlie functional deficiency of hearing impairment-associated connexin-31 mutants. *Protein Cell* **2010**, *1*, 935–943. [CrossRef] [PubMed]

320. Chen, S.; Zhang, D. Friend or foe: Endoplasmic reticulum protein 29 (ERP29) in epithelial cancer. *FEBS Open Biol.* **2015**, *5*, 91–98. [CrossRef] [PubMed]

321. Li, H.; Spagnol, G.; Pontifex, T.K.; Burt, J.M.; Sorgen, P.L. Chemical shift assignments of the connexin37 carboxyl terminal domain. *Biomol. NMR Assign.* **2017**, *11*, 137–141. [CrossRef] [PubMed]

322. Schlingmann, B.; Schadzek, P.; Hemmerling, F.; Schaarschmidt, F.; Heisterkamp, A.; Ngezahayo, A. The role of the C-terminus in functional expression and internalization of rat connexin46 (rcx46). *J. Bioenergy Biomembr.* **2013**, *45*, 59–70. [CrossRef] [PubMed]

323. Laing, J.G.; Koval, M.; Steinberg, T.H. Association with ZO-1 correlates with plasma membrane partitioning in truncated connexin45 mutants. *J. Membr. Biol.* **2005**, *207*, 45–53. [CrossRef] [PubMed]

324. Trease, A.J.; Capuccino, J.M.V.; Contreras, J.; Harris, A.L.; Sorgen, P.L. Intramolecular signaling in a cardiac connexin: Role of cytoplasmic domain dimerization. *J. Mol. Cell. Cardiol.* **2017**, *111*, 69–80. [CrossRef] [PubMed]

325. Calero, G.; Kanemitsu, M.; Taffet, S.M.; Lau, A.F.; Delmar, M. A 17MER peptide interferes with acidification-induced uncoupling of connexin43. *Circ. Res.* **1998**, *82*, 929–935. [CrossRef] [PubMed]

326. Stergiopoulos, K.; Alvarado, J.L.; Mastroianni, M.; Ek-Vitorin, J.F.; Taffet, S.M.; Delmar, M. Hetero-domain interactions as a mechanism for the regulation of connexin channels. *Circ. Res.* **1999**, *84*, 1144–1155. [CrossRef] [PubMed]

327. Banerjee, D.; Das, S.; Molina, S.A.; Madgwick, D.; Katz, M.R.; Jena, S.; Bossmann, L.K.; Pal, D.; Takemoto, D.J. Investigation of the reciprocal relationship between the expression of two gap junction connexin proteins, connexin46 and connexin43. *J. Biol. Chem.* **2011**, *286*, 24519–24533. [CrossRef] [PubMed]

328. Basheer, W.; Shaw, R. The "tail" of connexin43: An unexpected journey from alternative translation to trafficking. *Biochim. Biophys. Acta* **2016**, *1863*, 1848–1856. [CrossRef] [PubMed]

329. Wellen, K.E.; Thompson, C.B. Cellular metabolic stress: Considering how cells respond to nutrient excess. *Mol. Cell* **2010**, *40*, 323–332. [CrossRef] [PubMed]

330. James, C.C.; Zeitz, M.J.; Calhoun, P.J.; Lamouille, S.; Smyth, J.W. Altered translation initiation of GJA1 limits gap junction formation during epithelial-mesenchymal transition. *Mol. Biol. Cell* **2018**, *29*, 797–808. [CrossRef]

331. Smyth, J.W.; Vogan, J.M.; Buch, P.J.; Zhang, S.S.; Fong, T.S.; Hong, T.T.; Shaw, R.M. Actin cytoskeleton rest stops regulate anterograde traffic of connexin 43 vesicles to the plasma membrane. *Circ. Res.* **2012**, *110*, 978–989. [CrossRef] [PubMed]

332. Shaw, R.M.; Fay, A.J.; Puthenveedu, M.A.; von Zastrow, M.; Jan, Y.N.; Jan, L.Y. Microtubule plus-end-tracking proteins target gap junctions directly from the cell interior to adherens junctions. *Cell* **2007**, *128*, 547–560. [CrossRef] [PubMed]

333. Giepmans, B.N.; Verlaan, I.; Hengeveld, T.; Janssen, H.; Calafat, J.; Falk, M.M.; Moolenaar, W.H. Gap junction protein connexin-43 interacts directly with microtubules. *Curr. Biol.* **2001**, *11*, 1364–1368. [CrossRef]

334. Yamane, Y.; Shiga, H.; Asou, H.; Ito, E. Gap junctional channel inhibition alters actin organization and calcium propagation in rat cultured astrocytes. *Neuroscience* **2002**, *112*, 593–603. [CrossRef]

335. Theiss, C.; Meller, K. Microinjected anti-actin antibodies decrease gap junctional intercellular commmunication in cultured astrocytes. *Exp. Cell Res.* **2002**, *281*, 197–204. [CrossRef] [PubMed]

336. Cheng, C.; Nowak, R.B.; Gao, J.; Sun, X.; Biswas, S.K.; Lo, W.K.; Mathias, R.T.; Fowler, V.M. Lens ion homeostasis relies on the assembly and/or stability of large connexin 46 gap junction plaques on the broad sides of differentiating fiber cells. *Am. J. Physiol. Cell Physiol.* **2015**, *308*, C835–C847. [CrossRef] [PubMed]

337. Lauf, U.; Giepmans, B.N.; Lopez, P.; Braconnot, S.; Chen, S.C.; Falk, M.M. Dynamic trafficking and delivery of connexons to the plasma membrane and accretion to gap junctions in living cells. *Proc. Natl. Acad. Sci. USA* **2002**, *99*, 10446–10451. [CrossRef] [PubMed]

338. Rhett, J.M.; Jourdan, J.; Gourdie, R.G. Connexin 43 connexon to gap junction transition is regulated by zonula occludens-1. *Mol. Biol. Cell* **2011**, *22*, 1516–1528. [CrossRef] [PubMed]

339. Thevenin, A.F.; Margraf, R.A.; Fisher, C.G.; Kells-Andrews, R.M.; Falk, M.M. Phosphorylation regulates connexin43/zo-1 binding and release, an important step in gap junction turnover. *Mol. Biol. Cell* **2017**, *28*, 3595–3608. [CrossRef] [PubMed]

340. Van Itallie, C.M.; Fanning, A.S.; Bridges, A.; Anderson, J.M. Zo-1 stabilizes the tight junction solute barrier through coupling to the perijunctional cytoskeleton. *Mol. Biol. Cell* **2009**, *20*, 3930–3940. [CrossRef] [PubMed]

341. Fanning, A.S.; Van Itallie, C.M.; Anderson, J.M. Zonula occludens-1 and -2 regulate apical cell structure and the zonula adherens cytoskeleton in polarized epithelia. *Mol. Biol. Cell* **2012**, *23*, 577–590. [CrossRef] [PubMed]

342. Waxse, B.J.; Sengupta, P.; Hesketh, G.G.; Lippincott-Schwartz, J.; Buss, F. Myosin vi facilitates connexin 43 gap junction accretion. *J. Cell Sci.* **2017**, *130*, 827–840. [CrossRef] [PubMed]

343. Wang, H.Y.; Lin, Y.P.; Mitchell, C.K.; Ram, S.; O'Brien, J. Two-color fluorescent analysis of connexin 36 turnover: Relationship to functional plasticity. *J. Cell Sci.* **2015**, *128*, 3888–3897. [CrossRef] [PubMed]

344. Windoffer, R.; Beile, B.; Leibold, A.; Thomas, S.; Wilhelm, U.; Leube, R.E. Visualization of gap junction mobility in living cells. *Cell Tissue Res.* **2000**, *299*, 347–362. [CrossRef] [PubMed]

345. Wang, Y. Two-Color Fluorescent Analysis of Connexin 36 Turnover—Relationship to Functional plasticity. Ph.D. Dissertation, University of Texas, Houston, UT, USA, May 2015. GSBS Dissertations and Theses (Open Access).

346. Stahley, S.N.; Saito, M.; Faundez, V.; Koval, M.; Mattheyses, A.L.; Kowalczyk, A.P. Desmosome assembly and disassembly are membrane raft-dependent. *PLoS ONE* **2014**, *9*, e87809. [CrossRef] [PubMed]

347. Jennings, J.M.; Tucker, D.K.; Kottke, M.D.; Saito, M.; Delva, E.; Hanakawa, Y.; Amagai, M.; Kowalczyk, A.P. Desmosome disassembly in response to pemphigus vulgaris igg occurs in distinct phases and can be reversed by expression of exogenous DSG3. *J. Investig. Dermatol.* **2011**, *131*, 706–718. [CrossRef] [PubMed]

348. Schlingmann, B.; Overgaard, C.E.; Molina, S.A.; Lynn, K.S.; Mitchell, L.A.; Dorsainvil White, S.; Mattheyses, A.L.; Guidot, D.M.; Capaldo, C.T.; Koval, M. Regulation of claudin/zonula occludens-1 complexes by hetero-claudin interactions. *Nat. Commun.* **2016**, *7*, 12276. [CrossRef] [PubMed]

349. Ward, C.; Schlingmann, B.; Stecenko, A.A.; Guidot, D.M.; Koval, M. Nf-kb inhibitors impair lung epithelial tight junctions in the absence of inflammation. *Tissue Barriers* **2015**, *3*, e982424. [CrossRef] [PubMed]

350. Overgaard, C.E.; Schlingmann, B.; Dorsainvil White, S.; Ward, C.; Fan, X.; Swarnakar, S.; Brown, L.A.; Guidot, D.M.; Koval, M. The relative balance of GM-CSF and tgfbeta1 regulates lung epithelial barrier function. *Am. J. Physiol. Lung Cell. Mol. Physiol.* **2015**, *308*, L1212–L1223. [CrossRef] [PubMed]

351. Lafemina, M.J.; Rokkam, D.; Chandrasena, A.; Pan, J.; Bajaj, A.; Johnson, M.; Frank, J.A. Keratinocyte growth factor enhances barrier function without altering claudin expression in primary alveolar epithelial cells. *Am. J. Physiol. Lung Cell. Mol. Physiol.* **2010**, *299*, L724–L734. [CrossRef] [PubMed]

International Journal of
Molecular Sciences

MDPI

Review

Protein–Protein Interactions with Connexin 43: Regulation and Function

Paul L. Sorgen [1], Andrew J. Trease [1], Gaelle Spagnol [1], Mario Delmar [2] and Morten S. Nielsen [3,*]

[1] Department of Biochemistry and Molecular Biology, University of Nebraska Medical Center, Omaha, NE 68198, USA; psorgen@unmc.edu (P.L.S.); andrew.trease@unmc.edu (A.J.T.); gspagnol@unmc.edu (G.S.)
[2] Leon H Charney Division of Cardiology, NYU School of Medicine, New York, NY 10016, USA; Mario.Delmar@nyumc.org
[3] Department of Biomedical Sciences, Faculty of Health and Medical Sciences, University of Copenhagen, DK-2200 Copenhagen, Denmark
* Correspondence: schak@sund.ku.dk; Tel.: +45-28-757-427

Received: 20 April 2018; Accepted: 8 May 2018; Published: 10 May 2018

Abstract: Connexins are integral membrane building blocks that form gap junctions, enabling direct cytoplasmic exchange of ions and low-molecular-mass metabolites between adjacent cells. In the heart, gap junctions mediate the propagation of cardiac action potentials and the maintenance of a regular beating rhythm. A number of connexin interacting proteins have been described and are known gap junction regulators either through direct effects (e.g., kinases) or the formation of larger multifunctional complexes (e.g., cytoskeleton scaffold proteins). Most connexin partners can be categorized as either proteins promoting coupling by stimulating forward trafficking and channel opening or inhibiting coupling by inducing channel closure, internalization, and degradation. While some interactions have only been implied through co-localization using immunohistochemistry, others have been confirmed by biophysical methods that allow detection of a direct interaction. Our understanding of these interactions is, by far, most well developed for connexin 43 (Cx43) and the scope of this review is to summarize our current knowledge of their functional and regulatory roles. The significance of these interactions is further exemplified by demonstrating their importance at the intercalated disc, a major hub for Cx43 regulation and Cx43 mediated effects.

Keywords: gap junction; connexin; protein–protein interaction; intrinsically disordered protein; post-translational modification; intercalated disc

1. Introduction

The Cx43 carboxyl terminal (Cx43CT) domain plays a role in the trafficking, localization, and turnover of gap junction channels via numerous post-translational modifications and protein–protein interactions [1–5]. The Cx43CT is also important for regulating junctional conductance and voltage sensitivity [6–9]. Structural studies from our laboratory revealed that the Cx43CT as well as the CT domain from other connexins are predominately unstructured [10–13]. Intrinsically disordered domains are now well recognized to be loci for regulation of protein function because their conformation can be readily modulated by the local environment, phosphorylation, and interaction with proteins and small-molecules. We and others have shown that the Cx43CT binds multiple proteins, some of which have been shown to modulate channel function (for review see [14]). These data strongly suggest that protein–protein interactions mediated by any part of the CT are likely to have regulatory effects. Numerous excellent reviews have summarized the functional significance of these Cx43-interacting proteins [15–17]; here we provide a different perspective. We separated the proteins known to affect Cx43 function into three categories. The first are those proteins that directly interact with the CT and are associated with trafficking Cx43 to the gap junction plaque and open gap junction channels. Cx43-protein interactions identified from cell biology studies (e.g., immunoprecipitation and co-localization) that have been confirmed using different biophysical techniques (e.g., nuclear magnetic resonance, X-ray crystallography, and surface plasmon resonance) are considered a "direct" interaction. The second are those proteins that directly interact with the CT and are associated with channel closure, disassembly, and degradation. The third, which will not be a focus of this review, are those proteins that can affect all aspects of the Cx43 life cycle, but no evidence exists they directly interact with the Cx43CT (Table 1; albeit we realize a number of the proteins in Table 1 will eventually be shown to directly interact with Cx43 or may never be identified because binding requires a connexin embedded within the membrane or in context of a connexon, thus posing extreme challenges to performing in vitro assays). Additionally, we will not focus on those post-translational modifications such as ubiquitination, sumoylation, methylation, phosphorylation, and hydroxylation that form covalent bonds with connexins to modify function (for review see [18]). For the proteins that directly interact, we provide their location on the Cx43CT domain, residues (de)phosphorylated where necessary, and their diameter as estimated from their molecular weight (Available online: http://www.calctool.org/CALC/prof/bio/protein_size). Of note, these values are on the conservative side because proteins like ZO-1 and 14-3-3 have multiple modular domains and would have a larger diameter. For the Cx43CT, we combined the knowledge that the intrinsically disordered Cx43CT domain (length of 3.8 Å per residue; [19]) can contain as high as 35% α-helical structure (length of 1.50 Å per residue) depending on the level of phosphorylation [20]. The rationale for this perspective is to visually illustrate that only a small number of proteins can bind at any one time. The importance of Cx43 cellular localization (spatial), Cx43CT phosphorylation state, as well as the cellular condition (temporal) will help determine which proteins will bind the Cx43CT domain.

Table 1. Proteins suggested to interact with Cx43, but where no evidence currently exist for a direct protein–protein interaction. Abbreviations: IP, immunoprecipitation; co-Loc, co-localization; PLA, proximity ligation assay; TEM, transmission electron microscopy; PD, pull-down; IV, in vitro assay; FW, Far-Western.

Interacting Protein	Type of Detection	References
Actin	co-Loc	[21–23]
AGS8	IP, co-Loc	[24]
A-kinase anchoring protein 95	IP, co-Loc	[25]
Ankyrin G	IP	[26]
Apoptosis-inducing factor	IP, co-Loc, PLA	[27]
Atg16L/Atg14/Atg9/Vps34	IP, co-Loc	[28]
Bax	IP, co-Loc	[29]
β-arrestin	IP, co-Loc	[30]
β-subunit of the electron-transfer protein	IP, co-Loc, PLA	[27]
Brain-derived integrating factor-1	IP, co-Loc	[31]
CASK (LIN2)	IP, co-Loc	[32]
Caveolin-1,2,3	IP, co-Loc	[33–35]
Clathrin	IP, co-Loc	[36]
Claudin 5	IP, co-Loc	[37]
CIP85	IP, co-Loc	[38]
Consortin	IP, co-Loc	[39]
Cyclin E	IP, PLA, TEM	[40]
Desmocollin-2a	PD	[41]
Dlg	co-Loc	[42]
Dynamin	IP, co-Loc	[43]
EB1	IP	[44]
Eps15	IP, co-Loc	[45]

Table 1. *Cont.*

Interacting Protein	Type of Detection	References
ERp29	IP, co-Loc	[46]
Hrs	co-Loc	[47]
HSP70	IP, PD	[48]
HSP90	IP, co-Loc	[49]
Light chain 3	IP, co-Loc	[50,51]
Lin-7	PD	[52]
Myosin-VI	co-Loc	[53]
DMPK	IP, co-Loc	[54]
Na$_V$1.5	co-Loc	[55]
N-cadherin	co-Loc	[56]
NOV/CCN3	IP, PD	[57]
Occludin	IP, co-Loc	[37]
p120^{ctn}	co-Loc	[58]
P2X7	IP, co-Loc	[59,60]
P62	IP	[50]
PKG	IV	[61]
Plakophilin-2	co-Loc	[62]
PP1/PP2A	IP, co-Loc	[63]
RPTPμ	IP	[64]
Smurf2	IP, co-Loc	[65]
STAMBP (AMSH)	IP, co-Loc	[66]
TOM20	IP, co-Loc	[49]
TRIM21	IP, co-Loc	[67]
USP8	IP	[68]
Vinculin	IP, co-Loc	[60]
Wwp1	IP	[69]
ZO-2	IP, co-Loc, PD, FW	[52,70]

2. Direct Interactions with Cx43 and Their Functional Consequence

2.1. Interactions that Promote Synthesis, Trafficking to the Gap Junction Plaque, and Channel Opening

Intercellular coupling is eventually determined by the number of open channels in gap junction plaques, which is governed by the synthesis, forward trafficking, and channel open probability. A number of protein partners affect these processes (Figure 1).

Cx43 is translationally integrated into the endoplasmic reticulum (ER) and oligomerization occurs only after exit of the ER in the trans-Golgi network [71]. One of the first proteins likely to directly interact with Cx43 is the Connexin Interacting Protein of 75 kDa (CIP75). CIP75 interacts with Cx43CT residues K264-Q317 through its ubiquitin-associated (UBA) domain [72,73]. The importance of CIP75 is to mediate ER associated degradation of Cx43 for quality control and fine-tune the level of expression through dislocation of Cx43 from the ER and proteasomal degradation [73–76]. Use of cellular denaturants increased the association of CIP75 with Cx43, suggesting only pools of Cx43 lacking association with CIP75 escape ER dislocation and travel to the Golgi [75]. Upon exiting the trans-Golgi network, Cx43 containing vesicles are transported via the microtubular network to the plasma membrane [77].

Microtubular transport of connexons coincides with the recruitment of a number of protein interactors to the Cx43CT, a number of which have been implicated, however a direct interaction was not confirmed (Table 1; for review see [4,78,79]). In addition to microtubules, the actin cytoskeleton aids in connexon delivery to the gap junction plaque (for review see [80]). Curiously, regulation of Cx43 forward trafficking may in part be regulated by internally translated fragments of the Cx43CT [81]. One of these fragments, GJA1-20k, was recently shown to stabilize filamentous actin and suggested to help target microtubules to cell–cell junctions [82]. Full length Cx43 did not stabilize actin and the relation between the ability of GJA1-20k and Cx43 (see below) to target microtubules to the membrane remains to be established.

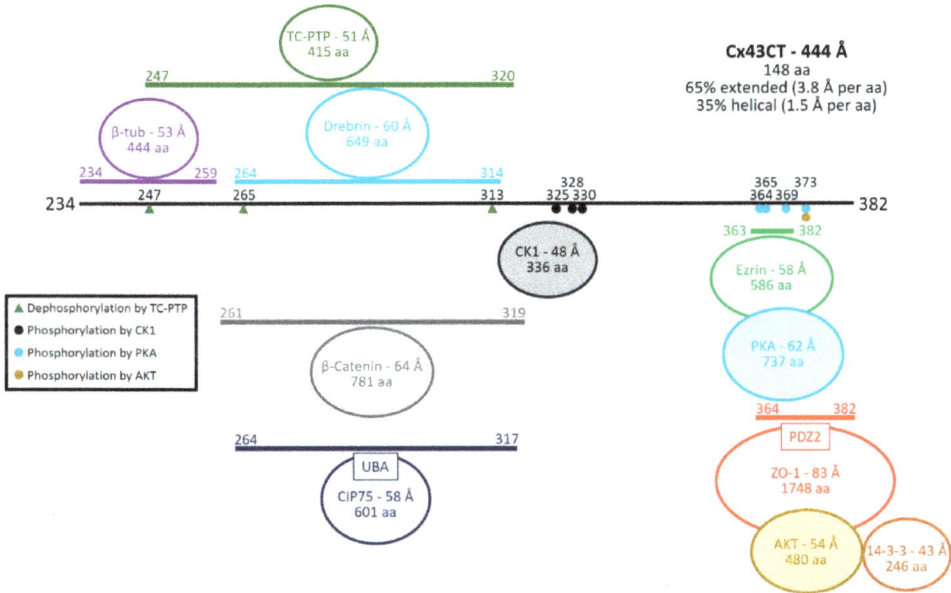

Figure 1. Protein partners that directly interact with the Cx43CT domain to promote intercellular communication. The black line represents Cx43CT domain residues 234–382. Provided for each Cx43CT protein partner (circle) is its diameter (in Å) as estimated from their molecular weight, and number of amino acids (aa), and the Cx43CT residues affected as a result of the interaction (lines). If the protein partner is a kinase or phosphatase, the Cx43CT residues affected are labeled on the Cx43CT (circle or triangle). Abbreviations are as follows: β-tubulin (β-tub), T-cell protein tyrosine phosphatase (TC-PTP), Connexin interacting protein 75 kDa (CIP75), Ubiquitin-associating domain (UBA), Casein kinase 1 (CK1), Protein kinase A (PKA), Zonula Occludens 1 (ZO-1), and Protein kinase B (AKT). Kinases have been highlighted (shaded circle).

In proximity of the plasma membrane, the actin- and protein kinase A (PKA)-binding protein Ezrin, binds the Cx43CT and enables PKA to phosphorylate Cx43CT serine residues. In particular, phosphorylation of S364 is a likely precursor to binding with the tight junction protein Zonula occludens 1 (ZO-1), another actin scaffolding protein [83]. Functional studies demonstrating increased gap junction intercellular communication following activation of PKA support this hypothesis [84,85]. Work by Pidoux et al. 2014, identified the minimal binding motif of Cx43CT for Ezrin as [366]RASSR[370] using a peptide screening approach [86]. Furthermore, PKA and ZO-1 interact with the Cx43CT over the same region as Ezrin (S364-I382), however phosphorylation by PKA (S365, S369) did not appear to alter binding of Ezrin to Cx43, nor binding of ZO-1 [86,87]. Work from Thévenin et al. 2017, and others have highlighted phosphorylation of S373 as a critical modulator of ZO-1 binding, a site phosphorylated by both PKA and protein kinase B (AKT) [83,87–89]. Association with ZO-1 is a critical mediator of gap junction plaque size; when bound to ZO-1 Cx43 is retained in the perinexal region "poised" for docking with apposing connexons, and upon release Cx43 is incorporated into the gap junction plaque proper [88–91]. Whether Ezrin and ZO-1 simultaneously bind the Cx43CT remains to be determined, but based on their size and location of binding on the Cx43CT, it seems unlikely.

Capture and incorporation of Cx43 containing vesicles at the plasma membrane (gap junction periphery) has been attributed to 14-3-3 [92–94]. Like Ezrin, 14-3-3 interacts with the Cx43CT in the same region as ZO-1, hovering over S373 [94]. Unlike the reduced binding of ZO-1, phosphorylation of S373 by PKA enhances 14-3-3 binding and likely serves as a switch of perinexal Cx43 to junctional Cx43 through tethering to integrins (specifically integrin α5; [88,89,94]). Taken together these studies highlight the intricacy of spatial-temporal and post-translational regulation of Cx43 trafficking to the gap junction plaque and suggest that association of Ezrin (and PKA) with the Cx43CT precedes association with ZO-1. This is further advanced by phosphorylation of S373 promoting the exchange of ZO-1 for 14-3-3 and incorporation into the gap junction plaque [88,89,94]. Of note AKT and 14-3-3 proteins are also involved in gap junction disassembly, a topic covered in the next section. Once incorporated into the plaque a number of interactions serve to stabilize and maintain Cx43 and control channel maturation (opening; for review see [95]).

Fully open channels require phosphorylation by casein kinase 1 (CK1) on residues S325, S328, S330 [96]. Interestingly, Cx43 knock-in mice in which Cx43CT residues S325, S328, and S330 were replaced with glutamic acids (phospho-mimicking) were immune to acute and chronic pathological gap junction remodeling and ventricular arrhythmias after transverse aortic constriction [97]. In addition to channel opening, stability of the gap junction plaque regulates gap junction intercellular communication. Direct protein interaction with microtubules via β-tubulin and association with the actin cytoskeleton through the scaffolding protein Developmentally Regulated Brain Protein 1 (Drebrin) are two key interactions, which stabilize gap junctions (for review see [98,99]). β-tubulin binds the Cx43CT over Y247, a known site of phosphorylation by Src kinase, and Drebrin binds over Y265 and Y313, two other substrates for Src phosphorylation [100–106]. Importantly, the interaction of β-tubulin with the Cx43CT likely occurs subsequent to plasma membrane incorporation as a direct interaction prior to plasma membrane incorporation would prevent Cx43 trafficking to the membrane (no motor proteins). This hypothesis is supported by data from Francis et al. 2011, indicating that Cx43 regulates microtubule dynamics at plasma membrane [107]. NMR and cell based work from our laboratory identified a phosphatase T-cell Protein Tyrosine Phosphatase (TC-PTP) which directly interacts with the Cx43CT and dephosphorylates the Y247 and Y265 reversing the down-regulating effects of Src kinase (described further in the next section; [108]).

Finally, β-catenin is another protein identified to interact with Cx43. In response to Wnt signaling, β-catenin can interact with the Cx43 gene to increase transcription as well as modulate gap junction stability at the plaque [109–112]. Works from several laboratories have shown indirect evidence of this interaction at the plaque by reciprocal co-immunoprecipitation as well as co-localization [109,113]. β-catenin was added in this section because we recently identified a direct interaction with the Cx43CT domain over three areas (residues G261-T275, S282-N295, and N302-R319) using a combination of surface plasmon resonance (SPR) and NMR experiments [114].

2.2. Interactions that Promote Channel Closure, Gap Junction Disassembly, Internalization and Degradation

Similarly, to facilitating coupling, down regulation of Cx43-mediated intercellular communication requires a number of direct protein interactions and phosphorylation events (Figure 2). Indeed, phosphorylation of Cx43 by Src is a key initiator of gap junction closure, internalization, and turnover [103,104,115–119]. Src-induced phosphorylation of Cx43 has been correlated with channel closure [101]. Current research suggests a "particle–receptor" mechanism for Src-mediated channel closure similar to that proposed for pH gating of Cx43 channels [7,104,120]. The impact of Src phosphorylation on channel activity is decreased electrical coupling by reducing open probability and changes in selectivity [121]. Work from our laboratory and others support an additional mechanism of Src to decrease gap junctional intercellular communication: the altering of Cx43 protein partners to enhance degradation. A commonality between the proteins that link Cx43 to the cytoskeleton is that Src can inhibit their interaction. For example, Cx43CT residues Y247 and Y265 phosphorylated by Src inhibit the binding of β-tubulin and Drebrin, respectively [122].

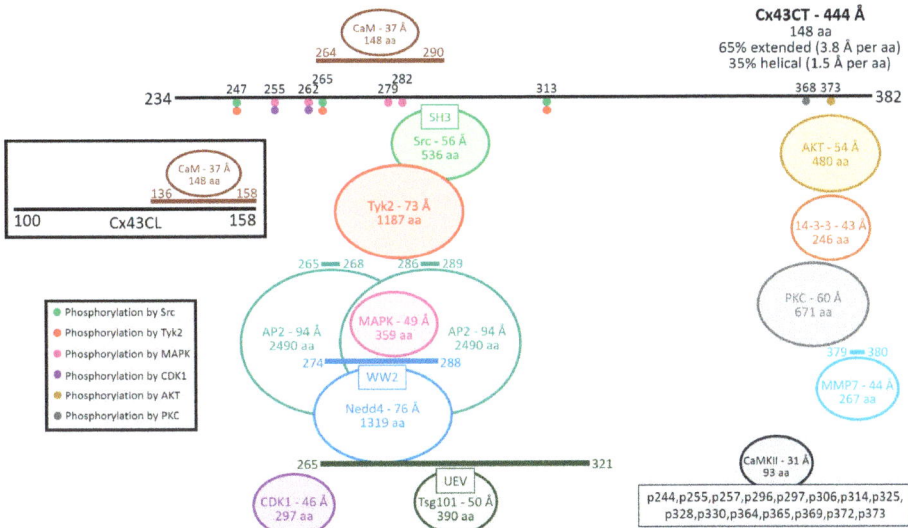

Figure 2. Protein partners that directly interact with the Cx43 CT and CL domains to impede intercellular communication. The black lines represents Cx43CT domain residues 234–382 and Cx43CL domain residues 100–158. Provided for each Cx43 CT and CL protein partner (circle) is its diameter (in Å) as estimated from their molecular weight, and number of amino acids (aa), and the Cx43CT residues affected as a result of the interaction (lines). If the protein partner is a kinase, the Cx43CT residues affected are labeled on the Cx43CT (circle). Abbreviations are as follows: Calmodulin (CaM), Src homology 3 domain (SH3), Tyrosine kinase 2 (Tyk2), Mitogen-activated protein kinase (MAPK), Neural precursor cell expressed developmentally down-regulated protein 4 (Nedd4), Cyclin-dependent kinase 1 (CDK1), Tumor susceptibility gene 101 protein (Tsg101), Ubiquitin E2 variant domain (UEV), Protein kinase B (AKT), Protein kinase C (PKC), matrix metalloproteinase-7 (MMP7), and Ca^{2+}/calmodulin-dependent protein kinase II (CaMKII). Kinases have been highlighted (shaded circle).

In the case of β-tubulin, at the gap junction plaque, this may be a mechanism in the disassembly process; at the trans-Golgi network, in cardiomyocytes this may re-route trafficking from the intercalated disc to lateral membranes; or inhibit trafficking to the plasma membrane altogether, leading to increased proteasomal and/or lysosomal degradation. For Drebrin, depletion in cells results in impaired cell–cell coupling, internalization of gap junctions, and targeting of Cx43 for degradation [123]. While phosphorylation of the Cx43CT by Src does not inhibit ZO-1 binding, we found that active c-Src can compete with Cx43 to directly bind ZO-1 [124]. Studies from the Gourdie and Lampe laboratories would suggest blocking these protein partners would transition Cx43 from the non-junctional plasma membrane into the gap junction plaque, and then through the degradation pathway(s) [91]. Finally, Src activation also indirectly leads to serine phosphorylation by AKT (S373), PKC (S368), and MAPK (S255, S279, and S282) that contributes to reduced Cx43 at the plasma membrane. AKT may act in a similar manner as Src in that phosphorylation of S373 inhibits the Cx43 interaction with ZO-1 [88]. In addition, phosphorylation of S373 enables the binding of 14-3-3 leading to gap junction ubiquitination, internalization, and degradation during acute cardiac ischemia [94]. Altogether, the data point to Src playing a significant role in inhibiting Cx43-mediated cell-to-cell communication by channel closure and enhanced degradation.

In addition to Src, another tyrosine kinase identified to directly interact with and phosphorylate the Cx43CT was the Janus kinase family member non-receptor tyrosine-protein kinase 2 (Tyk2; [125]).

Interestingly, Tyk2 can functionally substitute for Src as work from our laboratory identified that it phosphorylates Cx43CT residues Y247 and Y265 and results in concomitant loss of coupling and disassembly of gap junction plaques [125]. While phosphorylation of these sites by either Tyk2 or Src would result in disruption of the direct binding of β-tubulin and Drebrin, one difference is that Tyk2 unlikely disrupts the Cx43/ZO-1 interaction as Tyk2 does not contain a SH3 domain (for review see [126]). Whether Tyk2 binds to Cx43 via its SH2 domain or FERM domain remains to be determined [127–129]. It is becoming clear that overlap in the phosphorylated residues of Cx43 by a number of kinases provides the cell with a highly dynamic ability to alter gap junction function in response to various initial stimuli. In addition, like Src, activation of Tyk2 coincides with increased phosphorylation of S279/282 by MAPK and S368 by PKC [125]. MAPK also phosphorylates Cx43 residues S255 and S262, all of which alter the secondary structure of the Cx43CT to increase α-helical content, a mechanism which can promote or inhibit interactions with other protein partners [20].

One protein partner that undergoes recruitment following MAPK activation, is the E3 ubiquitin ligase Neural precursor cell expressed developmentally down-regulated 4 (Nedd4; [130]). Specifically, work by Leykauf et al. 2006, demonstrated that phosphorylation of S279/282 increased the affinity (K_D pS279/282 585 μM vs non-pS279/282 1064 μM) of Nedd4 for Cx43 [131]. Our laboratory confirmed this approximate 2-fold increase in the binding affinity for Nedd4 via NMR [132]. Furthermore, we determined that Nedd4 binds to the Cx43CT primarily through its WW2 domain via the PPXY motif (P283-Y286; [132]). Importantly, other proteins also interact with the Cx43CT in proximity to the PPXY motif, these are tumor susceptibility gene 101 (Tsg101) and the AP2 adaptor protein complex (AP2) both of which are involved in the endocytosis and retrograde trafficking of Cx43 [133,134]. In addition to MAPK, Src phosphorylation also primes Cx43 for phosphorylation by PKC at S368 [102]. A point worth noting is that phosphorylation of Cx43 S368 requires dephosphorylation of S365, as work from Solan et al. 2007, demonstrated that phosphorylation of these sites is mutually exclusive [135].

Phosphorylation of Cx43 by PKC occurs via indirect mechanisms following phosphorylation by Src [102]. Cx43 residue S368 is well established as a site for PKC phosphorylation and this site is correlated with a decrease in unitary conductance of approximately 50% (~100 pS down to ~50 pS; [136]). This decrease works together with phosphorylation by MAPK on S262 to close the channel completely. Since MAPK and PKC interact with and phosphorylate Cx43 over different regions it is likely, they can both interact simultaneously. Indeed, time course experiments following the changes in levels of site-specific phosphorylations (MAPK and PKC sites) following treatment of porcine aorta endothelial cells with vascular endothelial growth factor (VEGF) revealed a concomitant increase in phosphorylation on S255, S262, S279/282, and S368 [137]. However, the same study demonstrated that inhibition of PKC by GF109203X also resulted in a decrease in phosphorylation of S255, S279/282, and S368. The authors suggest it is likely the PKC phosphorylation may precede MAPK phosphorylation at least in VEGF activated cells to create a binding site for AP2 [137]. Similar phosphorylation patterns occur in a number of other cell types with different initiating stimuli suggesting this as a likely critical kinase program for the closure and internalization of Cx43 gap junctions [138–140]. Furthermore, in the same study the authors demonstrated that the phosphomimetic Cx43CT S365D mutation resulted in a significant change in structure of CT residues (T275-A276, G285-Y286, L356-S368, and R370-D379) as indicated by significant changes in chemical shift as observed in a heteronuclear single quantum coherence experiment [135]. Taken together these two lines of data suggest that phosphorylation of Cx43 by PKA on S365, induces a shift in structure which precludes binding of and phosphorylation by PKC. Finally, activation of PKC can halt the assembly of new gap junctions and its phosphorylation on S368 has been implicated in affecting gating and/or disassembly [141,142].

AP2 is one protein member of a family of five adaptor protein complexes (AP1-5) that are involved in both clathrin and non-clathrin (AP4/5) mediated trafficking events (for review see [143]). AP2 associates specifically with its cargo proteins via either two tyrosine based sorting motifs (YXXΦ or NPXY) or dileucine based sorting motifs ([D/E]XXXL[L/I]) (for review see [144]). The Cx43CT domain contains three tyrosine based sorting motifs (S1-Y^{230}VFF, S2-Y^{265}AYF, and S3-Y^{286}KLV; [134,145]). Only S2 and S3 interacted with AP2 to initiate clathrin-mediated internalization [134]. S1 was not involved due to its membrane juxtaposition. Furthermore, the study by Thomas et al. 2003, illustrated that the Cx43 AP2 S3 overlaps with the proline rich PPXY motif which Nedd4 recognizes [145]. This suggests that it is unlikely both Nedd4 and AP2 bind Cx43 at the same time, indicating potential diverging roles for ubiquitin and clathrin mediated internalization. The significance of Cx43 containing two tyrosine based sorting signal is unclear, however, work by Johnson et al. 2013, using yeast two-hybrid analysis indicated that the Cx43CT with a Y286A mutation (abolishing S3) did not function as bait for the μ2 subunit of the AP2 complex [140]. Although they suggest a requirement for post-translational modification [140], most likely, coordination of the tyrosine ring is important for binding AP2 as tyrosine phosphorylation within the Yxxϕ-type-binding motif of other proteins inhibits the interaction with AP2 (e.g., [146]).

Two additional proteins that directly interact with Cx43 are calmodulin (CaM) and CaM-dependent kinase 2 (CaMKII). Ca^{2+}/CaM activates CaMKII leading to autophosphorylation and subsequent phosphorylation of target proteins, including Cx43 [147–149]. In vitro work using mass spectroscopy identified extensive phosphorylation of the Cx43CT by CaMKII (15 Cx43CT residues; [147]). Whether all of these sites identified occur in vivo remains to be determined as this high degree of phosphorylation could be a result of non-specific binding under in vitro conditions as the only identified CaMKII consensus is R-X-X-S/T (only four in the Cx43CT domain; for review see [150]). However, of the sites identified by Huang et al. 2011, phosphorylation of S306 has been shown to increase rather than decrease coupling [148]. NMR experiments showed that CaM directly binds the Cx43 cytoplasmic loop residues K136-S158 [151]. This occurs in a Ca^{2+} dependent manner and leads to gap junction channel closure, perhaps via occlusion of the pore (for review see [152]). We recently identified that CaM also binds Cx43CT residues K264-T290 [153]. It is tempting to speculate that this may be the mechanism by which Cx43 channels close, but remain at the plasma membrane, unlike the effects of Src phosphorylation. Along with regular turnover, gap junctions disassemble during cell division as they serve as a source of cell–cell adhesion (for review see [154]). During mitosis Cx43 phosphorylation patterns change with phosphorylation detected on S255 and S262 [155]. These changes in phosphorylation correlate with reduced intercellular communication as well as increased concentration of Cx43 in intracellular structures [156–158]. Interestingly, a pool of this internalized Cx43 can be recycled to nucleate the formation of new gap junction channels [155]. Similar to phosphorylation of S255 and S262 by MAPK, cyclin-dependent kinase 1 (CDK1) phosphorylates these same residues to closes the gap junction channel [156,157].

In addition to the phosphorylation-mediated changes in protein partner associations described above, new studies have begun to illustrate Cx43 as a potential target for proteolytic cleavage in various pathologies [159–162]. Lindsey et al. 2006, using in vivo, in vitro, and in silico methods demonstrated that Cx43 is a substrate for matrix metalloproteinase-7 (MMP-7; [159]). The Cx43CT domain contains two putative MMP-7 cleavage sites (G350-R362 and R374-I382); however, biochemical analysis using epitope-mapped antibodies (antibody 1: 252-270, antibody 2: 363-382) suggested cleavage was occurring only at the R374-I382 site [159]. A direct MMP-7 interaction with Cx43 was shown by SPR, in proximity to S373, suggesting potential regulation by PKA/AKT [83,88,89,159].

3. The Intercalated Disc as a Hub of Cx43 Mediated Protein–Protein Interactions

Cx43 is expressed in a large variety of cells [5], where it may interact with the proteins discussed above as well as yet unidentified binding partners. The expression and localization of the interacting partners vary between cell types, which possibly underlie the bewildering number of contradictory findings on the role and regulation of Cx43. In the following, we will give examples from the current knowledge about interactions and regulation of Cx43 at the intercalated disc (ID) of cardiomyocytes. The ID is a region of particular interest since it contains large amounts of Cx43 in close contact with several known interaction partners [163]. Although we only have evidence of direct interaction with a few of the nearby proteins, the list of possible partners is growing. Using a proteomics approach, Girao and coworkers showed that 236 proteins precipitated with Cx43 isolated from rat hearts [164]. Even if a lot of these are not direct or may occur outside the ID, the number of potential partners is overwhelming.

3.1. Nedd4 Regulates the Cx43 Content of Cardiac Gap Junctions

The ubiquitin ligase Nedd4 interacts directly with Cx43 [132] and both proteins co-localize in cardiomyocytes [165,166]. Studies indicate that multiple pathways may induce Cx43 ubiquitination in cardiomyocytes, such as activation of G-protein coupled receptors [166] and cardiac ischemia [165], and that the underlying mechanism may differ between experimental models. In the case of G-protein-coupled receptor activation, ubiquitination was achieved via a depletion of PIP2 without a measurable change in Cx43-Nedd4 co-IP [166,167], whereas cardiac ischemia increased both co-localization at the ID and increased co-IP [165]. Rather than closing the channel per se, ubiquitination most likely targets Cx43 to internalization [131] that may involve binding to the adaptor protein Eps15 [45] followed by endocytosis and lysosomal degradation [168].

3.2. Cx43 Regulates the Forward Trafficking of the Cardiac Sodium Channel Na$_V$1.5

In contrast to the binding of Nedd4 that primarily regulates the Cx43-dependent coupling, other binding partners may be important for the regulation of nearby partners. This has proven particularly crucial at the ID, as evidenced by the fact that mutations in a number of ID components lead to wide spread dysregulation of ID function [169]. Although the exact nature of cross regulation remains obscure for many ID interactions, the interdependence of Cx43 and the cardiac sodium channel Na$_V$1.5 has recently been unraveled in some detail.

Knock out of Cx43 in the heart leads to severe arrhythmias [170,171], originally believed to rely solely on the lack of intercellular coupling. However, several lines of evidence suggested a co-regulation of Cx43 and Na$_V$1.5 [26,172]; and van Rijen and coworkers demonstrated that Cx43 knock out indeed reduces sodium channel expression in mice in vivo [173], a result that was reproduced in the cardiac HL-1 cell line, where Cx43 knock down reduces sodium current [173]. Intriguingly, the deletion of the last five amino acids of the Cx43CT (D378stop), which interact with the scaffolding protein ZO-1, also induced a highly arrhythmogenic phenotype in mice, despite an apparently normal intercellular coupling [174]. As for the complete loss of Cx43 described above, sodium current as well as Na$_V$1.5 expression were reduced in cardiomyocytes from D378stop mice [174], showing that an intact CT is needed for full Na$_V$1.5 expression at the membrane. The lack of Na$_V$1.5 at the ID suggested that forward trafficking of Na$_V$1.5 might be compromised. Using super resolution microscopy Agullo-Pascual et al. demonstrated that the plus end microtubule marker EB1 was partially dislocated from the ID in mice expressing Cx43-D378stop, which correlated with the presence of Na$_V$1.5 clusters that came very close to the ID membrane without reaching it properly [175]. This led to the hypothesis that Cx43 acts as an anchoring point for microtubules and thereby regulates the forward trafficking of other proteins to the ID. Such an anchoring function was already demonstrated by Lo and coworkers, who showed that KO of Cx43 reduces fibroblast motility and destabilizes the microtubular network [107]. Deletion of the tubulin binding domain between amino acids 234 and 243 in the Cx43-CT recapitulated the effect of removing Cx43 altogether [107], demonstrating the important functional role of the Cx43-tubulin interaction. The role of the Cx43-tubulin interaction was also demonstrated in the cardiac HL1 cell line. As mentioned above, knock down of Cx43 in HL1 cells reduces the sodium current by ~50% and re-transfection with Cx43 restores the sodium current [173]. In contrast, transfection of the same HL1 cells with Cx43 with the tubulin binding domain truncated, failed to restore sodium current [175], supporting a role for Cx43 as a microtubule anchoring point and thereby for guiding in sodium channels. Using the HL1 cells, it was also demonstrated that Cx43-D378stop channels were unable to restore the sodium current [175], indicating that both the tubulin- and ZO-1-binding domains are needed for proper transportation of sodium channels to the membrane.

3.3. Cx43, the Area Composita and the Connexome

There is overwhelming evidence indicating that the functions of Cx43 extend beyond that of forming gap junction channels. Studies from various laboratories indicate that in fact, Cx43 is not only localized at the gap junction or in the perinexus [176], but also as part of a molecular/structural conglomerate named the "area composita" [177]. This term was coined to describe the fact that in the heart cells, in addition to well-defined desmosomes, there are structures with features of both, desmosomes and adherens junctions. Work of Agullo-Pascual et al. 2014 showed that Cx43 can be localized to these structures [178]. Furthermore, loss of Cx43 can decrease intercellular adhesion strength [179]. Finally, changes in desmosomal molecules can affect the integrity of gap junctions [180]. All of these complex interactions have brought us to the conclusion that in the heart, desmosomes, gap junctions, and sodium channel complexes are not separated and apart from each other. Instead, they form a protein interacting network where molecules classically defined as belonging to one of these groups, interact with others and together bring about excitability, adhesion, and intercellular coupling in the heart. This protein interacting network (dubbed "the connexome" [178,181,182]) provides for a coordinated response between the different elements that are necessary for an integrated functional syncytium.

4. Conclusions

It has been over 30 years since the description by Beyer, Paul, and Goodenough of Cx43 as the major gap junction protein in the heart [183]. Since this description, there has been abundant research demonstrating that Cx43 is far from a lonely and aloof piece of the intercalated disc, geared for only one function. Rather, Cx43 is part of a complex interacting protein network, not only as a recipient of interactors that modify gap junctions, but also as a component of complexes that exert other functions. As such, the view of Cx43 as a single-function molecule (to make gap junctions) is now changed to that of a multi-tasking protein, webbed into other networks to synchronize cell coupling. The extent to which those functions are involved in disease remains a matter of controversy. Whether gap junctions, or Cx43, participate in arrhythmia syndromes, or in limiting the size of infarcts, or as good (or bad?) pharmacological targets, remains incompletely defined. These last 30 years have brought us a long way in understanding Cx43 as part of a molecular ecosystem. Hopefully, the next 30 years will help us improve our ability to forecast the storms that may result from Cx43 deficiency.

Acknowledgments: P.L.S. was supported by NIH Grants GM072631, GM319613, and GM103427. M.D. was supported by grants RO1 HL134328 and RO1 HL136179. M.S.N. was supported by grant 16-R107-A6812-22015 from the Danish Heart Foundation.

Conflicts of Interest: The authors declare no conflict of interest. The founding sponsors had no role in the writing of the manuscript, and in the decision to publish the results.

Abbreviations

AA	amino acids
AGS8	Activator of G protein signaling 8
AKT	protein kinase B
AMSH	associated molecule with the SH3 domain of STAM
AP2	adaptor protein 2
Atg	Autophagy-related protein
β-tub	β-tubulin
CaM	Calmodulin
CaMKII	Ca2+/calmodulin-dependent protein kinase II
CASK	Ca2+/calmodulin-activated serine kinase
CCN3	CYR61/CTGF/NOV
CDK1	Cyclin-dependent kinase 1
CIP75	connexin interacting protein 75 kDa

CIP85	Cx43-interacting protein of 85-kDa
CK1	Casein kinase 1
co-Loc	co-localization
CT	carboxyl terminal
Cx43	connexin 43
Cx43CT	Cx43 carboxyl terminal
Dlg	Discs-large
DMPK	dystrophia myotonica protein kinase
Drebrin	Developmentally Regulated Brain Protein 1
EB1	End binding 1
Eps15	Epidermal growth factor receptor substrate 15
ER	endoplasmic reticulum
ERp29	Endoplasmic reticulum protein 29
FERM domain	Domain found in 4.1 protein (F), Ezrin, Radixin and Moesin
FW	Far-Western
Hrs	hepatocyte growth factor-regulated tyrosine kinase substrate
HSP70	heat shock protein 70
HSP90	heat shock protein 90
ID	intercalated disc
IP	immunoprecipitation
IV	in vitro assay
Lin-7	linage-7
MAPK	Mitogen-activated protein kinase
MMP7	matrix metalloproteinase-7
Nedd4	Neural precursor cell expressed developmentally down-regulated protein 4
NMR	nuclear magnetic resonance
NOV	nephroblastoma overexpressed
p120ctn	p120-catenin
PD	pull-down
PIP2	Phosphatidylinositol-bisphosphate
PKA	protein kinase A
PKC	protein kinase C
PKG	protein kinase G
PLA	proximity ligation assay
PP	protein phosphatase
RPTPμ	receptor-like protein tyrosine phosphatase μ
SH3	Src homology 3 domain
Smurf2	Smad ubiquitination regulatory factor-2
SPR	surface plasmon resonance
STAMBP	Signal transducing adapter molecule 1 binding protein
TC-PTP	T-cell protein tyrosine phosphatase
TEM	transmission electron microscopy
TOM20	mitochondrial outer membrane receptor 20
TRIM21	Tripartite motif-containing protein 21
Tsg101	Tumor susceptibility gene 101 protein
Tyk2	Tyrosine kinase 2
UBA	Ubiquitin-associating domain
UEV	Ubiquitin E2 variant domain
USP8	Ubiquitin specific protease 8
VEGF	Vascular endothelial growth factor
ZO-1	Zonula occludens-1
ZO-2	Zonula occludens-2

References

1. Laird, D.W. The gap junction proteome and its relationship to disease. *Trends Cell Biol.* **2010**, *20*, 92–101. [CrossRef] [PubMed]
2. Lampe, P.D.; Lau, A.F. The effects of connexin phosphorylation on gap junctional communication. *Int. J. Biochem. Cell Biol.* **2004**, *36*, 1171–1186. [CrossRef]
3. Herve, J.; Bourmeyster, N.; Sarrouilhe, D.; Duffy, H. Gap junctional complexes: From partners to functions. *Progress. Biophys. Mol. Biol.* **2007**, *94*, 29–65. [CrossRef] [PubMed]
4. Thevenin, A.F.; Kowal, T.J.; Fong, J.T.; Kells, R.M.; Fisher, C.G.; Falk, M.M. Proteins and mechanisms regulating gap-junction assembly, internalization, and degradation. *Physiology (Bethesda)* **2013**, *28*, 93–116. [CrossRef] [PubMed]
5. Nielsen, M.S.; Axelsen, L.N.; Sorgen, P.L.; Verma, V.; Delmar, M.; Holstein-Rathlou, N.H. Gap junctions. *Compr. Physiol.* **2012**, *2*, 1981–2035. [PubMed]
6. Moreno, A.P.; Chanson, M.; Elenes, S.; Anumonwo, J.; Scerri, I.; Gu, H.; Taffet, S.M.; Delmar, M. Role of the carboxyl terminal of connexin43 in transjunctional fast voltage gating. *Circ. Res.* **2002**, *90*, 450–457. [CrossRef] [PubMed]
7. Morley, G.E.; Taffet, S.M.; Delmar, M. Intramolecular interactions mediate ph regulation of connexin43 channels. *Biophys. J.* **1996**, *70*, 1294–1302. [CrossRef]
8. Anumonwo, J.M.; Taffet, S.M.; Gu, H.; Chanson, M.; Moreno, A.P.; Delmar, M. The carboxyl terminal domain regulates the unitary conductance and voltage dependence of connexin40 gap junction channels. *Circ. Res.* **2001**, *88*, 666–673. [CrossRef] [PubMed]
9. Revilla, A.; Castro, C.; Barrio, L.C. Molecular dissection of transjunctional voltage dependence in the connexin-32 and connexin-43 junctions. *Biophys. J.* **1999**, *77*, 1374–1383. [CrossRef]
10. Sorgen, P.L.; Duffy, H.S.; Sahoo, P.; Coombs, W.; Delmar, M.; Spray, D.C. Structural changes in the carboxyl terminus of the gap junction protein connexin43 indicates signaling between binding domains for c-src and zonula occludens-1. *J. Biol. Chem.* **2004**, *279*, 54695–54701. [CrossRef] [PubMed]
11. Bouvier, D.; Kieken, F.; Kellezi, A.; Sorgen, P.L. Structural changes in the carboxyl terminus of the gap junction protein connexin 40 caused by the interaction with c-src and zonula occludens-1. *Cell Commun. Adhes.* **2008**, *15*, 107–118. [CrossRef] [PubMed]
12. Stauch, K.; Kieken, F.; Sorgen, P. Characterization of the structure and intermolecular interactions between the connexin 32 carboxyl-terminal domain and the protein partners synapse-associated protein 97 and calmodulin. *J. Biol. Chem.* **2012**, *287*, 27771–27788. [CrossRef] [PubMed]
13. Nelson, T.K.; Sorgen, P.L.; Burt, J.M. Carboxy terminus and pore-forming domain properties specific to cx37 are necessary for cx37-mediated suppression of insulinoma cell proliferation. *Am. J. Physiol. Cell Physiol.* **2013**, *305*, C1246–C1256. [CrossRef] [PubMed]
14. Gilleron, J.; Carette, D.; Chevallier, D.; Segretain, D.; Pointis, G. Molecular connexin partner remodeling orchestrates connexin traffic: From physiology to pathophysiology. *Crit. Rev. Biochem. Mol. Biol.* **2012**, *47*, 407–423. [CrossRef] [PubMed]
15. Leithe, E.; Mesnil, M.; Aasen, T. The connexin 43 c-terminus: A tail of many tales. *Biochim. Biophys. Acta* **2018**, *1860*, 48–64. [CrossRef] [PubMed]
16. Solan, J.L.; Lampe, P.D. Spatio-temporal regulation of connexin43 phosphorylation and gap junction dynamics. *Biochim. Biophys. Acta* **2018**, *1860*, 83–90. [CrossRef] [PubMed]
17. Falk, M.M.; Bell, C.L.; Kells Andrews, R.M.; Murray, S.A. Molecular mechanisms regulating formation, trafficking and processing of annular gap junctions. *BMC Cell Biol.* **2016**, *17* (Suppl. 1), 22. [CrossRef] [PubMed]
18. Axelsen, L.N.; Calloe, K.; Holstein-Rathlou, N.H.; Nielsen, M.S. Managing the complexity of communication: Regulation of gap junctions by post-translational modification. *Front. Pharmacol.* **2013**, *4*, 130. [CrossRef] [PubMed]
19. Xue, B.; Romero, P.R.; Noutsou, M.; Maurice, M.M.; Rudiger, S.G.; William, A.M., Jr.; Mizianty, M.J.; Kurgan, L.; Uversky, V.N.; Dunker, A.K. Stochastic machines as a colocalization mechanism for scaffold protein function. *FEBS Lett.* **2013**, *587*, 1587–1591. [CrossRef] [PubMed]

20. Grosely, R.; Kopanic, J.L.; Nabors, S.; Kieken, F.; Spagnol, G.; Al-Mugotir, M.; Zach, S.; Sorgen, P.L. Effects of phosphorylation on the structure and backbone dynamics of the intrinsically disordered connexin43 c-terminal domain. *J. Biol. Chem.* **2013**, *288*, 24857–24870. [CrossRef] [PubMed]

21. Yamane, Y.; Shiga, H.; Asou, H.; Haga, H.; Kawabata, K.; Abe, K.; Ito, E. Dynamics of astrocyte adhesion as analyzed by a combination of atomic force microscopy and immuno-cytochemistry: The involvement of actin filaments and connexin 43 in the early stage of adhesion. *Arch. Histol. Cytol.* **1999**, *62*, 355–361. [CrossRef] [PubMed]

22. Squecco, R.; Sassoli, C.; Nuti, F.; Martinesi, M.; Chellini, F.; Nosi, D.; Zecchi-Orlandini, S.; Francini, F.; Formigli, L.; Meacci, E. Sphingosine 1-phosphate induces myoblast differentiation through cx43 protein expression: A role for a gap junction-dependent and -independent function. *Mol. Biol. Cell* **2006**, *17*, 4896–4910. [CrossRef] [PubMed]

23. Wall, M.E.; Otey, C.; Qi, J.; Banes, A.J. Connexin 43 is localized with actin in tenocytes. *Cell Motil. Cytoskelet.* **2007**, *64*, 121–130. [CrossRef] [PubMed]

24. Sato, M.; Jiao, Q.; Honda, T.; Kurotani, R.; Toyota, E.; Okumura, S.; Takeya, T.; Minamisawa, S.; Lanier, S.M.; Ishikawa, Y. Activator of g protein signaling 8 (ags8) is required for hypoxia-induced apoptosis of cardiomyocytes: Role of g betagamma and connexin 43 (cx43). *J. Biol. Chem.* **2009**, *284*, 31431–31440. [CrossRef] [PubMed]

25. Chen, X.; Kong, X.; Zhuang, W.; Teng, B.; Yu, X.; Hua, S.; Wang, S.; Liang, F.; Ma, D.; Zhang, S.; et al. Dynamic changes in protein interaction between akap95 and cx43 during cell cycle progression of a549 cells. *Sci. Rep.* **2016**, *6*, 21224. [CrossRef] [PubMed]

26. Sato, P.Y.; Coombs, W.; Lin, X.; Nekrasova, O.; Green, K.J.; Isom, L.L.; Taffet, S.M.; Delmar, M. Interactions between ankyrin-g, plakophilin-2, and connexin43 at the cardiac intercalated disc. *Circ. Res.* **2011**, *109*, 193–201. [CrossRef] [PubMed]

27. Denuc, A.; Nunez, E.; Calvo, E.; Loureiro, M.; Miro-Casas, E.; Guaras, A.; Vazquez, J.; Garcia-Dorado, D. New protein-protein interactions of mitochondrial connexin 43 in mouse heart. *J. Cell. Mol. Med.* **2016**, *20*, 794–803. [CrossRef] [PubMed]

28. Bejarano, E.; Yuste, A.; Patel, B.; Stout, R.F., Jr.; Spray, D.C.; Cuervo, A.M. Connexins modulate autophagosome biogenesis. *Nat. Cell Biol.* **2014**, *16*, 401–414. [CrossRef] [PubMed]

29. Sun, Y.; Zhao, X.; Yao, Y.; Qi, X.; Yuan, Y.; Hu, Y. Connexin 43 interacts with bax to regulate apoptosis of pancreatic cancer through a gap junction-independent pathway. *Int. J. Oncol.* **2012**, *41*, 941–948. [CrossRef] [PubMed]

30. Bivi, N.; Lezcano, V.; Romanello, M.; Bellido, T.; Plotkin, L.I. Connexin43 interacts with betaarrestin: A pre-requisite for osteoblast survival induced by parathyroid hormone. *J. Cell. Biochem.* **2011**, *112*, 2920–2930. [CrossRef] [PubMed]

31. Ito, T.; Ueki, T.; Furukawa, H.; Sato, K. The identification of novel protein, brain-derived integrating factor-1 (bdif1), which interacts with astrocytic gap junctional protein. *Neurosci. Res.* **2011**, *70*, 330–333. [CrossRef] [PubMed]

32. Marquez-Rosado, L.; Singh, D.; Rincon-Arano, H.; Solan, J.L.; Lampe, P.D. Cask (lin2) interacts with cx43 in wounded skin and their coexpression affects cell migration. *J. Cell Sci.* **2012**, *125*, 695–702. [CrossRef] [PubMed]

33. Schubert, A.L.; Schubert, W.; Spray, D.C.; Lisanti, M.P. Connexin family members target to lipid raft domains and interact with caveolin-1. *Biochemistry* **2002**, *41*, 5754–5764. [CrossRef] [PubMed]

34. Langlois, S.; Cowan, K.N.; Shao, Q.; Cowan, B.J.; Laird, D.W. Caveolin-1 and -2 interact with connexin43 and regulate gap junctional intercellular communication in keratinocytes. *Mol. Biol. Cell* **2008**, *19*, 912–928. [CrossRef] [PubMed]

35. Liu, L.; Li, Y.; Lin, J.; Liang, Q.; Sheng, X.; Wu, J.; Huang, R.; Liu, S.; Li, Y. Connexin43 interacts with caveolin-3 in the heart. *Mol. Biol. Rep.* **2010**, *37*, 1685–1691. [CrossRef] [PubMed]

36. Huang, X.D.; Horackova, M.; Pressler, M.L. Changes in the expression and distribution of connexin 43 in isolated cultured adult guinea pig cardiomyocytes. *Exp. Cell Res.* **1996**, *228*, 254–261. [CrossRef] [PubMed]

37. Nagasawa, K.; Chiba, H.; Fujita, H.; Kojima, T.; Saito, T.; Endo, T.; Sawada, N. Possible involvement of gap junctions in the barrier function of tight junctions of brain and lung endothelial cells. *J. Cell. Physiol.* **2006**, *208*, 123–132. [CrossRef] [PubMed]

38. Lan, Z.; Kurata, W.E.; Martyn, K.D.; Jin, C.; Lau, A.F. Novel rab gap-like protein, cip85, interacts with connexin43 and induces its degradation. *Biochemistry* **2005**, *44*, 2385–2396. [CrossRef] [PubMed]

39. Del Castillo, F.J.; Cohen-Salmon, M.; Charollais, A.; Caille, D.; Lampe, P.D.; Chavrier, P.; Meda, P.; Petit, C. Consortin, a trans-golgi network cargo receptor for the plasma membrane targeting and recycling of connexins. *Hum. Mol. Genet.* **2010**, *19*, 262–275. [CrossRef] [PubMed]

40. Johnstone, S.R.; Kroncke, B.M.; Straub, A.C.; Best, A.K.; Dunn, C.A.; Mitchell, L.A.; Peskova, Y.; Nakamoto, R.K.; Koval, M.; Lo, C.W.; et al. Mapk phosphorylation of connexin 43 promotes binding of cyclin e and smooth muscle cell proliferation. *Circ. Res.* **2012**, *111*, 201–211. [CrossRef] [PubMed]

41. Gehmlich, K.; Lambiase, P.D.; Asimaki, A.; Ciaccio, E.J.; Ehler, E.; Syrris, P.; Saffitz, J.E.; McKenna, W.J. A novel desmocollin-2 mutation reveals insights into the molecular link between desmosomes and gap junctions. *Heart Rhythm* **2011**, *8*, 711–718. [CrossRef] [PubMed]

42. Macdonald, A.I.; Sun, P.; Hernandez-Lopez, H.; Aasen, T.; Hodgins, M.B.; Edward, M.; Roberts, S.; Massimi, P.; Thomas, M.; Banks, L.; et al. A functional interaction between the maguk protein hdlg and the gap junction protein connexin 43 in cervical tumour cells. *Biochem. J.* **2012**, *446*, 9–21. [CrossRef] [PubMed]

43. Gilleron, J.; Carette, D.; Fiorini, C.; Dompierre, J.; Macia, E.; Denizot, J.P.; Segretain, D.; Pointis, G. The large gtpase dynamin2: A new player in connexin 43 gap junction endocytosis, recycling and degradation. *Int. J. Biochem. Cell Biol.* **2011**, *43*, 1208–1217. [CrossRef] [PubMed]

44. Shaw, R.M.; Fay, A.J.; Puthenveedu, M.A.; von Zastrow, M.; Jan, Y.N.; Jan, L.Y. Microtubule plus-end-tracking proteins target gap junctions directly from the cell interior to adherens junctions. *Cell* **2007**, *128*, 547–560. [CrossRef] [PubMed]

45. Girao, H.; Catarino, S.; Pereira, P. Eps15 interacts with ubiquitinated cx43 and mediates its internalization. *Exp. Cell Res.* **2009**, *315*, 3587–3597. [CrossRef] [PubMed]

46. Das, S.; Smith, T.D.; Sarma, J.D.; Ritzenthaler, J.D.; Maza, J.; Kaplan, B.E.; Cunningham, L.A.; Suaud, L.; Hubbard, M.J.; Rubenstein, R.C.; et al. Erp29 restricts connexin43 oligomerization in the endoplasmic reticulum. *Mol. Biol. Cell* **2009**, *20*, 2593–2604. [CrossRef] [PubMed]

47. Leithe, E.; Kjenseth, A.; Sirnes, S.; Stenmark, H.; Brech, A.; Rivedal, E. Ubiquitylation of the gap junction protein connexin-43 signals its trafficking from early endosomes to lysosomes in a process mediated by hrs and tsg101. *J. Cell Sci.* **2009**, *122*, 3883–3893. [CrossRef] [PubMed]

48. Hatakeyama, T.; Dai, P.; Harada, Y.; Hino, H.; Tsukahara, F.; Maru, Y.; Otsuji, E.; Takamatsu, T. Connexin43 functions as a novel interacting partner of heat shock cognate protein 70. *Sci. Rep.* **2013**, *3*, 2719. [CrossRef] [PubMed]

49. Rodriguez-Sinovas, A.; Boengler, K.; Cabestrero, A.; Gres, P.; Morente, M.; Ruiz-Meana, M.; Konietzka, I.; Miro, E.; Totzeck, A.; Heusch, G.; et al. Translocation of connexin 43 to the inner mitochondrial membrane of cardiomyocytes through the heat shock protein 90-dependent tom pathway and its importance for cardioprotection. *Circ. Res.* **2006**, *99*, 93–101. [CrossRef] [PubMed]

50. Bejarano, E.; Girao, H.; Yuste, A.; Patel, B.; Marques, C.; Spray, D.C.; Pereira, P.; Cuervo, A.M. Autophagy modulates dynamics of connexins at the plasma membrane in a ubiquitin-dependent manner. *Mol. Biol. Cell* **2012**, *23*, 2156–2169. [CrossRef] [PubMed]

51. Martins-Marques, T.; Catarino, S.; Zuzarte, M.; Marques, C.; Matafome, P.; Pereira, P.; Girao, H. Ischaemia-induced autophagy leads to degradation of gap junction protein connexin43 in cardiomyocytes. *Biochem. J.* **2015**, *467*, 231–245. [CrossRef] [PubMed]

52. Singh, D.; Lampe, P.D. Identification of connexin-43 interacting proteins. *Cell Commun. Adhes.* **2003**, *10*, 215–220. [CrossRef] [PubMed]

53. Piehl, M.; Lehmann, C.; Gumpert, A.; Denizot, J.P.; Segretain, D.; Falk, M.M. Internalization of large double-membrane intercellular vesicles by a clathrin-dependent endocytic process. *Mol. Biol. Cell* **2007**, *18*, 337–347. [CrossRef] [PubMed]

54. Schiavon, G.; Furlan, S.; Marin, O.; Salvatori, S. Myotonic dystrophy protein kinase of the cardiac muscle: Evaluation using an immunochemical approach. *Microsc. Res. Tech.* **2002**, *58*, 404–411. [CrossRef] [PubMed]

55. Malhotra, J.D.; Thyagarajan, V.; Chen, C.; Isom, L.L. Tyrosine-phosphorylated and nonphosphorylated sodium channel beta1 subunits are differentially localized in cardiac myocytes. *J. Biol. Chem.* **2004**, *279*, 40748–40754. [CrossRef] [PubMed]

56. Akar, F.G.; Spragg, D.D.; Tunin, R.S.; Kass, D.A.; Tomaselli, G.F. Mechanisms underlying conduction slowing and arrhythmogenesis in nonischemic dilated cardiomyopathy. *Circ. Res.* **2004**, *95*, 717–725. [CrossRef] [PubMed]

57. Fu, C.T.; Bechberger, J.F.; Ozog, M.A.; Perbal, B.; Naus, C.C. Ccn3 (nov) interacts with connexin43 in c6 glioma cells: Possible mechanism of connexin-mediated growth suppression. *J. Biol. Chem.* **2004**, *279*, 36943–36950. [CrossRef] [PubMed]

58. Xu, X.; Li, W.E.; Huang, G.Y.; Meyer, R.; Chen, T.; Luo, Y.; Thomas, M.P.; Radice, G.L.; Lo, C.W. Modulation of mouse neural crest cell motility by n-cadherin and connexin 43 gap junctions. *J. Cell Biol.* **2001**, *154*, 217–230. [CrossRef] [PubMed]

59. Fortes, F.S.; Pecora, I.L.; Persechini, P.M.; Hurtado, S.; Costa, V.; Coutinho-Silva, R.; Braga, M.B.; Silva-Filho, F.C.; Bisaggio, R.C.; De Farias, F.P.; et al. Modulation of intercellular communication in macrophages: Possible interactions between gap junctions and p2 receptors. *J. Cell Sci.* **2004**, *117*, 4717–4726. [CrossRef] [PubMed]

60. Iacobas, D.A.; Suadicani, S.O.; Iacobas, S.; Chrisman, C.; Cohen, M.A.; Spray, D.C.; Scemes, E. Gap junction and purinergic p2 receptor proteins as a functional unit: Insights from transcriptomics. *J. Membr. Biol.* **2007**, *217*, 83–91. [CrossRef] [PubMed]

61. Kwak, B.R.; Saez, J.C.; Wilders, R.; Chanson, M.; Fishman, G.I.; Hertzberg, E.L.; Spray, D.C.; Jongsma, H.J. Effects of cgmp-dependent phosphorylation on rat and human connexin43 gap junction channels. *Pflugers Arch.* **1995**, *430*, 770–778. [CrossRef] [PubMed]

62. Li, M.W.; Mruk, D.D.; Lee, W.M.; Cheng, C.Y. Connexin 43 and plakophilin-2 as a protein complex that regulates blood-testis barrier dynamics. *Proc. Natl. Acad. Sci. USA* **2009**, *106*, 10213–10218. [CrossRef] [PubMed]

63. Ai, X.; Pogwizd, S.M. Connexin 43 downregulation and dephosphorylation in nonischemic heart failure is associated with enhanced colocalized protein phosphatase type 2a. *Circ. Res.* **2005**, *96*, 54–63. [CrossRef] [PubMed]

64. Lezcano, V.; Bellido, T.; Plotkin, L.I.; Boland, R.; Morelli, S. Osteoblastic protein tyrosine phosphatases inhibition and connexin 43 phosphorylation by alendronate. *Exp. Cell Res.* **2014**, *324*, 30–39. [CrossRef] [PubMed]

65. Fykerud, T.A.; Kjenseth, A.; Schink, K.O.; Sirnes, S.; Bruun, J.; Omori, Y.; Brech, A.; Rivedal, E.; Leithe, E. Smad ubiquitination regulatory factor-2 controls gap junction intercellular communication by modulating endocytosis and degradation of connexin43. *J. Cell Sci.* **2012**, *125*, 3966–3976. [CrossRef] [PubMed]

66. Ribeiro-Rodrigues, T.M.; Catarino, S.; Marques, C.; Ferreira, J.V.; Martins-Marques, T.; Pereira, P.; Girao, H. Amsh-mediated deubiquitination of cx43 regulates internalization and degradation of gap junctions. *FASEB J.* **2014**, *28*, 4629–4641. [CrossRef] [PubMed]

67. Chen, V.C.; Kristensen, A.R.; Foster, L.J.; Naus, C.C. Association of connexin43 with e3 ubiquitin ligase trim21 reveals a mechanism for gap junction phosphodegron control. *J. Proteome Res.* **2012**, *11*, 6134–6146. [CrossRef] [PubMed]

68. Sun, J.; Hu, Q.; Peng, H.; Peng, C.; Zhou, L.; Lu, J.; Huang, C. The ubiquitin-specific protease usp8 deubiquitinates and stabilizes cx43. *J. Biol. Chem.* **2018**. [CrossRef] [PubMed]

69. Basheer, W.A.; Harris, B.S.; Mentrup, H.L.; Abreha, M.; Thames, E.L.; Lea, J.B.; Swing, D.A.; Copeland, N.G.; Jenkins, N.A.; Price, R.L.; et al. Cardiomyocyte-specific overexpression of the ubiquitin ligase wwp1 contributes to reduction in connexin 43 and arrhythmogenesis. *J. Mol. Cell Cardiol.* **2015**, *88*, 1–13. [CrossRef] [PubMed]

70. Singh, D.; Solan, J.L.; Taffet, S.M.; Javier, R.; Lampe, P.D. Connexin 43 interacts with zona occludens-1 and -2 proteins in a cell cycle stage-specific manner. *J. Biol. Chem.* **2005**, *280*, 30416–30421. [CrossRef] [PubMed]

71. Musil, L.S.; Goodenough, D.A. Multisubunit assembly of an integral plasma membrane channel protein, gap junction connexin43, occurs after exit from the er. *Cell* **1993**, *74*, 1065–1077. [CrossRef]

72. Kieken, F.; Spagnol, G.; Su, V.; Lau, A.F.; Sorgen, P.L. Nmr structure note: Uba domain of cip75. *J. Biomol. NMR* **2010**, *46*, 245–250. [CrossRef] [PubMed]

73. Kopanic, J.L.; Schlingmann, B.; Koval, M.; Lau, A.F.; Sorgen, P.L.; Su, V.F. Degradation of gap junction connexins is regulated by the interaction with cx43-interacting protein of 75 kda (cip75). *Biochem. J.* **2015**, *466*, 571–585. [CrossRef] [PubMed]

74. Li, X.; Su, V.; Kurata, W.E.; Jin, C.; Lau, A.F. A novel connexin43-interacting protein, cip75, which belongs to the ubl-uba protein family, regulates the turnover of connexin43. *J. Biol. Chem.* **2008**, *283*, 5748–5759. [CrossRef] [PubMed]

75. Su, V.; Hoang, C.; Geerts, D.; Lau, A.F. Cip75 (connexin43-interacting protein of 75 kda) mediates the endoplasmic reticulum dislocation of connexin43. *Biochem. J.* **2014**, *458*, 57–67. [CrossRef] [PubMed]

76. Su, V.; Nakagawa, R.; Koval, M.; Lau, A.F. Ubiquitin-independent proteasomal degradation of endoplasmic reticulum-localized connexin43 mediated by cip75. *J. Biol. Chem.* **2010**, *285*, 40979–40990. [CrossRef] [PubMed]

77. Thomas, T.; Jordan, K.; Simek, J.; Shao, Q.; Jedeszko, C.; Walton, P.; Laird, D.W. Mechanisms of cx43 and cx26 transport to the plasma membrane and gap junction regeneration. *J. Cell Sci.* **2005**, *118*, 4451–4462. [CrossRef] [PubMed]

78. Akhmanova, A.; Steinmetz, M.O. Microtubule +tips at a glance. *J. Cell Sci.* **2010**, *123*, 3415–3419. [CrossRef] [PubMed]

79. Welte, M.A. Bidirectional transport along microtubules. *Curr. Biol.* **2004**, *14*, R525–R537. [CrossRef] [PubMed]

80. Zhang, S.S.; Shaw, R.M. Trafficking highways to the intercalated disc: New insights unlocking the specificity of connexin 43 localization. *Cell Commun. Adhes.* **2014**, *21*, 43–54. [CrossRef] [PubMed]

81. Smyth, J.W.; Shaw, R.M. Autoregulation of connexin43 gap junction formation by internally translated isoforms. *Cell Rep.* **2013**, *5*, 611–618. [CrossRef] [PubMed]

82. Basheer, W.A.; Xiao, S.; Epifantseva, I.; Fu, Y.; Kleber, A.G.; Hong, T.; Shaw, R.M. Gja1-20k arranges actin to guide cx43 delivery to cardiac intercalated discs. *Circ. Res.* **2017**, *121*, 1069–1080. [CrossRef] [PubMed]

83. Dukic, A.R.; Gerbaud, P.; Guibourdenche, J.; Thiede, B.; Tasken, K.; Pidoux, G. Ezrin-anchored pka phosphorylates serine 369 and 373 on connexin 43 to enhance gap junction assembly, communication, and cell fusion. *Biochem. J.* **2017**, *475*, 455–476. [CrossRef] [PubMed]

84. Atkinson, M.M.; Lampe, P.D.; Lin, H.H.; Kollander, R.; Li, X.R.; Kiang, D.T. Cyclic amp modifies the cellular distribution of connexin43 and induces a persistent increase in the junctional permeability of mouse mammary tumor cells. *J. Cell Sci.* **1995**, *108*, 3079–3090. [PubMed]

85. Spray, D.C.; Moreno, A.P.; Kessler, J.A.; Dermietzel, R. Characterization of gap junctions between cultured leptomeningeal cells. *Brain Res.* **1991**, *568*, 1–14. [CrossRef]

86. Pidoux, G.; Gerbaud, P.; Dompierre, J.; Lygren, B.; Solstad, T.; Evain-Brion, D.; Tasken, K. A pka-ezrin-cx43 signaling complex controls gap junction communication and thereby trophoblast cell fusion. *J. Cell Sci.* **2014**, *127*, 4172–4185. [CrossRef] [PubMed]

87. Thevenin, A.F.; Margraf, R.A.; Fisher, C.G.; Kells-Andrews, R.M.; Falk, M.M. Phosphorylation regulates connexin43/zo-1 binding and release, an important step in gap junction turnover. *Mol. Biol. Cell* **2017**, *28*, 3595–3608. [CrossRef] [PubMed]

88. Dunn, C.A.; Lampe, P.D. Injury-triggered akt phosphorylation of cx43: A zo-1-driven molecular switch that regulates gap junction size. *J. Cell Sci.* **2014**, *127*, 455–464. [CrossRef] [PubMed]

89. Dunn, C.A.; Su, V.; Lau, A.F.; Lampe, P.D. Activation of akt, not connexin 43 protein ubiquitination, regulates gap junction stability. *J. Biol. Chem.* **2012**, *287*, 2600–2607. [CrossRef] [PubMed]

90. Hunter, A.W.; Barker, R.J.; Zhu, C.; Gourdie, R.G. Zonula occludens-1 alters connexin43 gap junction size and organization by influencing channel accretion. *Mol. Biol. Cell* **2005**, *16*, 5686–5698. [CrossRef] [PubMed]

91. Rhett, J.M.; Jourdan, J.; Gourdie, R.G. Connexin 43 connexon to gap junction transition is regulated by zonula occludens-1. *Mol. Biol. Cell* **2011**, *22*, 1516–1528. [CrossRef] [PubMed]

92. Batra, N.; Riquelme, M.A.; Burra, S.; Jiang, J.X. 14-3-3theta facilitates plasma membrane delivery and function of mechanosensitive connexin 43 hemichannels. *J. Cell Sci.* **2014**, *127*, 137–146. [CrossRef] [PubMed]

93. Park, D.J.; Freitas, T.A.; Wallick, C.J.; Guyette, C.V.; Warn-Cramer, B.J. Molecular dynamics and in vitro analysis of connexin43: A new 14-3-3 mode-1 interacting protein. *Protein Sci.* **2006**, *15*, 2344–2355. [CrossRef] [PubMed]

94. Park, D.J.; Wallick, C.J.; Martyn, K.D.; Lau, A.F.; Jin, C.; Warn-Cramer, B.J. Akt phosphorylates connexin43 on ser373, a "mode-1" binding site for 14-3-3. *Cell Commun. Adhes.* **2007**, *14*, 211–226. [CrossRef] [PubMed]

95. Solan, J.L.; Lampe, P.D. Connexin43 phosphorylation: Structural changes and biological effects. *Biochem. J.* **2009**, *419*, 261–272. [CrossRef] [PubMed]

96. Cooper, C.D.; Lampe, P.D. Casein kinase 1 regulates connexin-43 gap junction assembly. *J. Biol. Chem.* **2002**, *277*, 44962–44968. [CrossRef] [PubMed]

97. Remo, B.F.; Qu, J.; Volpicelli, F.M.; Giovannone, S.; Shin, D.; Lader, J.; Liu, F.Y.; Zhang, J.; Lent, D.S.; Morley, G.E.; et al. Phosphatase-resistant gap junctions inhibit pathological remodeling and prevent arrhythmias. *Circ. Res.* **2011**, *108*, 1459–1466. [CrossRef] [PubMed]

98. Giepmans, B.N. Role of connexin43-interacting proteins at gap junctions. *Adv. Cardiol.* **2006**, *42*, 41–56. [PubMed]

99. Giepmans, B.N. Gap junctions and connexin-interacting proteins. *Cardiovasc. Res.* **2004**, *62*, 233–245. [CrossRef] [PubMed]

100. Kanemitsu, M.Y.; Loo, L.W.; Simon, S.; Lau, A.F.; Eckhart, W. Tyrosine phosphorylation of connexin 43 by v-src is mediated by sh2 and sh3 domain interactions. *J. Biol. Chem.* **1997**, *272*, 22824–22831. [CrossRef] [PubMed]

101. Lin, R.; Warn-Cramer, B.J.; Kurata, W.E.; Lau, A.F. V-src phosphorylation of connexin 43 on tyr247 and tyr265 disrupts gap junctional communication. *J. Cell Biol.* **2001**, *154*, 815–827. [CrossRef] [PubMed]

102. Solan, J.L.; Lampe, P.D. Connexin 43 in la-25 cells with active v-src is phosphorylated on y247, y265, s262, s279/282, and s368 via multiple signaling pathways. *Cell Commun. Adhes.* **2008**, *15*, 75–84. [CrossRef] [PubMed]

103. Swenson, K.I.; Piwnica-Worms, H.; McNamee, H.; Paul, D.L. Tyrosine phosphorylation of the gap junction protein connexin43 is required for the pp60v-src-induced inhibition of communication. *Cell Regul.* **1990**, *1*, 989–1002. [CrossRef] [PubMed]

104. Zhou, L.; Kasperek, E.M.; Nicholson, B.J. Dissection of the molecular basis of pp60(v-src) induced gating of connexin 43 gap junction channels. *J. Cell Biol.* **1999**, *144*, 1033–1045. [CrossRef] [PubMed]

105. Ambrosi, C.; Ren, C.; Spagnol, G.; Cavin, G.; Cone, A.; Grintsevich, E.E.; Sosinsky, G.E.; Sorgen, P.L. Connexin43 forms supramolecular complexes through non-overlapping binding sites for drebrin, tubulin, and zo-1. *PLoS ONE* **2016**, *11*, e0157073. [CrossRef] [PubMed]

106. Zheng, Li.; Li, H.; Spagnol, G.; Patel, K.; Sorgen, P.L. Src phosphorylation of Cx43 residue Y313 contributes to inhibiting the interaction with Drebrin and gap junction disassembly. *J. Mol. Cell. Cardiol.* (under review).

107. Francis, R.; Xu, X.; Park, H.; Wei, C.J.; Chang, S.; Chatterjee, B.; Lo, C. Connexin43 modulates cell polarity and directional cell migration by regulating microtubule dynamics. *PLoS ONE* **2011**, *6*, e26379. [CrossRef] [PubMed]

108. Li, H.; Spagnol, G.; Naslavsky, N.; Caplan, S.; Sorgen, P.L. Tc-ptp directly interacts with connexin43 to regulate gap junction intercellular communication. *J. Cell Sci.* **2014**, *127*, 3269–3279. [CrossRef] [PubMed]

109. Ai, Z.; Fischer, A.; Spray, D.C.; Brown, A.M.; Fishman, G.I. Wnt-1 regulation of connexin43 in cardiac myocytes. *J. Clin. Investig.* **2000**, *105*, 161–171. [CrossRef] [PubMed]

110. Nakashima, T.; Ohkusa, T.; Okamoto, Y.; Yoshida, M.; Lee, J.K.; Mizukami, Y.; Yano, M. Rapid electrical stimulation causes alterations in cardiac intercellular junction proteins of cardiomyocytes. *Am. J. Physiol. Heart Circ. Physiol.* **2014**, *306*, H1324–H1333. [CrossRef] [PubMed]

111. Swope, D.; Cheng, L.; Gao, E.; Li, J.; Radice, G.L. Loss of cadherin-binding proteins beta-catenin and plakoglobin in the heart leads to gap junction remodeling and arrhythmogenesis. *Mol. Cell. Biol.* **2012**, *32*, 1056–1067. [CrossRef] [PubMed]

112. Wang, H.X.; Gillio-Meina, C.; Chen, S.; Gong, X.Q.; Li, T.Y.; Bai, D.; Kidder, G.M. The canonical wnt2 pathway and fsh interact to regulate gap junction assembly in mouse granulosa cells. *Biol. Reprod.* **2013**, *89*, 39. [CrossRef] [PubMed]

113. Rinaldi, F.; Hartfield, E.M.; Crompton, L.A.; Badger, J.L.; Glover, C.P.; Kelly, C.M.; Rosser, A.E.; Uney, J.B.; Caldwell, M.A. Cross-regulation of connexin43 and beta-catenin influences differentiation of human neural progenitor cells. *Cell Death Dis.* **2014**, *5*, e1017. [CrossRef] [PubMed]

114. Spagnol, G.; Trease, A.J.; Zheng, Li.; Phillips, A.; Sorgen, P.L. Regulation of Connexin43 by the direct interaction with β-catenin. *Int. J. Mol. Sci.* (under review).

115. Giepmans, B.N.; Hengeveld, T.; Postma, F.R.; Moolenaar, W.H. Interaction of c-src with gap junction protein connexin-43. Role in the regulation of cell-cell communication. *J. Biol. Chem.* **2001**, *276*, 8544–8549. [CrossRef] [PubMed]

116. Gilleron, J.; Fiorini, C.; Carette, D.; Avondet, C.; Falk, M.M.; Segretain, D.; Pointis, G. Molecular reorganization of cx43, zo-1 and src complexes during the endocytosis of gap junction plaques in response to a non-genomic carcinogen. *J. Cell Sci.* **2008**, *121*, 4069–4078. [CrossRef] [PubMed]

117. Mitra, S.S.; Xu, J.; Nicholson, B.J. Coregulation of multiple signaling mechanisms in pp60v-src-induced closure of cx43 gap junction channels. *J. Membr. Biol.* **2012**, *245*, 495–506. [CrossRef] [PubMed]

118. Pahujaa, M.; Anikin, M.; Goldberg, G.S. Phosphorylation of connexin43 induced by src: Regulation of gap junctional communication between transformed cells. *Exp. Cell Res.* **2007**, *313*, 4083–4090. [CrossRef] [PubMed]

119. Toyofuku, T.; Akamatsu, Y.; Zhang, H.; Kuzuya, T.; Tada, M.; Hori, M. C-src regulates the interaction between connexin-43 and zo-1 in cardiac myocytes. *J. Biol. Chem.* **2001**, *276*, 1780–1788. [CrossRef] [PubMed]

120. Homma, N.; Alvarado, J.L.; Coombs, W.; Stergiopoulos, K.; Taffet, S.M.; Lau, A.F.; Delmar, M. A particle-receptor model for the insulin-induced closure of connexin43 channels. *Circ. Res.* **1998**, *83*, 27–32. [CrossRef] [PubMed]

121. Cottrell, G.T.; Lin, R.; Warn-Cramer, B.J.; Lau, A.F.; Burt, J.M. Mechanism of v-src- and mitogen-activated protein kinase-induced reduction of gap junction communication. *Am. J. Physiol. Cell Physiol.* **2003**, *284*, C511–C520. [CrossRef] [PubMed]

122. Saidi Brikci-Nigassa, A.; Clement, M.J.; Ha-Duong, T.; Adjadj, E.; Ziani, L.; Pastre, D.; Curmi, P.A.; Savarin, P. Phosphorylation controls the interaction of the connexin43 c-terminal domain with tubulin and microtubules. *Biochemistry* **2012**, *51*, 4331–4342. [CrossRef] [PubMed]

123. Butkevich, E.; Hulsmann, S.; Wenzel, D.; Shirao, T.; Duden, R.; Majoul, I. Drebrin is a novel connexin-43 binding partner that links gap junctions to the submembrane cytoskeleton. *Curr. Biol.* **2004**, *14*, 650–658. [CrossRef] [PubMed]

124. Kieken, F.; Mutsaers, N.; Dolmatova, E.; Virgil, K.; Wit, A.L.; Kellezi, A.; Hirst-Jensen, B.J.; Duffy, H.S.; Sorgen, P.L. Structural and molecular mechanisms of gap junction remodeling in epicardial border zone myocytes following myocardial infarction. *Circ. Res.* **2009**, *104*, 1103–1112. [CrossRef] [PubMed]

125. Li, H.; Spagnol, G.; Zheng, L.; Stauch, K.L.; Sorgen, P.L. Regulation of connexin43 function and expression by tyrosine kinase 2. *J. Biol. Chem.* **2016**, *291*, 15867–15880. [CrossRef] [PubMed]

126. Yamaoka, K.; Saharinen, P.; Pesu, M.; Holt, V.E., 3rd; Silvennoinen, O.; O'Shea, J.J. The janus kinases (jaks). *Genome Biol.* **2004**, *5*, 253. [CrossRef] [PubMed]

127. Girault, J.A.; Labesse, G.; Mornon, J.P.; Callebaut, I. The n-termini of fak and jaks contain divergent band 4.1 domains. *Trends Biochem. Sci.* **1999**, *24*, 54–57. [CrossRef]

128. Haan, C.; Kreis, S.; Margue, C.; Behrmann, I. Jaks and cytokine receptors—An intimate relationship. *Biochem. Pharmacol.* **2006**, *72*, 1538–1546. [CrossRef] [PubMed]

129. Wilks, A.F.; Harpur, A.G.; Kurban, R.R.; Ralph, S.J.; Zurcher, G.; Ziemiecki, A. Two novel protein-tyrosine kinases, each with a second phosphotransferase-related catalytic domain, define a new class of protein kinase. *Mol. Cell Biol.* **1991**, *11*, 2057–2065. [CrossRef] [PubMed]

130. Huibregtse, J.M.; Scheffner, M.; Beaudenon, S.; Howley, P.M. A family of proteins structurally and functionally related to the e6-ap ubiquitin-protein ligase. *Proc. Natl. Acad. Sci. USA* **1995**, *92*, 2563–2567. [CrossRef] [PubMed]

131. Leykauf, K.; Salek, M.; Bomke, J.; Frech, M.; Lehmann, W.D.; Durst, M.; Alonso, A. Ubiquitin protein ligase nedd4 binds to connexin43 by a phosphorylation-modulated process. *J. Cell Sci.* **2006**, *119*, 3634–3642. [CrossRef] [PubMed]

132. Spagnol, G.; Kieken, F.; Kopanic, J.L.; Li, H.; Zach, S.; Stauch, K.L.; Grosely, R.; Sorgen, P.L. Structural studies of the nedd4 ww domains and their selectivity for the connexin43 (cx43) carboxyl terminus. *J. Biol. Chem.* **2016**, *291*, 7637–7650. [CrossRef] [PubMed]

133. Auth, T.; Schluter, S.; Urschel, S.; Kussmann, P.; Sonntag, S.; Hoher, T.; Kreuzberg, M.M.; Dobrowolski, R.; Willecke, K. The tsg101 protein binds to connexins and is involved in connexin degradation. *Exp. Cell Res.* **2009**, *315*, 1053–1062. [CrossRef] [PubMed]

134. Fong, J.T.; Kells, R.M.; Falk, M.M. Two tyrosine-based sorting signals in the cx43 c-terminus cooperate to mediate gap junction endocytosis. *Mol. Biol. Cell* **2013**, *24*, 2834–2848. [CrossRef] [PubMed]

135. Solan, J.L.; Marquez-Rosado, L.; Sorgen, P.L.; Thornton, P.J.; Gafken, P.R.; Lampe, P.D. Phosphorylation at s365 is a gatekeeper event that changes the structure of cx43 and prevents down-regulation by pkc. *J. Cell Biol.* **2007**, *179*, 1301–1309. [CrossRef] [PubMed]

136. Ek-Vitorin, J.F.; King, T.J.; Heyman, N.S.; Lampe, P.D.; Burt, J.M. Selectivity of connexin 43 channels is regulated through protein kinase c-dependent phosphorylation. *Circ. Res.* **2006**, *98*, 1498–1505. [CrossRef] [PubMed]

137. Nimlamool, W.; Andrews, R.M.; Falk, M.M. Connexin43 phosphorylation by pkc and mapk signals vegf-mediated gap junction internalization. *Mol. Biol. Cell* **2015**, *26*, 2755–2768. [CrossRef] [PubMed]

138. Cone, A.C.; Cavin, G.; Ambrosi, C.; Hakozaki, H.; Wu-Zhang, A.X.; Kunkel, M.T.; Newton, A.C.; Sosinsky, G.E. Protein kinase cdelta-mediated phosphorylation of connexin43 gap junction channels causes movement within gap junctions followed by vesicle internalization and protein degradation. *J. Biol. Chem.* **2014**, *289*, 8781–8798. [CrossRef] [PubMed]

139. Fong, J.T.; Nimlamool, W.; Falk, M.M. Egf induces efficient cx43 gap junction endocytosis in mouse embryonic stem cell colonies via phosphorylation of ser262, ser279/282, and ser368. *FEBS Lett.* **2014**, *588*, 836–844. [CrossRef] [PubMed]

140. Johnson, K.E.; Mitra, S.; Katoch, P.; Kelsey, L.S.; Johnson, K.R.; Mehta, P.P. Phosphorylation on ser-279 and ser-282 of connexin43 regulates endocytosis and gap junction assembly in pancreatic cancer cells. *Mol. Biol. Cell* **2013**, *24*, 715–733. [CrossRef] [PubMed]

141. Lampe, P.D. Analyzing phorbol ester effects on gap junctional communication: A dramatic inhibition of assembly. *J. Cell Biol.* **1994**, *127*, 1895–1905. [CrossRef] [PubMed]

142. Lampe, P.D.; TenBroek, E.M.; Burt, J.M.; Kurata, W.E.; Johnson, R.G.; Lau, A.F. Phosphorylation of connexin43 on serine368 by protein kinase c regulates gap junctional communication. *J. Cell Biol.* **2000**, *149*, 1503–1512. [CrossRef] [PubMed]

143. Kirchhausen, T.; Owen, D.; Harrison, S.C. Molecular structure, function, and dynamics of clathrin-mediated membrane traffic. *Cold Spring Harb. Perspect. Biol.* **2014**, *6*, a016725. [CrossRef] [PubMed]

144. Bonifacino, J.S.; Traub, L.M. Signals for sorting of transmembrane proteins to endosomes and lysosomes. *Annu. Rev. Biochem.* **2003**, *72*, 395–447. [CrossRef] [PubMed]

145. Thomas, M.A.; Zosso, N.; Scerri, I.; Demaurex, N.; Chanson, M.; Staub, O. A tyrosine-based sorting signal is involved in connexin43 stability and gap junction turnover. *J. Cell Sci.* **2003**, *116*, 2213–2222. [CrossRef] [PubMed]

146. Kittler, J.T.; Chen, G.; Kukhtina, V.; Vahedi-Faridi, A.; Gu, Z.; Tretter, V.; Smith, K.R.; McAinsh, K.; Arancibia-Carcamo, I.L.; Saenger, W.; et al. Regulation of synaptic inhibition by phospho-dependent binding of the ap2 complex to a yecl motif in the gabaa receptor gamma2 subunit. *Proc. Natl. Acad. Sci. USA* **2008**, *105*, 3616–3621. [CrossRef] [PubMed]

147. Huang, R.Y.; Laing, J.G.; Kanter, E.M.; Berthoud, V.M.; Bao, M.; Rohrs, H.W.; Townsend, R.R.; Yamada, K.A. Identification of camkii phosphorylation sites in connexin43 by high-resolution mass spectrometry. *J. Proteome Res.* **2011**, *10*, 1098–1109. [CrossRef] [PubMed]

148. Procida, K.; Jorgensen, L.; Schmitt, N.; Delmar, M.; Taffet, S.M.; Holstein-Rathlou, N.H.; Nielsen, M.S.; Braunstein, T.H. Phosphorylation of connexin43 on serine 306 regulates electrical coupling. *Heart Rhythm* **2009**, *6*, 1632–1638. [CrossRef] [PubMed]

149. Shifman, J.M.; Choi, M.H.; Mihalas, S.; Mayo, S.L.; Kennedy, M.B. Ca2+/calmodulin-dependent protein kinase ii (camkii) is activated by calmodulin with two bound calciums. *Proc. Natl. Acad. Sci. USA* **2006**, *103*, 13968–13973. [CrossRef] [PubMed]

150. Braun, A.P.; Schulman, H. The multifunctional calcium/calmodulin-dependent protein kinase: From form to function. *Annu. Rev. Physiol.* **1995**, *57*, 417–445. [CrossRef] [PubMed]

151. Zhou, Y.; Yang, W.; Lurtz, M.M.; Ye, Y.; Huang, Y.; Lee, H.W.; Chen, Y.; Louis, C.F.; Yang, J.J. Identification of the calmodulin binding domain of connexin 43. *J. Biol. Chem.* **2007**, *282*, 35005–35017. [CrossRef] [PubMed]

152. Zou, J.; Salarian, M.; Chen, Y.; Veenstra, R.; Louis, C.F.; Yang, J.J. Gap junction regulation by calmodulin. *FEBS Lett.* **2014**, *588*, 1430–1438. [CrossRef] [PubMed]

153. Spagnol, G.; Chenavas, S.; Trease, A.; Li, H.; Kieken, F.; Brownell, S.; Sorgen, P.L. Characterizing the interaction between calmodulin and the Cx43 cytoplasmic domains (manuscript in preparation).

154. Dbouk, H.A.; Mroue, R.M.; El-Sabban, M.E.; Talhouk, R.S. Connexins: A myriad of functions extending beyond assembly of gap junction channels. *Cell Commun. Signal.* **2009**, *7*, 4. [CrossRef] [PubMed]

155. Boassa, D.; Solan, J.L.; Papas, A.; Thornton, P.; Lampe, P.D.; Sosinsky, G.E. Trafficking and recycling of the connexin43 gap junction protein during mitosis. *Traffic* **2010**, *11*, 1471–1486. [CrossRef] [PubMed]

156. Kanemitsu, M.Y.; Jiang, W.; Eckhart, W. Cdc2-mediated phosphorylation of the gap junction protein, connexin43, during mitosis. *Cell Growth Differ.* **1998**, *9*, 13–21. [PubMed]

157. Lampe, P.D.; Kurata, W.E.; Warn-Cramer, B.J.; Lau, A.F. Formation of a distinct connexin43 phosphoisoform in mitotic cells is dependent upon p34cdc2 kinase. *J. Cell Sci.* **1998**, *111*, 833–841. [PubMed]

158. Stein, L.S.; Boonstra, J.; Burghardt, R.C. Reduced cell-cell communication between mitotic and nonmitotic coupled cells. *Exp. Cell Res.* **1992**, *198*, 1–7. [CrossRef]

159. Lindsey, M.L.; Escobar, G.P.; Mukherjee, R.; Goshorn, D.K.; Sheats, N.J.; Bruce, J.A.; Mains, I.M.; Hendrick, J.K.; Hewett, K.W.; Gourdie, R.G.; et al. Matrix metalloproteinase-7 affects connexin-43 levels, electrical conduction, and survival after myocardial infarction. *Circulation* **2006**, *113*, 2919–2928. [CrossRef] [PubMed]

160. Kowluru, R.A.; Mohammad, G.; dos Santos, J.M.; Zhong, Q. Abrogation of mmp-9 gene protects against the development of retinopathy in diabetic mice by preventing mitochondrial damage. *Diabetes* **2011**, *60*, 3023–3033. [CrossRef] [PubMed]

161. Mohammad, G.; Kowluru, R.A. Novel role of mitochondrial matrix metalloproteinase-2 in the development of diabetic retinopathy. *Invest. Ophthalmol. Vis. Sci.* **2011**, *52*, 3832–3841. [CrossRef] [PubMed]

162. Wu, X.; Huang, W.; Luo, G.; Alain, L.A. Hypoxia induces connexin 43 dysregulation by modulating matrix metalloproteinases via mapk signaling. *Mol. Cell Biochem.* **2013**, *384*, 155–162. [CrossRef] [PubMed]

163. Vermij, S.H.; Abriel, H.; van Veen, T.A. Refining the molecular organization of the cardiac intercalated disc. *Cardiovasc. Res.* **2017**, *113*, 259–275. [CrossRef] [PubMed]

164. Martins-Marques, T.; Anjo, S.I.; Pereira, P.; Manadas, B.; Girao, H. Interacting network of the gap junction (gj) protein connexin43 (cx43) is modulated by ischemia and reperfusion in the heart. *Mol. Cell Proteom.* **2015**, *14*, 3040–3055. [CrossRef] [PubMed]

165. Martins-Marques, T.; Catarino, S.; Marques, C.; Matafome, P.; Ribeiro-Rodrigues, T.; Baptista, R.; Pereira, P.; Girao, H. Heart ischemia results in connexin43 ubiquitination localized at the intercalated discs. *Biochimie* **2015**, *112*, 196–201. [CrossRef] [PubMed]

166. Mollerup, S.; Hofgaard, J.P.; Braunstein, T.H.; Kjenseth, A.; Leithe, E.; Rivedal, E.; Holstein-Rathlou, N.H.; Nielsen, M.S. Norepinephrine inhibits intercellular coupling in rat cardiomyocytes by ubiquitination of connexin43 gap junctions. *Cell Commun. Adhes.* **2011**, *18*, 57–65. [CrossRef] [PubMed]

167. Hofgaard, J.P.; Banach, K.; Mollerup, S.; Jorgensen, H.K.; Olesen, S.P.; Holstein-Rathlou, N.H.; Nielsen, M.S. Phosphatidylinositol-bisphosphate regulates intercellular coupling in cardiac myocytes. *Pflugers Arch.* **2008**, *457*, 303–313. [CrossRef] [PubMed]

168. Totland, M.Z.; Bergsland, C.H.; Fykerud, T.A.; Knudsen, L.M.; Rasmussen, N.L.; Eide, P.W.; Yohannes, Z.; Sorensen, V.; Brech, A.; Lothe, R.A.; et al. The e3 ubiquitin ligase nedd4 induces endocytosis and lysosomal sorting of connexin 43 to promote loss of gap junctions. *J. Cell Sci.* **2017**, *130*, 2867–2882. [CrossRef] [PubMed]

169. Stroemlund, L.W.; Jensen, C.F.; Qvortrup, K.; Delmar, M.; Nielsen, M.S. Gap junctions-guards of excitability. *Biochem. Soc. Trans.* **2015**, *43*, 508–512. [CrossRef] [PubMed]

170. Gutstein, D.E.; Morley, G.E.; Tamaddon, H.; Vaidya, D.; Schneider, M.D.; Chen, J.; Chien, K.R.; Stuhlmann, H.; Fishman, G.I. Conduction slowing and sudden arrhythmic death in mice with cardiac-restricted inactivation of connexin43. *Circ. Res.* **2001**, *88*, 333–339. [CrossRef] [PubMed]

171. Van Rijen, H.V.; Eckardt, D.; Degen, J.; Theis, M.; Ott, T.; Willecke, K.; Jongsma, H.J.; Opthof, T.; de Bakker, J.M. Slow conduction and enhanced anisotropy increase the propensity for ventricular tachyarrhythmias in adult mice with induced deletion of connexin43. *Circulation* **2004**, *109*, 1048–1055. [CrossRef] [PubMed]

172. Desplantez, T.; McCain, M.L.; Beauchamp, P.; Rigoli, G.; Rothen-Rutishauser, B.; Parker, K.K.; Kleber, A.G. Connexin43 ablation in foetal atrial myocytes decreases electrical coupling, partner connexins, and sodium current. *Cardiovasc. Res.* **2012**, *94*, 58–65. [CrossRef] [PubMed]

173. Jansen, J.A.; Noorman, M.; Musa, H.; Stein, M.; de Jong, S.; van der Nagel, R.; Hund, T.J.; Mohler, P.J.; Vos, M.A.; van Veen, T.A.; et al. Reduced heterogeneous expression of cx43 results in decreased Na$_V$1.5 expression and reduced sodium current that accounts for arrhythmia vulnerability in conditional cx43 knockout mice. *Heart Rhythm* **2012**, *9*, 600–607. [CrossRef] [PubMed]

174. Lubkemeier, I.; Requardt, R.P.; Lin, X.; Sasse, P.; Andrie, R.; Schrickel, J.W.; Chkourko, H.; Bukauskas, F.F.; Kim, J.S.; Frank, M.; et al. Deletion of the last five c-terminal amino acid residues of connexin43 leads to lethal ventricular arrhythmias in mice without affecting coupling via gap junction channels. *Basic Res. Cardiol.* **2013**, *108*, 348. [CrossRef] [PubMed]

175. Agullo-Pascual, E.; Lin, X.; Leo-Macias, A.; Zhang, M.; Liang, F.X.; Li, Z.; Pfenniger, A.; Lubkemeier, I.; Keegan, S.; Fenyo, D.; et al. Super-resolution imaging reveals that loss of the c-terminus of connexin43 limits microtubule plus-end capture and Na$_V$1.5 localization at the intercalated disc. *Cardiovasc. Res.* **2014**, *104*, 371–381. [CrossRef] [PubMed]

176. Rhett, J.M.; Gourdie, R.G. The perinexus: A new feature of cx43 gap junction organization. *Heart Rhythm* **2012**, *9*, 619–623. [CrossRef] [PubMed]

177. Franke, W.W.; Borrmann, C.M.; Grund, C.; Pieperhoff, S. The area composita of adhering junctions connecting heart muscle cells of vertebrates. I. Molecular definition in intercalated disks of cardiomyocytes by immunoelectron microscopy of desmosomal proteins. *Eur. J. Cell Biol.* **2006**, *85*, 69–82. [CrossRef] [PubMed]

178. Agullo-Pascual, E.; Reid, D.A.; Keegan, S.; Sidhu, M.; Fenyo, D.; Rothenberg, E.; Delmar, M. Super-resolution fluorescence microscopy of the cardiac connexome reveals plakophilin-2 inside the connexin43 plaque. *Cardiovasc. Res.* **2013**, *100*, 231–240. [CrossRef] [PubMed]

179. Agullo-Pascual, E.; Delmar, M. The noncanonical functions of cx43 in the heart. *J. Membr. Biol.* **2012**, *245*, 477–482. [CrossRef] [PubMed]

180. Oxford, E.M.; Musa, H.; Maass, K.; Coombs, W.; Taffet, S.M.; Delmar, M. Connexin43 remodeling caused by inhibition of plakophilin-2 expression in cardiac cells. *Circ. Res.* **2007**, *101*, 703–711. [CrossRef] [PubMed]

181. Agullo-Pascual, E.; Cerrone, M.; Delmar, M. Arrhythmogenic cardiomyopathy and brugada syndrome: Diseases of the connexome. *FEBS Lett.* **2014**, *588*, 1322–1330. [CrossRef] [PubMed]

182. Leo-Macias, A.; Agullo-Pascual, E.; Delmar, M. The cardiac connexome: Non-canonical functions of connexin43 and their role in cardiac arrhythmias. *Semin. Cell Dev. Biol.* **2016**, *50*, 13–21. [CrossRef] [PubMed]

183. Beyer, E.C.; Paul, D.L.; Goodenough, D.A. Connexin43: A protein from rat heart homologous to a gap junction protein from liver. *J. Cell Biol.* **1987**, *105*, 2621–2629. [CrossRef] [PubMed]

International Journal of
Molecular Sciences

MDPI

Article

Cx43 Channel Gating and Permeation: Multiple Phosphorylation-Dependent Roles of the Carboxyl Terminus

José F. Ek-Vitorín *, Tasha K. Pontifex and Janis M. Burt

Department of Physiology, University of Arizona, P.O. Box 245051, Tucson, AZ 85724, USA;
tasha@email.arizona.edu (T.K.P.); jburt@email.arizona.edu (J.M.B.)
* Correspondence: ekvitori@email.arizona.edu; Tel.: +1-520-626-1351

Received: 25 April 2018; Accepted: 31 May 2018; Published: 4 June 2018

Abstract: Connexin 43 (Cx43), a gap junction protein seemingly fit to support cardiac impulse propagation and synchronic contraction, is phosphorylated in normoxia by casein kinase 1 (CK1). However, during cardiac ischemia or pressure overload hypertrophy, this phosphorylation fades, Cx43 abundance decreases at intercalated disks and increases at myocytes' lateral borders, and the risk of arrhythmia rises. Studies in wild-type and transgenic mice indicate that enhanced CK1-phosphorylation of Cx43 protects from arrhythmia, while dephosphorylation precedes arrhythmia vulnerability. The mechanistic bases of these Cx43 (de)phosphoform-linked cardiac phenotypes are unknown. We used patch-clamp and dye injection techniques to study the channel function (gating, permeability) of Cx43 mutants wherein CK1-targeted serines were replaced by aspartate (Cx43-CK1-D) or alanine (Cx43-CK1-A) to emulate phosphorylation and dephosphorylation, respectively. Cx43-CK1-D, but not Cx43-CK1-A, displayed high Voltage-sensitivity and variable permselectivity. Both mutants showed multiple channel open states with overall increased conductivity, resistance to acidification-induced junctional uncoupling, and hemichannel openings in normal external calcium. Modest differences in the mutant channels' function and regulation imply the involvement of dissimilar structural conformations of the interacting domains of Cx43 in electrical and chemical gating that may contribute to the divergent phenotypes of CK1-(de)phospho-mimicking Cx43 transgenic mice and that may bear significance in arrhythmogenesis.

Keywords: casein kinase 1; phosphorylation; channel gating; gap junction permeability; arrhythmia

1. Introduction

Gap junction channels made of Connexin (Cx) proteins support the propagation of electrical impulses from cell to cell, and are therefore indispensable for synchronic heart contractions. Cx43, the most widely distributed cardiac Cx and abundant in all four chambers [1], can compensate for atrial Cx40 deficiency and prevent fibrillation [2]. However, replacing or supplementing Cx43 with other Cx isotypes promotes arrhythmia vulnerability [3,4]. These data suggest that Cx43 is optimally endowed to support cardiac conduction. While such uniqueness may derive from Cx43 amenability to regulation by multiple kinases [5,6], the functional ramifications of such regulation remain poorly defined. To address this deficit, we aim to outline the operational profiles of Cx43 modified by specific phosphorylation events.

In normal heart, Casein Kinase 1 (CK1) phosphorylates serines 325, 328 and 330 of the Cx43 protein residing at gap junctions (GJs) of intercalated disks (ID) [7,8]. During acute ischemia or chronic pressure overload hypertrophy (induced by transverse aortic constriction, TAC), phosphorylation of these serine residues is greatly reduced, while total Cx43 protein decreases at the ID and increases at the lateral sides of cardiomyocytes [8,9]. This gap junction remodeling (GJR) alone might slow impulse propagation and contribute to arrhythmia. In accordance, spironolactone, an aldosterone antagonist with beneficial effects in patients with heart failure [10], prevents dephosphorylation of the CK1 sites and reduces or reverses pathological GJR [11]. These data suggest that reversible CK1-phosphorylation is a pivotal regulatory event that establishes the fate of Cx43 protein/channels during the transition from physiological to pathological states. In agreement, hearts from transgenic mice expressing a CK1-dephospho-mimicking mutant of Cx43 (substitutions S325,328,330A, named S3A) exhibit enhanced GJR and high propensity to arrhythmias after ischemia or TAC-induced hypertrophy. A converse CK1-phospho-mimicking mutant (substitutions S325,328,330E, named S3E) had opposite effects, that is, hearts were resistant to arrhythmia induction and to pathological GJR after ischemia or TAC [9]. Despite the clinical interest of such Cx43 (de)phosphoform-linked cardiac phenotypes their mechanistic (functional) bases remain unknown.

We hypothesized that differences in channel gating/permeability, linked to the phosphorylation status of Cx43, contribute to the cardiac phenotypes of S3A and S3E mice. Therefore, we assessed gating and permeability of mutant Cx43-S325,328,330D (here dubbed Cx43-CK1-D), where aspartates supply the charge effect of phosphorylation, and mutant Cx43-S325,328,330A (dubbed Cx43-CK1-A), where non-phosphorylatable alanines provide a stable "dephosphorylated" state. While both mutants displayed properties typical of wild-type Cx43 (Cx43WT), they also showed unexpected properties that challenge current models of channel function and may be significant for the cardiac phenotypes of the (S3A, S3E) transgenic mice, as well as for the pathophysiology of cardiac ischemia and arrhythmogenesis.

2. Results

2.1. Cx43-CK1-D Displays Stronger V_j-Sensitivity than Cx43-CK1-A

The decrease in junctional current (I_j) observed in GJs subjected to large transjunctional voltages (V_j) is called V_j-gating (Figure 1A,B) and is quantified by fitting the ratio of steady state and instantaneous junctional conductance (g_j^{ss}/g_j^{inst}) through a broad V_j range with a Boltzmann function [12,13], with half-maximal response reported as V_0 (see Section 4). When measured in rat insulinoma (Rin) cells, Cx43 V_j-gating (Figure S1, Table 1, Tables S1 and S2 and [14]) was similar to that reported for Cx43WT expressed in other cell types [15–17]. Cx43-CK1-A showed V_j-gating comparable to Cx43WT (Figure 1; Table 1, Tables S1 and S2). In contrast, Cx43-CK1-D showed a faster response to V_j (Figure 1A,B) and smaller V_0 values (Figure 1C,D, Table 1, Tables S1 and S2), revealing a more sensitive closure mechanism than Cx43-CK1-A. This suggests that Cx43 phosphorylation by CK1 enhances junctional V_j-gating, and that highly V_j-dependent gap junction channels (GJChs) populate the intercalated disks.

Table 1. Boltzmann parameters of Cx43 WT and mutants.

Mutant	$G_{j\,max}$	$G_{j\,min}$	V_0 (mV)	A
Cx43WT	0.98/0.99	0.32/0.34	−64/+70	4.8/4.1
Cx43-CK1-D	1.1/1.0	0.21/0.34	−49/+46	12.1/5.4
Cx43-CK1-A	1.0/0.98	0.35/0.40	−71/+70	8.4/4.1

Boltzmann fit parameters are shown as rounded values for both negative and positive polarities. Exact values are reported on Table S2. Cx43WT values obtained from experiments are displayed in Figure S1.

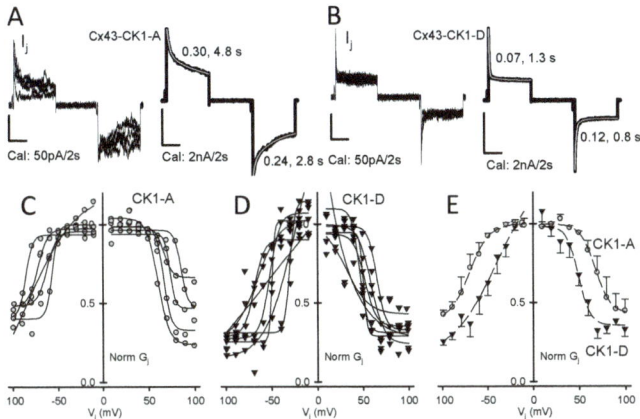

Figure 1. Cx43-CK1-D displays stronger V_j-sensitivity than Cx43-CK1-A. (**A**) Five individual I_j responses to $V_j = \pm 80$ mV pulses (left) and sum (right) of 38 similar traces from Cx43-CK1-A cell pairs; (**B**) Five individual I_j responses to $V_j = \pm 80$ mV pulses (left) and sum (right) of 30 similar traces from Cx43-CK1-D cell pairs. For (**A,B**), tau values of I_j inactivation (2nd order exponential decays) are shown and the fits (white) displayed over the corresponding sum traces; (**C**) V_j-dependence of G_j from individual experiments in Cx43-CK1-A ($g_j = 2.8 \pm 1.0$; $n = 5$) cell pairs; (**D**) V_j-dependence of G_j from individual experiments in Cx43-CK1-D ($g_j = 1.7 \pm 1.1$; $n = 7$) cell pairs. For (**C,D**), g_j from each experiment was normalized as described in the Methods and the Boltzmann fits are shown in solid black lines; (**E**) Average V_j-dependence for CK1-A (gray circles) and CK1-D (black triangles) and their corresponding Boltzmann fits (dashed lines). Fast inactivation and V_0 values were different between Cx43-CK1-D and Cx43-CK1-A. For fitting parameters, see Table 1, Tables S1 and S2.

2.2. Cx43-CK1-D and Cx43-CK1-A Display Highly Conductive, V_j-Sensitive, Channel Transition Amplitudes

As deduced from their unitary conductances (γ_j) at $V_j < V_0$, Cx43 GJChs exist in multiple states: closed (C), fully open (O), residually open (R), and several intermediate open states, or substates (S) [16,18–22]. Thus, while only one fully open state may exist, channels can occupy multiple less conductive configurations. These multiple configurations could result from variable structural conformation of the connexin molecules comprising each channel. Transitions between multiple channel states yield a broad range of apparent γ_j values that seem to vary with the cell type and with the Cx phosphorylation state (cf. [23,24]). In addition, the presence of other Cxs can modify both the γ_j profile and function of predominantly Cx43-comprised GJs [25,26]. It is thus important that recording conditions are standardized and that cells express only the protein of interest. In our current experimental settings [14,22,27,28] and at $V_j < V_0$, the γ_j values for Cx43WT channels are C = 0, O = 100–125, R = 17–35, and S = 55–70 pS. Transitions between these states would produce transition amplitudes of approximately (in pS) 100–125 (O↔C), 17–35 (R↔C), 55–70 (S↔C), 83–90 (O↔R), 45–50 (O↔S), and 35–40 (S↔R). Given the conductance ranges for each transition type, distinguishing each as a distinct peak is generally not possible. $V_j > V_0$ brings GJChs to the residual state (O↔R) from which full closure (R↔C) may occur. Hence, at $V_j < V_0$ Cx43WT displays many transition amplitudes, and at $V_j > V_0$ most likely O↔C, O↔R and R↔C transitions (Figure S2; cf. [16,18,28]). Because Cx43-CK1-D emulates a Cx43 phosphoform found in normoxic hearts (Cx43-p^{CK1}), we expected the distribution of channel transition amplitudes of this mutant to resemble that of Cx43WT. In comparison, Cx43-CK1-A emulates a Cx43 phosphoform (Cx43-dp^{CK1}) found during hypoxia and displays a decreased incidence of fully open channels when expressed in mesenchymal cells [8].

In Rin cells both mutants showed multiple transition amplitudes, including those consistent with a fully open channel (Figure 2A,B,D,E). Differences in transition amplitudes and distribution with respect to each other and to Cx43WT were documented as follows (Figure 2C,F; cf. Figure S1D): At $V_j = 40$ mV

($<<V_0$), main transitions in Cx43-CK1-D were 90–150 pS, and in Cx43 CK1-A, 75–140 pS. In both mutants, R↔C transitions (≤30 pS) were rare. Transitions >150 pS were sporadically observed in Cx43-CK1-D, extending the range of apparent channel amplitudes to values not present in Cx43-CK1-A and unreported for Cx43WT. At $V_j = 80$ mV ($>>V_0$), Cx43-CK1-D mostly displayed 35–55 and 75–120 pS transitions, and a few >150 pS transitions. In comparison, Cx43 CK1-A displayed 25–40 and 60–105 pS transitions, but essentially no >150 pS transitions. Note that while both mutants showed transition amplitudes above the typical 120 pS fully open state, they also displayed intermediate transitions compatible with a substate (80 > S > 40 pS), particularly at $V_j = 80$ mV. Thus, in contrast to Cx43WT [14] both mutants preferentially displayed O↔C and O↔R transitions at small V_j gradients, and O↔R and R↔C transitions at large V_j gradients. On the whole, the data indicate that in the absence of V_j gradients, mutant (and WT) channels may favor a fully open state. In addition, transition amplitude distributions from both mutants suggest a shift toward higher channel conductivity than Cx43WT.

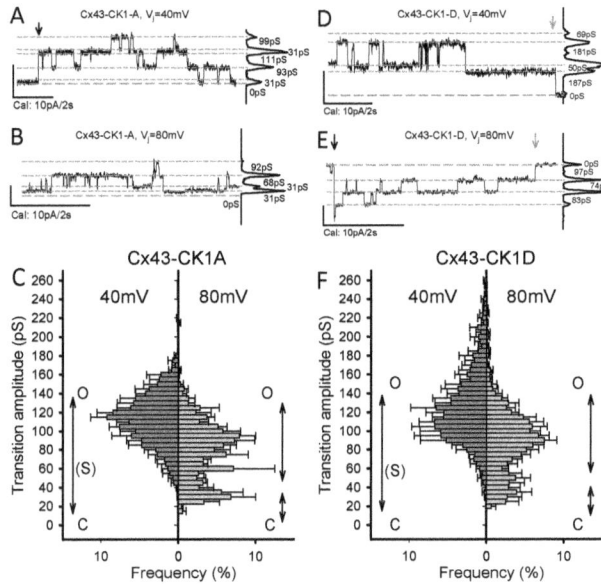

Figure 2. Cx43-CK1-A and Cx43-CK1-D display fully open, V_j-sensitive gap junction channels. (**A,B,D,E**) Illustrative traces of channel activity from Cx43-CK1-A (**A,B**) and Cx43-CK1-D (**D,E**) expressing cell pairs, at 40 mV (**A,D**) and 80 mV (**B,E**) transjunctional gradients. For all traces: zero current (long-dashed line) and the most evident I_j levels (short-dashed lines) are indicated; when present, downward arrows mark the beginning (black) and end (gray) of pulses; plots at right are the all-points histogram for the displayed record segment, showing the fraction of time at each I_j level; numbers indicate the conductance change between current levels. Notice that channel transitions often occur between the identified levels. (**C,F**) Average transition amplitude histograms at 40 and 80 mV V_j values from Cx43-CK1-A (**C**) and Cx43-CK1-D (**F**). Peak fits indicated by solid black lines. Likely transitions between channel states (see main text for further explanation) are indicated by double arrowed vertical lines. Transition amplitude distributions of Cx43-CK1-D and Cx43-CK1-A differed from each other and from Cx43WT, at both *Vj* values of 40 and 80 mV. However, at $V_j = 40$ mV, both mutants displayed transitions amplitudes compatible with O↔C and O↔R transitions (if O = 150 pS and R = 30 pS). Transitions larger than 150 pS were documented for Cx43-CK1-D. At $V_j = 80$ mV, O-R and R-C transitions were more evident for both mutants. However, at both 40 and 80 mV, transitions between closed and levels smaller than fully open states were observed, suggesting substates (S). For each group, the number of experiments (*n*) and measured transitions (N) were respectively, as follows: For CK1-A, 6 and 1867 at 40 mV, 5 and 1080 at 80 mV. For CK1-D, 4 and 1369 at 40 mV, 6 and 1032 at 80 mV.

2.3. Cx43-CK1-A Displayed Lower Permselectivity than Cx43-CK1-D

The appearance of highly conductive channels in the CK1-(de)phospho-mimicking mutants raised the possibility of these mutants displaying higher GJ permselectivity (molecular/atomic permeability: $P_{j\text{-NBD}}/g_j$) than Cx43WT [14,28]. Thus, NBD transjunctional diffusion (illustrated in Figure S3) and g_j were measured in cell pairs expressing either mutant (see Section 4). Cx43-CK1-D expressing cells displayed (~three-fold) larger electrical coupling ($g_j = 30.86 \pm 7.81$ nS; $n = 21$) than those expressing Cx43WT ($g_j = 10.49 \pm 1.48$ nS; $n = 53$); however, permselectivity (Figure 3) of Cx43-CK1-D (0.074 ± 0.022 and Cx43WT (0.094 ± 0.018; [28]) expressing cells was similar. Cx43-CK1-A cells also displayed (~2 fold) greater electrical coupling ($g_j = 24.56 \pm 8.79$ nS; $n = 12$) than Cx43WT cells, but showed lower permselectivity (0.017 ± 0.001) than Cx43-CK1-D or Cx43WT (Figure 3), in agreement with previous studies [8]. It is noteworthy that the variability in permselectivity of Cx43-CK1-A was far less than either Cx43WT or Cx43-CK1-D (6% vs. 19% and 30%, respectively). These data suggest a link between Cx43 phosphorylation by CK1 and variable permselectivity; in contrast, low Cx43 permselectivity may follow dephosphorylation of the CK1 target serines.

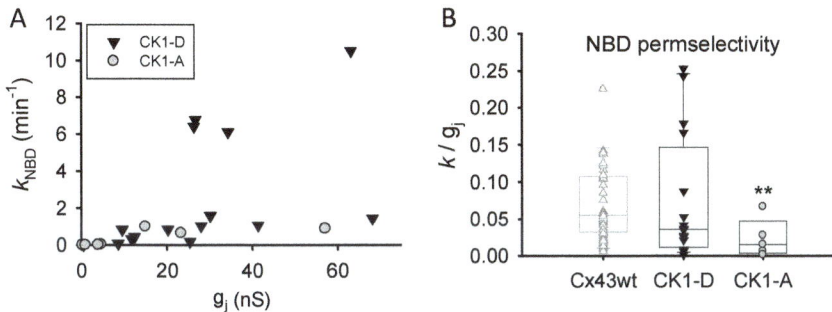

Figure 3. Permselectivity of dephospho-mimicking Cx43-CK1-A GJs is lower and less variable than that of phospho-mimicking Cx43-CK1-D GJs. (**A**) Rate constant of transjunctional dye diffusion vs. junctional conductance for the indicated mutants. Each symbol represents a single experiment; (**B**) Distribution (box plots) of collected permselectivity values for Cx43-CK1-D (0.074 ± 0.022; $n = 16$) and Cx43-CK1-A (0.017 ± 0.001; $n = 7$); Cx43WT data (from [14]) shown in light gray for comparison. Note that permselectivity values of Cx43-CK1-D (and Cx43WT) do not display a normal distribution. ** Median and variance of Cx43-CK1-A are different from Cx43-CK1-D and Cx43-WT ($p < 0.05$). See text for further explanation.

2.4. Cx43-CK1-D and Cx43-CK1-A GJs are Resistant to Intracellular Acidification

In ischemic tissue, internal pH (pH$_i$) falls to very acidic values (pH ~6.0) [29–31], and low pH-induced uncoupling can be arrhythmogenic [32–35]. To determine whether decreased pH-sensitivity of the Cx43 CK1-phosphoform confers the arrhythmia-resistant phenotype of S3E transgenic mice, we documented the response of Cx43-CK1-D g_j to acidification. Unlike Cx43WT GJs, which typically close within ~5 min of exposure to low external pH (pH$_o$), Cx43-CK1-D GJs remained open for > 5 min (Figure 4): ~54% and ~40% of initial g_j lingered at 15 and 20 min, respectively, after the start of acidification. The overall delayed uncoupling reveals a major reduction of pH-sensitivity, suggesting that phosphorylation influences pH-gating of Cx43. In particular, phosphorylation by CK1 may decrease Cx43 susceptibility to low pH-induced uncoupling. Unexpectedly, however, the same acidification protocol also failed to uncouple Cx43-CK1-A GJs (Figure 4). Indeed, Cx43-CK1-A g_j was ~93 and ~90% of initial at 15 and 20 min, respectively, and remained essentially unchanged following 25 min of low pH exposure, suggesting that dephosphorylation of CK1 targeted serines of

Cx43 renders GJChs unable to close upon acidification. These data show that resistance of GJ coupling to acidification alone cannot explain the divergent cardiac phenotypes of transgenic S3A and S3E mice.

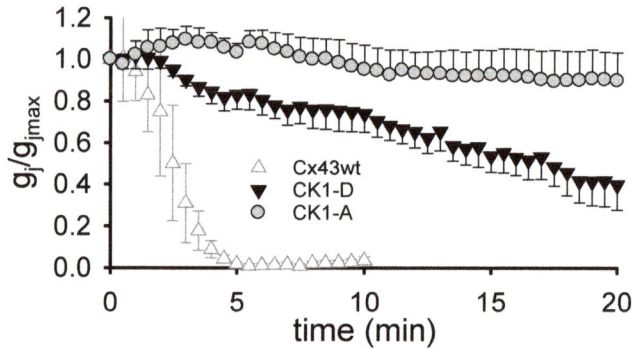

Figure 4. Cx43-CK1-D and Cx43-CK1-A mutants form gap junctions resistant to closure by low pH. Superfusion (starting at time 0) with a bicarbonate solution buffered at pH = 6.0–6.2 caused slow g_j decrease in Cx43-CK1-D (black triangles; $n = 6$), and no g_j decrease in Cx43-CK1-A (gray circles; $n = 4$) during an observation period of $\geq$ 20 min. The response of Cx43WT junctions to similar treatment (white triangles; $n = 5$) is reproduced from [14] for comparison.

2.5. Cx43-CK1-D and -CK1-A Hemichannels Open Frequently

In Cx43WT-expressing cells at normal external calcium ($[Ca^{2+}]_o$) = 1–2 mM (and internal calcium, ($[Ca^{2+}]_i$) from nominally 0 to ~40 nM), connexin hemichannel (HCh) openings are rare and may require depolarization >+60 mV [36–39]. In Rin cells under our experimental conditions (see Section 4), this rarity is exacerbated, and few HCh-like transitions from Cx43WT have been recorded at V_m = 80–100 mV (Figure S4). In contrast, membrane current (I_m) transitions suggestive of HCh openings were seen in several, but not all cells transfected with Cx43-CK1-D (43 out of 127 explored, or ~33%) or Cx43-CK1-A (11 out of 62 explored, or ~18%). The amplitude of presumptive HCh transitions varied for both Cx43-CK1-D (140–280 pS; Figure 5A–C) and for Cx43-CK1-A (100–240 pS; Figure 5D–F). HCh activity of mutant Cx43-CK1-D varied from only a few events per second to profuse transitions (Figure 5A,B), not inactivated during >50 s pulses. In contrast, Cx43-CK1-A displayed less organized, harder to detect HCh activity (Figure 5D,E). Clear step-wise openings could be found in Cx43-CK1-D (Figure 5C), but less often in Cx43-CK1-A expressing cells. However, in three Cx43-CK1-A expressing cells, sudden I_m increases suggesting multiple HCh openings, were seen during prolonged depolarizing pulses. One such I_m increase (and recovery) occurred before the trace shown in Figure 5F, under well controlled V_m clamp. The variability of HCh activities, in particular the infrequent appearance of Cx43-CK1-A events, encumbers wider comparisons between the mutants. Nevertheless, the records suggest that HChs formed by Cx43-CK1-D or Cx43-CK1-A can open under physiological conditions of internal and external Ca^{2+}.

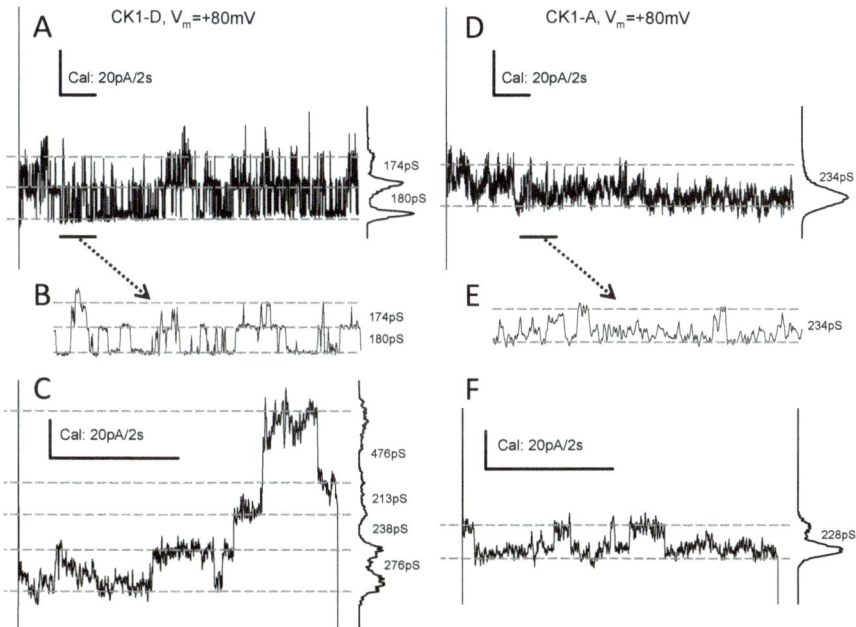

Figure 5. Cx43-CK1-D and Cx43-CK1-A expressing cells display frequent connexin HCh activity. Plasma membrane channel transitions compatible with opening of undocked connexons from single cells expressing Cx43-CK1-D (**A–C**) or Cx43-CK1-A (**D–F**). Lines, marks, plots, and numbers as in Figure 2. (**A,D**) are 20 s samples of longer recordings; (**B,E**) are expanded displays of the 2 s interval marked by black lines at the bottom left of (**A,C**); (**C,F**) Further examples of transitions recorded during 5-s pulses. Hemichannel activity from Cx43-CK1-D is usually well defined (as in **A–C**), in contrast to Cx43-CK1-A (**D–F**).

2.6. Cx43-CK1-D and -CK1-A are Phosphorylated at S368

Our published data suggest PKC phosphorylation at residue S368 increases the dye permselectivity ($P_{j\text{-NBD}}/g_j$) of Cx43 GJs [28]. Hence, we pondered whether the discrepancy between Cx43-CK1-D and Cx43-CK1-A permselectivity might be consequential to differences in S368 phosphorylation. To address this possibility, we used phospho-specific antibodies to determine the presence and location of Cx43 phosphorylated at S368 (Cx43-pS368) in Cx43WT- and mutant-expressing cells. Representative images of this scrutiny show that the Cx43-pS368 epitope was present both at junctional (mainly) and non-junctional membranes, abundantly and in multiple spots in Cx43WT cells (Figure 6A,B). Although fewer cells in the Cx43-CK1-A clone were Cx43 positive, Cx43-pS368 signal was also present in this mutant (Figure 6D,E), with distribution comparable to Cx43WT. In Cx43-CK1-D expressing cells, Cx43-pS368 signal was readily found decorating long cell contacts (Figure 6G,H). Total Cx43 was profusely expressed, with a significant fraction located to junctional plaques, in all cell groups (Figure 6C,F,I, green). In addition, an antibody against CK1-phosphorylated Cx43 demonstrated the presence of this phosphoform in Cx43WT plaques, confirming that Rin cells are able to properly process and place this junctional protein (Figure 6C). These data demonstrate that phosphorylation of S368 occurs in both Cx43 mutants, but the data do not preclude the possibility that high permselectivity (in Cx43-CK1-D) is due to higher phosphorylation of S368 by PKC, and in turn, low permselectivity (in Cx43-CK1-A) results from lower pS368-Cx43.

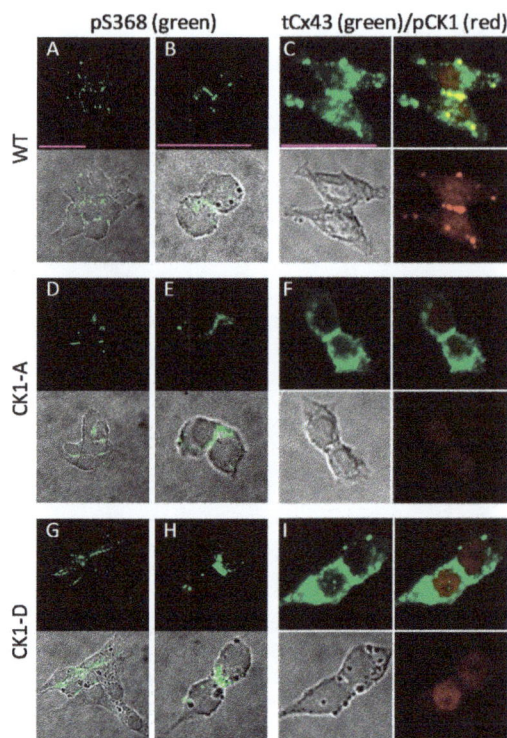

Figure 6. Residue S368 of Cx43 is phosphorylated in Rin cells expressing Cx43-WT, Cx43-CK1-A or Cx43-CK1-D. (**A,B,D,E,G,H**) Fluorescent (upper) and fluorescent merged with the corresponding differential interference contrast (DIC, lower) images of cells stained with a Cx43-pS368 phospho-specific antibody (pS368). Cx43-pS368 (green) was found in junctional plaques in groups and isolated pairs of Cx43WT (**A,B**), Cx43-CK1-A (**D,E**) and Cx43-CK1-D (**G,H**) cells. (**C,F,I**) DIC and fluorescent images of cells stained simultaneously with a polyclonal Cx43 antibody ("total" Cx43, tCx43) and a phospho-specific antibody against Cx43-p^{CK1} (pCK1). In all groups, total Cx43 (green) was found in junctional plaques and other cell areas (**C,F,I**, upper left panels). Bona fide CK1-phosphorylated Cx43 (red) was found only in Cx43WT cells (**C**, bottom right) colocalized with a membrane fraction of total Cx43 at junctional plaques (**C**, upper right, yellow). The pCK1 antibody labels a non-specific nuclear signal in the dephospho- (**F**) and phospho-mimicking (**I**) mutant expressing cells that is also found in parental Rin cells, devoid of connexins (Figure S5). Calibration bars (pink lines, 25 µm) apply to each column.

3. Discussion

3.1. On the Role Cx43 of Phosphorylation

The cytoplasmic carboxyl terminus (CT) domain of Cx43 harbors several kinase consensus sites [5,40], but whether phosphorylation at these sites occurs independently or in tandem, transiently or continuously, alternately or progressively, remains unclear. Nevertheless, (de)phosphorylation at specific sites has been linked to specific Cx43 cellular locations [41–47]. How these phosphorylation differences and their associated structural modifications translate into functional changes in the context of cellular phenomena where Cx43 plays a role (e.g., cell cycle progression and proliferation [48–51], healing [52–54], migration [55–58]; differentiation [59–61] and electrical coupling [34,35,41,62,63]), remains undetermined. Here, we studied the electrical/chemical gating and the permselectivity of Cx43 (de)phospho-mimicking mutants associated to disparate cardiac phenotypes in transgenic mice (arrhythmia susceptibility vs. arrhythmia resistance). We found differences in channel function/regulation that may contribute to these cardiac phenotypes. Our data may be relevant to the pathophysiology of cardiac ischemia.

3.2. What the Data Suggest

3.2.1. Voltage Gating and Channel States

Cx43-CK1-D and Cx43-CK1-A display differences in GJ V_j-sensitivity and channel behavior. The fast exponential component of V_j-gating may involve an interaction of the Carboxyl Terminus (CT) domain with a receptor-like structure near the pore mouth, formed by a region of the Cytoplasmic Loop (CL) [19,64,65]. Our V_j-gating data suggest that CK1-phosphorylation enhances V_j-sensitivity of Cx43 residing at the ID, and that the residual state would be more readily occupied by this phosphoform (Cx43-p^{CK1}) when subjected to high V_j gradients. In turn, the channel data suggest that at low V_j gradients (the most likely environment of GJs), both the dephospho-(Cx43-dp^{CK1}) and the phosphoform exist in either fully open or closed states, and at high V_j gradients, both transit frequently between fully open and residual states, from which they can close.

Because at low V_j and despite their charge differences, both mutants show larger channel transitions than Cx43WT, it is possible that high conductivity results from structural modifications of the mutated region rather than different charge polarity/density. Alternatively, the large unitary conductance (γ_j) of mutant channels could reflect the "homomeric" nature of these channels (at least in regard to the CK1 targets) compared to Cx43WT, where differences in phosphorylation state between Cx subunits certainly must exist. Large γ_j values in mutants representing the extremes of Cx43 phosphorylation by CK1 cannot readily explain their opposite cardiac phenotypes. However, the Cx43-CK1-D data suggest that phosphorylation by CK1 enhances conductivity of Cx43 channels, as evinced by the larger O↔C and O↔R transitions (at low V_j), and the wider residual state range (at high V_j), than either Cx43-CK1-A or Cx43WT. Of note, oversized transitions were not observed in all Cx43-CK1-D cell pairs and did not establish a predominant γ_j peak, suggesting that CK1-phosphorylation alone is not sufficient to yield this highly conductive channel configuration. Another possibility is that large transitions are double channel events brought about by the high V_j-sensitivity of the mutant, which cannot be discriminated under our recording conditions.

It is important to consider here the possible meaning of multiple channel transition amplitudes and their relative abundance. Such multiplicity is observed at $V_j < V_0$ as well as $> V_0$, in Cx43WT as well as Cx43-CK1-D and Cx43-CK1-A mutants. This multiplicity of transition amplitudes indicates that, in the absence of V_j gradients, Cx43 channels reside stably ($\geq$50 ms) in multiple open configurations, each of which can display V_j sensitivity. If the only function of GJChs were to establish an electrical pathway between cells, the only relevant measure would be the absolute value of g_j, irrespective of the multiple channel states that inhabit the junctional plaque. What, then, is the benefit of supporting multiple open states? It has been shown that electrical coupling and dye coupling are not linearly correlated [28,66]; perhaps we should ask whether and how multiple channel states regulate the intercellular flux of the "metabolites and second messengers" so often mentioned in the literature.

3.2.2. pH-Gating

Susceptibility to acidification-induced uncoupling is a common trait of GJs, but pH-sensitivity varies among connexins [67–70]. As with V_j-gating, a CT-CL interaction has also been proposed as the pH-gating mechanism [64]. Uncoupling of Cx43 GJs is quickly induced by superfusing cells with weak acid solutions buffered at pH$_o$ ~6.0–6.2 [68,71], which pH$_i$ readily follows [72]. In ischemic tissue, pH$_i$ may fall to similar low values [29–31,73]. Therefore, during myocardial ischemia low pH$_i$ could contribute to the decreased electrical coupling in damaged tissue, setting the stage for arrhythmogenesis [32–35]. In this scenario, delayed closure upon acidification and transient preservation of electrical coupling at the cells' end-to-end GJs could temporarily protect impulse conduction, cardiac anisotropy, and metabolic rescue of failing cells, all favoring preservation of function early in an ischemic episode. However, during continuous ischemia, complete GJ uncoupling could limit tissue damage, to the detriment of the still living cells in the infarct area, but to the benefit of the organ. Thus, the transient resistance to acidification of Cx43-CK1-D would be consistent with the arrhythmia-resilience of S3E mice. On the other hand, persistent junctional coupling during prolonged acidification would negate the advantages of a transient preservation of coupling, by allowing the steady diffusion of noxious substances from the damaged area to surrounding tissue [32,33]. If this were the case, the imperviousness of Cx43-CK1-A to acidification-induced closure would be consistent with the arrhythmia-vulnerability of S3A mice, where increasingly larger lesions may facilitate arrhythmia. This possibility does not necessarily exclude TAC-induced arrhythmia susceptibility, a chronic condition where acute acidification is not a known factor, but where junctional permselectivity to regulatory molecules may play an important role.

3.2.3. Channel Selectivity

The permselective variability of homomeric, homotypic Cx43 GJs has been linked to variable phosphorylation of the Cxs comprising the channels [28]. Results from Cx43-CK1-D and Cx43-CK1-A support the notion that permselectivity is an inherent property of Cx43 channels, not directly linked to the level of electrical coupling, but to the phosphorylation-induced functional states of the channels comprising the junction. Explicitly, CK1 phosphorylation may provide part of, or be permissive to, the permselective variability of Cx43 junctions. In turn, dephosphorylation of CK1 sites may be less permissive to, or limit (but not completely prevent) such variability. It was also proposed that decreasing permselectivity with unchanging g_j was due to a decreasing proportion of highly permeable channels within the junction [74]. Our results from Cx43-CK1-A and previous data showing low dye (LY) coupling in mesenchymal cells expressing this mutant [8], are in agreement with that idea.

At this point, we should consider possible differences between the transgenic models and the fate of Cx43WT phosphoforms during pathological events. First, Cx43-CK1-A showed lower and less variable permselectivity ($P_{j\text{-NBD}}/g_j$) than Cx43WT, suggesting that despite its likely contribution to persistent electrical coupling (in the mouse model), this mutant's ability to support transjunctional diffusion of organic ions (e.g., signaling molecules, nutrients, and waste products) may be limited. Whether this reduced permselectivity would make Cx43-CK1-A (or Cx43-dp^{CK1}) in the infarct border zone more liable to propagating damage (bystander effect) or efficient in protecting non-injured tissue (metabolic rescue) requires further study. Interestingly, in transgenic mice with cardiac-selective Cx45 overexpression, increased arrhythmia vulnerability was linked to altered permselectivity (lower LY and higher biotin coupling) between myocytes [4], suggesting that positively charged junctional permeants in the molecular size range of biotin may be of interest in arrhythmogenesis.

Second, during ischemia (or pressure overload hypertrophy), Cx43WT dephosphorylated at the CK1 sites may be phosphorylated or dephosphorylated at other sites as it moves away from the IDs. Judging by its electrophoretic mobility alone, the Cx43-CK1-A mutant would appear overall as dephosphorylated [9]; but despite being a "CK1-dephosphoform" this mutant generates (electrically) stable GJs, even in conditions of low pH$_i$, and therefore, may not accurately represent the Cx43-dp^{CK1}. Possible functional effects of kinases targeting other Cx43 sites (see below) during ischemia and GJR, and the fate of Cx43-dp^{CK1} remain subjects of study.

3.2.4. Hemichannels

Cardiac ischemia- and TAC-induced GJR involve an apparent mobilization of Cx43 from IDs to lateral membranes of myocytes [8,11]. It is uncertain whether this lateralized protein pool makes functional channels (as docked connexons or undocked HChs), but reports suggest that massive HCh opening is a pathophysiological occurrence in the aftermath of myocardial ischemia [75–79]. Cx43-CK1-D and Cx43-CK1-A HChs can display high activity in the presence of normal [Ca^{2+}]$_o$.

To date, only one other Cx43 construct, Cx43*NT37 [14], has displayed robust HCh activity in our experimental settings, but this chimera, with exceptional inability to respond to several gating triggers, is an unnatural construct made to explore the interaction of Cx domains. In contrast, Cx43-CK1-D and Cx43-CK1-A emulate natural states of the protein. This raises the questions of whether the massive HCh opening during metabolic inhibition or ischemia is indeed linked to stable CK1-dependent Cx43 phosphoforms or, alternatively, the mutants' HChs are artifactual. As shown previously [8, 11], Cx43-p^{CK1} disappears from IDs during GJR. In contrast, CK1-A and CK1-D mutants seem to exist (and stay) in junctional and in non-junctional membranes ([9] and our current data), even in pathological conditions.

If the Cx43-CK1-D and Cx43-CK1-A mutants replicate phosphorylated states of Cx43WT, then the corresponding alternate Cx43 phosphoforms make HChs that open at different cellular locales. Thus, Cx43-p^{CK1} HChs would open toward the "gap" (the virtual space containing the extracellular regions of Cx43) or the perinexus, but it is unclear whether their opening would be consequential (e.g., modify the ionic composition of perinexus or cytoplasm; cf. [80]). In contrast, HChs made by Cx43-dp^{CK1} may open at the lateral membranes, where they would contribute to arrhythmia propensity by facilitating collapse of the membrane potential, decreasing (Na$^+$ channel) excitability and mediating leak of substances into and out of the cells [75,81,82].

3.2.5. Further Cx43 Phosphorylation

The cardioprotective S3E phenotype may be linked to a stable presence of Cx43 at the ID, and/or to a permissive/protective role of the mutation itself for phosphorylation at other Cx43 sites ([9], their Figures 1–4). Without conflict with these interpretations, arrhythmia resilience might be partly due to the temporary resistance of the CK1-phospho-mimicking Cx43 mutant to low pH-induced gating.

Slowly migrating electrophoretic Cx43 bands, perhaps representing phosphorylated isoforms, are preserved in the S3E hearts [9]. This observation poses an interesting line of thought: In Cx43WT, dephosphorylated S365 (Cx43-dpS365) is permissive to S368 phosphorylation by PKC [83], an event linked to smaller γ_j and higher permselectivity [15,18,28] of the remaining active GJChs [50,84]. Because spironolactone and the S3E mutation yield parallel effects on Cx43 distribution and arrhythmia susceptibility [9,11], one can speculate that the Cx43-pS365 isoform, protected by the hormone inhibitor, is also protected in the transgenic S3E hearts. Were this the case, the rise of Cx43-pS368 and the shift toward smaller γ_j and higher GJ permselectivity would be deterred in the S3E hearts and in the CK1-D expressing cells. Our results partially concur with this possibility, as the CK1-D mutant displayed lower frequency of small GJCh transitions and no higher permselectivity values than Cx43WT. However, Cx43-pS368 was readily found in cells expressing Cx43-CK1-D, and less abundantly in Cx43-CK1-A expressing cells; high permselectivity was observed in some Cx43-CK1-D, but not in Cx43-CK1-A cell pairs. These data demonstrate that S368 remains a target of PKC in both Cx43-CK1-D and Cx43-CK1-A, but leaves the possibility open that differences between these mutants may be due to differences in the level of S368 phosphorylation. To address this issue, one possible strategy would involve Cx43 constructs carrying both the CK1- and PKC-(de)phospho- mimicking mutations.

Despite their possible drawbacks, and because of their relative ease, mutational techniques contribute importantly to the study of phosphorylation [85,86]. However, alternate mutant (mimicking) forms of single phosphorylation sites with outcomes equally differing from Cx43WT have been reported [87]. These and some of our data (e.g., single channel transitions, pH-gating) suggest that (de)phospho-mimicking amino acid substitutions may not recapitulate all features of biological (de)phosphorylation. It is also possible that similar outcomes indicate that the mutated residues are not absolutely essential, or that complementary or sequential changes at other sites are necessary, for the explored/expected outcomes. Thus, in addition to S368, PKC-phosphorylation of S262 was linked to cardioprotection and prevention of Cx43 lateralization in the context of preconditioning and reperfusion injury [88–91]. Also during ischemia, Cx43 can be phosphorylated at S373, which activates a 14-3-3 mode-1 binding domain in Cx43; pS373 may disrupt Cx43/ZO-1 interaction, while the complex Cx43/14-3-3 facilitates ubiquitination, internalization and degradation of Cx43 [84]. The role of phosphorylation in Cx43/ZO1 binding/release was recently further explored using (de)phospho-mimicking amino acid substitutions [87]. Results suggest that Cx43/ZO-1 interaction is set within a series of sequential and hierarchical phosphorylation/dephosphorylation steps that involve not only residue S373, but also S365, S368, S279/S282, S255, and S262. Specific phospho-antibodies against these and other important sites (e.g., S262, S279, S365) are not widely available. However, it would be interesting to explore the channel from (de)phospho-mimicking amino acid substitutions of all these sites.

3.2.6. Implications of CK1-Phospho-Mimicking Mutants for a Channel Gating Model

Overall, the data shown here strongly suggest that the gating of gap junction channels cannot be explained with a simple/unique closure mechanism. The data also confirm that the CT domain plays a role in channel closure, and demonstrate that such a role is modified by phosphorylation. To illustrate these points, let us assume that the various triggers that cause GJCh closure shared a single gating mechanism. Because modifications to the mechanism itself must affect its response to any trigger, then, weak pH-gating, for instance, would be expected to match weak V_j-gating. We showed strong V_j-gating paired with weak pH-gating (opposite), and "WT-like" V_j-gating paired with absence of pH-gating (opposite). Thus, in contrast to the chimera Cx43*NT37, which inhibited both pH- and V_j-gating in the presence of an intact Cx43CT [14], the CK1-(de)phospho-mimicking mutations differentially modulate the response of GJs to gating triggers through Cx43CT. These observations can only be understood if the (CT-CL) interactions involved in V_j- and pH-induced channel closure are dissimilar, at least for Cx43. In other words, while different triggers may share common elements, the specific molecular structures involved in electrical and chemical gating may not always be identical. Moreover, the availability and readiness of the gating elements are amenable to modification by phosphorylation, and thus by the changing cellular conditions.

4. Materials and Methods

4.1. Plasmid Construction

pcDNA3 neo plasmid (Thermo Fisher Scientific, Waltham, MA, USA) containing the rat Cx43 sequence was mutated at the N341 site to convert the amino acid sequence to mouse using the Stratagene's QuikChange Lightning kit (Stratagene, San Diego, CA, USA) according to the manufacturer instructions and the following primers (mutations shown in bold): 341S F 5′ gatttccccgacgac**agc**cagaatgccaaaaag3′ and N341S R 5′ cttttggcattctg**gct**gtcgtcggggaaatcg 3′. After verifying the sequence at the University of Arizona's UAGC sequencing facility, mutations S325,328,330D (dubbed CK1-D) and S325,328,330A (dubbed CK1-A) were introduced also with the QuikChange kit (Stratagene, San Diego, CA, USA) and the primers (mutations shown in bold): CKIDx3 F 5′ catggggcaggccgga**gaca**ccatc**gac**aac**gat**cacgcccagccgttcg 3′ and CKIDx3 R 5′ cgaacggctgggcgtg**atc**gttg**tc**gatggt**gtc**tccggcct gccccatg 3′; CK1-Ax3 F 5′ catggggcaggccgga**gcc**accatc**gcc**aac**gca**cacgcccagccgttcg 3′ and CKIAx3 R 5′ cgaacggctgggcgtg**tgc**gttg**gc**gatggt**ggc**tccggcctgccccatg 3′. pcDNA containing each mutation was amplified with the QIAgen Maxiprep kit (QIAgen, Hilden, Germany)) as per manufacturer instructions to produce material for transfection. Sequence was confirmed at the University of Arizona UAGC Sequencing Facility (Local University Services).

4.2. Cell Culture, Transfections and Protein Expression

Rat insulinoma (Rin) cells [51] were transfected with the Lipofectamine 2000 (Life Technologies, Grand Island, NY, USA) and pcDNA3 plasmid containing Cx43-CK1-D. Subclones were isolated by dilution cloning and tested for gene expression by Western blotting. GJ coupling was examined in subclones or recently transfected cells. Control experiments were performed in cells stably transfected with Cx43WT (Rin43). Protein expression of Cx43-CK1-D and CK1-A was confirmed by immunofluorescence and Western blot. Immunocytochemistry (ICC) was performed using primary antibodies against total Cx43 (Sigma-Aldrich, St. Louis, MO, USA), pS368-Cx43 (AbCam, Cambridge, MA, USA) and pS325/328/330-Cx43 (generous gift from Paul Lampe). Secondary antibodies (Jackson ImmunoResearch, West Grove, PA, USA) labeled with Alexa488 or Alexa647 were used for fluorescence detection.

4.3. Electrophysiology

Electrical recordings were performed as described [14]. Briefly, junctional (I_j) or membrane (I_m) currents were recorded with square voltage pulses using dual or single whole-cell (WC) patch clamp and osmotically matched (300–330 mOsm) external (in mM: 142.5 NaCl, 4 KCl, 1 MgCl$_2$, 5 Glucose, 2 Na-pyruvate, 10 HEPES, 15 CsCL, 10 TEA-Cl, 1 CaCl$_2$), and pipette solutions (in mM: 124 KCl, 3 MgCl$_2$, 5 Glucose, 9 HEPES, 9 EGTA, 14 CsCl, 9 TEA-Cl, 5 Na$_2$-ATP, 0.5 CaCl$_2$; calculated [92] free cytosolic Ca^{2+} $\leq$ 20 nM). Macroscopic junctional conductance (g_j) was measured with repeated, 2-second transjunctional voltage (V_j) pulses of $\pm$10 mV. V_j-gating was assessed with 5-s V_j pulses of increasing magnitude from 0 to $\pm$100 mV in 10 mV increments (step protocol) or with repeated pulses of $\pm$80 mV (for V_j-dependent fast I_j inactivation). To quantify V_j-gating, normalized g_j (Gj = steady state/instantaneous g_j (g_j^{ss}/g_j^{inst})) values obtained with the V_j step protocol were fit with a Boltzmann function, which depicts distribution of two channel states: maximally/minimally open, over the V_j range [12,13,16]; from these G$_j$/V_j relationships, the following parameters were obtained: Gj$_{max}$ and Gj$_{min}$ (maximum and minimum normalized g_j^{ss}), V_0 (V_j at which G$_j^{ss}$ is halfway between Gj$_{max}$ and Gj$_{min}$) and z, a value representing gating charges [12,13]. To quantify fast I_j inactivation, the decrease of composite I_j during repeated $\pm$80 mV pulses were fit with exponential decays of 1st or 2nd order (cf. [13]). Channel transition amplitudes (γ_j) were documented with $V_j = \pm$40 or 80 mV in poorly-coupled cell pairs with or without halothane treatment. Transitions (see Supplemental Figure S1) were considered to occur between fully open (O), residual (R) and closed (C) states; transitions with values intermediate between O↔R and R↔C point to the existence of a substate (S). To reveal the presence of hemichannels, repeated, 5-second or longer depolarizing pulses ($V_m = 80$ mV, WC configuration) were applied to non-coupled or single cells. All-points histograms were made from short (5–12 s) fragments of extended recordings. As intracellular pH (pH$_i$) closely follows the extracellular pH (pH$_o$) when buffered with weak acid solutions [72], g_j uncoupling was achieved by superfusing cells with bicarbonate-containing solution adjusted to pH = 6.0 to 6.4 (when bubbled with 95% CO$_2$/5% O$_2$).

4.4. Permselectivity

Measurements of the ratio of permeation to dye vs. current carrying ions (P$_{j\text{-dye}}$/g_j) were fully described [14,27]. Briefly, dyes NBD-m-TMA (NBD, a junctional permeant) and rhodamine-labeled 3000Da dextran (rhodex3000, unable to permeate junctions, Molecular Probes, Eugene, OR, USA) were delivered through a patch pipette in Whole Cell Voltage Clamp (WCVC) mode to one cell of a pair; total NBD fluorescence was timely imaged for up to 17 min or until NBD equilibrated in both cells; the second cell was then accessed in WCVC mode to document g_j. For each pair, a rate constant ($k \equiv$ junctional permeability to dye, P$_{j\text{-dye}}$) representing the speed of transjunctional dye diffusion, was calculated and plotted vs. the associated g_j (thus, permselectivity $\equiv$ P$_{j\text{-NBD}}$/g_j). Rhodex3000 images (Molecular Probes, Eugene, OR, USA) and/or halothane-induced uncoupling helped to discard dye diffusion through cytoplasmic bridges.

4.5. Statistical Analysis

Analyses were performed in Excel (Office 2010, Microsoft Corp., Redmond, WA, USA), Origin (Version 7, OriginLab Corp., Northampton, MA, USA) and GraphPad Prism (Version 7, GraphPad Software, La Jolla, CA, USA). Values are reported as Mean $\pm$ SEM. Comparisons were performed with ANOVA and unpaired *t*-Test (significance at $p < 0.05$). For permselectivity and channel distribution,

Kruskal-Wallis (ANOVA on ranks) and Mann-Whitney U test (significance at $p < 0.05$) were used. Graphics were created with SigmaPlot 2001 (Version 7.101, Systat Software, San Jose, CA, USA).

5. Conclusions

Our data suggest that phosphorylation of Cx43 by CK1 (normoxia) yields GJChs with strong V_j-gating, large γ_j values, variable permselectivity, and resistance to low pH-induced uncoupling, all of which may be compatible with arrhythmia-resistance. In contrast, dephosphorylation of the CK1 sites (hypoxia) yields GJChs with near "wild type" (as seen in Rin cells) V_j-gating, large γ_j values, low permselectivity and imperviousness to low pH-induced gating. These data suggest that persistently open GJChs may be deleterious during ischemia, despite (or because of) their low permselectivity. In addition, Cx43-dp^{CK1} HCh openings at the lateral membranes (rather than the ID) may worsen the arrhythmic propensity of the afflicted myocardium. While phosphorylation of S368 was shown in Cx43WT and in the CK1-(de)phospho-mimicking mutants, a role for pS368 levels in determining permselectivity remains possible. The data offer possible explanations for the cardiac phenotypes of S3A and S3E mice, and for the pathophysiological events that attend the development of ischemia- or TAC-induced GJR and arrhythmias. Gap junction channel gating cannot be described with a single closure mechanism; instead, the role played by the CT in channel closure can be modified by phosphorylation.

Supplementary Materials: Supplementary materials can be found at http://www.mdpi.com/1422-0067/19/6/1659/s1.

Author Contributions: J.M.B. and J.F.E.-V. conceived and designed the experiments. J.F.E.-V. and T.K.P. performed the experiments. J.F.E.-V. and T.K.P. analyzed the data. J.F.E.-V. and J.M.B. wrote the paper that was critically proof read by T.K.P.

Funding: This research was funded by the National Institutes of Health (Heart, Lung and Blood Institute) Grants HL058732 and HL131712 (to J.M.B.).

Acknowledgments: We are grateful for a Faculty Stipend Award provided to J.F.E.-V. by the Office of the Diversity and Inclusion (ODI) and the Arizona Center of Excellence (AZ-COE) of the University of Arizona. We thank Paul Lampe for his gracious gift of phospho-specific antibodies against Cx43.

Conflicts of Interest: The authors declare no conflict of interest.

Abbreviations

CK1	Casein Kinase 1
Cx	Connexin
Cx43	Connexin 43
CT	Carboxyl Terminus
CL	Cytoplasmic Loop
S3A	Transgenic mouse line expressing mutant Cx43-S325,328,330A
S3E	Transgenic mouse line expressing mutant Cx43-S325,328,330E
TAC	Transverse Aortic Constriction
GJR	Gap Junction Remodelling
ID(s)	Intercalated disc(s)
GJCh(s)	Gap Junction Channel(s)
HCh(s)	Hemichannel(s)
Rin	Rat insulinoma
WCVC	Whole Cell Voltage Clamp
pS	PicoSiemens
nS	nanoSiemens
pH_i	internal pH
pH_o	external pH
DIC	Differential Interference Contrast
CT	Carboxyl Terminus
CL	Cytoplasmic Loop
ZO-1	Zonula Occludens 1

References

1. Vozzi, C.; Dupont, E.; Coppen, S.R.; Yeh, H.I.; Severs, N.J. Chamber-related differences in connexin expression in the human heart. *J. Mol. Cell. Cardiol.* **1999**, *31*, 991–1003. [CrossRef] [PubMed]

2. Igarashi, T.; Finet, J.E.; Takeuchi, A.; Fujino, Y.; Strom, M.; Greener, I.D.; Rosenbaum, D.S.; Donahue, J.K. Connexin gene transfer preserves conduction velocity and prevents atrial fibrillation. *Circulation* **2012**, *125*, 216–225. [CrossRef] [PubMed]

3. Plum, A.; Hallas, G.; Magin, T.; Dombrowski, F.; Hagendorff, A.; Schumacher, B.; Wolpert, C.; Kim, J.; Lamers, W.H.; Evert, M.; et al. Unique and shared functions of different connexins in mice. *Curr. Biol.* **2000**, *10*, 1083–1091. [CrossRef]

4. Betsuyaku, T.; Nnebe, N.S.; Sundset, R.; Patibandla, S.; Krueger, C.M.; Yamada, K.A. Overexpression of cardiac connexin45 increases susceptibility to ventricular tachyarrhythmias in vivo. *Am. J. Physiol. Heart Circ. Physiol.* **2006**, *290*, H163–H171. [CrossRef] [PubMed]

5. Lampe, P.D.; Lau, A.F. The effects of connexin phosphorylation on gap junctional communication. *Int. J. Biochem. Cell Biol.* **2004**, *36*, 1171–1186. [CrossRef]

6. Pogoda, K.; Kameritsch, P.; Retamal, M.A.; Vega, J.L. Regulation of gap junction channels and hemichannels by phosphorylation and redox changes: A revision. *BMC Cell Biol.* **2016**, *17* (Suppl. 1), 137–150. [CrossRef] [PubMed]

7. Cooper, C.D.; Lampe, P.D. Casein kinase 1 regulates connexin-43 gap junction assembly. *J. Biol. Chem.* **2002**, *277*, 44962–44968. [CrossRef] [PubMed]

8. Lampe, P.D.; Cooper, C.D.; King, T.J.; Burt, J.M. Analysis of Connexin43 phosphorylated at S325, S328 and S330 in normoxic and ischemic heart. *J. Cell Sci.* **2006**, *119*, 3435–3442. [CrossRef] [PubMed]

9. Remo, B.F.; Qu, J.; Volpicelli, F.M.; Giovannone, S.; Shin, D.; Lader, J.; Liu, F.Y.; Zhang, J.; Lent, D.S.; Morley, G.E.; et al. Phosphatase-resistant gap junctions inhibit pathological remodeling and prevent arrhythmias. *Circ. Res.* **2011**, *108*, 1459–1466. [CrossRef] [PubMed]

10. Pitt, B.; Zannad, F.; Remme, W.J.; Cody, R.; Castaigne, A.; Perez, A.; Palensky, J.; Wittes, J. The effect of spironolactone on morbidity and mortality in patients with severe heart failure. *N. Engl. J. Med.* **1999**, *341*, 709–717. [CrossRef] [PubMed]

11. Qu, J.; Volpicelli, F.M.; Garcia, L.I.; Sandeep, N.; Zhang, J.; Marquez-Rosado, L.; Lampe, P.D.; Fishman, G.I. Gap junction remodeling and spironolactone-dependent reverse remodeling in the hypertrophied heart. *Circ. Res.* **2009**, *104*, 365–371. [CrossRef] [PubMed]

12. Wang, H.-Z.; Li, J.; Lemanski, L.F.; Veenstra, R.D. Gating of mammalian cardiac gap junction channels by transjunctional voltage. *Biophys. J.* **1992**, *63*, 139–151. [CrossRef]

13. Srinivas, M.; Costa, M.; Gao, Y.; Fort, A.; Fishman, G.I.; Spray, D.C. Voltage dependence of macroscopic and unitary currents of gap junction channels formed by mouse connexin50 expressed in rat neuroblastoma cells. *J. Physiol.* **1999**, *517*, 673–689. [CrossRef] [PubMed]

14. Ek Vitorin, J.F.; Pontifex, T.K.; Burt, J.M. Determinants of Cx43 Channel Gating and Permeation: The Amino Terminus. *Biophys. J.* **2016**, *110*, 127–140. [CrossRef] [PubMed]

15. Moreno, A.P.; Fishman, G.I.; Spray, D.C. Phosphorylation shifts unitary conductance and modifies voltage dependent kinetics of human connexin43 gap junction channels. *Biophys. J.* **1992**, *62*, 51–53. [CrossRef]

16. Moreno, A.P.; Rook, M.B.; Fishman, G.I.; Spray, D.C. Gap junction channels: Distinct voltage-sensitivie and -insensitive conductance states. *Biophys. J.* **1994**, *67*, 113–119. [CrossRef]

17. Gonzalez, D.; Gomez-Hernandez, J.M.; Barrio, L.C. Molecular basis of voltage dependence of connexin channels: An integrative appraisal. *Prog. Biophys. Mol. Biol.* **2007**, *94*, 66–106. [CrossRef] [PubMed]

18. Moreno, A.P.; Saez, J.C.; Fishman, G.I.; Spray, D.C. Human Connexin43 gap junction channels: Regulation of unitary conductances by phosphorylation. *Circ. Res.* **1994**, *74*, 1050–1057. [CrossRef] [PubMed]

19. Moreno, A.P.; Chanson, M.; Elenes, S.; Anumonwo, J.; Scerri, I.; Gu, H.; Taffet, S.M.; Delmar, M. Role of the carboxyl terminal of connexin43 in transjunctional fast voltage gating. *Circ. Res.* **2002**, *90*, 450–457. [CrossRef] [PubMed]

20. Moore, L.K.; Beyer, E.C.; Burt, J.M. Characterization of gap junction channels in A7r5 vascular smooth muscle cells. *Am. J. Physiol.* **1991**, *260*, C975–C981. [CrossRef] [PubMed]

21. Lampe, P.D.; Tenbroek, E.M.; Burt, J.M.; Kurata, W.E.; Johnson, R.G.; Lau, A.F. Phosphorylation of connexin43 on serine368 by protein kinase C regulates gap junctional communication. *J. Cell Biol.* **2000**, *149*, 1503–1512. [CrossRef] [PubMed]

22. Cottrell, G.T.; Lin, R.; Warn-Cramer, B.J.; Lau, A.F.; Burt, J.M. Mechanism of v-Src- and mitogen-activated protein kinase-induced reduction of gap junction communication. *Am. J. Physiol. Cell Physiol.* **2003**, *284*, C511–C520. [CrossRef] [PubMed]

23. Nelson, T.K.; Sorgen, P.L.; Burt, J.M. Carboxy terminus and pore-forming domain properties specific to Cx37 are necessary for Cx37-mediated suppression of insulinoma cell proliferation. *Am. J. Physiol. Cell Physiol.* **2013**, *305*, C1246–C1256. [CrossRef] [PubMed]

24. Jacobsen, N.L.; Pontifex, T.K.; Li, H.; Solan, J.L.; Lampe, P.D.; Sorgen, P.L.; Burt, J.M. Regulation of Cx37 channel and growth-suppressive properties by phosphorylation. *J. Cell Sci.* **2017**, *130*, 3308–3321. [CrossRef] [PubMed]

25. Gemel, J.; Nelson, T.K.; Burt, J.M.; Beyer, E.C. Inducible coexpression of connexin37 or connexin40 with connexin43 selectively affects intercellular molecular transfer. *J. Membr. Biol.* **2012**, *245*, 231–241. [CrossRef] [PubMed]

26. Gu, H.; Ek-Vitorin, J.F.; Taffet, S.M.; Delmar, M. UltraRapid communication: Coexpression of connexins 40 and 43 enhances the pH sensitivityof gap junctions: A model for synergistic interactions among connexins. *Circ. Res.* **2000**, *86*, e98–e103. [CrossRef] [PubMed]

27. Ek-Vitorin, J.F.; Burt, J.M. Quantification of Gap Junction Selectivity. *Am. J. Physiol. Cell Physiol.* **2005**, *289*, C1535–C1546. [CrossRef] [PubMed]

28. Ek-Vitorin, J.F.; King, T.J.; Heyman, N.S.; Lampe, P.D.; Burt, J.M. Selectivity of connexin 43 channels is regulated through protein kinase C-dependent phosphorylation. *Circ. Res.* **2006**, *98*, 1498–1505. [CrossRef] [PubMed]

29. Gadian, D.G.; Hoult, D.I.; Radda, G.K.; Seeley, P.J.; Chance, B.; Barlow, C. Phosphorus nuclear magnetic resonance studies on normoxic and ischemic cardiac tissue. *Proc. Natl. Acad. Sci. USA* **1976**, *73*, 4446–4448. [CrossRef] [PubMed]

30. Chen, W.; Wetsel, W.; Steenbergen, C.; Murphy, E. Effect of ischemic preconditioning and PKC activation on acidification during ischemia in rat heart. *J. Mol. Cell. Cardiol.* **1996**, *28*, 871–880. [CrossRef] [PubMed]

31. Schaefer, M.; Gross, W.; Gebhard, M.M. Hearts during ischemia with or without HTK-protection analysed by dielectric spectroscopy. *Physiol. Meas.* **2018**, *39*, 025002. [CrossRef] [PubMed]

32. Ruiz-Meana, M.; Garcia-Dorado, D.; Lane, S.; Pina, P.; Inserte, J.; Mirabet, M.; Soler-Soler, J. Persistence of gap junction communication during myocardial ischemia. *Am. J. Physiol. Heart Circ. Physiol.* **2001**, *280*, H2563–H2571. [CrossRef] [PubMed]

33. De Groot, J.R. Ischaemia-induced cellular electrical uncoupling and ventricular fibrillation. *Neth. Heart J.* **2002**, *10*, 360–365. [PubMed]

34. Jain, S.K.; Schuessler, R.B.; Saffitz, J.E. Mechanisms of delayed electrical uncoupling induced by ischemic preconditioning. *Circ. Res.* **2003**, *92*, 1138–1144. [CrossRef] [PubMed]

35. Cascio, W.E.; Yang, H.; Muller-Borer, B.J.; Johnson, T.A. Ischemia-induced arrhythmia: The role of connexins, gap junctions, and attendant changes in impulse propagation. *J. Electrocardiol.* **2005**, *38*, 55–59. [CrossRef] [PubMed]

36. Li, H.; Liu, T.-F.; Lazrak, A.; Peracchia, C.; Goldberg, G.S.; Lampe, P.D.; Johnson, C.M. Properties and regulation of gap junctional hemichannels in the plasma membranes of cultured cells. *J. Cell Biol.* **1996**, *134*, 1019–1030. [CrossRef] [PubMed]

37. John, S.A.; Kondo, R.; Wang, S.Y.; Goldhaber, J.I.; Weiss, J.N. Connexin-43 hemichannels opened by metabolic inhibition. *J. Biol. Chem.* **1999**, *274*, 236–240. [CrossRef] [PubMed]

38. Shahidullah, M.; Delamere, N.A. Connexins form functional hemichannels in porcine ciliary epithelium. *Exp. Eye Res.* **2014**, *118*, 20–29. [CrossRef] [PubMed]

39. Contreras, J.E.; Saez, J.C.; Bukauskas, F.F.; Bennett, M.V. Gating and regulation of connexin 43 (Cx43) hemichannels. *Proc. Natl. Acad. Sci. USA* **2003**, *100*, 11388–11393. [CrossRef] [PubMed]

40. Lampe, P.D.; Lau, A.F. Regulation of gap junctions by phosphorylation of connexins. *Arch. Biochem. Biophys.* **2000**, *384*, 205–215. [CrossRef] [PubMed]

41. Beardslee, M.A.; Lerner, D.L.; Tadros, P.N.; Laing, J.G.; Beyer, E.C.; Yamada, K.A.; Kleber, A.G.; Schuessler, R.B.; Saffitz, J.E. Dephosphorylation and intracellular redistribution of ventricular connexin43 during electrical uncoupling induced by ischemia. *Circ. Res.* **2000**, *87*, 656–662. [CrossRef] [PubMed]

42. Solan, J.L.; Lampe, P.D. Connexin phosphorylation as a regulatory event linked to gap junction channel assembly. *Biochim. Biophys. Acta* **2005**, *1711*, 154–163. [CrossRef] [PubMed]

43. King, T.J.; Lampe, P.D. Temporal regulation of connexin phosphorylation in embryonic and adult tissues. *Biochim. Biophys. Acta* **2005**, *1719*, 24–35. [CrossRef] [PubMed]

44. Solan, J.L.; Lampe, P.D. Connexin43 phosphorylation: Structural changes and biological effects. *Biochem. J.* **2009**, *419*, 261–272. [CrossRef] [PubMed]

45. Marquez-Rosado, L.; Solan, J.L.; Dunn, C.A.; Norris, R.P.; Lampe, P.D. Connexin43 phosphorylation in brain, cardiac, endothelial and epithelial tissues. *Biochim. Biophys. Acta* **2012**, *1818*, 1985–1992. [CrossRef] [PubMed]

46. Solan, J.L.; Lampe, P.D. Specific Cx43 phosphorylation events regulate gap junction turnover in vivo. *FEBS Lett.* **2014**, *588*, 1423–1429. [CrossRef] [PubMed]

47. Solan, J.L.; Lampe, P.D. Kinase programs spatiotemporally regulate gap junction assembly and disassembly: Effects on wound repair. *Semin. Cell Dev. Biol.* **2015**. [CrossRef] [PubMed]

48. Reynhout, J.K.; Lampe, P.D.; Johnson, R.G. An activator of protein kinase C inhibits gap junction communication between cultured bovine lens cells. *Exp. Cell Res.* **1992**, *198*, 337–342. [CrossRef]

49. Stein, L.S.; Boonstra, J.; Burghardt, R.C. Reduced cell-cell communication between mitotic and nonmitotic coupled cells. *Exp. Cell Res.* **1992**, *198*, 1–7. [CrossRef]

50. Solan, J.L.; Fry, M.D.; Tenbroek, E.M.; Lampe, P.D. Connexin43 phosphorylation at S368 is acute during S and G2/M and in response to protein kinase C activation. *J. Cell Sci.* **2003**, *116*, 2203–2211. [CrossRef] [PubMed]

51. Burt, J.M.; Nelson, T.K.; Simon, A.M.; Fang, J.S. Connexin 37 profoundly slows cell cycle progression in rat insulinoma cells. *Am. J. Physiol. Cell Physiol.* **2008**, *295*, C1103–C1112. [CrossRef] [PubMed]

52. Richards, T.S.; Dunn, C.A.; Carter, W.G.; Usui, M.L.; Olerud, J.E.; Lampe, P.D. Protein kinase C spatially and temporally regulates gap junctional communication during human wound repair via phosphorylation of connexin43 on serine368. *J. Cell Biol.* **2004**, *167*, 555–562. [CrossRef] [PubMed]

53. Fang, J.S.; Angelov, S.N.; Simon, A.M.; Burt, J.M. Cx37 deletion enhances vascular growth and facilitates ischemic limb recovery. *Am. J. Physiol. Heart Circ. Physiol.* **2011**, *301*, H1872–H1881. [CrossRef] [PubMed]

54. Fang, J.S.; Angelov, S.N.; Simon, A.M.; Burt, J.M. Cx40 is required for, and Cx37 limits, postischemic hindlimb perfusion, survival and recovery. *J. Vasc. Res.* **2012**, *49*, 2–12. [CrossRef] [PubMed]

55. Huang, G.Y.; Cooper, E.S.; Waldo, K.; Kirby, M.L.; Gilula, N.B.; Lo, C.W. Gap junction-mediated cell-cell communication modulates mouse neural crest migration. *J. Cell Biol.* **1998**, *143*, 1725–1734. [CrossRef] [PubMed]

56. Van Rijen, H.V.; Van Kempen, M.J.; Postma, S.; Jongsma, H.J. Tumour necrosis factor alpha alters the expression of connexin43, connexin40, and connexin37 in human umbilical vein endothelial cells. *Cytokine* **1998**, *10*, 258–264. [CrossRef] [PubMed]

57. Kwak, B.R.; Pepper, M.S.; Gros, D.B.; Meda, P. Inhibition of endothelial wound repair by dominant negative connexin inhibitors. *Mol. Biol. Cell* **2001**, *12*, 831–845. [CrossRef] [PubMed]

58. Okamoto, T.; Akita, N.; Kawamoto, E.; Hayashi, T.; Suzuki, K.; Shimaoka, M. Endothelial connexin32 enhances angiogenesis by positively regulating tube formation and cell migration. *Exp. Cell Res.* **2014**, *321*, 133–141. [CrossRef] [PubMed]

59. Sawey, M.J.; Goldschmidt, M.H.; Risek, B.; Gilula, N.B.; Lo, C.W. Perturbation in connexin 43 and connexin 26 gap-junction expression in mouse skin hyperplasia and neoplasia. *Mol. Carcinog.* **1996**, *17*, 49–61. [CrossRef]

60. Hirschi, K.K.; Burt, J.M.; Hirschi, K.D.; Dai, C. Gap junction communication mediates transforming growth factor-β activation and endothelial-induced mural cell differentiation. *Circ. Res.* **2003**, *93*, 429–437. [CrossRef] [PubMed]

61. Fang, J.S.; Dai, C.; Kurjiaka, D.T.; Burt, J.M.; Hirschi, K.K. Connexin45 regulates endothelial-induced mesenchymal cell differentiation toward a mural cell phenotype. *Arterioscler. Thromb. Vasc. Biol.* **2013**, *33*, 362–368. [CrossRef] [PubMed]

62. Lerner, D.L.; Yamada, K.A.; Schuessler, R.B.; Saffitz, J.E. Accelerated onset and increased incidence of ventricular arrhythmias induced by ischemia in Cx43-deficient mice. *Circulation* **2000**, *101*, 547–552. [CrossRef] [PubMed]

63. Procida, K.; Jorgensen, L.; Schmitt, N.; Delmar, M.; Taffet, S.M.; Holstein-Rathlou, N.H.; Nielsen, M.S.; Braunstein, T.H. Phosphorylation of connexin43 on serine 306 regulates electrical coupling. *Heart Rhythm* **2009**, *6*, 1632–1638. [CrossRef] [PubMed]

64. Morley, G.E.; Ek-Vitorin, J.F.; Taffet, S.M.; Delmar, M. Structure of connexin43 and its regulation by pHi. *J. Cardiovasc. Electrophysiol.* **1997**, *8*, 939–951. [CrossRef] [PubMed]

65. Duffy, H.S.; Sorgen, P.L.; Girvin, M.E.; O'Donnell, P.; Coombs, W.; Taffet, S.M.; Delmar, M.; Spray, D.C. pH-dependent intramolecular binding and structure involving Cx43 cytoplasmic domains. *J. Biol. Chem.* **2002**, *277*, 36706–36714. [CrossRef] [PubMed]

66. Eckert, R. Gap-junctional single channel permeability for fluorescent tracers in mammalian cell cultures. *Biophys. J.* **2006**, *91*, 565–579. [CrossRef] [PubMed]

67. Wang, X.G.; Peracchia, C. Connexin 32/38 chimeras suggest a role for the second half of inner loop in gap junction gating by low pH. *Am. J. Physiol.* **1996**, *271*, C1743–C1749. [CrossRef] [PubMed]

68. Bukauskas, F.F.; Peracchia, C. Two distinct gating mechanisms in gap junction channels: CO_2-sensitive and voltage-sensitive. *Biophys. J.* **1997**, *72*, 2137–2142. [CrossRef]

69. Stergiopoulos, K.; Alvarado, J.L.; Mastroianni, M.; Ek-Vitorin, J.F.; Taffet, S.M.; Delmar, M. Hetero-domain interactions as a mechanism for the regulation of connexin channels. *Circ. Res.* **1999**, *84*, 1144–1155. [CrossRef] [PubMed]

70. Francis, D.; Stergiopoulos, K.; Ek-Vitorin, J.F.; Cao, F.L.; Taffet, S.M.; Delmar, M. Connexin diversity and gap junction regulation by pHi. *Dev. Genet.* **1999**, *24*, 123–136. [CrossRef]

71. Liu, S.; Taffet, S.; Stoner, L.; Delmar, M.; Vallano, M.L.; Jalife, J. A structural basis for the unequal sensitivity of the major cardiac and liver gap junctions to intracellular acidification: The carboxyl tail length. *Biophys. J.* **1993**, *64*, 1422–1433. [CrossRef]

72. Morley, G.E.; Taffet, S.M.; Delmar, M. Intramolecular interactions mediate pH regulation of connexin43 channels. *Biophys. J.* **1996**, *70*, 1294–1302. [CrossRef]

73. Inserte, J.; Barba, I.; Hernando, V.; Abellan, A.; Ruiz-Meana, M.; Rodriguez-Sinovas, A.; Garcia-Dorado, D. Effect of acidic reperfusion on prolongation of intracellular acidosis and myocardial salvage. *Cardiovasc. Res.* **2008**, *77*, 782–790. [CrossRef] [PubMed]

74. Heyman, N.S.; Burt, J.M. Hindered diffusion through an aqueous pore describes invariant dye selectivity of Cx43 junctions. *Biophys. J.* **2008**, *94*, 840–854. [CrossRef] [PubMed]

75. Kondo, R.P.; Wang, S.Y.; John, S.A.; Weiss, J.N.; Goldhaber, J.I. Metabolic inhibition activates a non-selective current through connexin hemichannels in isolated ventricular myocytes. *J. Mol. Cell Cardiol.* **2000**, *32*, 1859–1872. [CrossRef] [PubMed]

76. Shintani-Ishida, K.; Uemura, K.; Yoshida, K. Hemichannels in cardiomyocytes open transiently during ischemia and contribute to reperfusion injury following brief ischemia. *Am. J. Physiol. Heart Circ. Physiol.* **2007**, *293*, H1714–H1720. [CrossRef] [PubMed]

77. Clarke, T.C.; Williams, O.J.; Martin, P.E.; Evans, W.H. ATP release by cardiac myocytes in a simulated ischaemia model: Inhibition by a connexin mimetic and enhancement by an antiarrhythmic peptide. *Eur. J. Pharmacol.* **2009**, *605*, 9–14. [CrossRef] [PubMed]

78. Hawat, G.; Benderdour, M.; Rousseau, G.; Baroudi, G. Connexin 43 mimetic peptide Gap26 confers protection to intact heart against myocardial ischemia injury. *Pflugers Arch.* **2010**, *460*, 583–592. [CrossRef] [PubMed]

79. Wang, N.; De Vuyst, E.; Ponsaerts, R.; Boengler, K.; Palacios-Prado, N.; Wauman, J.; Lai, C.P.; De Bock, M.; Decrock, E.; Bol, M.; et al. Selective inhibition of Cx43 hemichannels by Gap19 and its impact on myocardial ischemia/reperfusion injury. *Basic Res. Cardiol.* **2013**, *108*, 309. [CrossRef] [PubMed]

80. Veeraraghavan, R.; Lin, J.; Hoeker, G.S.; Keener, J.P.; Gourdie, R.G.; Poelzing, S. Sodium channels in the Cx43 gap junction perinexus may constitute a cardiac ephapse: An experimental and modeling study. *Pflugers Arch.* **2015**. [CrossRef] [PubMed]

81. Li, F.; Sugishita, K.; Su, Z.; Ueda, I.; Barry, W.H. Activation of connexin-43 hemichannels can elevate $[Ca^{2+}]_i$ and $[Na^+]_i$ in rabbit ventricular myocytes during metabolic inhibition. *J. Mol. Cell Cardiol.* **2001**, *33*, 2145–2155. [CrossRef] [PubMed]

82. Ye, Z.C.; Wyeth, M.S.; Baltan-Tekkok, S.; Ransom, B.R. Functional hemichannels in astrocytes: A novel mechanism of glutamate release. *J. Neurosci.* **2003**, *23*, 3588–3596. [CrossRef] [PubMed]

83. Solan, J.L.; Marquez-Rosado, L.; Sorgen, P.L.; Thornton, P.J.; Gafken, P.R.; Lampe, P.D. Phosphorylation at S365 is a gatekeeper event that changes the structure of Cx43 and prevents down-regulation by PKC. *J. Cell Biol.* **2007**, *179*, 1301–1309. [CrossRef] [PubMed]

84. Smyth, J.W.; Zhang, S.S.; Sanchez, J.M.; Lamouille, S.; Vogan, J.M.; Hesketh, G.G.; Hong, T.; Tomaselli, G.F.; Shaw, R.M. A 14-3-3 mode-1 binding motif initiates gap junction internalization during acute cardiac ischemia. *Traffic* **2014**, *15*, 684–699. [CrossRef] [PubMed]

85. Chen, Z.; Cole, P.A. Synthetic approaches to protein phosphorylation. *Curr. Opin. Chem. Biol.* **2015**, *28*, 115–122. [CrossRef] [PubMed]

86. Dissmeyer, N.; Schnittger, A. Use of phospho-site substitutions to analyze the biological relevance of phosphorylation events in regulatory networks. *Methods Mol. Biol.* **2011**, *779*, 93–138. [CrossRef] [PubMed]

87. Thevenin, A.F.; Margraf, R.A.; Fisher, C.G.; Kells-Andrews, R.M.; Falk, M.M. Phosphorylation regulates connexin43/ZO-1 binding and release, an important step in gap junction turnover. *Mol. Biol. Cell* **2017**, *28*, 3595–3608. [CrossRef] [PubMed]

88. Doble, B.W.; Ping, P.; Kardami, E. The epsilon subtype of protein kinase C is required for cardiomyocyte connexin-43 phosphorylation. *Circ. Res.* **2000**, *86*, 293–301. [CrossRef] [PubMed]

89. Cross, H.R.; Murphy, E.; Bolli, R.; Ping, P.; Steenbergen, C. Expression of activated PKC epsilon (PKC epsilon) protects the ischemic heart, without attenuating ischemic H⁺ production. *J. Mol. Cell Cardiol.* **2002**, *34*, 361–367. [CrossRef] [PubMed]

90. Doble, B.W.; Dang, X.; Ping, P.; Fandrich, R.R.; Nickel, B.E.; Jin, Y.; Cattini, P.A.; Kardami, E. Phosphorylation of serine 262 in the gap junction protein connexin-43 regulates DNA synthesis in cell-cell contact forming cardiomyocytes. *J. Cell Sci.* **2004**, *117*, 507–514. [CrossRef] [PubMed]

91. Srisakuldee, W.; Jeyaraman, M.M.; Nickel, B.E.; Tanguy, S.; Jiang, Z.S.; Kardami, E. Phosphorylation of connexin-43 at serine 262 promotes a cardiac injury-resistant state. *Cardiovasc. Res.* **2009**, *83*, 672–681. [CrossRef] [PubMed]

92. Patton, C. 2018. 2018. Available online: https://web.stanford.edu/~cpatton/CaEGTA-TS.htm (accessed on 17 April 2018).

International Journal of
Molecular Sciences

MDPI

Article

Connexin43 Carboxyl-Terminal Domain Directly Interacts with β-Catenin

Gaelle Spagnol [†], Andrew J. Trease [†], Li Zheng, Mirtha Gutierrez, Ishika Basu, Cleofes Sarmiento, Gabriella Moore, Matthew Cervantes and Paul L. Sorgen *

Department of Biochemistry and Molecular Biology, University of Nebraska Medical Center, Omaha, NE 68198, USA; gspagnol@unmc.edu (G.S.); andrew.trease@unmc.edu (A.J.T.); li.zheng@unmc.edu (L.Z.); MGutierrez4413@CSM.edu (M.G.); ishika.basu@unmc.edu (I.B.); cleofes.sarmiento@unmc.edu (C.S.); gabriella.moore@unmc.edu (G.M.); Matthew.R.Cervantes.9@nd.edu (M.C.)
* Correspondence: psorgen@unmc.edu; Tel.: +1-(402)-559-7557; Fax: +1-(402)-559-6650
† These authors contributed equally to this work.

Received: 25 April 2018; Accepted: 22 May 2018; Published: 24 May 2018

Abstract: Activation of Wnt signaling induces Connexin43 (Cx43) expression via the transcriptional activity of β-catenin, and results in the enhanced accumulation of the Cx43 protein and the formation of gap junction channels. In response to Wnt signaling, β-catenin co-localizes with the Cx43 protein itself as part of a complex at the gap junction plaque. Work from several labs have also shown indirect evidence of this interaction via reciprocal co-immunoprecipitation. Our goal for the current study was to identify whether β-catenin directly interacts with Cx43, and if so, the location of that direct interaction. Identifying residues involved in direct protein–protein interaction is of importance when they are correlated to the phosphorylation of Cx43, as phosphorylation can modify the binding affinities of Cx43 regulatory protein partners. Therefore, combining the location of a protein partner interaction on Cx43 along with the phosphorylation pattern under different homeostatic and pathological conditions will be crucial information for any potential therapeutic intervention. Here, we identified that β-catenin directly interacts with the Cx43 carboxyl-terminal domain, and that this interaction would be inhibited by the Src phosphorylation of Cx43CT residues Y265 and Y313.

Keywords: Cx43; β-catenin; phosphorylation

1. Introduction

Gap junctions are intercellular channels that permit the passage of ions, small metabolites, and signaling molecules between neighboring cells [1]. They are important in a number of physiological processes, including cellular development, growth, and differentiation. In the heart, gap junctions mediate the propagation of cardiac action potentials and the maintenance of a regular beating rhythm [2]. Dysfunctional intercellular communication via gap junctions has been implicated in causing many human diseases [3]. Gap junctions are formed by the apposition of connexons from adjacent cells, where six connexin proteins form each connexon. Although the 21-connexin isoforms (e.g., 43-kDa isoform, Cx43) share significant sequence homology, differences in the amino acid sequence occur in the cytoplasmic loop and carboxyl terminal (CT) domains.

The CT domain is involved in regulating the trafficking of connexons to and from the plasma membrane, as well as the level of gap junction intercellular communication via a number of post-translational modifications and interactions with protein partners [4–11]. The CT domain is predominately unstructured (i.e., intrinsically disordered), making it an ideal substrate for the regulation of intercellular signaling by facilitating both high specificity and low affinity interaction with many different binding partners to allow the rapid feedback to cytoplasmic signals [12–15]. Over 20 protein partners have been identified to directly interact with Cx43, which can be categorized according to those that can promote or inhibit intercellular communication (for review see [16,17]). For example, connexin localization and stability at the gap junction plaques are strongly determined by interaction with cytoskeletal-associated proteins; these interactions are modulated by specific phosphorylation events [18,19]. The demonstration that Src phosphorylates Y247 and Y265 of Cx43 [20,21] enabled subsequent findings that pY247 inhibits the Cx43 interaction with β-tubulin [22] and that pY265 inhibits the Cx43 interaction with drebrin [23]. At the gap junction plaque, inhibiting the β-tubulin interaction could be a mechanism involved in the disassembly process. If this inhibition occurs after connexon formation in the trans-Golgi network, trafficking may be re-routed for degradation or to the lateral membrane [24]. The depletion of drebrin results in impaired gap junction intercellular communication, also by targeting Cx43 for degradation [25]. Further, while Cx43CT phosphorylation by Src does not inhibit Zonula occludens-1 (ZO-1) binding, active Src directly competes with Cx43 to bind ZO-1 [26]. Studies from the Gourdie and Lampe labs suggest that interaction with ZO-1 transitions Cx43 from the non-junctional membrane into the gap junction plaque, and that Src inhibits this process [27,28]. Thus, even though our knowledge is no doubt incomplete, it is clear that for Cx43, there exists a network of integrated processes involving phosphorylations and binding partners that control junctional Cx43 [19]. Another protein that has been identified to modulate both Cx43 expression and gap junction intercellular communication is β-catenin [29].

β-catenin is a critical protein in the canonical Wnt signaling transduction cascade. β-catenin (781 amino acids) consists of a well-structured central region made up of 12 armadillo repeats that are flanked by intrinsically disordered N-terminal and C-terminal domains [30,31]. β-catenin is a multi-functional protein whose activity depends on its subcellular localization. β-catenin at the plasma membrane is a component of cell adhesion junctions, while cytosolic accumulation leads to increased nuclear localization and transcriptional activity [32]. The activation of Wnt signaling induces Cx43 expression via the transcriptional activity of β-catenin, which leads to the increased formation of gap junction channels [29,33–38]. In response to Wnt signaling, β-catenin co-localizes with the Cx43 protein itself as part of a complex at the gap junction plaque [29]. Unfortunately, none of these studies was able to determine whether the β-catenin and Cx43 interaction is direct, or instead requires other protein partners.

2. Results

2.1. β-Catenin CT Domain Directly Interacts with the Cx43CT

The domain architecture of β-catenin includes a disordered N-terminal domain (~150 residues; binds α-catenin, glycogen synthase kinase 3β (GSK3β), and is phosphorylated by casein kinase I), a well-structured central armadillo repeat domain (~530 residues; major protein partner binding domain; e.g., binds axin, adenomatous polyposis coli protein, 14-3-3ζ, and E-cadherin), and a disordered C-terminal domain (~100 residues; binds several transcriptional coactivators) [39,40]. To determine whether β-catenin directly interacts with the Cx43CT domain, we performed a ^{15}N-heteronuclear single quantum coherence (HSQC) nuclear magnetic resonance (NMR) experiment using purified ^{15}N-lableled Cx43CT (V236–I382) and unlabeled full-length β-catenin (Figure 1A, top). Each chemical shift (or peak) in this two-dimensional experiment corresponds to one amide group; thus, the number of peaks should correspond to the number of Cx43CT residues (except proline). We have previously published the ^{15}N-HSQC assignment for the Cx43CT domain [41]. These chemical shifts are sensitive to their environment and small changes in structure and/or dynamics, such as those that would occur from a direct protein–protein interaction, can influence the chemical shift (i.e., change the location or broaden beyond detection) of an amino acid. The advantage of using NMR to study direct protein–protein interactions over cellular assays such as immunoprecipitation is its specificity. As only two proteins are present in the solution, any detected interaction is the result of a direct interaction, as opposed to the possibility that both are part of a larger molecular complex (limits of immunoprecipitation). Moreover, because the chemical shifts of the affected amino acids drift or diminish, the specific residues involved in the interaction can be determined. The addition of β-catenin affected a subset of Cx43CT residues (Figure 1A, top). When mapped onto the Cx43CT sequence, they were located within three areas: K259–T275, S282–N295, and N302–R319 (Figure 1A, bottom). The data indicates that the Cx43CT domain and β-catenin directly interact.

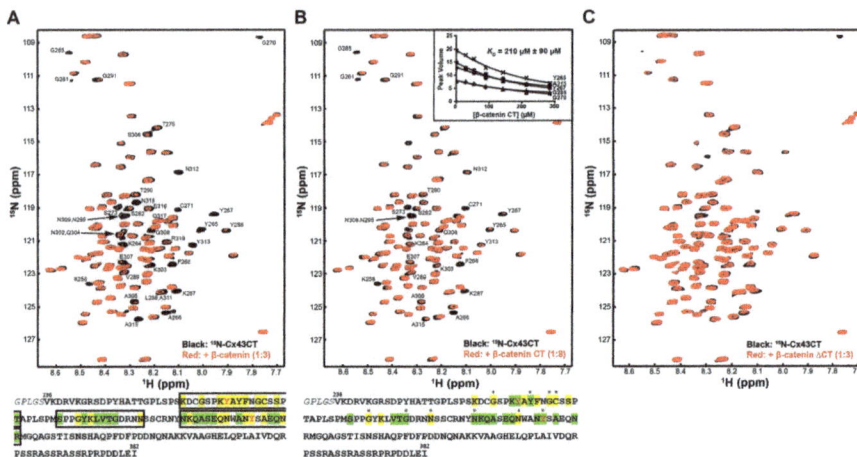

Figure 1. Nuclear magnetic resonance spectra showing the direct interaction between the Cx43CT and β-catenin. ^{15}N- heteronuclear single quantum coherence (HSQC) spectra of Cx43CT alone (black) and in the presence of (**A**) full-length β-catenin (red); (**B**) the β-catenin carboxyl-terminal (CT) domain (red); or (**C**) the β-catenin ΔCT domains (red). Molar ratio for each experiment is indicated in the figure. In panel B, provided is a subset of residues used to calculate the K_D of the interaction by fitting their decrease in signal intensity according to β-catenin CT concentration. Below each ^{15}N-HSQC spectra is the Cx43CT amino acid sequence. Highlighted are the affected residues (yellow—peaks broadened beyond detection; green—peaks decreased in intensity). Black boxes delimit the three areas of interaction with β-catenin. Asterisks denote that amino acids that were used to calculate the binding affinity for the β-catenin CT. Two of the residues phosphorylated by Src and affected by β-catenin are also highlighted (red letters).

To identify the β-catenin domain mediating the direct interaction with Cx43CT, we initially focused on the β-catenin CT domain (S681–L781). The N-terminal domain is the primary locus for Wnt signaling (GSK3β/E3 ubiquitin ligase), and armadillo repeat domains have been well characterized for binding partners involved in cell adhesion. Moreover, most of the binding partners to the C-terminal domain occur in the nucleus, thus potentially leaving the CT domain free to associate with a different set of proteins in the cytoplasm [42]. Upon purification, circular dichroism was used to determine whether the β-catenin CT domain contains any secondary structure (Figure 2). Analysis of the circular dichroism spectrum by Dichroweb (London, UK) determined that the protein is predominately intrinsically disordered with between 18–21% α-helical content [43,44]. The low amount of secondary structure is consistent with the low peak dispersion (<1 ^{1}H ppm) that was previously seen in the β-catenin CT domain ^{15}N-HSQC [32]. The addition of the unlabeled β-catenin CT domain affected the same subset of ^{15}N-Cx43CT residues as full-length β-catenin (Figure 1B, top). These have been highlighted on the Cx43CT sequence (Figure 1B, bottom). Next, a titration of the unlabeled β-catenin CT domain was performed to determine the binding affinity (K_D). The decrease in signal intensity caused by increasing the β-catenin CT domain concentration was fit according to the nonlinear least-square method (Figure 1B, inset). The K_D was determined to be 210 μM (±90 μM). A surface plasmon resonance (SPR) experiment confirmed the NMR results (Figure 3). When the Cx43CT domain was immobilized onto a carboxymethyl-dextran 5 chip, the addition of the β-catenin CT domain resulted in a direct interaction. A peptide to the Cx43 first extracellular loop (EL1, residues G38–R76) served as a negative control, and the Cx43CT domain itself served as a positive control (specific areas of dimerization include M281–N295, R299–Q304, S314–I327, and Q342–A348, [45]). To ensure that the β-catenin CT is the only β-catenin domain interacting with the Cx43CT, we purified a β-catenin construct containing the N-terminal and armadillo repeat domains (β-catenin ΔCT, i.e., deleted the CT domain). The addition of the unlabeled β-catenin ΔCT had no effect on the ^{15}N-Cx43CT residues (Figure 1C). Of note, the N-terminal domain of β-catenin has poor expression and degraded during purification, which prevented any attempt to test this domain only. Altogether, the data indicate that the β-catenin CT is the domain that directly interacts with Cx43CT.

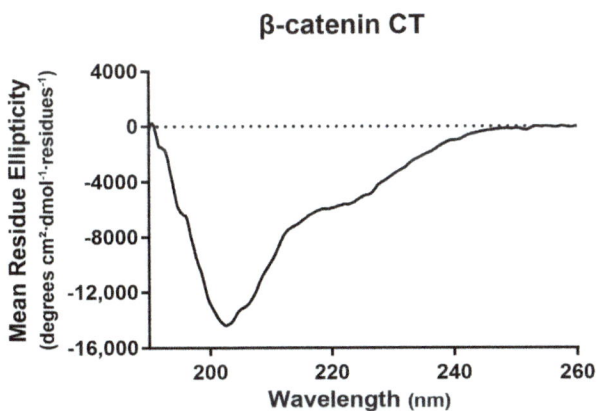

Figure 2. Circular dichroism spectrum showing the secondary structure of the β-catenin CT domain. The spectrum is represented as mean residue ellipticity as a function of wavelength. Data were analyzed using Dichroweb.

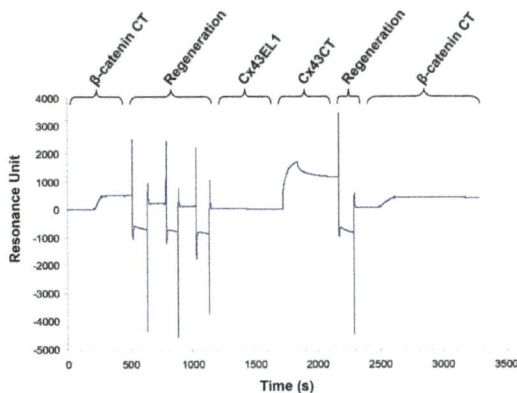

Figure 3. Surface plasmon resonance spectra showing direct interaction between the Cx43CT and β-catenin CT domains. Cx43CT was immobilized onto a CM5 chip by amine coupling (Biacore; GE Healthcare, Uppsala, Sweden) and either β-catenin CT (500 response units), Cx43EL1 (negative control), or Cx43CT (residues S255–I382, positive control) were flown over the chip as indicated on the top of the graph. The chip was regenerated after an interaction was observed. Repeat of an injection of β-catenin CT was performed to confirm the interaction.

2.2. Phosphorylation of Y265 and Y313 Inhibits Cx43 Binding with β-Catenin

Previous studies have identified that Src phosphorylates Cx43CT residues Y247 and Y265 [21,46]. Additional studies have identified that Src also phosphorylates Cx43CT residues Y313 (Li et al., 2018, Journal of Molecular and Cellular Cardiology, publication under revision; PhosphoSitePlus, [47]). Since the β-catenin CT domain affected Cx43CT residues Y265 and Y313, we addressed whether the phosphorylation of both sites could dissociate β-catenin from Cx43. A number of gap junction studies have determined that an aspartic acid can mimic a phosphate for Cx43 [41,48,49] and responds to the need to purify enough protein for biophysical studies. Therefore, NMR titration experiments were performed using purified soluble ^{15}N-labeled Cx43CT single or double phosphomimetics (Y313D or Y265,313D) and different concentrations of either an unlabeled β-catenin CT domain (Figure 4A,B) or full-length β-catenin (Figure 4C). The ^{15}N-HSQC spectrum of each control (no β-catenin, black) has been overlaid with spectra when either the β-catenin CT domain or full-length β-catenin (red) were added at a single molar ratio. For the single phosphomimetic construct, among the three areas where the β-catenin CT domain interacted with Cx43CT wild type (WT), binding was completely inhibited for area three (N302–R319), significantly reduced in area two (S282–N295), and mostly preserved in area one (K259–T275). A titration of the unlabeled β-catenin CT domain was performed to determine the K_D. The interaction with β-catenin decreased by approximately two-fold compared to WT (K_D = 341 μM vs. 202 μM) (Figure 4A, inset). When both Cx43CT Y265 and Y313 sites were mutated to mimic Src phosphorylation, the Cx43CT interaction with β-catenin was completely inhibited. These results confirm the direct interaction between Cx43CT and CT portion of β-catenin, and strongly suggest that Src phosphorylation of Cx43CT regulates this interaction.

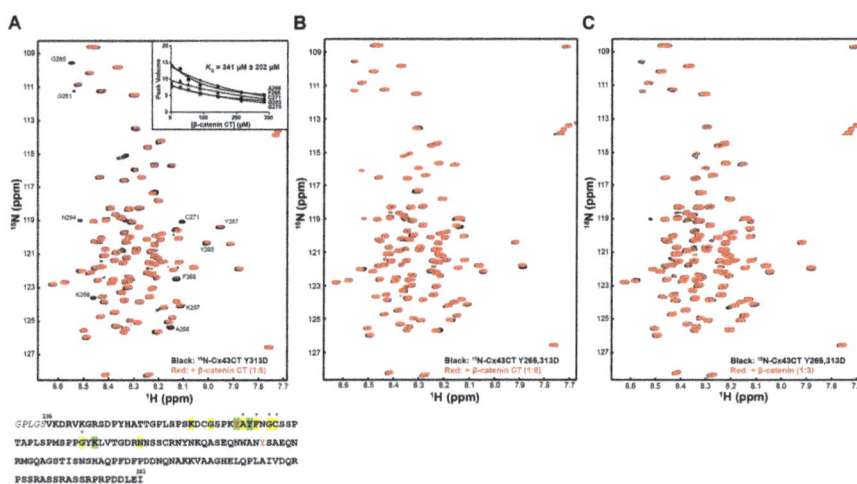

GPLGSVKDRVKGRSDFYHATTGPLSPSKDCGSPKYAYFNGCSSP
TAPLSPMSPPGYKLVTGDRNNSSCRNYNKQASEQNWANYSAEQN
RMGQAGSTISNSHAQPFDFPDDNQNAKKVAAGHELQFLAIVDQR
PSSRASSRASSRPRPDDLEI

Figure 4. Mimicking phosphorylation of Cx43CT residues Y265 and Y313 inhibits the interaction with the β-catenin CT domain. ^{15}N-HSQC spectra of (**A**) Cx43CT Y313D alone (black) and in the presence of the β-catenin CT domain (red) or Y265,313D alone (black), and in the presence of the (**B**) β-catenin CT domain and (**C**) full-length β-catenin (red). The molar ratio for each experiment is indicated in the figure. In panel A, provided is a subset of residues used to calculate the K_D of the interaction by fitting their decrease in signal intensity according to β-catenin CT concentration. Also represented below the ^{15}N-HSQC spectra is the Cx43CT amino acid sequence. Highlighted are the affected residues (yellow—peaks broadened beyond detection; green—peaks decreased in intensity). Asterisks denote amino acids used to calculate the binding affinity for the β-catenin CT interaction with the Cx43CT. Tyrosine residues 265 and 313 phosphorylated by Src are indicated in red.

3. Discussion

The Cx43CT domain binds multiple proteins, many of which have been shown to regulate Cx43 function through altering trafficking to the gap junction plaque, opening/closing gap junction channels, disassembly, and degradation (for review see [16,50]). Numerous excellent reviews have summarized the functional significance of these Cx43-interacting proteins [50–53]. These proteins can be partitioned into those that directly interact with the Cx43CT, and those proteins that can affect all aspects of the Cx43 life cycle, but no current evidence exists that they directly interact with the Cx43CT. One of the proteins in the latter category is β-catenin. Work from several labs has shown indirect evidence of this interaction, including reciprocal co-immunoprecipitation as well as co-localization of Cx43 with β-catenin [29,35]. Additionally, β-catenin segregates in triton insoluble fractions with Cx43 [29]. Our goal here was to identify whether β-catenin directly interacts with Cx43, and the location of that direct interaction.

β-catenin is an intracellular signal transducer in the Wnt signaling pathway that is involved in the regulation and coordination of cell–cell adhesion and gene transcription [54]. Ai et al. first identified that in response to Wnt signaling by the addition of Li⁺, β-catenin interacts with the Cx43 gene *GJA1* (contains three transcription factor 4 (TCF)/lymphoid enhancer binding factor binding sites; [55]) to increase transcription expression [29]. The accumulating Cx43 in the junctional membrane increased neonatal rat cardiomyocyte cell-to-cell coupling and co-localization with β-catenin [29]. Cx43 and β-catenin co-localization also occurs at the intercalated discs of adult rat cardiomyocytes [56]. A similar response to Li⁺ elicited Wnt/β-catenin signaling, which increased Cx43 expression and gap junction intercellular communication in skeletal myoblasts [57]. Additionally, the rapid electrical stimulation of cardiomyocytes had a similar effect as Li⁺, leading to the increased nuclear localization of β-catenin and subsequent Cx43 expression [58]. The suggested sequestering of β-catenin by Cx43 would serve to reduce the transactivation potential of β-catenin [29]. This observation is consistent with the increase in the active form of β-catenin with the knockdown of Cx43 seen in human neural progenitor cells [35], and the decrease in the nuclear localization of β-catenin with Cx43 overexpression in breast adenocarcinoma cell lines [36]. Conversely, Moorer et al. observed in a Cx43CT truncation (K258stop) mouse model that osteoblasts had reduced active β-catenin (along with protein kinase C δ and extracellular signal-regulated kinase 1/2), leading to altered proliferation, differentiation, collagen processing, and organization [34]. While the phenotype observed from loss of the Cx43CT matches that from the complete loss of Cx43 in bone cells, the same is not true in the cardiovascular system [34,59]. This suggests the influence of Cx43 on the activity and cellular localization of β-catenin may be tissue specific [34,59].

Shaw et al. put forth a model for connexin trafficking to the plasma membrane [56]. Cx43 oligomerizes into connexons in the trans-Golgi network [60]. Upon exiting, they use microtubules to travel to adherens junctions, which capture the microtubules allowing for connexon offloading to the plasma membrane [56,61–64]. Based upon the co-localization of Cx43 and β-catenin at the gap junction plaque, at some point in the trafficking, either to or at the adherens junctions, the interaction occurs. The β-catenin interaction occurs at Cx43 residues K259–T275, S282–N295, and N302–R319. Interestingly, similar residues directly interact with drebrin [23]. A commonality with these proteins is they would both help Cx43 indirectly interact with F-actin (β-catenin indirectly through α-catenin; drebrin directly) and stabilize gap junctions to favor intercellular communication. Conversely, they both cannot interact at the same time. Therefore, the available data would suggest that β-catenin binds first, and then at some point in the maturation of the gap junction plaque, Cx43CT switches to interact with drebrin. Since the phosphorylation of Y265 and Y313 also inhibits the Cx43 interaction with drebrin, this would not be the mechanism [23]. The possibility exists that regulation from the β-catenin perspective inhibits the interaction with Cx43. Consistent with this is that: (1) the phosphorylation of β-catenin at S552 by protein kinase B increases the association between β-catenin and 14-3-3ζ, leading to β-catenin translocation into the cytosol and nucleus [65]; (2) the phosphorylation of β-catenin by casein kinase 1 is necessary for subsequent glycogen synthase kinase-3 phosphorylation and then degradation [66]; (3)

the protein kinase A phosphorylation of β-catenin leads to the nuclear localization of β-catenin [67]. Interestingly, protein kinase B, casein kinase 1, and protein kinase A also phosphorylate Cx43 to promote synthesis, trafficking to the gap junction plaque, and channel opening.

The importance of identifying if and where a direct protein interaction occurs is in relationship to the phosphorylation of Cx43, because phosphorylation modifies the binding affinities of the Cx43 protein partners that regulate assembly, disassembly, and channel function [68]. For example, we demonstrated that mitogen-activated protein kinase phosphorylation of Cx43 increases the binding affinity for the E3 ubiquitin ligase neural precursor cell expressed, developmentally down-regulated 4 [69], which leads to Cx43 degradation [70,71]. Therefore, combining the location of a protein partner interaction on Cx43 along with the phosphorylation pattern under different homeostatic and pathological conditions will be crucial information for any potential therapeutic intervention. Here, we identified that β-catenin directly interacts with the Cx43CT domain, and that this interaction would be inhibited by Src phosphorylation of Cx43CT residues Y265 and Y313.

4. Material and Methods

4.1. Expression and Purification of Recombinant Proteins

The rat Cx43CT (V236–I382) (or (S255–I382) for the SPR study) polypeptide (unlabeled or ^{15}N-labeled) cloned into the bacterial expression vector pGEX-6P-2 (GST-tagged; Amersham Biosciences, Little Chalfont, UK) was expressed and purified in 1× phosphate-buffered saline (PBS), as previously described [72,73]. Y313D and Y265,313D mutations in the Cx43CT plasmid were incorporated using the Quick Change Lightning kit (Qiagen, Hilden, Germany). Human β-catenin in pET-28a (+) was purchased from Addgene, expressed (unlabeled), and purified by Nickel affinity column (Buffer A: 50 mM Tris, 150 mM NaCl, 10 mM Imidazole, 2 mM β-mercaptoethanol (BME), pH 8.0; Buffer B: 50 mM Tris, 150 mM NaCl, 600 mM Imidazole, 2 mM BME, pH 8.0) followed by Anion exchange (Buffer A: 50 mM Tris, 2 mM BME, pH 8; Buffer B: 50 mM Tris, 1 M NaCl, 2 mM BME, pH 8). β-catenin ΔCT (M1-S680) was obtained by introducing a stop codon after serine 680 using the Quick Change Lightning kit (Agilent, Santa Clara, CA, USA). Purification was identical to the full-length β-catenin. β-catenin CT (S681–L781) was subcloned into the pET-16b vector, expressed (unlabeled), and purified by Nickel affinity column similarly to the full-length β-catenin. Purity and analysis for degradation was assessed by SDS-PAGE, and all of the polypeptides were equilibrated by dialysis in Slide-A-Lyzer G2 Dialysis Cassettes (Thermo Scientific, Waltham, MA, USA) in 1× PBS at pH 7.8 in presence of 2 mM dithiothreitol.

4.2. Nuclear Magnetic Resonance (NMR)

NMR data were acquired at 7 °C using a 600-MHz Varian INOVA spectrometer (Agilent, Palo Alto, CA, USA) upgraded with a Bruker Avance-III HD console (Bruker, Billerica, MA, USA) and outfitted with a Bruker z-axis PFG "inverse" triple-resonance cryogenic (cold) probe (Bruker). Gradient-enhanced two-dimensional ^{15}N-HSQC experiments were used to obtain the binding isotherms of the ^{15}N-labeled Cx43CT WT, Y313D, and Y265,313D at a constant concentration (35 μM) in the absence or presence of increasing amounts (up to 285 μM) of β-catenin, β-catenin CT, or β-catenin ΔCT. Data acquisition, processing, and analysis, including calculation of the dissociation constants (K_D), have been previously described [14,46,69].

4.3. Circular dichroism (CD)

The CD experiment was performed on a JASCO J-815 CD spectrometer (JASCO, Mary's Court, Easton, MD, USA) at 7 °C in the far UV (260–190 nm). Spectra of the β-catenin CT were collected in 1× PBS at pH 7.4, with a 0.1-mm path length quartz cell, using a bandwidth of 1 nm, an integration time of 1 s, and a scan rate of 50 nm/min. The final spectrum was obtained from the average of five scans. All of the spectra were corrected by subtracting the solvent spectrum acquired under identical conditions. CD data were processed and converted to mean residue ellipticity using Spectra Analysis from the Jasco Spectra Manager software, Version 2.05.01 (JASCO, Mary's Court, Easton, MD, USA).

4.4. Surface plasmon resonance (SPR)

The SPR experiments were performed on a Biacore (GE Healthcare) 1000 at 25 °C. The Cx43CT (S255–I382) was immobilized onto a CM5 sensor chip by amine coupling, and the flow cell was equilibrated with the reaction buffer at a flow rate of 5 µL/min (213 mM phosphate buffer, pH 7.1). Then, 5 µL of either the β-catenin CT (4 µM), the Cx43EL1 (residues G38–R76, 10 µM, negative control), or the Cx43CT (10 µM, positive control) were injected over the chip, and the responses were recorded as resonance units (RU).

Author Contributions: P.L.S. and G.S. conceived and designed the experiments; G.S., A.J.T., L.Z., M.G., I.B., C.S., G.M. and M.C. performed the experiments; G.S., A.J.T. and L.Z. analyzed the data; P.L.S., G.S. and A.J.T. wrote the paper.

Acknowledgments: P.L.S. was supported by National Institutes of Health Grants GM072631, GM319613, and GM103427.

Conflicts of Interest: The authors declare no conflict of interest.

References

1. Goodenough, D.A.; Goliger, J.A.; Paul, D.L. Connexins, connexons, and intercellular communication. *Annu. Rev. Biochem.* **1996**, *65*, 475–502. [CrossRef] [PubMed]
2. Severs, N.J.; Bruce, A.F.; Dupont, E.; Rothery, S. Remodelling of gap junctions and connexin expression in diseased myocardium. *Cardiovasc. Res.* **2008**, *80*, 9–19. [CrossRef] [PubMed]
3. Evans, W.H.; Martin, P.E. Gap junctions: Structure and function. *Mol. Membr. Biol.* **2002**, *19*, 121–136. [CrossRef] [PubMed]
4. Laird, D.W. The gap junction proteome and its relationship to disease. *Trends Cell Biol.* **2010**, *20*, 92–101. [CrossRef] [PubMed]
5. Lampe, P.D.; Lau, A.F. The effects of connexin phosphorylation on gap junctional communication. *Int. J. Biochem. Cell Biol.* **2004**, *36*, 1171–1186. [CrossRef]
6. Herve, J.; Bourmeyster, N.; Sarrouilhe, D.; Duffy, H. Gap junctional complexes: From partners to functions. *Prog. Biophys. Mol. Biol.* **2007**, *94*, 29–65. [CrossRef] [PubMed]
7. Thevenin, A.F.; Kowal, T.J.; Fong, J.T.; Kells, R.M.; Fisher, C.G.; Falk, M.M. Proteins and mechanisms regulating gap-junction assembly, internalization, and degradation. *Physiology* **2013**, *28*, 93–116. [CrossRef] [PubMed]
8. Moreno, A.P.; Chanson, M.; Elenes, S.; Anumonwo, J.; Scerri, I.; Gu, H.; Taffet, S.M.; Delmar, M. Role of the carboxyl terminal of connexin43 in transjunctional fast voltage gating. *Circ. Res.* **2002**, *90*, 450–457. [CrossRef] [PubMed]
9. Morley, G.E.; Taffet, S.M.; Delmar, M. Intramolecular interactions mediate pH regulation of connexin43 channels. *Biophys. J.* **1996**, *70*, 1294–1302. [CrossRef]
10. Anumonwo, J.M.; Taffet, S.M.; Gu, H.; Chanson, M.; Moreno, A.P.; Delmar, M. The carboxyl terminal domain regulates the unitary conductance and voltage dependence of connexin40 gap junction channels. *Circ. Res.* **2001**, *88*, 666–673. [CrossRef] [PubMed]
11. Revilla, A.; Castro, C.; Barrio, L.C. Molecular dissection of transjunctional voltage dependence in the connexin-32 and connexin-43 junctions. *Biophys. J.* **1999**, *77*, 1374–1383. [CrossRef]
12. Sorgen, P.L.; Duffy, H.S.; Sahoo, P.; Coombs, W.; Delmar, M.; Spray, D.C. Structural changes in the carboxyl terminus of the gap junction protein connexin43 indicates signaling between binding domains for c-Src and zonula occludens-1. *J. Biol. Chem.* **2004**, *279*, 54695–54701. [CrossRef] [PubMed]
13. Bouvier, D.; Kieken, F.; Kellezi, A.; Sorgen, P.L. Structural changes in the carboxyl terminus of the gap junction protein connexin 40 caused by the interaction with c-Src and zonula occludens-1. *Cell Commun. Adhes.* **2008**, *15*, 107–118. [CrossRef] [PubMed]

14. Stauch, K.; Kieken, F.; Sorgen, P. Characterization of the structure and intermolecular interactions between the connexin 32 carboxyl-terminal domain and the protein partners synapse-associated protein 97 and calmodulin. *J. Biol. Chem.* **2012**, *287*, 27771–27788. [CrossRef] [PubMed]

15. Nelson, T.K.; Sorgen, P.L.; Burt, J.M. Carboxy terminus and pore-forming domain properties specific to Cx37 are necessary for Cx37-mediated suppression of insulinoma cell proliferation. *Am. J. Physiol. Cell Physiol.* **2013**, *305*, C1246–C1256. [CrossRef] [PubMed]

16. Gilleron, J.; Carette, D.; Chevallier, D.; Segretain, D.; Pointis, G. Molecular connexin partner remodeling orchestrates connexin traffic: From physiology to pathophysiology. *Crit. Rev. Biochem. Mol. Biol.* **2012**, *47*, 407–423. [CrossRef] [PubMed]

17. Sorgen, P.L.; Trease, A.J.; Spagnol, G.; Delmar, M.; Nielsen, M.S. Protein–Protein Interactions with Connexin 43: Regulation and Function. *Int. J. Mol. Sci.* **2018**, *19*. [CrossRef] [PubMed]

18. Basheer, W.; Shaw, R. The "tail" of Connexin43: An unexpected journey from alternative translation to trafficking. *Biochim. Biophys. Acta* **2016**, *1863*, 1848–1856. [CrossRef] [PubMed]

19. Epifantseva, I.; Shaw, R.M. Intracellular trafficking pathways of Cx43 gap junction channels. *Biochim. Biophys. Acta* **2017**, *1860*, 40–47. [CrossRef] [PubMed]

20. Lau, A.F.; Kurata, W.E.; Kanemitsu, M.Y.; Loo, L.W.; Warn-Cramer, B.J.; Eckhart, W.; Lampe, P.D. Regulation of connexin43 function by activated tyrosine protein kinases. *J. Bioenerg. Biomembr.* **1996**, *28*, 359–368. [CrossRef] [PubMed]

21. Lin, R.; Warn-Cramer, B.J.; Kurata, W.E.; Lau, A.F. v-Src phosphorylation of connexin 43 on Tyr247 and Tyr265 disrupts gap junctional communication. *J. Cell Biol.* **2001**, *154*, 815–827. [CrossRef] [PubMed]

22. Saidi Brikci-Nigassa, A.; Clement, M.J.; Ha-Duong, T.; Adjadj, E.; Ziani, L.; Pastre, D.; Curmi, P.A.; Savarin, P. Phosphorylation controls the interaction of the connexin43 C-terminal domain with tubulin and microtubules. *Biochemistry* **2012**, *51*, 4331–4342. [CrossRef] [PubMed]

23. Ambrosi, C.; Ren, C.; Spagnol, G.; Cavin, G.; Cone, A.; Grintsevich, E.E.; Sosinsky, G.E.; Sorgen, P.L. Connexin43 Forms Supramolecular Complexes through Non-Overlapping Binding Sites for Drebrin, Tubulin, and ZO-1. *PLoS ONE* **2016**, *11*, e0157073. [CrossRef] [PubMed]

24. Alberts, B.; Johnson, A.; Lewis, J.; Raff, M.; Roberts, K.; Walter, P. Transport from the Trans Golgi Network to Lysosomes. In *Molecular Biology of the Cell*; Garland Science: New York, NY, USA, 2002.

25. Butkevich, E.; Hulsmann, S.; Wenzel, D.; Shirao, T.; Duden, R.; Majoul, I. Drebrin is a novel connexin-43 binding partner that links gap junctions to the submembrane cytoskeleton. *Curr. Biol.* **2004**, *14*, 650–658. [CrossRef] [PubMed]

26. Kieken, F.; Mutsaers, N.; Dolmatova, E.; Virgil, K.; Wit, A.L.; Kellezi, A.; Hirst-Jensen, B.J.; Duffy, H.S.; Sorgen, P.L. Structural and molecular mechanisms of gap junction remodeling in epicardial border zone myocytes following myocardial infarction. *Circ. Res.* **2009**, *104*, 1103–1112. [CrossRef] [PubMed]

27. Rhett, J.M.; Jourdan, J.; Gourdie, R.G. Connexin 43 connexon to gap junction transition is regulated by zonula occludens-1. *Mol. Biol. Cell* **2011**, *22*, 1516–1528. [CrossRef] [PubMed]

28. Solan, J.L.; Lampe, P.D. Connexin 43 in LA-25 cells with active v-src is phosphorylated on Y247, Y265, S262, S279/282, and S368 via multiple signaling pathways. *Cell Commun. Adhes.* **2008**, *15*, 75–84. [CrossRef] [PubMed]

29. Ai, Z.; Fischer, A.; Spray, D.C.; Brown, A.M.; Fishman, G.I. Wnt-1 regulation of connexin43 in cardiac myocytes. *J. Clin. Investig.* **2000**, *105*, 161–171. [CrossRef] [PubMed]

30. Prakash, S.; Swaminathan, U.; Nagamalini, B.R.; Krishnamurthy, A.B. β-catenin in disease. *J. Oral Maxillofacc. Pathol.* **2016**, *20*, 289–299. [CrossRef] [PubMed]

31. Lorenzon, A.; Calore, M.; Poloni, G.; De Windt, L.J.; Braghetta, P.; Rampazzo, A. Wnt/β-catenin pathway in arrhythmogenic cardiomyopathy. *Oncotarget* **2017**, *8*, 60640–60655. [CrossRef] [PubMed]

32. Xing, Y.; Takemaru, K.; Liu, J.; Berndt, J.D.; Zheng, J.J.; Moon, R.T.; Xu, W. Crystal structure of a full-length β-catenin. *Structure* **2008**, *16*, 478–487. [CrossRef] [PubMed]

33. Olson, D.J.; Christian, J.L.; Moon, R.T. Effect of wnt-1 and related proteins on gap junctional communication in Xenopus embryos. *Science* **1991**, *252*, 1173–1176. [CrossRef] [PubMed]

34. Moorer, M.C.; Hebert, C.; Tomlinson, R.E.; Iyer, S.R.; Chason, M.; Stains, J.P. Defective signaling, osteoblastogenesis and bone remodeling in a mouse model of connexin 43 C-terminal truncation. *J. Cell Sci.* **2017**, *130*, 531–540. [CrossRef] [PubMed]

35. Rinaldi, F.; Hartfield, E.M.; Crompton, L.A.; Badger, J.L.; Glover, C.P.; Kelly, C.M.; Rosser, A.E.; Uney, J.B.; Caldwell, M.A. Cross-regulation of Connexin43 and β-catenin influences differentiation of human neural progenitor cells. *Cell Death Dis.* **2014**, *5*, e1017. [CrossRef] [PubMed]

36. Talhouk, R.S.; Fares, M.B.; Rahme, G.J.; Hariri, H.H.; Rayess, T.; Dbouk, H.A.; Bazzoun, D.; Al-Labban, D.; El-Sabban, M.E. Context dependent reversion of tumor phenotype by connexin-43 expression in MDA-MB231 cells and MCF-7 cells: Role of β-catenin/connexin43 association. *Exp. Cell Res.* **2013**, *319*, 3065–3080. [CrossRef] [PubMed]

37. Swope, D.; Cheng, L.; Gao, E.; Li, J.; Radice, G.L. Loss of cadherin-binding proteins β-catenin and plakoglobin in the heart leads to gap junction remodeling and arrhythmogenesis. *Mol. Cell. Biol.* **2012**, *32*, 1056–1067. [CrossRef] [PubMed]

38. Du, W.J.; Li, J.K.; Wang, Q.Y.; Hou, J.B.; Yu, B. Lithium chloride preconditioning optimizes skeletal myoblast functions for cellular cardiomyoplasty in vitro via glycogen synthase kinase-3β/β-catenin signaling. *Cells Tissues Organs* **2009**, *190*, 11–19. [CrossRef] [PubMed]

39. Xu, W.; Kimelman, D. Mechanistic insights from structural studies of β-catenin and its binding partners. *J. Cell Sci.* **2007**, *120*, 3337–3344. [CrossRef] [PubMed]

40. Kim, W.; Kim, M.; Jho, E.H. Wnt/β-catenin signalling: From plasma membrane to nucleus. *Biochem. J.* **2013**, *450*, 9–21. [CrossRef] [PubMed]

41. Grosely, R.; Kopanic, J.L.; Nabors, S.; Kieken, F.; Spagnol, G.; Al-Mugotir, M.; Zach, S.; Sorgen, P.L. Effects of phosphorylation on the structure and backbone dynamics of the intrinsically disordered connexin43 C-terminal domain. *J. Biol. Chem.* **2013**, *288*, 24857–24870. [CrossRef] [PubMed]

42. Valenta, T.; Hausmann, G.; Basler, K. The many faces and functions of β-catenin. *EMBO J.* **2012**, *31*, 2714–2736. [CrossRef] [PubMed]

43. Whitmore, L.; Wallace, B.A. DICHROWEB, an online server for protein secondary structure analyses from circular dichroism spectroscopic data. *Nucleic Acids Res.* **2004**, *32*, W668–W673. [CrossRef] [PubMed]

44. Whitmore, L.; Wallace, B.A. Protein secondary structure analyses from circular dichroism spectroscopy: Methods and reference databases. *Biopolymers* **2008**, *89*, 392–400. [CrossRef] [PubMed]

45. Sorgen, P.L.; Duffy, H.S.; Spray, D.C.; Delmar, M. pH-dependent dimerization of the carboxyl terminal domain of Cx43. *Biophys. J.* **2004**, *87*, 574–581. [CrossRef] [PubMed]

46. Li, H.; Spagnol, G.; Naslavsky, N.; Caplan, S.; Sorgen, P.L. TC-PTP directly interacts with connexin43 to regulate gap junction intercellular communication. *J. Cell Sci.* **2014**, *127*, 3269–3279. [CrossRef] [PubMed]

47. Hornbeck, P.V.; Kornhauser, J.M.; Tkachev, S.; Zhang, B.; Skrzypek, E.; Murray, B.; Latham, V.; Sullivan, M. PhosphoSitePlus: A comprehensive resource for investigating the structure and function of experimentally determined post-translational modifications in man and mouse. *Nucleic Acids Res.* **2012**, *40*, D261–D270. [CrossRef] [PubMed]

48. Solan, J.L.; Marquez-Rosado, L.; Sorgen, P.L.; Thornton, P.J.; Gafken, P.R.; Lampe, P.D. Phosphorylation at S365 is a gatekeeper event that changes the structure of Cx43 and prevents down-regulation by PKC. *J. Cell Biol.* **2007**, *179*, 1301–1309. [CrossRef] [PubMed]

49. Remo, B.F.; Qu, J.; Volpicelli, F.M.; Giovannone, S.; Shin, D.; Lader, J.; Liu, F.Y.; Zhang, J.; Lent, D.S.; Morley, G.E.; et al. Phosphatase-resistant gap junctions inhibit pathological remodeling and prevent arrhythmias. *Circ. Res.* **2011**, *108*, 1459–1466. [CrossRef] [PubMed]

50. Trease, A.J.; Capuccino, J.M.V.; Contreras, J.; Harris, A.L.; Sorgen, P.L. Intramolecular signaling in a cardiac connexin: Role of cytoplasmic domain dimerization. *J. Mol. Cell. Cardiol.* **2017**, *111*, 69–80. [CrossRef] [PubMed]

51. Leithe, E.; Mesnil, M.; Aasen, T. The connexin 43 C-terminus: A tail of many tales. *Biochim. Biophys. Acta* **2018**, *1860*, 48–64. [CrossRef] [PubMed]

52. Solan, J.L.; Lampe, P.D. Spatio-temporal regulation of connexin43 phosphorylation and gap junction dynamics. *Biochim. Biophys. Acta* **2018**, *1860*, 83–90. [CrossRef] [PubMed]

53. Falk, M.M.; Bell, C.L.; Kells Andrews, R.M.; Murray, S.A. Molecular mechanisms regulating formation, trafficking and processing of annular gap junctions. *BMC Cell Biol.* **2016**, *17*, S22. [CrossRef] [PubMed]

54. Deb, A. Cell-cell interaction in the heart via Wnt/β-catenin pathway after cardiac injury. *Cardiovasc. Res.* **2014**, *102*, 214–223. [CrossRef] [PubMed]

55. Van der Heyden, M.A.; Rook, M.B.; Hermans, M.M.; Rijksen, G.; Boonstra, J.; Defize, L.H.; Destree, O.H. Identification of connexin43 as a functional target for Wnt signalling. *J. Cell Sci.* **1998**, *111 Pt 12*, 1741–1749. [PubMed]

56. Shaw, R.M.; Fay, A.J.; Puthenveedu, M.A.; von Zastrow, M.; Jan, Y.N.; Jan, L.Y. Microtubule plus-end-tracking proteins target gap junctions directly from the cell interior to adherens junctions. *Cell* **2007**, *128*, 547–560. [CrossRef] [PubMed]

57. Du, W.J.; Li, J.K.; Wang, Q.Y.; Hou, J.B.; Yu, B. Lithium chloride regulates connexin43 in skeletal myoblasts in vitro: Possible involvement in Wnt/β-catenin signaling. *Cell Commun. Adhes.* **2008**, *15*, 261–271. [CrossRef] [PubMed]

58. Nakashima, T.; Ohkusa, T.; Okamoto, Y.; Yoshida, M.; Lee, J.K.; Mizukami, Y.; Yano, M. Rapid electrical stimulation causes alterations in cardiac intercellular junction proteins of cardiomyocytes. *Am. J. Physiol. Heart Circ. Physiol.* **2014**, *306*, H1324–H1333. [CrossRef] [PubMed]

59. Reaume, A.G.; de Sousa, P.A.; Kulkarni, S.; Langille, B.L.; Zhu, D.; Davies, T.C.; Juneja, S.C.; Kidder, G.M.; Rossant, J. Cardiac malformation in neonatal mice lacking connexin43. *Science* **1995**, *267*, 1831–1834. [CrossRef] [PubMed]

60. Musil, L.S.; Goodenough, D.A. Multisubunit assembly of an integral plasma membrane channel protein, gap junction connexin43, occurs after exit from the ER. *Cell* **1993**, *74*, 1065–1077. [CrossRef]

61. Lauf, U.; Giepmans, B.N.; Lopez, P.; Braconnot, S.; Chen, S.C.; Falk, M.M. Dynamic trafficking and delivery of connexons to the plasma membrane and accretion to gap junctions in living cells. *Proc. Natl. Acad. Sci. USA* **2002**, *99*, 10446–10451. [CrossRef] [PubMed]

62. Giepmans, B.N.; Verlaan, I.; Hengeveld, T.; Janssen, H.; Calafat, J.; Falk, M.M.; Moolenaar, W.H. Gap junction protein connexin-43 interacts directly with microtubules. *Curr. Biol.* **2001**, *11*, 1364–1368. [CrossRef]

63. Fort, A.G.; Murray, J.W.; Dandachi, N.; Davidson, M.W.; Dermietzel, R.; Wolkoff, A.W.; Spray, D.C. In vitro motility of liver connexin vesicles along microtubules utilizes kinesin motors. *J. Biol. Chem.* **2011**, *286*, 22875–22885. [CrossRef] [PubMed]

64. Smyth, J.W.; Vogan, J.M.; Buch, P.J.; Zhang, S.S.; Fong, T.S.; Hong, T.T.; Shaw, R.M. Actin cytoskeleton rest stops regulate anterograde traffic of connexin 43 vesicles to the plasma membrane. *Circ. Res.* **2012**, *110*, 978–989. [CrossRef] [PubMed]

65. Fang, D.; Hawke, D.; Zheng, Y.; Xia, Y.; Meisenhelder, J.; Nika, H.; Mills, G.B.; Kobayashi, R.; Hunter, T.; Lu, Z. Phosphorylation of β-catenin by AKT promotes β-catenin transcriptional activity. *J. Biol. Chem.* **2007**, *282*, 11221–11229. [CrossRef] [PubMed]

66. Liu, C.; Li, Y.; Semenov, M.; Han, C.; Baeg, G.H.; Tan, Y.; Zhang, Z.; Lin, X.; He, X. Control of β-catenin phosphorylation/degradation by a dual-kinase mechanism. *Cell* **2002**, *108*, 837–847. [CrossRef]

67. Zhang, M.; Mahoney, E.; Zuo, T.; Manchanda, P.K.; Davuluri, R.V.; Kirschner, L.S. Protein kinase A activation enhances β-catenin transcriptional activity through nuclear localization to PML bodies. *PLoS ONE* **2014**, *9*, e109523. [CrossRef] [PubMed]

68. Solan, J.L.; Lampe, P.D. Specific Cx43 phosphorylation events regulate gap junction turnover in vivo. *FEBS Lett.* **2014**, *588*, 1423–1429. [CrossRef] [PubMed]

69. Spagnol, G.; Kieken, F.; Kopanic, J.L.; Li, H.; Zach, S.; Stauch, K.L.; Grosely, R.; Sorgen, P.L. Structural Studies of the Nedd4 WW Domains and Their Selectivity for the Connexin43 (Cx43) Carboxyl Terminus. *J. Biol. Chem.* **2016**, *291*, 7637–7650. [CrossRef] [PubMed]

70. Girao, H.; Catarino, S.; Pereira, P. Eps15 interacts with ubiquitinated Cx43 and mediates its internalization. *Exp. Cell Res.* **2009**, *315*, 3587–3597. [CrossRef] [PubMed]

71. Leykauf, K.; Salek, M.; Bomke, J.; Frech, M.; Lehmann, W.D.; Durst, M.; Alonso, A. Ubiquitin protein ligase Nedd4 binds to connexin43 by a phosphorylation-modulated process. *J. Cell Sci.* **2006**, *119*, 3634–3642. [CrossRef] [PubMed]

72. Duffy, H.S.; Sorgen, P.L.; Girvin, M.E.; O'Donnell, P.; Coombs, W.; Taffet, S.M.; Delmar, M.; Spray, D.C. pH-dependent intramolecular binding and structure involving Cx43 cytoplasmic domains. *J. Biol. Chem.* **2002**, *277*, 36706–36714. [CrossRef] [PubMed]

73. Hirst-Jensen, B.J.; Sahoo, P.; Kieken, F.; Delmar, M.; Sorgen, P.L. Characterization of the pH-dependent interaction between the gap junction protein connexin43 carboxyl terminus and cytoplasmic loop domains. *J. Biol. Chem.* **2007**, *282*, 5801–5813. [CrossRef] [PubMed]

International Journal of
Molecular Sciences

MDPI

Article

N-Glycosylation Regulates Pannexin 2 Localization but Is Not Required for Interacting with Pannexin 1

Rafael E. Sanchez-Pupo, Danielle Johnston and Silvia Penuela *

Department of Anatomy and Cell Biology, Schulich School of Medicine & Dentistry,
University of Western Ontario, London, ON N6A5C1, Canada; rsnchezp@uwo.ca (R.E.S.-P.);
Danielle.Johnston@schulich.uwo.ca (D.J.)
* Correspondence: spenuela@uwo.ca; Tel.: +1-519-661-2111 (ext. 84735)

Received: 27 April 2018; Accepted: 20 June 2018; Published: 22 June 2018

Abstract: Pannexins (Panx1, 2, 3) are channel-forming glycoproteins expressed in mammalian tissues. We previously reported that N-glycosylation acts as a regulator of the localization and intermixing of Panx1 and Panx3, but its effects on Panx2 are currently unknown. Panx1 and Panx2 intermixing can regulate channel properties, and both pannexins have been implicated in neuronal cell death after ischemia. Our objectives were to validate the predicted N-glycosylation site of Panx2 and to study the effects of Panx2 glycosylation on localization and its capacity to interact with Panx1. We used site-directed mutagenesis, enzymatic de-glycosylation, cell-surface biotinylation, co-immunoprecipitation, and confocal microscopy. Our results showed that N86 is the only N-glycosylation site of Panx2. Panx2 and the N86Q mutant are predominantly localized to the endoplasmic reticulum (ER) and cis-Golgi matrix with limited cell surface localization was seen only in the presence of Panx1. The Panx2 N86Q mutant is glycosylation-deficient and tends to aggregate in the ER reducing its cell surface trafficking but it can still interact with Panx1. Our study indicates that N-glycosylation may be important for folding and trafficking of Panx2. We found that the un-glycosylated forms of Panx1 and 2 can readily interact, regulating their localization and potentially their channel function in cells where they are co-expressed.

Keywords: pannexin; Panx1; Panx2; post-translational modification; traffic; N-glycosylation; channels; subcellular localization

---

1. Introduction

Pannexins (Panx1, Panx2, and Panx3) are membrane-spanning glycoproteins capable of forming large pore channels that allow the passage of ions and macromolecules involved in paracrine and autocrine signaling [1,2]. Panx1 is the most widely expressed pannexin and the most studied, with evidence supporting that its channels act as ATP and Ca^{2+} conduits [3–6] and are implicated in critical cellular processes such as cell death after brain ischemia [7] and inflammation [8]. On the other hand, Panx2 is the largest member of the family and its expression was thought to be restricted to the central nervous system (CNS). Recently, it has been reported that Panx2 can also be expressed in other tissues such as skin, kidney and liver [9,10], while Panx3 is predominantly expressed in skin, cartilage and bone [4,5,11–13]. In contrast to the hexameric type of channels formed by Panx1, it has been suggested that Panx2 can form octameric or heptameric channels [14] and it is still unclear whether its channel function would be similar to the other family members. Interestingly, Panx1 and Panx2 expression have been found to overlap in adult rodent brains although they are inversely regulated throughout development, with Panx1 being more abundant in neonatal and young tissues, whereas Panx2 is more abundant in the adult [15–18]. Under ischemic conditions, both Panx1 and Panx2 are expressed in the brain and their overlapping channel functions contribute to neurodegeneration. In fact, the deletion of Panx1 and Panx2 in a double knockout mouse model was necessary to observe a reduction in cell death after ischemia, perhaps due to their redundant and/or complementary functions [19].

N-glycosylation is a posttranslational modification that occurs in the endoplasmic reticulum (ER) and is recognized to have profound effects on protein folding and trafficking of membrane-bound proteins [20]. The prediction of putative N-linked glycosylation sites for pannexins has been done in the past, and it has been demonstrated that Panx1 and Panx3 have sites for N-linked glycosylation at Asn (N) 254 and N71, respectively. These studies comprised further characterization using enzymatic digestion with endoglycosidases that confirmed that all three members of the pannexin family are differentially N-glycosylated but not O-glycosylated [11,21]. However, for Panx2 the predicted site of N-glycosylation at residue N86 remains to be validated [22].

Unlike other pannexins, Panx2 is only modified to a high-mannose glycosylation species (termed as Gly-1) and presents a predominantly intracellular localization that has been associated with this lower level of N-glycosylation [22–26]. Previous evidence supports the concept that complex glycoprotein species (Gly-2) (further processed at Golgi) present in Panx1 and Panx3 traffic readily to the cell surface [22]. However, our group and others have stated that under certain circumstances Panx2 can also translocate to the plasma membrane [14,17,22], but it is still unclear whether glycosylation plays a role in regulating Panx2 trafficking.

Panx2 has been shown to interact with Panx1, and when co-expressed together they can form heteromeric channels with reduced channel properties compared to homomeric ones [22,27]. Although this has only been tested in ectopic expression systems it has been suggested that it might function as a mode of regulation in cells that endogenously express both proteins. Interestingly, the Panx1/Panx2 interaction only occurs with the Gly-0 and Gly-1 species of Panx1 [22] and it is unknown whether Panx2 glycosylation has any impact on the formation of these intermixed channels. Since previous studies have shown that Panx1 and Panx2 are often co-expressed in the same cells under normal conditions [15,18,27–30], and that their co-expression modulates important processes such as ischemia-induced neurodegeneration and brain damage in vivo [31], it is important to understand how these channels may be regulated by post-translational modifications (PTMs) such as N-glycosylation, and how this PTM may regulate their interaction.

The present study aimed to validate the predicted N-linked-glycosylation site of Panx2 and determine its role in the regulation of the subcellular localization and intermixing of Panx2 with Panx1. We generated a Panx2 mutant protein completely devoid of N-glycosylation that, when overexpressed, exhibits a high level of intracellular aggregation with decreased traffic to the plasma membrane compared with wild-type Panx2. We found that N-glycosylation of Panx2 is not required for Panx1/Panx2 intermixing but facilitates Panx2 trafficking and localization at the plasma membrane when co-expressed with Panx1. The intracellular localization of un-glycosylated Panx2 reduces its co-localization with Panx1 at the cell surface and may impact their channel function in cells that co-express both glycoproteins, such as neurons. Collectively, we propose that N-glycosylation may be necessary for proper processing of Panx2 at the endoplasmic reticulum, regulating its intracellular distribution, but it is not required for interacting with Panx1.

2. Results

2.1. Characterization of the Panx2 N-Glycosylation Site at Asparagine 86

Previous research has shown that ectopic expression of full-length mouse Panx2 presented N-glycosylated species with high-mannose modification [22]. Based on the analysis of the canonical sequence of Panx2 (UniProt ID: Q6IMP4-1), this modification might occur at asparagine 86 (Asn86 or N86) that is located in the first extracellular loop of Panx2 [20] (Figure 1A). To validate this prediction, we generated a full-length Panx2 mutant (Panx2^{N86Q}, referred here as N86Q) with the substitution of the N86 with a glutamine (Gln, Q) that prevents the attachment of N-linked glycans to this specific site. As shown in Figure 1B, Western blotting (WB) of overexpressed constructs of N86Q and wild-type

Panx2 (as glycosylated control) indicated that the electrophoretic band corresponding to N86Q mutant migrated faster than its wildtype counterpart, characteristic of a reduction in molecular weight.

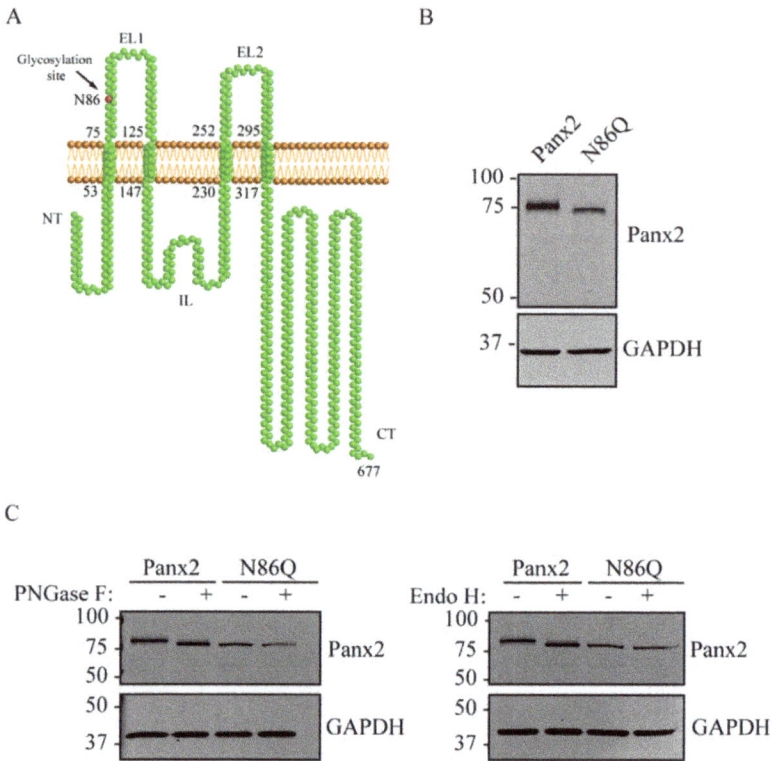

Figure 1. Asn86 is the N-glycosylation site of Panx2. (**A**) Based on sequence analysis, Panx2 (Uniprot ID: Q6IMP4-1) is predicted to contain four transmembrane domains, one intracellular (IL) and two extracellular loops (EL). The predicted N-glycosylation site is located at Asn86 in the first extracellular loop (EL1) (red residue). (**B**) Western blot (WB) comparing wildtype Panx2 and mutant N86Q, the latter shows a faster migrating band than the wildtype counterpart, indicative of decreased molecular weight. (**C**) Cell lysates of HEK293T transiently expressing Panx2 and N86Q mutant were subjected to enzymatic digestions with PNGase F and EndoH N-glycosidases. WB analysis confirmed that N86 is the only glycosylation site for Panx2 since only the wildtype protein exhibited a band shift after treatment with both glycosidases, and the de-glycosylated Panx2 band ran to the same position as the N86Q mutant. GAPDH was used as loading control. Molecular weights are noted in kDa.

To analyze whether N-glycosylation was prevented in the N86Q mutant, enzymatic de-glycosylation with endoglycosidase H (Endo H) and Peptide-N-Glycosidase F (PNGase F) were applied to the protein lysates and analyzed by WB (Figure 1C). Both de-glycosylation treatments did not cause any shift in migration of the N86Q band. Yet, wild-type Panx2 exhibited a small shift after treatment which relocated the migration of this band to the same position as that of the faster migrating N86Q mutant. Thus, the differences in the electrophoretic migration of the N86Q compared with the wild-type (WT) Panx2 can be explained by differences in molecular weight due to the absence of N-glycosylation. These findings validate N86 as the only N-glycosylation site present in Panx2 as was predicted by Penuela and others [22].

2.2. N86Q Forms Intracellular Aggregates

To determine whether glycosylation of Panx2 has any effect on its subcellular localization, we ectopically expressed Panx2 WT and N86Q in different reference cell lines and evaluated whether the intracellular distribution observed was dependent on the cell type as has been reported for Panx1 and Panx3 [32]. Because our group has demonstrated before [21] that Panx1 influences Panx2 trafficking, we selected Normal Rat Kidney cells (NRK), that in our experience have very low expression of endogenous Panx1 (Figure 2D), and "adherent" human embryonic kidney cells (AD293) that endogenously express the human ortholog PANX1 (Figure 3D). Likewise, these cell lines were selected as they are suitable transfection hosts with a relatively large cytoplasm which facilitates the visualization of the subcellular localization of proteins.

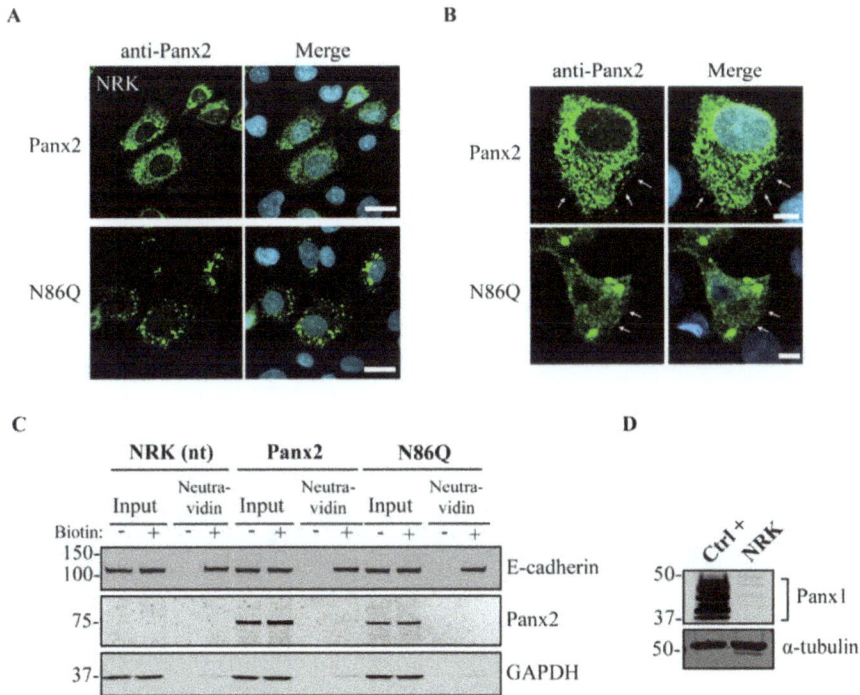

Figure 2. Panx2 has a predominantly intracellular localization, and its N-glycosylation-deficient mutant (N86Q) forms aggregates in NRK cells with low endogenous Panx1 protein expression. (**A**) Confocal micrographs of Panx2 and N86Q ectopically expressed in NRK cells, immunolabeled with anti-Panx2 antibody (green), revealed that Panx2 is predominantly localized intracellularly. The mutant N86Q is also localized intracellularly but appeared mostly as punctate aggregates. Nuclei (blue) were counterstained with Hoechst 33342. Scale bars = 20 μm. (**B**) Representative images of a small subpopulation of NRK cells expressing Panx2 or N86Q that showed minimal apparent localization at the cell surface (indicated with white arrows). Scale bars = 5 μm. (**C**) Western blot analysis of cell-surface-biotinylated proteins with EZ-Link™ Sulfo-NHS-SS-Biotin pulled down with NeutrAvidin® beads showed very low detection of Panx2 but not the N86Q mutant at the cell surface of NRK cells. E-cadherin was used as a positive control of cell surface protein labeling and GAPDH was used as a negative control of intracellular proteins (no biotin internalization). Non-transfected NRKs (nt) were used as a negative control. (**D**) Western blot of protein lysate from NRK cells indicated that these cells have very low detectable levels of endogenous Panx1. Overexpressed Panx1 was used as positive control (Ctrl +) and endogenous α-tubulin was used as protein loading control. Molecular weights noted in kDa.

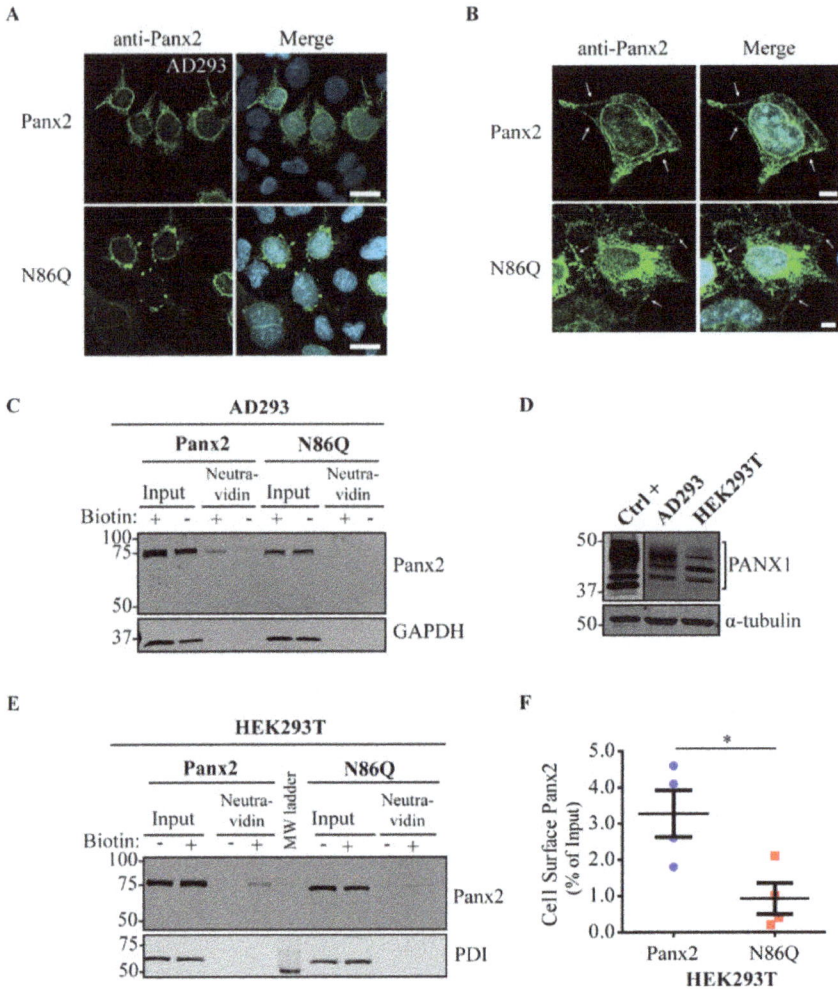

Figure 3. Analysis of Panx2 and N86Q localization in AD293 and HEK293T cells expressing endogenous PANX1. (**A**) Immunolabeling of Panx2 and N86Q mutant (green) showed that both localized mostly intracellularly, but the N86Q mutant aggregated intracellularly in AD293 cells. Scale bars = 20 μm. (**B**) A subpopulation of AD293 cells displayed Panx2 localization at the cell surface, less evident with the N86Q mutant (indicated with white arrows) Scale bars = 5 μm. Nuclei (blue). (**C**) Cell Surface Biotinylation Assays on AD293 cells showed a weak detection of the Panx2 wildtype but not N86Q mutant in surface-labeled fractions. GAPDH was used as a control for biotin internalization. (**D**) Immunoblots of AD293 and HEK293T cells confirmed that both cell lines express endogenous PANX1. Overexpressed human PANX1 served as positive control (Ctrl +) and endogenous α-tubulin was used as loading control. Line dividing upper panel of PANX1 WB indicates differences in exposure of the same blot to show a better detection of endogenous PANX1 compared to the overexpressed positive control. (**E**) Cell surface biotinylation experiments performed on HEK293T cells showed that overexpressed Panx2 and the mutant N86Q are detectable at the cell surface. Protein disulfide-isomerase (PDI) was used as a control for biotin internalization. (**F**) Densitometric analysis and quantification of cell surface biotinylation experiments performed in HEK293T cells revealed a significant reduction of N86Q cell surface detection compared to Panx2. Cell surface detection was calculated relative to the total protein in input lanes. Statistical significance was considered when $p < 0.05$ (* $p = 0.0286$, $N = 4$ independent experiments), Mann Whitney U test. Error bars denote mean ± S.E.M. Molecular weights are noted in kDa.

Confocal immunofluorescence imaging revealed an intracellular localization of ectopic Panx2 (Figures 2A and 3A, top panel), with perinuclear distribution and spread in intracellular compartments. Interestingly, in both cell types assayed, the mutant N86Q localized intracellularly with large subpopulations forming punctate aggregates (Figures 2A and 3A, bottom panel).

2.3. Limited Panx2 Localization at the Cell Surface Is Reliant on N-Glycosylation Status and the Level of Panx1

Despite the predominant intracellular localization, in a subpopulation of cells, Panx2 and the N86Q mutant (to a lesser extent) were apparent in limited regions of the cell surface (Figures 2B and 3B, arrows). To corroborate these results, cell surface biotinylation assays followed by immunoblotting were conducted using a cell-impermeable biotinylation reagent (Sulfo-NHS-SS-Biotin). Cell-surface biotinylation experiments in NRKs (low endogenous Panx1, Figure 2D) showed a faint band of Panx2 and no detection of N86Q at the cell surface in the neutravidin pull-downs (Figure 2C). Also, AD293 cells (with a higher level of endogenous PANX1, Figure 3D) exhibited low Panx2 and no detectable N86Q mutant protein at the cell surface (Figure 3C). However, in subsequent experiments we also used human embryonic kidney (HEK293T) cells (Figure 3D) that have been used in previous studies [11,22], because of their increased transfection efficiency and enhanced protein expression due to the SV40 T-antigen [33]. After ectopic expression in HEK293T cells, we performed the same cell-surface biotinylation assays and noticed that Panx2 WT protein was detected (approximately 4% of the total Panx2 expression) at the cell surface (Figure 3E). Under these overexpression conditions, the N86Q mutant was also detected in the biotinylated-protein fractions (Figure 3E) but there was a significant ($p = 0.0286$, $N = 4$) reduction (to ~1% of its total amount) in the cell surface protein pool of the mutant (Figure 3F). Therefore, although the Panx2 cell membrane trafficking is reduced when Panx2 is not N-glycosylated at N86, its cell surface localization is not completely abrogated when overexpressed in HEK293T cells.

2.4. Panx2 and N86Q Aggregates Localize to the Endoplasmic Reticulum and Golgi Apparatus

Because of the prominent intracellular localization of both Panx2 and the mutant N86Q, we were interested in determining the subcellular compartments to which these proteins could be trafficking. We transiently expressed these proteins in AD293 cells and used immunolabeling with different organelle markers to assess their intracellular location by confocal microscopy (Figures 4 and 5).

As shown in Figure 4A, Panx2 immunolabeling exhibits a broad cytoplasmic distribution highly overlapping (Pearson's Colocalization Coefficient (PCC)$_{Panx2-PDI}$ = 0.49 ± 0.02; $n = 41$, $N = 3$) with the chaperone protein disulfide-isomerase (PDI), a known marker of the endoplasmic reticulum (ER). The mutant N86Q had significantly ($p < 0.0001$) less overlap (PCC$_{N86Q-PDI}$ = 0.34 ± 0.03, $n = 57$, $N = 3$) with PDI although there was also a more punctate distribution of PDI that co-localized with N86Q. These observations suggest that N86Q is being confined to some punctate regions enriched in PDI, along with other subcellular organelles. We also observed an apparent alteration of the ER morphology when N86Q was expressed.

Quantitation of colocalization with cis-Golgi marker (GM130) (Figure 4B) showed a significantly ($p = 0.0001$) higher colocalization with the N86Q mutant (PCC$_{N86Q-GM130}$ = 0.43 ± 0.04; $n = 41$, $N = 3$) than with Panx2 WT (PCC$_{Panx2-GM130}$ = 0.23 ± 0.03; $n = 30$, $N = 3$). In this case, cells that overexpressed Panx2 WT exhibited changes in the distribution of GM130 compared to un-transfected ones (in the same field of view, Figure 4B). These changes were more pronounced in AD293 cells overexpressing the mutant N86Q, in which not only the cis-Golgi morphology changed, but GM130 also appeared to accumulate within N86Q aggregates.

On the other hand, lysosome-associated membrane protein 2 (Lamp-2) and the fixable and cell permeant Mitotracker™ Red CMXRos were assayed to label lysosomes/late endosomes, and active mitochondria, respectively. We did not observe colocalization with Lamp-2 (PCC$_{Panx2-Lamp-2}$ = −0.08 ± 0.01; $n = 59$, $N = 3$; PCC$_{N86Q-Lamp-2}$ = −0.05 ± 0.02; $n = 49$, $N = 3$) or Mitotracker (PCC$_{Panx2-Mitotracker}$ = −0.08 ± 0.03; $n = 25$, $N = 3$; PCC$_{N86Q-Mitotracker}$ = −0.13 ± 0.06;

$n = 31$, $N = 3$), and there was no difference in the distribution of both markers upon overexpression of Panx2 WT or N86Q.

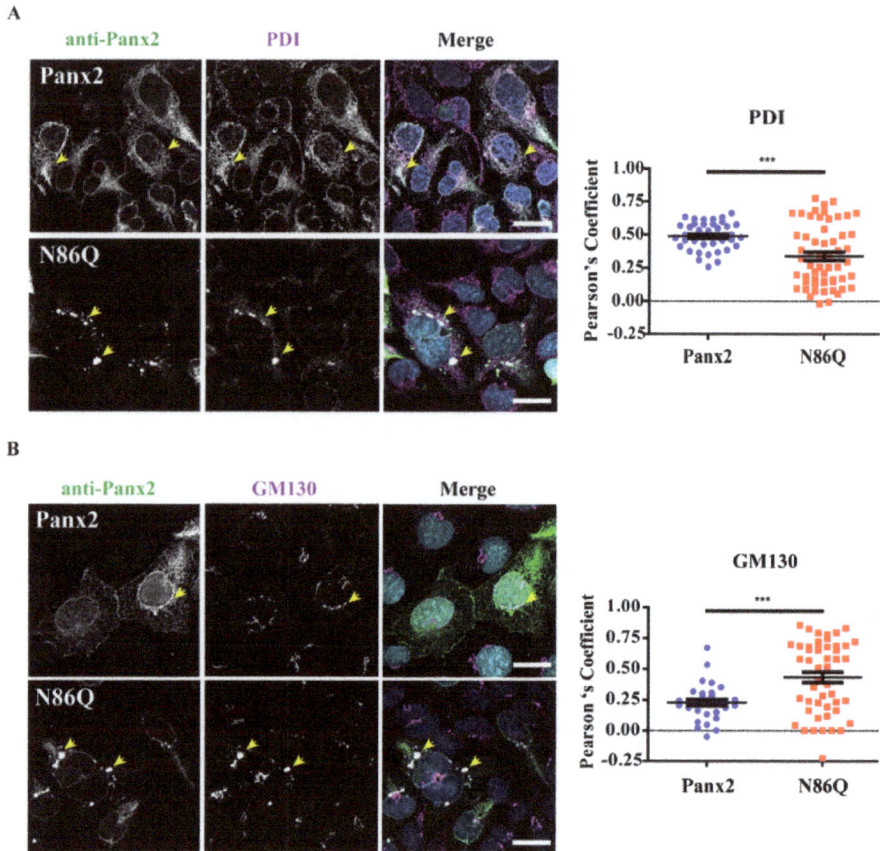

Figure 4. Panx2 and N86Q colocalize with markers of the endoplasmic reticulum and Golgi. Representative confocal micrographs of Panx2 and N86Q ectopically expressed in AD293 cells. Co-immunolabeling with anti-Panx2 antibody (green) and organelle markers (magenta): (**A**) PDI, endoplasmic reticulum (ER); (**B**) GM-130, cis-Golgi matrix. Panx2 has a perinuclear localization and is spread intracellularly in the cytoplasm partially colocalizing with markers of the endoplasmic reticulum and Golgi. N86Q aggregates also overlap with ER and Golgi markers and disrupt their distribution. Yellow arrowheads indicate representative regions of colocalization. Nuclei (blue) were counterstained with Hoechst 33342. Scale bars = 20 μm. Pearson Correlation Coefficients (right) were calculated for multiple regions of interest (ROI)s corresponding to double-labeled cells. Statistical significance was considered when $p < 0.05$ (*** $p \leq 0.0001$, $N = 3$ independent experiments), using unpaired two-tailed t test with Welch's correction. Error bars denote mean $\pm$ S.E.M.

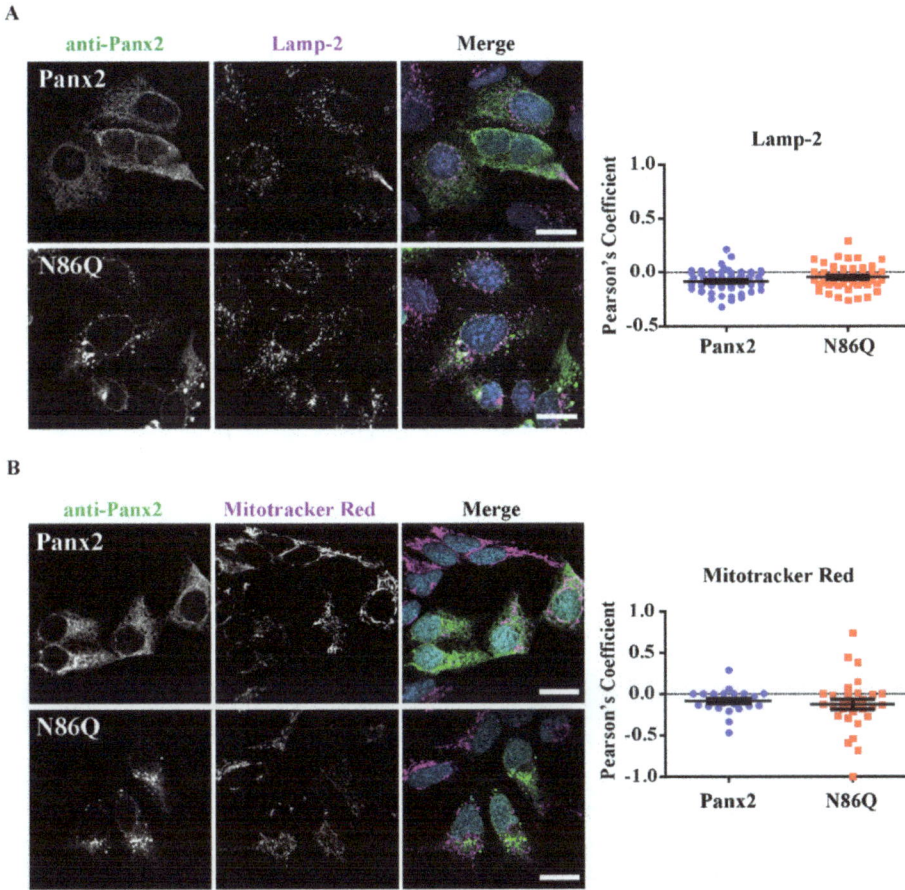

Figure 5. Panx2 and N86Q do not localize to late endosomes/lysosomes or mitochondria. Confocal micrographs of Panx2 and N86Q ectopically expressed in AD293 cells. Co-immunolabeling with anti-Panx2 antibody (green) and organelle markers (magenta): (**A**) Lamp-2, lysosomes and late endosomes, and (**B**) Mitotracker® Red, mitochondria showed that neither Panx2 or N86Q mutant exhibited significant overlap with the markers. Nuclei (blue) were counterstained with Hoechst 33342. Nuclei (blue) were counterstained with Hoechst 33342. Scale bars = 20 μm. Pearson Correlation Coefficients (right) were calculated for multiples ROIs corresponding to double-labeled cells. There was no statistical significance ($p > 0.05$, $N = 3$, unpaired two-tailed t test with Welch's correction) in the degree of colocalization between Panx2 and N86Q with the markers. Error bars denote mean ± S.E.M.

2.5. Panx2 N-Glycosylation Is Not Required for the Interaction with Panx1

Due to our previous report [22] in which we showed that glycosylation regulates intermixing of pannexins and that Panx2 interacts only with the core (non-glycosylated, Gly-0) and high-mannose species (Gly-1) of Panx1, we were interested in determining what would be the outcome with the un-glycosylated species of Panx2. For these experiments, Panx1 was ectopically co-expressed with either Panx2 or non-glycosylated N86Q mutant in HEK293T cells. Using co-immunoprecipitation assays (co-IP, Figure 6A), we observed that both Panx2 and N86Q can co-IP in a complex with Panx1 and occasionally, although not statistically significant ($p > 0.05$, $N = 4$) (Figure 6B,C), N86Q pulled down more Panx1 than the WT Panx2. In addition, consistently with what was reported before by

Penuela et al. (2009), we noticed that only Gly-0 and Gly-1 Panx1 species interacted with both variants of Panx2.

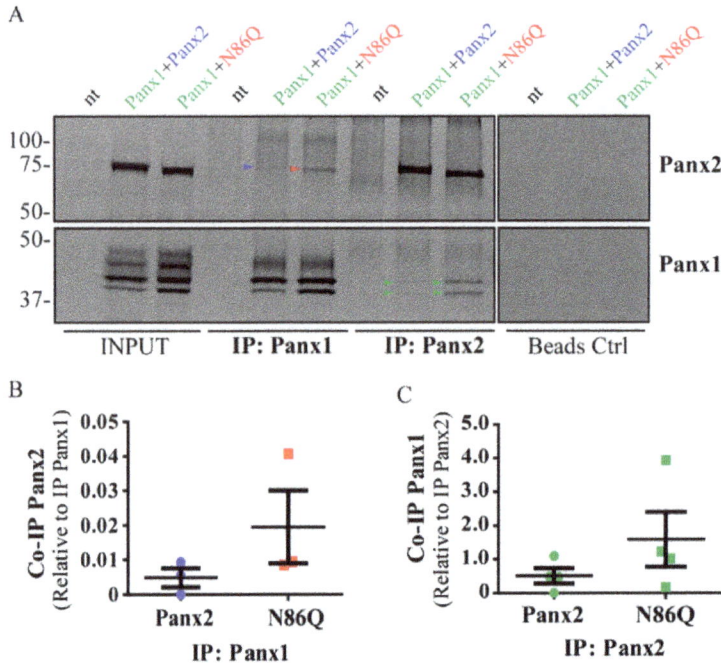

Figure 6. Panx2 glycosylation is not required for the interaction with Panx1 by immunoprecipitation. (**A**) Reciprocal co-immunoprecipitation (co-IP) experiments showed that both Panx2 and N86Q co-IP in a complex with overexpressed Panx1 in HEK293T cells. Colored arrowheads denote bands of co-IP proteins detected in WB. (**B,C**) Quantitative analysis (see Materials and Methods Section 4.8) of co-IP shows that the interaction of N86Q with Panx1 is not significantly different ($p > 0.05$, $N = 4$) than with Panx2, and in both cases the complexes only involved the lower glycosylated species of Panx1 (Gly-0 and Gly-1). Beads Ctrl denote control IPs done in parallel without antibodies. Protein sizes in kDa.

Due to the lack of available antibodies from different species to perform double immunolabeling of both pannexins, we used C-terminal FLAG-tagged Panx2 and N86Q that were co-expressed with Panx1 in HEK293T cells. Confocal imaging of Panx2-FLAG or N86Q-FLAG co-expressed with Panx1 (Figure 7A) showed that Panx2-FLAG exhibits both an intracellular and cell surface localization overlap with Panx1 (see Linescan analysis, Figure 7A). N86Q-FLAG formed mostly intracellular aggregates like its untagged counterpart (Figure 3A). A small subpopulation of N86Q-FLAG could still be seen at the cell surface colocalizing with Panx1 (see Linescan analysis, Figure 7B), but to a lesser degree than Panx2-FLAG. Taken together, these findings suggest that the glycosylation of Panx2 is not required for the interaction of Panx1/Panx2 but can determine the differential localization of glycosylated and un-glycosylated species.

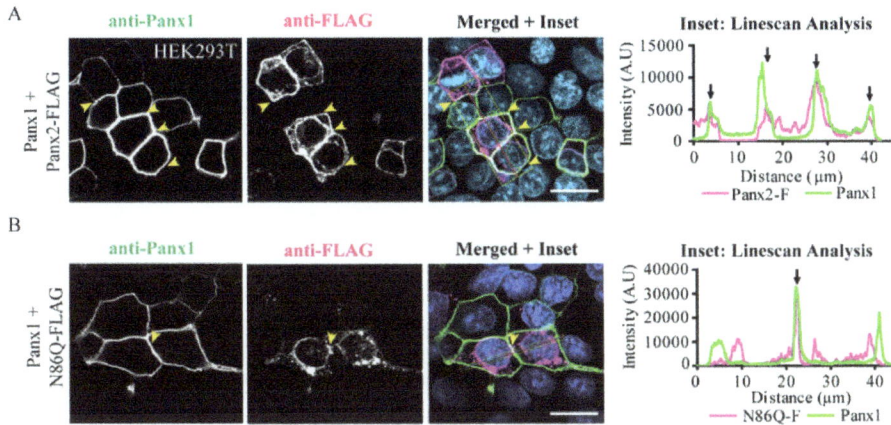

Figure 7. N-glycosylation may enhance plasma membrane localization of Panx2 when co-expressed with Panx1. Confocal micrographs of (**A**) Panx2-FLAG or (**B**) N86Q-FLAG (magenta) ectopically co-expressed with mouse Panx1 (green) in HEK293T cells. 72 h post-transfection Panx2-FLAG partially colocalized with Panx1 at the cell membrane (see black arrows in Linescan, **panel A**) with a subpopulation still in intracellular compartments. N86Q-FLAG formed intracellular aggregates and showed limited colocalization with Panx1 at the plasma membrane (see black arrows in Linescan, **panel B**). Yellow arrowheads denote regions of colocalization of Panx1 and FLAG labeling also depicted with black arrows in the corresponding linescans. Insets: Linescans showing the overlapping (black arrows) between fluorescence peaks to denote colocalization. Nuclei (blue, Hoechst 33342). Scale bars = 20 μm.

3. Discussion

Pannexins are a family of channel proteins implicated in important physiological and pathological functions and most of the current research has been conducted to analyze their level of expression and distribution within mammalian tissues and their role in diverse diseases [34]. However, there is still a need to understand the biophysical properties of these channels and the different ways of regulation that prevent the detrimental effects of their exacerbated channel activity. Pannexins are N-glycosylated and as integral membrane proteins, this modification seems to be essential in regulating their trafficking, as was demonstrated formerly for Panx1 and Panx3 [11,22]. Unlike the other pannexins, Panx2 is modified only to a high-mannose glycosylation (Gly-1) which is known to be an early post-translational modification occurring in the ER lumen. For many other glycoproteins, this step is generally followed by further oligosaccharide editing in the Golgi (complex glycosylation). To date, there is no evidence showing further processing of Panx2 in Golgi and only studies in Panx1 and Panx3 showed that trafficking of these two pannexins to the plasma membrane is mediated by Sar1-dependent COPII vesicles [35] with N-glycosylation affecting their final delivery [11,36,37]. This suggests that N-glycosylation regulates the route of pannexin trafficking and modifies their localization and channel formation in immortalized culture cells. A previous report of Panx2 localization pointed to a predominantly intracellular distribution that can also be modified by other PTMs like palmytoilation, which can determine its subcellular localization in neurons [17].

Panx2 seems to be mostly intracellularly localized, but in some instances, it can also translocate to the plasma membrane [14,17]. However, the exact site of N-glycosylation and whether this modification affects Panx2 trafficking was unknown. In this study, we sought to determine the N-glycosylation site by generating a glycosylation-deficient mutant (N86Q) based on the predicted N-linked glycosylation site reported by Penuela et al. [22]. For our experiments we transfected constructs encoding mouse Panx2 and an N86Q mutant into HEK293T cells. Our results showed that N86Q substitution generated a Panx2 mutant with a faster-migrating electrophoretic band compared to Panx2 WT, that does not shift after specific N-glycosidase digestion with endoglycosidases, thus confirming that N86 is the only N-glycosylation site for Panx2. In the other cell lines assayed (NRK and AD293) we also observed that overexpression of the same Panx2 WT and N86Q constructs exhibited the same electrophoretic properties seen with HEK293T (Figures 2C and 3C).

Independently of the cell-type used for ectopic expression, the N86Q mutant appeared to form punctate aggregates that were localized to intracellular compartments along with the ER-chaperone PDI, and the cis-Golgi matrix marker GM130. Compared to the N-glycosylation-deficient mutants of Panx1 and Panx3 [11], un-glycosylated Panx2 exhibited an exacerbated abnormal intracellular aggregation. This raises the possibility that the lack of N-glycosylation may have affected proper protein folding of Panx2, which is the largest member of the pannexin family. We cannot rule out that the intracellular accumulation of N86Q could be an artifact of overexpression, that concomitant with the lack of glycosylation of the Panx2 mutant, might have induced misfolding and ER-stress [38].

A previous study in murine postnatal hippocampal neural progenitor cells (NPC)s [17] showed that treatment with glycosidases had no effect on endogenous Panx2 electrophoretic mobility, suggesting that they were un-glycosylated. That study relied on antibody detection of endogenous Panx2, and the protein bands identified were of lower molecular weight than the predicted full-length Panx2 used in this study (677 aa, [39]). Further studies are needed to determine first, if the expression of certain un-glycosylated endogenous isoforms of Panx2 is cell-type specific, and second, whether glycosylation has measurable effects on the Panx2 cellular function. To date, there are no reports of mutations in the *Panx2* gene, but our results would predict significant changes in Panx2 behavior if its glycosylation is affected. It is also possible that the un-glycosylated form of Panx2 may be preferentially expressed in some cells and tissues determining the primary function of the Panx2 channel and its subcellular localization.

In our study, we detected large intracellular subpopulations of Panx2 likely localized in the ER, and partially colocalized with cis-Golgi marker. Interestingly, we found that overexpression of the N86Q mutant changed the distribution of PDI and GM130 compared to Panx2 WT, suggesting that the formation of aggregates may disrupt the morphology of ER and cis-Golgi. GM-130 is a peripheral membrane protein in *cis*-Golgi matrix that is important for maintenance of Golgi structure [40], and the regulation of ER-to-Golgi transport of proteins and glycosylation [41]. It is possible that the aggregation of N86Q may interfere with the mutant protein transport from the ER causing accumulation of GM130.

As reviewed by Boyce et al. [42], pannexins contain putative recognition sequences for endocytic and endo-lysosomal targeting which could account for the control of pannexin trafficking. Interestingly, recently published work by Boassa et al., described the localization of a recombinant Panx2 fused to mini-SOG tag that was transiently expressed in HeLa cells [26]. These authors used correlated light and electron microscopy imaging to detect Panx2 localization at cytoplasmic protrusions. Also, immuno-colocalization in HEK293T cells using assorted vesicular markers displayed Panx2 WT (untagged) localized to early or recycling endosomes rather than ER. Although, they mentioned in the manuscript that when the ER marker calnexin was used they detected colocalization with overexpressed Panx2 in HEK293T cells. Our findings are consistent with these reports [24], since we found Panx2 primarily in the ER (Figure 4A). However, we did not find conclusive evidence of Panx2 in endo-lysosome compartments (Figure 5A) as others have reported [25,26]. Based on our results, it is possible that Panx2 may have an intracellular channel function in the ER, similar to the proposed calcium-leak channels formed by Panx1 and Panx3 [6].

In some instances, Panx2 distribution exhibited limited cell surface localization that was more apparent when higher endogenous or ectopic Panx1 protein was expressed in the studied cell lines. This is consistent with our previous observation of increased Panx2 at the cell surface when co-expressed with Panx1 under overexpression conditions [22]. Here, we used three different cells lines with varying levels of endogenous Panx1 and a different capacity of protein production. We noticed that in cells with low endogenous Panx1 (e.g., NRKs), there was barely any Panx2 at the cell surface based on immunolabeling and cell surface biotinylation pull-downs. In the case of AD293 and HEK293T cells, it was possible to detect low levels of overexpressed Panx2 at the cell surface, while the mutant Panx2 (N86Q) had a detectable but significantly decreased presence only at the cell membrane of HEK293T cells. This result could be attributed to the increased protein expression in HEK293T cells that contains the SV40 T-antigen and have high transfection efficiency. This feature might have allowed the overexpressed Panx2 to bypass mechanisms of protein quality control resulting in more Panx2 trafficking to the plasma membrane [43].

Further research is needed to evaluate endogenous Panx2 in terms of N-glycosylation and subcellular localization of its isoforms, and to examine if Panx2 can form channels at the plasma membrane under physiological conditions. To date, limited studies have attempted to evaluate the Panx2 channel function [14,15] and several factors make it difficult to test Panx2 channel activity, such as its intracellular localization, the lack of evidence of in vivo functional channel formation [27] and the unknown mechanisms of activation [14].

Work done by Bruzzone et al. [15], showed that Panx1 and Panx2 were abundantly expressed in the CNS, and co-injection of both pannexin RNAs in paired *Xenopus* oocytes resulted in the formation of heteromeric channels with functional characteristics different from those formed by Panx1 monomers but with similar pharmacological sensitivity [27]. Ambrosi et al. [14] suggested that Panx1/Panx2 heteromeric channels tend to be unstable and they attributed that to differences in monomers size and oligomeric symmetry between these two pannexins. We have previously shown that Panx1 and Panx2 do form a complex as determined by co-IP experiments. Interestingly, when both pannexins are co-expressed, the level of interaction between Panx2 and glycosylated-species of Panx1 is dependent of the glycosylation of the latter [22]. Here, we showed in vitro, that the Panx2 glycosylation-deficient mutant can readily form complexes with Panx1, thus Panx2 glycosylation is not required for intermixing of the two pannexins. In fact, although it was not statistically significant, N86Q seemed to pull-down Panx1 more efficiently than the WT Panx2. However, in confocal images the N86Q aggregates did not show higher overlap with Panx1-immunolabeling than Panx2 WT. Consistent with our previous report [22], complex N-glycosylation of Panx1 hinders their interaction, since both Panx2 and N86Q interacted only with the Gly-0 and Gly-1 forms of Panx1. Taken together, these results suggest that in an ectopic expression system, glycosylation of Panx2 is not required for Panx1/Panx2 intermixing, but it does help with the transport of Panx2 to the cell surface, which is also increased by the presence of Panx1.

Finally, we propose that N-glycosylation of pannexins is an important post-translational modification that partially regulates their subcellular localization. Whether N-glycosylation represents a post-translational mechanism that regulates trafficking and Panx1/Panx2 interactions in vivo would be important questions to address in future studies. In cells where both pannexins are co-expressed, glycosylation may act as a form of regulation defining whether these channels will serve as intracellular or plasma membrane channels with different physiological and pathological functions.

4. Materials and Methods

4.1. Cell Lines, Constructs and Transient Transfections

Media, supplements and reagents were obtained from GIBCO® and Invitrogen™ (Carlsbad, CA, USA). Normal rat kidney (NRK) (ATCC® CRL-6509™) and human embryonic kidney cells (HEK293T) (ATCC® CRL-3216™) were obtained from ATCC (Manassas, VA, USA). Adherent HEK293 cells (AD293, Cat# 240085) were obtained from Agilent Technologies, Inc. (Santa Clara, CA, USA). Cell cultures were grown in high-glucose DMEM supplemented with 10% fetal bovine serum (FBS), 100 U/mL penicillin, 100 μg/mL streptomycin and 2mM L-Glutamine. At ~50% of confluency, cells were transfected adding Lipofectamine 3000 (Invitrogen™) following manufacturer directions. 2 μg of pcDNA3.1 (Invitrogen™) plasmids encoding mouse Panx2 [22] or Panx2^{N86Q} or their respective FLAG-tagged versions were used for transfections in 35 mm culture plates. After 48 h for single transfections and 72 h for co-transfections, proteins were extracted, and expression levels were determined by Western blot. For co-transfections experiments with mPanx1 plasmid [11], levels were reduced to 0.5 μg of DNA.

4.2. Mutagenesis and Cloning of New FLAG-Tagged Panx2 Constructs

As described previously [11], Panx2 has a predicted N-glycosylation consensus site located at asparagine (N) 86 on the first extracellular loop. Site-directed mutagenesis service (NorClone Biotech Labs, London, ON, Canada) was used to generate a new expression Panx2 construct encoding a replacement of asparagine by glutamine at position 86, referred to as Panx2 N86Q. FLAG-tagged Panx2 was obtained by inserting a single FLAG sequence with In-Fusion HD Cloning Kit (Clontech Laboratories, Inc., Takara Bio, Inc., Mountain View, CA, USA) at the end of the coding region of the Panx2 C-termini. Primers used for FLAG insertion were Forward: 5′-GTTTAAACTTAAGCTTCATGCACCACCTCCTGGAG-3′ and Reverse: 5′-GCCCTCTAGACTCGAGCTCACTTGTCATCGTCGTCCTTGTAATCAAACTCCACAGTACT-3′. All the constructs were verified by sequencing.

4.3. Protein Extractions and Western Blots

For co-immunoprecipitation (Co-IP) assays, cell lysates were obtained using a Triton-based extraction buffer (IP buffer) (1% Triton X-100, 150 mM NaCl, 10 mM Tris, 1 mM EDTA, 1 mM EGTA, 0.5% NP-40). The rest of the protein extractions were performed with SDS-based buffer (RIPA buffer) (0.1% SDS, 50 mM Tris-HCl, pH 8.0, 150 mM NaCl, 1% NP-40 and 0.5% Sodium Deoxycholate). In each case, lysis buffers were complemented with a final concentration of 1 mM NaF, 1 mM Na_3VO_4, and one tablet of cOmplete™-mini, EDTA-free Protease Inhibitor Cocktail (Roche, Mannheim, Germany). Total protein concentrations were quantitated with Pierce™ BCA Protein Assay Kit (Thermo Scientific, Rockford, IL, USA). For Western Blots, 50 μg of total protein were resolved in 8% SDS-PAGE and transferred onto nitrocellulose membranes using an iBlot™ Blotting System (Invitrogen, Carlsbad, CA, USA). Membranes were blocked with 3% bovine serum albumin (BSA, Burlington, ON, Canada) and 0.05% Tween 20-Phosphate Buffer Saline (T-PBS) for 45 min at room temperature and probed overnight with a 1:1000 dilution of the rabbit affinity-purified antibodies anti-Panx2-CT-523 [22]. Mouse monoclonal anti-FLAG® M2 (Sigma, St. Louis, MO, USA, Cat# F3165), monoclonal mouse anti-GAPDH (Millipore, Burlington, MA, USA, Cat# MAB374, RRID: AB_2107445), and anti-α-Tubulin (Millipore-Sigma Cat# 05-829, clone DM1A) antibodies were used at 1:2000, 1:1000 and 1:5000 dilutions, respectively. For detection, IRDye-800 and -680RD (Life Technologies™, Carlsbad, CA, USA) were used as secondary antibodies at 1:10,000 dilution and the membranes were scanned on a Li-Cor Odyssey infrared system (Li-Cor, Lincoln, NL, USA). In most cases, GAPDH was used as a loading control.

4.4. Immunofluorescence, Confocal Imaging, Linescans and Colocalization Analysis

Cells were grown on coverslips at ~70% of confluency and were transfected as described previously in Section 4.1. After 48 h of transfection, coverslips were washed with D-PBS (Gibco®) and fixed with ice-cold 80% methanol and 20% acetone for 15 min at 4 °C. Coverslips were blocked with 2% BSA-PBS for 1h and primary antibodies were used diluted in blocking buffer as follows: polyclonal anti-Panx2-CT (2 mg/mL, 1:100 dilution), polyclonal anti-Panx1-CT (1 mg/mL, 1:100 dilution), anti-PDI monoclonal antibody (1 mg/mL, 1:400 dilution) (1D3, Enzo® Life Sciences, Burlington, ON, Canada, ADI-SPA-891-D), cis-Golgi marker anti-GM130 (1 mg/mL, 1:300) (Abcam, Toronto, ON, Canada, Prod#: ab169276), anti-FLAG® M2 (1 mg/mL, 1:500) (F3165, Sigma, St. Louis, MO, USA), mouse monoclonal anti-Lamp-2 (1 mg/mL, 1:300) (DHSB, clone H4B4). Coverslips were incubated with primary antibodies for 1 h at room temperature, then washed with PBS and incubated with the secondary antibodies: Alexa Fluor 488 goat anti-rabbit IgG (2 mg/mL, 1:700) or goat anti-mouse antibody Alexa Fluor 647 (2 mg/mL, 1:400) (Life Technologies), that were selected to avoid bleed-through between dyes. Mitochondrial labeling was performed by using MitoTracker™ Red CMXRos (M7512, Thermofisher, Life Technologies, Eugene, OR, USA) as per manufacturer directions, and cells were fixed with freshly prepared paraformaldehyde (4%) and then permeabilized with 0.1% Triton X-100 before the blocking step. Coverslips were rinsed with PBS followed by water once and counterstained with Hoechst 33342 (H3570, Life Technologies™, Eugene, OR, USA) (1:1000, in water) for 5 min to stain nuclei and then were mounted with the custom-made Airvol mounting medium. Imaging was performed with an LSM 800 Confocal Microscope (Carl Zeiss, Oberkochen, Germany) using a Plan-Apochromat 63x/1.40 Oil DIC objective (Carl Zeiss, Oberkochen, Germany). Image acquisition for colocalization was performed with sequential laser scanning and with the multitracking feature of the Zeiss software with settings to avoid wrong excitation-crosstalk and bleed-through of the channels. Colocalization was quantitated with the colocalization plug-in of the Zeiss software (ZEN, version 2.3, blue edition). Regions of interest (ROI) were drawn in dual-labeled cells selecting individual cells expressing Panx2 or the mutant and co-stained with organelle markers. Controls of single-labeled cells were used to determine thresholds of intensities for each single channel and a manual thresholding was used to determine the region of pixels colocalized in the intensities scatterplots. Pearson correlation coefficient was determined for each cell as a measure of colocalization and was expressed as means $\pm$ S.E.M., representative of at least three independent transfections. Linescans using Zeiss software tool were used to detect overlaps of fluorescence peaks in cells co-transfected with Panx1.

4.5. Cell Surface Biotinylation Assays

Cell surface biotinylation assays were performed as described in Reference [22]. Briefly, cells were grown in 60 mm plates and used for biotinylation 72 h following transfection with Panx2 or N86Q constructs. After, culture media was aspirated, the cell monolayer was rinsed twice with ice-cold D-PBS supplemented with Ca^{2+} and Mg^{2+} (Gibco®). Then, cells were incubated only with D-PBS (non-labeling as a negative control) or with a solution of 1.5 mg/mL EZ-Link™ Sulfo-NHS-SS-Biotin (Thermo Scientific, Rockford, IL, USA) in D-PBS for 30 min on ice and covered from light. Plates were washed once again with D-PBS and then incubated with 100 mM glycine dissolved in D-PBS for 30 min to quench the remaining labeling biotin washed once more with D-PBS. Lysates were prepared using RIPA buffer as described before. For pull-down of cell surface biotinylated proteins, 250 μg of total protein was incubated overnight with 50% slurry of 50 μL NeutrAvidin agarose beads (Thermo Scientific, Rockford, IL, USA). Samples from lysates and initial flow-through wash were collected, the beads were spun down (at $500\times g$, 4 °C) and then washed three times with RIPA buffer. The samples and the beads were then mixed with 2X Laemmli buffer and 10% β-mercaptoethanol and placed at 95 °C in heat block for 5 min. 50 μg of total protein from lysates and the beads supernatant were resolved in parallel with 8% SDS-PAGE gel and then transferred to nitrocellulose membranes as previously described. PDI or GAPDH were used as controls of non-specific biotinylation of intracellular/cytoplasmic proteins and E-cadherin was used as positive control of cell surface protein.

4.6. Co-Immunoprecipitation (Co-IP) Assays

Co-immunoprecipitation (Co-IP) of protein complexes was performed at 4 °C by incubating overnight 1 mg of total protein from pre-cleared (with Protein A/G beads alone) lysates with 5 μg/mL of rabbit polyclonal anti-Panx2-CT or anti-Panx1-CT affinity-purified antibody crosslinked to Pierce Protein A/G-Agarose beads (Thermo Scientific); the same amount of beads (~30 μL) were used for IP in each case. Control experiments to evaluate unspecific binding to the beads were performed in parallel using beads with no antibody. To remove un-bound proteins, four washes with 500 μL of ice-cold IP buffer were performed. Then, beads were dried by aspiration and re-suspended in 2X Laemmli buffer (10% (*v:v*) β-mercaptoethanol), boiled for 5 min, spun down and the supernatants (IP samples) were used for WB. For WB analysis, 50 μg of protein of each lysate was loaded into the INPUT lanes and ran along with the IP samples. The intensities of the bands in each lane were obtained by densitometry and were used for quantitation.

4.7. De-Glycosylation Assays

Lysates from HEK293T cells ectopically expressing mouse Panx2 and the N86Q were used for validation of N-glycosylation site of the mouse Panx2 construct. Enzymatic de-glycosylation with Peptide-N-glycosidase F (PNGase F) and Endoglycosidase H (EndoH) were used to detect the presence of all complex forms of N-glycosylation and high-mannose modification, respectively. PNGase F (Roche, Indianapolis, IN, USA) and EndoH (New England Biolabs Ipswich, MA, USA) digestions were performed according to their manufacturer's instructions. Briefly, at least 35 μg of total protein was denatured at $100\,^{\circ}$C for 5 min in denaturing buffer (0.1% (v/v) SDS, 0.05 M 2-mercaptoethanol, 50 mM phosphate buffer, pH 7.5) and subsequently incubated for 1h at $37\,^{\circ}$C with 10 units of the PNGase F, 0.7% (v/v) of Triton X-100 or 2500 U of Endo H in supplier's digestion buffer. In the parallel control, samples were assayed without endoglycosidases. Protein samples were separated on an 8% SDS-polyacrylamide gel electrophoresis gel (PAGE) and transferred to nitrocellulose membranes for WB.

4.8. Densitometric Analysis of Western Blots

Densitometry analysis was performed in the Odyssey Application Software Version 3.0.16 (LI-COR Biosciences, Lincoln, NL, USA) as follows: For cell surface biotinylation experiments, the fraction of biotinylated-protein detected at the cell surface was calculated using the integrated intensity (I.I.) of protein bands detected in the Neutravidin lanes divided by the I.I. of protein bands detected in their corresponding input-lysate lanes. For Co-IP experiments, quantitative analysis was performed by calculating the ratio of the I.I detected for Co-IP protein divided by the I.I of the IP-target protein. Quantitation results were expressed as means $\pm$ S.E.M. representative of at least three independent experiments.

4.9. Statistics

Statistical analysis was performed using the statistical package of GraphPad Prism® Ver. 5.03 (GraphPad Software, Inc., San Diego, CA, USA). Cell surface biotinylation data were analyzed with non-parametric Mann-Whitney U test for unpaired data. Data derived from quantitation of Co-IP assays and colocalization were analyzed with a two-tailed unpaired t test with Welch's correction. A probability of $p < 0.05$ was considered as statistically significant.

Author Contributions: R.E.S.-P. devised and performed the experiments, analysis of data. R.E.S.-P. and S.P. wrote the manuscript. D.J., assisted on the design of mutagenesis, and provided technical help in immunostaining and imaging and revised the manuscript. S.P. designed the original project, devised experiments, served as scientific advisor and provided critical revision of results and manuscript.

Funding: This work and the cost of publication was funded by the Natural Sciences and Engineering Research Council of Canada with an NSERC Discovery Grant RGPIN-2015-06794 to Silvia Penuela.

Conflicts of Interest: The authors declare no conflict of interest.

Abbreviations

AD293	Adherent human embryonic kidney cells
ATP	Adenosine triphosphate nucleotide
CNS	Central nervous system
Co-IP	Co-immunoprecipitation
Endo H	Endoglycosidase H
ER	Endoplasmic reticulum
GAPDH	Glyceraldehyde 3-phosphate dehydrogenase
GM130	Golgi matrix protein 130 kDa
HEK293T	Human embryonic kidney cells with SV40 T-antigen
IP	Immunoprecipitation
LAMP2	Lysosome-associated membrane glycoprotein 2
NPC	Neural progenitor cells
NRK	Normal rat kidney cells
PDI	Protein disulfide isomerase
PNGase F	Peptide-N-glycosidase F
PTM	Posttranslational modification
SV40	Simian vacuolating virus 40

References

1. Sosinsky, G.E.; Boassa, D.; Dermietzel, R.; Duffy, H.S.; Laird, D.W.; MacVicar, B.; Naus, C.C.; Penuela, S.; Scemes, E.; Spray, D.C.; et al. Pannexin channels are not gap junction hemichannels. *Channels* **2011**, *5*, 193–197. [CrossRef] [PubMed]

2. Panchin, Y.; Kelmanson, I.; Matz, M.; Lukyanov, K.; Usman, N.; Lukyanov, S. A ubiquitous family of putative gap junction molecules. *Curr. Biol.* **2000**, *10*, R473–R474. [CrossRef]

3. Bao, L.; Locovei, S.; Dahl, G. Pannexin membrane channels are mechanosensitive conduits for atp. *FEBS Lett.* **2004**, *572*, 65–68. [CrossRef] [PubMed]

4. Iwamoto, T.; Nakamura, T.; Doyle, A.; Ishikawa, M.; de Vega, S.; Fukumoto, S.; Yamada, Y. Pannexin 3 regulates intracellular atp/camp levels and promotes chondrocyte differentiation. *J. Biol. Chem.* **2010**, *285*, 18948–18958. [CrossRef] [PubMed]

5. Ishikawa, M.; Iwamoto, T.; Nakamura, T.; Doyle, A.; Fukumoto, S.; Yamada, Y. Pannexin 3 functions as an er Ca$^{(2+)}$ channel, hemichannel, and gap junction to promote osteoblast differentiation. *J. Cell Biol.* **2011**, *193*, 1257–1274. [CrossRef] [PubMed]

6. Vanden Abeele, F.; Bidaux, G.; Gordienko, D.; Beck, B.; Panchin, Y.V.; Baranova, A.V.; Ivanov, D.V.; Skryma, R.; Prevarskaya, N. Functional implications of calcium permeability of the channel formed by pannexin 1. *J. Cell Boil.* **2006**, *174*, 535–546. [CrossRef] [PubMed]

7. Thompson, R.J. Pannexin channels and ischaemia. *J. Physiol.* **2015**, *593*, 3463–3470. [CrossRef] [PubMed]

8. Pelegrin, P.; Surprenant, A. Pannexin-1 couples to maitotoxin- and nigericin-induced interleukin-1beta release through a dye uptake-independent pathway. *J. Biol. Chem.* **2007**, *282*, 2386–2394. [CrossRef] [PubMed]

9. Baranova, A.; Ivanov, D.; Petrash, N.; Pestova, A.; Skoblov, M.; Kelmanson, I.; Shagin, D.; Nazarenko, S.; Geraymovych, E.; Litvin, O.; et al. The mammalian pannexin family is homologous to the invertebrate innexin gap junction proteins. *Genomics* **2004**, *83*, 706–716. [CrossRef] [PubMed]

10. Le Vasseur, M.; Lelowski, J.; Bechberger, J.F.; Sin, W.C.; Naus, C.C. Pannexin 2 protein expression is not restricted to the cns. *Front. Cell. Neurosci.* **2014**, *8*, 392. [CrossRef] [PubMed]

11. Penuela, S.; Bhalla, R.; Gong, X.-Q.; Cowan, K.N.; Celetti, S.J.; Cowan, B.J.; Bai, D.; Shao, Q.; Laird, D.W. Pannexin 1 and pannexin 3 are glycoproteins that exhibit many distinct characteristics from the connexin family of gap junction proteins. *J. Cell Sci.* **2007**, *120*, 3772–3783. [CrossRef] [PubMed]

12. Celetti, S.J.; Cowan, K.N.; Penuela, S.; Shao, Q.; Churko, J.; Laird, D.W. Implications of pannexin 1 and pannexin 3 for keratinocyte differentiation. *J. Cell Sci.* **2010**, *123*, 1363–1372. [CrossRef] [PubMed]

13. Bond, S.R.; Lau, A.; Penuela, S.; Sampaio, A.V.; Underhill, T.M.; Laird, D.W.; Naus, C.C. Pannexin 3 is a novel target for runx2, expressed by osteoblasts and mature growth plate chondrocytes. *J. Bone Min. Res.* **2011**, *26*, 2911–2922. [CrossRef] [PubMed]

14. Ambrosi, C.; Gassmann, O.; Pranskevich, J.N.; Boassa, D.; Smock, A.; Wang, J.; Dahl, G.; Steinem, C.; Sosinsky, G.E. Pannexin1 and pannexin2 channels show quaternary similarities to connexons and different oligomerization numbers from each other. *J. Biol. Chem.* **2010**, *285*, 24420–24431. [CrossRef] [PubMed]

15. Bruzzone, R.; Hormuzdi, S.G.; Barbe, M.T.; Herb, A.; Monyer, H. Pannexins, a family of gap junction proteins expressed in brain. *Proc. Natl. Acad. Sci. USA* **2003**, *100*, 13644–13649. [CrossRef] [PubMed]

16. Swayne, L.A.; Bennett, S.A. Connexins and pannexins in neuronal development and adult neurogenesis. *BMC Cell Biol.* **2016**, *17* (Suppl. S1), 10. [CrossRef] [PubMed]

17. Swayne, L.A.; Sorbara, C.D.; Bennett, S.A. Pannexin 2 is expressed by postnatal hippocampal neural progenitors and modulates neuronal commitment. *J. Biol. Chem.* **2010**, *285*, 24977–24986. [CrossRef] [PubMed]

18. Vogt, A.; Hormuzdi, S.G.; Monyer, H. Pannexin1 and pannexin2 expression in the developing and mature rat brain. *Brain Res. Mol. Brain Res.* **2005**, *141*, 113–120. [CrossRef] [PubMed]

19. Bargiotas, P.; Krenz, A.; Monyer, H.; Schwaninger, M. Functional outcome of pannexin-deficient mice after cerebral ischemia. *Channels* **2012**, *6*, 453–456. [CrossRef] [PubMed]

20. Blom, N.; Sicheritz-Ponten, T.; Gupta, R.; Gammeltoft, S.; Brunak, S. Prediction of post-translational glycosylation and phosphorylation of proteins from the amino acid sequence. *Proteomics* **2004**, *4*, 1633–1649. [CrossRef] [PubMed]

21. Penuela, S.; Lohman, A.W.; Lai, W.; Gyenis, L.; Litchfield, D.W.; Isakson, B.E.; Laird, D.W. Diverse post-translational modifications of the pannexin family of channel-forming proteins. *Channels* **2014**, *8*, 124–130. [CrossRef] [PubMed]

22. Penuela, S.; Bhalla, R.; Nag, K.; Laird, D.W. Glycosylation regulates pannexin intermixing and cellular localization. *Mol. Boil. Cell* **2009**, *20*, 4313–4323. [CrossRef] [PubMed]

23. Zappalà, A.; Li Volti, G.; Serapide, M.F.; Pellitteri, R.; Falchi, M.; La Delia, F.; Cicirata, V.; Cicirata, F. Expression of pannexin2 protein in healthy and ischemized brain of adult rats. *Neuroscience* **2007**, *148*, 653–667. [CrossRef] [PubMed]

24. Lai, C.P.K.; Bechberger, J.F.; Naus, C.C. Pannexin2 as a novel growth regulator in c6 glioma cells. *Oncogene* **2009**, *28*, 4402–4408. [CrossRef] [PubMed]

25. Wicki-Stordeur, L.E.; Boyce, A.K.J.; Swayne, L.A. Analysis of a pannexin 2-pannexin 1 chimeric protein supports divergent roles for pannexin c-termini in cellular localization. *Cell Commun. Adhes.* **2013**, *20*, 73–79. [CrossRef] [PubMed]

26. Boassa, D.; Nguyen, P.; Hu, J.; Ellisman, M.H.; Sosinsky, G.E. Pannexin2 oligomers localize in the membranes of endosomal vesicles in mammalian cells while pannexin1 channels traffic to the plasma membrane. *Front. Cell. Neurosci.* **2015**, *8*, 15. [CrossRef] [PubMed]

27. Bruzzone, R.; Barbe, M.T.; Jakob, N.J.; Monyer, H. Pharmacological properties of homomeric and heteromeric pannexin hemichannels expressed in xenopus oocytes. *J. Neurochem.* **2005**, *92*, 1033–1043. [CrossRef] [PubMed]

28. Bruzzone, R.; Dermietzel, R. Structure and function of gap junctions in the developing brain. *Cell. Tissue Res.* **2006**, *326*, 239–248. [CrossRef] [PubMed]

29. Dvoriantchikova, G.; Ivanov, D.; Panchin, Y.; Shestopalov, V.I. Expression of pannexin family of proteins in the retina. *FEBS Lett.* **2006**, *580*, 2178–2182. [CrossRef] [PubMed]

30. Tang, W.; Ahmad, S.; Shestopalov, V.I.; Lin, X. Pannexins are new molecular candidates for assembling gap junctions in the cochlea. *Neuroreport* **2008**, *19*, 1253–1257. [CrossRef] [PubMed]

31. Bargiotas, P.; Krenz, A.; Hormuzdi, S.G.; Ridder, D.A.; Herb, A.; Barakat, W.; Penuela, S.; von Engelhardt, J.; Monyer, H.; Schwaninger, M. Pannexins in ischemia-induced neurodegeneration. *Proc. Natl. Acad. Sci. USA* **2011**, *108*, 20772–20777. [CrossRef] [PubMed]

32. Penuela, S.; Celetti, S.J.; Bhalla, R.; Shao, Q.; Laird, D.W. Diverse subcellular distribution profiles of pannexin 1 and pannexin 3. *Cell. Commun. Adhes.* **2008**, *15*, 133–142. [CrossRef] [PubMed]

33. DuBridge, R.B.; Tang, P.; Hsia, H.C.; Leong, P.M.; Miller, J.H.; Calos, M.P. Analysis of mutation in human cells by using an epstein-barr virus shuttle system. *Mol. Cell. Biol.* **1987**, *7*, 379–387. [CrossRef] [PubMed]

34. Penuela, S.; Harland, L.; Simek, J.; Laird, D.W. Pannexin channels and their links to human disease. *Biochem. J.* **2014**, *461*, 371–381. [CrossRef] [PubMed]

35. Bhalla-Gehi, R.; Penuela, S.; Churko, J.M.; Shao, Q.; Laird, D.W. Pannexin1 and pannexin3 delivery, cell surface dynamics, and cytoskeletal interactions. *J. Biol. Chem.* **2010**, *285*, 9147–9160. [CrossRef] [PubMed]

36. Boassa, D.; Ambrosi, C.; Qiu, F.; Dahl, G.; Gaietta, G.; Sosinsky, G. Pannexin1 channels contain a glycosylation site that targets the hexamer to the plasma membrane. *J. Boil. Chem.* **2007**, *282*, 31733–31743. [CrossRef] [PubMed]

37. Boassa, D.; Qiu, F.; Dahl, G.; Sosinsky, G. Trafficking dynamics of glycosylated pannexin 1 proteins. *Cell Commun. Adhes.* **2008**, *15*, 119–132. [CrossRef] [PubMed]

38. Hiramatsu, N.; Joseph, V.T.; Lin, J.H. Monitoring and manipulating mammalian unfolded protein response. *Methods Enzymol.* **2011**, *491*, 183–198. [PubMed]

39. Penuela, S.; Laird, D.W. The cellular life of pannexins. *Wiley Interdiscip. Rev. Membr. Transp. Signal. Banner* **2012**, *1*, 621–632. [CrossRef]

40. Nakamura, N.; Rabouille, C.; Watson, R.; Nilsson, T.; Hui, N.; Slusarewicz, P.; Kreis, T.E.; Warren, G. Characterization of a cis-golgi matrix protein, gm130. *J. Cell Biol.* **1995**, *131*, 1715–1726. [CrossRef] [PubMed]

41. Nakamura, N. Emerging new roles of gm130, a cis-golgi matrix protein, in higher order cell functions. *J. Pharmacol. Sci.* **2010**, *112*, 255–264. [CrossRef] [PubMed]

42. Boyce, A.K.; Prager, R.T.; Wicki-Stordeur, L.E.; Swayne, L.A. Pore positioning: Current concepts in pannexin channel trafficking. *Channels* **2014**, *8*, 110–117. [CrossRef] [PubMed]

43. Chen, B.; Retzlaff, M.; Roos, T.; Frydman, J. Cellular strategies of protein quality control. *Cold Spring Harb. Perspect. Biol.* **2011**, *3*, a004374. [CrossRef] [PubMed]

International Journal of
Molecular Sciences

MDPI

Communication

A Cell Junctional Protein Network Associated with Connexin-26

Ana C. Batissoco [1,2,*], Rodrigo Salazar-Silva [1], Jeanne Oiticica [2], Ricardo F. Bento [2], Regina C. Mingroni-Netto [1] and Luciana A. Haddad [1]

[1] Human Genome and Stem Cell Research Center, Department of Genetics and Evolutionary Biology, Instituto de Biociências, Universidade de São Paulo, 05508-090 São Paulo, Brazil; rodrigo.salazar.silva@usp.br (R.S.-S.); renetto@ib.usp.br (R.C.M.-N.); haddadL@usp.br (L.A.H.)
[2] Laboratório de Otorrinolaringologia/LIM32, Hospital das Clínicas, Faculdade de Medicina, Universidade de São Paulo, 01246-903 São Paulo, Brazil; jeanneoiticica@bioear.com.br (J.O.); rbento@gmail.com (R.F.B.)
* Correspondence: anacarlabatissoco@usp.br; Tel.: +55-11-30617166

Received: 17 July 2018; Accepted: 21 August 2018; Published: 27 August 2018

Abstract: *GJB2* mutations are the leading cause of non-syndromic inherited hearing loss. *GJB2* encodes connexin-26 (CX26), which is a connexin (CX) family protein expressed in cochlea, skin, liver, and brain, displaying short cytoplasmic N-termini and C-termini. We searched for CX26 C-terminus binding partners by affinity capture and identified 12 unique proteins associated with cell junctions or cytoskeleton (CGN, DAAM1, FLNB, GAPDH, HOMER2, MAP7, MAPRE2 (EB2), JUP, PTK2B, RAI14, TJP1, and VCL) by using mass spectrometry. We show that, similar to other CX family members, CX26 co-fractionates with TJP1, VCL, and EB2 (EB1 paralogue) as well as the membrane-associated protein ASS1. The adaptor protein CGN (cingulin) co-immuno-precipitates with CX26, ASS1, and TJP1. In addition, CGN co-immunoprecipitation with CX30, CX31, and CX43 indicates that CX association is independent on the CX C-terminus length or sequence. CX26, CGN, FLNB, and DAMM1 were shown to distribute to the organ of Corti and hepatocyte plasma membrane. In the mouse liver, CX26 and TJP1 co-localized at the plasma membrane. In conclusion, CX26 associates with components of other membrane junctions that integrate with the cytoskeleton.

Keywords: connexin; connexin 26; *GJB2* gene; deafness; protein-protein interaction; TJP1; CGN; FLNB; DAAM1; organ of Corti

1. Introduction

The *GJB2* gene encodes connexin 26 (CX26), which is a protein that plays central roles in hearing, promoting cochlear development, and sustaining auditory function in the mature cochlea [1–4]. In addition to the cochlea, expression of CX26 is also observed in the skin, the liver, the brain, the mammary gland, the salivary gland, the uterus, testis, the pancreas, lungs, the stomach, the thyroid, and the parathyroid [5]. *GJB2* mutations are the most frequent cause of non-syndromic recessive hearing loss across diverse populations [6–9]. In addition, some heterozygous *GJB2* mutations behave in a dominant fashion, which leads to non-syndromic autosomal dominant hearing loss or to the keratitis-ichthyosis-deafness syndrome [10].

In vertebrates, connexins (CX) assemble intercellular gap junctions (GJ), which result from the interaction between two distinct hemi-channels from adjacent cells with each composed of six CX units. GJ directly allows the passage of various small (<2 kDa) molecules between two cells such as ions, secondary messengers, nucleotides, amino acids, and short RNAs [11]. GJ are highly organized structures in which CX interact among themselves as well as with a number of other cellular components including cytoskeleton-associated elements and adhesion and signaling molecules [12–14].

While, among CX family members, the C-termini are dissimilar and present unique binding partners and signaling, they may share common protein interactors [15–17]. The C-terminus from CX26 is strikingly different from that of other CX [18]. Among mouse CX family members, CX26 has the second lowest molecular mass due to shorter segments outside the four transmembrane domains (the extracellular and intracellular loops as well as N-termini and C-termini). Due to its limited length, few binding partners have been identified for CX26 cytosolic segments, e.g., amino-termini and carboxyl-termini and the loop between the second and third transmembrane domains [19–21].

The aim of this study was to search for proteins that interact with the cytoplasmic ten-residue carboxyl-terminal tail of CX26. Employing two distinct biochemical approaches, we disclosed a cytoskeleton and membrane junction-associated protein network that co-fractionates with CX26. CX26 interaction with the molecular complex depends on its C-terminus. Additionally, our results revealed that proteins from this macromolecular complex may also associate with CX30, CX31, or CX43, which indicates that assembly of CX in the macromolecular complex is independent of the CX C-terminus length or sequence.

2. Results

We employed affinity precipitation assays to search for proteins that interact with the cytoplasmic carboxyl-terminal tail of CX26. To that end, the portion of the *GJB2* mouse gene coding for the 10 most C-terminal amino acids of Cx26 was cloned and expressed in *Escherichia coli* as a peptide in fusion with the glutathione-*S*-transferase (GST) C-terminus (GST–CX26). The purified fusion protein or GST was submitted to affinity capture assays. Mass spectrometry analyses identified 447 proteins from the mouse brain or liver that precipitated in sepharose beads conjugated to glutathione and bound by affinity to the GST–CX26 fusion protein or only GST. After exclusion of potential contaminants, 39 proteins were found to co-fractionate in the GST–CX26 assay but not in the negative control (GST-only assay). The number of peptides identified by mass spectrometry for each of the 39 proteins varied from two to seven and the protein coverage by peptides ranged from 1% to 15%. The number of unique interactor candidates was reduced from 39 to 26 proteins when the following exclusion criteria were applied: redundancy of representation within the GST–CX26 group, discrepancy between the observed and expected molecular weights, and inconsistency in tissue/cell spatial distribution. For instance, biglycan, canstatin, and fibronectin were excluded because, as secreted fibrous proteins, the interaction results would probably be false-positive due to unspecific precipitation or a transient association during synthesis and trafficking in the secretory pathway. As a result, we retrieved a total of 26 candidate proteins to interact with the cytosolic C-terminus of CX26.

Gene ontology and scientific literature searches allowed us to classify the 26 interactor candidates in the following groups: (i) 12 proteins with direct or indirect association with cell junctions and/or the cytoskeleton (Table 1); (ii) seven proteins from the secretory pathway; (iii) four mitochondrial proteins; (iv) two chaperone proteins; and (v) one nucleus-cytoplasm shuttling protein. Two proteins from the mitochondrion group and one protein from the secretory pathway have been detected at the plasma membrane. Therefore, 15 out of 26 proteins may be directly or indirectly associated with the plasma membrane (12 from cell junctions or the cytoskeleton, two from the mitochondrial group, and one from the secretory group). A general subcellular classification of the 26 proteins is illustrated in Figure 1A.

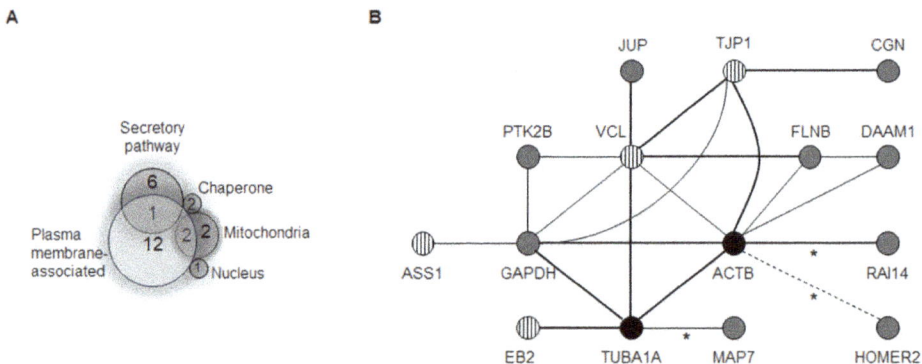

Figure 1. (**A**) Venn diagram representing the distribution of 26 candidate proteins to interact with CX26. The subcellular classification and the number of proteins in each of the five distinct groups are indicated as well as their overlapping sorting; (**B**) The network of protein-protein interaction (PPI) is predicted at http://string-db.org. Acronyms are according to the genes listed on Table 1 in addition to actin B (ACTB), α-tubulin (TUBA1A), and arginine succinate synthase 1 (ASS1). Black circles represent proteins absent from the list of interactors fed by the user to the software input list. Asterisks indicate protein interactions manually added to the PPI network according to the literature (MAP7, RAI14, Homer2). Striped circles denote the four CX26 interactor candidates previously demonstrated to associate with other CX. Thicker lines indicate experimentally demonstrated associations that may be either more reproducible or a direct interaction. A dashed line indicates inference of an intermediate interactor.

Table 1. Candidate proteins to interact with CX26 are classified as a cell junction or cytoskeleton-associated proteins.

Human Gene Acronym	Human Gene	Protein Name, Aliases, and Acronyms
CGN	Cingulin	Cingulin
DAAM1	Disheveled-associated activator of morphogenesis 1	DAAM1
FLNB	Filamin B	Filamin-B, Filamin-3, β-filamin
GAPDH	Glyceraldehyde-3-phosphate dehydrogenase	GAPDH
HOMER2	Homer scaffold protein 2	Homer-2, cupidin
JUP	Junction plakoglobin	Plakoglobin, γ-catenin, desmoplakin-3
MAP7	Microtubule-associated protein 7	MAP7, EMAP-115, ensconsin
MAPRE2	Microtubule-associated RP/EB family member 2	EB2
PTK2B	Protein tyrosine kinase 2β	Focal adhesion kinase 2, FAK2, PYK2, PTK2B
RAI14	Retinoic acid-induced 14	RAI14, ankycorbin, NORPEG
TJP1	Tight Junction Protein 1	*Zonula occludens* 1, ZO-1, TJP1
VCL	Vinculin	Vinculin, VCL

2.1. A Membrane-Cytoskeleton Protein Network Associated with CX26

We continued the in silico analyses by verifying the potential ability of the 26 proteins to assemble protein-protein interaction (PPI) networks according to data from literature reports on the interaction and co-expression experiments [22]. Nine out of 26 proteins comprised a single PPI network (ASS1, CGN, DAAM1, FLNB, GAPDH, JUP, PTK2B, VCL, and TJP1). When cytoskeleton proteins α-tubulin and actin were included on the protein list, the PPI network encompassed 10 out of the 26 proteins (EB2 was included). A further individual literature data search allowed for the manual inclusion of three additional proteins in the PPI network (HOMER2, MAP7, and RAI14) (Figure 1B) [23–27]. Consequently, among the 26 proteins, 13 constitute a single PPI network along with α-tubulin and actin (Figure 1B). The PPI network contains all 12 proteins classified as cell junction/cytoskeleton (Table 1, Figure 1B) and one from the mitochondrion group (argininosuccinate synthase 1—ASS1, Figure 1A), which has been reported to interact with the plasma membrane in addition to the mitochondrial outer membrane [28]. In the present study, we report the analysis of the group of 13 proteins present in the PPI network.

2.2. Known CX Binding Partners in the CX26 Molecular Complex

Four of the 13 proteins present in the PPI network (Figure 1B, striped circles) have been described to interact with other members from the CX family. They are ASS1, microtubule-associated RP/EB family member 2 (EB2), tight junction protein 1/zonula occludens protein 1 (TJP1), and vinculin (VCL) [29–34]. While the former classifies as mitochondria-associated and plasma membrane-associated, the latter three proteins are cell junction or cytoskeleton proteins. As CX interaction with TJP1 appears to be direct for the majority of family members studied [29–31,35–40], we adopted the yeast two-hybrid split-ubiquitin system to search for direct, pairwise interaction between full-length CX26 individually with TJP1, ASS1, EB2, and VCL.

In the yeast split-ubiquitin system, the interaction is expected to take place at the membrane and cleavage of the fusion protein by a ubiquitin-specific processing protease and then releases the transcription factor lexA-VP16. The reporter genes *lacZ*, *HIS3*, and *ADE2* were employed in this study as they are responsive to lexA-VP16 binding after its nuclear translocation. As presented in Figure 2A, no specific activation of the reporter genes was observed for any test bait-prey pair (CX26–TJP1, CX26–VCL, CX26–EB2, or CX26–ASS1). Leaky activation was observed for the *lacZ* gene expression for all pairs and the *ADE2* gene was activated by the preys themselves. No test pair allowed for yeast growth in minimal medium without histidine when compared to the positive control (Figure 2A). Therefore, we concluded that, under these conditions, we did not obtain data indicating direct interaction between full-length CX26 and TJP1, VCL, EB2, or ASS1.

Antibodies that recognize each of the four CX interactors including TJP1, VCL, EB2, and ASS1 were employed in immunofluorescence assays of adult mouse liver with double staining for CX26 in all four experiments. CX26, TJP1, VCL, and EB2 all presented staining patterns consistent with plasma membrane distribution in hepatocytes (Figure 2B). TJP1 and CX26 co-localized in different points at the plasma membrane (Figure 2B, arrows). On the other hand, although in double staining assays, there were sporadic merged signals between CX26 and either EB2 or VCL (Figure 2B, arrows), we considered that VCL and EB2 do not co-localize with CX26 in the mouse liver under our assay conditions. ASS1 disclosed a cytoplasmic staining pattern and no co-localization with CX26 under the conditions employed (Figure 2B). However, in a control experiment, ASS1 did not co-localize with the mitochondrial cell marker mtHSP70 (data not shown), which suggests that the anti-ASS1 antibody employed should be unspecific in our immunofluorescence assays of mouse liver sections.

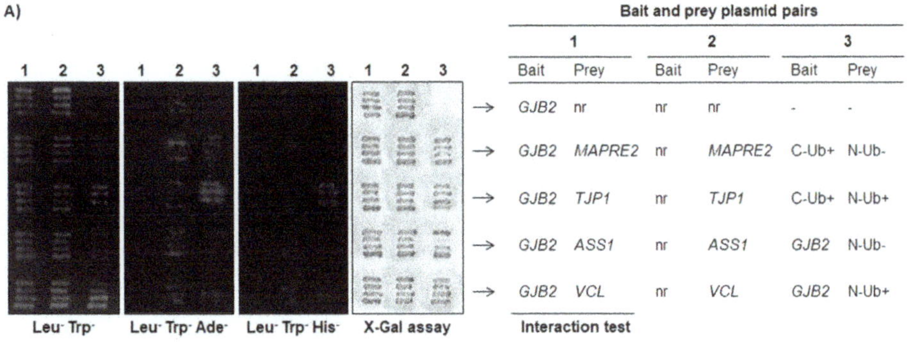

Figure 2. *Cont.*

B)

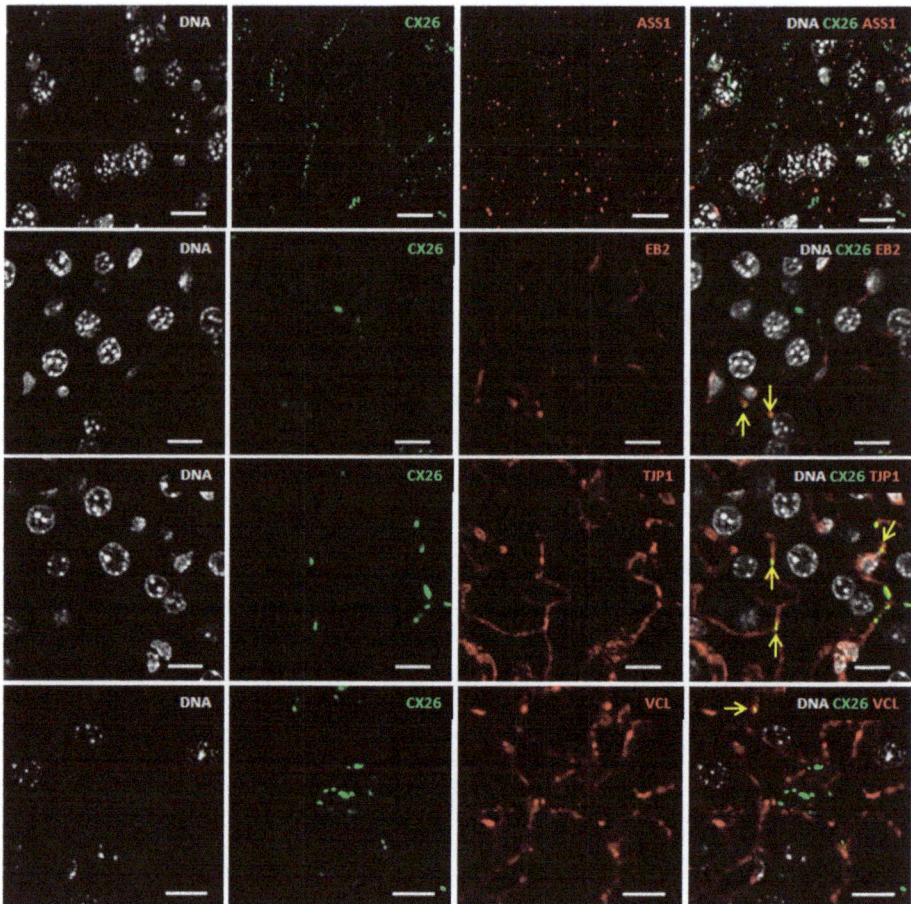

Figure 2. (**A**) Results of the yeast two-hybrid split ubiquitin assay has, in column 1, interaction tests with *GJB2* full-length sequence as bait and the preys non-recombinant vector (nr, control), EB2, TJP1, ASS1, or VCL. Column 2 has the non-recombinant (nr) vector as bait. Therefore, it leads to testing for leaky activation of the reporter gene by each prey fusion protein. Column 3 presents different controls: untransformed, positive, and negative controls for each vector (coding for either N- (N-Ub) or C-terminal (C-Ub) ubiquitin moieties in fusion with known strong (+) or negative (−) interactors. Different media respectively select for the presence of both vectors (SD, Leu⁻, Trp⁻), activation of the reporter gene *ADE2* (SD, Leu⁻, Trp⁻, Ade⁻), *HIS3* (SD, Leu⁻, Trp⁻, His⁻), or β-galactosidase (X-Gal test); (**B**) Indirect immunofluorescence of adult (P60) mouse liver cryosections with anti-CX26 antibody (green) and ASS1, VCL, EB2, and TJP1 (all in red). DNA is in a pseudo white color, according to DAPI staining. The analysis was performed at confocal microscopy with z-sections of 0.5 µm (LSM880, Carl Zeiss, Oberkochen, Germany). Arrows: merged signals. Scale bar: 10 µm.

2.3. Expanding the Protein Interaction Network with Cx26

Since some CX26 interactor candidates retrieved in our affinity capture assay are common to other CX and do not appear to be in direct association with CX26, we looked for adaptor proteins in the PPI network that could play roles in bringing TJP1, VCL, ASS1, or EB2 to proximity to CX26.

Among the 13 proteins from the PPI network, there are four adaptor/scaffolding proteins: cingulin (CGN), filaminB (FLNB), dishevelled-associated activator of morphogenesis 1 (DAAM1), and homer scaffolding protein 2 (HOMER2). As the N-terminal globular domain of the adaptor protein CGN associates with TJP1 PDZ domain [41], we performed immunoprecipitation experiments with anti-CGN antibodies in RIPA or EGTA buffer lysates of neonate mouse liver tissue and confirmed that CGN, as expected, co-immunoprecipitates with TJP1. We confirmed co-immunoprecipitation of CGN and CX26 in both buffer compositions employed, RIPA buffer (Figure 3), and EGTA buffer (data not shown). ASS1 and EB2 were disclosed to co-immunoprecipitate with CGN. Data presented in Figure 3 are clear evidence for CGN co-immunoprecipitation with TJP1 and ASS1. The faint band observed for EB2 in a CGN immunoprecipitation lane led us to consider it may be due to weak and indirect association with proteins from any CGN protein complex. We did not observe VCL in CGN immunoprecipitates in either buffer condition (Figure 3 and data not shown) even though it may still bind CX26 through an additional intermediate interactor. Although CX43-specific blot band migrates very close to an unspecific band observed in the negative control, its co-immunoprecipitation with CGN is demonstrated (Figure 3). Lastly, we show that CGN co-immunoprecipitates with CX30 and CX31 (Figure 3).

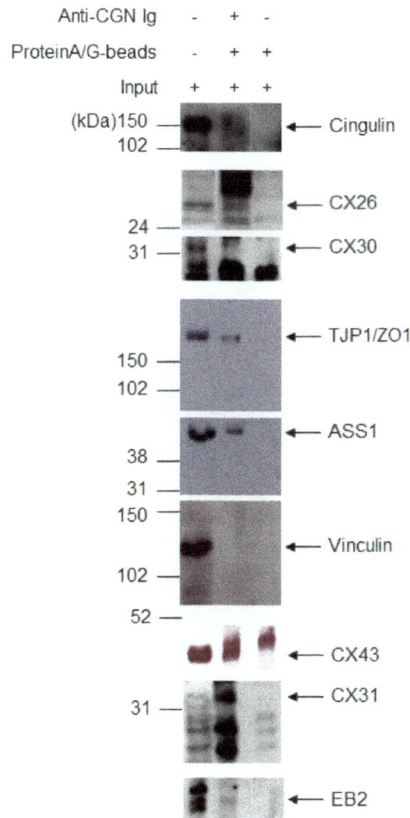

Figure 3. Immunoprecipitation of cingulin from adult mouse liver and its co-immunoprecipitation with CX26, CX30, CX31, CX43, EB2, TJP1, and ASS1, which is indicated by the respective arrows. No co-immunoprecipitation is observed with VCL. The protein molecular mass is indicated in kiloDaltons (kDa).

The adult mouse liver was co-labeled for Cx26 and three adaptor proteins known as CGN, FLNB, or DAAM1. The four proteins were detected in hepatocytes and in patterns consistent with plasma membrane localization (Figure 4, arrowheads) as well as cytoplasm organelles (Figure 4). Few overlapping signals were observed in hepatocytes for CX26 and DAAM1 at the plasma membrane (Figure 4, arrows). The paucity of these observations led us to exclude co-localization of those proteins. CX26 co-localization with CGN or FLNB was not observed under those conditions. Moreover, we show that all three adaptor proteins including CGN, FLNB, and DAAM1 are expressed in key regions of P14 mouse cochlea such as the organ of Corti (OC), *stria vascularis*, spiral ligament, spiral limbus, and external sulcus cells (Figure 5). DAAM1 is also present in spiral ganglion cells (Figure 5). Likewise, in *stria vascularis* DAAM1, staining appeared more evident than CGN and FLNB staining. On P14, CX26 distributed to the OC, *stria vascularis*, spiral ligament, spiral limbus, and the spiral ganglion. The tectorial membrane is acellular and known to yield unspecific staining due to spontaneous fluorescence. CX26 and all three adaptor proteins seem to be co-expressed in the OC even though OC cell identity could not be precisely defined since distinct cell type markers have not been employed.

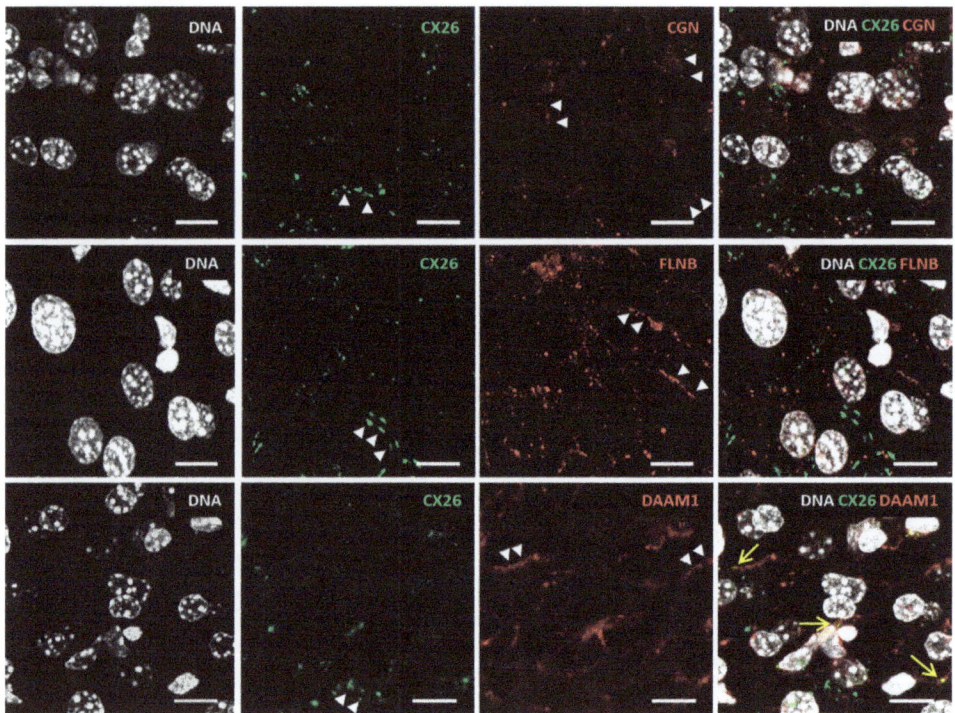

Figure 4. Indirect immunofluorescence of adult (P60) mouse liver cryosections with anti-Cx26 antibody (green) and CGN, FLNB, and DAAM1 antibodies (all in red). DNA is in a white pseudo color, according to DAPI staining. The analysis was performed at confocal microscopy with z-sections of 0.5 μm (LSM880, Carl Zeiss, Oberkochen, Germany). Each image consists of the maximum intensity projection of all z-sections obtained. Arrowheads: signal at the plasma membrane. Arrows: merged signals. Scale bar: 10 μm.

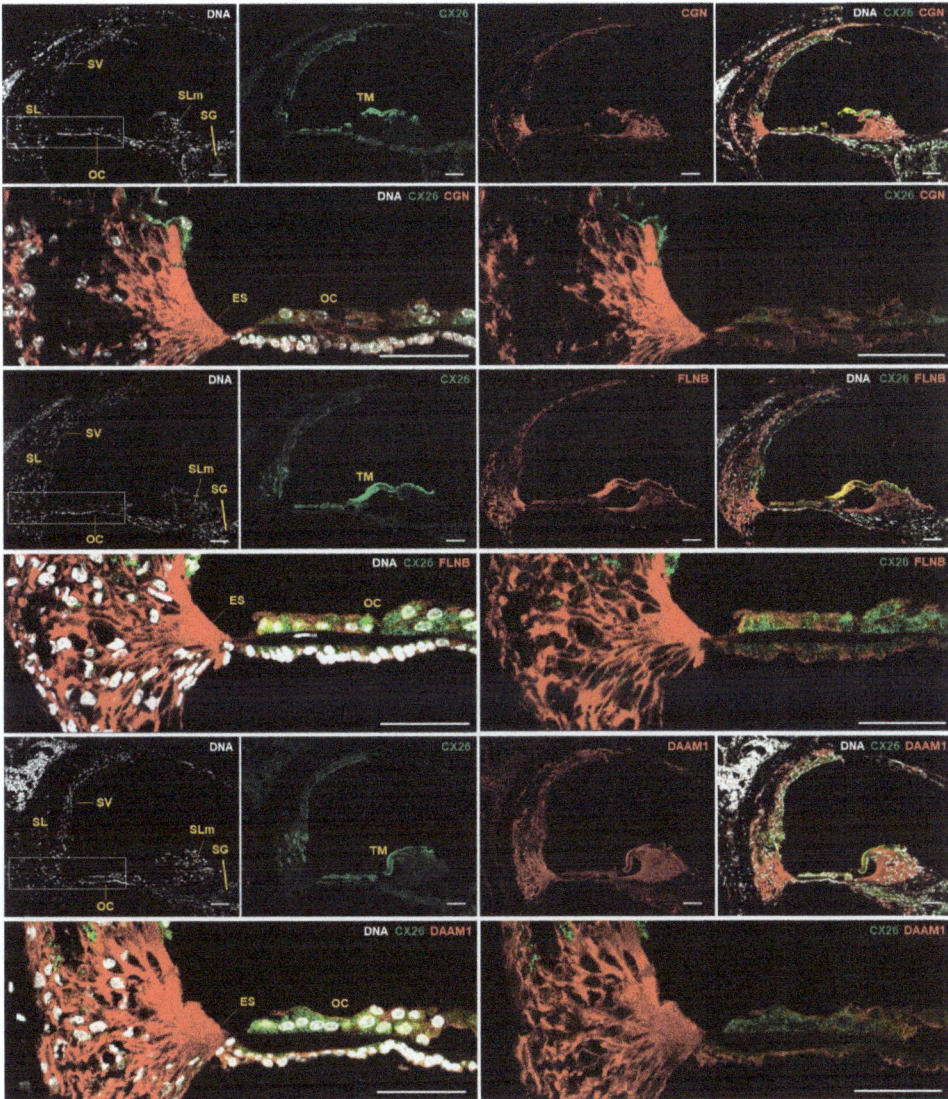

Figure 5. Indirect immunofluorescence of P14 mouse cochlea cryosections and zooming-in at the organ of Corti (OC) indicated by a rectangle. The sections highlight the spiral ligament (SL), *stria vascularis* (SV), spiral limbus (SLm), spiral ganglion (SG), external sulcus cells (ES), OC, and tectorial membrane (TM). Labeling by anti-Cx26 antibody (green) and antibodies for CGN, FLNB, and DAAM1 (all in red) are presented. DNA is in a white pseudo color, according to DAPI staining. The analysis was performed at confocal microscopy with z-sections of 1 μm (LSM880, Carl Zeiss, Oberkochen, Germany). Each image consists of maximum intensity projection of all z-sections obtained. Scale bar: 50 μm.

3. Discussion

CX26 assembly as heteromeric hemi-channels and heterotypical gap junctions has been demonstrated in particular through its association with CX30 [42]. CX26 association with other

CX has also been disclosed by global interactome analyses [33]. CX26 physical interaction with paralogues is, therefore, a common feature since it is for other family members [43]. Few additional binding partners have been reported for CX26. The trans-Golgi network protein consortin interacts with CX26 in the secretory pathway [21]. At the plasma membrane, CX26 binding to caveolin-1 is necessary for its localization in caveolae from lipid rafts [44]. Lastly, CX26 association with dynamin-2 has been implicated in its turnover by endocytosis [45]. The finding of CX26 interaction with the SCF E3 ubiquitin ligase component known as the F-box protein OCP1 has also contributed to clarify its turnover mechanism [46]. As seen, few proteins are known as binding partners of CX26. Therefore, we employed the CX26 C-terminus as bait and sought for interacting proteins from the adult mouse brain or liver.

In this paper, we presented 13 proteins that have been identified by mass spectrometry analysis of the CX26 C-terminus affinity precipitation assays with 12 of them having been classified as cell junction and cytoskeleton-associated proteins (Table 1). Four proteins have previously been identified as other CX interactors (ASS1, EB2, TJP1, VCL, Figure 1B). Three proteins from this subgroup are part of cell junctions and the cytoskeleton (EB2, TJP1, and VCL). TJP1 directly interacts with the C-termini of CX30, CX31.9, CX32, CX35, CX36, CX43, CX45, CX46, CX47, and CX50 [29–31,35–40] as well as with VCL (Figure 1B) [34] and is important to stabilize CX43 gap junctions [34]. Moreover, EB1, which is a paralogue of EB2, has been shown to be necessary for targeting CX26 and CX43 to the plasma membrane and co-immunoprecipitates with CX43 [32].

TJP1 is a large protein with three tandem N-terminal PDZ domains, which mediate its interaction with CX. TJP1 binding to CX C-terminus is an important regulatory step in a gap junction assembly, internalization, and degradation [47]. Apparently, TJP1 binding needs a CX C-terminus to be anchored at the membrane or protein complex. For affinity capture, we employed CX26 C-terminus in fusion with the GST C-terminus. This configuration may have contributed to in vitro binding of TJP1 to the GST–CX26 C-terminus. However, contrary to other connexins such as CX43, CX26 does not have a PDZ-binding motif in its C-terminus (data not shown). In fact, PDZ-binding motifs need to be internal at the C-terminus to properly mediate protein interaction [48] and the CX26 C-terminus is only 11-amino acids long. Therefore, it was not surprising that the yeast two-hybrid system did not reveal a direct interaction between TJP1 and CX26 (Figure 2A). We then looked at indirect interactions between these proteins and employed CGN in immunoprecipitation assays. There is a direct association between the N-terminal globular domain of the adaptor protein CGN and the TJP1 PDZ domain [41] and we demonstrated that CX26 co-immunoprecipitates with CGN (Figure 3). No extensive co-localization was observed between CGN and CX26 in the liver (Figure 4) or cochlea (Figure 5). Therefore, a direct interaction between these proteins is not suggested by our data.

Due to the limited length of the CX26 C-terminus, other scenarios could be considered to speculate how CX26 interactions described here would take place. Each hemi-channel displays six CX N-termini, six intracellular loops, and six C-termini at the cytosolic face of the plasma membrane. Hence, it is plausible that the CX26 C-termini combine in specific configurations that allow for protein interactions, which individually would not be possible. Those segments could be stabilized through an interaction with other membrane proteins or the lipid content from the plasma membrane in a stoichiometric ratio likely different from one-to-one.

Multiple sequence alignments among CX C-termini display limited similarity between the short cytosolic C-termini of CX26 and other B-group CX paralogues (data not shown). On the other hand, manual alignments between human and mouse CX26 11-residue C-terminus and up to 42 residues of the C-termini from the A-group and the B-group CX disclosed enrichment (30% to 50%) in basic amino acids in the first 21 cytosolic amino acids of all CX analyzed (Figure 6). Functional roles in the gap junction sensitivity have been proposed for the CX32 C-terminus basic residues [49]. Although basic residues are commonly observed in cytosolic protein segments contiguous with transmembrane domains and by balancing the net negative charge of the lipid bilayer, they may also mediate common interactions with scaffolding proteins [50]. Therefore, an additional scenario would be that the basic amino acids of at least two CX26 C-termini help to stabilize molecular interactions in the PPI network in close adjacency to the membrane. This network could be assembled on the cytoplasmic face of membranes at any point along the secretory pathway.

Gene	Protein	Last 14 residues of TM4	Residues 1 to 21 of the cytosolic C-terminus	Residues 22 to 42 of the cytosolic C-terminus	Number of basic residues Segment 1	Segment 2
GJB2	hCX26	CILLNVTELCYLLI	RYCSGKSKKPV*		4/11	
	mCx26	CILLNITELCYLFV	RYCSGKSKRPV*		4/11	
			******** +**			
GJB1	hCX32	CIILNVAEVVYLII	RACARRAQRRSNPPSRKGSGF	GHRLSPEYKQNEINKLLSEQD	7/21	3/21
	mCx32	CIILNVAEVVYLII	RACARRAQRRSNPPSRKGSGF	GHRLSPEYKQNEINKLLSEQD	7/21	3/21
		************** **	*********************			
GJB3	hCX31	CIVLTICELCYLIC	HRVLRGLHKDKPRGGCSPSSS	ASRASTCRCHHKLVEAGEVDP	7/21	5/21
	mCx31	CIILTICEICYLIF	HRIMRGISKGKSTKSISSPKS	SSRASTCRCHHKLLESGDPEA	7/21	5/21
		** ** * *	******** * *			
GJB4	hCX30.3	CILLNLSEVFYLVG	KRCMEIFGPRHRRPRCRECLP	DTCPPYVLSQGGHPEDGNSVL	8/21	1/21
	mCx30.3	CILLNLSEVVYLVG	KRCMEVFRPRRRKASRRHQLP	DTCPPYVISKGGHPQDESVIL	10/21	2/21
		***** * **+*+	** ******** * **** *			
GJB6	hCX30	CMLLNVAELCYLLL	KVCFRRSKRAQTQKNHPNHAL	KESKQNEMNELISDSGQNAIT	8/21	2/21
	mCx30	CMLLNVAELCYLLL	KLCFRRSKRTQAQRNHPNHAL	KESKQNEMNELISDSGQNAIT	8/21	2/21
		* ******* * *+	**********************			
GJA1	hCX43	SLALNIIELFYVFF	KGVKDRVKGRSDPYHATSGALS	PAKDCGSQKYAYFNGCSSPTA	6/21	2/21
	mCx43	SLALNIIELFYVFF	KGVKDRVKGRSDPYHATTGPLS	PSKDCGSPKYAYFNGCSSPTA	6/21	2/21
		**************** *	*** *****************			
GJA3	hCX46	SLLLNMLEIYHLGW	KKLKQGVTSRLGPDA-EAPLG	TADPPPLPPSSRPPAVAIGFPP	4/21	1/21
	mCx46	SLVLNMLEIYHLGW	KKLKQGVTNHFNPDASEARHK	PLDPLPTATSSGPPSVSIGFPP	6/21	0/21
		******** *** **	** * ** ** * *****			
GJA4	hCX37	SLVLNLLELVHLLC	RCLSRGMRARQGQDAPPTQGTS	SDPYTDQVFFYLPVGQGPSSP	4/21	0/21
	mCx37	SLVLNLLELVHLLC	RCVSREIKARRDHDARPAQGSA	SDPYPEQVFFYLPMGEGPSSP	7/21	0/21
		** ** ** ** * **	*** *** ********** *			
GJA5	hCX40	SLLLSLAELYHLGW	KKIRQRFVKPRQHMAKCQLSGP	SVGIVQSCTPPPDFNQCLENG	8/21	0/21
	mCx40	SLFLSLAELYHLGW	KKIRQRFGKSRQGVDKHQLPGP	PTSLVQSLTPPPDFNQCLKNS	7/21	1/21
		******* * ** * **	*** *** ********** *			
GJA8	hCX50	SLFLNVMELGHLGL	KGIRSALKRPVEQPLGEIPEKS	LHSIAVSSIQKAKGYQLLEEE	5/21	3/21
	mCx50	SLFLNIMEMSHLGM	KGIRSAFKRPVEQPLGEIAEKS	LHSIAVSSIQKAKGYQLLEEE	5/21	3/21
		***** ***************** *				

Figure 6. Pairwise alignments of human (h) and mouse (m) CX orthologues as indicated by comprehending part of the last transmembrane domain (TM4) and 42 amino acids that follow that domain. Basic amino acids are underlined and their counting indicated on the two right columns, respectively, for the first and second 21-residue initial segments of C-termini. (*) identical residues; (+) amino acids with lateral chains from the same biochemical group.

Our results do not indicate a specific interactor that could be a direct binding partner for CX26. Hence, the macromolecular assembly hypothesis is corroborated and it contributes to a protein platform at the cytosolic face of CX26 hemi-channels associated with the membrane and other junction proteins. While interaction between tight junctions and CX has previously been observed solely for CX32 [51], we could detect co-localization of CX26 and TJP1 at the plasma membrane of hepatocytes from the mouse liver (Figure 2B). It is also a possibility that CX26 hemi-channels would associate in vivo with tight junction proteins if composed of heteromeric assemblies. Alternatively, the tight junction proteins could associate with CX26 C-terminus during trafficking at the Golgi cytoplasmic face.

EB2 and VCL disclosed no convincing co-localization with CX26 in the mouse liver (Figure 2B). Specifically, microtubule plus end-binding proteins, EB1 and EB3, have been implicated in microtubule dynamics promoting microtubule growth and inhibiting its catastrophe [52]. In microtubule-assisted disassembly of focal adhesions, microtubule growth is believed to take place on underlying actin microfilaments and associated proteins. It has been demonstrated that EB2 knocking-down decreases cell motility and causes aberrant focal adhesion dynamics. EB2 has been shown to be essential for focal adhesion disassembly as a direct microtubule interactor and through its interaction with MAP4K4 (mitogen-activated protein kinase 4) [53]. Moreover, EB1 plays roles in CX43 trafficking to regions of the plasma membrane where adherens junctions had already been formed [32,54,55].

VCL is a membrane-cytoskeletal protein in focal adhesion plaques involved in the linkage of integrin adhesion molecules to the actin cytoskeleton. It is a cytoskeletal protein associated with cell-cell and cell-matrix junctions where it is thought to function as one of several interacting proteins involved in anchoring F-actin to the membrane [56]. VCL binding to CX43 has already been demonstrated by in vivo and in vitro studies including co-immunoprecipitation and co-localization [57]. Therefore, focal adhesions are cytoskeleton-membrane association sites where CX26 interaction with VCL and EB2 could be investigated.

ASS1 is the fourth protein from the CX26 PPI network that has previously been identified as a CX interactor since it has been detected in the CX32 interactome [33]. Although ASS1 is not part of the cell junction or the cytoskeleton, it has been analyzed in this report since it distributes to the plasma membrane of endothelial cells. More specifically, associated with endothelial nitric oxide synthase in caveolae from lipid rafts [28], where CX26 has also been identified [44]. However, most commonly, ASS1 is described in the vicinity of the mitochondria outer membrane [58]. ASS1 and other enzymes from the urea cycle are believed to form a macromolecular complex that facilitates and concentrates arginine metabolism components near mitochondria. Since ASS1 gene expression and ASS1 protein localization have been demonstrated to be regulated by hormones and amino acids [58], it is logical to assume that, when driven to caveolae, ASS1 association to cell membrane junction proteins such as CX and paralogues would be more pronounced. On the other hand, a few reports have implicated CX in mitochondrial functions. CX43 has been shown to localize in the mitochondria inner membrane [59] where it is assembled as a hemi-channel and functions in homeostasis and cell death [33,60]. Therefore, on the one hand, plasma membrane caveolae are a likely address for interaction between CX and ASS1. On the other hand, although CX43 and ASS1 have been reported in different mitochondria compartments including mitochondrial inner and outer membranes, indirect interaction could take place between CX and ASS1 during the transport to mitochondria. This is probably a considerable alternative since we did not observe co-localization of CX26 and ASS1 at the plasma membrane (Figure 2B).

We showed that CGN, DAAM1, and FLNB distribute to the organ of Corti (Figure 5). Moreover, all 13 proteins from the CX26 PPI network have been reported to be expressed in the inner ear, according to databases [61–65]. Among the 13 genes that encode CX26 interactors, the HOMER2 gene has been related to autosomal dominant hearing loss in humans with the description of a missense mutation [66]. In addition, tricellulin, which is a protein encoded by the *TRIC* gene, presents in its C-terminal region a domain for binding to occludin, which is known as a TJP1 direct binding partner [67]. Protein-truncating mutations in the *TRIC* gene led to the loss of the occludin-binding domain and autosomal recessive hearing loss in humans [68]. In the inner ear, tricellulin is in cell junctions of supporting and ciliated cells. These data corroborate our results on TJP1 as part of the CX26 interactome. Lastly, the localization of CX26 and adaptor proteins belonging to its cell junctional network in the cochlea confirms their potential for physiological roles in hearing. Their corresponding genes are unveiled as good candidates to be explored in hearing loss studies. Among other functions, they may participate in the mechano-electrical transduction of sound vibrations in the organ of Corti [69] or in the maintenance of cochlear ion homeostasis regulated through *stria vascularis* [70].

4. Materials and Methods

4.1. Animals and Tissue

The experimental protocol was previously approved by the Internal Review Board on Ethics in Animal Research from the Institute of Biosciences of the University of São Paulo (CEP 062/2007, 3 April 2007). All experiments were conducted in accordance with the guidelines for the care and use of laboratory animals established by the American National Research Council. Postnatal day 3 (P3) Balbc mice (*Mus musculus*), postnatal day 14 (P14), and 60 (P60) CBL57/6 mice (*Mus musculus*) were obtained from specialized breeders from the University of São Paulo (São Paulo, Brazil). Animals presenting acute or chronic ear infection or congenital malformations were excluded from the study.

The P60 CBL57/6 mouse brain and liver were obtained for affinity precipitation assays. P3 mouse liver was employed in immunoprecipitation assays. To obtain the biological material of interest, the animals were submitted to profound anesthesia with ketamine and xylazine (0.17 and 0.03 mg/g of body mass, respectively) followed by decapitation. The heads and abdomen were bathed in 70% ethanol, which was followed by the removal of the liver and sagittal incision of the skull to remove all brain tissue. Tissues were snap-frozen in liquid nitrogen and later stored at $-80\,^{\circ}$C. P14 CBL57/6 mice provided cochlea for immunofluorescence assays. The animals were sacrificed as described above. The heads were bathed in 70% ethanol, temporal bones were removed, labyrinth dissected, and cochleae was harvested with micro tweezers (Dumont #5 and #54, Electron Microscopy Sciences, Hatfield, PA, USA), under a trinocular stereomicroscope (Discovery V12, Carl Zeiss, Oberkochen, Germany) and incubated in 4% paraformaldehyde. For immunofluorescence assays, P60 CBL57/6 mice were submitted to profound anesthesia with ketamine and xylazine (0.17 and 0.03 mg/g of body mass, respectively), which was followed by trans-cardiac perfusion with 4% paraformaldehyde in phosphate-buffered saline solution (PBS; 137 mM NaCl, 2.7 mM KCl, 10 mM Na_2HPO_4, and 1.76 mM KH_2PO_4, pH 7.4) at $4\,^{\circ}$C.

4.2. DNA Clones

To obtain a recombinant pGEX-4T-1 clone containing the DNA coding sequence for glutathione-*S*-transferase (GST) in fusion with that encoding CX26 C-terminus, genomic DNA from mouse liver was used as a template for a polymerase chain reaction (PCR) amplification of human *GJB2* gene DNA from base chr13:2018852 to 20188933 (GRCh37/hg19) with the following pair of primers: (1S) 5′ CGA GGC CCG GGT TAT TGC TCA GGA AAG TCC A 3′ and (1AS) 5′ CGA GGG CGG CCG CTG GGT TCC TCT CTC CTG TC 3′, with the 5′-end having restriction sites respectively for *Sma*I and *Not*I (underlined). The undigested, purified PCR product was cloned in a pCR2.1-TOPO vector (Invitrogen, Carlsbad, CA, USA) in DH5α *E. coli*. Plasmid DNA from one recombinant clone with its insert sequence confirmed by Sanger sequencing was digested with *Sma*I and *Not*I (Invitrogen). A released insert was subcloned in the pGEX-4T-1 vector (GE Healthcare, Little Chalfont, UK) and the final recombinant clone was named PGEX-GST–CX26. For the yeast two-hybrid interaction test, the bait construct was obtained after a cloning mouse *GJB2* full-length coding sequence (present in a single exon) into the vector pBT3-N (MoBiTec, Göttingen, Germany) while prey constructs (full-length coding sequence for mouse VCL, TJP1, ASS1, or MAPRE2) were sub-cloned into the vector pPR3-N (MoBiTec).

4.3. Antibodies

The antibodies from Santa Cruz Biotechnology (Santa Cruz, CA, USA) used to detect the following proteins were: CX26 (*GJB2*, goat polyclonal antibody N-19, and rabbit polyclonal antibody O-24), CX30 (GJB6, rabbit polyclonal antibody C20), CX31 (GJB1, rabbit polyclonal antibody H-43), CX43 (GJA1, mouse monoclonal antibody F-7), ASS1 (rabbit monoclonal antibody H-231), CGN (cingulin, mouse monoclonal antibody G-6), DAAM1 (disheveled-associated activator of morphogenesis 1, mouse monoclonal antibody WW-3), FLNB (filamin B, mouse monoclonal antibody F-8), and TJP1 (zonula occludens 1 protein, ZO-1, rat polyclonal antibody R40.76). Antibodies from Abcam

(Cambridge, MA, USA) were as follows: MAPRE2 (EB2, rat polyclonal antibody K52), VCL (vinculin, mouse monoclonal antibody SPM227), and TJP1 (rabbit polyclonal antibody ab59720). An additional anti-CX26 (Zymed mouse monoclonal antibody, Thermo Fisher Scientific, Waltham, MA, USA) was employed in immunofluorescence assays. Western blot secondary antibodies were conjugated to horseradish peroxidase (GE Healthcare, Wauwatosa, WI, USA). Immunofluorescence secondary antibodies were conjugated to Alexa488 or Alexa594 (Jackson ImmunoResearch Laboratories, West Grove, PA, USA).

4.4. Bacteria Expression of Fusion Protein

For fusion protein expression, pGEX-GST–CX26 or pGEX-GST plasmid DNA was used to transform BL21 *E. coli*. Recombinant clones were grown at 37 °C in liquid LB medium supplemented with 50 μg/mL ampicillin (TEUTO, Anapólis, Brazil) and 0.5 mM IPTG (Invitrogen) until optic density (600 nm) reached values between 0.4 and 0.6. For soluble protein isolation, a bacteria pellet was suspended in PBS with a protease inhibitor (Pefabloc, Roche Applied Science, Indianapolis, IN, USA), 10 mg/mL lysozyme (Sigma-Aldrich, St Louis, MO, USA), and incubated on ice for 15 min. After two quick cycles of freezing and thawing, lysates were centrifuged at 10,000× *g*, for 15 min at 4 °C. One-hundred μL of the supernatant were mixed with 40 μL of GST-Bind™ resin (glutathione (GSH)-sepharose, Novagen, Darmstadt, Germany) under agitation at 4 °C for 30 min. After washing the pellet, samples of sepharose beads containing glutathione-bound proteins were boiled and submitted to SDS-PAGE for protein quantification in comparison to bovine serum albumin (BSA) standards. Bacterium soluble protein fractions containing 600 pmols of GST–CX26 or GST were aliquoted and stored at −80 °C.

4.5. Affinity Capture Assay

Three different lysis buffers named EDTA, EGTA, or PHEM were employed for obtaining mouse tissue lysates. For each one, 10 to 20 mg of mouse liver or brain were homogenized in 1 mL of the lysis buffer using a Douncer homogenizer (40 slow strokes). The basic composition of EDTA and EGTA buffers was the same: 50 mM Tris-HCl pH 7.4, 150 mM NaCl, 0.75% triton X-100, 2 mM Na_3VO_4, 10 mM NaF, 1× protease inhibitor (cOmplete, EDTA-free, Sigma-Aldrich, St Louis, MO, USA). EDTA and EGTA buffers differed on a divalent cation composition with the former having 1 mM EDTA and the latter 10 mM EGTA and 2 mM $MgCl_2$. PHEM buffer was composed of 60 mM piperazine-*N*,*N'*-bis [2-ethane-sulfonic acid] (PIPES) pH 6.9, 25 mM *N*-2-hydroxyethylpiperazine-*N'*-2-ethane-sulfonic acid (HEPES), 2 mM $MgCl_2$, 10 mM EGTA, 0.75% triton X-100, 5 μM phallacidin (Sigma-Aldrich, St Louis, MO, USA), 2 mM Na_3VO_4, 10 mM NaF, 1× protease inhibitor (complete, EDTA-free, Sigma-Aldrich). After tissue homogenization in EDTA or EGTA buffers, the suspension was incubated on ice for 30 min and then centrifuged for 30 min at a temperature of 4 °C at 14,000× *g*. The supernatant was transferred to a new tube, the protein was quantified, and the lysate was aliquoted and stored at −80 °C. After Douncer homogenization of tissue in PHEM, the suspension was incubated for 2 min at 37 °C and centrifuged for 5 min at a temperature of 4 °C at 14,000× *g*. The pellet was re-suspended in PHEM buffer and, after a novel centrifugation, both supernatants containing soluble proteins were pooled.

For affinity precipitation, 600 pmoles of GST–CX26 or GST were bound to 200 μL of 50% GST™Bind Resin sepharose beads (Novagen) and incubated at RT for 30 min under agitation. After three washes with PBS and protease inhibitor (Pefabloc, Roche Applied Science), lysates were added to the beads and the samples were maintained under rocking at 4 °C for 16 h. The samples were washed in a lysis buffer without triton-X-100, centrifuged and submitted to SDS-PAGE, and followed by Coomassie blue staining, according to standard procedures.

4.6. Mass Spectrometry Analyses

Gel bands identified after Coomassie blue staining were compared between the two lanes of SDS-polyacrylamide in which precipitates from GST–CX26 or GST-only had been electrophoresed side by side. Lanes with apparently similar total protein loading had specific gel bands visually compared. Bands with discrepant intensities between the two lanes were excised from the whole gel, which led to a total of 11 gel band pairs. The in-gel digest and mass spectrometry experiments were performed by the Proteomics platform of the Research Center at the Quebec University Hospital Center (CHUQ, Laval University, QC, Canada). Protein from the excised gel bands were digested with trypsin on a MassPrep liquid handling robot (Waters, Milford, CT, USA), according to the manufacturer's specifications and to the protocol of Shevchenko et al. [71] with the modifications suggested by Havlis et al. [72]. Proteins were reduced with 10 mM DTT and alkylated with 55 mM iodoacetamide. Trypsin digestion was performed using 126 nM of modified porcine trypsin (Sequencing grade, Promega, Madison, WI, USA) at 58 °C for 1 h. Digestion products were extracted using 1% formic acid, 2% acetonitrile followed by 1% formic acid, and 50% acetonitrile. The recovered extracts were pooled, vacuum centrifuge-dried, and then suspended into 7 µL of 0.1% formic acid and 2 µL were analyzed by mass spectrometry.

The resulting peptides were separated by online reversed-phase (RP) nanoscale capillary liquid chromatography (nanoLC) and analyzed by electrospray mass spectrometry (ES MS/MS). The experiments were performed with a Thermo Surveyor MS pump connected to a LTQ linear ion trap mass spectrometer (Thermo Fisher, San Jose, CA, USA) equipped with a nano-electrospray ion source (Thermo Fisher). Peptide separation took place on a self-packed PicoFrit column (New Objective, Woburn, MA, USA) packed with a C18 Jupiter HPLC column (5-µm particle size, 300-Å pore size; Phenomenex, Torrance, CA, USA). Peptides were eluted with a linear gradient of 2–50% acetonitrile, 0.1% formic acid in 30 min at 200 nL/min (obtained by flow-splitting). Mass spectra were acquired using a data-dependent acquisition mode using Xcalibur software (Version 2.0, Thermo Fisher Scientific). Each full scan mass spectrum (400 to 2000 m/z) was followed by collision-induced dissociation of the seven most intense ions. The dynamic exclusion (30-s duration) function was enabled and the relative collisional fragmentation energy was set to 35%.

All MS/MS samples were analyzed using Mascot (Version 2.3.0; Matrix Science, London, UK). The Mascot was set up to search the Uniref100 mouse database (release of June 2010; 80,419 entries) when assuming the digestion by trypsin. Mascot was searched with a fragment ion mass tolerance of 0.50 Da and a parent ion tolerance of 2.0 Da. Iodoacetamide derivative of cysteine was specified as a fixed modification and oxidation of methionine was specified as a variable modification. Two missed cleavages were allowed.

Scaffold (Version 3.6.2; Proteome Software Inc., Portland, OR, USA) was used to validate MS/MS based peptide and protein identifications. Protein identifications were accepted if they could be established at greater than 95% probability and contained at least two identified peptides, which is specified by the Peptide Prophet algorithm [73]. Proteins that contained similar peptides and could not be differentiated based on the MS/MS analysis alone were grouped to satisfy the principles of parsimony.

4.7. NCBI Protein Reference Sequence Accession Numbers

Initial protein identification was completed according to Uniprot (http://www.uniprot.org/) accession numbers. Corresponding NCBI mouse protein RefSeq accession numbers were as follows: ASS1 (argininosuccinate synthase, NP_031520.1), CGN (cingulin, NP_001032800.2), DAAM1 (disheveled-associated activator of morphogenesis, NP_766052.2), FLNB (filamin-B, NP_001074896.1), GAPDH (glyceraldehyde-3-phosphate dehydrogenase, NP_001276675.1), Homer2 (homer protein homolog 2 isoform 1: NP_036113.1), JUP (junction plakoglobin, NP_034723.1), MAP7 (ensconsin, NP_032661.2), MAPRE2 (microtubule-associated protein RP/EB family member 2, NP_694698.3), PTK2B (protein tyrosine kinase 2β, NP_001155838), RAI14 (ankycorbin, NP_001159880.1), TJP1 (tight junction protein, NP_033412.2), and VCL (vinculin, NP_033528.3). RefSeq accession numbers for human and mouse CX protein paralogues employed in multiple sequence alignments were: connexin 26 (hCX26 NP_003995.2, mCX26 NP_032151.1), connexin 30 (hCX30 NP_001103689.1, mCX30 NP_001258592.1), connexin 30.3 (hCX30.3 NP_694944.1, mCX30.3 NP_032153.1), connexin

31 CX31 (hCX31 NP_001005752.1, mCX31 NP_001153484.1), connexin 32 (hCX32 NP_000157.1, mCX32 NP_032150.2), connexin 37 (hCX37 NP_002051.2, mCX37 NP_032146.1), connexin 40 (hCX40 NP_859054.1, mCX40 NP_001258557.1), connexin 43 (hCX43 NP_000156.1, mCX43NP_034418.1), connexin 46 (hCX46 NP_068773.2, mCX46, NP_001258552.1), and connexin 50 (hCX50 NP_005258.2, mCX50 NP_032149.1).

4.8. In Silico Analyses

Computational analyses of nucleic acids and proteins were performed at the following web sites, according to their recommendations: EMBL-EBI [74], ExPASy Proteomics server [75], Hereditary Hearing Loss Homepage [76], Kyte Doolittle Hydropathy [77], Mouse Genome Informatics [64], NCBI [62], PSORT [78], The Connexin-Deafness Homepage [79], SWISSPROT [80], STRING [22], BioGRID [81], and WebGestalt [82].

The Connexin family (pfam00029) comprises 215 members (updating of August, 2016, https: //www.ncbi.nlm.nih.gov/Structure/cdd/cddsrv.cgi?uid=pfam00029). RefSeq accession numbers from protein full-length sequences of five human and five mouse paralogues from each CX group A or B were retrieved and used in multiple sequence alignments at Clustal Omega [83]. The last transmembrane domain of each CX was identified at pfam00029. Alignment of the following 42 amino acids as available was manually finalized and basic residues were highlighted.

4.9. Yeast Two-Hybrid Assay

The sequence-verified bait or prey constructs were used in self-activation testing by individually transforming the strain NMY51 (MATa his3Δ200 trp1-901 leu2-3,112 ade2 LYS2::(lexAop)4-HIS3 ura3::(lexAop)8-lacZ ade2::(lexAop)8-ADE2 GAL4) using standard procedures. For the yeast two-hybrid interaction test, bait and prey were employed in co-transformation of the yeast strain NMY51. Interaction was verified by testing for His and Ade activation. Lastly, both bait and prey plasmids were used to co-transform yeast Y2HGold. In the case of bait-prey interaction, the reporter genes (*HIS3* and *ADE2*) were activated and yeast was able to grow on SD–Leu⁻ Trp–His⁻ medium and activate the β-galactosidase expression in the X-gal assay (Creative BioLabs, Shirley, NY, USA).

4.10. Immunoprecipitation and Western Blotting

Whole livers from P2–P3 mice were lysed in EGTA buffer as described above or in RIPA buffer [50 mM Tris-HCl, pH 7.4, 150 mM NaCl, 50 mM NaF, 5 mM Na_3VO_4, 2 mM EGTA, 1% NP-40, 0.1% SDS, 0.5% sodium deoxycholate, 1X protease inhibitor (cOmplete, EDTA-free, Sigma-Aldrich)]. The protein quantity was estimated using a Bradford reagent at 595-nm absorbance. For immunoprecipitation, lysates were precleared in a 1:1 mixture of protein-A and protein-G conjugated to sepharose beads (GE Healthcare) and 1:50 volume of normal mouse serum. Nearly 500 μg of precleared lysates were submitted to incubation with 2 μg of anti-CGN specific antibodies or normal mouse serum for 16 h at 4 °C under rocking. The antibody-lysate mix was then transferred to a microtube containing a 1:1 mixture of protein-A plus protein-G beads (GE Healthcare). Further agitation was at 4 °C for two hours. Beads were pelleted at 8000× *g* for three minutes at 4 °C, washed twice in 50 mM Tris-HCl, pH 7.4, 150 mM NaCl, 50 mM NaF, 5 mM Na_3VO_4, and suspended in sample buffer (2% SDS, 100 mM dithiothreitol, 10% glycerol). Western blotting was performed by submitting samples to electrophoresis (6% or 14% SDS-PAGE) and electro-transferring proteins to a 45-μm nitrocellulose filter (BioRad, Hercules, CA, USA) for 16 h at 25 Volts. Transfer efficiency was observed after 1.5% Ponceau S staining. Proteins were blocked for one hour in 1% casein (Novagen) and was followed by 10 min in 3% hydrogen peroxide. Blots were incubated with primary antibody followed by secondary antibody for 1 h each at room temperature. Antibody dilutions were in 2% immunoglobulin-free bovine serum albumin (BSA, Jackson ImmunoResearch Laboratories) in TBS-T (20 mM Tris pH 7.6, 135 mM NaCl, 0.05% Tween-20). All washes were in TBS-T. The filter was incubated in ECL™ Plus (GE Healthcare) and exposed to Amersham Hyperfilm™ ECL film (GE Healthcare).

4.11. Indirect Immunofluorescence

Isolated P60 mouse livers were post-fixed in 4% paraformaldehyde for 6 h at 4 °C and then incubated in increasing concentrations of sucrose up to 30% sucrose for 48 h at 4 °C. After inclusion immersion in Jung Tissue Freezing Medium (Leica Biosystems, Buffalo Grove, IL, USA) for two hours, the tissue was frozen and cryo-sectioned in 12-μm histological sections. Isolated cochleae were punctured in the apex for perfusion of 4% paraformaldehyde and was followed by fixation in 4% paraformaldehyde at 4 °C for 16 h. The tissues were then immersed in decalcification solution (10% EDTA and 1% paraformaldehyde) at 4 °C for four days. After washing in PBS, all cochleae were immersed in 10% sucrose solution until the tissue sunk and then all cochleae were incubated in 20% sucrose solution at 4 °C for 20 h. Lastly, the cochleae were immersed in Jung Tissue Freezing Medium (Leica Biosystems) for two hours. The tissue was frozen and cryo-sectioned in 12-μm histological sections (Leica CM-1850, Leica Biosystems).

The histological slides were incubated in acetone at −20 °C for ten minutes and permeabilized in 0.3% triton-X-100 for 20 min, which was followed by 20 min in 0.1 M glycine. Blocking was performed in 5% normal donkey serum in PBS, supplemented with 1× donkey anti-mouse Fab' (Jackson ImmunoResearch Laboratories). Primary antibody incubation was for 16 h and secondary antibody incubation followed for three hours. All washes were in PBS and antibody dilutions were in 5% normal donkey serum and 0.1% triton-X-100. After the final wash, slides were mounted in DAPI-containing mounting medium (Invitrogen) and sealed after a coverslip was placed. Images were captured in a LSM 880 confocal microscope (Carl Zeiss) after a background fluorescence intensity subtraction from a negative control section (omission of primary antibody) obtained in the same experiment. Pseudo colors and zooming-in were obtained by the ZEN software (Version 2.5, Zeiss Efficient Navigation, Oberkochen, Germany).

Author Contributions: Conceptualization, L.A.H. Formal analysis, A.C.B. Funding acquisition, R.C.M.-N. Investigation, A.C.B., R.S.-S., J.O., and L.A.H. Methodology, A.C.B. and L.A.H. Project administration, A.C.B. and R.C.M.-N. Supervision, J.O., R.F.B., R.C.M.-N., and L.A.H. Validation, A.C.B., R.S.-S. and L.A.H. Visualization, A.C.B., R.S.-S., and L.A.H. Writing-original draft, A.C.B. Writing-review & editing, R.S.-S., R.C.M.-N., and L.A.H.

Funding: Fundação de Amparo à Pesquisa do Estado de São Paulo (FAPESP)—CEPID—Centro de Pesquisa sobre o Genoma Humano e Células-Tronco, (13/08028-1), FAPESP 12/50154-1 and Conselho Nacional de Desenvolvimento Científico e Tecnológico/CNPq (140009/2007-8) fellowships to A.C.B.

Acknowledgments: The authors thank Luiz Roberto G. Britto and laboratory technologist Adilson S. Alves (Institute of Biomedical Sciences, University of São Paulo, São Paulo, Brazil) for fixation of the mouse liver, Alberto A.G., Ribeiro F.C. and laboratory staff, Sheila Schuindt-do-Carmo and Waldir Caldeira (Laboratory of Cell Biology and Electron Microscopy, University of São Paulo, São Paulo, Brazil) for tissue sectioning and immunofluorescence image capturing, Juliana C. Correa (Chemistry Institute, University of São Paulo, São Paulo, Brazil) and Thiago G. Alegria (University of São Paulo, São Paulo, Brazil) for technical assistance, and Karina Lezirovitz and biologists Claudia R M L Hemza and Gleiciele A V Silva (Otolaryngology Lab/LIM32- University of São Paulo School of Medicine, São Paulo, Brazil) for animal management.

Conflicts of Interest: The authors declare no conflict of interest.

References

1. Zhu, Y.; Zong, L.; Mei, L.; Zhao, H.-B. Connexin26 Gap Junction Mediates MiRNA Intercellular Genetic Communication in the Cochlea and Is Required for Inner Ear Development. *Sci. Rep.* **2015**, *5*, 15647. [CrossRef] [PubMed]
2. Johnson, S.L.; Ceriani, F.; Houston, O.; Polishchuk, R.; Polishchuk, E.; Crispino, G.; Zorzi, V.; Mammano, F.; Marcotti, W. Connexin-Mediated Signaling in Nonsensory Cells Is Crucial for the Development of Sensory Inner Hair Cells in the Mouse Cochlea. *J. Neurosci.* **2017**, *37*, 258–268. [CrossRef] [PubMed]
3. Beltramello, M.; Piazza, V.; Bukauskas, F.F.; Pozzan, T.; Mammano, F. Impaired Permeability to Ins(1,4,5)P$_3$ in a Mutant Connexin Underlies Recessive Hereditary Deafness. *Nat. Cell Biol.* **2005**, *7*, 63–69. [CrossRef] [PubMed]

4. Zhang, Y.; Tang, W.; Ahmad, S.; Sipp, J.A.; Chen, P.; Lin, X. Gap Junction-Mediated Intercellular Biochemical Coupling in Cochlear Supporting Cells Is Required for Normal Cochlear Functions. *Proc. Natl. Acad. Sci. USA* **2005**, *102*, 15201–15206. [CrossRef] [PubMed]

5. The Human Protein Atlas. Available online: https://www.proteinatlas.org/ (accessed on 12 April 2018).

6. Kelsell, D.P.; Dunlop, J.; Stevens, H.P.; Lench, N.J.; Liang, J.N.; Parry, G.; Mueller, R.F.; Leigh, I.M. Connexin 26 Mutations in Hereditary Non-Syndromic Sensorineural Deafness. *Nature* **1997**, *387*, 80–83. [CrossRef] [PubMed]

7. Gasparini, P.; Estivill, X.; Volpini, V.; Totaro, A.; Castellvi-Bel, S.; Govea, N.; Mila, M.; Della, M.M.; Ventruto, V.; De, M.B.; et al. Linkage of DFNB1 to Non-Syndromic Neurosensory Autosomal-Recessive Deafness in Mediterranean Families. *Eur. J. Hum. Genet.* **1997**, *5*, 83–88. [PubMed]

8. Angeli, S.; Lin, X.; Liu, X.Z. Genetics of Hearing and Deafness. *Anat. Rec.* **2012**, *295*, 1812–1829. [CrossRef] [PubMed]

9. Smith, R.J.; Jones, M.-K.N. Nonsyndromic Hearing Loss and Deafness, DFNB1. In *GeneReviews*®; Adam, M.P., Ardinger, H.H., Pagon, R.A., Wallace, S.E., Bean, L.J., Stephens, K., Amemiya, A., Eds.; University of Washington: Seattle, WA, USA, 1993.

10. Shearer, A.E.; Hildebrand, M.S.; Smith, R.J. Hereditary Hearing Loss and Deafness Overview. In *GeneReviews*®; Adam, M.P., Ardinger, H.H., Pagon, R.A., Wallace, S.E., Bean, L.J., Stephens, K., Amemiya, A., Eds.; University of Washington: Seattle, WA, USA, 1993.

11. Donahue, H.J.; Qu, R.W.; Genetos, D.C. Joint Diseases: From Connexins to Gap Junctions. *Nat. Rev. Rheumatol.* **2018**, *14*, 42–51. [CrossRef] [PubMed]

12. Derangeon, M.; Spray, D.C.; Bourmeyster, N.; Sarrouilhe, D.; Hervé, J.-C. Reciprocal Influence of Connexins and Apical Junction Proteins on Their Expressions and Functions. *Biochim. Biophys. Acta* **2009**, *1788*, 768–778. [CrossRef] [PubMed]

13. Dbouk, H.A.; Mroue, R.M.; El-Sabban, M.E.; Talhouk, R.S. Connexins: A Myriad of Functions Extending beyond Assembly of Gap Junction Channels. *Cell Commun. Signal.* **2009**, *7*, 4. [CrossRef] [PubMed]

14. Laird, D.W. The Gap Junction Proteome and Its Relationship to Disease. *Trends Cell Biol.* **2010**, *20*, 92–101. [CrossRef] [PubMed]

15. Hervé, J.-C.; Derangeon, M.; Sarrouilhe, D.; Giepmans, B.N.G.; Bourmeyster, N. Gap Junctional Channels Are Parts of Multiprotein Complexes. *Biochim. Biophys. Acta* **2012**, *1818*, 1844–1865. [CrossRef] [PubMed]

16. Stains, J.P.; Civitelli, R. Connexins in the Skeleton. *Semin. Cell Dev. Biol.* **2016**, *50*, 31–39. [CrossRef] [PubMed]

17. Leithe, E.; Mesnil, M.; Aasen, T. The Connexin 43 C-Terminus: A Tail of Many Tales. *Biochim. Biophys. Acta* **2018**, *1860*, 48–64. [CrossRef] [PubMed]

18. Maeda, S.; Nakagawa, S.; Suga, M.; Yamashita, E.; Oshima, A.; Fujiyoshi, Y.; Tsukihara, T. Structure of the Connexin 26 Gap Junction Channel at 3.5 Å Resolution. *Nature* **2009**, *458*, 597–602. [CrossRef] [PubMed]

19. Nusrat, A.; Chen, J.A.; Foley, C.S.; Liang, T.W.; Tom, J.; Cromwell, M.; Quan, C.; Mrsny, R.J. The Coiled-Coil Domain of Occludin Can Act to Organize Structural and Functional Elements of the Epithelial Tight Junction. *J. Biol. Chem.* **2000**, *275*, 29816–29822. [CrossRef] [PubMed]

20. Nelson, R.F.; Glenn, K.A.; Zhang, Y.; Wen, H.; Knutson, T.; Gouvion, C.M.; Robinson, B.K.; Zhou, Z.; Yang, B.; Smith, R.J.H.; et al. Selective Cochlear Degeneration in Mice Lacking the F-Box Protein, Fbx2, a Glycoprotein-Specific Ubiquitin Ligase Subunit. *J. Neurosci.* **2007**, *27*, 5163–5171. [CrossRef] [PubMed]

21. Castillo, D.J.F.; Cohen-Salmon, M.; Charollais, A.; Caille, D.; Lampe, P.D.; Chavrier, P.; Meda, P.; Petit, C. Consortin, a Trans-Golgi Network Cargo Receptor for the Plasma Membrane Targeting and Recycling of Connexins. *Hum. Mol. Genet.* **2010**, *19*, 262–275. [CrossRef] [PubMed]

22. STRING: Functional Protein Association Networks. Available online: https://string-db.org/cgi/input.pl (accessed on 15 July 2018).

23. Masson, D.; Kreis, T.E. Binding of E-MAP-115 to Microtubules Is Regulated by Cell Cycle-Dependent Phosphorylation. *J. Cell Biol.* **1995**, *131*, 1015–1024. [CrossRef] [PubMed]

24. Peng, Y.F.; Mandai, K.; Sakisaka, T.; Okabe, N.; Yamamoto, Y.; Yokoyama, S.; Mizoguchi, A.; Shiozaki, H.; Monden, M.; Takai, Y. Ankycorbin: A Novel Actin Cytoskeleton-associated Protein. *Genes Cells* **2001**, *5*, 1001–1008. [CrossRef]

25. Ajima, R.; Kajiya, K.; Inoue, T.; Tani, M.; Shiraishi-Yamaguchi, Y.; Maeda, M.; Segawa, T.; Furuichi, T.; Sutoh, K.; Yokota, J. HOMER2 Binds MYO18B and Enhances Its Activity to Suppress Anchorage Independent Growth. *Biochem. Biophys. Res. Commun.* **2007**, *356*, 851–856. [CrossRef] [PubMed]

26. Shiraishi-Yamaguchi, Y.; Sato, Y.; Sakai, R.; Mizutani, A.; Knöpfel, T.; Mori, N.; Mikoshiba, K.; Furuichi, T. Interaction of Cupidin/Homer2 with Two Actin Cytoskeletal Regulators, Cdc42 Small GTPase and Drebrin, in Dendritic Spines. *BMC Neurosci.* **2009**, *10*, 25. [CrossRef] [PubMed]

27. Metzger, T.; Gache, V.; Xu, M.; Cadot, B.; Folker, E.S.; Richardson, B.E.; Gomes, E.R.; Baylies, M.K. MAP and Kinesin-Dependent Nuclear Positioning Is Required for Skeletal Muscle Function. *Nature* **2012**, *484*, 120–124. [CrossRef] [PubMed]

28. Flam, B.R.; Hartmann, P.J.; Harrell-Booth, M.; Solomonson, L.P.; Eichler, D.C. Caveolar Localization of Arginine Regeneration Enzymes, Argininosuccinate Synthase, and Lyase, with Endothelial Nitric Oxide Synthase. *Nitric Oxide* **2001**, *5*, 187–197. [CrossRef] [PubMed]

29. Giepmans, B.N.G.; Moolenaar, W.H. The Gap Junction Protein Connexin43 Interacts with the Second PDZ Domain of the Zona Occludens-1 Protein. *Curr. Biol.* **1998**, *8*, 931–934. [CrossRef]

30. Toyofuku, T.; Yabuki, M.; Otsu, K.; Kuzuya, T.; Hori, M.; Tada, M. Direct Association of the Gap Junction Protein Connexin-43 with ZO-1 in Cardiac Myocytes. *J. Biol. Chem.* **1998**, *273*, 12725–12731. [CrossRef] [PubMed]

31. Kojima, T.; Sawada, N.; Chiba, H.; Kokai, Y.; Yamamoto, M.; Urban, M.; Lee, G.-H.; Hertzberg, E.L.; Mochizuki, Y.; Spray, D.C. Induction of Tight Junctions in Human Connexin 32 (HCx32)-Transfected Mouse Hepatocytes: Connexin 32 Interacts with Occludin. *Biochem. Biophys. Res. Commun.* **1999**, *266*, 222–229. [CrossRef] [PubMed]

32. Shaw, R.M.; Fay, A.J.; Puthenveedu, M.A.; von Zastrow, M.; Jan, Y.-N.; Jan, L.Y. Microtubule Plus-End-Tracking Proteins Target Gap Junctions Directly from the Cell Interior to Adherens Junctions. *Cell* **2007**, *128*, 547–560. [CrossRef] [PubMed]

33. Fowler, S.L.; Akins, M.; Zhou, H.; Figeys, D.; Bennett, S.A.L. The Liver Connexin32 Interactome Is a Novel Plasma Membrane-Mitochondrial Signaling Nexus. *J. Proteome Res.* **2013**, *12*, 2597–2610. [CrossRef] [PubMed]

34. Zemljic-Harpf, A.E.; Godoy, J.C.; Platoshyn, O.; Asfaw, E.K.; Busija, A.R.; Domenighetti, A.A.; Ross, R.S. Vinculin Directly Binds Zonula Occludens-1 and Is Essential for Stabilizing Connexin-43-Containing Gap Junctions in Cardiac Myocytes. *J. Cell Sci.* **2014**, *127*, 1104–1116. [CrossRef] [PubMed]

35. Kausalya, P.J.; Reichert, M.; Hunziker, W. Connexin45 Directly Binds to ZO-1 and Localizes to the Tight Junction Region in Epithelial MDCK Cells. *FEBS Lett.* **2001**, *505*, 92–96. [CrossRef]

36. Nielsen, P.A.; Baruch, A.; Shestopalov, V.I.; Giepmans, B.N.G.; Dunia, I.; Benedetti, E.L.; Kumar, N.M. Lens Connexins A3Cx46 and A8Cx50 Interact with Zonula Occludens Protein-1 (ZO-1). *Mol. Biol. Cell* **2003**, *14*, 2470–2481. [CrossRef] [PubMed]

37. Li, X.; Olson, C.; Lu, S.; Kamasawa, N.; Yasumura, T.; Rash, J.E.; Nagy, J.I. Neuronal Connexin36 Association with Zonula Occludens-1 Protein (ZO-1) in Mouse Brain and Interaction with the First PDZ Domain of ZO-1. *Eur. J. Neurosci.* **2004**, *19*, 2132–2146. [CrossRef] [PubMed]

38. Singh, D.; Solan, J.L.; Taffet, S.M.; Javier, R.; Lampe, P.D. Connexin 43 Interacts with Zona Occludens-1 and -2 Proteins in a Cell Cycle Stage-Specific Manner. *J. Biol. Chem.* **2005**, *280*, 30416–30421. [CrossRef] [PubMed]

39. Penes, M.C.; Li, X.; Nagy, J.I. Expression of Zonula Occludens-1 (ZO-1) and the Transcription Factor ZO-1-associated Nucleic Acid-binding Protein (ZONAB)–MsY3 in Glial Cells and Colocalization at Oligodendrocyte and Astrocyte Gap Junctions in Mouse Brain. *Eur. J. Neurosci.* **2005**, *22*, 404–418. [CrossRef] [PubMed]

40. Flores, C.E.; Li, X.; Bennett, M.V.L.; Nagy, J.I.; Pereda, A.E. Interaction between Connexin35 and Zonula Occludens-1 and Its Potential Role in the Regulation of Electrical Synapses. *Proc. Natl. Acad. Sci. USA* **2008**, *105*, 12545–12550. [CrossRef] [PubMed]

41. Cordenonsi, M.; D'Atri, F.; Hammar, E.; Parry, D.A.; Kendrick-Jones, J.; Shore, D.; Citi, S. Cingulin Contains Globular and Coiled-Coil Domains and Interacts with ZO-1, ZO-2, ZO-3, and Myosin. *J. Cell Biol.* **1999**, *147*, 1569–1582. [CrossRef] [PubMed]

42. Yum, S.W.; Zhang, J.; Valiunas, V.; Kanaporis, G.; Brink, P.R.; White, T.W.; Scherer, S.S. Human Connexin26 and Connexin30 Form Functional Heteromeric and Heterotypic Channels. *Am. J. Physiol. Cell Physiol.* **2007**, *293*, 1032–1048. [CrossRef] [PubMed]

43. Karademir, L.B.; Aoyama, H.; Yue, B.; Chen, H.; Bai, D. Engineered Cx26 Variants Established Functional Heterotypic Cx26/Cx43 and Cx26/Cx40 Gap Junction Channels. *Biochem. J.* **2016**, *473*, 1391–1403. [CrossRef] [PubMed]

44. Schubert, A.-L.; Schubert, W.; Spray, D.C.; Lisanti, M.P. Connexin Family Members Target to Lipid Raft Domains and Interact with Caveolin-1. *Biochemistry* **2002**, *41*, 5754–5764. [CrossRef] [PubMed]

45. Xiao, D.; Chen, S.; Shao, Q.; Chen, J.; Bijian, K.; Laird, D.W.; Alaoui-Jamali, M.A. Dynamin 2 Interacts with Connexin 26 to Regulate Its Degradation and Function in Gap Junction Formation. *Int. J. Biochem. Cell Biol.* **2014**, *55*, 288–297. [CrossRef] [PubMed]

46. Henzl, M.T.; Thalmann, I.; Larson, J.D.; Ignatova, E.G.; Thalmann, R. The Cochlear F-Box Protein OCP1 Associates with OCP2 and Connexin 26. *Hear. Res.* **2004**, *191*, 101–109. [CrossRef] [PubMed]

47. Sorgen, P.L.; Trease, A.J.; Spagnol, G.; Delmar, M.; Nielsen, M.S. Protein–Protein Interactions with Connexin 43: Regulation and Function. *Int. J. Mol. Sci.* **2018**, *19*, 1428. [CrossRef] [PubMed]

48. Thévenin, A.F.; Kowal, T.J.; Fong, J.T.; Kells, R.M.; Fisher, C.G.; Falk, M.M. Proteins and Mechanisms Regulating Gap-Junction Assembly, Internalization, and Degradation. *Physiology* **2013**, *28*, 93–116. [CrossRef] [PubMed]

49. Wang, X.G.; Peracchia, C. Chemical Gating of Heteromeric and Heterotypic Gap Junction Channels. *J. Membr. Biol.* **1998**, *162*, 169–176. [CrossRef] [PubMed]

50. Baker, J.A.; Wong, W.-C.; Eisenhaber, B.; Warwicker, J.; Eisenhaber, F. Charged Residues next to Transmembrane Regions Revisited: "Positive-inside Rule" Is Complemented by the "Negative inside Depletion/Outside Enrichment Rule". *BMC Biol.* **2017**, *15*, 66. [CrossRef] [PubMed]

51. Kojima, T.; Kokai, Y.; Chiba, H.; Yamamoto, M.; Mochizuki, Y.; Sawada, N. Cx32 but Not Cx26 Is Associated with Tight Junctions in Primary Cultures of Rat Hepatocytes. *Exp. Cell Res.* **2001**, *263*, 193–201. [CrossRef] [PubMed]

52. Komarova, Y.; Lansbergen, G.; Galjart, N.; Grosveld, F.; Borisy, G.G.; Akhmanova, A. EB1 and EB3 Control CLIP Dissociation from the Ends of Growing Microtubules. *Mol. Biol. Cell* **2005**, *16*, 5334–5345. [CrossRef] [PubMed]

53. Yue, J.; Xie, M.; Gou, X.; Lee, P.; Schneider, M.D.; Wu, X. Microtubules Regulate Focal Adhesion Dynamics through MAP4K4. *Dev. Cell* **2014**, *31*, 572–585. [CrossRef] [PubMed]

54. Chkourko, H.S.; Guerrero-Serna, G.; Lin, X.; Darwish, N.; Pohlmann, J.R.; Cook, K.E.; Martens, J.R.; Rothenberg, E.; Musa, H.; Delmar, M. Remodeling of Mechanical Junctions and of Microtubule-Associated Proteins Accompany Cardiac Connexin43 Lateralization. *Heart Rhythm* **2012**, *9*, 1133–1140.e6. [CrossRef] [PubMed]

55. Agullo-Pascual, E.; Lin, X.; Leo-Macias, A.; Zhang, M.; Liang, F.-X.; Li, Z.; Pfenniger, A.; Lübkemeier, I.; Keegan, S.; Fenyö, D.; et al. Super-Resolution Imaging Reveals That Loss of the C-Terminus of Connexin43 Limits Microtubule plus-End Capture and NaV1.5 Localization at the Intercalated Disc. *Cardiovasc. Res.* **2014**, *104*, 371–381. [CrossRef] [PubMed]

56. Ziegler, W.H.; Gingras, A.R.; Critchley, D.R.; Emsley, J. Integrin Connections to the Cytoskeleton through Talin and Vinculin. *Biochem. Soc. Trans.* **2008**, *36*, 235–239. [CrossRef] [PubMed]

57. Xu, X.; Francis, R.; Wei, C.J.; Linask, K.L.; Lo, C.W. Connexin 43-Mediated Modulation of Polarized Cell Movement and the Directional Migration of Cardiac Neural Crest Cells. *Dev. Camb. Engl.* **2006**, *133*, 3629–3639. [CrossRef] [PubMed]

58. Haines, R.J.; Pendleton, L.C.; Eichler, D.C. Argininosuccinate Synthase: At the Center of Arginine Metabolism. *Int. J. Biochem. Mol. Biol.* **2011**, *2*, 8–23. [PubMed]

59. Rodriguez-Sinovas, A.; Boengler, K.; Cabestrero, A.; Gres, P.; Morente, M.; Ruiz-Meana, M.; Konietzka, I.; Miró, E.; Totzeck, A.; Heusch, G.; et al. Translocation of Connexin 43 to the Inner Mitochondrial Membrane of Cardiomyocytes through the Heat Shock Protein 90-Dependent TOM Pathway and Its Importance for Cardioprotection. *Circ. Res.* **2006**, *99*, 93–101. [CrossRef] [PubMed]

60. Peixoto, P.M.; Ryu, S.-Y.; Pruzansky, D.P.; Kuriakose, M.; Gilmore, A.; Kinnally, K.W. Mitochondrial Apoptosis Is Amplified through Gap Junctions. *Biochem. Biophys. Res. Commun.* **2009**, *390*, 38–43. [CrossRef] [PubMed]

61. Sajan, S.A.; Warchol, M.E.; Lovett, M. Toward a Systems Biology of Mouse Inner Ear Organogenesis: Gene Expression Pathways, Patterns and Network Analysis. *Genetics* **2007**, *177*, 631–653. [CrossRef] [PubMed]

62. Home-UniGene-NCBI. Available online: https://www.ncbi.nlm.nih.gov/unigene (accessed on 24 April 2018).

63. Shen, J.; Scheffer, D.I.; Kwan, K.Y.; Corey, D.P. SHIELD: An Integrative Gene Expression Database for Inner Ear Research. *Database* **2015**, *2015*, bav071. [CrossRef] [PubMed]

64. MGI-Mouse Genome Informatics. The International Database Resource for the Laboratory Mouse. Available online: http://www.informatics.jax.org/ (accessed on 24 April 2018).

65. International Mouse Phenotyping Consortium. Available online: http://www.mousephenotype.org/ (accessed on 24 April 2018).

66. Azaiez, H.; Decker, A.R.; Booth, K.T.; Simpson, A.C.; Shearer, A.E.; Huygen, P.L.M.; Bu, F.; Hildebrand, M.S.; Ranum, P.T.; Shibata, S.B.; et al. HOMER2, a Stereociliary Scaffolding Protein, Is Essential for Normal Hearing in Humans and Mice. *PLOS Genet.* **2015**, *11*, e1005137. [CrossRef] [PubMed]

67. Furuse, M.; Fujita, K.; Hiiragi, T.; Fujimoto, K.; Tsukita, S. Claudin-1 and -2: Novel Integral Membrane Proteins Localizing at Tight Junctions with No Sequence Similarity to Occludin. *J. Cell Biol.* **1998**, *141*, 1539–1550. [CrossRef] [PubMed]

68. Riazuddin, S.; Ahmed, Z.M.; Fanning, A.S.; Lagziel, A.; Kitajiri, S.; Ramzan, K.; Khan, S.N.; Chattaraj, P.; Friedman, P.L.; Anderson, J.M.; et al. Tricellulin Is a Tight-Junction Protein Necessary for Hearing. *Am. J. Hum. Genet.* **2006**, *79*, 1040–1051. [CrossRef] [PubMed]

69. Dror, A.A.; Avraham, K.B. Hearing Loss: Mechanisms Revealed by Genetics and Cell Biology. *Annu. Rev. Genet.* **2009**, *43*, 411–437. [CrossRef] [PubMed]

70. Wangemann, P. K+ Cycling and Its Regulation in the Cochlea and the Vestibular Labyrinth. *Audiol. Neurotol.* **2002**, *7*, 199–205. [CrossRef] [PubMed]

71. Shevchenko, A.; Wilm, M.; Vorm, O.; Mann, M. Mass Spectrometric Sequencing of Proteins from Silver-Stained Polyacrylamide Gels. *Anal. Chem.* **1996**, *68*, 850–858. [CrossRef] [PubMed]

72. Havlis, J.; Thomas, H.; Sebela, M.; Shevchenko, A. Fast-Response Proteomics by Accelerated in-Gel Digestion of Proteins. *Anal. Chem.* **2003**, *75*, 1300–1306. [CrossRef] [PubMed]

73. Keller, A.; Nesvizhskii, A.I.; Kolker, E.; Aebersold, R. Empirical Statistical Model to Estimate the Accuracy of Peptide Identifications Made by MS/MS and Database Search. *Anal. Chem.* **2002**, *74*, 5383–5392. [CrossRef] [PubMed]

74. The European Bioinformatics Institute < EMBL-EBI. Available online: https://www.ebi.ac.uk/ (accessed on 15 July 2018).

75. ExPASy: SIB Bioinformatics Resource Portal—Home. Available online: https://www.expasy.org/ (accessed on 15 July 2018).

76. Hereditary Hearing Loss. Hereditary Hearing Loss Homepage. Available online: http://hereditaryhearingloss.org/ (accessed on 15 July 2018).

77. Genomics and Bioinformatics @ Davidson College. Available online: http://gcat.davidson.edu/ (accessed on 15 July 2018).

78. PSORT WWW Server. Available online: https://psort.hgc.jp/ (accessed on 15 July 2018).

79. The Connexin-Deafness Homepage. Available online: http://davinci.crg.es/deafness/ (accessed on 15 July 2018).

80. Uniprot < EMBL-EBI. Available online: https://www.ebi.ac.uk/uniprot (accessed on 15 July 2018).

81. BioGRID. Database of Protein, Chemical, and Genetic Interactions. Available online: https://thebiogrid.org/ (accessed on 15 July 2018).

82. WebGestalt GSAT. Available online: http://www.webgestalt.org/option.php (accessed on 15 July 2018).

83. Clustal Omega < Multiple Sequence Alignment < EMBL-EBI. Available online: https://www.ebi.ac.uk/Tools/msa/clustalo/ (accessed on 15 July 2018).

International Journal of
Molecular Sciences

MDPI

Article

Concatenation of Human Connexin26 (hCx26) and Human Connexin46 (hCx46) for the Analysis of Heteromeric Gap Junction Hemichannels and Heterotypic Gap Junction Channels

Patrik Schadzek [1], Doris Hermes [1,2], Yannick Stahl [1], Nadine Dilger [1] and Anaclet Ngezahayo [1,3,*]

[1] Institut für Biophysik, Leibniz Universität Hannover, Herrenhäuser Straße 2, 30419 Hannover, Germany; p.schadzek@biophysik.uni-hannover.de (P.S.); hermes@em.mpg.de (D.H.); yannick.stahl1@googlemail.com (Y.S.); n.dilger@biophysik.uni-hannover.de (N.D.)
[2] Department of Clinical Neurophysiology, University of Göttingen, Robert-Koch Str. 40, D-37075 Göttingen, Germany
[3] Zentrum für Systemische Neurowissenschaften Stiftung Tierärztliche Hochschule Hannover, Bünteweg 2, 30559 Hannover, Germany
* Correspondence: ngezahayo@biophysik.uni-hannover.de; Tel.: +49-511-762-4568

Received: 20 August 2018; Accepted: 11 September 2018; Published: 13 September 2018

Abstract: Gap junction channels and hemichannels formed by concatenated connexins were analyzed. Monomeric (hCx26, hCx46), homodimeric (hCx46-hCx46, hCx26-hCx26), and heterodimeric (hCx26-hCx46, hCx46-hCx26) constructs, coupled to GFP, were expressed in HeLa cells. Confocal microscopy showed that the tandems formed gap junction plaques with a reduced plaque area compared to monomeric hCx26 or hCx46. Dye transfer experiments showed that concatenation allows metabolic transfer. Expressed in *Xenopus* oocytes, the inside-out patch-clamp configuration showed single channels with a conductance of about 46 pS and 39 pS for hemichannels composed of hCx46 and hCx26 monomers, respectively, when chloride was replaced by gluconate on both membrane sides. The conductance was reduced for hCx46-hCx46 and hCx26-hCx26 homodimers, probably due to the concatenation. Heteromerized hemichannels, depending on the connexin-order, were characterized by substates at 26 pS and 16 pS for hCx46-hCx26 and 31 pS and 20 pS for hCx26-hCx46. Because of the linker between the connexins, the properties of the formed hemichannels and gap junction channels (e.g., single channel conductance) may not represent the properties of hetero-oligomerized channels. However, should the removal of the linker be successful, this method could be used to analyze the electrical and metabolic selectivity of such channels and the physiological consequences for a tissue.

Keywords: oligomerization; concatenated connexins; gap junction; channel stoichiometry; heteromeric connexons; human connexin46; human connexin26; inside-out patch-clamp configuration; dual whole-cell patch-clamp; dye transfer

---

1. Introduction

Gap junction channels (GJC) are formed between adjacent cells by docking of two hemichannels. The hemichannels are formed by oligomerization of connexins (Cx). Connexins are membrane proteins encoded by a gene family which in the human genome comprises 21 members [1]. In many tissues, different connexins are concurrently expressed and the formation of heteromeric connexons and heterotypic gap junction channels has been postulated. It is assumed that by forming heteromeric connexons and heterotypic gap junction channels, the cells in a tissue could achieve a rectification of gap junctions as it is observed in tissue [2,3]. Using expression systems such as the *Xenopus* oocytes, the formation of heterotypic gap junction channels could be unequivocally demonstrated by coupling two oocytes expressing two different connexins [4]. With such experiments, it was possible to specify which connexins were able to form heterotypic gap junction channels with each other. In combination with molecular biology and protein modeling, it was possible to classify the connexins into two different groups depending on some residues found in the second extracellular loop (EL2). One group is represented by Cx26 and named K-N group with the sequence φ(K/R)CXXXPCPNXVDCΩψS and a second group is represented by Cx43 called H group with the sequence φXCXXXPCPHXVDCΩψS. In the sequences, φ represents a hydrophobic residue, X is any residue, Ω an aromatic residue, and ψ indicates a residue with a large aliphatic side chain [5–10]. Within a group the connexins formed compatible connexons. An asparagine residue in position 168 of Cx26 or in a homologous position in other connexins belonging to the K-N group was shown to form hydrogen bonds and was therefore essential for the docking between hemichannels of this group [11–13]. The analysis of hCx26, for which a crystal structure was generated, revealed that each asparagine residue at position 176 (N176) in a hemichannel formed three hydrogen bonds with a lysine residue at position 168 (K168), a threonine residue at position 177 (T177), and an aspartic acid residue at position 179 (D179) in the E2 domain of the counterpart hCx26 in the hemichannel of the adjacent cell [6–9,11,14–18]. For hCx32 and hCx46, homologous N residues to N176 were described [12,13,19,20]. For hCx32, the central N residue was N175, which interacted with K167, T176, and D178. For hCx46, the N188 formed corresponding hydrogen bonds with R180, T189, and D191. The importance of N176 and K168 for the docking interaction was recently demonstrated by Karademir et al. (2016) [21]. The authors showed that by adapting the homologous residues, heterotypic docking between Cx26 and Cx40 connexons could be achieved.

With respect to oligomerization in connexons, the first transmembrane domain (TM1) and the transition between the cytoplasmic loop (CL) and the third transmembrane domain (TM3) were identified as the critical regions [10,22,23]. Analyzing Cx26 mutants, the sequence V37-A40 (VVAA) of the wild type Cx26 was identified as an important motive for Cx26 oligomerization. However, as stated in Jara et al. (2012), the motive did not determine hetero-oligomerization of connexins [10,22,23]. Concerning the hetero-oligomerization of different connexin types within a connexon, the compatibility between connexins was mostly related to the amino acid residues in the region of the transition between the cytoplasmic loop (CL) and the third transmembrane domain (TM3) [10]. According to the sequence of these regions, the connexins were classified into the R-type connexins, which contain a conserved arginine or lysine motif in this region, and the W-type connexins with a di-tryptophan motif. In compliance to this classification, only connexins belonging to the same type can hetero-oligomerize. However, even if the motif in this region is important for the oligomerization, it was proposed that indirect mechanisms associated with the motif were necessary to achieve the discrimination between connexins that do not belong to the same type. Cx43 proteins, for example, are maintained as monomers in the endoplasmic reticulum (ER) by the chaperone protein ERp29, which is associated with the second extracellular loop, thereby avoiding oligomerization. After the transport to the trans-Golgi network, the Cx43 proteins are separated from ERp29 for oligomerization [24]. Therefore, it was suggested that the sequence did not per se hinder the oligomerization between connexins belonging to different types. The sequence was rather an element involved in recruiting quality control proteins like ERp29 that in turn regulated the oligomerization.

Concatenation of subunits of the GABA, Ach, and $P_{2\times7}$ ionotropic receptors has been used to analyse the architecture of these membrane channels [25–28]. In these studies, it was shown that concatenated proteins were completely inserted in the cell membrane. In the present report, we concatenated hCx26 and hCx46 (Figure 1). Cx26 and Cx46 are concurrently expressed in trachea and alveolar epithelium type 1 and 2 cells [29,30]. With respect to the docking they belong to the K-N group. Therefore, they form hemichannels that can dock to each other. But with respect to the oligomerization behavior, Cx26 belongs to the W-type while Cx46 belongs to the R-type so that they are not supposed to oligomerize [10]. The analysis of the formed gap junction channels with imaging methods and functional assays, as well as electrophysiological characterization, showed that the concatenated proteins were able to form functional hemichannels and gap junction channels. Although aspects of the channels e.g., single channel conductance might be affected by concatenation, aspects such as docking of hetero-oligomerized connexins in variable and clearly determined stoichiometry could be analyzed using concatenation of connexins. For an accurate analysis of the biophysical properties of the concatenated channels, it would be desirable to cleave the linkers of the concatenated connexins in a connexon to reintroduce the full C- and N-terminal flexibility (and thus the natural functionality), as it was done for the concatenated pannexin1 [31].

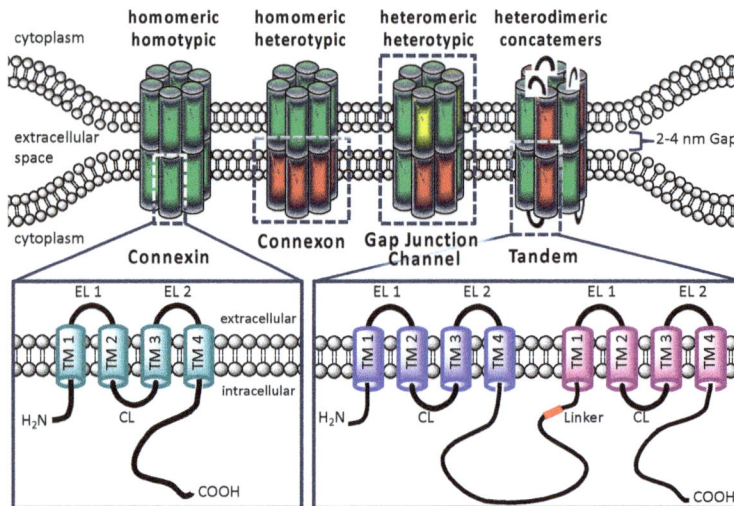

Figure 1. Structural organization of connexins and gap junctions. A gap junction channel is composed of two hemichannels or connexons, which consist of six connexins. Connexins have four transmembrane domains (TM), two extracellular loops (EL), one cytoplasmatic loop (CL), as well an intracellularly localized N- and C-terminus. Homomeric connexons and homotypic gap junction channels are formed by a single connexin isoform. In the postulated heteromeric connexons and heterotypic gap junction channels, different connexin isoforms are expected. The lower right pictogram shows the constructed concatemeric connexins as heterodimeric tandem.

2. Results

2.1. Gap Junction Plaques Formed by Concatenated Variants of hCx26 and hCx46

Molecular cloning was used to generate concatenated hCx26-hCx26-GFP (green fluorescent protein), hCx46-hCx46-GFP, hCx26-hCx46-GFP, and hCx46-hCx26-GFP. In order to concatenate two neighboring connexins, a 19-amino acid long linker was inserted into the sequence. The linker, which was used between the connexin and the GFP tag, was 23 amino acids long (Table 1).

Table 1. The amino acid linkers (one-letter code) between the two concatenated connexins and between the connexin and the GFP tag. The GFP tag was always located at the C-terminus.

Linker	Amino Acid Sequence
Cx-Cx	GGNLQSTVPR ATTLYTKVV
Cx-GFP	GGNLQSTVPR AHPAFLYKVV RSR

The concatenated constructs, as well as hCx26-GFP and hCx46-GFP, were expressed in HeLa cells to analyze their capacity to form gap junction plaques. Confocal laser scanning microscopy showed that the GFP labeled constructs formed gap junction plaques between neighboring cells (Figure 2A and Figure S1 in the Supplemental Materials). Compared to the expressed monomeric hCx26-GFP or hCx46-GFP, the dimeric connexins formed gap junction plaques with a reduced surface, suggesting a possible reduction of the protein synthesis for both heterodimeric and homodimeric tandems (Figure 2B). Moreover, western blotting of the hCx46 monomer and hCx46-hCx46 homodimer seems to corroborate the suggestion (Figure 2C). When quantifying the protein amount of the hCx46 monomer and the hCx46-hCx46 homodimer using an anti-Cx46 antibody, cells expressing the homodimer showed a reduction of about 40% of the relative hCx46 amount compared to cells expressing the monomeric hCx46.

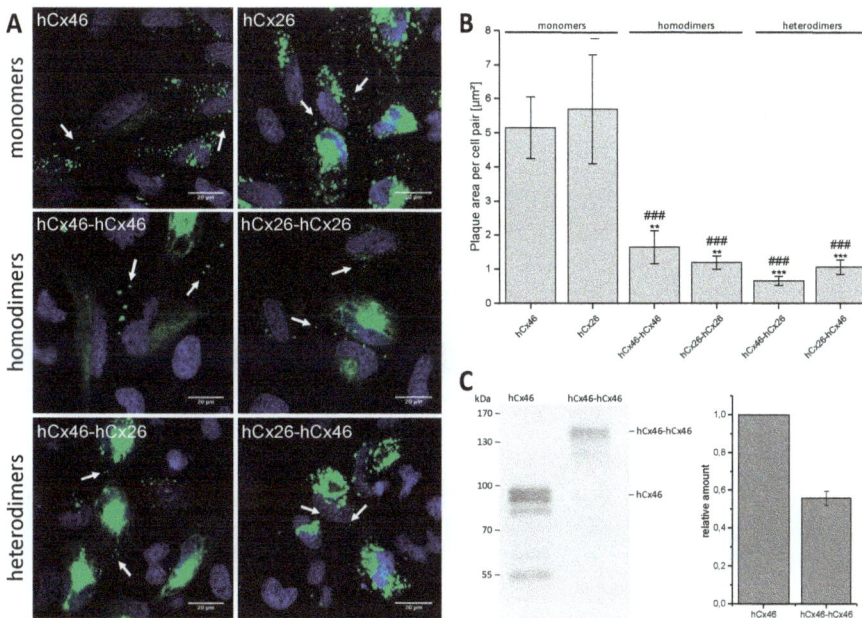

Figure 2. Expression of the GFP-labeled monomeric and concatemeric connexins in HeLa cells. (**A**) Representative micrographs of cell pairs expressing GFP-labeled hCx46, hCx26, hCx46-hCx46, hCx26-hCx26, hCx46-hCx26, and hCx26-hCx46 are shown. The cells were imaged 24 h after transfection using a confocal laser scanning microscope. The nuclei (blue) were stained with Hoechst 33342. Gap junction plaques are indicated by arrows. Gap junction plaques were found in HeLa cells expressing hCx46, hCx26, and the four different tandems. In cells expressing the tandems, a trend to accumulate the proteins in intracellular organelles was observed. (**B**) Quantification of the gap junction plaque area formed by the monomers and the four different tandems in HeLa cells. The plaque area was calculated using the particle analyzer of ImageJ and normalized to the number of transfected cell pairs. At least three transfections were performed per construct. The results are given as average plaque area per cell pair [μm^2]. Error bars represent the SEM. The data were evaluated by a one-way ANOVA and a post-hoc Tukey test (** $p \leq 0.01$, *** $p \leq 0.001$) in comparison to hCx46 and hCx26 (### $p \leq 0.001$). (**C**) Quantification of the relative protein amount in HeLa cells expressing the monomeric hCx46-GFP or the homodimeric hCx46-hCx46-GFP. An anti-Cx46 antibody was used for the western blotting. For the quantification, four independent replicates were analyzed by using the gel analyzer tool of the FiJi software [32]. The data was normalized to the intensity of the hCx46 monomer.

To test the physiological functionality of the built gap junction channels, dye transfer experiments with the monomeric and the tandem connexins in all variations were performed in HeLa and N2A cells (Figure 3). HeLa cells, transfected with either a GFP-labelled monomeric connexin or a homo- or heterodimeric tandem, showed a degree of Lucifer Yellow dye coupling of about 45%, while mock-transfected HeLa cells showed a significantly reduced degree of dye coupling (11%). When using AMCA (7-amino-4-methyl-3-coumarinylacetic acid) as tracer dye in Neuro2A (N2A) cells, expressing the untagged homo- or heterodimeric tandems or the monomeric variants together with soluble GFP (IRES-GFP plasmids), a dye transfer rate between 32% and 46% could be observed, while N2A cells expressing only the soluble GFP showed a significantly reduced dye coupling ability of about 5%. Interestingly, the concatenation did not alter the formation of the functionally coupled gap junction channels.

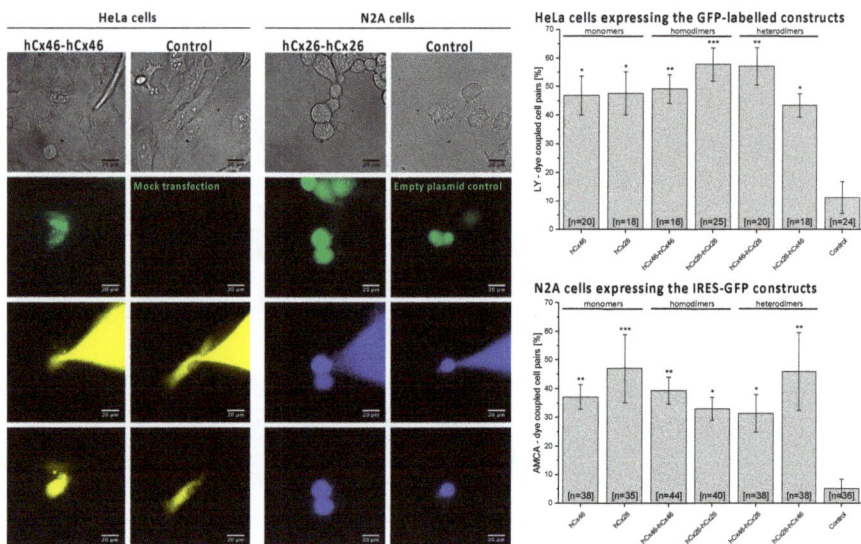

Figure 3. Analyzing the gap junction functionality by dye transfer experiments. A whole-cell patch-clamp configuration with a pipette solution containing 1 mg/mL Lucifer yellow was established on a HeLa cell pair expressing GFP-labelled monomeric hCx26 or hCx46 or one of the concatemeric variants. Mock transfected cells were used as control. The first row of the micrographs shows the phase contrast images of example experiments. In the second row, the GFP fluorescence signal before a dye transfer experiment is shown. The third and fourth rows show the fluorescence signal of the tracer dye 5 min and 10 min after establishment of the whole cell configuration. For the sake of clarity, the image in the fourth row was taken after removal of the dye filled capillary. Likewise, the experiments were performed with N2A cells. The cells were transfected with IRES-GFP constructs resulting in the expression of untagged constructs in the membrane and GFP in the cytosol. As control, cells expressing only GFP were used. The experiments were performed with a pipette solution containing 1 mg/mL AMCA, which could easily be distinguished from GFP under the fluorescence microscope. The cells were considered as coupled if the fluorescence intensity, which was measured in the unpatched cell of a cell pair after 10 min, was at least twice as bright as the background, which was measured at the beginning of the experiment. The probability of coupling (bar diagrams) was estimated as ratio of the sum of coupled cell pairs per the sum of tested pairs. The results are given as average. Error bars represent the SEM. The data were evaluated by a one-way ANOVA and post-hoc Tukey test (* $p \leq 0.05$, ** $p \leq 0.01$, *** $p \leq 0.01$) in comparison to the control cells.

2.2. Single Channel Activity of Connexons

Cx46 is known to form gap junction hemichannels when expressed in *Xenopus* oocytes [33–35]. Therefore, *Xenopus* oocytes were used as expression system to measure the single channel activity of connexons formed by the single connexins, as well as the variant tandems. In inside-out patch-clamp experiments, we found that the open probability p of channels composed of the monomeric connexins, as well as the homodimers and heterodimers, was increased by suppression of Ca^{2+} on both side of the channels (Figure 4). Moreover, we observed that all configurations were sensitive to the gap junction channel inhibitor carbenoxolone (CBX). With respect to CBX, the Cx26 connexons were less sensitive to the agent than the hCx46 connexons (Figure 4). The insensitivity to CBX was even more pronounced for the hCx26-hCx26 homodimer, while the hCx46-hCx46 heterodimeric hemichannels were almost completely closed by CBX (Figure 4B). For the hemichannels formed by the heterodimers in either configuration, the sensitivity to CBX was more similar to that observed for hCx46 connexons than that of hCx26 connexons. For a further characterization of the hemichannels, we analyzed the single channel conductance in absence of Ca^{2+} (Figure 4A). Under our experimental conditions, in which the chloride was completely replaced by gluconate at both sides of the membrane, a single channel conductance of 46.0 ± 5.3 pS and 39.2 ± 5.5 pS was found for hCx46 and hCx26, respectively. For the homodimers, the conductance was reduced to 24.5 ± 3.3 pS and 32.9 ± 6.3 pS for hCx46-hCx46 and hCx26-hCx26, respectively. For the heterodimers, two substates were observed for each configuration: A conductance of 15.7 ± 0.8 pS and 26.2 ± 1.8 pS for hCx46-hCx26, and 20.1 ± 1.4 pS and 31.2 ± 3.0 pS for the hCx26-hCx46 configuration (Table 2, Figure 4). The changes in conductance are also related to the concatenation. A successful cleavage of the linker to separate the connexins in the channels is needed in order to evaluate changes solely caused by hetero-oligomerization of the connexins within a channel.

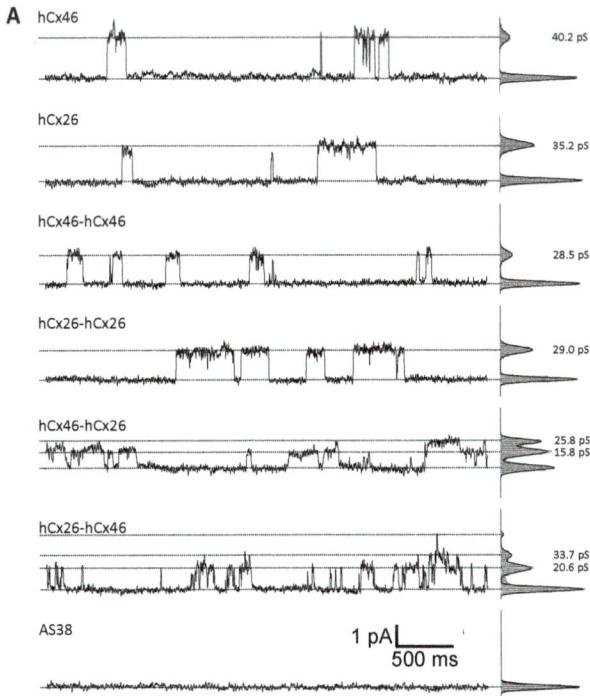

Figure 4. *Cont.*

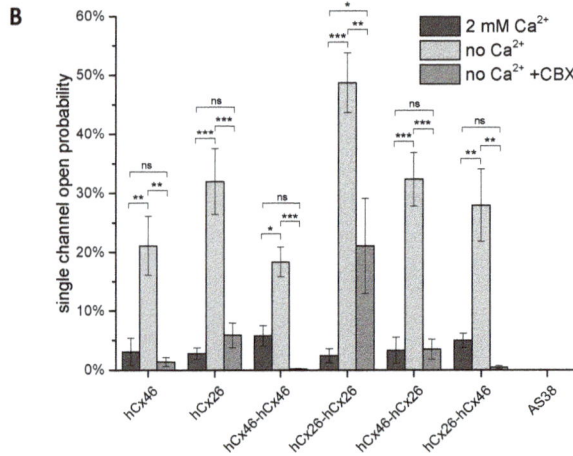

Figure 4. Analysis of single hemichannels formed by concatemeric connexins. The stripped membrane of *Xenopus* oocytes, which were injected with hCx46, hCx26, or the four different concatemeric constructs, as well as the AS38 (control) cRNA 24 h before, was used to perform the inside-out patch-clamp recordings. The measurements were performed in presence of Cs^+ and in absence of Cl^- on both sides of the membrane. (**A**) Examples of single channel currents elicited by a depolarizing voltage pulse of +50 mV in absence of Ca^{2+} in the bath solution are shown. (**B**) The open probability of all measured single channels was analyzed. The error bars represent the SEM. The data were evaluated by a one-way ANOVA followed by a Tukey test (* $p \leq 0.05$, ** $p \leq 0.01$, *** $p \leq 0.001$, ns: not significant). The statistical comparison showed that the presence of Ca^{2+} or carbenoxolone (CBX) in the bath solution significantly reduced the open probability of all tested variants.

Table 2. Conductance states of the hemichannels expressed in *Xenopus* oocytes as measured in inside out patch-clamp configuration. n gives the number of analyzed oocytes.

Injected cRNA	Large Substates ± SEM	Small Substates ± SEM
hCx46 (n = 5)	46.0 ± 5.3 pS	
hCx26 (n = 5)	39.2 ± 5.5 pS	
hCx46-hCx46 (n = 4)	24.5 ± 3.3 pS	
hCx26-hCx26 (n = 7)	32.9 ± 6.3 pS	
hCx46-hCx26 (n = 4)	26.2 ± 1.8 pS	15.7 ± 0.8 pS
hCx26-hCx46 (n = 5)	31.2 ± 3.0 pS	20.1 ± 1.4 pS
AS38 (n = 6)	0 pS	

2.3. Dye Uptake through Hemichannels

Using ethidium bromide (Etd), we observed the dye uptake by HeLa cells expressing hemichannels formed by the different variants in order to clarify how the constructed tandems could affect the function of the channels. First, we found that the cells expressing the monomers or the tandems in different variations did not differ from mock cells in their capacity to absorb ethidium bromide as long as external Ca^{2+} was present (Figure 5). The dye uptake in presence of external Ca^{2+} was therefore considered as background uptake. Specific hemichannel uptake of the dye was initialized when the cells were superfused with a Ca^{2+}-free external solution (Figure 5). To determine a possible mechanical effect on the dye uptake, the cells were first superfused with a 2 mM Ca^{2+}-containing solution. In cells expressing the homodimeric hCx26-hCx26, a tendency to increase the rate of dye uptake during the perfusion with the Ca^{2+}-containing solution was observed. However, this mechanical sensitivity of the channel was not statistically significant when analyzed by a two-way ANOVA and a post-hoc Tukey test. For cells expressing the GFP control or the other variants, the tendency to respond to a mechanical stimulus was not observed (Figure 5). For all variants, the ethidium bromide uptake was accelerated by the superfusion of the cells with Ca^{2+}-free external solution compared to cells expressing the GFP control (Figure 5A,B and Figure S2: Table in the Supplemental Materials). In the context of low external Ca^{2+}, only the hCx26-hCx26 homodimer showed an increased rate of dye uptake compared to the monomeric hCx26 and hCx46, as well as the hCx46-hCx46 homodimer and both heterodimers (S2). These changes in the ethidium bromide dye uptake rate seems to be an artifact due to the concatenation. For hCx46 however, the hCx46-hCx46 homodimer as well as the hCx46 monomer showed a comparable rate of dye uptake, and this uptake rate of hCx46 was also measured for the heterodimers in either order. Additionally, dye uptake by hemichannels in all variants except for the hCx26 monomer (and the GFP control cells) was significantly reduced by the superfusion with a La^{3+}-containing medium. When using a two-way ANOVA and a post-hoc Tukey test for the comparison of the different variants, neither the perfusion with the Ca^{2+}-containing solution nor the perfusion with a Ca^{2+}-free and La^{3+}-containing solution showed a significant difference to another variant or the GFP expressing control cells. Only the perfusion with the Ca^{2+}-free solution led to significant differences, as described above (S2).

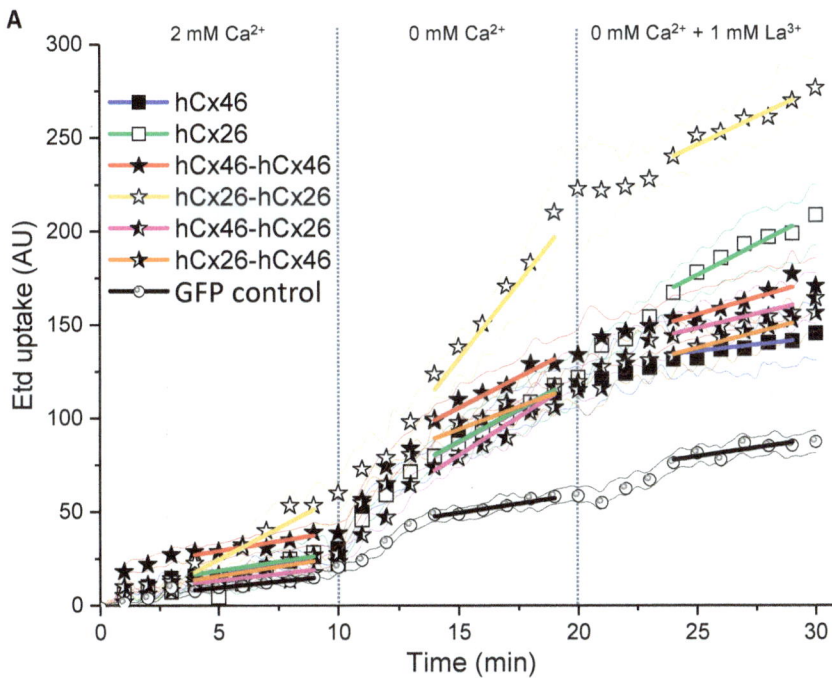

Figure 5. *Cont.*

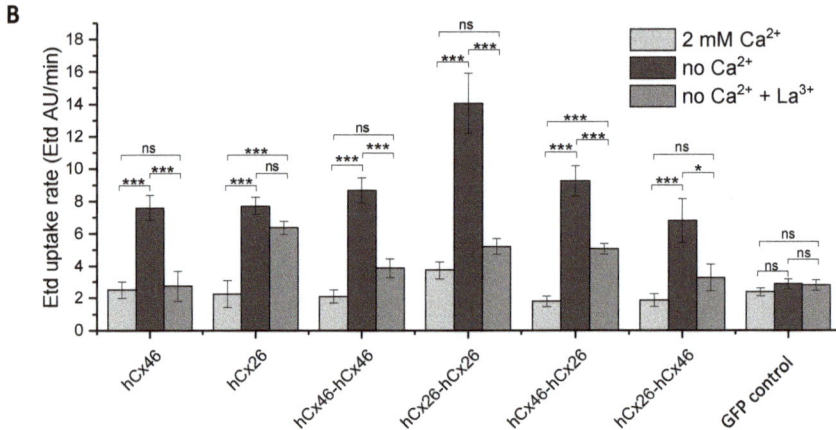

Figure 5. Dye uptake through hemichannels using ethidium bromide. HeLa cells were grown on coverslips to a confluency of about 40–50% and transfected with IRES-GFP-plasmids. The GFP allowed the identification of transfected cells by fluorescent microscopy. (**A**) Time course of dye uptake experiments by cells expressing the different variants when perfused with bath solutions containing 2 mM Ca^{2+}, no Ca^{2+}, and no Ca^{2+} but 1 mM La^{3+}. The fine lines show the SEM spread for all measured points. The symbols indicate the average for data points measured every 1 min. The solid lines indicate the part of the curves that was used to estimate the dye uptake rate. (**B**) Quantification of the dye uptake rate (Etd AU/min) for all tested variants and the backbone control in absence or presence of Ca^{2+} or La^{3+}. The error bars represent the SEM. The data were evaluated by a one-way ANOVA and a post-hoc Tukey test (* $p \leq 0.05$, *** $p \leq 0.001$, ns: not significant).

2.4. Activity of Single Gap Junction Channels

The activity of single gap junction channels was analyzed by dual whole-cell patch-clamp experiments [36,37] applied on N2A cells expressing hCx26 and hCx46, as well as the different dimeric variants. Compared to HeLa cells, which were used for the analysis of gap junction plaques, N2A cells did not form a monolayer, which offered the advantage of an easy identification of unambiguous cell pairs suitable for the dual whole-cell patch-clamp experiments. Moreover, probably due to their round morphology, the cells formed small gap junction plaques with less gap junction channels allowing a better observation of single gap junction channel activity (Figure 6).

Cells expressing the different variants were found to form gap junction channels with a total macroscopic conductance of up to 1000 pS. In some of these cell pairs, it was possible to follow the closing and opening of single gap junction channels at different transjunctional voltages, and to estimate the conductance of the single gap junction channels in dependency of the expressed variant. Although multiple simultaneously opened channels were recorded, it was possible to follow the opening and closing of single gap junction channels (Figure 6).

Considering the clear opening and closing of the single channels, maximal conductance levels of 202 pS, 198 pS, 138 pS, 184 pS, 137 pS, and 371 pS were estimated for the Cx46 monomer, Cx46-Cx46 homodimer, Cx26 monomer, Cx26-Cx26 homodimer, Cx46-Cx26 heterodimer, and Cx26-Cx46 heterodimer, respectively. The mean of the large conductance ± SEM of these variants is given in Table 3. Besides these large conductance levels, low subconductance states were observed (Figure 6 and Figure S3 in the Supplemental Materials). However, because of the rapid flickering, a clear estimation of the numeric values was not possible.

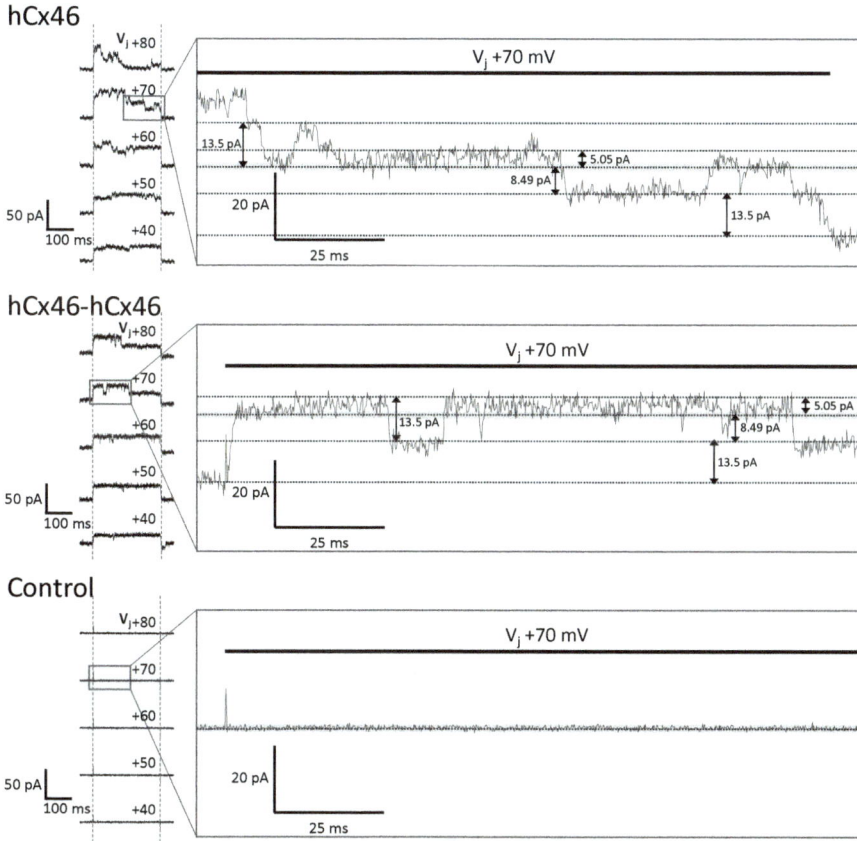

Figure 6. Recordings of dual whole-cell patch-clamp experiments. N2A cells were cultured and transfected with the different IRES-GFP-plasmids. As control, cells expressing only soluble GFP in the cytosol were used. Twenty-four hours post transfection, the dual whole-cell patch-clamp experiments were performed. The resting membrane potential was set to −40 mV for both cells. One cell of a cell pair was alternatingly stepped from −120 mV to +60 mV, while the junctional currents were recorded in the other cell. The junctional currents (ΔI2) recorded during the 250 ms-long voltage pulses at different transjunctional potentials are shown above the current responses. Magnification of the Vj +70 mV traces of the hCx46 monomer, as well as of the hCx46-hCx46 homodimer, showed several simultaneously open channels, with a large conductance of ~193 pS (13.5 pA step), and low conductance of ~72 pS (5.05 pA step) and ~121 pS (8.49 pA step). Similar steps were observed for Cx46 by other authors [38–41]. The control cells showed only the background noise, which was below 2 pA (grey band), indicating that the fluctuations of about 5 pA were conducting substates.

Table 3. Large conductance states of the gap junction channels expressed in transfected N2A cell pairs. n gives the number of analyzed cell pairs for each variant.

Expressed Variant	Large Conductance $\pm$ SEM
hCx46 (n = 7)	175.5 $\pm$ 5.7 pS
hCx26 (n = 7)	182.8 $\pm$ 1.0 pS
hCx46-hCx46 (n = 14)	193.5 $\pm$ 2.4 pS
hCx26-hCx26 (n = 9)	125.1 $\pm$ 4.0 pS
hCx46-hCx26 (n = 18)	110.5 $\pm$ 8.5 pS
hCx26-hCx46 (n = 11)	281.2 $\pm$ 24.7 pS
GFP control (n = 8)	0 pS

3. Discussion

In cells, different connexin isoforms are concurrently expressed. The formation of heteromeric connexons and heterotypic channels with a variable stoichiometry offers the cell a mode for fine tuning the gap junction coupling, thereby producing a flux rectification and selectivity [3,4]. The goal of the present report was to test whether concatenated connexins form heteromeric connexons and heterotypic gap junction channels with clearly determined stoichiometry (Figure 1). This could be a promising method to study the physiological consequences of the hetero-oligomerization of connexins in hemichannels and gap junction channels.

By expressing GFP labeled hCx46 and hCx26 monomers, hCx46-hCx46 and hCx26-hCx26 homodimers, as well as hCx46-hCx26 and hCx26-hCx46 heterodimers in HeLa cells, we found that the monomers as well as the homodimers and heterodimers were transported to the cell membrane and were able to form gap junction plaques (Figure 2A). However, as shown for the hCx46 monomer and the hCx46-hCx46 homodimer, the concatenation reduced the quantity of the produced protein. This might contribute to the reduction of the gap junction plaque area formed by the tandems (Figure 2B). The results suggest that concatenation might induce changes in the synthesis or the trafficking of the proteins to the membrane. Furthermore, a tendency to retain the proteins in the ER and Golgi apparatus was observed in cells expressing the hCx46-hCx46 tandem compared to cells expressing the hCx46 monomer (Figure 2; cells expressing the hCx26 per se showed a higher intracellular signal, which limited the possibility to compare trafficking of homomers and dimers). Although this trend was not statistically significant, it could indicate a possible concatenation-related problem with protein trafficking. Nevertheless, the different tandem proteins as well as the monomers formed functional gap junction channels, as demonstrated by the dye transfer experiments (Figure 3). Consequently, the concatenation of connexins might be helpful for the analysis of hetero-oligomerization of connexins and, inter alia, opens the possibility to prospectively study metabolite selectivity, which could not be predicted so far. Oligomerization, trafficking to the membrane, and the assembly to gap junction plaques were not affected when GFP was coupled to the N-terminus. However, the fusion of a fluorescent protein tag to the N-terminus of a connexin has been shown to block the channel conductance of hemichannels and gap junction channels [42,43]. The mechanism responsible for this is still a matter of speculation. The crystal structure of Cx26 led to the prediction that the N-termini folded in the pore of the channel [11]. We can assume that the hydrophilic GFP subunits coupled to the N-termini might stay in the cytosol. They might hinder the correct folding of the N-terminus or form a lid on the cytosolic mouth of the channel. In case of the concatenated connexins, the hydrophobic transmembrane domains of the two linked connexins led to the formation of a protein with eight transmembrane domains. How the connexin C-terminus is structured is not fully understood. However, experimental data showed that the C-terminus is highly flexible [44] and interacts with cytoskeleton-associated proteins to structure the gap junction plaques in the cell membrane, suggesting that the C-terminus does not form a barrel like GFP [45]. We therefore assume that, by interacting with these associated proteins and in combination with the high flexibility, the C-terminus of the first connexin of the concatemer could be close enough to the membrane to allow a correct structure formation of the linked N-terminus of the following connexin. However, the NMR (nuclear magnetic resonance) data also showed disorders in the C-termini that can be affected by binding to partners. Additionally, the NMR data showed the capability of dimerization in some parts of the C-termini [44]. Consequently, it is possible that the concatenation, even if it is compatible with the formation of hemichannels and gap junction channels, may affect the properties of the channels e.g., single channel conductance. Therefore, concatenation offers the possibility to form hemichannels with a determined stoichiometry and gap junction channels with manageable variabilities. However, a tool for the removal of the linker between the connexins after formation of the hemichannels is still needed to allow an exploitation of the potentialities of the method.

Considering the single channel conductance, the inside-out patch-clamp analysis of connexons formed in the membrane of *Xenopus* oocytes showed conductance values of about 46 pS and 39 pS for hCx46 and hCx26 monomers, respectively. In isolated lens fibers (Cx46) of rodents, a conductance of about 240 pS was measured [46]. Similarly, for Cx46 hemichannels expressed in *Xenopus* oocytes, HeLa, or N2A cells, main conductance levels of 250–300 pS could be identified [47–50]. For Cx26 expressed in *Xenopus* oocytes, HeLa, or N2A cells, a main conductance of 320 pS [51] and even a higher conductance above 400 pS was found [52]. In our inside-out patch-clamp experiments, we could not identify these large conductance values. However, the conductance values presented in this report of about 39 pS and 46 pS for hemichannels composed of hCx26 and hCx46 monomers are similar to measured subconductance states of these channels that were published by other authors [38–41]. For Cx26, Gaßmann et al. (2009) reported three conductance states with $G_1 = 34 \pm 8$ pS, $G_2 = 70 \pm 8$ pS, and $G_3 = 165 \pm 19$ pS. G_1 represents the vast majority of the detected events [39]. Moreover, the conductance values were measured after the replacement of chloride by the less mobile gluconate and acetate ions on both sides of the membrane. This allowed silencing the background currents and thereby isolating specific currents that were only observed in patches from oocytes expressing connexons (Figure 4). It is known that the replacement of chloride by less mobile ions leads to lower conductance values [53,54]. We therefore assume that the measurement of hemichannels formed by hCx46 and hCx26 monomers in our experimental conditions, where chloride was replaced by gluconate and acetate ions, only allowed the recording of substates with a low conductance.

The conductance values measured for the hCx46-hCx46 (24.5 $\pm$ 3.3 pS) and hCx26-hCx26 (32.9 $\pm$ 6.3 pS) homodimers were slightly lower than the conductance values found for the hemichannels formed by the respective monomers. Although the shown differences are comparably slight, concatenation-related artifacts might need to be suppressed by successfully removing the linker between the proteins within the hemichannels to reveal the properties of heteromerization-related changes. However, by allowing the generation of channels with a defined stoichiometry, concatenation could allow studying the aspects of heterodimerization of connexins. Correspondingly, we found that heterodimerization of hCx26 and hCx46 introduced two new substates with ~16 pS and ~26 pS for hCx46-hCx26, and ~20 pS and ~31 pS for hCx26-hCx46. The conductance of hemichannels formed by hCx46-hCx46 or hCx26-hCx26 homodimers was clearly different to that of hemichannels formed by hCx46-hCx26 or hCx26-hCx46 heterodimers, respectively. This result suggests that both parts of the concatenated connexins participated in the hemichannels. Further non-published data showed that concatenation of the hCx46 and the hCx46N188T mutant, which did not form gap junction plaques [13], reduced the formation of gap junction plaques compared to the homodimer hCx46-hCx46, suggesting that in concatenated form the hCx46N188T participated to formation of gap junction channels. The observation that concatenated connexins are inserted in the membrane as whole is an agreement with other experiments in which subunits of membrane proteins such as Ach, GABA, or ATP ionotropic receptors channels were concatenated [25–28]. As for the changes in conductance observed in the present report, the results indicate that the formation of heterodimeric hemichannels would change the conductance and opening properties of the channels in comparison to the respective homomeric hemichannels. A separation of the two connexins in a concatemer is desirable for the analysis of the biophysical properties of the channel.

Dual whole-cell patch-clamp experiments were performed to analyze the corresponding gap junction channels. The hCx46 and hCx26 homomers formed gap junction channels with a maximal conductance of ~200 pS for hCx46 and ~140 pS for hCx26. In other expression systems, similar conductance values were found for the gap junction channels formed by these connexins [41,55]. Beside these main conductance levels, other open substates and other residual substates with conductance values of about 20 pS for Cx46 and 17 pS for hCx26 were observed, suggesting that the conductance values (40 pS for hCx46 and 35 pS for hCx26) observed in the membrane of oocytes might represent the residual substates of the hemichannels. The concatenation of hCx46 did not change the maximal conductance of the channels (Figure 6). For hCx26, the concatenation resulted in channels with a slightly increased maximal conductance of about 180 pS. Similar to the hemichannels, the analysis of gap junction channels formed by the heterodimers revealed changes in the channel properties compared to the homodimers. Therefore, these changes, such as the strong increase of the maximal conductance, might be more related to heteromerization than to the concatenation. To confirm this hypothesis, a cleavage of the linker between the connexins in a concatemer is necessary.

The comparison of the electrophysiological data obtained from hCx46-hCx26 and hCx26-hCx46 heterodimers showed some unexpected variability with respect to hemichannels and gap junction channel activity, as well as to conductance states. The hCx26-hCx46 showed a higher activity as hemichannels and as gap junction channels in comparison to homodimers. In the case of gap junction channels, we found that channels formed by hCx26-hCx46 had a more elevated conductance (371 pS) than the channels composed of hCx46-hCx26 (137 pS). It shows that the concatenation might affect some properties of the channels, especially the electrical conductance, in a way that we cannot explain. We presume that this might be related to the short C-terminus of Cx26, which might affect the following N-terminus of Cx46. In their model, Maeda et al. (2009) suggested that the N-termini form a voltage-sensitive funnel in the cytoplasmic mouth of the pore [11]. In our experiments with the tandems, three N-termini were free while the other three termini were linked to the C-termini of the preceding molecule. If the C-terminus of Cx46 is linked, the length of this terminus may allow more flexibility to the linked N-terminus than if the C-terminus is given by Cx26, which is very short. The observation that the homodimer hCx26-hCx26 formed hemichannels and gap junction channels with more activity, as well as gap junction channels with higher conductance levels than those of the Cx26 homomers, is compatible with this presumption. Additionally, the presumed model could explain the difference in the trend of insensitivity to the inhibition by La^{3+}. Using ethidium bromide, Jara et al. (2012) already showed a degree of insensitivity of Cx26 hemichannels to La^{3+} [22]. This insensitivity was transferred to the hemichannels formed by hCx46-hCx26 heteromers and did not affect the hCx26-hCx46 heteromers (Figure 5). This trend might be related to the number of free N-termini. Because the C-terminus of Cx46 in hCx46-hCx26 is long, the N-terminus of the following hCx26 has more freedom to move. As result, we have three N-termini of hCx26, which are almost free to interact with the completely free N-termini of hCx46. In hCx26-hCx46, three N-termini of hCx46 are linked to the short C-termini of hCx26, thereby limiting the interaction of these N-termini with the free N-termini of hCx26. At that point, the concatenation limits the bearing of the information shown in this report. However, it will be a valuable method if the cleavage of the linker is successful. An increase of the conductance would be a good indication of a successful removal of the linker.

Using different methods, in this report we showed that concatenation could be a technique to understand the consequences of formation of heteromeric gap junction hemichannels, as well as of heterotypic gap junction channels built between cells. Concerning the oligomerization of connexins, two critical motifs, which were not mutually exclusive, have been identified: The end of the first transmembrane domain (TM1) and the transition between the cytoplasmic loop (CL) and the third transmembrane domain (TM3) [10,22,23]. The motif in TM1 which was found to be critical for the oligomerization of Cx26 was not important for hetero-oligomerization [10,22,23]. On the other side, the amino acid dendrogram showed that the compatibility of different connexin isoforms to hetero-oligomerize into a connexon was related to a motif situated at the transition region between CL and TM3 [10]. According to the sequence of these regions, the connexins were classified between the R-type connexins, which contain a conserved arginine or lysine residue, and the W-type connexins with a di-tryptophan motif. According to this classification, connexins belonging to different types do not oligomerize.

In our experiments, we showed that Cx26 (W-type) and Cx46 (R-type), which are also different in the TM1 region, could form heteromeric connexons with each other. Thereby, our results support the assumption that the control of hetero-oligomerization is not regulate by the TM1 [10,22,23]. Moreover, the residues R and W of the connexins in the transition between the cytoplasmic loop (CL) and the third transmembrane domain (TM3) are indirect control and not an intrinsic property of the connexins that regulate the hetero-oligomerization. Our results show the importance of indirect controls for hetero-oligomerization above the sequences in the different parts of the connexins.

4. Materials and Methods

4.1. Molecular Cloning

In order to express various concatemers in HeLa cells, as well as in N2A cells and in *Xenopus laevis* oocytes, the multisite gateway cloning system with three different destination plasmids was used.

For the transfection of the cell lines, the destination plasmids pEF-I-GFP GX, which has an IRES element between the gateway cassette and the reporter GFP, and the psDEST47 were used. pEF-I-GFP GX [56] was a gift from John Brigande (Addgene plasmid # 45443). psDEST47 was created by using a "reverse" BP-cloning reaction with the expression clone pcDNA-DEST47-GFP-GFP and the pDONR221 linearized with EcoNI. pcDNA-DEST47-GFP-GFP [57] was a gift from Patrick Van Oostveldt (Addgene plasmid # 36139). The psDEST47 was transformed into ccdB survival *Escherichia coli* BD3.1 cells and selected on ampicillin- and chloramphenicol-containing LB-Agar plates. The purified psDEST47, which is similar to the commercially available pcDNA™-DEST47 vector (#12281010, Thermo Fisher Scientific, Waltham, MA, USA), was used to create a C-terminally GFP-labeled fusion protein via the LR-cloning reaction.

The vector psGEMHE-GW was used for in vitro transcription to produce the cRNA for the *Xenopus* oocytes. The vector was created by restriction enzyme cloning with XbaI and HindIII using the pGEMHE [35] as backbone. As insert the gateway cassette was amplified with the attR1-ccdB-attR2 XbaI F and attR1-ccdB-attR2 HindIII R primers (see Table 4). *Escherichia coli* BD3.1 cells (invitrogen, Carlsbad, CA, USA) were used to host the three different destination vectors.

Table 4. Primers used for restriction enzyme cloning to produce the destination vector psGEMHE-GW and for the BP-cloning to generate the various Entry clones.

Primer	5′-3′ Sequence
attR1-ccdB-attR2 XbaI F	CTTCATCTAGACACGCTCGAGATCACAAGTTTGTAC
attR1-ccdB-attR2 HindIII R	CTTCGAAGCTTTTACATCTCGAGCACCACTTTGTACAAG
GW_BP-cloning hCx46 attB1 F	GGGGACAAGTTTGTACAAAAAAGCAGGCTCCATGGGCGACTGGAGCTTTCTGG
GW_BP-cloning hCx46 attB2 R	GGGGACCACTTTGTACAAGAAAGCTGGGTGGGCCCGCGGTACCGTCGAC
GW_BP-cl. hCx46 stop attB2 R	GGGGACCACTTTGTACAAGAAAGCTGGGTTCTAGATGGCCAAGTCCTCCGGT
GW_BP-cloning hCx46 attB5r R	GGGGACAACTTTTGTATACAAAGTTGTGGCCCGCGGTACCGTCG
GW_BP-cloning hCx46 attB5 F	GGGGACAACTTTGTATACAAAAGTTGTAATGGGCGACTGGAGCTTTCTGG
GW_BP-cloning hCx26 attB1 F	GGGGACAAGTTTGTACAAAAAAGCAGGCTTAATGGATTGGGGCACGCT
GW_BP-cloning hCx26 attB2 R	GGGGACCACTTTGTACAAGAAAGCTGGGTTGGCCCGCGGTACCG
GW_BP-cl. hCx26 stop attB2 R	GGGGACCACTTTGTACAAGAAAGCTGGGTTCTAAACTGGCTTTTTTGACTTCCCAGAAC
GW_BP-cloning hCx26 attB5r R	GGGGACAACTTTTGTATACAAAGTTGTGGCCCGCGGTACCG
GW_BP-cloning hCx26 attB5 F	GGGGACAACTTTGTATACAAAAGTTGTAATGGATTGGGGCACGCT

The various Entry vectors were built by amplifying hCx46 or hCx26 with the primers listed in Table 4 with a proofreading DNA polymerase (Phusion, Thermo Fisher Scientific, Waltham, MA, USA) followed by the BP-clonase reaction (Thermo Fisher Scientific, Waltham, MA, USA) with the donor plasmids pDONR™221, pDONR™221 P1-P5r, and pDONR™221 P5-P2 (Thermo Fisher Scientific, Waltham, MA, USA). For the psGEMHE-GW and the pEF-I-GFP GX plasmids, the stop attB2 R primers were used. *Escherichia coli* MachI (Thermo Fisher Scientific, Waltham, MA, USA) was used to host the ten different Entry plasmids. The purified Entry and destination plasmids were used to perform the multisite LR reaction (LR clonase II plus, Thermo Fisher Scientific, Waltham, MA, USA). *Escherichia coli* MachI was used to host the 18 different expression clones (three different destination plasmids, each with monomeric hCx46 and hCx26, homodimeric hCx46-hCx46 and hCx26-hCx26, as well as the heterodimeric hCx46-hCx26 and hCx26-hCx46). The BP-clonase II and LR-clonase II plus reactions were successfully performed in a total volume of only 2.5 µL. Restriction enzyme cloning and gateway cloning were verified by sequencing (Seqlab, Göttingen, Germany).

4.2. Cell Culture

HeLa cells (DSMZ no.: ACC 57, Leibniz Institute DSMZ-German Collection of Microorganisms and Cell Cultures, Braunschweig, Germany) were cultured in DMEM/Ham's F12 (1:1) medium (FG 4815, Biochrom, Berlin, Germany) supplemented with 10% fetal calf serum (Biochrom), 1 mg/mL penicillin, and 0.1 mg/mL streptomycin (Biochrom). The mouse neuroblastoma cells N2A, abbreviation for Neuro-2A (DSMZ no.: ACC 148), were cultured in DMEM with 1.0 g/L D-glucose (FG 0415, Biochrom) supplemented with 10% heat inactivated fetal calf serum (Biochrom), 1x non-essential amino acids (Biochrom), 1 mg/mL penicillin, and 0.1 mg/mL streptomycin (Biochrom). The cells were cultured in a humidified atmosphere with 5% CO_2 at 37 °C. Every two to three days the cell culture medium was renewed.

4.3. Quantification of the Expression Behavior

To analyze the formation of gap junctions, 7×10^4 HeLa cells were seeded on collagen I-coated glass coverslips with a diameter of 1 cm into a well of a 24-well plate 24 h before transfection to reach a confluency of about 70–80%. Prior to the transfection, the cell culture medium was replaced by 500 µL OptiMEM I medium (Thermo Fisher Scientific). The transfection was performed as described before [13,58,59]. In brief, per well, 500 ng purified plasmid and 1.5 µL FuGene HD (Promega, Mannheim, Germany) transfection reagent were incubated in 25 µL OptiMEM I medium for 15 min at room temperature and added to the prepared cells. After 4–6 h, the transfection medium was exchanged to the penicillin- and streptomycin-free culture medium.

For the quantification of the expression behavior, the psDEST47 constructs were used, which resulted in C-terminally labeled GFP fusion proteins. The cells were fixed 24 h after transfection with 3.7% formaldehyde. The nuclei of the cells were stained with Hoechst 33342 (1 µg/mL; Sigma Aldrich, St. Louis, MO, USA) and the cell membranes were stained with Alexa 555-conjugated Wheat Germ Agglutinin (5 µg/mL; Molecular Probes, Eugene, OR, USA) to improve the visibility of the cell-cell contact regions. The cells were imaged with a confocal Nikon Eclipse TE2000-E C1 laser scanning microscope (Nikon, Düsseldorf, Germany) as described previously [13,58,59]. For each variant, at least five different transfections and coverslips were evaluated. Four images were taken of different regions of each coverslip.

To analyze the plaque area per cell pair, the micrographs were evaluated using FiJi [32]. The resulting plaque areas per cell pair of the concatemers were evaluated in comparison to the monomeric hCx46 and hCx26 by using a one-way ANOVA, followed by a Tukey test and are given as mean ± SEM.

4.4. Western Blot

HeLa cells were grown to about 80% confluence in a 100 mm diameter cell culture plate. The cells were transfected with the psDEST47 hCx46 or the psDEST47 hCx46-hCx46 plasmids and cultivated for further 24 h. For the transfection of a 100 mm diameter cell culture plate, 5 µg plasmid DNA and 15 µL FuGene HD were used (details are described above in Section 4.3). For the protein isolation, the cells were washed twice with ice-cold PBS and were removed from the culture plate with a cell scraper in presence of 1 mL ice-cold PBS. After a centrifugation step at $750\times g$ for 3 min at 4 °C, the pellet was resuspended in 50 μL RIPA buffer containing 25 mM Tris HCl pH 7.6, 150 mM NaCl, 1% nonidet P-40, 1% sodium desoxycholate, 0.1% SDS, freshly added 0.5% protease inhibitor cocktail (Roche, Waiblingen, Germany), 10 mM NaF, 1 mM PMSF, and 1 mM Na_3VO_4. After an incubation for 15 min on ice, a centrifugation at $14,000 \times g$ for 15 min at 4 °C was used to separate the protein solution from the cell debris. A Bradford assay (Sigma Aldrich) was used to determine the protein concentration of the supernatant using BSA as standard. $1 \times$ Laemmli buffer (13 mM Tris HCl, 10 mM DTT, 2% glycerol, 0.4% SDS, 0.002% Bromphenol Blue, pH 6.8) was added to the protein solution and incubated for 10 min at 70 °C. Next, 100 μg protein per lane were separated in a 5% SDS-polyacrylamide stacking gel and a 10% separation gel. The proteins were transferred to a nitrocellulose membrane using a semi-dry blot (transfer buffer: 25 mM Tris HCL, pH 8.3, 192 mM glycine, 0.1% SDS, and 20% methanol). Afterwards, the membrane was blocked with 5% non-fat dry milk powder in PBS containing 0.1% Tween 20 (PBS-T) for 2 h at room temperature. Anti-Cx46 antibody (sc-365394, Santa Cruz Biotechnology, Heidelberg, Germany) was diluted 1:1000 in PBS-T and applied to the membranes for an overnight incubation at 4 °C. After washing, the secondary anti-mouse antibody (A9044, Sigma Aldrich) was diluted 1:100,000 and applied for 1 h at room temperature. For the detection, the SuperSignal West chemiluminescent reagent (Thermo Fisher Scientific) was used. The blot was imaged with a CCD camera system (Intas Science Imaging, Göttingen, Germany). For the quantification, four independent replicates were analyzed by using the gel analyzer tool of the FiJi software [32]. Data are displayed normalized to the intensity of the hCx46 monomer.

4.5. Dye Transfer Experiments

To test the functionality of the formed gap junction channels, dye transfer experiments with Lucifer Yellow and 7-amino-4-methyl-3-coumarinylacetic acid (AMCA) in HeLa and N2A cells were performed, respectively. HeLa cells were prepared and transfected with the different psDEST47-plasmids as described above in Section 4.3. As control, mock transfected HeLa cells were used. For the dye transfer experiments with the N2A cells, the cells were transfected with the different pEF-I-GFP GX-plasmids. For control experiments, the N2A cells were transfected with the empty destination vector pEF-I-GFP GX. Coverslips with transfected cells were transferred to a perfusion chamber containing 400 μL of a bath medium consisting of (in mM) 140 NaCl, 5 KCl, 10 HEPES, 10 glucose, 1 $MgCl_2$, and 2 $CaCl_2$ at pH 7.4 and osmolarity (π) of 295 mosmol/L. The chamber was mounted on an inverted fluorescence microscope (Ti-E, Nikon GmbH, Duesseldorf, Germany) equipped with a Polychrome V monochromator (T.I.L.L. Photonics GmbH, Planegg, Germany), a CCD Orca-Flash 4.0 camera (Hamamatsu Photonics Deutschland GmbH, Herrsching, Germany), and the NIS-Elements AR 4.4 software (Nikon GmbH).

For the dye transfer experiments, a whole-cell patch-clamp configuration was established on one cell of a transfected cell pair using an EPC 10 USB double patch-clamp amplifier (HEKA Elektronik Dr. Schulze GmbH, Lambrecht/Pfalz, Germany) coupled to the PatchMaster 2.9 software (HEKA Elektronik Dr. Schulze GmbH). For the pipette filling solution used for the HeLa cells, 1 mg/mL Lucifer Yellow (LY) lithium salt (Biotium, Hayward, CA, USA) was diluted in a pipette medium containing (in mM) 145 K gluconate, 5 KCl, 10 HEPES, 2.5 MgATP, 5 glucose, 0.5 Na_2ATP, 1 EGTA, and 0.2 $CaCl_2$ at pH 7.4 and π 295 mosmol/L. For the experiments with N2A cells, the LY was replaced by 1 mg/mL AMCA (Sigma Aldrich, St. Louis, MO, USA). The Polychrome V was used to excite the GFP-labeled connexin variants at 488 nm, LY at 410 nm, and AMCA at 350 nm. For each dye transfer experiment, micrographs of the GFP and LY or AMCA fluorescence were taken before the whole-cell configuration was established, during the experiment, and after 10 min with prior removal of the LY or AMCA containing pipette. For each variant, the degree of dye coupling was estimated as the ratio of the number of coupled pairs to the total number of tested pairs expressing the particular variant. The results are given as mean values $\pm$ SEM. The significance of the difference was evaluated by a one-way ANOVA and a post-hoc Tukey test (*** for $p \leq 0.001$, ** for $p \leq 0.01$, and * for $p \leq 0.05$).

4.6. Expression in Xenopus Oocytes

For the in vitro transcription, the mMESSAGE mMACHINE® T7 (Thermo Fisher Scientific, Waltham, MA, USA) and the PeaI-linearized (Thermo Fisher Scientific) psGEMHE vectors were used to generate the artificial cRNA. The cRNA was purified by a phenol/chloroform extraction and an isopropanol precipitation.

Xenopus laevis oocytes were harvested from an anaesthetized female frog. After mechanical disruption of the tissue, the oocytes were separated by an incubation (1 h at room temperature) in 190–240 U/mL collagenase type II (Worthington, Berlin, Germany) containing oocyte control medium composed of (in mM) 88 NaCl, 1 KCl, 10 Tris-HCl, and 0.82 $MgCl_2$ (pH 7.4 and π 180 mosmol/L). During the tissue digestion with collagenase, the tissue was shaken at 100 rpm. After the collagenase treatment, the oocytes were washed with oocyte control medium supplemented with 2 mM $CaCl_2$. The oocytes were stored and used for injection for up to three days after isolation.

Stage V and stage VI oocytes were injected with 23 nL of an aqueous solution containing 1 µg/µL connexin mRNA and 400 ng/µL antisense to the endogenous Cx38 (AS38) using the Nanoliter Injector (World Precision Instruments, Berlin, Germany). The antisense DNA (AS38) with the sequence C*T*GACTGCTCGTCTGTCCACAC*A*G* (* indicates phosphorothioate modification) was purchased from Microsynth AG (Balgach, Switzerland). The injected oocytes were incubated at 16 °C in the Ca^{2+}-containing oocyte medium and used for the measurement of single channels 18 to 48 h post injection.

4.7. Single Channel Recordings of the Connexons

For the recording of the single channels, the vitelline membrane of a connexin expressing oocyte was mechanically removed. The oocyte was incubated for at least 3 min in a stripping solution containing (in mM) 88 NaCl, 1 KCl, 10 Tris-HCl, 0.82 $MgCl_2$, 2 $CaCl_2$, and 200 D-mannitol (pH 7.4 and π 444 mosmol/L) to release the vitelline membrane from the oocyte membrane. The released vitelline membrane was removed using two Dumont no. 5 forceps (Manufactures D'Outils Dumont SA, Montignez, Switzerland) under a LEICA GZ4 binocular (Leica Mikrossysteme Vertrieb GmbH, Wetzlar, Germany). The stripped oocyte was transferred into a perfusion chamber containing 400 µL of a control bath solution consisting of (in mM) 88 Na gluconate, 1 K gluconate, 10 Tris, 2 Ca acetate, 0.82 Mg acetate, and 20 Cs acetate (pH 7.4 and π 220 mosmol/L) and mounted on an inverted fluorescence microscope described in Section 4.4. The patch-clamp experiments were performed using an EPC 10 USB double patch-clamp amplifier (HEKA Elektronik Dr. Schulze GmbH). The data were recorded with filter 1 (Bessel) at 10 kHz and filter 2 (I_Bessel) at 1 kHz. The patch pipettes were made from PG150T-7.5 glass capillaries (Clark Electromedical Instruments, Pangbourne, UK). Filled with the pipette filling solution composed of (in mM) 80 Na gluconate, 20 Cs acetate, and 10 HEPES (pH 7.4 and π 220 mosmol/L) the pipettes had an electrical resistance of about 5 MΩ. The reference electrode was filled with K gluconate solution to avoid problems related to Ag/AgCl junction [60].

Three minutes after the formation of a gigaseal, the inside-out patch-clamp configuration [61] was established. To measure the single channels, test voltage pulses between −70 mV and 60 mV were applied for 20 s in 10 mV steps. Between the voltage pulses, the membrane was clamped at 0 mV for 30 s. Thereafter, the bath solution was changed to a Ca^{2+}-free solution composed of (in mM) 88 Na gluconate, 1 K gluconate, 10 Tris, 0.82 Mg acetate, and 20 Cs acetate at pH 7.4 and π 220 mosmol/L. After application of the test voltages, the membrane was perfused with a Ca^{2+}-free solution supplied with 100 µM carbenoxolone (CBX) and the voltage pulses were applied again.

For the data analysis and the example curves, the data were filtered with the digital filter at 100 Hz in the FitMaster 2.90 software (HEKA Elektronik Dr. Schulze GmbH). The software was also used to generate an amplitude histogram, which was fitted using a multi Gaussian fit to calculate the single channel conductance states. The single channel open probability was estimated by using the single channel event detection tool of the FitMaster software. Simultaneous opening of channels was rarely observed. If it was observed this event was excluded in the calculation of the single channel open probability. For the calculation, only the +40 mV to +60 mV traces, which had an unambiguous signal-to-noise ratio, were used. A measurement time of at least 4 min for at least four injected oocytes of each variant was used for the estimation. For the comparison of the data a one-way ANOVA followed by a Tukey test was used.

4.8. Dye Uptake through Hemichannels

The hemichannel activity was analyzed by measuring the ethidium bromide (Etd) uptake slightly modified from Schalper et al. (2008) [62]. A day before the experiment, subconfluent HeLa cells grown on collagen I-coated coverslips were transfected with the pEF-I-GFP variants. For the control group experiments, the empty destination plasmid pEF-I-GFP GX was used to transfect the cells. A coverslip with transfected cells was placed in a perfusion chamber with a chamber volume of approximately 400 µL mounted on an inverse Nikon Ti-E fluorescence microscope, as described in Section 4.4. The ISMATEC REGIO ICC peristaltic pump (Cole-Parmer GmbH, Wertheim, Germany) controlled by the software ISMATEC® Pump Control (Cole-Parmer GmbH) allowed the constant exchange of the medium with a flow rate of 1 mL/min. Prior to the experiment, the GFP fluorescence of the transfected cells was used to define the regions of interest (ROIs) for the Etd uptake measurement. During the first 10 min of a 30-min long dye uptake experiment, the cells in the chamber were perfused with a prewarmed (37 °C) bath solution composed of (in mM) 121 NaCl, 5.4 KCl, 25 HEPES, 0.8 $MgCl_2$, 5.5 glucose, 6 $NaHCO_3$, 2 $CaCl_2$, and 5 µM ethidium bromide (pH 7.4 and π 296 mosmol/L). After 10 min, the medium was exchanged for additional 10 min to a Ca^{2+}- and Mg^{2+}-free solution, which was consisting of (in mM) 121 NaCl, 5.4 KCl, 25 HEPES, 5.5 glucose, 6 $NaHCO_3$, and 5 µM ethidium bromide (pH 7.4 and π 295 mosmol/L). In the last 10 min of a dye uptake experiment, 1 mM La^{3+} was added to the Ca^{2+}/Mg^{2+}-free solution. Before starting an experiment, regions of interest (ROIs) were selected in a fluorescent micrograph of the cells taken by an Orca flash 4.0 CCD camera (Hamamatsu Photonics Germany, Herrsching am Ammersee, Germany). During the entire experiment, fluorescent images were taken every 15 s with an exposure time of 700 ms. The images were used to assess the changes of the fluorescence intensity of the ROIs during an experiment. For the recording of the images and the measurement of fluorescence intensity in the ROIs, the NIS-Elements AR 4.4 software (Nikon GmbH) was used. The dye uptake rate (Etd AU/min) was calculated with OriginPro 2017 (OriginLab Corporation, Northampton, MA, USA) from minute 4–9, 14–19, and 24–29, respectively (stationary rate). The results are given as mean values ± SEM. The significance of the difference was evaluated by a one-way ANOVA and a post-hoc Tukey test (*** for $p \leq 0.001$, ** for $p \leq 0.01$, and * for $p \leq 0.05$). For the comparison between the groups a two-way ANOVA followed by a Tukey test was used.

4.9. Dual Whole-Cell Patch-Clamp Experiments

For the dual whole-cell patch-clamp experiments, approximately 3×10^4 N2A cells were cultured on a collagen I-coated coverslip in a well of a 24-well plate and were transfected as described above with the pEF-I-GFP variants, which resulted in the separate expression of the different connexin-variants and GFP as the reporter. For the control experiments, the N2A cells were transfected with the empty pEF-I-GFP destination vector. The coverslip with the transfected cells was transferred into a perfusion chamber filled with 400 µL of a bath solution composed of (in mM) 121 NaCl, 5.4 KCl, 25 HEPES, 0.8 $MgCl_2$, 5.5 glucose, 6 $NaHCO_3$, 2 $CaCl_2$ and mounted on the inverse Nikon Ti-E fluorescence microscope described in Section 4.4. The patch pipettes were filled with a pipette solution containing (in mM) 125 K gluconate, 15 CsCl, 0.2 $CaCl_2$, 2.5 $MgCl_2$, 1 MgATP, 5 glucose, 0.5 EGTA, 4 Na_2ATP, 0.1 cAMP, and 10 HEPES (pH 7.4 and π 295 mosmol/L). The patch pipettes were made from 40A502 glass capillaries (Kimble Chase Life Science and Research Products, Rockwood, TN, USA). Filled with the pipette solution the pipettes had an electrical resistance of 2–5 MΩ.

Dual whole-cell patch-clamp experiments were performed with the EPC 10 USB double patch-clamp amplifier described in Section 4.6. After establishing a whole-cell configuration on both cells of a cell pair, both cells were clamped at −40 mV. For the measurements, one cell of the cell pair (cell 1) was alternatingly stepped from −120 mV to +60 mV (V_1) for a duration of 250 ms, while the junctional currents were recorded in the other cell (cell 2), which was maintained at −40 mV (V_2). The transjunctional voltage gradient ($V_j = V_2 − V_1$) was calculated.

5. Conclusions

In summary, the present paper shows that the expression of concatenated connexins leads to a reduced plaque area between cells. However, concatenation of connexins was compatible with trafficking of the hemichannels to the membrane and the formation of functional gap junction hemichannels in the cell membrane and gap junction channels between cells. It could be used to generate hemichannels and gap junction channels with a determined stoichiometry. Because of the linker between the connexins, the properties of the formed hemichannels and gap junction channels (e.g., single channel conductance) do not represent the properties of hetero-oligomerized channels. However, should the removal of the linker be successful, this method could be used to analyze the electrical and metabolic selectivity of such channels and the physiological consequences for a tissue.

Supplementary Materials: The following are available online at http://www.mdpi.com/1422-0067/19/9/2742/s1.

Author Contributions: Conceptualization, P.S. and A.N.; Data curation, P.S.; Formal analysis, P.S.; Funding acquisition, A.N.; Investigation, P.S., D.H. and Y.S.; Methodology, P.S. and A.N.; Project administration, A.N.; Supervision, A.N.; Visualization, P.S. and N.D.; Writing—original draft, P.S. and A.N.; Writing—review & editing, P.S. and A.N. The paper was proofread by all authors.

Funding: This research was partly funded by TransRegio TR37. The publication of this article was funded by the Open Access Fund of the Leibniz Universität Hannover.

Acknowledgments: Nadine Dilger was partly supported by the DFG project Elektrodenoptimierung für Neuroprothesen NG 4/10-1. We thank Viviana Berthoud for the hCx46 clone.

Conflicts of Interest: The authors declare no conflict of interest.

Abbreviations

ACh	acetylcholine
AMCA	7-amino-4-methyl-3-coumarinylacetic acid
ANOVA	analysis of variance
AS38	antisence38 oligonucleotide, against *Xenopus laevis* Cx38
CBX	carbenoxolone
CL	cytoplasmatic loop
Cx	connexin
EL1	first extracellular loop
EL2	second extracellular loop
ER	endoplasmatic reticulum
Etd	ethidium bromide
GABA	γ-aminobutyric acid
GJC	gap junction channel
hCx	human connexin
LY	lucifer Yellow
N2A	neuro-2A, mouse neuroblastoma cell line
ns	not significant
ROIs	regions of interest
SEM	standard error of the mean
TM	transmembrane domain

References

1. Söhl, G.; Willecke, K. Gap junctions and the connexin protein family. *Cardiovasc. Res.* **2004**, *62*, 228–232. [CrossRef] [PubMed]
2. Desplantez, T.; Grikscheit, K.; Thomas, N.M.; Peters, N.S.; Severs, N.J.; Dupont, E. Relating specific connexin co-expression ratio to connexon composition and gap junction function. *J. Mol. Cell. Cardiol.* **2015**, *89*, 195–202. [CrossRef] [PubMed]
3. Oh, S.; Bargiello, T.A. Voltage regulation of connexin channel conductance. *Yonsei Med. J.* **2015**, *56*, 1–15. [CrossRef] [PubMed]
4. White, T.W.; Paul, D.L.; Goodenough, D.A.; Bruzzone, R. Functional analysis of selective interactions among rodent connexins. *Mol. Biol. Cell* **1995**, *6*, 459–470. [CrossRef] [PubMed]
5. Bai, D.; Wang, A.H. Extracellular domains play different roles in gap junction formation and docking compatibility. *Biochem. J.* **2014**, *458*, 1–10. [CrossRef] [PubMed]
6. Neijssen, J.; Herberts, C.; Drijfhout, J.W.; Reits, E.; Janssen, L.; Neefjes, J. Cross-presentation by intercellular peptide transfer through gap junctions. *Nature* **2005**, *434*, 83–88. [CrossRef] [PubMed]
7. Bedner, P.; Niessen, H.; Odermatt, B.; Kretz, M.; Willecke, K.; Harz, H. Selective permeability of different connexin channels to the second messenger cyclic AMP. *J. Biol. Chem.* **2006**, *281*, 6673–6681. [CrossRef] [PubMed]
8. Bennett, M.V.M.; Verselis, V.K.V. Biophysics of gap junctions. *Semin. Cell Biol.* **1992**, *3*, 29–47. [CrossRef]
9. Niessen, H.; Harz, H.; Bedner, P.; Krämer, K.; Willecke, K. Selective permeability of different connexin channels to the second messenger inositol 1,4,5-trisphosphate. *J. Cell Sci.* **2000**, *113*, 1365–1372. [PubMed]
10. Koval, M.; Molina, S.A.; Burt, J.M. Mix and match: Investigating heteromeric and heterotypic gap junction channels in model systems and native tissues. *FBBS Lett.* **2014**, *588*, 1193–1204. [CrossRef] [PubMed]
11. Maeda, S.; Nakagawa, S.; Suga, M.; Yamashita, E.; Oshima, A.; Fujiyoshi, Y.; Tsukihara, T. Structure of the connexin 26 gap junction channel at 3.5 A resolution. *Nature* **2009**, *458*, 597–602. [CrossRef] [PubMed]
12. Nakagawa, S.; Gong, X.-Q.; Maeda, S.; Dong, Y.; Misumi, Y.; Tsukihara, T.; Bai, D. Asparagine 175 of connexin32 is a critical residue for docking and forming functional heterotypic gap junction channels with connexin26. *J. Biol. Chem.* **2011**, *286*, 19672–19681. [CrossRef] [PubMed]
13. Schadzek, P.; Schlingmann, B.; Schaarschmidt, F.; Lindner, J.; Koval, M.; Heisterkamp, A.; Preller, M.; Ngezahayo, A. The cataract related mutation N188T in human connexin46 (hCx46) revealed a critical role for residue N188 in the docking process of gap junction channels. *Biochim. Biophys. Acta* **2016**, *1858*, 57–66. [CrossRef] [PubMed]
14. Laird, D.W. The gap junction proteome and its relationship to disease. *Trends Cell Biol.* **2009**, *20*, 92–101. [CrossRef] [PubMed]
15. Diez, J.A.; Ahmad, S.; Evans, W.H. Assembly of heteromeric connexons in guinea-pig liver en route to the Golgi apparatus, plasma membrane and gap junctions. *Eur. J. Biochem.* **1999**, *262*, 142–148. [CrossRef] [PubMed]
16. Koval, M.; Harley, J.E.; Hick, E.; Steinberg, T.H. Connexin46 is retained as monomers in a trans-Golgi compartment of osteoblastic cells. *J. Cell Biol.* **1997**, *137*, 847–857. [CrossRef] [PubMed]
17. Musil, L.S.; Goodenough, D.A. Multisubunit assembly of an integral plasma membrane channel protein, gap junction connexin43, occurs after exit from the ER. *Cell* **1993**, *74*, 1065–1077. [CrossRef]
18. Maeda, S.; Tsukihara, T. Structure of the gap junction channel and its implications for its biological functions. *Cell. Mol. Life Sci.* **2011**, *68*, 1115–1129. [CrossRef] [PubMed]
19. Schadzek, P.; Schlingmann, B.; Schaarschmidt, F.; Lindner, J.; Koval, M.; Heisterkamp, A.; Ngezahayo, A.; Preller, M. Data of the molecular dynamics simulations of mutations in the human connexin46 docking interface. *Data Brief* **2016**, *7*, 93–99. [CrossRef] [PubMed]
20. Bai, D.; Yue, B.; Aoyama, H. Crucial motifs and residues in the extracellular loops influence the formation and specificity of connexin docking. *Biochim. Biophys. Acta* **2018**, *1860*, 9–21. [CrossRef] [PubMed]
21. Karademir, L.B.; Aoyama, H.; Yue, B.; Chen, H.; Bai, D. Engineered Cx26 variants established functional heterotypic Cx26/Cx43 and Cx26/Cx40 gap junction channels. *Biochem. J.* **2016**, *473*, 1391–1403. [CrossRef] [PubMed]

22. Jara, O.; Acuña, R.; García, I.E.; Maripillán, J.; Figueroa, V.; Sáez, J.C.; Araya-Secchi, R.; Lagos, C.F.; Perez-Acle, T.; Berthoud, V.M.; et al. Critical role of the first transmembrane domain of Cx26 in regulating oligomerization and function. *Mol. Biol. Cell* **2012**, *23*, 3299–3311. [CrossRef] [PubMed]

23. Martínez, A.D.; Maripillán, J.; Acuña, R.; Minogue, P.J.; Berthoud, V.M.; Beyer, E.C. Different domains are critical for oligomerization compatibility of different connexins. *Biochem. J.* **2011**, *436*, 35–43. [CrossRef] [PubMed]

24. Das, S.; Smith, T.D.; Sarma, J.D.; Ritzenthaler, J.D.; Maza, J.; Kaplan, B.E.; Cunningham, L.A.; Suaud, L.; Hubbard, M.J.; Rubenstein, R.C.; et al. ERp29 restricts Connexin43 oligomerization in the endoplasmic reticulum. *Mol. Biol. Cell* **2009**, *20*, 2593–2604. [CrossRef] [PubMed]

25. Ahring, P.K.; Liao, V.W.Y.; Balle, T. Concatenated nicotinic acetylcholine receptors: A gift or a curse? *J. Gen. Physiol.* **2018**, *150*, 453–473. [CrossRef] [PubMed]

26. Baumann, S.W.; Baur, R.; Sigel, E. Subunit arrangement of gamma-aminobutyric acid type A receptors. *J. Biol. Chem.* **2001**, *276*, 36275–36280. [CrossRef] [PubMed]

27. Sigel, E.; Kaur, K.H.; Lüscher, B.P.; Baur, R. Use of concatamers to study GABAA receptor architecture and function: Application to delta-subunit-containing receptors and possible pitfalls. *Biochem. Soc. Trans.* **2009**, *37*, 1338–1342. [CrossRef] [PubMed]

28. Stoop, R.; Thomas, S.; Rassendren, F.; Kawashima, E.; Buell, G.; Surprenant, A.; North, R.A. Contribution of individual subunits to the multimeric P2X(2) receptor: Estimates based on methanethiosulfonate block at T336C. *Mol. Pharmacol.* **1999**, *56*, 973–981. [CrossRef] [PubMed]

29. Isakson, B.E.; Olsen, C.E.; Boitano, S. Laminin-332 alters connexin profile, dye coupling and intercellular Ca^{2+} waves in ciliated tracheal epithelial cells. *Respir. Res.* **2006**, *7*, 105. [CrossRef] [PubMed]

30. Oviedo-Orta, E.; Kwak, B.R.; Evans, W.H. *Connexin Cell Communication Channels*; CRC Press: Boca Raton, FL, USA, 2013.

31. Chiu, Y.-H.; Jin, X.; Medina, C.B.; Leonhardt, S.A.; Kiessling, V.; Bennett, B.C.; Shu, S.; Tamm, L.K.; Yeager, M.; Ravichandran, K.S.; et al. A quantized mechanism for activation of pannexin channels. *Nat. Commun.* **2017**, *8*, 14324. [CrossRef] [PubMed]

32. Schindelin, J.; Arganda-Carreras, I.; Frise, E.; Kaynig, V.; Longair, M.; Pietzsch, T.; Preibisch, S.; Rueden, C.; Saalfeld, S.; Schmid, B.; et al. Fiji: An open-source platform for biological-image analysis. *Nat. Methods* **2012**, *9*, 676–682. [CrossRef] [PubMed]

33. Paul, D.L.; Ebihara, L.; Takemoto, L.J.; Swenson, K.I.; Goodenough, D.A. Connexin46, a novel lens gap junction protein, induces voltage-gated currents in nonjunctional plasma membrane of Xenopus oocytes. *J. Cell Biol.* **1991**, *115*, 1077–1089. [CrossRef] [PubMed]

34. Ngezahayo, A.; Zeilinger, C.; Todt, I.; Marten, I.; Kolb, H.A. Inactivation of expressed and conducting rCx46 hemichannels by phosphorylation. *Pflugers Arch.* **1998**, *436*, 627–629. [CrossRef] [PubMed]

35. Walter, W.J.; Zeilinger, C.; Bintig, W.; Kolb, H.-A.; Ngezahayo, A. Phosphorylation in the C-terminus of the rat connexin46 (rCx46) and regulation of the conducting activity of the formed connexons. *J. Bioenerg. Biomembr.* **2008**, *40*, 397–405. [CrossRef] [PubMed]

36. Neyton, J.; Trautmann, A. Single-channel currents of an intercellular junction. *Nature* **1985**, *317*, 331–335. [CrossRef] [PubMed]

37. Ngezahayo, A.; Altmann, B.; Kolb, H.A. Regulation of ion fluxes, cell volume and gap junctional coupling by cGMP in GFSHR-17 granulosa cells. *J. Membr. Biol.* **2003**, *194*, 165–176. [CrossRef] [PubMed]

38. Bao, L.; Sachs, F.; Dahl, G. Connexins are mechanosensitive. *Am. J. Physiol. Cell Physiol.* **2004**, *287*, 1389–1395. [CrossRef] [PubMed]

39. Gaßmann, O.; Kreir, M.; Ambrosi, C.; Pranskevich, J.; Oshima, A.; Röling, C.; Sosinsky, G.; Fertig, N.; Steinem, C. The M34A mutant of Connexin26 reveals active conductance states in pore-suspending membranes. *J. Struct. Biol.* **2009**, *168*, 168–176. [CrossRef] [PubMed]

40. Hopperstad, M.G.; Srinivas, M.; Spray, D.C. Properties of gap junction channels formed by Cx46 alone and in combination with Cx50. *Biophys. J.* **2000**, *79*, 1954–1966. [CrossRef]

41. Oh, S.; Rubin, J.B.; Bennett, M.V.; Verselis, V.K.; Bargiello, T.A. Molecular determinants of electrical rectification of single channel conductance in gap junctions formed by connexins 26 and 32. *J. Gen. Physiol.* **1999**, *114*, 339–364. [CrossRef] [PubMed]

42. Contreras, J.E.; Sáez, J.C.; Bukauskas, F.F.; Bennett, M.V.L. Gating and regulation of connexin 43 (Cx43) hemichannels. *Proc. Natl. Acad. Sci. USA* **2003**, *100*, 11388–11393. [CrossRef] [PubMed]

43. Laird, D.W.; Jordan, K.; Thomas, T.; Qin, H.; Fistouris, P.; Shao, Q. Comparative analysis and application of fluorescent protein-tagged connexins. *Microsc. Res. Tech.* **2001**, *52*, 263–272. [CrossRef]

44. Sorgen, P.L.; Duffy, H.S.; Sahoo, P.; Coombs, W.; Delmar, M.; Spray, D.C. Structural changes in the carboxyl terminus of the gap junction protein connexin43 indicates signaling between binding domains for c-Src and zonula occludens-1. *J. Biol. Chem.* **2004**, *279*, 54695–54701. [CrossRef] [PubMed]

45. Leithe, E.; Mesnil, M.; Aasen, T. The connexin 43 C-terminus: A tail of many tales. *Biochim. Biophys. Acta* **2018**, *1860*, 48–64. [CrossRef] [PubMed]

46. Ebihara, L.; Tong, J.-J.; Vertel, B.; White, T.W.; Chen, T.-L. Properties of connexin 46 hemichannels in dissociated lens fiber cells. *Investig. Ophthalmol. Vis. Sci.* **2011**, *52*, 882–889. [CrossRef] [PubMed]

47. Trexler, E.B.; Bennett, M.V.; Bargiello, T.A.; Verselis, V.K. Voltage gating and permeation in a gap junction hemichannel. *Proc. Natl. Acad. Sci. USA* **1996**, *93*, 5836–5841. [CrossRef] [PubMed]

48. Trexler, E.B.; Bukauskas, F.F.; Kronengold, J.; Bargiello, T.A.; Verselis, V.K. The first extracellular loop domain is a major determinant of charge selectivity in connexin46 channels. *Biophys. J.* **2000**, *79*, 3036–3051. [CrossRef]

49. Hu, X.; Dahl, G. Exchange of conductance and gating properties between gap junction hemichannels. *FEBS Lett.* **1999**, *451*, 113–117. [CrossRef]

50. Srinivas, M.; Kronengold, J.; Bukauskas, F.F.; Bargiello, T.A.; Verselis, V.K. Correlative studies of gating in Cx46 and Cx50 hemichannels and gap junction channels. *Biophys. J.* **2005**, *88*, 1725–1739. [CrossRef] [PubMed]

51. Mese, G.; Sellitto, C.; Li, L.; Wang, H.-Z.; Valiunas, V.; Richard, G.; Brink, P.R.; White, T.W. The Cx26-G45E mutation displays increased hemichannel activity in a mouse model of the lethal form of keratitis-ichthyosis-deafness syndrome. *Mol. Biol. Cell* **2011**, *22*, 4776–4786. [CrossRef] [PubMed]

52. Sánchez, H.A.; Mese, G.; Srinivas, M.; White, T.W.; Verselis, V.K. Differentially altered Ca^{2+} regulation and Ca^{2+} permeability in Cx26 hemichannels formed by the A40V and G45E mutations that cause keratitis ichthyosis deafness syndrome. *J. Gen. Physiol.* **2010**, *136*, 47–62. [CrossRef] [PubMed]

53. Slavi, N.; Rubinos, C.; Li, L.; Sellitto, C.; White, T.W.; Mathias, R.; Srinivas, M. Cx46 Gap Junctions Provide a Pathway for the Delivery of Glutathione to the Lens Nucleus. *J. Biol. Chem.* **2014**, *289*, 32694–32702. [CrossRef] [PubMed]

54. Suchyna, T.M.; Nitsche, J.M.; Chilton, M.; Harris, A.L.; Veenstra, R.D.; Nicholson, B.J. Different ionic selectivities for connexins 26 and 32 produce rectifying gap junction channels. *Biophys. J.* **1999**, *77*, 2968–2987. [CrossRef]

55. Slavi, N.; Wang, Z.; Harvey, L.; Schey, K.L.; Srinivas, M. Identification and Functional Assessment of Age-Dependent Truncations to Cx46 and Cx50 in the Human Lens. *Investig. Ophthalmol. Vis. Sci.* **2016**, *57*, 5714–5722. [CrossRef] [PubMed]

56. Gubbels, S.P.; Woessner, D.W.; Mitchell, J.C.; Ricci, A.J.; Brigande, J.V. Functional auditory hair cells produced in the mammalian cochlea by in utero gene transfer. *Nature* **2008**, *455*, 537–541. [CrossRef] [PubMed]

57. Dieriks, B.; Van Oostveldt, P. Spatiotemporal behavior of nuclear cyclophilin B indicates a role in RNA transcription. *Int. J. Mol. Med.* **2012**, *29*, 1031–1038. [CrossRef] [PubMed]

58. Schlingmann, B.; Schadzek, P.; Busko, S.; Heisterkamp, A.; Ngezahayo, A. Cataract-associated D3Y mutation of human connexin46 (hCx46) increases the dye coupling of gap junction channels and suppresses the voltage sensitivity of hemichannels. *J. Bioenerg. Biomembr.* **2012**, *44*, 607–614. [CrossRef] [PubMed]

59. Schlingmann, B.; Schadzek, P.; Hemmerling, F.; Schaarschmidt, F.; Heisterkamp, A.; Ngezahayo, A. The role of the C-terminus in functional expression and internalization of rat connexin46 (rCx46). *J. Bioenerg. Biomembr.* **2013**, *45*, 59–70. [CrossRef] [PubMed]

60. Raynauld, J.P.; Laviolette, J.R. The silver-silver chloride electrode: A possible generator of offset voltages and currents. *J. Neurosci. Methods* **1987**, *19*, 249–255. [CrossRef]

61. Hamill, O.P.; Marty, A.; Neher, E.; Sakmann, B.; Sigworth, F.J. Improved patch-clamp techniques for high-resolution current recording from cells and cell-free membrane patches. *Pflugers Arch.* **1981**, *391*, 85–100. [CrossRef] [PubMed]

62. Schalper, K.A.; Palacios-Prado, N.; Retamal, M.A.; Shoji, K.F.; Martínez, A.D.; Sáez, J.C. Connexin hemichannel composition determines the FGF-1-induced membrane permeability and free [Ca^{2+}]i responses. *Mol. Biol. Cell* **2008**, *19*, 3501–3513. [CrossRef] [PubMed]

International Journal of
Molecular Sciences

MDPI

Review

Connexins and Pannexins in Vascular Function and Disease

Filippo Molica [1], Xavier F. Figueroa [2], Brenda R. Kwak [1,*,†], Brant E. Isakson [3,4,†] and Jonathan M. Gibbins [5,†]

1   Department of Pathology and Immunology, University of Geneva, CH-1211 Geneva, Switzerland; filippo.Molica@unige.ch
2   Departamento de Fisiología, Facultad de Ciencias Biológicas, Pontifica Universidad Católica de Chile, Santiago 8330025, Chile; xfigueroa@bio.puc.cl
3   Robert M. Berne Cardiovascular Research Center, University of Virginia School of Medicine, Charlottesville, VA 22908, USA; bei6n@virginia.edu
4   Department of Molecular Physiology and Biophysics, University of Virginia School of Medicine, Charlottesville, VA 22908, USA
5   Institute for Cardiovascular & Metabolic Research, School of Biological Sciences, Harborne Building, University of Reading, Reading RG6 6AS, UK; j.m.gibbins@reading.ac.uk
*   Correspondence: brenda.kwakchanson@unige.ch; Tel.: +41-22-379-5737
†   These authors contributed equally to this work.

Received: 27 April 2018; Accepted: 31 May 2018; Published: 5 June 2018

Abstract: Connexins (Cxs) and pannexins (Panxs) are ubiquitous membrane channel forming proteins that are critically involved in many aspects of vascular physiology and pathology. The permeation of ions and small metabolites through Panx channels, Cx hemichannels and gap junction channels confers a crucial role to these proteins in intercellular communication and in maintaining tissue homeostasis. This review provides an overview of current knowledge with respect to the pathophysiological role of these channels in large arteries, the microcirculation, veins, the lymphatic system and platelet function. The essential nature of these membrane proteins in vascular homeostasis is further emphasized by the pathologies that are linked to mutations and polymorphisms in Cx and Panx genes.

Keywords: connexin; pannexin; vascular physiology; vascular disease

1. Introduction

The cardiovascular system consists of the heart pumping the blood to a closed circuit of interconnected blood vessels, allowing for the indispensable and constant supply of O_2 and vital nutriments to every single tissue throughout the human body. The blood contains components, such as coagulation factors and platelets, that are essential to keeping the cardiovascular circuit closed after injury by initiating haemostasis and formation of a platelet clot. The systemic circulation is composed of large elastic arteries such as the aorta, serving as high-pressure conduits for the blood to smaller muscular arteries and arterioles. Arterioles are resistance arteries controlling blood flow into capillary beds by their high vasodilatory and vasoconstrictive ability. In the capillary beds, the actual exchange of O_2/CO_2, nutriments, catabolites and fluid takes place between the blood and the surrounding tissue. The blood returns to the heart via venules and veins. The excess of interstitial fluid returns to the systemic circulation via lymphatics, a blind-ended system of lymphatic capillaries converging into collecting vessels and ending into the subclavian vein. Much evidence has demonstrated an important role for connexins (Cxs) and pannexins (Panxs) in many aspects of vascular physiology and pathology. In this review, we will focus on the role of Cxs and Panxs in the physiology/pathophysiology of the vascular system by systematically following the route of the blood from the left ventricle through the systemic circulation to its way back to heart.

2. Connexins and Pannexins

Cxs belong to a family of 20 to 21 proteins expressed in a wide variety of tissues [1]. Cx genes are separated into 5 subfamilies according to their sequence homologies [2]; most cardiovascular Cxs are found in the α subfamily (for instance, *GJA4*). The names of Cx proteins, on the other hand, are determined by their specific molecular weight in kDa (for instance, Cx37). Structurally, Cxs comprise 4 α-helical transmembrane domains (TM1–TM4) and two extracellular loops (EL1 and EL2) that are highly conserved among the family members. Substantial differences among Cxs, both in length and composition, are found in their cytoplasmic amino-terminal (NT) and carboxy-terminal (CT) parts, as well as in the intracellular loop (IL). The synthesis of Cxs occurs in the endoplasmic reticulum (ER) and their oligomerization in the ER/Golgi or trans-Golgi network results in the formation of hexameric connexons [3,4]. Then, connexons traffic to the plasma membrane along microtubules. When the membranes of two cells are in close proximity, connexons from one cell can connect with their corresponding parts in the adjacent cell and form gap junction channels, which permit the intercellular exchange of ions and metabolites up to ~1kDa. Connexons are normally closed but may operate in a pathological setting as hemi-channels enabling the transmembrane passage of Ca^{2+}, ATP and glutamate for instance [5,6]. Cx channel gating is critically regulated by a number of factors, including voltage, pH and Ca^{2+} and post-translational modifications such as phosphorylation [7,8]. The "connexin interactome", a protein interacting network with the Cx as central mediator [9,10], has been receiving increasing attention in recent years. For example, an interacting complex of gap junctions, desmosomes and Na^+ channels that cooperate to control excitability, electrical coupling and intercellular adhesion are found at intercalated discs in the heart [10,11]. The plethora of diseases associated with mutations and polymorphisms in Cx genes further underlines the crucial role of these structures in tissue homeostasis [12,13].

Pannexins (Panxs) represent a smaller family of 3 transmembrane proteins (Panx1–3) exhibiting a topology similar to Cxs but no sequence homology [14–16]. The glycosylation of specific sites in the ELs of Panx1 and Panx3 likely prevents docking of pannexons [16,17], and it is presumed that pannexons act as single-membrane channels connecting the cytoplasm to the extracellular compartment. While Panx2 and Panx3 display a rather limited expression pattern (central nervous system for Panx2 and bones and skin for Panx3), Panx1 shows a ubiquitous expression pattern and is thus also found in vascular cells. Similar to Cx hemi-channels, pannexons serve as "communication channels" by permitting the release of small molecules, for instance purines, that subsequently signal via the activation of membrane receptors in neighboring cells or even at distance. An important difference between Cx hemi-channels and Panx channels is that the latter can be opened at physiological membrane potential and physiological intra- and extra-cellular Ca^{2+} concentration by, for example, mechanical stretching or upon activation of purinergic P2 receptors [18]. Instead, Cx hemi-channels only become functional under conditions associated with pathologies such as hypoxia or ischemia and will be only briefly mentioned in this review. Excellent reviews on Cx hemi-channels have been published recently [19–21].

3. Role of Cxs and Panxs in Distributing Arteries and Atherosclerosis

The largest distributing arteries are elastic vessels, which allows them to receive a high and pulsatile pressure from the heart. The elastic properties of the distributing vessels further contribute to the so-called Windkessel effect, transforming a pulsatile flow at the entry into a constant flow at the level of the capillaries. The aorta, pulmonary trunk, carotids as well as the illiac and subclavian arteries are all examples of elastic arteries. As elastic arteries have such fundamental roles in the vascular physiology, any pathology affecting the function of these vessels by inducing a stiffening of their wall may exert dramatic effects on the supply of vital substances to organs. Atherosclerosis principally impacts on large and medium-sized arteries and is the leading cause of mortality worldwide [22]. In brief, the pathogenesis of the disease initiates with the dysfunction of endothelial cells (ECs) characterized by expression of adhesion molecules, secretion of chemokines and increased permeability to low-density lipoproteins (LDL), which will subsequently accumulate in the sub-endothelial space where they get oxidized. The expression of adhesion molecules and secretion of chemokines promotes the entry of inflammatory cells such as monocytes, T lymphocytes and neutrophils into the intimal layer [23,24]. After their infiltration into the intima, monocytes differentiate into macrophages that will take up oxidized LDL and convert it into foam cells. Inflammatory cells secrete metalloproteinases that degrade extracellular matrix (ECM), as well as growth factors stimulating the proliferation and migration of smooth muscle cells (SMCs) from the media to the intima. Intimal SMCs synthesize collagen and their further proliferation eventually leads to the formation of a fibrous cap segregating the necrotic core of the plaque from the luminal blood flow. Plaques with a large necrotic core, a thin fibrous cap, many inflammatory cells and a few SMCs display increased propensity to rupture [25]. Upon rupture, ECM and tissue factor (TF) present in atherosclerotic lesions are exposed to the bloodstream, which initiates a coagulation cascade leading to the formation of a fibrin monolayer covering the site of injury [26,27]. In parallel, the activation of platelet receptors by atherosclerotic plaque components leads to platelet activation and aggregation [26]. If platelet aggregation is not limited, thrombus formation may compromise the arterial lumen and provoke acute ischemic events such as myocardial infarction and stroke.

3.1. Connexins and Atherosclerotic Disease

The integrity of the endothelial barrier is warranted by various types of endothelial junctions including gap junctions [28]. During atherosclerotic plaque development pro-inflammatory molecules induce a progressive deterioration of EC junctions and an increase in endothelial permeability. ECs are, for instance, responsive to TNF-α, which induces the expression of adhesion molecules and inflammatory cell recruitment. It has been shown that treating ECs with TNF-α dampens the expression of some Cxs, in particular Cx37 and Cx40, suggesting a possible implication of Cxs in the pathogenesis of atherosclerosis [29]. Moreover, increased vascular permeability is associated with an elevation of Cx43 expression in ECs [30,31]. The first support for the hypothesis that Cxs may participate in the development of atherosclerotic disease came from studies analyzing atherosclerotic lesions at different disease stages in specimen of human, rabbit or mouse origin. In summary, it has been reported that Cx43 was generally absent in ECs of large arteries, but that its expression was induced in ECs at the shoulder region of advanced atherosclerotic plaques, a localization known to experience disturbed blood flow [32]. In addition, high expression of Cx43 was found in macrophages and SMCs of young atherosclerotic lesions, whereas Cx43 levels were downregulated in SMCs of more mature plaques [32–35]. Interestingly, the oxidized phospholipid derivative 1-palmitoyl-2-(5′-oxo-valeroyl)-*sn*-glycero-3-phosphocholine (POVPC) has been shown to decrease Cx43 expression in SMCs, to increase its phosphorylation, and to promote SMC proliferation in vitro and in vivo in a mouse model of atherosclerosis [36]. Besides Cx43, the expression patterns of Cx37 and Cx40 have also been reported to be affected during atherosclerotic plaque development in humans and in mice [32]. In fact, Cx40 and Cx37 expression was abolished in ECs covering advanced plaques, while Cx37 levels were increased in foam cells. Moreover, long-term hypercholesterolemia in mice decreased Cx37 and Cx40 expression in aortic ECs. Interestingly, this outcome could be reversed exclusively for Cx37 by a one-week treatment with simvastatin, a well-known lipid-lowering drug [37]. Collectively, these observations support the idea that Cxs expression or their post-translational modifications might evolve in atherosclerotic plaques over time, depending on the stage of the lesion, and might thus affect atherogenesis.

As the ubiquitous deletion of Cx43 is lethal [38], *Cx43*$^{+/-}$ mice were crossed with atherosclerosis-prone LDL receptor-deficient (*Ldlr*$^{-/-}$) mice and fed a high cholesterol diet to study atheroma formation. These initial studies revealed that Cx43 has an overall atherogenic effect, and that reducing Cx43 might be beneficial by both reducing plaque burden as well as stabilizing the lesions [39].

However, the exact scenario by which global reduction in Cx43 ultimately led to this dual benefit was unclear, due to Cx43 expression in multiple atheroma-associated cell types. To examine specifically the role of Cx43 in immune cells, $Ldlr^{-/-}$ mice were lethally irradiated and reconstituted with $Cx43^{+/+}$, $Cx43^{+/-}$ or $Cx43^{-/-}$ hematopoietic fetal liver cells [40]. Intriguingly, the progression of atherosclerosis was lower in $Cx43^{+/-}$ chimeras compared with $Cx43^{+/+}$ and $Cx43^{-/-}$ chimeras, and their plaques contained fewer neutrophils. It turned out that chemoattraction of neutrophils, which did not themselves express Cx43, was reduced in response to supernatant secreted by $Cx43^{+/-}$ macrophages in comparison with the ones of $Cx43^{+/+}$ and $Cx43^{-/-}$ macrophages. Thus, titration of Cx43 levels in macrophages might regulate their chemoattractant secretion, leading to reduced atherosclerosis [40]. Recently, it was shown that an upregulation of Cx43 expression in human umbilical vein ECs resulted in enhanced adhesion of monocytes via a mechanism involving increased vascular adhesion molecule-1 and intercellular cell adhesion-1. This effect was independent from the expression of other Cxs such as Cx37 and Cx40 [41].

In contrast to Cx43, Cx40 expression has been reported to protect against atherosclerosis in mice by synchronizing endothelial anti-inflammatory signaling thus inhibiting leukocyte recruitment to the atherosclerotic lesion [42]. Interestingly, Cx40 expression is induced in arterial ECs by high laminar shear stress, as normally observed in straight parts of arteries that are known to be protected from atherosclerosis [43]. IκBα, a member of a protein complex inhibiting the activation of the transcription factor NFκB, was recently identified as a binding partner of Cx40-CT. The Cx40 interactome may be relevant for the control of NFκB activation in arterial ECs and the initiation of atherogenesis [43].

Deletion of Cx37 has been shown to promote atheroma formation in atherosclerosis-susceptible apolipoprotein E-deficient ($Apoe^{-/-}$) mice. Mechanistically it was demonstrated that Cx37 hemichannels in monocytes modulate the initial steps of atherosclerosis by regulating their adhesion to the endothelium [44]. Even in later stages of the disease, Cx37 deletion also reduced the stability of shear stress-induced atherosclerotic plaques in $Apoe^{-/-}$ mice by increasing macrophage contents of the advanced plaques [45]. As the Cx37-CT directly binds to the NO reductase domain of endothelial nitric oxide synthase (eNOS), thereby influencing the function of the enzyme and NO production [46], absence of Cx37 in ECs covering the atherosclerotic lesion may contribute to the dysfunctionality of these cells. Of note, a single nucleotide polymorphism (SNP) in the human Cx37 gene (*Cx37 1019C > T*) associates with an increased risk for coronary artery disease, myocardial infarction, stroke and peripheral artery disease [47]. This *Cx37 1019C > T* SNP results in a non-conservative Proline-to-Serine substitution in the CT of Cx37 and appeared to have a significant impact on channel function under basal and phosphorylating conditions [46,48,49]. When transfected in HeLa or N2A cells, both polymorphic channels are efficiently transported to the cell membrane, where they may function both as hemi-channels and gap junction channels; however, the unitary conductance of channels formed by the Cx37-Proline isoform appeared 1.5 times larger than the one of the Cx37-Serine isoform [48]. In addition, it was shown that monocytic cells expressing Cx37-319P were markedly less adhesive than cells expressing Cx37-319S. Thus, Cx37-319P polymorphic hemi-channels may function as a protective genetic variant by specifically retarding recruitment of monocytes to human atherosclerotic lesions [44].

Altogether, these studies revealed important and diverse contribution of vascular Cxs to the development of atherosclerosis. Before we may consider Cx-based strategies to fight atherosclerotic disease, more work is needed to discriminate between beneficial effects of reduction of (hemi-) channel function and alteration of the Cx interactome of atherogenesis. Moreover, it remains to be determined whether Cxs may play a role in the mechanisms linked to plaque regression.

3.2. Panx1 and Atherosclerosis

As illustrated in the next section, Panx1 channels are important regulators of microvascular physiology, mostly through their capacity to release purines, including ATP [50,51]. As such, Panx1 channels were long time hypothesized to play a role in atherosclerotic disease via their effects on inflammasome activation, neutrophil and macrophage chemotaxis and the activation of T cells [52]. Moreover, Panx1 may play a potential role in macrophage apoptosis and clearance from atherosclerotic lesions by allowing the release of "find me" signals from apoptotic cells to recruit phagocytes at the initial steps of programmed cell death [53–55]. Examination of Panx1 expression in carotid arteries of *Apoe*$^{-/-}$ mice fed with high cholesterol diet revealed Panx1 in the arterial endothelium and in macrophage foam cells in atherosclerotic lesions, and confirmed its absence in the SMCs of the media in these large arteries [56] (Figure 1).

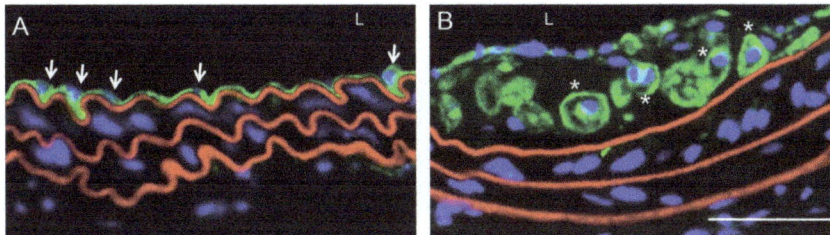

Figure 1. Panx1 expression in healthy and atherosclerotic arteries. (**A**) Panx1 (in green) is expressed in ECs (arrowheads) separating the arterial wall from the lumen (L) of a healthy mouse carotid artery; (**B**) Panx1 is found in lipid-laden macrophages (asterisks) present in atherosclerotic lesions. Of note, Panx1 is absent from the SMC-rich media of non-diseased and diseased conduit arteries. Nuclei are stained with DAPI (in blue) and elastic laminae are counterstained with Evans Blue (in red). Scale bar represents 25 μm.

To investigate the potential contribution of Panx1 in endothelial and monocytic cells to atherosclerosis, mice with a conditional deletion of Panx1 were generated. Atherosclerotic lesion development in response to high cholesterol diet was enhanced in *Tie2-Cre*^{Tg}*Panx1*$^{fl/fl}$*Apoe*$^{-/-}$ mice as compared to *Panx1*$^{fl/fl}$*Apoe*$^{-/-}$ controls, pointing to a protective role for Panx1 in endothelial and/or monocytic cells in atherosclerosis. Unexpectedly, atherogenesis was not altered in mice with ubiquitous Panx1 deletion (*Panx1*$^{-/-}$*Apoe*$^{-/-}$), but these mice displayed reduced body weight, serum cholesterol, triglycerides (TG) and free fatty acids (FFA), suggesting altered lipid metabolism in mice with ubiquitous Panx1 deletion. As it is well known that lowering serum cholesterol and TG levels protects against atherosclerosis in human, it was hypothesized that the lack of effect of ubiquitous deletion of Panx1 on the extent of atherosclerosis may be explained by simultaneous opposite effects of Panx1 on lipid metabolism and inflammation. Interestingly, Panx1-deficient mice show impaired lymphatic function [56] (see Section 6). Future work should unravel the mechanisms linking the lymphatic system, lipid metabolism and atherosclerosis.

4. Coordination of Microvascular Function by Gap Junctions

The arterial vascular system supplies oxygen and nutrients to peripheral tissues by controlling blood flow distribution through a complex network of vessels. Resistance to blood flow is a function of the lumen diameter of the vessels, which depends on the degree of vascular smooth muscle constriction (i.e., vasomotor tone). Most of the total resistance to blood flow resides on feed arteries and arterioles; therefore, coordination of changes in vasomotor tone in the microvascular network plays a central role in the regulation of blood flow distribution and arterial blood pressure [57].

The endothelium plays an essential role in the tonic control of vascular function by Ca^{2+}-dependent production of vasodilator signals such as NO and prostaglandins [58–60]. Although NO is the primary endothelium-dependent vasodilator signal in large conduit vessels, the inhibition of NO or prostaglandin production only attenuates the relaxation initiated by endothelium-dependent vasodilators in small resistance arteries [61–63]. The NO- and prostaglandin-independent response observed in these arteries is associated with the hyperpolarization of SMCs, which leads to smooth muscle relaxation by the consequent reduction in the open probability of L-type voltage-dependent Ca^{2+} channels. In addition to the complex EC signaling, the appropriate control of blood flow distribution also relies on the direct cell-to-cell communication via gap junctions, which has emerged as a key pathway to coordinate vascular wall function in resistance arteries by radial (among ECs and SMCs) and longitudinal (along the vessel length) conduction of vasomotor signals [62,64–66].

4.1. Radial Conduction in the Vascular Wall

Gap junctions play a central role in the intercellular communication of the endothelium-generated vasodilator signals. Although ECs and SMCs are physically separated by the internal elastic lamina in resistance arteries, these cells can make contact through cell projections that penetrate the internal elastic lamina and reach the other cell type at discrete points known as myoendothelial junctions [62,67–69]. These points of contact appear to constitute highly specialized subcellular signaling microdomains and gap junctions located at myoendothelial junctions (i.e., myoendothelial gap junctions) provide a critical pathway for fine regulation of vasomotor responses through the radial transmission of current, Ca^{2+} and small signaling molecules such as IP_3 [68,70–72].

The endothelium-mediated NO-independent smooth muscle hyperpolarization was first attributed to a diffusible factor released by ECs and, in consequence, this vasodilator signal was termed endothelium-derived hyperpolarizing factor (EDHF) [61,62]. Several EDHF candidates have been proposed, such as K^+ ions [73], epoxyeicosatrienoic acids [74,75], hydrogen peroxide [76], and C-type natriuretic peptide [77,78]. Although the NO-independent smooth muscle hyperpolarization is likely to rely on a combination of these signals, depending on the vascular territory [61,79] and experimental preparation used in the study [80], this vasodilator component is however typically paralleled by the hyperpolarization of ECs [61,62]. In addition, it has been consistently observed that the endothelium-dependent smooth muscle hyperpolarization is sensitive to simultaneous inhibition of Ca^{2+}-activated K^+ channels (K_{Ca}) of small (SK_{Ca}) and intermediate (IK_{Ca}) conductance [61,63,81]. In the vessel wall, these K^+ channels are only expressed in ECs [81,82], which prompted the proposal that a prominent component of the EDHF signaling is the simple direct electrotonic transmission from ECs to SMCs via myoendothelial gap junctions of a hyperpolarizing current initiated by SK_{Ca} and IK_{Ca} activation [61,81,83–85]. In which case, the release of a diffusible factor is not consistent with this signaling mechanism, which led to replacing the term abbreviated as EDHF with the expression endothelium-derived hyperpolarization (EDH) [86]. Consistent with this notion, the contribution of the EDH-mediated responses and the expression of myoendothelial gap junctions increase as the vessel size decreases [87,88] and the EDH-associated vasodilator signaling has been shown to be attenuated or abolished by the Cx-mimetic peptides ^{37,40}Gap26, ^{40}Gap27 and ^{37,43}Gap27 [89,90]. These peptides are homologous to specific domains of EL1 (Gap26) or EL2 (Gap27) and were designed to block channels formed by Cx37 or Cx40 in the case of ^{37,40}Gap26, Cx40 in the case of ^{40}Gap27, and Cx37 or Cx43 in the case of ^{37,43}Gap27. In addition to these findings, EC-selective loading with antibodies directed against the carboxyl-terminal region of Cx40 [91] or deletion of Cx40 specifically in ECs also leads to a reduction in the EDH pathway [80], which highlight the functional relevance of this Cx in the endothelial cell signaling and in the control of vasomotor tone.

Interestingly, a pool of eNOS is also found at myoendothelial junctions [92], which provide a subcellular location that is coherent, not only with the vasodilator function of the enzyme, but also with the intercellular signaling pathway of NO. Although the biophysical properties of NO are compatible with the assumption that it can diffuse freely across cell membranes, blockade of gap junction communication in mesenteric resistance vessels with 18β-glycyrrhetinic acid was shown to prevent the NO transfer from ECs to SMCs and the associated NO-dependent vasodilation observed in response to acetylcholine (ACh) [93], suggesting that myoendothelial gap junctions provide a directional pathway for effective NO signaling in the wall of small arteries. The Cx isoforms involved in the gap junction-mediated NO signaling have not been identified, but, as NO-induced relaxation is mediated by a reduction in the Ca^{2+} sensitivity of smooth muscle contractile machinery [94,95] and EDH signaling decreases the intracellular Ca^{2+} concentration of SMCs [61,94], regulation of myoendothelial gap junctions may play a pivotal role in the balance of these two complementary vasodilator components.

4.2. Longitudinal Conduction of Vasomotor Responses

Control of peripheral vascular resistance and blood flow distribution is a dynamic process that depends on coordination of changes in diameter between different segments and cellular elements of the vascular resistance network [57,96]. Vasomotor signals generated in a short arteriolar segment (100 μm) rapidly spread (<1 s) several millimeters along the vessel length without apparent delay, demonstrating functional coupling between distal and proximal segments of the vasculature [97,98]. Therefore, longitudinal conduction of vasomotor signals endows the microvascular network with a mechanism that is most likely to contribute to integrate function within the arteriolar network and between arterioles and feed arteries [96,99,100]. Direct measurements of membrane potential indicate that conducted vasomotor responses are associated with changes in the membrane potential of cells of the vessel wall [97,101,102]. As gap junctions provide a low-resistance intercellular pathway between ECs and SMCs, the conduction of vasomotor responses along the vessel length is thought to be the result of electrotonic spread of changes in membrane potential generated at the stimulation site through gap junctions connecting cells of the vessel wall [103,104]. Then, in the case of endothelium-dependent vasodilators, such as ACh, the conduction of the vasodilation is thought to be the result of the electrotonic spread along the vessel length of an EDH-initiated vasodilation [62,64,105]. In contrast, in the case of vasoconstrictor signals, such as those activated by phenylephrine (PE), a depolarization is conducted [101,106].

The cellular pathway of conducted vasomotor signals seems to depend on the cell type that initiates the response, and vasoconstrictor responses activated by the stimulation of SMCs are consistently conducted by SMCs [107,108]. In contrast, vasodilator signals have been shown to spread either exclusively by the endothelium in feed arteries [97,109] or by both SMCs and ECs in arterioles [107,108], which led to the proposal that the cellular pathway for conduction of vasodilations depends on the functional location of the vessel in the microvascular network [66]. However, the cellular pathway of vasodilator signals may also depend on the stimulus that initiated the response, because, in contrast to ACh, selective damage of the endothelium blocked the vasodilation induced by bradykinin in arterioles [104,108].

Conduction of vasomotor responses may be mediated by interaction of one or more of the five Cx isoforms that are expressed in the vascular system: Cx32, Cx37, Cx40, Cx43, and Cx45 [100,110–112]. Although the contribution of each of these Cxs to the longitudinal coordination of the changes in diameter has not been clearly determined, it has been consistently observed that global deletion of Cx40 results in the development of an irregular arteriolar vasomotion and in a reduced spread of vasodilator signals activated by ACh or bradykinin in feed arteries as well as in arterioles of the cremaster muscle microcirculation [98,113,114]. In blood vessels of the mouse, the expression of Cx40 is restricted to the endothelium [98,115,116], which raises an apparent disagreement with the participation of SMCs in the conducted vasodilation in arterioles. However, the involvement of Cx40 in the transmission of the EDH signaling may explain the detriment of the alternative conduction through SMCs observed previously in response to ACh [80]. In addition, ablation of Cx40 is also associated with a decrease in Cx37 expression and the development of a hypertension caused by a dysregulation of renin production [98,99,114,116,117]. As with Cx40, the expression of Cx37 is also confined to ECs in the vessel wall of mice [98,116], and then, the decline in Cx37-mediated communication in the absence of Cx40 and the development of hypertension may contribute to the reduction in the conduction of vasodilator signals observed in Cx40 knockout animals. Nevertheless, conducted vasodilator responses are intact in Cx37 knockout mice [98] and in animals with an angiotensin-dependent hypertension evoked by deletion of Cx40 in the renin-producing cells [116]. Furthermore, the disruption in the propagation of the response to endothelium-mediated vasodilators attained after global deletion of Cx40 was also observed in EC-specific Cx40 knockout mice [116] and in animals expressing a mutated Cx40 (Cx40A96S) that exhibits a substantially lower junctional conductance [116,118–120]. Although the mutation Cx40A96S causes a renin-dependent hypertension, as that observed with global deletion of Cx40, the endothelial Cx37 levels are normal in these mice [116]. Therefore, these findings in conjunction confirm the critical role of Cx40 in the control and coordination of microvascular function by ECs.

5. Coordination of Microvascular Function by Pannexins

There are three different pannexin isoforms (Panx1, Panx2 and Panx3), with Panx1 being the most ubiquitously expressed throughout the vasculature [121]. There are organ-specific circulations where it appears that other Panx isoforms have been described, but their function has not yet been described [121]. In general, Panx1 is expressed in endothelium throughout conduit and microcirculation, whereas Panx1 is restricted to smooth muscle of resistance arteries, and is not found in conduit smooth muscle [56,121]. Overall, much less is known about the Panxs (compared with Cxs) in the microcirculation, likely due to their more recent discovery, the inherent problems associated with the global Panx1 knockout mouse (e.g., compensation with up regulation of Panx3 throughout the vasculature [122], as well as other cell types [123]), and specific inhibitors for Panx1 that do not also block connexin-built gap junctions (e.g., [124]). However, there are exciting pieces of data emerging using inducible cell type specific Panx1 knockout mice that have revealed phenotypes that are fundamental to the microcirculation.

For example, multiple groups have now demonstrated that Panx1 and the α1-adrenergic receptor (AR) are uniquely coupled in a signaling axis that can regulate vasoconstriction [125–129]. Either SMC-specific Panx1 deletion, or use of multiple Panx inhibitors, blunts noradrenaline and PE mediated vasoconstriction of resistance arteries, but leaves other vasoconstriction pathways intact [125,127–129]. This translates to a hypotensive blood pressure response by the mouse at periods of highest sympathetic nerve activity (evening) [130]. Importantly, the Panx1-α1-AR signaling axis is not observed in large conduit arteries (e.g., aorta or carotid), which is likely because Panx1 is absent from conduit vessel smooth muscle [56,121], and sympathetic nerve innervation is very low.

The Panx1-α1-AR signaling axis also highlights a potent link between sympathetic nerves and vasoconstriction that may be directly druggable for treatment of hypertension in humans (e.g., [127,129]). Indeed, this was recently highlighted by the discovery of trovafloxacin and spironolactone being able to work directly on Panx1 channels [127,131], as evidenced by electrophysiology, inhibiting ATP release, and blunting of vasoconstriction. Spironolactone in particular has been used for decades as a potent anti-hypertensive whose primary effect had been thought to be due to mineralcorticoid antagonism. Other more specific mineralcorticoid antagonists failed to block the Panx1 channel, indicating that the potent effect of spironolactone may be due to blocking both mineralcorticoids and Panx1.

The mechanism of α1-AR activation of Panx1 is still under investigation, although based on previous work it is thought that Panx1 may be selectively regulated by receptor stimulation at the intracellular loop [130]. The use of both peptides and amino acid mutagenesis have confirmed the importance of this region [130]. However, there are likely other regions where Panx1 can be regulated that are especially important in the vasculature. For example, NO potently inhibits Panx1 channels by S-nitrosylating amino acids cysteine 40 and cysteine 346 to prevent channel opening and ATP release [132]. This could be an important mechanism for feedback on sympathetic nerve vasoconstriction. How the cross-talk of several different post-translational modifications fit together to regulate Panx1 channel gating properties will be important moving forward.

There are other more specific regions where Panx1 may play a role in the microcirculation. There is no identified role yet for Panx1 in regulation of endothelial-mediated dilation, except in large conduit vessels, which do have augmented responses to endothelial-induced vasodilation in global Panx1 knockout animals [133]. However, this effect is not seen in resistance arteries, and endothelial specific deletion of Panx1 has no effect on blood pressure [134]. Thus, it is not clear what exactly the augmented endothelial mediated dilation in conduit arteries may mean physiologically at this point.

Panx1 utilization could also be considered vascular bed specific. For example, it has recently been demonstrated that myogenic tone is attenuated in the cerebral circulation of EC-specific Panx1 knockout animals, but is not altered in the mesenteric circulation of the same animals [134]. The EC-specific Panx1 knockout mice also had resistance to middle cerebral artery occlusion (stroke model). SMC-specific Panx1 knockout animals did not have an attenuation of myogenic tone in the cerebral or mesenteric circulation, and were not resistant to induction of stroke [134]. Thus, different vascular beds may utilize Panx1 differently. It also highlights the importance of using cell type-specific Panx1 knockouts in order to properly identify phenotypes.

Besides regulation of blood pressure, among other important aspects of the microcirculation, it is an important regulator of the acute inflammatory response. It was recently demonstrated that TNFα stimulation activates Panx1 in the venous, but not the arterial microcirculation [135]. Interestingly, TNFα (but not IL-1β) induced ATP release via Panx1, as demonstrated in cultured venous ECs, as well as isolated murine veins, but not in any arterial EC or isolated arteries [135]. The effect of the increased ATP release caused an increase in leukocytes that was Panx1 dependent as shown by genetic deletion [135]. The effect of EC deletion of Panx1 has also recently been shown in ischemia, with deletion of Panx1 inducing a significant decrease in leukocytes after occlusion of the middle cerebral artery, lessening the overall impact of the ischemic response [136]. Recently, this work was even further expanded to include ischemic models in the lung and kidney [137,138]. These exciting observations point to a central role for Panx1 in ECs regulating ischemia and the acute inflammatory response.

Also, in the microcirculation, there has been significant attention paid to the possible role of purinergic signaling from red blood cells (RBCs) to endothelium to induce vasodilation, especially during hypoxia. It had been hypothesized that Panx1 on RBCs was the mechanism by which ATP (or other purinergic signals) could leave the RBC, bind to purinergic receptors on endothelium, and induce vasodilation. However, although Panx1 can be localized to RBCs [136], the role for ATP coming from RBCs has recently been called into question, especially via activation of the channel by cAMP/PKA [136,139]. Indeed, the mechanism for increased ATP may be lysis of the RBCs [136,139]. This highlights the need to be careful with measurement of ATP, which is an inconsistent and difficult methodological technique. However, further questions that arise based on this potential heterocellular communication between RBCs and endothelium include what the possible role of Panx1 on RBCs would be if it was present, or if other signaling mechanisms besides cAMP could induce Panx1 channel opening on RBCs.

6. Role of Connexins and Pannexins in Venous and Lymphatic Function

Apart from their role in inflammatory cell recruitment at the level of venules (see Section 5), the function of Panxs and Cxs in larger veins is much less studied. Venous valves play a crucial role in the systemic circulation, promoting the one-way movement of blood from peripheral veins towards the heart and augmenting venous return. In humans, valvular dysfunction or (congenital) absence of valves in large veins typically result in common venous disorders such a varicose veins and edema in the legs. Three gap junction proteins, i.e., Cx37, Cx43, and Cx47, are expressed in ECs, covering venous valves in a highly polarized fashion, with Cx43 on the upstream side of the valve leaflet and Cx37 on the downstream side. Cx47 seems more restricted to a small subset of ECs in the venous valves [140]. Similar to earlier observations in the lymphatic vasculature [141,142] veins from Cx37-deficient mice lack valves [140]. As Cx37 seems a crucial regulator of valve development in both veins and lymphatic vessels, there may be common molecular pathways controlling valve development in these distinct vessel types. Mechanistically, it has been shown in lymphatic valves that the transcription factors Prox1, Foxc2, as well as lymphatic flow, coordinately control the expression of Cx37 and activation of calcineurin/NFAT signaling. Indeed, Cx37 and calcineurin are required for the assembly and delimitation of lymphatic valve territory during development and for its postnatal maintenance [142]. Interestingly, the development of venous valves, but not the formation of lymphatic valves, is affected in Cx47-deficient mice [143,144]. Accordingly, Cx47 null mice also display normal lymphatic vascular function [145]. Mutations in the Cx47 gene are associated with reduced venous valve number and length, a crucial finding for understanding how some Cx47 mutations cause inherited (lymph) edema in humans [144]. Recently, both Cx47 and Cx43 have been added to the limited repertoire of primary lymphedema-associated genes (such as Foxc2, Vegfr3 and Sox18) [146–149]. Furthermore, Cx47 and Cx37 mutations have been associated with increased risk for secondary lymphedema following breast cancer treatment [150,151].

The lymphatic system regulates tissue fluid homeostasis, trafficking of immune cells to draining lymph nodes and absorption of dietary fat. To investigate whether Panx1 affects lymphatic flow, drainage of interstitial fluids following injection of Evans Blue in the footpad was recently compared between $Panx1^{-/-} Apoe^{-/-}$ and $Apoe^{-/-}$ mice. The dye progressively spread throughout the lymphatic system to successive draining lymph nodes and finally the systemic circulation. The dye transport was considerably smaller in $Panx1^{-/-} Apoe^{-/-}$ mice than in control $Apoe^{-/-}$ mice. Moreover, tails of $Panx1^{-/-} Apoe^{-/-}$ mice showed increased diameters and increased interstitial fluid content than control $Apoe^{-/-}$ mice, suggesting that lymphatic flow is impaired in mice with ubiquitous deletion of Panx1. Finally, $Panx1^{-/-} Apoe^{-/-}$ mice had reduced dietary fat absorption with control animals. Collectively, these findings suggest a pivotal role for Panx1 in lymphatic function [56], and it will be exciting to learn more on the cell type and molecular mechanism involved in this regulation.

7. Connexins and Pannexins in the Control of Platelet Function, Haemostasis and Thrombosis

Cx hemichannels and gap junctions have been studied widely in various cell types where sustained cell interactions occur. Some reports, however, indicate the presence of Cxs on the surface of some circulating cells, such as monocytes and T-cells, where gap junction and hemichannel functions control cellular functions [152–154]. In recent studies, fundamental roles for these proteins in platelets have emerged.

7.1. Platelets: Mediators of Haemostasis and Thrombosis

Platelets provide a front line of defense in response to injury, triggering haemostasis at sites of injury, and are increasingly recognized for their involvement in a range of other (patho)physiological processes, including inflammation, atherogenesis, and cancer cell metastasis. Platelets adhere to collagens that are exposed at sites of arterial damage, initially via an indirect interaction with plasma von Willebrand factor (VWF), which through binding to the platelet glycoprotein (GP) Ib-V-IX receptor complex, a short-lived interaction that therefore serves to slow platelets, allows subsequent direct binding to platelet collagen receptors GPVI and integrin $\alpha_2\beta_1$ (Figure 2A) [155,156]. Integrin $\alpha_2\beta_1$ functions principally as an adhesive receptor for collagen [157], while collagen binding to GPVI stimulates platelet activation.

Collagen binding to GPVI causes receptor clustering and the tyrosine phosphorylation of the Fc receptor γ-chain [158,159] by Src-family kinases [160]. The tyrosine kinase Syk is then recruited and initiates the first step in a complex and branching signaling pathways incorporating, among others, the linker for activation of T cells, phosphatidylinositol 3-kinase, protein kinase B, Bruton's tyrosine kinase, phospholipase Cγ2, integrin-linked kinase, and the mobilization of intracellular calcium stores [161,162]. This culminates, via the GTP binding protein Rap1b, in an increase in affinity of integrins that enhance adhesion to collagen ($\alpha_2\beta_1$), and causes aggregation through the binding of plasma fibrinogen to integrin $\alpha_{IIb}\beta_3$ [161,163].

A rapid and full platelet response is ensured through the autocrine and paracrine actions of factors that are released by activated platelets such as ADP and thromboxane A_2. Following the binding of fibrinogen to integrin $\alpha_{IIb}\beta_3$ and collagen to integrin $\alpha_2\beta_1$, outside-in signaling through these integrins also contributes to sustained platelet activation and irreversible thrombus formation [164,165].

Thrombin, which is generated on the surface of activated platelets within a thrombus, is also a potent platelet agonist that is important for effective haemostasis.

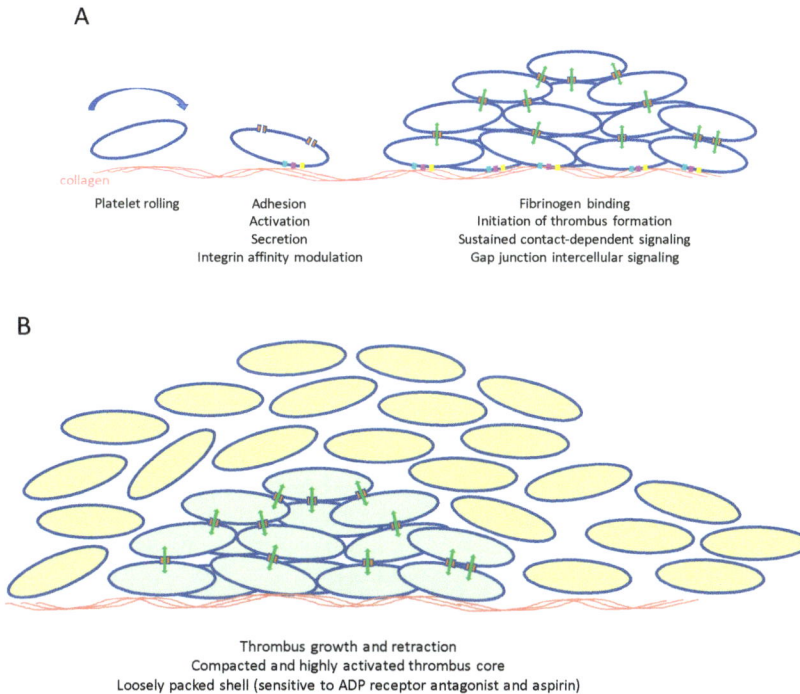

Figure 2. Gap junction intercellular communication between platelets: a working model. (**A**) Blood vessel injury leads to the exposure of subendothelial collagens. Through interaction with von Willebrand factor, which also binds to collagens, platelets roll along the surface, slowing their movement and allowing direct binding of collagen to the cell surface receptors integrin, $\alpha_2\beta_1$, and GPVI, initiating platelet intracellular signaling. This results in the secretion or release of prothrombotic factors such as ADP and TXA_2 that further propagate platelet activation. This culminates in an increase in affinity of integrin $\alpha_{IIb}\beta_3$ which then binds fibrinogen that supports platelet-platelet adhesion and thrombus formation. Close contact between platelets allows the formation of gap junctions that permit intercellular signaling during thrombus formation and stabilization; (**B**) Intercellular signaling controls thrombus contraction by enabling the formation of a core of platelets that are highlight activated and tightly packed. A more loosely packed shell of platelets develops, although this is inhibited in the presence of ADP receptor antagonists or aspirin (to prevent TXA_2 formation). Whether gap junctional intercellular communication controls platelet thrombus core and shell assembly has yet to be formally established.

A reactive system such as this, which incorporates many positive feedback mechanisms, requires precise control to prevent un-needed platelet activation. The healthy endothelium produces molecular signals, NO and prostaglandin I_2, short-lived molecules that exert powerful inhibitory effects on platelets through the stimulation of cyclic nucleotide-dependent intracellular signaling [166].

Inappropriate platelet activation, for example, at the site of atherosclerotic plaque formation or rupture, results in the exposure of platelets to activatory substances, including collagens, resulting in thrombosis and the occlusion of blood flow. As the principle cause of myocardial infarction and ischemic stroke, platelets represent an important therapeutic target [163].

In the last 20 years, substantial progress has been made in understanding the molecular mechanisms that control platelets, and this is beginning to impact in the development of new therapeutic approaches. These advances have largely involved study of traditional intra-cellular signaling, the identification of key activatory or inhibitory signals, cell surface receptors required to respond to these signals and the intracellular signaling pathways or networks that these control. While platelets are singular circulating cells, activation and thrombus formation bring them into close proximity for prolonged periods, and increasing evidence supports the importance of sustained inter-platelet communications within the thrombus derived largely through integrin outside-in signaling, with additional contributions from, for example, Eph family kinases and ephrin counter-ligands [167–169]. Sustained signaling within the thrombus enables platelets to cooperate to regulate thrombus compaction, structure and stability and subsequently clot retraction, a step believed to be important for wound repair [170]. The ability of platelets to coordinate their functions within a developing thrombus, particularly in the control of calcium signaling, led to early experimental evidence that this may be mediated by intercellular communication (Figure 2B), although this was not initially attributed to gap junctions [171].

7.2. Platelets Possess Connexins

Messenger RNA profiling of megakaryocytes, the precursor bone marrow cells from which platelets form, revealed that these cells likely possess notable levels of Cx37, with additional expression of Cx40 and Cx62. Indeed, Cx37 mRNA and protein were first reported to be expressed in human and mouse platelets, and were found to be present at the cell surface [172,173]. Scanning electron microscopy of sections of human platelet thrombi revealed regions of apposite membrane structures with a typical appearance of gap-junction-like structures [173].

7.3. Platelet Gap Junction Formation—Orchestration of Intercellular Signaling within Arterial Thrombi

The transport of dye between platelets was first demonstrated following micro-injection of neurobiotin, and transfer into surrounding platelets [172]. Gap junctional intercellular coupling between platelets was confirmed by fluorescence recovery after photo-bleaching (FRAP) analysis of thrombi preformed under arterial flow conditions using blood reconstituted with platelet that were labelled preloaded with cytosolic calcein [173]. Transfer of dye was inhibited by selective or non-selective inhibition of Cx37 (^{37,43}Gap27). Clot retraction responses were also inhibited in the presence of inhibitor, or the absence of Cx37 (using blood from Cx37-deficient mice) suggesting that gap junctional coupling mediates physiological responses within platelets.

There currently exists a difference of experimental conclusions drawn from the study of Angelillo-Scherrer, who reported modestly elevated platelet aggregation responses on Cx37-deficient mouse platelets and following the use of inhibitory peptides with human platelets [172], while Vaiyapuri reported substantially diminished responses [173]. Vaiyapuri also reported similar outcomes following the inhibition or deletion of Cx40 [173]. These differences are likely to be explained by differences in experimental conditions, although both studies indicate the potential importance of Cxs in the control of platelet function.

The use of flow cytometry gated to examine the function of individual platelets revealed that platelet activation, prior to platelet-platelet contact is inhibited in the absence of functional Cx37, which is suggestive of important roles for Cx hemichannels in the initiation of platelet responses [173,174]. Whether this is due to channel function or through interaction with other cell surface proteins has yet to be established, although inhibition of Cx37 or Cx40 is associated with diminished intracellular mobilization of calcium from stores, indicating a fundamental role in the propagation of platelet cell signaling [173,174]. Consistent with this, inhibition of Cx37 in whole blood results in diminished thrombus formation on a collagen-coated micro-fluidic flow cells under arterial flow conditions. Infusion of ^{37,43}Gap27 [173] or ^{40}Gap27 (unpublished observation) into mice prior analysis of laser induced thrombosis in cremaster muscle arterioles was found to result in diminished thrombotic responses. Diminished thrombus formation appears to be at the expense of haemostatic control, since bleeding times were extended modestly. It is interesting to note the Cx37-deficient mice were found to exhibit reduced survival time using a thrombo-embolism model, induced by intravenous injection of collagen and adrenalin [172], which may be a consequence of reduced thrombus stability that results in increased thrombus fragmentation and lung occlusion.

7.4. Panx1 Contributes to Platelet Function at Low Agonist Concentrations

Recent analysis of the megakaryocyte (and therefore the platelet) channelome revealed that among a range of cell surface channel proteins, Panx1 is also likely to be expressed. A series of pharmacological approaches using selective mimetic peptide inhibitor [175], subsequently confirmed using Panx1-deficient mouse platelets [176], demonstrated this protein to contribute to calcium responses to low concentrations of various platelet agonists. The ability of Panxs to facilitate the release of ATP from cells is a property that has also been observed in platelets. Panx1 mediated ATP release results in subsequent stimulation of the ATP-gated calcium channel P2X1, which causes enhanced calcium influx and therefore propagation of platelet functional responses [175,176]. At higher concentrations of collagen or other platelets agonists P2X1 makes little contribution to cell responses, and therefore the effects of Panx1 are restricted to conditions where agonist concentrations are limited.

7.5. Cx37 and Panx1 Polymorphisms

It is fair to ask, are the effects of platelet Cxs likely to be physiologically important, or do they contribute to cardiovascular disease risk? The clearest indication that gap junction and hemichannel function may be important stem from studies that have explored the effects of common gene SNPs in the human population. As discussed before (Section 3), a SNP in the coding sequence of Cx37 (P319S) influences the gating of Cx37 channels [46,48]. When transfected into HeLa cells, the 319P polymorphism is associated with reduced diffusion of dye between cells [172]. 96 Caucasian men were genotyped to explore the relationship between this polymorphism and platelet function. A relationship was observed between the number of 1019C alleles of the *Cx* gene possessed by volunteers, with the CC genotype associated with modestly increased platelet function [172]. These data suggest that platelet reactivity levels may be determined by connexin-mediated platelet function.

Three variants exist in the human Panx1 coding sequence. These result in a Q5H variant at the N-terminus, a I272V variant within the 4th transmembrane domain and deletion of amino acids 401 to 404 at the CT due a splice variant. In a population of 96 male Caucasian volunteers the splice variant was not detected [176]. Two thirds of subjects possessed the allele coding for the histidine variant at position 5 in the protein sequence, and the associated with a small increase in platelet aggregation in response to collagen, although responses to other agonists were unaltered. Comparison between the Panx1 alleles responsible for variability at position 272 showed no relationship with platelet reactivity.

The numbers of subjects included in these studies were relatively small, and more detailed analysis of the relationship between platelet Cx and Panx polymorphisms and variability in platelet responsiveness would be required to confirm these observations. Current data are, however, consistent with a role of Cxs and Panxs in the regulation of haemostasis and potentially thrombosis, allowing the potential development of new strategies for the prevention of thrombosis, or other conditions in which these cells are implicated.

8. Conclusions and Perspectives

It is now well established that Cxs have an important function in the control of blood flow distribution and tissue homeostasis, as well as in pathologies that involve a tight regulation and coordination between cells in the blood vessel wall and circulating blood cells such as atherosclerosis and hypertension. Strong evidence now also supports an important role for gap junctions and Cx hemichannels in the control of haemostasis and thrombosis, although many questions remain to be addressed in this field. More insight into the nature of molecular signals that are transported through gap junctions and hemichannels will be required in order to tease apart the basis of their ability to modulate diverse aspects of vascular function. Systems biology approaches have revealed exquisite detail regarding the architecture of platelet thrombi, which are organized into a densely packed core, surrounded by a less densely packed shell, which is sensitive to the actions of anti-platelet drugs [177]. It is possible that gap junctional intercellular communication mediates the organization of thrombus architecture, function and sensitivity to anti-thrombotic medication. The ability of gap junctions to support interactions between different cell types that are implicated in the stimulation of localized inflammatory responses and atherogenesis [44,178,179], further supports the notion that gap junctional coupling between different vascular cell types may impact on a wide(r) range of (patho)physiological processes.

The release of ATP or Ca^{2+} are generally assumed to be the most relevant signaling mechanisms mediated by both Cx hemichannels and Panx1 channels in vascular (patho)physiology. While recent years have shown great progress in the knowledge on Cx43 protein domains involved in gap junction channel vs. hemichannel gating [180], one of the inherit problems with Panxs is that the biophysical properties of the channels are still being worked out, and thus technically there remains a significant number of unknowns. For example, does ATP always come out of a channel? There is no reason that this needs to be the case, as Panxs in general are large pore channels. Also, being able to distinguish between receptor-mediated and caspase cleavage of the channel has become an important technical differential. The physiological effects described in Section 5 would be considered receptor-mediated Panx1 channel opening, which is uniquely different than caspase cleavage-mediated Panx1 opening that occurs during apoptosis. Key physiological parameters such as electrophysiology and ATP release can be observed in both receptor- and caspase cleavage-mediated Panx1 channel opening. The difference being that receptor mediated Panx1-channel opening is transient, and caspase cleavage produces a permanently "open" channel [126]. Perhaps differential dye uptake could help differentiate these two events? Whatever the case, there is still a significant amount to learn about the relatively recently discovered Panx1 channel and how it may affect vascular and platelet function.

Author Contributions: F.M., X.F.F., B.R.K., B.E.I. and J.M.G. contributed all to the writing of the paper.

Acknowledgments: This work was supported by grants from the Swiss National Science Foundation (310030_162579 and 310030E_176050 to BRK), Fondo Nacional de Desarrollo Científico y Tecnológico – FONDECYT (1150530 to XFF) and the British Heart Foundation (PG/11/125/29320, RG/15/2/31224, PG/17/76/33082 to JMG).

Conflicts of Interest: The authors declare no conflict of interest.

Abbreviations

Cx	Connexin
Panx	Pannexin
TM	Transmembrane
EL	Extracellular loop
CT	Carboxy-terminus
NT	Amino-terminus
IL	Intracellular loop
ER	Endoplasmic reticulum
EC	Endothelial cell
LDL	Low-density lipoprotein
ECM	Extracellular matrix
SMC	Smooth muscle cell
TF	Tissue factor
POPVC	1-palmitoyl-2-(5′-oxo-valeroyl)-*sn*-glycero-3-phosphocholine
eNOS	Endothelial NO synthase
SNP	Single nucleotide polymorphism
TG	Triglycerides
FFA	Free fatty acids

EDHF	Endothelium-derived hyperpolarizing factor
EDH	Endothelium-derived hyperpolarization
ACh	Acetylcholine
PE	Phenylephrine
AR	Adrenergic receptor
RBC	Red blood cells
VWF	von Willebrand factor
GP	Glycoprotein
FRAP	Fluorescence recovery after photo-bleaching

References

1. Molica, F.; Meens, M.J.; Morel, S.; Kwak, B.R. Mutations in cardiovascular connexin genes. *Biol. Cell* **2014**, *106*, 269–293. [CrossRef] [PubMed]
2. Abascal, F.; Zardoya, R. Evolutionary analyses of gap junction protein families. *Biochim. Biophys. Acta* **2013**, *1828*, 4–14. [CrossRef] [PubMed]
3. Laird, D.W. Life cycle of connexins in health and disease. *Biochem. J.* **2006**, *394 Pt 3*, 527–543. [CrossRef] [PubMed]
4. Thevenin, A.F.; Kowal, T.J.; Fong, J.T.; Kells, R.M.; Fisher, C.G.; Falk, M.M. Proteins and mechanisms regulating gap-junction assembly, internalization, and degradation. *Physiology* **2013**, *28*, 93–116. [CrossRef] [PubMed]
5. Patel, D.; Zhang, X.; Veenstra, R.D. Connexin hemichannel and pannexin channel electrophysiology: How do they differ? *FEBS Lett.* **2014**, *588*, 1372–1378. [CrossRef] [PubMed]
6. Saez, J.C.; Leybaert, L. Hunting for connexin hemichannels. *FEBS Lett.* **2014**, *588*, 1205–1211. [CrossRef] [PubMed]
7. Nielsen, M.S.; Axelsen, L.N.; Sorgen, P.L.; Verma, V.; Delmar, M.; Holstein-Rathlou, N.H. Gap junctions. *Compr. Physiol.* **2012**, *2*, 1981–2035. [PubMed]
8. Solan, J.L.; Lampe, P.D. Connexin43 phosphorylation: Structural changes and biological effects. *Biochem. J.* **2009**, *419*, 261–272. [CrossRef] [PubMed]
9. Laird, D.W. The gap junction proteome and its relationship to disease. *Trends Cell Biol.* **2010**, *20*, 92–101. [CrossRef] [PubMed]
10. Agullo-Pascual, E.; Cerrone, M.; Delmar, M. Arrhythmogenic cardiomyopathy and Brugada syndrome: Diseases of the connexome. *FEBS Lett.* **2014**, *588*, 1322–1330. [CrossRef] [PubMed]
11. Veeraraghavan, R.; Poelzing, S.; Gourdie, R.G. Old cogs, new tricks: A scaffolding role for connexin43 and a junctional role for sodium channels? *FEBS Lett.* **2014**, *588*, 1244–1248. [CrossRef] [PubMed]
12. Kelly, J.J.; Simek, J.; Laird, D.W. Mechanisms linking connexin mutations to human diseases. *Cell Tissue Res.* **2015**, *360*, 701–721. [CrossRef] [PubMed]
13. Pfenniger, A.; Wohlwend, A.; Kwak, B.R. Mutations in connexin genes and disease. *Eur. J. Clin. Investig.* **2011**, *41*, 103–116. [CrossRef] [PubMed]
14. Bond, S.R.; Naus, C.C. The pannexins: Past and present. *Front. Physiol.* **2014**, *5*, 58. [CrossRef] [PubMed]
15. Dahl, G.; Muller, K.J. Innexin and pannexin channels and their signaling. *FEBS Lett.* **2014**, *588*, 1396–1402. [CrossRef] [PubMed]
16. Penuela, S.; Harland, L.; Simek, J.; Laird, D.W. Pannexin channels and their links to human disease. *Biochem. J.* **2014**, *461*, 371–381. [CrossRef] [PubMed]
17. Boassa, D.; Ambrosi, C.; Qiu, F.; Dahl, G.; Gaietta, G.; Sosinsky, G. Pannexin1 channels contain a glycosylation site that targets the hexamer to the plasma membrane. *J. Biol. Chem.* **2007**, *282*, 31733–31743. [CrossRef] [PubMed]
18. Scemes, E. Nature of plasmalemmal functional "hemichannels". *Biochim. Biophys. Acta* **2012**, *1818*, 1880–1883. [CrossRef] [PubMed]
19. De Bock, M.; Leybaert, L.; Giaume, C. Connexin Channels at the Glio-Vascular Interface: Gatekeepers of the Brain. *Neurochem. Res.* **2017**, *42*, 2519–2536. [CrossRef] [PubMed]
20. Danesh-Meyer, H.V.; Zhang, J.; Acosta, M.L.; Rupenthal, I.D.; Green, C.R. Connexin43 in retinal injury and disease. *Prog. Retin. Eye Res.* **2016**, *51*, 41–68. [CrossRef] [PubMed]

21. Kim, Y.; Davidson, J.O.; Gunn, K.C.; Phillips, A.R.; Green, C.R.; Gunn, A.J. Role of Hemichannels in CNS Inflammation and the Inflammasome Pathway. *Adv. Protein Chem. Struct. Biol.* **2016**, *104*, 1–37. [PubMed]

22. Roth, G.A.; Johnson, C.; Abajobir, A.; Abd-Allah, F.; Abera, S.F.; Abyu, G.; Ahmed, M.; Aksut, B.; Alam, T.; Alam, K.; et al. Global, Regional, and National Burden of Cardiovascular Diseases for 10 Causes, 1990 to 2015. *J. Am. Coll. Cardiol.* **2017**, *70*, 1–25. [CrossRef] [PubMed]

23. Libby, P.; Lichtman, A.H.; Hansson, G.K. Immune effector mechanisms implicated in atherosclerosis: From mice to humans. *Immunity* **2013**, *38*, 1092–1104. [CrossRef] [PubMed]

24. Weber, C.; Noels, H. Atherosclerosis: Current pathogenesis and therapeutic options. *Nat. Med.* **2011**, *17*, 1410–1422. [CrossRef] [PubMed]

25. Newby, A.C. Metalloproteinases promote plaque rupture and myocardial infarction: A persuasive concept waiting for clinical translation. *Matrix Biol.* **2015**, *44–46*, 157–166. [CrossRef] [PubMed]

26. Badimon, L.; Vilahur, G. Thrombosis formation on atherosclerotic lesions and plaque rupture. *J. Intern. Med.* **2014**, *276*, 618–632. [CrossRef] [PubMed]

27. Owens, A.P., 3rd; Mackman, N. Role of tissue factor in atherothrombosis. *Curr. Atheroscler. Rep.* **2012**, *14*, 394–401. [CrossRef] [PubMed]

28. Bazzoni, G.; Dejana, E. Endothelial cell-to-cell junctions: Molecular organization and role in vascular homeostasis. *Physiol. Rev.* **2004**, *84*, 869–901. [CrossRef] [PubMed]

29. Van Rijen, H.V.; van Kempen, M.J.; Postma, S.; Jongsma, H.J. Tumour necrosis factor alpha alters the expression of connexin43, connexin40, and connexin37 in human umbilical vein endothelial cells. *Cytokine* **1998**, *10*, 258–264. [CrossRef] [PubMed]

30. Danesh-Meyer, H.V.; Huang, R.; Nicholson, L.F.; Green, C.R. Connexin43 antisense oligodeoxynucleotide treatment down-regulates the inflammatory response in an in vitro interphase organotypic culture model of optic nerve ischaemia. *J. Clin. Neurosci.* **2008**, *15*, 1253–1263. [CrossRef] [PubMed]

31. De Bock, M.; Culot, M.; Wang, N.; Bol, M.; Decrock, E.; De Vuyst, E.; da Costa, A.; Dauwe, I.; Vinken, M.; Simon, A.M.; et al. Connexin channels provide a target to manipulate brain endothelial calcium dynamics and blood-brain barrier permeability. *J. Cereb. Blood Flow Metab.* **2011**, *31*, 1942–1957. [CrossRef] [PubMed]

32. Kwak, B.R.; Mulhaupt, F.; Veillard, N.; Gros, D.B.; Mach, F. Altered pattern of vascular connexin expression in atherosclerotic plaques. *Arterioscler. Thromb. Vasc. Biol.* **2002**, *22*, 225–230. [CrossRef] [PubMed]

33. Blackburn, J.P.; Peters, N.S.; Yeh, H.I.; Rothery, S.; Green, C.R.; Severs, N.J. Upregulation of connexin43 gap junctions during early stages of human coronary atherosclerosis. *Arterioscler. Thromb. Vasc. Biol.* **1995**, *15*, 1219–1228. [CrossRef] [PubMed]

34. Polacek, D.; Bech, F.; McKinsey, J.F.; Davies, P.F. Connexin43 gene expression in the rabbit arterial wall: Effects of hypercholesterolemia, balloon injury and their combination. *J. Vasc. Res.* **1997**, *34*, 19–30. [CrossRef] [PubMed]

35. Polacek, D.; Lal, R.; Volin, M.V.; Davies, P.F. Gap junctional communication between vascular cells. Induction of connexin43 messenger RNA in macrophage foam cells of atherosclerotic lesions. *Am. J. Pathol.* **1993**, *142*, 593–606. [PubMed]

36. Johnstone, S.R.; Ross, J.; Rizzo, M.J.; Straub, A.C.; Lampe, P.D.; Leitinger, N.; Isakson, B.E. Oxidized phospholipid species promote in vivo differential cx43 phosphorylation and vascular smooth muscle cell proliferation. *Am. J. Pathol.* **2009**, *175*, 916–924. [CrossRef] [PubMed]

37. Yeh, H.I.; Lu, C.S.; Wu, Y.J.; Chen, C.C.; Hong, R.C.; Ko, Y.S.; Shiao, M.S.; Severs, N.J.; Tsai, C.H. Reduced expression of endothelial connexin37 and connexin40 in hyperlipidemic mice: Recovery of connexin37 after 7-day simvastatin treatment. *Arterioscler. Thromb. Vasc. Biol.* **2003**, *23*, 1391–1397. [CrossRef] [PubMed]

38. Reaume, A.G.; de Sousa, P.A.; Kulkarni, S.; Langille, B.L.; Zhu, D.; Davies, T.C.; Juneja, S.C.; Kidder, G.M.; Rossant, J. Cardiac malformation in neonatal mice lacking connexin43. *Science* **1995**, *267*, 1831–1834. [CrossRef] [PubMed]

39. Kwak, B.R.; Veillard, N.; Pelli, G.; Mulhaupt, F.; James, R.W.; Chanson, M.; Mach, F. Reduced connexin43 expression inhibits atherosclerotic lesion formation in low-density lipoprotein receptor-deficient mice. *Circulation* **2003**, *107*, 1033–1039. [CrossRef] [PubMed]

40. Morel, S.; Chanson, M.; Nguyen, T.D.; Glass, A.M.; Richani Sarieddine, M.Z.; Meens, M.J.; Burnier, L.; Kwak, B.R.; Taffet, S.M. Titration of the gap junction protein Connexin43 reduces atherogenesis. *Thromb. Haemost.* **2014**, *112*, 390–401. [CrossRef] [PubMed]

41. Yuan, D.; Sun, G.; Zhang, R.; Luo, C.; Ge, M.; Luo, G.; Hei, Z. Connexin 43 expressed in endothelial cells modulates monocyteendothelial adhesion by regulating cell adhesion proteins. *Mol. Med. Rep.* **2015**, *12*, 7146–7152. [CrossRef] [PubMed]

42. Chadjichristos, C.E.; Scheckenbach, K.E.; van Veen, T.A.; Richani Sarieddine, M.Z.; de Wit, C.; Yang, Z.; Roth, I.; Bacchetta, M.; Viswambharan, H.; Foglia, B.; et al. Endothelial-specific deletion of connexin40 promotes atherosclerosis by increasing CD73-dependent leukocyte adhesion. *Circulation* **2010**, *121*, 123–131. [CrossRef] [PubMed]

43. Denis, J.F.; Scheckenbach, K.E.L.; Pfenniger, A.; Meens, M.J.; Krams, R.; Miquerol, L.; Taffet, S.; Chanson, M.; Delmar, M.; Kwak, B.R. Connexin40 controls endothelial activation by dampening NFkappaB activation. *Oncotarget* **2017**, *8*, 50972–50986. [CrossRef] [PubMed]

44. Wong, C.W.; Christen, T.; Roth, I.; Chadjichristos, C.E.; Derouette, J.P.; Foglia, B.F.; Chanson, M.; Goodenough, D.A.; Kwak, B.R. Connexin37 protects against atherosclerosis by regulating monocyte adhesion. *Nat. Med.* **2006**, *12*, 950–954. [CrossRef] [PubMed]

45. Pfenniger, A.; Meens, M.J.; Pedrigi, R.M.; Foglia, B.; Sutter, E.; Pelli, G.; Rochemont, V.; Petrova, T.V.; Krams, R.; Kwak, B.R. Shear stress-induced atherosclerotic plaque composition in ApoE(−/−) mice is modulated by connexin37. *Atherosclerosis* **2015**, *243*, 1–10. [CrossRef] [PubMed]

46. Pfenniger, A.; Derouette, J.P.; Verma, V.; Lin, X.; Foglia, B.; Coombs, W.; Roth, I.; Satta, N.; Dunoyer-Geindre, S.; Sorgen, P.; et al. Gap junction protein Cx37 interacts with endothelial nitric oxide synthase in endothelial cells. *Arterioscler. Thromb. Vasc. Biol.* **2010**, *30*, 827–834. [CrossRef] [PubMed]

47. Meens, M.J.; Pfenniger, A.; Kwak, B.R. Risky communication in atherosclerosis and thrombus formation. *Swiss Med. Wkly.* **2012**, *142*, w13553. [CrossRef] [PubMed]

48. Derouette, J.P.; Desplantez, T.; Wong, C.W.; Roth, I.; Kwak, B.R.; Weingart, R. Functional differences between human Cx37 polymorphic hemichannels. *J. Mol. Cell. Cardiol.* **2009**, *46*, 499–507. [CrossRef] [PubMed]

49. Morel, S.; Burnier, L.; Roatti, A.; Chassot, A.; Roth, I.; Sutter, E.; Galan, K.; Pfenniger, A.; Chanson, M.; Kwak, B.R. Unexpected role for the human Cx37 C1019T polymorphism in tumour cell proliferation. *Carcinogenesis* **2010**, *31*, 1922–1931. [CrossRef] [PubMed]

50. Isakson, B.E.; Thompson, R.J. Pannexin-1 as a potentiator of ligand-gated receptor signaling. *Channels* **2014**, *8*, 118–123. [CrossRef] [PubMed]

51. Meens, M.J.; Kwak, B.R.; Duffy, H.S. Role of connexins and pannexins in cardiovascular physiology. *Cell. Mol. Life Sci.* **2015**, *72*, 2779–2792. [CrossRef] [PubMed]

52. Adamson, S.E.; Leitinger, N. The role of pannexin1 in the induction and resolution of inflammation. *FEBS Lett.* **2014**, *588*, 1416–1422. [CrossRef] [PubMed]

53. Sandilos, J.K.; Chiu, Y.H.; Chekeni, F.B.; Armstrong, A.J.; Walk, S.F.; Ravichandran, K.S.; Bayliss, D.A. Pannexin 1, an ATP release channel, is activated by caspase cleavage of its pore-associated C-terminal autoinhibitory region. *J. Biol. Chem.* **2012**, *287*, 11303–11311. [CrossRef] [PubMed]

54. Chekeni, F.B.; Elliott, M.R.; Sandilos, J.K.; Walk, S.F.; Kinchen, J.M.; Lazarowski, E.R.; Armstrong, A.J.; Penuela, S.; Laird, D.W.; Salvesen, G.S.; et al. Pannexin 1 channels mediate 'find-me' signal release and membrane permeability during apoptosis. *Nature* **2010**, *467*, 863–867. [CrossRef] [PubMed]

55. Qu, Y.; Misaghi, S.; Newton, K.; Gilmour, L.L.; Louie, S.; Cupp, J.E.; Dubyak, G.R.; Hackos, D.; Dixit, V.M. Pannexin-1 is required for ATP release during apoptosis but not for inflammasome activation. *J. Immunol.* **2011**, *186*, 6553–6561. [CrossRef] [PubMed]

56. Molica, F.; Meens, M.J.; Dubrot, J.; Ehrlich, A.; Roth, C.L.; Morel, S.; Pelli, G.; Vinet, L.; Braunersreuther, V.; Ratib, O.; et al. Pannexin1 links lymphatic function to lipid metabolism and atherosclerosis. *Sci. Rep.* **2017**, *7*, 13706. [CrossRef] [PubMed]

57. Segal, S.S. Integration and Modulation of Intercellular Signaling Underlying Blood Flow Control. *J. Vasc. Res.* **2015**, *52*, 136–157. [CrossRef] [PubMed]

58. Feletou, M.; Huang, Y.; Vanhoutte, P.M. Endothelium-mediated control of vascular tone: COX-1 and COX-2 products. *Br. J. Pharmacol.* **2011**, *164*, 894–912. [CrossRef] [PubMed]

59. Moncada, S.; Higgs, E.A. The discovery of nitric oxide and its role in vascular biology. *Br. J. Pharmacol.* **2006**, *147* (Suppl. 1), S193–S201. [CrossRef] [PubMed]

60. Vanhoutte, P.M. COX-1 and vascular disease. *Clin. Pharmacol. Ther.* **2009**, *86*, 212–215. [CrossRef] [PubMed]

61. Busse, R.; Edwards, G.; Feletou, M.; Fleming, I.; Vanhoutte, P.M.; Weston, A.H. EDHF: Bringing the concepts together. *Trends Pharmacol. Sci.* **2002**, *23*, 374–380. [CrossRef]

62. Figueroa, X.F.; Duling, B.R. Gap junctions in the control of vascular function. *Antioxid. Redox Signal.* **2009**, *11*, 251–266. [CrossRef] [PubMed]

63. Gaete, P.S.; Lillo, M.A.; Ardiles, N.M.; Perez, F.R.; Figueroa, X.F. Ca^{2+}-activated K$^+$ channels of small and intermediate conductance control eNOS activation through NAD(P)H oxidase. *Free Radic. Biol. Med.* **2012**, *52*, 860–870. [CrossRef] [PubMed]

64. Figueroa, X.F.; Isakson, B.E.; Duling, B.R. Connexins: Gaps in our knowledge of vascular function. *Physiology* **2004**, *19*, 277–284. [CrossRef] [PubMed]

65. Schmidt, V.J.; Wolfle, S.E.; Boettcher, M.; de Wit, C. Gap junctions synchronize vascular tone within the microcirculation. *Pharmacol. Rep.* **2008**, *60*, 68–74. [PubMed]

66. Segal, S.S. Regulation of blood flow in the microcirculation. *Microcirculation* **2005**, *12*, 33–45. [CrossRef] [PubMed]

67. Beny, J.L.; Koenigsberger, M.; Sauser, R. Role of myoendothelial communication on arterial vasomotion. *Am. J. Physiol. Heart Circ. Physiol.* **2006**, *291*, H2036–H2038. [CrossRef] [PubMed]

68. Emerson, G.G.; Segal, S.S. Electrical coupling between endothelial cells and smooth muscle cells in hamster feed arteries: Role in vasomotor control. *Circ. Res.* **2000**, *87*, 474–479. [CrossRef] [PubMed]

69. Sandow, S.L.; Looft-Wilson, R.; Doran, B.; Grayson, T.H.; Segal, S.S.; Hill, C.E. Expression of homocellular and heterocellular gap junctions in hamster arterioles and feed arteries. *Cardiovasc. Res.* **2003**, *60*, 643–653. [CrossRef] [PubMed]

70. Dora, K.A.; Doyle, M.P.; Duling, B.R. Elevation of intracellular calcium in smooth muscle causes endothelial cell generation of NO in arterioles. *Proc. Natl. Acad. Sci. USA* **1997**, *94*, 6529–6534. [CrossRef] [PubMed]

71. Isakson, B.E.; Ramos, S.I.; Duling, B.R. Ca^{2+} and inositol 1,4,5-trisphosphate-mediated signaling across the myoendothelial junction. *Circ. Res.* **2007**, *100*, 246–254. [CrossRef] [PubMed]

72. Nausch, L.W.; Bonev, A.D.; Heppner, T.J.; Tallini, Y.; Kotlikoff, M.I.; Nelson, M.T. Sympathetic nerve stimulation induces local endothelial Ca^{2+} signals to oppose vasoconstriction of mouse mesenteric arteries. *Am. J. Physiol. Heart Circ. Physiol.* **2012**, *302*, H594–H602. [CrossRef] [PubMed]

73. Edwards, G.; Dora, K.A.; Gardener, M.J.; Garland, C.J.; Weston, A.H. K$^+$ is an endothelium-derived hyperpolarizing factor in rat arteries. *Nature* **1998**, *396*, 269–272. [CrossRef] [PubMed]

74. Archer, S.L.; Gragasin, F.S.; Wu, X.; Wang, S.; McMurtry, S.; Kim, D.H.; Platonov, M.; Koshal, A.; Hashimoto, K.; Campbell, W.B.; et al. Endothelium-derived hyperpolarizing factor in human internal mammary artery is 11,12-epoxyeicosatrienoic acid and causes relaxation by activating smooth muscle BK(Ca) channels. *Circulation* **2003**, *107*, 769–776. [CrossRef] [PubMed]

75. Fleming, I. Cytochrome P450 epoxygenases as EDHF synthase(s). *Pharmacol. Res.* **2004**, *49*, 525–533. [CrossRef] [PubMed]

76. Shimokawa, H.; Morikawa, K. Hydrogen peroxide is an endothelium-derived hyperpolarizing factor in animals and humans. *J. Mol. Cell. Cardiol.* **2005**, *39*, 725–732. [CrossRef] [PubMed]

77. Ahluwalia, A.; Hobbs, A.J. Endothelium-derived C-type natriuretic peptide: More than just a hyperpolarizing factor. *Trends Pharmacol. Sci.* **2005**, *26*, 162–167. [CrossRef] [PubMed]

78. Chauhan, S.D.; Nilsson, H.; Ahluwalia, A.; Hobbs, A.J. Release of C-type natriuretic peptide accounts for the biological activity of endothelium-derived hyperpolarizing factor. *Proc. Natl. Acad. Sci. USA* **2003**, *100*, 1426–1431. [CrossRef] [PubMed]

79. Garland, C.J.; Dora, K.A. EDH: Endothelium-dependent hyperpolarization and microvascular signalling. *Acta Physiol.* **2017**, *219*, 152–161. [CrossRef] [PubMed]

80. Boettcher, M.; de Wit, C. Distinct endothelium-derived hyperpolarizing factors emerge in vitro and in vivo and are mediated in part via connexin 40-dependent myoendothelial coupling. *Hypertension* **2011**, *57*, 802–808. [CrossRef] [PubMed]

81. Eichler, I.; Wibawa, J.; Grgic, I.; Knorr, A.; Brakemeier, S.; Pries, A.R.; Hoyer, J.; Kohler, R. Selective blockade of endothelial Ca^{2+}-activated small- and intermediate-conductance K$^+$-channels suppresses EDHF-mediated vasodilation. *Br. J. Pharmacol.* **2003**, *138*, 594–601. [CrossRef] [PubMed]

82. Doughty, J.M.; Plane, F.; Langton, P.D. Charybdotoxin and apamin block EDHF in rat mesenteric artery if selectively applied to the endothelium. *Am. J. Physiol.* **1999**, *276 Pt 2*, H1107–H1112. [CrossRef] [PubMed]

83. Dora, K.A.; Sandow, S.L.; Gallagher, N.T.; Takano, H.; Rummery, N.M.; Hill, C.E.; Garland, C.J. Myoendothelial gap junctions may provide the pathway for EDHF in mouse mesenteric artery. *J. Vasc. Res.* **2003**, *40*, 480–490. [CrossRef] [PubMed]

84. Feletou, M.; Vanhoutte, P.M.; Weston, A.H.; Edwards, G. EDHF and endothelial potassiun channels: IK$_{Ca}$ and SK$_{Ca}$. *Br. J. Pharmacol.* **2003**, *140*, 225. [CrossRef] [PubMed]

85. Griffith, T.M. Endothelium-dependent smooth muscle hyperpolarization: Do gap junctions provide a unifying hypothesis? *Br. J. Pharmacol.* **2004**, *141*, 881–903. [CrossRef] [PubMed]

86. Feletou, M.; Vanhoutte, P.M. Endothelium-dependent hyperpolarization: No longer an f-word! *J. Cardiovasc. Pharmacol.* **2013**, *61*, 91–92. [CrossRef] [PubMed]

87. Sandow, S.L.; Hill, C.E. Incidence of myoendothelial gap junctions in the proximal and distal mesenteric arteries of the rat is suggestive of a role in endothelium-derived hyperpolarizing factor-mediated responses. *Circ. Res.* **2000**, *86*, 341–346. [CrossRef] [PubMed]

88. Shimokawa, H.; Yasutake, H.; Fujii, K.; Owada, M.K.; Nakaike, R.; Fukumoto, Y.; Takayanagi, T.; Nagao, T.; Egashira, K.; Fujishima, M.; et al. The importance of the hyperpolarizing mechanism increases as the vessel size decreases in endothelium-dependent relaxations in rat mesenteric circulation. *J. Cardiovasc. Pharmacol.* **1996**, *28*, 703–711. [CrossRef] [PubMed]

89. Chaytor, A.T.; Bakker, L.M.; Edwards, D.H.; Griffith, T.M. Connexin-mimetic peptides dissociate electrotonic EDHF-type signalling via myoendothelial and smooth muscle gap junctions in the rabbit iliac artery. *Br. J. Pharmacol.* **2005**, *144*, 108–114. [CrossRef] [PubMed]

90. De Vriese, A.S.; Van de Voorde, J.; Lameire, N.H. Effects of connexin-mimetic peptides on nitric oxide synthase- and cyclooxygenase-independent renal vasodilation. *Kidney Int.* **2002**, *61*, 177–185. [CrossRef] [PubMed]

91. Mather, S.; Dora, K.A.; Sandow, S.L.; Winter, P.; Garland, C.J. Rapid endothelial cell-selective loading of connexin 40 antibody blocks endothelium-derived hyperpolarizing factor dilation in rat small mesenteric arteries. *Circ. Res.* **2005**, *97*, 399–407. [CrossRef] [PubMed]

92. Straub, A.C.; Lohman, A.W.; Billaud, M.; Johnstone, S.R.; Dwyer, S.T.; Lee, M.Y.; Bortz, P.S.; Best, A.K.; Columbus, L.; Gaston, B.; et al. Endothelial cell expression of haemoglobin alpha regulates nitric oxide signalling. *Nature* **2012**, *491*, 473–477. [CrossRef] [PubMed]

93. Figueroa, X.F.; Lillo, M.A.; Gaete, P.S.; Riquelme, M.A.; Saez, J.C. Diffusion of nitric oxide across cell membranes of the vascular wall requires specific connexin-based channels. *Neuropharmacology* **2013**, *75*, 471–478. [CrossRef] [PubMed]

94. Bolz, S.S.; de Wit, C.; Pohl, U. Endothelium-derived hyperpolarizing factor but not NO reduces smooth muscle Ca^{2+} during acetylcholine-induced dilation of microvessels. *Br. J. Pharmacol.* **1999**, *128*, 124–134. [CrossRef] [PubMed]

95. Bolz, S.S.; Vogel, L.; Sollinger, D.; Derwand, R.; de Wit, C.; Loirand, G.; Pohl, U. Nitric oxide-induced decrease in calcium sensitivity of resistance arteries is attributable to activation of the myosin light chain phosphatase and antagonized by the RhoA/Rho kinase pathway. *Circulation* **2003**, *107*, 3081–3087. [CrossRef] [PubMed]

96. Bagher, P.; Segal, S.S. Regulation of blood flow in the microcirculation: Role of conducted vasodilation. *Acta Physiol.* **2011**, *202*, 271–284. [CrossRef] [PubMed]

97. Emerson, G.G.; Segal, S.S. Endothelial cell pathway for conduction of hyperpolarization and vasodilation along hamster feed artery. *Circ. Res.* **2000**, *86*, 94–100. [CrossRef] [PubMed]

98. Figueroa, X.F.; Duling, B.R. Dissection of two Cx37-independent conducted vasodilator mechanisms by deletion of Cx40: Electrotonic versus regenerative conduction. *Am. J. Physiol. Heart Circ. Physiol.* **2008**, *295*, H2001–H2007. [CrossRef] [PubMed]

99. De Wit, C.; Griffith, T.M. Connexins and gap junctions in the EDHF phenomenon and conducted vasomotor responses. *Pflugers Arch.* **2010**, *459*, 897–914. [CrossRef] [PubMed]

100. Haefliger, J.A.; Nicod, P.; Meda, P. Contribution of connexins to the function of the vascular wall. *Cardiovasc. Res.* **2004**, *62*, 345–356. [CrossRef] [PubMed]

101. Xia, J.; Duling, B.R. Electromechanical coupling and the conducted vasomotor response. *Am. J. Physiol.* **1995**, *269 Pt 2*, H2022–H2030. [CrossRef] [PubMed]

102. Xia, J.; Little, T.L.; Duling, B.R. Cellular pathways of the conducted electrical response in arterioles of hamster cheek pouch in vitro. *Am. J. Physiol.* **1995**, *269 Pt 2*, H2031–H2038. [CrossRef] [PubMed]

103. Gustafsson, F.; Holstein-Rathlou, N. Conducted vasomotor responses in arterioles: Characteristics, mechanisms and physiological significance. *Acta Physiol. Scand.* **1999**, *167*, 11–21. [CrossRef] [PubMed]

104. Welsh, D.G.; Segal, S.S. Endothelial and smooth muscle cell conduction in arterioles controlling blood flow. *Am. J. Physiol.* **1998**, *274 Pt 2*, H178–H186. [CrossRef] [PubMed]

105. Welsh, D.G.; Segal, S.S. Role of EDHF in conduction of vasodilation along hamster cheek pouch arterioles in vivo. *Am. J. Physiol. Heart Circ. Physiol.* **2000**, *278*, H1832–H1839. [CrossRef] [PubMed]

106. Segal, S.S.; Welsh, D.G.; Kurjiaka, D.T. Spread of vasodilatation and vasoconstriction along feed arteries and arterioles of hamster skeletal muscle. *J. Physiol.* **1999**, *516 Pt 1*, 283–291. [CrossRef] [PubMed]

107. Bartlett, I.S.; Segal, S.S. Resolution of smooth muscle and endothelial pathways for conduction along hamster cheek pouch arterioles. *Am. J. Physiol. Heart Circ. Physiol.* **2000**, *278*, H604–H612. [CrossRef] [PubMed]

108. Budel, S.; Bartlett, I.S.; Segal, S.S. Homocellular conduction along endothelium and smooth muscle of arterioles in hamster cheek pouch: Unmasking an NO wave. *Circ. Res.* **2003**, *93*, 61–68. [CrossRef] [PubMed]

109. Segal, S.S.; Jacobs, T.L. Role for endothelial cell conduction in ascending vasodilatation and exercise hyperaemia in hamster skeletal muscle. *J. Physiol.* **2001**, *536 Pt 3*, 937–946. [CrossRef] [PubMed]

110. Okamoto, T.; Akiyama, M.; Takeda, M.; Gabazza, E.C.; Hayashi, T.; Suzuki, K. Connexin32 is expressed in vascular endothelial cells and participates in gap-junction intercellular communication. *Biochem. Biophys. Res. Commun.* **2009**, *382*, 264–268. [CrossRef] [PubMed]

111. Severs, N.J.; Rothery, S.; Dupont, E.; Coppen, S.R.; Yeh, H.I.; Ko, Y.S.; Matsushita, T.; Kaba, R.; Halliday, D. Immunocytochemical analysis of connexin expression in the healthy and diseased cardiovascular system. *Microsc. Res. Tech.* **2001**, *52*, 301–322. [CrossRef]

112. Van Kempen, M.J.; Jongsma, H.J. Distribution of connexin37, connexin40 and connexin43 in the aorta and coronary artery of several mammals. *Histochem. Cell Biol.* **1999**, *112*, 479–486. [CrossRef] [PubMed]

113. De Wit, C.; Roos, F.; Bolz, S.S.; Kirchhoff, S.; Kruger, O.; Willecke, K.; Pohl, U. Impaired conduction of vasodilation along arterioles in connexin40-deficient mice. *Circ. Res.* **2000**, *86*, 649–655. [CrossRef] [PubMed]

114. De Wit, C.; Roos, F.; Bolz, S.S.; Pohl, U. Lack of vascular connexin 40 is associated with hypertension and irregular arteriolar vasomotion. *Physiol. Genom.* **2003**, *13*, 169–177. [CrossRef] [PubMed]

115. Figueroa, X.F.; Paul, D.L.; Simon, A.M.; Goodenough, D.A.; Day, K.H.; Damon, D.N.; Duling, B.R. Central role of connexin40 in the propagation of electrically activated vasodilation in mouse cremasteric arterioles in vivo. *Circ. Res.* **2003**, *92*, 793–800. [CrossRef] [PubMed]

116. Jobs, A.; Schmidt, K.; Schmidt, V.J.; Lubkemeier, I.; van Veen, T.A.; Kurtz, A.; Willecke, K.; de Wit, C. Defective Cx40 maintains Cx37 expression but intact Cx40 is crucial for conducted dilations irrespective of hypertension. *Hypertension* **2012**, *60*, 1422–1429. [CrossRef] [PubMed]

117. Wagner, C.; de Wit, C.; Kurtz, L.; Grunberger, C.; Kurtz, A.; Schweda, F. Connexin40 is essential for the pressure control of renin synthesis and secretion. *Circ. Res.* **2007**, *100*, 556–563. [CrossRef] [PubMed]

118. Gollob, M.H.; Jones, D.L.; Krahn, A.D.; Danis, L.; Gong, X.Q.; Shao, Q.; Liu, X.; Veinot, J.P.; Tang, A.S.; Stewart, A.F.; et al. Somatic mutations in the connexin 40 gene (GJA5) in atrial fibrillation. *N. Engl. J. Med.* **2006**, *354*, 2677–2688. [CrossRef] [PubMed]

119. Lubkemeier, I.; Andrie, R.; Lickfett, L.; Bosen, F.; Stockigt, F.; Dobrowolski, R.; Draffehn, A.M.; Fregeac, J.; Schultze, J.L.; Bukauskas, F.F.; et al. The Connexin40A96S mutation from a patient with atrial fibrillation causes decreased atrial conduction velocities and sustained episodes of induced atrial fibrillation in mice. *J. Mol. Cell. Cardiol.* **2013**, *65*, 19–32. [CrossRef] [PubMed]

120. Santa Cruz, A.; Mese, G.; Valiuniene, L.; Brink, P.R.; White, T.W.; Valiunas, V. Altered conductance and permeability of Cx40 mutations associated with atrial fibrillation. *J. Gen. Physiol.* **2015**, *146*, 387–398. [CrossRef] [PubMed]

121. Lohman, A.W.; Billaud, M.; Straub, A.C.; Johnstone, S.R.; Best, A.K.; Lee, M.; Barr, K.; Penuela, S.; Laird, D.W.; Isakson, B.E. Expression of pannexin isoforms in the systemic murine arterial network. *J. Vasc. Res.* **2012**, *49*, 405–416. [CrossRef] [PubMed]

122. Lohman, A.W.; Isakson, B.E. Differentiating connexin hemichannels and pannexin channels in cellular ATP release. *FEBS Lett.* **2014**, *588*, 1379–1388. [CrossRef] [PubMed]

123. Penuela, S.; Kelly, J.J.; Churko, J.M.; Barr, K.J.; Berger, A.C.; Laird, D.W. Panx1 regulates cellular properties of keratinocytes and dermal fibroblasts in skin development and wound healing. *J. Investig. Dermatol.* **2014**, *134*, 2026–2035. [CrossRef] [PubMed]

124. Billaud, M.; Lohman, A.W.; Johnstone, S.R.; Biwer, L.A.; Mutchler, S.; Isakson, B.E. Regulation of cellular communication by signaling microdomains in the blood vessel wall. *Pharmacol. Rev.* **2014**, *66*, 513–569. [CrossRef] [PubMed]

125. Angus, J.A.; Wright, C.E. Novel alpha1-adrenoceptor antagonism by the fluroquinolone antibiotic trovafloxacin. *Eur. J. Pharmacol.* **2016**, *791*, 179–184. [CrossRef] [PubMed]

126. Chiu, Y.H.; Jin, X.; Medina, C.B.; Leonhardt, S.A.; Kiessling, V.; Bennett, B.C.; Shu, S.; Tamm, L.K.; Yeager, M.; Ravichandran, K.S.; et al. A quantized mechanism for activation of pannexin channels. *Nat. Commun.* **2017**, *8*, 14324. [CrossRef] [PubMed]

127. Good, M.E.; Chiu, Y.H.; Poon, I.K.H.; Medina, C.B.; Butcher, J.T.; Mendu, S.K.; DeLalio, L.J.; Lohman, A.W.; Leitinger, N.; Barrett, E.; et al. Pannexin 1 Channels as an Unexpected New Target of the Anti-Hypertensive Drug Spironolactone. *Circ. Res.* **2018**, *122*, 606–615. [CrossRef] [PubMed]

128. Kauffenstein, G.; Tamareille, S.; Prunier, F.; Roy, C.; Ayer, A.; Toutain, B.; Billaud, M.; Isakson, B.E.; Grimaud, L.; Loufrani, L.; et al. Central Role of P2Y6 UDP Receptor in Arteriolar Myogenic Tone. *Arterioscler. Thromb. Vasc. Biol.* **2016**, *36*, 1598–1606. [CrossRef] [PubMed]

129. Nyberg, M.; Piil, P.; Kiehn, O.T.; Maagaard, C.; Jorgensen, T.S.; Egelund, J.; Isakson, B.E.; Nielsen, M.S.; Gliemann, L.; Hellsten, Y. Probenecid Inhibits alpha-Adrenergic Receptor-Mediated Vasoconstriction in the Human Leg Vasculature. *Hypertension* **2018**, *71*, 151–159. [CrossRef] [PubMed]

130. Billaud, M.; Chiu, Y.H.; Lohman, A.W.; Parpaite, T.; Butcher, J.T.; Mutchler, S.M.; DeLalio, L.J.; Artamonov, M.V.; Sandilos, J.K.; Best, A.K.; et al. A molecular signature in the pannexin1 intracellular loop confers channel activation by the alpha1 adrenoreceptor in smooth muscle cells. *Sci. Signal.* **2015**, *8*, ra17. [CrossRef] [PubMed]

131. Poon, I.K.; Chiu, Y.H.; Armstrong, A.J.; Kinchen, J.M.; Juncadella, I.J.; Bayliss, D.A.; Ravichandran, K.S. Unexpected link between an antibiotic, pannexin channels and apoptosis. *Nature* **2014**, *507*, 329–334. [CrossRef] [PubMed]

132. Lohman, A.W.; Weaver, J.L.; Billaud, M.; Sandilos, J.K.; Griffiths, R.; Straub, A.C.; Penuela, S.; Leitinger, N.; Laird, D.W.; Bayliss, D.A.; et al. S-nitrosylation inhibits pannexin 1 channel function. *J. Biol. Chem.* **2012**, *287*, 39602–39612. [CrossRef] [PubMed]

133. Gaynullina, D.; Tarasova, O.S.; Kiryukhina, O.O.; Shestopalov, V.I.; Panchin, Y. Endothelial function is impaired in conduit arteries of pannexin1 knockout mice. *Biol. Direct* **2014**, *9*, 8. [CrossRef] [PubMed]

134. Good, M.E.; Eucker, S.A.; Li, J.; Bacon, H.M.; Lang, S.M.; Butcher, J.T.; Johnson, T.J.; Gaykema, R.P.; Patel, M.K.; Zuo, Z.; et al. Endothelial cell Pannexin1 modulates severity of ischemic stroke by regulating cerebral inflammation and myogenic tone. *JCI Insight* **2018**, *3*. [CrossRef] [PubMed]

135. Lohman, A.W.; Leskov, I.L.; Butcher, J.T.; Johnstone, S.R.; Stokes, T.A.; Begandt, D.; DeLalio, L.J.; Best, A.K.; Penuela, S.; Leitinger, N.; et al. Pannexin 1 channels regulate leukocyte emigration through the venous endothelium during acute inflammation. *Nat. Commun.* **2015**, *6*, 7965. [CrossRef] [PubMed]

136. Keller, A.S.; Diederich, L.; Panknin, C.; DeLalio, L.J.; Drake, J.C.; Sherman, R.; Jackson, E.K.; Yan, Z.; Kelm, M.; Cortese-Krott, M.M.; et al. Possible roles for ATP release from RBCs exclude the cAMP-mediated Panx1 pathway. *Am. J. Physiol. Cell Physiol.* **2017**, *313*, C593–C603. [CrossRef] [PubMed]

137. Sharma, A.K.; Charles, E.J.; Zhao, Y.; Narahari, A.K.; Baderdinni, P.K.; Good, M.E.; Lorenz, U.M.; Kron, I.L.; Bayliss, D.A.; Ravichandran, K.S.; et al. Pannexin 1 channels on endothelial cells mediate vascular inflammation during lung ischemia-reperfusion injury. *Am. J. Physiol. Lung Cell. Mol. Physiol.* **2018**. [CrossRef] [PubMed]

138. Jankowski, J.; Perry, H.M.; Medina, C.B.; Huang, L.; Yao, Y.; Bajwa, A.; Lorenz, U.M.; Rosin, D.L.; Ravichandran, K.S.; Isakson, B.E.; et al. Epithelial and Endothelial Pannexin 1 Channels Mediate AKI. *J. Am. Soc. Nephrol.* **2018**, in press.

139. Sikora, J.; Orlov, S.N.; Furuya, K.; Grygorczyk, R. Hemolysis is a primary ATP-release mechanism in human erythrocytes. *Blood* **2014**, *124*, 2150–2157. [CrossRef] [PubMed]

140. Munger, S.J.; Kanady, J.D.; Simon, A.M. Absence of venous valves in mice lacking Connexin37. *Dev. Biol.* **2013**, *373*, 338–348. [CrossRef] [PubMed]

141. Kanady, J.D.; Dellinger, M.T.; Munger, S.J.; Witte, M.H.; Simon, A.M. Connexin37 and Connexin43 deficiencies in mice disrupt lymphatic valve development and result in lymphatic disorders including lymphedema and chylothorax. *Dev. Biol.* **2011**, *354*, 253–266. [CrossRef] [PubMed]

142. Sabine, A.; Agalarov, Y.; Maby-El Hajjami, H.; Jaquet, M.; Hagerling, R.; Pollmann, C.; Bebber, D.; Pfenniger, A.; Miura, N.; Dormond, O.; et al. Mechanotransduction, PROX1, and FOXC2 cooperate to control connexin37 and calcineurin during lymphatic-valve formation. *Dev. Cell.* **2012**, *22*, 430–445. [CrossRef] [PubMed]

143. Munger, S.J.; Geng, X.; Srinivasan, R.S.; Witte, M.H.; Paul, D.L.; Simon, A.M. Segregated Foxc2, NFATc1 and Connexin expression at normal developing venous valves, and Connexin-specific differences in the valve phenotypes of Cx37, Cx43, and Cx47 knockout mice. *Dev. Biol.* **2016**, *412*, 173–190. [CrossRef] [PubMed]

144. Lyons, O.; Saha, P.; Seet, C.; Kuchta, A.; Arnold, A.; Grover, S.; Rashbrook, V.; Sabine, A.; Vizcay-Barrena, G.; Patel, A.; et al. Human venous valve disease caused by mutations in FOXC2 and GJC2. *J. Exp. Med.* **2017**. [CrossRef] [PubMed]

145. Meens, M.J.; Kutkut, I.; Rochemont, V.; Dubrot, J.; Kaladji, F.R.; Sabine, A.; Lyons, O.; Hendrikx, S.; Bernier-Latmani, J.; Kiefer, F.; et al. Cx47 fine-tunes the handling of serum lipids but is dispensable for lymphatic vascular function. *PLoS ONE* **2017**, *12*, e0181476. [CrossRef] [PubMed]

146. Brice, G.; Ostergaard, P.; Jeffery, S.; Gordon, K.; Mortimer, P.S.; Mansour, S. A novel mutation in GJA1 causing oculodentodigital syndrome and primary lymphoedema in a three generation family. *Clin. Genet.* **2013**, *84*, 378–381. [CrossRef] [PubMed]

147. Brouillard, P.; Boon, L.; Vikkula, M. Genetics of lymphatic anomalies. *J. Clin. Investig.* **2014**, *124*, 898–904. [CrossRef] [PubMed]

148. Ferrell, R.E.; Baty, C.J.; Kimak, M.A.; Karlsson, J.M.; Lawrence, E.C.; Franke-Snyder, M.; Meriney, S.D.; Feingold, E.; Finegold, D.N. GJC2 missense mutations cause human lymphedema. *Am. J. Hum. Genet.* **2010**, *86*, 943–948. [CrossRef] [PubMed]

149. Ostergaard, E.; Rodenburg, R.J.; van den Brand, M.; Thomsen, L.L.; Duno, M.; Batbayli, M.; Wibrand, F.; Nijtmans, L. Respiratory chain complex I deficiency due to NDUFA12 mutations as a new cause of Leigh syndrome. *J. Med. Genet.* **2011**, *48*, 737–740. [CrossRef] [PubMed]

150. Finegold, D.N.; Baty, C.J.; Knickelbein, K.Z.; Perschke, S.; Noon, S.E.; Campbell, D.; Karlsson, J.M.; Huang, D.; Kimak, M.A.; Lawrence, E.C.; et al. Connexin 47 mutations increase risk for secondary lymphedema following breast cancer treatment. *Clin. Cancer Res.* **2012**, *18*, 2382–2890. [CrossRef] [PubMed]

151. Hadizadeh, M.; Mohaddes Ardebili, S.M.; Salehi, M.; Young, C.; Mokarian, F.; McClellan, J.; Xu, Q.; Kazemi, M.; Moazam, E.; Mahaki, B.; et al. GJA4/Connexin 37 Mutations Correlate with Secondary Lymphedema Following Surgery in Breast Cancer Patients. *Biomedicines* **2018**, *6*, 23. [CrossRef] [PubMed]

152. Kuczma, M.; Lee, J.R.; Kraj, P. Connexin 43 signaling enhances the generation of Foxp3+ regulatory T cells. *J. Immunol.* **2011**, *187*, 248–257. [CrossRef] [PubMed]

153. Oviedo-Orta, E.; Perreau, M.; Evans, W.H.; Potolicchio, I. Control of the proliferation of activated CD4+ T cells by connexins. *J. Leukoc. Biol.* **2010**, *88*, 79–86. [CrossRef] [PubMed]

154. Pfenniger, A.; Chanson, M.; Kwak, B.R. Connexins in atherosclerosis. *Biochim. Biophys. Acta* **2013**, *1828*, 157–166. [CrossRef] [PubMed]

155. Kato, K.; Kanaji, T.; Russell, S.; Kunicki, T.J.; Furihata, K.; Kanaji, S.; Marchese, P.; Reininger, A.; Ruggeri, Z.M.; Ware, J. The contribution of glycoprotein VI to stable platelet adhesion and thrombus formation illustrated by targeted gene deletion. *Blood* **2003**, *102*, 1701–1707. [CrossRef] [PubMed]

156. Massberg, S.; Gawaz, M.; Gruner, S.; Schulte, V.; Konrad, I.; Zohlnhofer, D.; Heinzmann, U.; Nieswandt, B. A crucial role of glycoprotein VI for platelet recruitment to the injured arterial wall in vivo. *J. Exp. Med.* **2003**, *197*, 41–49. [CrossRef] [PubMed]

157. Inoue, O.; Suzuki-Inoue, K.; Dean, W.L.; Frampton, J.; Watson, S.P. Integrin alpha2beta1 mediates outside-in regulation of platelet spreading on collagen through activation of Src kinases and PLCgamma2. *J. Cell Biol.* **2003**, *160*, 769–780. [CrossRef] [PubMed]

158. Gibbins, J.; Asselin, J.; Farndale, R.; Barnes, M.; Law, C.L.; Watson, S.P. Tyrosine phosphorylation of the Fc receptor gamma-chain in collagen-stimulated platelets. *J. Biol. Chem.* **1996**, *271*, 18095–18099. [CrossRef] [PubMed]

159. Gibbins, J.M.; Okuma, M.; Farndale, R.; Barnes, M.; Watson, S.P. Glycoprotein VI is the collagen receptor in platelets which underlies tyrosine phosphorylation of the Fc receptor gamma-chain. *FEBS Lett.* **1997**, *413*, 255–259. [CrossRef]

160. Suzuki-Inoue, K.; Tulasne, D.; Shen, Y.; Bori-Sanz, T.; Inoue, O.; Jung, S.M.; Moroi, M.; Andrews, R.K.; Berndt, M.C.; Watson, S.P. Association of Fyn and Lyn with the proline-rich domain of glycoprotein VI regulates intracellular signaling. *J. Biol. Chem.* **2002**, *277*, 21561–21566. [CrossRef] [PubMed]

161. Gibbins, J.M. Platelet adhesion signalling and the regulation of thrombus formation. *J. Cell Sci.* **2004**, *117 Pt 16*, 3415–3425. [CrossRef] [PubMed]

162. Watson, S.P.; Auger, J.M.; McCarty, O.J.; Pearce, A.C. GPVI and integrin alphaIIb beta3 signaling in platelets. *J. Thromb. Haemost.* **2005**, *3*, 1752–1762. [CrossRef] [PubMed]

163. Ruggeri, Z.M. Platelets in atherothrombosis. *Nat. Med.* **2002**, *8*, 1227–1234. [CrossRef] [PubMed]

164. Shattil, S.J. Signaling through platelet integrin alpha IIb beta 3: Inside-out, outside-in, and sideways. *Thromb. Haemost.* **1999**, *82*, 318–325. [PubMed]

165. Shattil, S.J.; Kashiwagi, H.; Pampori, N. Integrin signaling: The platelet paradigm. *Blood* **1998**, *91*, 2645–2657. [PubMed]

166. Jones, C.I.; Barrett, N.E.; Moraes, L.A.; Gibbins, J.M.; Jackson, D.E. Endogenous inhibitory mechanisms and the regulation of platelet function. *Methods Mol. Biol.* **2012**, *788*, 341–366. [PubMed]

167. Prevost, N.; Woulfe, D.; Tanaka, T.; Brass, L.F. Interactions between Eph kinases and ephrins provide a mechanism to support platelet aggregation once cell-to-cell contact has occurred. *Proc. Natl. Acad. Sci. USA* **2002**, *99*, 9219–9224. [CrossRef] [PubMed]

168. Prevost, N.; Woulfe, D.S.; Jiang, H.; Stalker, T.J.; Marchese, P.; Ruggeri, Z.M.; Brass, L.F. Eph kinases and ephrins support thrombus growth and stability by regulating integrin outside-in signaling in platelets. *Proc. Natl. Acad. Sci. USA* **2005**, *102*, 9820–9825. [CrossRef] [PubMed]

169. Vaiyapuri, S.; Sage, T.; Rana, R.H.; Schenk, M.P.; Ali, M.S.; Unsworth, A.J.; Jones, C.I.; Stainer, A.R.; Kriek, N.; Moraes, L.A.; et al. EphB2 regulates contact-dependent and contact-independent signaling to control platelet function. *Blood* **2015**, *125*, 720–730. [CrossRef] [PubMed]

170. Prevost, N.; Woulfe, D.; Tognolini, M.; Brass, L.F. Contact-dependent signaling during the late events of platelet activation. *J. Thromb. Haemost.* **2003**, *1*, 1613–1627. [CrossRef] [PubMed]

171. Nesbitt, W.S.; Giuliano, S.; Kulkarni, S.; Dopheide, S.M.; Harper, I.S.; Jackson, S.P. Intercellular calcium communication regulates platelet aggregation and thrombus growth. *J. Cell Biol.* **2003**, *160*, 1151–1161. [CrossRef] [PubMed]

172. Angellilo-Scherrer, A.; Fontana, P.; Burnier, L.; Roth, I.; Sugamele, R.; Brisset, A.; Morel, S.; Nolli, S.; Sutter, E.; Chassot, A.; et al. Connexin 37 limits thrombus propensity by downregulating platelet reactivity. *Circulation* **2011**, *124*, 930–939. [CrossRef] [PubMed]

173. Vaiyapuri, S.; Jones, C.I.; Sasikumar, P.; Moraes, L.A.; Munger, S.J.; Wright, J.R.; Ali, M.S.; Sage, T.; Kaiser, W.J.; Tucker, K.L.; et al. Gap junctions and connexin hemichannels underpin hemostasis and thrombosis. *Circulation* **2012**, *125*, 2479–2491. [CrossRef] [PubMed]

174. Vaiyapuri, S.; Moraes, L.A.; Sage, T.; Ali, M.S.; Lewis, K.R.; Mahaut-Smith, M.P.; Oviedo-Orta, E.; Simon, A.M.; Gibbins, J.M. Connexin40 regulates platelet function. *Nat. Commun.* **2013**, *4*, 2564. [CrossRef] [PubMed]

175. Taylor, K.A.; Wright, J.R.; Vial, C.; Evans, R.J.; Mahaut-Smith, M.P. Amplification of human platelet activation by surface pannexin-1 channels. *J. Thromb. Haemost.* **2014**, *12*, 987–998. [CrossRef] [PubMed]

176. Molica, F.; Stierlin, F.B.; Fontana, P.; Kwak, B.R. Pannexin- and Connexin-Mediated Intercellular Communication in Platelet Function. *Int. J. Mol. Sci.* **2017**, *18*, 850. [CrossRef] [PubMed]

177. Stalker, T.J.; Traxler, E.A.; Wu, J.; Wannemacher, K.M.; Cermignano, S.L.; Voronov, R.; Diamond, S.L.; Brass, L.F. Hierarchical organization in the hemostatic response and its relationship to the platelet-signaling network. *Blood* **2013**, *121*, 1875–1885. [CrossRef] [PubMed]

178. Morel, S.; Burnier, L.; Kwak, B.R. Connexins participate in the initiation and progression of atherosclerosis. *Semin. Immunopathol.* **2009**, *31*, 49–61. [CrossRef] [PubMed]

179. Wong, C.W.; Burger, F.; Pelli, G.; Mach, F.; Kwak, B.R. Dual benefit of reduced Cx43 on atherosclerosis in LDL receptor-deficient mice. *Cell Commun. Adhes.* **2003**, *10*, 395–400. [CrossRef] [PubMed]

180. Leybaert, L.; Lampe, P.D.; Dhein, S.; Kwak, B.R.; Ferdinandy, P.; Beyer, E.C.; Laird, D.W.; Naus, C.C.; Green, C.R.; Schulz, R. Connexins in Cardiovascular and Neurovascular Health and Disease: Pharmacological Implications. *Pharmacol. Rev.* **2017**, *69*, 396–478. [CrossRef] [PubMed]

International Journal of
Molecular Sciences

MDPI

Article

Differential Association of Cx37 and Cx40 Genetic Variants in Atrial Fibrillation with and without Underlying Structural Heart Disease

Sebastian Carballo [1,*], Anna Pfenniger [2], David Carballo [3], Nicolas Garin [1], Richard W James [4], François Mach [3], Dipen Shah [3,†] and Brenda R Kwak [2,†]

[1]   Service of General Internal medicine, University Hospitals of Geneva, 1211 Geneva, Switzerland;
      nicolas.garin@hcuge.ch
[2]   Department of Pathology and Immunology, University of Geneva, 1211 Geneva, Switzerland;
      anna.pfenniger@gmail.com (A.P.); Brenda.KwakChanson@unige.ch (B.R.K.)
[3]   Service of Cardiology, University Hospitals of Geneva, 1211 Geneva, Switzerland;
      david.carballo@hcuge.ch (D.C.); francois.mach@hcuge.ch (F.M.); dipen.shah@hcuge.ch (D.S.)
[4]   Service of Endocrinology and Diabetes, University Hospitals of Geneva, 1211 Geneva, Switzerland;
      richard.james@hcuge.ch
*   Correspondence: sebastian.carballo@hcuge.ch; Tel.: +41-22-372-9216
†   These authors contributed equally to this work.

Received: 19 December 2017; Accepted: 16 January 2018; Published: 19 January 2018

Abstract: Atrial fibrillation (AF) appears in the presence or absence of structural heart disease. The majority of foci causing AF are located near the ostia of pulmonary veins (PVs), where cardiomyocytes and vascular smooth muscle cells interdigitate. Connexins (Cx) form gap junction channels and participate in action potential propagation. Genetic variants in genes encoding Cx40 and Cx37 affect their expression or function and may contribute to PV arrhythmogenicity. DNA was obtained from 196 patients with drug-resistant, symptomatic AF with and without structural heart disease, who were referred for percutaneous catheter ablation. Eighty-nine controls were matched for age, gender, hypertension, and BMI. Genotyping of the Cx40 $-44G > A$, Cx40 $+71A > G$, Cx40 $-26A > G$, and Cx37 $1019C > T$ polymorphisms was performed. The promoter A Cx40 polymorphisms ($-44G > A$ and $+71A > G$) showed no association with non-structural or structural AF. Distribution of the Cx40 promoter B polymorphism ($-26A > G$) was different in structural AF when compared to controls ($p = 0.03$). There was no significant difference with non-structural AF ($p = 0.50$). The distribution of the Cx37 $1019C > T$ polymorphism was different in non-structural AF ($p = 0.03$) but not in structural AF ($p = 0.08$) when compared to controls. Our study describes for the first time an association of drug-resistant non-structural heart disease AF with the Cx37 $1019C > T$ gene polymorphism. We also confirmed the association of the Cx40 $- 26G > A$ polymorphism in patients with AF and structural disease.

Keywords: atrial fibrillation; connexin; polymorphism; genetic variant

---

1. Introduction

Atrial fibrillation (AF) is the most common sustained cardiac arrhythmia, its incidence increasing with age to reach up to 35 new cases per 1000 population in the elderly [1,2]. In the general population, its prevalence is estimated at 1.5–2%. AF is associated with increased risk of stroke, incidence of heart failure, and an overall higher mortality [3,4]. Diagnosis of AF is based on electrocardiographic characteristics such as rapid, irregular, fibrillatory waves, varying in size, shape, and timing [1]. Five clinical types of AF are recognized, mainly based on duration [2–5]: paroxysmal AF, persistent AF, long-standing persistent AF, and permanent AF. Whilst controversial, the term idiopathic or lone AF is used in reference to AF appearing in the absence of structural disease such as atrial left enlargement or valvular heart disease [6]. This historical concept and term should probably and progressively be avoided as understanding of the multiple complex etiologies of AF is improved [7].

When underlying structural heart disease is present, such as in patients with hypertensive heart disease, valvulopathy, or ischemic heart disease, this may trigger progressive remodeling of both the ventricles and atria [3,4]. Structural abnormalities will induce remodeling of the myocardium by affecting the extracellular matrix and cardiomyocytes through inflammatory processes and increased fibrosis [3,4]. Electrophysiological alterations and remodeling are instrumental in maintaining AF after its onset [8]. Initiation of AF is thought to depend on a combination of abnormal impulse formation, conduction and a propensity for reentry at the ostia of pulmonary veins (PVs), where sheets of cardiomyocytes and vascular smooth muscle cells (VSMCs) interdigitate [9,10]. However, the molecular mechanism underlying this regional specificity remains to be discovered.

Variants in genes encoding for membrane channels have been associated with both familial and other forms of AF, including genes encoding potassium channels such as *KCNQ1* and *KCNA5* [11–13]. Interestingly, atrial vulnerability to AF has been associated with polymorphisms in the Connexin40 (Cx40) gene, *GJA5* [14–17]. In addition, specific somatic mutations in the same gene have been discovered in idiopathic AF [18]. Other clinical AF presentations have been described with mosaicism in another connexin gene, *GJA1* encoding for Cx43 [19].

Connexins are gap junction proteins playing an essential role in direct cell–cell communication in the vast majority of tissues in the body, including in electrical propagation in the heart [20]. There are 20 different connexins in humans, each forming channels with distinct properties and with specific expression patterns [21]. Cx40, encoded by *GJA5*, is expressed by endothelial cells (ECs), by coronary VSMCs, by atrial cardiomyocytes, and in the cardiac conduction system. Cx37, encoded by *GJA4*, is found in ECs, pulmonary vein VSMCs, monocyte/macrophages, and platelets. Cx43, encoded by *GJA1*, is expressed by ventricular and atrial cardiomyocytes, VSMCs, ECs, and monocyte/macrophages [22]. Several polymorphisms have been described in *GJA4* and *GJA5*, which affect either their connexin expression level or gap junction channel function, some of which may contribute to the arrhythmogenicity of the PVs [15]. In animal studies, downregulation of either Cx40 or Cx43 expression increases susceptibility to AF [23,24].

Variants in the Cx37 (*GJA4*) gene have been associated with atherosclerosis and coronary heart disease, most likely through their influence on monocyte adhesion, thereby regulating the extent of local inflammation [25]. Both systemic and local inflammation appear to influence the development of AF in various situations, in patients having undergone cardiac surgery as well as non-post-operative patients [26].

Here, we investigated the relation between polymorphisms in *GJA5* and *GJA4* in defined AF populations. Within the AF population, a distinction was made between patients with no underlying structural cardiomyopathy, for the purpose of clarity referred to as non-structural AF, and those with clearly documented structural disease, referred to as structural AF. All AF patients that were included had undergone a trial of anti-arrhythmic drugs and were referred for catheter ablation because of poor response.

2. Results

A total of 285 individuals were included in this study; 92 patients in the non-structural AF group, 104 patients with structural AF, and 89 controls. Characteristics of the study population are summarized in Table 1.

Table 1. Characteristics and cardiovascular risk factors in the three cohorts.

Characteristics	Controls $n = 89$	Non-Structural AF $n = 92$	Structural AF $n = 104$
Age, years (SD)	61.79 (7.81)	64.09 (8.87)	60.04 (9.96)
Males (%)	74.2	84.8	76.9
Smokers (%)	66.3	56.5	32.7 *
Diabetes (%)	10.1	3.3	13.4
Dyslipidemia (%)	17.0	26.1	42.3 *
Hypertension (%)	42.5	30.4	42.3
BMI (Kg/m^2) (SD)	27.54 (4.48)	27.15 (5.70)	27.65 (5.77)

Controls were matched against the atrial fibrillation groups for age, gender, hypertension, and BMI. SD: standard deviation; BMI: body mass index. Qualitative values compared with Fisher's exact test. Quantitative values compared with the Student's *t*-test. * $p < 0.05$.

The type of cardiomyopathy present in the structural AF group is detailed in Table 2. Valvular disease and dilated atria were the main structural abnormalities.

Table 2. Characteristics of heart disease in the structural AF group. (More than one type of cardiomyopathy was possible in each patient).

Characteristics	Numbers
Type of Heart Disease	**Total n (%) ($n = 104$)**
Valvular heart disease	
- Mitral regurgitation	71 (68.3)
- Mitral stenosis	2 (1.9)
- Aortic regurgitation	29 (27.9)
- Aortic stenosis	4 (3.8)
Ischemic heart disease	9 (8.6)
Other (HCM, interatrial communication, etc.)	34 (32.7)
Dilated left atrium (>20 cm^2)	61 (58.7)
Echocardiography values	
Average LVEF (%)	56
Average left atrial surface, cm^2 (SD)	23 (6.9)
Average left ventricular ejection fraction, % (SD)	56.2 (10.1)

HCM: hypertrophic cardiomyopathy, LVEF: Left Ventricular Ejection Fraction, SD: standard deviation.

2.1. Cx40 − 44G > A/+71A > G Polymorphisms

Analysis of the two promoter A Cx40 polymorphisms (−44G > A and +71A > G) showed an almost complete linkage disequilibrium of the −44G/+71A and −44A/+71G alleles. The overall genotype distribution in the control group was −44GG/+71AA = 67.4%, −44AG/+71AG = 28.1% and −44AA/+71GG = 4.5%. This genotype distribution was not significantly different in either of the AF groups (non-structural AF: $p = 0.846$; structural AF: $p = 0.132$) (Table 3). There was also no significant difference in between the two AF groups ($p = 0.275$).

2.2. Cx40 − 26G > A Polymorphism

The specific genotype distribution of the Cx40 − 26G > A polymorphism in the control group was −26GG = 25.8%, −26AG = 47.2%, and −26AA = 27.0%. This distribution was significantly different in the structural AF group: −26GG = 26.0%, −26AG = 61.5%, and −26AA = 12.5% ($p = 0.029$). There was no significant difference between the control and the non-structural AF group ($p = 0.511$) (Table 3) or between the two AF groups ($p = 0.3$).

2.3. Cx37 1019C > T Polymorphism

The specific genotype distribution of the Cx37 *1019C > T* polymorphism in the control group was *1019CC* = 42.7%, *1019CT* = 34.8%, and *1019TT* = 22.5%. This distribution was significantly different in the non-structural AF group: *1019CC* = 46.7%, *1019CT* = 44.6%, and *1019TT* = 8.7% (p = 0.034). There was no significant difference between the control and the structural AF group (p = 0.080) (Table 3) or between the two AF groups (p = 0.85).

Table 3. Allele distributions of the Cx40 (*GJA5*) and Cx37 (*GJA4*) polymorphisms in the control, non-structural AF, and structural AF groups. Proportions were compared with Fisher's exact test.

Genotype	Gene				p Value vs. Control Group
Cx40 −44G > A	*GG (n, %)*	*AG (n, %)*	*AA (n, %)*	*Total*	
Control	60 (67.4)	25 (28.1)	4 (4.5)	89	
Non-Structural AF	59 (64.1)	27 (29.3)	6 (6.5)	92	*p = 0.846*
Structural AF	55 (52.9)	41 (39.5)	8 (7.7)	104	*p = 0.132*
Cx40 +71A > G	*AA (n, %)*	*AG (n, %)*	*GG (n, %)*	*Total*	
Control	60 (67.4)	25 (28.1)	4 (4.5)	89	
Non-Structural AF	61 (66.3)	26 (28.3)	5 (5.4)	92	*p = 1.000*
Structural AF	55 (52.9)	41 (39.4)	8 (7.7)	104	*p = 0.133*
Cx40 − 26G > A	*GG (n, %)*	*AG (n, %)*	*AA (n, %)*	*Total*	
Control	23 (25.8)	42 (47.2)	24 (27.0)	89	
Non-Structural AF	26 (28.3)	48 (52.2)	18 (19.6)	92	*p = 0.511*
Structural AF	27 (26.0)	64 (61.5)	13 (12.5)	104	**p = 0.029**
Cx37 1019C > T	*CC (n, %)*	*CT (n, %)*	*TT (n, %)*	*Total*	
Control	38 (42.7)	31 (34.8)	20 (22.5)	89	
Non-Structural AF	43 (46.7)	41 (44.6)	8 (8.7)	92	**p = 0.034**
Structural AF	50 (48.1)	43 (41.3)	11 (10.6)	104	*p = 0.080*

In bold the genes and genotypes.

3. Discussion

Our study describes for the first time an association of the Cx37 *1019C > T* gene polymorphism with drug-resistant non-structural AF, but not with structural AF. The potential differences in gene susceptibility between non-structural and structural AF were further underlined by the fact that we found an association of the Cx40 − *26G > A* polymorphism with structural AF and not with non-structural AF.

These results suggest that there are potential differences in the genetic susceptibility to AF in given populations. Patients who develop AF in whom there is no underlying structural cardiac disease appear to have an increased frequency of Cx37 *1019C* alleles than patients who do not develop AF. This genetic polymorphism has previously been described in populations with coronary heart disease and myocardial infarction [27,28]. The *GJA4* genotype may predict survival after an acute coronary syndrome [29]. Cx37 is principally expressed in endothelial cells, VSMCs, monocyte/macrophages, and platelets. The *1019C > T* polymorphism in *GJA4* causes a Proline-to-Serine substitution at amino acid 319 (P319S) in the cytoplasmic tail of Cx37, which in turn alters channel conductance and permeability [25,30]. The Cx37 polymorphism has been shown to affect monocyte adhesion as well as platelet aggregation [25,31]. Both mechanisms are thought to underlie the development of CAD [25]. Inflammation and its associated immune response are increasingly recognized to be involved in the initiation and maintenance of atrial fibrillation [32]. Thus, the Cx37 *1019C > T* polymorphism might affect the susceptibility to non-structural AF by virtue of its effect on monocyte adhesion. Alternatively, initiation of non-structural AF is thought to depend on a combination of abnormal impulse formation and conduction [9]. As Cx37 is also expressed in VSMCs at the ostia of pulmonary veins (PVs), where sheets of cardiomyocytes and VSMCs interdigitate, the differences in gap junction channel electrophysiological properties associated with the Cx37 *1019C > T* polymorphism may accentuate the latter mechanism [33].

Our results also show an association of the Cx40 − *26G > A* polymorphism in patients with AF with underlying cardiomyopathy, with the −26G allele being more frequent than in control patients. This single nucleotide polymorphism (SNP) in the promoter B region of *GJA5*, the *Cx40* gene, which alters the TATA box sequence and modulates mRNA expression levels of Cx40, has previously been associated with AF [34]. Of note, our study did not show any association of AF (either structural or non-structural) with the promoter A polymorphisms, Cx40 − *44/+71*; these polymorphisms were also shown to reduce *GJA5* expression in vitro by about 50%; however, Wirka et al. showed no effect on *GJA5* expression in atrial tissue, suggesting that promoter A is not necessary to Cx40 expression in atrial myocytes [34]. In patients with structural AF, atrial remodeling is often present at a structural, electrophysiological, and molecular level with changes in the refractory period and cardiomyocyte contractility [8,10]. Some of these atrial cardiomyocyte characteristics can be explained by changes in potassium channel properties and important changes to calcium handling [8]. Inconsistent observations have been made with respect to Cx40 expression levels in patients with AF, with both higher and lower levels being reported [35–37]. However, only few patients were included in these studies—10 and 12, respectively—and almost exclusively with mitral valve disease. Complete deletion of Cx40 in murine models leads to decreased atrial conduction velocity and increased atrial stability [8,38]. Somatic mutations in *GJA5* have been shown to predispose patients to idiopathic AF by either impairing the assembly of Cx40 into gap junctions or by affecting the electrical coupling itself [14,18]. Importantly, a recent study revealed that one of the reported Cx40 mutations (A96S) caused decreased atrial conduction velocities and sustained episodes of induced atrial fibrillation in mice and rat [39,40].

The strength of our study is the distinction between patients with or without underlying cardiomyopathy. This enables a more precise association study and a dissection of possible different pathophysiological mechanism leading to AF. We chose to include patients in whom anti-arrhythmic treatment was not successful in order to have long-standing or permanent AF. It may be interesting in the future to include all AF. The main limitation of our study is nevertheless the relatively small sample size, which increases the risk of falsely positive results. Furthermore, we acknowledge that non-structural AF is a misnomer and an exhaustive search of underlying etiology reduces the number of instances when there is truly no explanation for AF [7]. We also acknowledge that the link between the severity of valvular disease, in particular mitral valve disease, and AF may be complex. This may be an important aspect to further investigate. Associations between smoking or serum lipid levels and Cx40 have never been reported; it is, however, quite unlikely that one of these confounding factors would induce a significant association of the Cx40 − *26G > A* polymorphism with structural AF.

4. Materials and Methods

4.1. Patients

Three patient cohorts were studied, namely patients with AF and no underlying structural cardiac disease (non-structural AF), patients with AF and documented structural cardiac disease (structural AF), and a control population with no AF or underlying cardiac disease. This study protocol conforms to the ethical guidelines of the 1975 Declaration of Helsinki, as reflected in prior approval by the institution's human research committee, namely the Internal Medicine Departmental Ethics Committee, Geneva University Hospital (Protocol 03-167, 3 April 2010). Informed consent was obtained from all patients.

The non-structural AF group consisted of patients undergoing catheter ablation of AF by radiofrequency ablation through pulmonary vein isolation (RFAPVI). Inclusion criteria were an age > 18 years, documented paroxysmal or persistent AF (for <2 years) with or without typical flutter (ECG or Holter) after failure of at least one anti-arrhythmic treatment of Class I or III, with planned and clinically indicated RFAPVI. Exclusion criteria were reversible AF, clinical heart failure, increased left atrial size (>20 cm^2 four chamber view), decreased left ventricular ejection fraction, any valvulopathy, or a history of prior AF or flutter ablation. The structural AF group consisted of patients undergoing RFAPVI but in whom there was underlying cardiomyopathy, be it structural or valvular disease, or documented ischemic heart disease (history of acute or chronic coronary disease, angiographically proven disease). Inclusion criteria were an age > 18 years, documented paroxysmal or persistent AF (for <2 years) with or without typical flutter (ECG or Holter) and with planned and clinically indicated RFAPVI. Inclusion and exclusion criteria for these groups are detailed in Table 4. Demographics and characteristics of each group are detailed in Table 1. The type of cardiomyopathy present in the structural AF group is detailed in Table 1. The control group was considered as a whole and variables were matched to the AF groups with priority given to age, gender, hypertension, and BMI, these being major risk factors associated with AF development. These individuals were from the same population basin as AF patients, seen in the same clinic, and were undergoing elective coronary angiography at the Geneva University Hospital due to symptoms of suspected CAD or unrelated conditions requiring angiographic evaluation [41]. Participants gave written informed consent for a blood draw at the time of angiography for use in confidential studies approved by the hospital's institutional review board. Exclusion criteria for this control group were AF or a history of AF, or arrhythmia on 12 lead ECG, a history of coronary heart disease, any significant angiographic lesions on coronary angiography, or any known heart disease including valvulopathy.

Table 4. Inclusion and exclusion criteria for the non-structural and structural atrial fibrillation and control groups.

Criteria	Non-Structural AF	Structural AF	Control Group
Inclusion criteria	Age > 18 years Paroxysmal or persistent AF Failure of at least 1 anti-arrhythmic drug of class I or III	Age > 18 years Any valvulopathy Left ventricular or atrial dilatation Documented ischemic heart disease Hypertrophic cardiomyopathy Any structural disease	Age > 18 years
Exclusion criteria	Reversible AF Clinical heart failure Decrease left ejection fraction Any valvulopathy		History or documented AF Any heart disease including coronary heart disease or significant coronary lesion on angiography Any valvulopathy

AF: atrial fibrillation; NYHA: New York Heart Association heart failure classification.

4.2. Echocardiography

Echocardiography was performed on a HP/Philips 5500, Philips Electronics North American Corporation, Andover, MA, USA, with digitalization of images and analyzed according to a standardized protocol in a core lab.

4.3. DNA Extraction and Genotyping

DNA was extracted from the peripheral blood and genotyping of the Cx40 − 44G > A, Cx40 + 71A > G, Cx40 − 26A > G, and Cx37 1019C > T polymorphisms was performed by polymerase chain reaction (PCR) and restriction fragment length polymorphism (RFLP) assays using previously established methods [34,42,43].

4.4. Statistical Analysis

Clinical, biological, and echocardiographic characteristics are presented using descriptive statistics and a number of cases with a percentage in parentheses for categorical variables. In view of skewed allele distribution and small numbers of certain genotypes, we used non-parametric tests. Proportions/categorical data were compared with the Fisher's exact test.

5. Conclusions

Our study suggests that there may be a different genetic predisposition to non-structural AF as compared to AF appearing in the context of underlying heart disease. We report here for the first time that the Cx37 *1019C > T* variant is associated with non-structural AF. Mechanistically, this might involve effects of altered Cx37 channel function on the inflammatory response (monocyte adhesion) or on abnormal impulse conduction (communication between VSMCs and atrial cardiomyocytes). In addition, we confirmed that the Cx40 − *26G > A* polymorphism affecting the expression level of this protein in atrial cardiomyocytes is associated with structural AF.

Acknowledgments: The authors thank Katia Galan and Bernard Foglia for technical assistance with DNA extraction and genotyping. This work was partly supported by the Research Fund of the Department of Internal Medicine of the University Hospital and the Faculty of Medicine of Geneva; this Fund receives an unrestricted grant from AstraZeneca Switzerland. Anna Pfenniger was supported by a joint grant from the Swiss National Science Foundation, the Swiss Academy of Medical Sciences, and the Velux Foundation (323630-123735), and Brenda R Kwak by a grant of the Swiss National Science Foundation (310030_162579).

Author Contributions: Sebastian Carballo, David Carballo, and Brenda R Kwak conceived and designed the experiments; Anna Pfenniger performed the experiments; Anna Pfenniger and Sebastian Carballo analyzed the data; Dipen Shah and Richard W James contributed to the recruitment of the patients and controls; Sebastian Carballo, Brenda R Kwak, David Carballo, Nicolas Garin, and François Mach wrote the manuscript.

Conflicts of Interest: The authors declare no conflict of interest.

References

1. Chugh, S.S.; Havmoeller, R.; Narayanan, K.; Singh, D.; Rienstra, M.; Benjamin, E.J.; Gillum, R.F.; Kim, Y.H.; McAnulty, J.H., Jr.; Zheng, Z.J.; et al. Worldwide epidemiology of atrial fibrillation: A Global Burden of Disease 2010 Study. *Circulation* **2014**, *129*, 837–847. [CrossRef] [PubMed]

2. January, C.T.; Wann, L.S.; Alpert, J.S.; Calkins, H.; Cigarroa, J.E.; Cleveland, J.C., Jr.; Conti, J.B.; Ellinor, P.T.; Ezekowitz, M.D.; Field, M.E.; et al. 2014 AHA/ACC/HRS guideline for the management of patients with atrial fibrillation: A report of the American College of Cardiology/American Heart Association Task Force on Practice Guidelines and the Heart Rhythm Society. *J. Am. Coll. Cardiol.* **2014**, *64*, e1–e76. [PubMed]

3. Camm, A.J.; Kirchhof, P.; Lip, G.Y.; Schotten, U.; Savelieva, I.; Ernst, S.; Van Gelder, I.C.; Al-Attar, N.; Hindricks, G.; Prendergast, B.; et al. Guidelines for the management of atrial fibrillation: The Task Force for the Management of Atrial Fibrillation of the European Society of Cardiology (ESC). *Eur. Heart J.* **2010**, *31*, 2369–2429. [PubMed]

4. Kirchhof, P.; Benussi, S.; Kotecha, D.; Ahlsson, A.; Atar, D.; Casadei, B.; Castella, M.; Diener, H.C.; Heidbuchel, H.; Hendriks, J.; et al. 2016 ESC Guidelines for the management of atrial fibrillation developed in collaboration with EACTS. *Eur. Heart J.* **2016**, *37*, 2893–2962. [CrossRef] [PubMed]

5. Camm, A.J.; Lip, G.Y.; De Caterina, R.; Savelieva, I.; Atar, D.; Hohnloser, S.H.; Hindricks, G.; Kirchhof, P. 2012 focused update of the ESC Guidelines for the management of atrial fibrillation: An update of the 2010 ESC Guidelines for the management of atrial fibrillation. Developed with the special contribution of the European Heart Rhythm Association. *Eur. Heart J.* **2012**, *33*, 2719–2747. [CrossRef] [PubMed]

6. Weijs, B.; Pisters, R.; Nieuwlaat, R.; Breithardt, G.; Le Heuzey, J.Y.; Vardas, P.E.; Limantoro, I.; Schotten, U.; Lip, G.Y.; Crijns, H.J. Idiopathic atrial fibrillation revisited in a large longitudinal clinical cohort. *Europace* **2012**, *14*, 184–190. [CrossRef] [PubMed]

7. Wyse, D.G.; Van Gelder, I.C.; Ellinor, P.T.; Go, A.S.; Kalman, J.M.; Narayan, S.M.; Nattel, S.; Schotten, U.; Rienstra, M. Lone atrial fibrillation: Does it exist? *J. Am. Coll. Cardiol.* **2014**, *63*, 1715–1723. [CrossRef] [PubMed]

8. Schotten, U.; Verheule, S.; Kirchhof, P.; Goette, A. Pathophysiological mechanisms of atrial fibrillation: A translational appraisal. *Physiol. Rev.* **2011**, *91*, 265–325. [CrossRef] [PubMed]

9. Haissaguerre, M.; Jais, P.; Shah, D.C.; Takahashi, A.; Hocini, M.; Quiniou, G.; Garrigue, S.; Le Mouroux, A.; Le Metayer, P.; Clementy, J. Spontaneous initiation of atrial fibrillation by ectopic beats originating in the pulmonary veins. *N. Engl. J. Med.* **1998**, *339*, 659–666. [CrossRef] [PubMed]

10. Wijffels, M.C.; Kirchhof, C.J.; Dorland, R.; Allessie, M.A. Atrial fibrillation begets atrial fibrillation. A study in awake chronically instrumented goats. *Circulation* **1995**, *92*, 1954–1968. [CrossRef] [PubMed]

11. Chen, Y.H.; Xu, S.J.; Bendahhou, S.; Wang, X.L.; Wang, Y.; Xu, W.Y.; Jin, H.W.; Sun, H.; Su, X.Y.; Zhuang, Q.N.; et al. KCNQ1 gain-of-function mutation in familial atrial fibrillation. *Science* **2003**, *299*, 251–254. [CrossRef] [PubMed]

12. Fatkin, D.; Otway, R.; Vandenberg, J.I. Genes and atrial fibrillation: A new look at an old problem. *Circulation* **2007**, *116*, 782–792. [CrossRef] [PubMed]

13. Olson, T.M.; Alekseev, A.E.; Liu, X.K.; Park, S.; Zingman, L.V.; Bienengraeber, M.; Sattiraju, S.; Ballew, J.D.; Jahangir, A.; Terzic, A. Kv1.5 channelopathy due to KCNA5 loss-of-function mutation causes human atrial fibrillation. *Hum. Mol. Genet.* **2006**, *15*, 2185–2191. [CrossRef] [PubMed]

14. Christophersen, I.E.; Holmegard, H.N.; Jabbari, J.; Sajadieh, A.; Haunso, S.; Tveit, A.; Svendsen, J.H.; Olesen, M.S. Rare variants in GJA5 are associated with early-onset lone atrial fibrillation. *Can. J. Cardiol.* **2013**, *29*, 111–116. [CrossRef] [PubMed]

15. Firouzi, M.; Ramanna, H.; Kok, B.; Jongsma, H.J.; Koeleman, B.P.; Doevendans, P.A.; Groenewegen, W.A.; Hauer, R.N. Association of human connexin40 gene polymorphisms with atrial vulnerability as a risk factor for idiopathic atrial fibrillation. *Circ. Res.* **2004**, *95*, e29–e33. [CrossRef] [PubMed]

16. Juang, J.M.; Chern, Y.R.; Tsai, C.T.; Chiang, F.T.; Lin, J.L.; Hwang, J.J.; Hsu, K.L.; Tseng, C.D.; Tseng, Y.Z.; Lai, L.P. The association of human connexin 40 genetic polymorphisms with atrial fibrillation. *Int. J. Cardiol.* **2007**, *116*, 107–112. [CrossRef] [PubMed]

17. Hailati, J.; Yang, Y.C.; Zhang, L.; He, P.Y.; Baikeyi, M.; Muhuyati, W.; Liu, Z.Q. Association between −44G/A and +71A/G polymorphisms in the connexin 40 gene and atrial fibrillation in Uyghur and Han populations in Xinjiang, China. *Genet. Mol. Res.* **2016**, *15*. [CrossRef] [PubMed]

18. Gollob, M.H.; Jones, D.L.; Krahn, A.D.; Danis, L.; Gong, X.Q.; Shao, Q.; Liu, X.; Veinot, J.P.; Tang, A.S.; Stewart, A.F.; et al. Somatic mutations in the connexin 40 gene (GJA5) in atrial fibrillation. *N. Engl. J. Med.* **2006**, *354*, 2677–2688. [CrossRef] [PubMed]

19. Thibodeau, I.L.; Xu, J.; Li, Q.; Liu, G.; Lam, K.; Veinot, J.P.; Birnie, D.H.; Jones, D.L.; Krahn, A.D.; Lemery, R.; et al. Paradigm of genetic mosaicism and lone atrial fibrillation: Physiological characterization of a connexin 43-deletion mutant identified from atrial tissue. *Circulation* **2010**, *122*, 236–244. [CrossRef] [PubMed]

20. Jansen, J.A.; van Veen, T.A.; de Bakker, J.M.; van Rijen, H.V. Cardiac connexins and impulse propagation. *J. Mol. Cell. Cardiol.* **2010**, *48*, 76–82. [CrossRef] [PubMed]

21. Pfenniger, A.; Wohlwend, A.; Kwak, B.R. Mutations in connexin genes and disease. *Eur. J. Clin. Investig.* **2011**, *41*, 103–116. [CrossRef] [PubMed]

22. Pfenniger, A.; Chanson, M.; Kwak, B.R. Connexins in atherosclerosis. *Biochim. Biophys. Acta* **2013**, *1828*, 157–166. [CrossRef] [PubMed]

23. Chaldoupi, S.M.; Loh, P.; Hauer, R.N.; de Bakker, J.M.; van Rijen, H.V. The role of connexin40 in atrial fibrillation. *Cardiovasc. Res.* **2009**, *84*, 15–23. [CrossRef] [PubMed]

24. Fontes, M.S.; van Veen, T.A.; de Bakker, J.M.; van Rijen, H.V. Functional consequences of abnormal Cx43 expression in the heart. *Biochim. Biophys. Acta* **2012**, *1818*, 2020–2029. [CrossRef] [PubMed]

25. Wong, C.W.; Christen, T.; Roth, I.; Chadjichristos, C.E.; Derouette, J.P.; Foglia, B.F.; Chanson, M.; Goodenough, D.A.; Kwak, B.R. Connexin37 protects against atherosclerosis by regulating monocyte adhesion. *Nat. Med.* **2006**, *12*, 950–954. [CrossRef] [PubMed]

26. Aranki, S.F.; Shaw, D.P.; Adams, D.H.; Rizzo, R.J.; Couper, G.S.; VanderVliet, M.; Collins, J.J., Jr.; Cohn, L.H.; Burstin, H.R. Predictors of atrial fibrillation after coronary artery surgery. Current trends and impact on hospital resources. *Circulation* **1996**, *94*, 390–397. [CrossRef] [PubMed]

27. Wen, D.; Du, X.; Nie, S.P.; Dong, J.Z.; Ma, C.S. Association of Connexin37 C1019T with myocardial infarction and coronary artery disease: A meta-analysis. *Exp. Gerontol.* **2014**, *58*, 203–207. [CrossRef] [PubMed]

28. Listi, F.; Candore, G.; Lio, D.; Russo, M.; Colonna-Romano, G.; Caruso, M.; Hoffmann, E.; Caruso, C. Association between C1019T polymorphism of connexin37 and acute myocardial infarction: A study in patients from Sicily. *Int. J. Cardiol.* **2005**, *102*, 269–271. [CrossRef] [PubMed]

29. Lanfear, D.E.; Jones, P.G.; Marsh, S.; Cresci, S.; Spertus, J.A.; McLeod, H.L. Connexin37 (GJA4) genotype predicts survival after an acute coronary syndrome. *Am. Heart J.* **2007**, *154*, 561–566. [CrossRef] [PubMed]

30. Derouette, J.P.; Desplantez, T.; Wong, C.W.; Roth, I.; Kwak, B.R.; Weingart, R. Functional differences between human Cx37 polymorphic hemichannels. *J. Mol. Cell. Cardiol.* **2009**, *46*, 499–507. [CrossRef] [PubMed]

31. Angelillo-Scherrer, A.; Fontana, P.; Burnier, L.; Roth, I.; Sugamele, R.; Brisset, A.; Morel, S.; Nolli, S.; Sutter, E.; Chassot, A.; et al. Connexin 37 limits thrombus propensity by downregulating platelet reactivity. *Circulation* **2011**, *124*, 930–939. [CrossRef] [PubMed]

32. Hu, Y.F.; Chen, Y.J.; Lin, Y.J.; Chen, S.A. Inflammation and the pathogenesis of atrial fibrillation. *Nat. Rev. Cardiol.* **2015**, *12*, 230–243. [CrossRef] [PubMed]

33. Verheule, S.; Wilson, E.E.; Arora, R.; Engle, S.K.; Scott, L.R.; Olgin, J.E. Tissue structure and connexin expression of canine pulmonary veins. *Cardiovasc. Res.* **2002**, *55*, 727–738. [CrossRef]

34. Wirka, R.C.; Gore, S.; Van Wagoner, D.R.; Arking, D.E.; Lubitz, S.A.; Lunetta, K.L.; Benjamin, E.J.; Alonso, A.; Ellinor, P.T.; Barnard, J.; et al. A common connexin-40 gene promoter variant affects connexin-40 expression in human atria and is associated with atrial fibrillation. *Circ. Arrhythm. Electrophysiol.* **2011**, *4*, 87–93. [CrossRef] [PubMed]

35. Nao, T.; Ohkusa, T.; Hisamatsu, Y.; Inoue, N.; Matsumoto, T.; Yamada, J.; Shimizu, A.; Yoshiga, Y.; Yamagata, T.; Kobayashi, S.; et al. Comparison of expression of connexin in right atrial myocardium in patients with chronic atrial fibrillation versus those in sinus rhythm. *Am. J. Cardiol.* **2003**, *91*, 678–683. [CrossRef]

36. Polontchouk, L.; Haefliger, J.A.; Ebelt, B.; Schaefer, T.; Stuhlmann, D.; Mehlhorn, U.; Kuhn-Regnier, F.; De Vivie, E.R.; Dhein, S. Effects of chronic atrial fibrillation on gap junction distribution in human and rat atria. *J. Am. Coll. Cardiol.* **2001**, *38*, 883–891. [CrossRef]

37. Gemel, J.; Levy, A.E.; Simon, A.R.; Bennett, K.B.; Ai, X.; Akhter, S.; Beyer, E.C. Connexin40 abnormalities and atrial fibrillation in the human heart. *J. Mol. Cell. Cardiol.* **2014**, *76*, 159–168. [CrossRef] [PubMed]

38. Hagendorff, A.; Schumacher, B.; Kirchhoff, S.; Luderitz, B.; Willecke, K. Conduction disturbances and increased atrial vulnerability in Connexin40-deficient mice analyzed by transesophageal stimulation. *Circulation* **1999**, *99*, 1508–1515. [CrossRef] [PubMed]

39. Lubkemeier, I.; Andrie, R.; Lickfett, L.; Bosen, F.; Stockigt, F.; Dobrowolski, R.; Draffehn, A.M.; Fregeac, J.; Schultze, J.L.; Bukauskas, F.F.; et al. The Connexin40A96S mutation from a patient with atrial fibrillation causes decreased atrial conduction velocities and sustained episodes of induced atrial fibrillation in mice. *J. Mol. Cell. Cardiol.* **2013**, *65*, 19–32. [CrossRef] [PubMed]

40. Kanthan, A.; Fahmy, P.; Rao, R.; Pouliopoulos, J.; Alexander, I.E.; Thomas, S.P.; Kizana, E. Human Connexin40 Mutations Slow Conduction and Increase Propensity for Atrial Fibrillation. *Heart Lung Circ.* **2018**, *27*, 114–121. [CrossRef] [PubMed]

41. James, R.W.; Leviev, I.; Righetti, A. Smoking is associated with reduced serum paraoxonase activity and concentration in patients with coronary artery disease. *Circulation* **2000**, *101*, 2252–2257. [CrossRef] [PubMed]

42. Wong, C.W.; Christen, T.; Pfenniger, A.; James, R.W.; Kwak, B.R. Do allelic variants of the connexin37 1019 gene polymorphism differentially predict for coronary artery disease and myocardial infarction? *Atherosclerosis* **2007**, *191*, 355–361. [CrossRef] [PubMed]

43. Pfenniger, A.; van der Laan, S.W.; Foglia, B.; Dunoyer-Geindre, S.; Haefliger, J.A.; Winnik, S.; Mach, F.; Pasterkamp, G.; James, R.W.; Kwak, B.R. Lack of association between connexin40 polymorphisms and coronary artery disease. *Atherosclerosis* **2012**, *222*, 148–153. [CrossRef] [PubMed]

International Journal of
Molecular Sciences

MDPI

Article

Functional Characterization of Novel Atrial Fibrillation-Linked *GJA5* (Cx40) Mutants

Mahmoud Noureldin, Honghong Chen and Donglin Bai *

Department of Physiology and Pharmacology, University of Western Ontario, London, ON, N6A 5C1 Canada;
mnourel@uwo.ca (M.N.); hchen38@uwo.ca (H.C.)
* Correspondence: donglin.bai@schulich.uwo.ca; Tel.: +1-519-850-2569

Received: 23 February 2018; Accepted: 21 March 2018; Published: 25 March 2018

Abstract: Atrial fibrillation (AF) is the most common form of cardiac arrhythmia. Recently, four novel heterozygous Cx40 mutations—K107R, L223M, Q236H, and I257L—were identified in 4 of 310 unrelated AF patients and a followup genetic analysis of the mutant carriers' families showed that the mutants were present in all the affected members. To study possible alterations associated with these Cx40 mutants, including their cellular localization and gap junction (GJ) function, we expressed GFP-tagged and untagged mutants in connexin-deficient model cells. All four Cx40 mutants showed clustered localization at cell–cell junctions similar to that observed of wildtype Cx40. However, cell pairs expressing Cx40 Q236H, but not the other individual mutants, displayed a significantly lower GJ coupling conductance (G_j) than wildtype Cx40. Similarly, co-expression of Cx40 Q236H with Cx43 resulted in a significantly lower G_j. Transjunctional voltage-dependent gating (V_j gating) properties were also altered in the GJs formed by Q236H. Reduced GJ function and altered V_j gating may play a role in promoting the Q236H carriers to AF.

Keywords: atrial fibrillation; gap junction channel; connexin40; V_j gating; patch clamp

1. Introduction

Atrial fibrillation (AF) is the most common sustained cardiac arrhythmia affecting millions of people worldwide [1,2]. With an overall prevalence of 1%, AF increases with age, starting from 0.1% in individuals younger than 55 years and reaching 9% in those over 80 years [3]. AF prevalence is expected to increase substantially due to an aging population [3]. AF is characterized by a fast sporadic beating of the atria, which causes substantial morbidity including a much higher risk of stroke [2,4]. Often, AF exists as a secondary disease to a wide range of other diseases, such as hypertension, diabetes, and coronary artery disease [5]. However, AF is the primary disease in about 30% of AF patients who are categorized as AF with genetic predisposition [1,6,7]. This group of AF patients has been linked to multiple genetic mutations including genes encoding ion channels, such as potassium channels, sodium channels, and gap junction (GJ) channels [8–14].

GJ channels are composed of connexins. In humans, there are 21 different connexins and all of them share a similar topological structure of four transmembrane domains (M1–M4), two extracellular domains (E1 and E2), one cytoplasmic loop (CL), and both the amino terminus (NT) and carboxyl terminus (CT) lie within the cytoplasm [15,16]. Six connexins oligomerize to form a hemichannel (also known as connexon) that could function as a channel on the plasma membrane [17]. Two hemichannels from adjacent cells could dock head-to-head at their extracellular domains to form a GJ channel [18]. Human heart expresses three different types of connexins—Cx40, Cx43, and Cx45 [19,20]—allowing for the possible formation of homomeric or heteromeric hemichannels and homotypic or heterotypic GJ channels. Cx45 is dominantly expressed in the sinoatrial (SA) and atrioventricular (AV) nodes while Cx43 and Cx40 are both expressed in the atrial myocardium and are often found to be co-localized at the intercalated discs between atrial myocytes [21,22]. Cx43 is the main connexin in the ventricles [23]. A much lower level of Cx45 is expressed in the atria and ventricles [24]. These connexins form GJ channels between cardiomyocytes to mediate rapid propagation of action potentials (APs) in the heart [24,25].

The importance of Cx40 and Cx43 in the heart has been highlighted in animal models and genetic mutation studies. Mice with an ablation of Cx43 in the heart develop ventricular arrhythmias leading to sudden cardiac death [26]. An in vitro study using cultured atrial synthetic strands from Cx43-deficient mice showed a decrease in conduction velocity [27]. Moreover, an early onset of AF is associated with a somatic Cx43 mutant, which exhibits GJ impairment [28]. Interestingly, viral expression of the exogenous wildtype Cx43 in the atria was found to prevent AF in pig models [29,30]. For Cx40, earlier studies reported that mice lacking the Cx40 gene exhibit a slower action potential propagation [31,32] and are more susceptible to inducible atrial arrhythmias [32,33]. Recent studies reported an increased conduction velocity and a decrease in the conduction heterogeneity in Cx40 knockout mice or cells derived from these mice [27,34,35]. Furthermore, Cx40 promoter polymorphisms result in lower levels of Cx40 mRNA and have been linked to an early onset of AF [36]. Somatic and germline mutations within the coding regions of human Cx40 gene (*GJA5*) have been linked to AF patients and families [11–14]. A recent genetic study identified four novel germline mutants in *GJA5* in four of 310 unrelated AF patients, resulting in heterozygous missense mutants in Cx40 protein: K107R, L223M, Q236H, and I257L [37]. Further testing on available relatives of the mutant carriers revealed that these mutants presented in all affected family members and were absent in 400 reference alleles [37]. Functional consequences of these AF-linked Cx40 mutants have not been studied. We hypothesize that these AF-linked Cx40 mutants impair GJ and/or hemichannel function, which may predispose the mutant carriers to AF.

2. Results

2.1. AF-Linked Cx40 Mutants Formed GJ Plaque-Like Structures at the Cell–Cell Interface

Expression of AF-linked Cx40 mutants (K107R, L223M, Q236H, and I257L, all tagged with YFP at the carboxyl terminus) was used to study their localization in live HeLa cells. As shown in Figure 1A, each of the mutants was localized in intracellular compartments and displayed GJ plaque-like clusters at the cell–cell interfaces similar to that of Cx40–YFP. The percentage of successful mutant-expressing cell pairs displaying GJ plaque-like structures at cell–cell interfaces was calculated and was found to be similar to that of cells expressing wildtype Cx40 (Figure 1B).

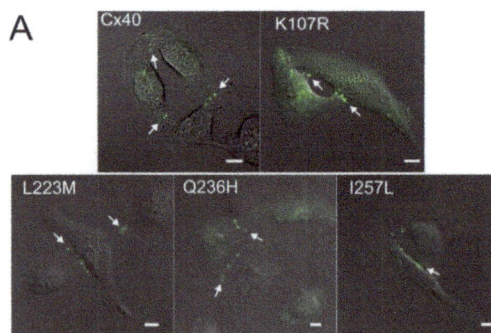

Figure 1. *Cont.*

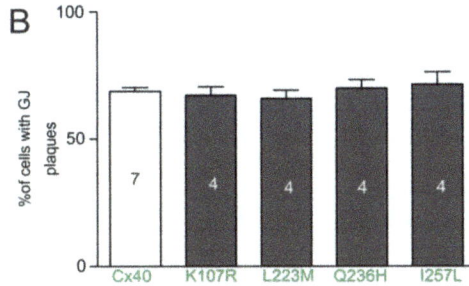

Figure 1. Localization of atrial fibrillation (AF)-linked Cx40 mutants. (**A**) Fluorescent images of HeLa cell clusters or pairs expressing YFP-tagged Cx40, K107R, L223M, Q236H, or I257L superimposed on their respective differential interference contrast (DIC) images. Cells expressing each of the AF-linked Cx40 mutants were able to form GJ plaque-like structures at the cell–cell interface similar to that of wildtype Cx40 (white arrows). Scale bars = 10 μm; (**B**) the bar graph summarizes the percentage of cell pairs showing GJ plaque-like structures at the cell–cell interface for each mutant. No statistical difference was observed between any of the mutants and wildtype Cx40. Approximately 100 positively-transfected cell pairs were examined for each transfection. The total number of transfections is indicated on each bar.

2.2. Coupling Conductance of GJs Formed by AF-Linked Mutants

Dual whole-cell patch clamp was used to study the functionality of untagged AF-linked Cx40 mutants in N2A cell pairs. Representative junctional currents (I_js) of cell pairs expressing each of the Cx40 mutants and wildtype Cx40 are presented (Figure 2A). The averaged coupling percentage of each Cx40 mutant in several transfections, plotted as a bar graph, was not different from that of wildtype Cx40 (Figure 2B, $p > 0.05$ for each of the mutants). The coupling conductance (G_j) of cell pairs expressing K107R, L223M, or I257L was also not different from that of wildtype Cx40 (Figure 2C). However, a significant reduction in G_j was observed in cell pairs expressing Q236H (Figure 2C, $p < 0.05$).

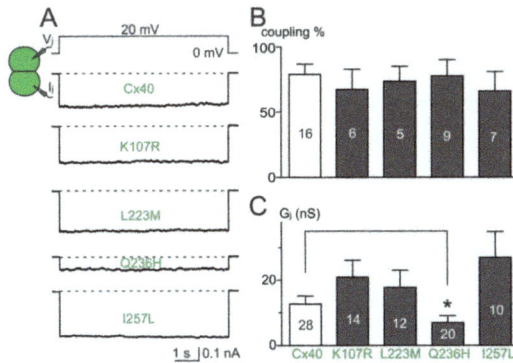

Figure 2. Coupling percentage and G_j of AF-linked mutants. (**A**) Dual whole-cell patch clamp technique was used to measure junctional current (I_j) from N2A cell pairs expressing untagged Cx40, K107R, L223M, Q236H, or I257L at 20 mV V_j; (**B**) bar graph summarizes the coupling percentages of cell pairs expressing the AF-linked Cx40 mutants. No statistical difference was observed between each of the mutants and the wildtype Cx40. The number of transfections is indicated on each bar; (**C**) bar graph illustrates the coupling conductance (G_j) of coupled cell pairs expressing Cx40, K107R, L223M, Q236H, or I257L. Cell pairs expressing Q236H showed a significantly lower G_j than those of wildtype Cx40 (* $p < 0.05$). The number of cell pairs is indicated on each bar.

2.3. Homotypic Cx40 Q236H GJs Showed an Altered V_j Gating

To investigate the transjunctional voltage-dependent gating (V_j gating) of AF-linked Cx40 mutants, we measured I_js in cell pairs in response to a series of V_j pulses (± 20 to ± 100 mV, Figure 3A). The I_js from cell pairs expressing untagged Cx40 mutant K107R, L233M, Q236H, or I257L showed similar symmetrical V_j-dependent deactivation (sometimes also called inactivation) when V_js $\geq$ 40 mV (Figure 3). The normalized steady state conductance ($G_{j,ss}$) of each mutant (filled circles) or wildtype Cx40 (open grey circles) was plotted at different V_js (Figure 3B). The smooth black lines are Boltzmann fitting curves for each of the mutants (Figure 3B). Boltzmann fittings of wildtype Cx40 (smooth grey dashed lines) are plotted and superimposed onto each mutant $G_{j,ss}$–V_j plot for comparison (Figure 3B). Compared to the wildtype Cx40, GJ channels formed by these mutants showed nearly identical Boltzmann fitting curves, except Q236H, which showed a significant reduction in V_0 for both V_j polarities (Figure 3B, Table 1).

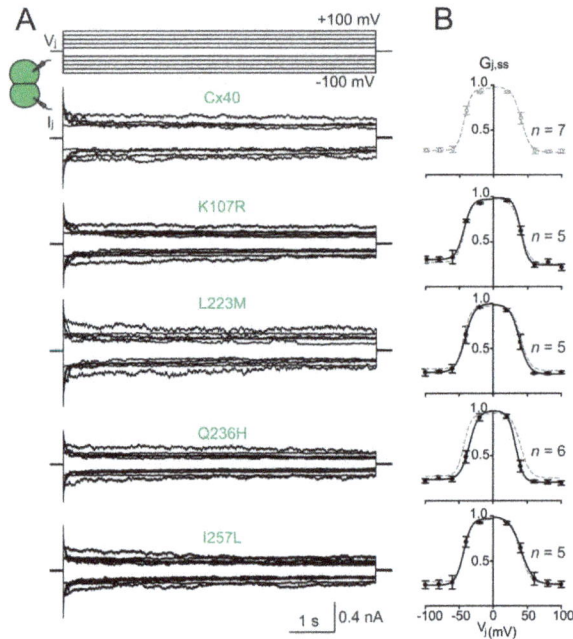

Figure 3. V_j gating of AF-linked mutant GJs. (**A**) Dual whole-cell patch clamp was used to measure I_js in N2A cell pairs expressing Cx40, K107R, L223M, Q236H, or I257L in response to a series of V_j pulses as indicated. Superimposed I_js for each mutant is shown; (**B**) normalized steady state junctional conductance, $G_{j,ss}$, of the Cx40 mutants (black filled circles) and wildtype Cx40 (grey open circles) were plotted at different V_js. The $G_{j,ss}$–V_j plot of each mutant was fitted with a two-state Boltzmann equation at each V_j polarity (smooth black lines). Boltzmann fittings of $G_{j,ss}$–V_j plot of wildtype Cx40 (smooth grey dashed lines) were obtained and superimposed on each plot for comparison. The number of cell pairs is indicated.

Table 1. Boltzmann fitting parameters for V_j gating of AF-linked mutants.

Cells Expressing	V_j Polarity	G_{min}	V_0 (mV)	A
Cx40	+	0.25 ± 0.02	40.2 ± 1.4	0.15 ± 0.05
(n = 7)	−	0.27 ± 0.02	42.9 ± 1.4	0.19 ± 0.07
K107R	+	0.21 ± 0.03	38.6 ± 1.6	0.15 ± 0.05
(n = 5)	−	0.23 ± 0.03	41.1 ± 1.9	0.15 ± 0.05

Table 1. *Cont.*

Cells Expressing	V_j Polarity	G_{min}	V_0 (mV)	A
L223M	+	0.24 ± 0.02	40.2 ± 1.0	0.20 ± 0.10
($n = 5$)	−	0.31 ± 0.03	43.2 ± 1.8	0.17 ± 0.06
Q236H	+	0.20 ± 0.02	33.3 ± 1.7 *,[1]	0.19 ± 0.04
($n = 6$)	−	0.24 ± 0.02	35.2 ± 2.0 *	0.15 ± 0.04
I257L	+	0.24 ± 0.03	40.9 ± 2.2	0.12 ± 0.03
($n = 5$)	−	0.24 ± 0.03	43.4 ± 2.2	0.17 ± 0.08

[1] One-way ANOVA followed by Tukey *post-hoc* test was used to compare the Boltzmann fitting parameters of each mutant and wildtype Cx40 at the corresponding V_j polarity. GJ channels formed by the mutant Q236H showed a significantly lower V_0 for both V_j polarities than those of wildtype Cx40 (* $p < 0.05$).

To analyze V_j-gating kinetics, we fitted the I_j deactivation by a single exponential process at V_js of ± 60 to ± 100 mV. As shown in Figure 4A, I_j deactivation of wildtype Cx40 GJs fitted well with a single exponential process (with a time constant, τ) at each of the tested V_js (Figure 4A). The averaged time constants (τs) showed a decrease with the increase of V_js (Figure 4B, open grey circles). The I_j deactivations of the GJs formed by AF-linked mutants could be fitted by a single exponential process and the τ–V_j plots were not statistically different from those of wildtype Cx40 GJs (filled black circles), except Q236H GJ that showed consistently lower τs at all tested V_js (Figure 4B, $p < 0.05$).

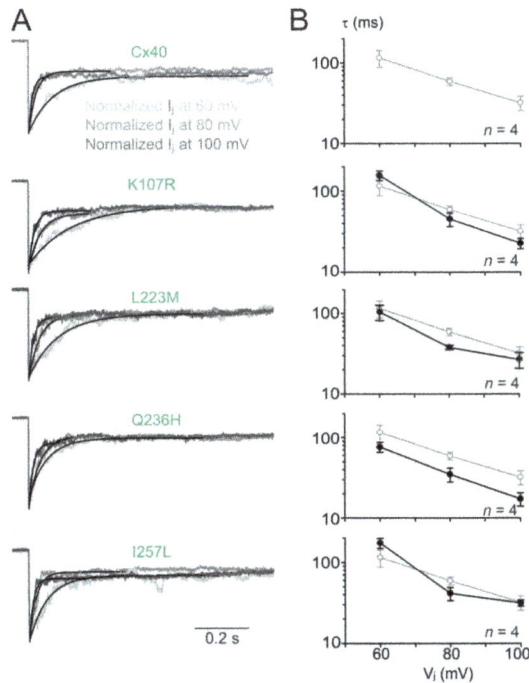

Figure 4. V_j-gating kinetics of AF-linked Cx40 mutants. (**A**) I_js induced at different V_js (60 mV light grey, 80 mV medium grey, 100 mV dark grey) were normalized and superimposed for each of the mutant or Cx40 GJs. I_j deactivations under different V_js were all fitted well with a single exponential process (smooth black lines); (**B**) The time constants (τs) were plotted on a semi logarithmic scale against different V_js. When the V_js increased, the averaged τs of the mutant GJs (black filled circles) decreased similar to those observed for the wildtype Cx40 (grey open circles). No consistent statistical difference was found between most of the mutant τs and the τs of wildtype Cx40, except the τs of Q236H was consistently lower than those of wildtype Cx40 (two-way ANOVA). The number of cell pairs are indicated.

2.4. Co-Expression of AF-Linked Mutants with Cx43

To investigate if AF-linked Cx40 mutants had a trans-dominant negative effect on wildtype Cx43, each of the mutants was co-expressed with Cx43 (with an untagged DsRed). Cell pairs successfully expressing both connexins were selected for dual whole-cell patch clamp. Cell pairs successfully co-expressing K107R:Cx43, L223M:Cx43, or I257L:Cx43 showed coupling percentages and G_js that were not statistically different from those of wildtype Cx40:Cx43 (Figure 5B,C). The coupling percentage of Q236H:Cx43 was also not statistically different from that of wildtype Cx40:Cx43. However, the G_j of cell pairs co-expressing Q236H:Cx43 was significantly lower than that of wildtype Cx40:Cx43 (Figure 5C, $p < 0.05$).

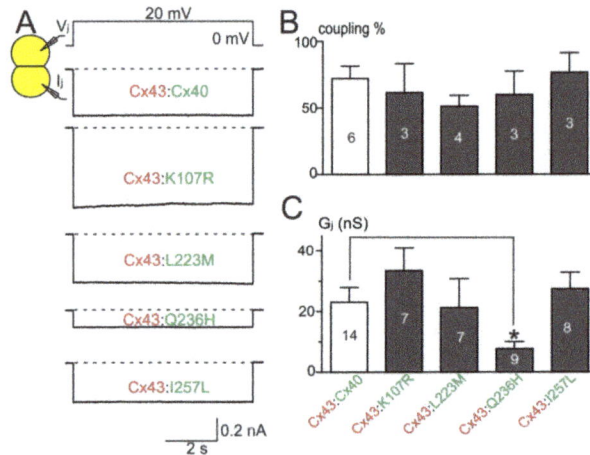

Figure 5. Coupling percentage and G_j of co-expressing AF-linked mutants with wildtype Cx43. (**A**) Representative I_js are shown from N2A cell pairs co-expressing Cx40, K107R, L223M, Q236H, or I257L (with an untagged reporter GFP) with wildtype Cx43 (with an untagged reporter DsRed); (**B**) bar graph illustrates coupling percentages of N2A cell pairs expressing each combination. The number of transfections is indicated; (**C**) bar graph illustrates the G_j of cell pairs co-expressing one of the Cx40 mutants (K107R, L223M, Q236H, or I257L) with Cx43. The G_j of cell pairs co-expressing Q236H:Cx43 was significantly lower than that of wildtype Cx40:Cx43 (* $p < 0.05$). The number of cell pairs is indicated.

2.5. Function of Heterotypic Mutant/Cx40 GJ Channels

The above results showed that Cx40 Q236H had a significantly lower G_js than wildtype Cx40 when expressed alone or co-expressed with Cx43. To further test whether Q236H could affect the function of heterotypic Q236H/Cx40 GJs, we mixed cells expressing Q236H (with untagged GFP) with cells expressing Cx40 (with untagged DsRed) and performed dual patch clamp on heterotypic cell pairs (one GFP+ and the other DsRed+, Figure 6A). The coupling percentages and G_js of heterotypic Q236H/Cx40 cell pairs were not statistically different from those of the control (Cx40/Cx40, Figure 6B,C). Similar results were also obtained for L223M (Figure 6).

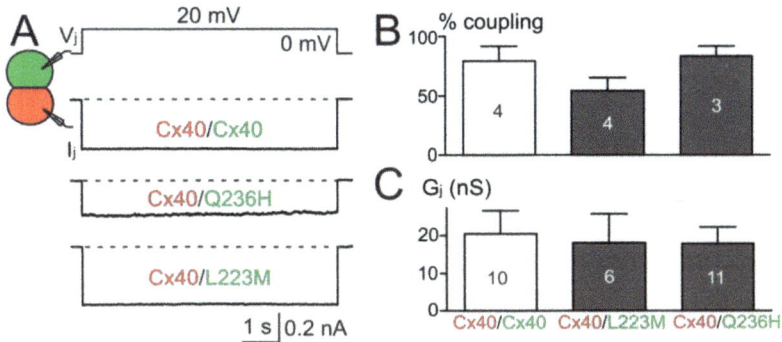

Figure 6. Functional test on heterotypic mutant/Cx40 GJs. (**A**) I_js were obtained from heterotypic Q236H/Cx40, L223M/Cx40, or Cx40/Cx40 (in all cases with untagged GFP or DsRed, respectively) N2A cell pairs; (**B**) Bar graph summarizes the coupling percentages of heterotypic cell pairs. No statistical difference was found between the coupling percentage of any of the mutant heterotypic GJs and that of Cx40/Cx40 GJs. The number of transfections is indicated; (**C**) G_js of cell pairs expressing L223M/Cx40 or Q236H/Cx40 were not statistically different from those of Cx40/Cx40. The number of cell pairs is indicated.

2.6. Propidium Iodide Uptake by AF-Linked Cx40 Mutant-Expressing Cells

Propidium iodide (PI) uptake assay was used to investigate the hemichannel function of AF-linked Cx40 mutants as elevated PI uptake was observed in AF-linked Cx40 mutants, including L221I [38,39]. Figure 7A shows the fluorescent images of individual HeLa cells expressing GFP alone, Cx40, or one of the mutants (L221I, K107R, L223M, Q236H, or I257L) in divalent cation-free solution. The percentage of individual cells showing PI uptake in cells expressing K107R, L223M, Q236H or I257L was not significantly different from either the wildtype Cx40 or the negative control (expressing GFP alone) but was statistically lower than the positive control L221I (69%, $p < 0.001$) (Figure 7B). These results suggest that the Cx40 mutants and the wildtype Cx40 failed to show PI uptake in divalent cation-free solution.

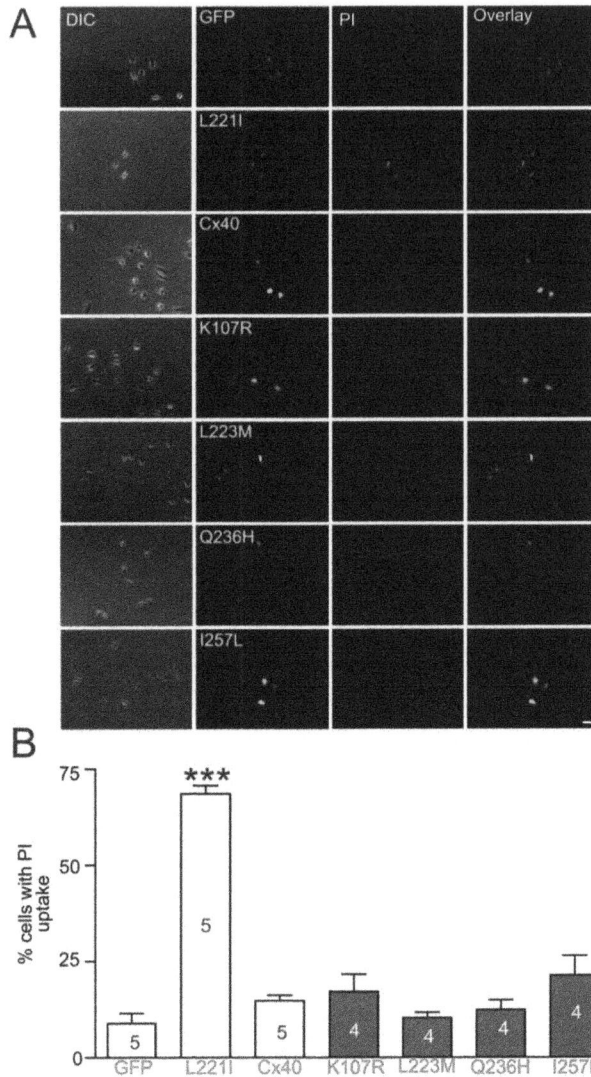

Figure 7. Propidium iodide uptake of AF-linked Cx40 mutants. (**A**) HeLa cells transfected with Cx40 mutants, empty vector GFP, or Cx40 are shown: column 1 (under DIC), column 2 (GFP fluorescence to show successful expression of respective vector), column 3 (propidium iodide [PI] uptake in red), column 4 (an overlay of images of column2 and 3). Only cells expressing L221I showed PI uptake. The scale bar = 50 μm; (**B**) bar graph summarizes PI uptake percentage of isolated individual cells expressing Cx40 mutants, Cx40, or GFP. PI uptake percentage for each of the AF-linked mutants was not statistically different from that of wildtype Cx40 or the empty vector (GFP), except L221I (*** $p < 0.001$). The number transfection is indicated with observations of over 60 isolated cells for each transfection.

3. Discussion

In this study, we examined morphological and functional characteristics of four recently identified AF-linked Cx40 mutations (K107R, L223M, Q236H and I257L) in vitro. Our localization experiments showed that YFP-tagged K107R, L223M, Q236H, and I257L were able to form GJ plaque-like structures at the cell–cell interface in HeLa cells, similar to that of wildtype Cx40. PI uptake by each of these mutant (untagged)-expressing cells showed no significant increase from that of Cx40, indicating no increase in hemichannel activity in any of the mutant-expressing cells. Dual patch clamp experiments revealed that the Cx40 mutants (K107R, L223M, and I257L) showed no apparent change in coupling conductance (G_j) when expressed alone or together with Cx43. The GJs formed by each of these mutants also failed to show any obvious change in the V_j gating properties. In contrast, Q236H GJs exhibited a significantly reduced G_j when expressed alone or together with Cx43. In addition, Q236H GJs also showed altered V_j gating, specifically a reduction in the V_j required to close the channel (V_0) and an increase in V_j-gating kinetics. These defects associated with Q236H might play a role in the pathogenesis of AF in the mutant carriers.

3.1. AF-Linked Cx40 Mutants Showed Multiple Defects in GJ or Hemichannel Function

So far, a total of ten germlines and three somatic mutations in the coding region of the *GJA5* gene (encoding Cx40) have been identified in AF patients with genetic predisposition [11–14,37,40]. In vitro studies on these AF-linked Cx40 mutants have revealed that these mutants display either a loss of GJ function or a gain of hemichannel function. The detailed molecular and cellular mechanisms leading to GJ or hemichannel functional changes appear to be quite different. (1) A Cx40 missense mutation P88S failed to localize to cell–cell interfaces to form GJ plaque-like structures [11]. Similarly, a nonsense Cx40 mutation (Q49X) was found to be retained in the endoplasmic reticulum and unable to reach cell–cell junctions [41]. Functional impairment of GJs in these two mutants were anticipated and confirmed experimentally, but interestingly both mutants showed dominant negative and transdominant negative actions on GJ function when co-expressed with wildtype Cx40 or Cx43, respectively [11,41]. (2) Other AF-linked Cx40 missense mutants (G38D, I75F, V85I, A96S, M163V, L221I, L229M) and those in the present study (K107R, L223M, Q236H, and I257L) showed GJ plaques at the cell–cell interfaces [11,14,38,39]. However, eliminated or significantly reduced macroscopic coupling conductance (G_j) was observed in G38D, I75F, A96S, and Q236H GJs, probably due to impairment at the GJ channel [11,42]. Some of the mutants in this category were also found to show dominant negative action on Cx40 (I75F and A96S) and/or transdominant negative action on Cx43 (I75F, A96S, L229M, and Q236H) [11,42]. Although some isolated disagreements on the G_j levels of G38D and A96S GJs have been reported [39,43], the majority of these studies agree on the GJ functional impairments in most of the AF-linked Cx40 mutants [11,14,39,41,43]. (3) Detailed characterizations of mutant-containing GJ channels revealed additional defects, including reduced G_j of heterotypic mutant/wildtype GJs, altered V_j gating properties of homotypic (G38D, A96S, M163V, Q236H) or heterotypic (I75F/Cx40, A96S/Cx40) V_j gating properties, a substantially reduced open probability without changing unitary channel conductance (I75F), or elevated unitary channel conductance (G38D and M163V) [14,39,42,43]. (4) Only a limited number of AF-linked mutants (G38D, M163V, and A96S) have been studied for GJ permeability changes but significant permeability change to anionic dye (Lucifer yellow) or cationic dye (ethidium bromide) was observed [43]. (5) The PI uptake assay was used to study hemichannel function in isolated cells expressing AF-linked mutants. Among the tested Cx40 mutants, only V85I- and L221I-expressing cells showed an elevated PI uptake compared to that of wildtype Cx40, indicating a gain of hemichannel function in these mutants [38]. Patch clamp on cells expressing Cx40 G38D showed unitary hemichannel currents [39]. Similar elevated hemichannel function was also observed in a few other disease-linked mutants in Cx26, Cx43, and Cx50 [44–47].

3.2. AF-Linked Cx40 Mutants and Their Possible Role in AF Pathogenesis

As discussed above, there is a variety of molecular/cellular changes associated with AF-linked Cx40 mutants. Whether these molecular/cellular changes play a role in the pathogenesis of AF is not clear. Several theoretical possibilities exist. First, a reduced macroscopic coupling conductance (G_j) of AF-linked Cx40 mutants due to either an impaired localization or GJ channel function is known to reduce the action potential conduction velocity [48,49], which could be an important contributing factor in promoting re-entrant atrial arrhythmias [50]. Consistent with this model, about half of AF-linked Cx40 mutants identified so far have shown G_j reduction not only in the mutant GJs but also when they are co-expressed with wildtype Cx40 and/or Cx43 [51]. Our present study showed that Cx40 Q236H also reduced G_j when expressed alone and together with Cx43. We did not perform co-expression of this mutant with wildtype Cx40 because our untagged Cx40 construct (Cx40-IRES-DsRed) had a very low transfection efficiency. Second, enhanced V_j gating by lower V_js and faster gating kinetics by Cx40 mutants could also dynamically down-regulate G_j when sufficient junctional delays exist [52,53]. Cx40 showed a pronounced V_j gating with a minimum conductance level reaching a quarter of the maximum G_j [54,55]. A reduction in the Boltzmann fitting parameter, V_0, and faster V_j-dependent deactivation kinetics of Q236H mutant GJs predict an increased V_j gating when sufficient junctional delay exists. It is not clear whether V_j gating of Cx40 or Q236H GJs could dynamically down-regulate G_j as observed for Cx45 GJs [56]. Third, AF-linked Cx40 mutants have been shown to alter their GJ permeability [43], which could alter intercellular exchanges of small signaling molecules, including second messengers. This altered permeability of GJs could restrict/enhance signaling molecules necessary for intercellular communication between atrial myocytes, which might be important for atrial function. Fourth, three of the AF-linked mutants showed elevated PI uptake and/or hemichannel current, indicating enhanced hemichannel activity under reduced divalent cations in the extracellular medium [38,39]. Physiological and/or pathological stresses, such as large repetitive membrane depolarizations, mechanical stretch, reduced extracellular divalent cations, reduced oxygen/glucose during ischemia, have all been shown to enhance several other connexin hemichannels. Whether these stress factors also promote the opening of Cx40 hemichannel remains to be determined. In summary, it is not clear how these defects in Cx40 mutants in model cells link to the pathogenesis of AF in the mutant carriers. Among the changes associated with AF-linked Cx40 mutants, the most consistent is a reduced or eliminated GJ coupling (G_j) in different possible atrial GJs. At present, we cannot rule out other changes, such as biosynthesis, and turnover of the mutant Cx40 protein may also change the abundance and function of Cx40 at the intercalated discs, which may need genetically modified animal models to assess fully.

3.3. AF-Linked Cx40 Mutants without Apparent Defects In Vitro

In our present study, we did not detect any obvious defects in GJ distribution, function, or hemichannel activities in three out of four AF-linked Cx40 mutants (K107R, L223M, and I257L). It is not clear how these mutants relate to AF. Here are some possibilities. (1) These mutants are located in the CL (K107R), M4 (L223M), and CT (I257L) domains of Cx40. The CL and CT domains of Cx40 show a lot more residue variation in different vertebrate species than the E1, E2, and M1–4 domains. Following this general trend, the conservation percentage at these residue positions in Cx40 are L223 (in M4 domain) 85%, K107 (in CL domain) 74%, and I257 (in CT domain) 28% across 47 different species (accessible online: https://omabrowser.org/oma/home/ OMA Group 752281). It is possible that one or more of these mutants are benign mutants that do not necessarily play a role in promoting AF, especially at the least conserved position, I257. (2) Both of our model cells, HeLa and N2A cells, are convenient model systems to study localization, function of GJs, and hemichannels. They are GJ deficient, easily transfected with connexin mutant constructs, and easily accessible for morphological, dual patch clamp, or dye-uptake experiments. However, these cells are not cardiomyocytes and may not recapitulate all aspects of GJs at the intercalated discs of cardiomyocytes and, therefore, some defects might go undetected. Future studies on cell models that are closer to atrial myocytes will likely help to resolve the role of these Cx40 mutants linking to AF. (3) We have good rationales to focus our study on the morphology and functional changes in GJs and hemichannel activities. However, there are other unconventional functions of connexins, including but not limited to aggregation of protein complexes at the cell–cell junctions, adhesion, cell growth control, and differentiation, etc., that require specific biological assays to evaluate.

3.4. Other AF-Linked Genetic Factors

It is well known that genetic factors play a role in AF. A lot of research is focused on genes responsible for inherited AF cases as these genes are putative independent AF risk factors [7,57]. The first genetic mutation linked to familial AF was identified in the *KCNQ1* gene, which encodes a potassium channel subunit [9]. Since then, more AF-linked mutations in different potassium channel subunits have been identified and are now extended into genes encoding sodium channels, transcription factors, Ca^{2+} handling proteins, nucleoporins, and atrial natriuretic peptide [8,58]. Our present study is consistent with several previous studies showing that atrial GJ impairments represent an independent risk factor for AF [11,14,28,38,40,41,51].

4. Materials and Methods

4.1. Plasmid Construction

The C-terminal fusion fluorescent protein tagged human Cx40-YFP and the untagged constructs (Cx40-IRES-GFP, Cx43-IRES-DsRed, Cx40-IRES-DsRed) were created as previously described [14,59]. The novel AF-linked tagged and untagged Cx40 mutants were generated by site-directed mutagenesis on the corresponding tagged/untagged Cx40 as templates with the following primers.

K107R	Forward: 5′ CAGGAGAAGCGCAGGCTACGGGAGGCC 3′
	Reverse: 5′ GGCCTCCCGTAGCCTGCGCTTCTCCTG 3′
L223M	Forward: 5′ CTCCTCCTTAGCATGGCTGAACTCT 3′
	Reverse: 5′ AGAGTTCAGCCATGCTAAGGAGG 3′
I257L	Forward: 5′ CCCTCTGTGGGCCTAGTCCAGAGCTGC3′
	Reverse: 5′ GCAGCTCTGGACTAGGCCCACAGAGGG3′
Q236H	Forward: 5′ GGAAGAAGATCAGACACCGATTTGTCAAACC3′
	Reverse: 5′ GGTTTGACAAATCGGTGTCTGATCTTCTTCC3′

All these Cx40 mutant constructs were sequenced to confirm the accuracy of the nucleotide sequence.

4.2. Cell Culture and Transfection

Connexin-deficient mouse neuroblastoma (N2A) and human cervical carcinoma (HeLa) cells (American Type Culture Collection, Manassas, VA, USA) were cultured at 37 °C with 5% CO_2. Cells were grown in Dulbecco's modified Eagle's medium (DMEM) (Cat# 10313-021, Thermo Fisher Scientific, Waltham, MA, USA) containing 10% fetal bovine serum, 1% penicillin, 1% streptomycin, 4.5 g/L D-(+)-glucose, 584 mg/L L-glutamine, and 110 mg/L sodium pyruvate. Twenty-four hours before cell transfection, N2A or HeLa cells were replated into a 35-mm dish at 60% confluency. Transfection was performed the next day by adding 0.8–1 µg of DNA with 2 µL of the transfection reagent X-tremeGENE HP DNA (Roche Applied Sciences, Indianapolis, IN, USA). To assess the effect of Cx40 mutants on wildtype Cx43, N2A cells were transfected in a 1:1 ratio of Cx40 mutants-IRES-GFP and Cx43-IRES-DsRed. Cell pairs successfully co-expressing both GFP and DsRed were selected for measuring coupling conductance with dual whole-cell patch clamp (see below).

4.3. Localization

HeLa cells were transiently transfected with YFP-tagged Cx40 mutants. One day after transfection, cells were replated on 10 mm glass coverslips and incubated overnight. The number of successfully transfected cell pairs forming GJ plaque-like structures at the cell–cell interface were counted. A confocal microscope (Zeiss LSM800 with Airyscan) (Zeiss, Oberkochen, Germany) was used to observe mutant-YFP and wildtype Cx40-YFP localizations as described earlier [59].

4.4. Electrophysiology

On the experimental day, transfected N2A cells were replated onto glass coverslips and incubated for 1.5 to 3 h. Coverslips with cells were transferred into a recording chamber and bathed in extracellular solution (ECS) containing 135 mM NaCl, 5 mM KCl, 10 mM Hepes, 1 mM MgCl$_2$, 2 mM CaCl$_2$, 1 mM BaCl$_2$, 2 mM CsCl, 2 mM Na Pyruvate, and 5 mM D-glucose with pH and osmolarity of 7.4 and 310–320 mOsm, respectively. The recording chamber was placed on an upright fluorescent microscope (BX51WI, Olympus, Center Valley, PA, USA) to visualize reporter (GFP)-positive cell pairs. Patch pipette was filled with intracellular solution (ICS) containing 130 mM CsCl, 10 mM EGTA, 0.5 mM CaCl2, 4 mM Na2ATP, and 10 mM Hepes with pH 7.2 and osmolarity of 290–300 mOsm. Dual whole-cell patch clamp technique was performed on isolated cell pairs expressing the Cx40 mutant. Initially, cell pairs were both voltage clamped at 0 mV. Then, a 20 mV voltage pulse was applied to one of the cell pairs (pulsing cell) while keeping the other clamped at 0 mV (the recording cell). If functional GJ channels exist between the cell pairs then a transjunctional current (I_j) can be measured at the recording cell via MultiClamp 700A (Molecular Devices, Sunnyvale, CA, USA) and stored in a PC via an AD/DA interface (Digidata 1322A) and pClamp9.2 software (Molecular Devices, Sunnyvale, CA, USA). G_j was calculated ($G_j = I_j/V_j$). V_j gating properties were studied by applying a series of voltage pulses (± 20 to ± 100 mV in 20 mV increment) as described in our previous studies [14,56].

4.5. Dye Uptake Assay

AF-linked Cx40 mutants hemichannel function were assessed using propidium iodide (PI) uptake assay. HeLa cells were transiently transfected with each of the Cx40 mutants in an IRES-GFP vector [38]. We used a previously characterized AF-linked Cx40 mutant L221I-IRES-GFP as the positive control and empty IRES-GFP vector as the negative control for these experiments [38]. Divalent cation containing extracellular solution (DCC-ECS) was composed of 142 mM NaCl, 5.4 mM KCl, 1.4 mM MgCl$_2$, 2 mM CaCl$_2$, 10 mM HEPES, and 25 mM D-(+)-glucose adjusted to pH 7.35 and osmolarity of ~298 mOsm. HeLa cells were incubated in a divalent cation-free extracellular solution (DCF-ECS) containing PI (150 µM). Removal of both Ca^{2+} and Mg^{2+} ions, as well as including EGTA (2 mM) in the DCF-ECS, were used to facilitate PI uptake via GJ hemichannels. After incubation at 37 °C for 15 min, cells were washed three times with DCC-ECS at room temperature prior to observation with a fluorescent microscope (DMIRE2, Leica, Cridersville, OH, USA). The number of transfected HeLa cells with or without PI uptake was counted and the percentage of cells with PI uptake was calculated. Only isolated individual HeLa cells were counted to prevent errors caused by GJ channels in cell clusters.

4.6. Data Analysis

Mann–Whitney U test was used to compare each of the mutants against wildtype Cx40 for the percentage of cell pairs with morphological GJ plaques, the coupling percentage and conductance (G_j) using dual patch clamp, and the percentage of cells displaying PI uptake for the hemichannel study. One-way ANOVA followed by Tukey post-hoc test was used to compare the Boltzmann fitting parameters of each mutant and wildtype Cx40 at the corresponding V_j polarity. Statistical significance is denoted with different levels of significance (* $p < 0.05$; ** $p < 0.01$; or *** $p < 0.001$).

5. Conclusions

In summary, our results indicate that the AF-linked Cx40 Q236H mutation exhibited GJ function impairment by reducing the overall G_j when expressed alone or with the wildtype Cx43, and altered V_j-gating kinetics and the V_0, which might play a role in AF pathogenesis. The other mutants—K107R, L223M, and I257L—did not exhibit any apparent GJ or hemichannel functional impairments in our cell models.

Acknowledgments: This work was supported by the Heart and Stroke Foundation of Canada (G-13-0003066 to Donglin Bai) and the Canadian Institutes of Health Research (153415 to Donglin Bai).

Author Contributions: Mahmoud Noureldin conceived, designed, and performed all patch clamp experiments and analyzed the data and wrote an early draft of the manuscript; Honghong Chen designed and generated all the AF-linked Cx40 mutants and designed and performed some localization experiments; Donglin Bai designed the project, supervised data analysis, provided funding support and critically revised the manuscript.

Conflicts of Interest: The authors declare no conflict of interest.

Abbreviations

Cx40	connexin40
$G_{j,ss}$	normalized steady-state junction conductance
I_j	macroscopic junctional current
V_j	transjunctional voltage

References

1. Fuster, V.; Ryden, L.E.; Cannom, D.S.; Crijns, H.J.; Curtis, A.B.; Ellenbogen, K.A.; Halperin, J.L.; Kay, G.N.; Le Huezey, J.Y.; Lowe, J.E.; et al. 2011 ACCF/AHA/HRS focused updates incorporated into the ACC/AHA/ESC 2006 Guidelines for the management of patients with atrial fibrillation: A report of the American College of Cardiology Foundation/American Heart Association Task Force on Practice Guidelines developed in partnership with the European Society of Cardiology and in collaboration with the European Heart Rhythm Association and the Heart Rhythm Society. *J. Am. Coll. Cardiol.* **2011**, *57*, e101–e198. [PubMed]
2. Wakili, R.; Voigt, N.; Kaab, S.; Dobrev, D.; Nattel, S. Recent advances in the molecular pathophysiology of atrial fibrillation. *J. Clin. Investig.* **2011**, *121*, 2955–2968. [CrossRef] [PubMed]
3. Go, A.S.; Hylek, E.M.; Phillips, K.A.; Chang, Y.; Henault, L.E.; Selby, J.V.; Singer, D.E. Prevalence of diagnosed atrial fibrillation in adults: National implications for rhythm management and stroke prevention: The AnTicoagulation and Risk Factors in Atrial Fibrillation (ATRIA) Study. *JAMA* **2001**, *285*, 2370–2375. [CrossRef] [PubMed]
4. Wolf, P.A.; Abbott, R.D.; Kannel, W.B. Atrial fibrillation as an independent risk factor for stroke: The Framingham Study. *Stroke* **1991**, *22*, 983–988. [CrossRef] [PubMed]
5. Saffitz, J.E.; Corradi, D. The electrical heart: 25 years of discovery in cardiac electrophysiology, arrhythmias and sudden death. *Cardiovasc. Pathol.* **2016**, *25*, 149–157. [CrossRef] [PubMed]
6. Levy, S.; Maarek, M.; Coumel, P.; Guize, L.; Lekieffre, J.; Medvedowsky, J.L.; Sebaoun, A. Characterization of different subsets of atrial fibrillation in general practice in France: The ALFA study. The College of French Cardiologists. *Circulation* **1999**, *99*, 3028–3035. [CrossRef] [PubMed]
7. Fox, C.S.; Parise, H.; D'Agostino, R.B., Sr.; Lloyd-Jones, D.M.; Vasan, R.S.; Wang, T.J.; Levy, D.; Wolf, P.A.; Benjamin, E.J. Parental atrial fibrillation as a risk factor for atrial fibrillation in offspring. *JAMA* **2004**, *291*, 2851–2855. [CrossRef] [PubMed]
8. Xiao, J.; Liang, D.; Chen, Y.H. The genetics of atrial fibrillation: From the bench to the bedside. *Annu. Rev. Genom. Hum. Genet.* **2011**, *12*, 73–96. [CrossRef] [PubMed]
9. Chen, Y.H.; Xu, S.J.; Bendahhou, S.; Wang, X.L.; Wang, Y.; Xu, W.Y.; Jin, H.W.; Sun, H.; Su, X.Y.; Zhuang, Q.N.; et al. KCNQ1 gain-of-function mutation in familial atrial fibrillation. *Science* **2003**, *299*, 251–254. [CrossRef] [PubMed]
10. Makiyama, T.; Akao, M.; Shizuta, S.; Doi, T.; Nishiyama, K.; Oka, Y.; Ohno, S.; Nishio, Y.; Tsuji, K.; Itoh, H.; et al. A novel SCN5A gain-of-function mutation M1875T associated with familial atrial fibrillation. *J. Am. Coll. Cardiol.* **2008**, *52*, 1326–1334. [CrossRef] [PubMed]
11. Gollob, M.H.; Jones, D.L.; Krahn, A.D.; Danis, L.; Gong, X.Q.; Shao, Q.; Liu, X.; Veinot, J.P.; Tang, A.S.; Stewart, A.F.; et al. Somatic mutations in the connexin 40 gene (GJA5) in atrial fibrillation. *N. Engl. J. Med.* **2006**, *354*, 2677–2688. [CrossRef] [PubMed]
12. Yang, Y.Q.; Zhang, X.L.; Wang, X.H.; Tan, H.W.; Shi, H.F.; Jiang, W.F.; Fang, W.Y.; Liu, X. Connexin40 nonsense mutation in familial atrial fibrillation. *Int. J. Mol. Med.* **2010**, *26*, 605–610. [CrossRef] [PubMed]
13. Yang, Y.Q.; Liu, X.; Zhang, X.L.; Wang, X.H.; Tan, H.W.; Shi, H.F.; Jiang, W.F.; Fang, W.Y. Novel connexin40 missense mutations in patients with familial atrial fibrillation. *Europace* **2010**, *12*, 1421–1427. [CrossRef] [PubMed]
14. Sun, Y.; Yang, Y.Q.; Gong, X.Q.; Wang, X.H.; Li, R.G.; Tan, H.W.; Liu, X.; Fang, W.Y.; Bai, D. Novel germline GJA5/connexin40 mutations associated with lone atrial fibrillation impair gap junctional intercellular communication. *Hum. Mutat.* **2013**, *34*, 603–609. [PubMed]

15. Sohl, G.; Willecke, K. Gap junctions and the connexin protein family. *Cardiovasc. Res.* **2004**, *62*, 228–232. [CrossRef] [PubMed]

16. Zimmer, D.B.; Green, C.R.; Evans, W.H.; Gilula, N.B. Topological analysis of the major protein in isolated intact rat liver gap junctions and gap junction-derived single membrane structures. *J. Biol. Chem.* **1987**, *262*, 7751–7763. [PubMed]

17. Evans, W.H.; De Vuyst, E.; Leybaert, L. The gap junction cellular internet: Connexin hemichannels enter the signalling limelight. *Biochem. J.* **2006**, *397*, 1–14. [CrossRef] [PubMed]

18. Bai, D.; Wang, A.H. Extracellular domains play different roles in gap junction formation and docking compatibility. *Biochem. J.* **2014**, *458*, 1–10. [CrossRef] [PubMed]

19. Jansen, J.A.; van Veen, T.A.; de Bakker, J.M.; van Rijen, H.V. Cardiac connexins and impulse propagation. *J. Mol. Cell. Cardiol.* **2010**, *48*, 76–82. [CrossRef] [PubMed]

20. Verheule, S.; Kaese, S. Connexin diversity in the heart: Insights from transgenic mouse models. *Front. Pharmacol.* **2013**, *4*, 81. [CrossRef] [PubMed]

21. Vozzi, C.; Dupont, E.; Coppen, S.R.; Yeh, H.I.; Severs, N.J. Chamber-related differences in connexin expression in the human heart. *J. Mol. Cell. Cardiol.* **1999**, *31*, 991–1003. [CrossRef] [PubMed]

22. Severs, N.J.; Bruce, A.F.; Dupont, E.; Rothery, S. Remodelling of gap junctions and connexin expression in diseased myocardium. *Cardiovasc. Res.* **2008**, *80*, 9–19. [CrossRef] [PubMed]

23. Fishman, G.I.; Spray, D.C.; Leinwand, L.A. Molecular characterization and functional expression of the human cardiac gap junction channel. *J. Cell Biol.* **1990**, *111*, 589–598. [CrossRef] [PubMed]

24. Desplantez, T.; Dupont, E.; Severs, N.J.; Weingart, R. Gap junction channels and cardiac impulse propagation. *J. Membr. Biol.* **2007**, *218*, 13–28. [CrossRef] [PubMed]

25. Kanno, S.; Saffitz, J.E. The role of myocardial gap junctions in electrical conduction and arrhythmogenesis. *Cardiovasc. Pathol.* **2001**, *10*, 169–177. [CrossRef]

26. Gutstein, D.E.; Morley, G.E.; Tamaddon, H.; Vaidya, D.; Schneider, M.D.; Chen, J.; Chien, K.R.; Stuhlmann, H.; Fishman, G.I. Conduction slowing and sudden arrhythmic death in mice with cardiac-restricted inactivation of connexin43. *Circ. Res.* **2001**, *88*, 333–339. [CrossRef] [PubMed]

27. Beauchamp, P.; Yamada, K.A.; Baertschi, A.J.; Green, K.; Kanter, E.M.; Saffitz, J.E.; Kleber, A.G. Relative contributions of connexins 40 and 43 to atrial impulse propagation in synthetic strands of neonatal and fetal murine cardiomyocytes. *Circ. Res.* **2006**, *99*, 1216–1224. [CrossRef] [PubMed]

28. Thibodeau, I.L.; Xu, J.; Li, Q.; Liu, G.; Lam, K.; Veinot, J.P.; Birnie, D.H.; Jones, D.L.; Krahn, A.D.; Lemery, R.; et al. Paradigm of genetic mosaicism and lone atrial fibrillation: Physiological characterization of a connexin 43-deletion mutant identified from atrial tissue. *Circulation* **2010**, *122*, 236–244. [CrossRef] [PubMed]

29. Igarashi, T.; Finet, J.E.; Takeuchi, A.; Fujino, Y.; Strom, M.; Greener, I.D.; Rosenbaum, D.S.; Donahue, J.K. Connexin gene transfer preserves conduction velocity and prevents atrial fibrillation. *Circulation* **2012**, *125*, 216–225. [CrossRef] [PubMed]

30. Bikou, O.; Thomas, D.; Trappe, K.; Lugenbiel, P.; Kelemen, K.; Koch, M.; Soucek, R.; Voss, F.; Becker, R.; Katus, H.A.; et al. Connexin 43 gene therapy prevents persistent atrial fibrillation in a porcine model. *Cardiovasc. Res.* **2011**, *92*, 218–225. [CrossRef] [PubMed]

31. Simon, A.M.; Goodenough, D.A. Diverse functions of vertebrate gap junctions. *Trends Cell Biol.* **1998**, *8*, 477–483. [CrossRef]

32. Kirchhoff, S.; Nelles, E.; Hagendorff, A.; Kruger, O.; Traub, O.; Willecke, K. Reduced cardiac conduction velocity and predisposition to arrhythmias in connexin40-deficient mice. *Curr. Biol.* **1998**, *8*, 299–302. [CrossRef]

33. Hagendorff, A.; Schumacher, B.; Kirchhoff, S.; Luderitz, B.; Willecke, K. Conduction disturbances and increased atrial vulnerability in Connexin40-deficient mice analyzed by transesophageal stimulation. *Circulation* **1999**, *99*, 1508–1515. [CrossRef] [PubMed]

34. Leaf, D.E.; Feig, J.E.; Vasquez, C.; Riva, P.L.; Yu, C.; Lader, J.M.; Kontogeorgis, A.; Baron, E.L.; Peters, N.S.; Fisher, E.A.; et al. Connexin40 imparts conduction heterogeneity to atrial tissue. *Circ. Res.* **2008**, *103*, 1001–1008. [CrossRef] [PubMed]

35. Bagwe, S.; Berenfeld, O.; Vaidya, D.; Morley, G.E.; Jalife, J. Altered right atrial excitation and propagation in connexin40 knockout mice. *Circulation* **2005**, *112*, 2245–2253. [CrossRef] [PubMed]

36. Firouzi, M.; Ramanna, H.; Kok, B.; Jongsma, H.J.; Koeleman, B.P.; Doevendans, P.A.; Groenewegen, W.A.; Hauer, R.N. Association of human connexin40 gene polymorphisms with atrial vulnerability as a risk factor for idiopathic atrial fibrillation. *Circ. Res.* **2004**, *95*, e29–e33. [CrossRef] [PubMed]

37. Shi, H.F.; Yang, J.F.; Wang, Q.; Li, R.G.; Xu, Y.J.; Qu, X.K.; Fang, W.Y.; Liu, X.; Yang, Y.Q. Prevalence and spectrum of GJA5 mutations associated with lone atrial fibrillation. *Mol. Med. Rep.* **2013**, *7*, 767–774. [CrossRef] [PubMed]

38. Sun, Y.; Hills, M.D.; Ye, W.G.; Tong, X.; Bai, D. Atrial fibrillation-linked germline GJA5/connexin40 mutants showed an increased hemichannel function. *PLoS ONE* **2014**, *9*, e95125. [CrossRef] [PubMed]

39. Patel, D.; Gemel, J.; Xu, Q.; Simon, A.R.; Lin, X.; Matiukas, A.; Beyer, E.C.; Veenstra, R.D. Atrial fibrillation-associated connexin40 mutants make hemichannels and synergistically form gap junction channels with novel properties. *FEBS Lett.* **2014**, *588*, 1458–1464. [CrossRef] [PubMed]

40. Christophersen, I.E.; Holmegard, H.N.; Jabbari, J.; Sajadieh, A.; Haunso, S.; Tveit, A.; Svendsen, J.H.; Olesen, M.S. Rare variants in GJA5 are associated with early-onset lone atrial fibrillation. *Can. J. Cardiol.* **2013**, *29*, 111–116. [CrossRef] [PubMed]

41. Sun, Y.; Tong, X.; Chen, H.; Huang, T.; Shao, Q.; Huang, W.; Laird, D.W.; Bai, D. An atrial-fibrillation-linked connexin40 mutant is retained in the endoplasmic reticulum and impairs the function of atrial gap-junction channels. *Dis. Models Mech.* **2014**, *7*, 561–569. [CrossRef] [PubMed]

42. Lubkemeier, I.; Andrie, R.; Lickfett, L.; Bosen, F.; Stockigt, F.; Dobrowolski, R.; Draffehn, A.M.; Fregeac, J.; Schultze, J.L.; Bukauskas, F.F.; et al. The Connexin40A96S mutation from a patient with atrial fibrillation causes decreased atrial conduction velocities and sustained episodes of induced atrial fibrillation in mice. *J. Mol. Cell. Cardiol.* **2013**, *65*, 19–32. [CrossRef] [PubMed]

43. Santa Cruz, A.; Mese, G.; Valiuniene, L.; Brink, P.R.; White, T.W.; Valiunas, V. Altered conductance and permeability of Cx40 mutations associated with atrial fibrillation. *J. Gen. Physiol.* **2015**, *146*, 387–398. [CrossRef] [PubMed]

44. Dobrowolski, R.; Sommershof, A.; Willecke, K. Some oculodentodigital dysplasia-associated Cx43 mutations cause increased hemichannel activity in addition to deficient gap junction channels. *J. Membr. Biol.* **2007**, *219*, 9–17. [CrossRef] [PubMed]

45. Lai, A.; Le, D.N.; Paznekas, W.A.; Gifford, W.D.; Jabs, E.W.; Charles, A.C. Oculodentodigital dysplasia connexin43 mutations result in non-functional connexin hemichannels and gap junctions in C6 glioma cells. *J. Cell Sci.* **2006**, *119 Pt 3*, 532–541. [CrossRef] [PubMed]

46. Mese, G.; Sellitto, C.; Li, L.; Wang, H.Z.; Valiunas, V.; Richard, G.; Brink, P.R.; White, T.W. The Cx26-G45E mutation displays increased hemichannel activity in a mouse model of the lethal form of keratitis-ichthyosis-deafness syndrome. *Mol. Biol. Cell* **2011**, *22*, 4776–4786. [CrossRef] [PubMed]

47. Minogue, P.J.; Tong, J.J.; Arora, A.; Russell-Eggitt, I.; Hunt, D.M.; Moore, A.T.; Ebihara, L.; Beyer, E.C.; Berthoud, V.M. A mutant connexin50 with enhanced hemichannel function leads to cell death. *Investig. Ophthalmol. Vis. Sci.* **2009**, *50*, 5837–5845. [CrossRef] [PubMed]

48. Rohr, S. Role of gap junctions in the propagation of the cardiac action potential. *Cardiovasc. Res.* **2004**, *62*, 309–322. [CrossRef] [PubMed]

49. Shaw, R.M.; Rudy, Y. Ionic mechanisms of propagation in cardiac tissue. Roles of the sodium and L-type calcium currents during reduced excitability and decreased gap junction coupling. *Circ. Res.* **1997**, *81*, 727–741. [CrossRef] [PubMed]

50. Zipes, D.P. Mechanisms of clinical arrhythmias. *J. Cardiovasc. Electrophysiol.* **2003**, *14*, 902–912. [CrossRef] [PubMed]

51. Bai, D. Atrial fibrillation-linked GJA5/connexin40 mutants impaired gap junctions via different mechanisms. *FEBS Lett.* **2014**, *588*, 1238–1243. [CrossRef] [PubMed]

52. Lin, X.; Gemel, J.; Beyer, E.C.; Veenstra, R.D. Dynamic model for ventricular junctional conductance during the cardiac action potential. *Am. J. Physiol. Heart Circ. Physiol.* **2005**, *288*, H1113–H1123. [CrossRef] [PubMed]

53. Verheule, S.; van Kempen, M.J.A.; Postma, S.; Rook, M.B.; Jongsma, H.J. Gap junctions in the rabbit sinoatrial node. *Am. J. Physiol.-Heart Circul. Physiol.* **2001**, *280*, H2103–H2115. [CrossRef] [PubMed]

54. Hennemann, H.; Suchyna, T.; Lichtenberg-Frate, H.; Jungbluth, S.; Dahl, E.; Schwarz, J.; Nicholson, B.J.; Willecke, K. Molecular cloning and functional expression of mouse connexin40, a second gap junction gene preferentially expressed in lung. *J. Cell Biol.* **1992**, *117*, 1299–1310. [CrossRef] [PubMed]

55. Bruzzone, R.; Haefliger, J.A.; Gimlich, R.L.; Paul, D.L. Connexin40, a component of gap junctions in vascular endothelium, is restricted in its ability to interact with other connexins. *Mol. Biol. Cell* **1993**, *4*, 7–20. [CrossRef] [PubMed]

56. Ye, W.G.; Yue, B.; Aoyama, H.; Kim, N.K.; Cameron, J.A.; Chen, H.; Bai, D. Junctional delay, frequency, and direction-dependent uncoupling of human heterotypic Cx45/Cx43 gap junction channels. *J. Mol. Cell. Cardiol.* **2017**, *111*, 17–26. [CrossRef] [PubMed]

57. Saffitz, J.E. Connexins, conduction, and atrial fibrillation. *N. Engl. J. Med.* **2006**, *354*, 2712–2714. [CrossRef] [PubMed]

58. Darbar, D.; Kannankeril, P.J.; Donahue, B.S.; Kucera, G.; Stubblefield, T.; Haines, J.L.; George, A.L., Jr.; Roden, D.M. Cardiac sodium channel (SCN5A) variants associated with atrial fibrillation. *Circulation* **2008**, *117*, 1927–1935. [CrossRef] [PubMed]

59. Jassim, A.; Aoyama, H.; Ye, W.G.; Chen, H.; Bai, D. Engineered Cx40 variants increased docking and function of heterotypic Cx40/Cx43 gap junction channels. *J. Mol. Cell. Cardiol.* **2016**, *90*, 11–20. [CrossRef] [PubMed]

International Journal of
Molecular Sciences

MDPI

Article

Irradiation-Induced Cardiac Connexin-43 and miR-21 Responses Are Hampered by Treatment with Atorvastatin and Aspirin

Csilla Viczenczova [1], Branislav Kura [1], Tamara Egan Benova [1], Chang Yin [2], Rakesh C. Kukreja [2], Jan Slezak [1], Narcis Tribulova [1] and Barbara Szeiffova Bacova [1,*]

[1] Institute for Heart Research, Center of Experimental Medicine, Slovak Academy of Sciences, Bratislava 841 04, Slovak Republic; viczencz.csilla@gmail.com (C.V.); branislav.kura@savba.sk (B.K.); tamara.benova@savba.sk (T.E.B.); jan.slezak@savba.sk (J.S.); narcisa.tribulova@savba.sk (N.T.)
[2] Division of Cardiology, Medical College of Virginia, Virginia Commonwealth University, Richmond, VA 23298, USA; rakesh@vcu.edu (C.Y.); rakesh.kukreja@vcuhealth.org (R.C.K.)
*   Correspondence: barbara.bacova@savba.sk; Tel.: +421-2-3229-5419

Received: 9 March 2018; Accepted: 5 April 2018; Published: 10 April 2018

Abstract: Radiation of the chest during cancer therapy is deleterious to the heart, mostly due to oxidative stress and inflammation related injury. A single sub-lethal dose of irradiation has been shown to result in compensatory up-regulation of the myocardial connexin-43 (Cx43), activation of the protein kinase C (PKC) signaling along with the decline of microRNA (miR)-1 and an increase of miR-21 levels in the left ventricle (LV). We investigated whether drugs with antioxidant, anti-inflammatory or vasodilating properties, such as aspirin, atorvastatin, and sildenafil, may affect myocardial response in the LV and right ventricle (RV) following chest irradiation. Adult, male Wistar rats were subjected to a single sub-lethal dose of chest radiation at 25 Gy and treated with aspirin (3 mg/day), atorvastatin (0.25 mg/day), and sildenafil (0.3 mg/day) for six weeks. Cx43, PKCε and PKCδ proteins expression and levels of miR-1 as well as miR-21 were determined in the LV and RV. Results showed that the suppression of miR-1 was associated with an increase of total and phosphorylated forms of Cx43 as well as PKCε expression in the LV while having no effect in the RV post-irradiation as compared to the non-irradiated rats. Treatment with aspirin and atorvastatin prevented an increase in the expression of Cx43 and PKCε without change in the miR-1 levels. Furthermore, treatment with aspirin, atorvastatin, and sildenafil completely prevented an increase of miR-21 in the LV while having partial effect in the RV post irradiation. The increase in pro-apoptotic PKCδ was not affected by any of the used treatment. In conclusion, irradiation and drug-induced changes were less pronounced in the RV as compared to the LV. Treatment with aspirin and atorvastatin interfered with irradiation-induced compensatory changes in myocardial Cx43 protein and miR-21 by preventing their elevation, possibly via amelioration of oxidative stress and inflammation.

Keywords: irradiation; heart; connexin-43; miR-1; miR-21; atorvastatin; aspirin

---

1. Introduction

Cardiovascular injury due to radiation is the most common cause of adverse events among cancer survivors [1,2]. Key factors responsible for the establishment of cardiovascular injury, i.e., oxidative stress, inflammation, and epigenetic modifications, have been linked to potential treatments and been recently described [1,2]. Ionizing radiation induces oxidative stress and causes changes in the expression of several microRNAs (miRNA)s, including miR-1 and miR-21. An increase of miR-21 is involved in myocardial hypertrophy [3,4] and fibrosis [5]. An increase in miR-21 has also been associated with the up-regulation of the protein kinase C (PKC) δ [6], which is also implicated in tissue remodeling. Fibrosis and necrosis were reduced by the treatment with a free radical-scavenging component such as melatonin [7]. However, molecular mechanisms of the irradiation induced injury are unknown and there is currently a lack of treatment strategies.

Cardiac connexin-43 (Cx43) channels are essential for coordinated heart function because they ensure electrical coupling and direct intercellular communication. We and others [6,8] have shown that a single sub-lethal dose of irradiation results in up-regulation of Cx43, which has been associated with the protection of the heart against malignant arrhythmias [9] and infarction [10]. An increase in myocardial Cx43 expression and its active phosphorylated forms has been shown to be associated with the suppression of miR-1 (which regulates GJA1 gene transcription for Cx43) and the enhancement of PKCε (which phosphorylates Cx43) [6]. It appears that these early post-irradiation related myocardial alterations, including up-regulation of Cx43, are most likely compensatory responses of the heart to maintain its normal function [10–12]. Since irradiation induces inflammation and oxidative stress [12], we hypothesized that compounds exerting anti-inflammatory and antioxidant actions would interfere with irradiation-induced compensatory responses. To this context, we considered several drugs, which included acetylsalicylic acid (aspirin), a non-selective inhibitor of cyclooxygenase-1 and cyclooxygenase-2, which prevents formation of pro-inflammatory prostaglandins and thromboxanes. Antioxidant properties of acetylsalicylic acid are attributed to its ability of inhibiting lipid peroxidation and DNA damage [13]. Atorvastatin is widely used for treatment of human dislipidemia due to inhibition of the 3-hydroxy-3-methylglutaryl coenzyme A reductase. In addition, its pleiotropic effects are associated with anti-inflammatory and antioxidative actions [14] as well as promoting the availability of vascular nitric oxide [15]. Sildenafil, the inhibitor of phosphodiestherase 5, is used for its vasodilatation and activation of nitric oxide [16].

Our goal was to demonstrate whether treatment with the drugs targeting oxidative stress and inflammation might result in attenuation of irradiation induced myocardial compensatory responses previously reported [6,10]. In particular, we examined myocardial changes in Cx43, PKC, miR-1 and miR-21 in rats exposed to single chest irradiation.

2. Results

2.1. Main Characteristics of Experimental Rats

Comparing to non-irradiated rats, the body, heart, and left ventricular weight were significantly decreased in irradiated animals after six weeks. On the other hand, irradiation did not alter the right ventricular weight. Treatment with selected drugs also had no effect on these biometric parameters in any treated group, except for the heart weight in post-irradiated + sildenafil group. Data are summarized in the Table 1.

Table 1. Biometric parameters registered in control and irradiated Wistar rats.

GROUP	BW (g)	HW (g)	LVW (g)	RVW (g)
C	344.22 ± 28.40	0.89 ± 0.11	0.36 ± 0.07	0.09 ± 0.02
C-A	382.61 ± 32.19	0.92 ± 0.09	0.39 ± 0.04	0.12 ± 0.01
C-AT	371.01 ± 20.45	0.87 ± 0.07	0.37 ± 0.04	0.10 ± 0.02
C-S	363.63 ± 23.93	0.90 ± 0.10	0.39 ± 0.04	0.11 ± 0.02
I	252.24 ± 14.31 [a]	0.83 ± 0.05	0.25 ± 0.04 [a]	0.10 ± 0.04
I-A	227.41 ± 26.47 [c]	0.73 ± 0.09 [c]	0.24 ± 0.02 [c]	0.11 ± 0.03
I-AT	253.23 ± 31.82 [c]	0.77 ± 0.08	0.27 ± 0.03 [c]	0.13 ± 0.03
I-S	233.62 ± 14.50 [c]	0.67 ± 0.05 [b,c]	0.23 ± 0.02 [c]	0.12 ± 0.03

BW—body weight; HW—heart weight; LVW—left ventricular weight; RVW—right ventricular weight; C—control non-irradiated; C-A—control + aspirin; C-AT—control + atorvastatin; C-S—control + sildenafil; I—irradiated rats; I-A—irradiated + aspirin; I-AT—irradiated + atorvastatin; I-S—irradiated + sildenafil. Results are the mean ± SD of 6 hearts. [a] $p < 0.05$ vs. C (C vs. C-A, C-AT, C-S, I); [b] $p < 0.05$ vs. I (I vs. I-A, I-AT, I-S); [c] $p < 0.05$ treated C vs. treated I (C-A vs. I-A; C-AT vs. I-AT; C-S vs. I-S).

2.2. Protein Expression of Myocardial Cx43 in Control and Irradiated Wistar Rats

Western blot analysis showed that total Cx43 protein was increased in the LV ($p < 0.05$) and to a lesser extent in RV post-irradiation as compared to the non-irradiated group (Figure 1A,B,D,E). In parallel, the active phosphorylated forms of Cx43 were significantly increased in the LV and RV of irradiated rats versus the non-irradiated controls (Figure 1A,C,D,F). Treatment with aspirin and atorvastatin for six weeks (starting one day before irradiation) suppressed the elevation of the total as well as the phosphorylated forms of Cx43, significantly in the LV (Figure 1A–C) while having no effect in RV (Figure 1D–F) of post-irradiated animals. The administration of sildenafil had no significant effect on the irradiation-induced increase in myocardial Cx43, i.e., either on total levels or its phosphorylated forms after six weeks (Figure 1).

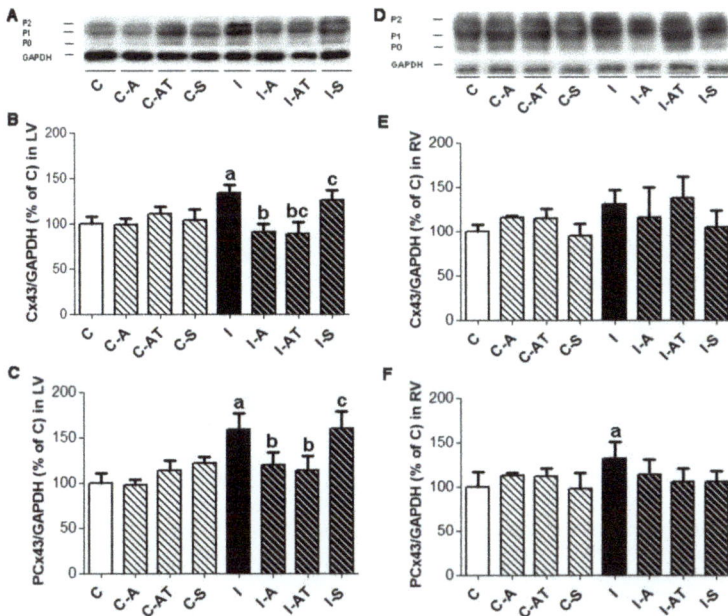

Figure 1. Representative immunoblots showing three forms of Cx43 (**A,D**) and densitometric quantification of total Cx43 expression (**B,E**), and its phosphorylated forms (**C,F**) normalized to GAPDH in the LV (left panel) and RV (right panel) of non-irradiated and post-irradiated Wistar rats with and without treatment. Abbreviations—P0: unphosphorylated form of Cx43; P1 and P2: phosphorylated forms of Cx43; GAPDH: housekeeper; C: control non-irradiated rats; C-A: control + aspirin; C-AT: control + atorvastatin; C-S: control + sildenafil; I: post-irradiated rats; I-A: post-irradiated + aspirin; I-AT: post-irradiated + atorvastatin; I-S: post-irradiated + sildenafil. Data are means ± SD of 6 hearts. [a] $p < 0.05$ vs. C (C vs. C-A, C-AT, C-S, I); [b] $p < 0.05$ vs. I (I vs. I-A, I-AT, I-S); [c] $p < 0.05$ treated C vs. treated I (C-A vs. I-A; C-AT vs. I-AT; C-S vs. I-S).

2.3. Expression of Protein Kinase Cε in Control and Irradiated Wistar Rats

PKCε was significantly increased in the LV (Figure 2A,B) but not in the RV (Figure 2C,D) of rats following irradiation when compared to the non-irradiated controls. Six weeks of treatment with aspirin and atorvastatin following irradiation normalized PKCε in the LV (Figure 2A,B) while having no effect in the RV (Figure 2C,D). Treatment with sildenafil had no effect on the PKCε expression of the irradiated rats. There was no change in the expression of PKCε in non-irradiated rats following treatment with drugs (Figure 2).

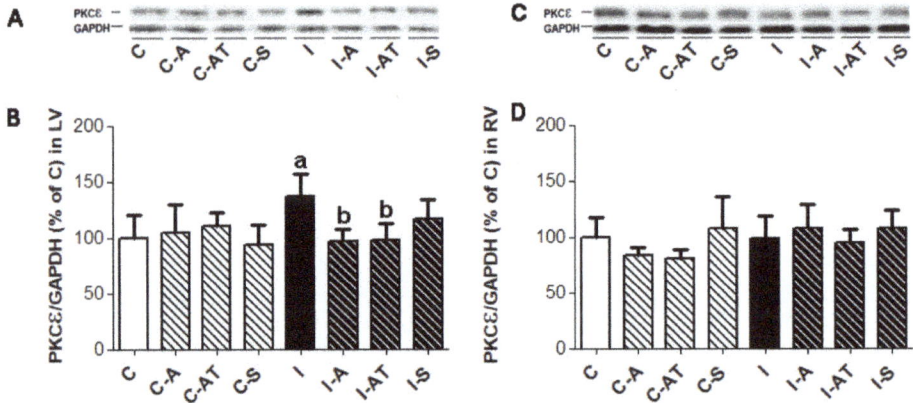

Figure 2. Representative immunoblots of PKCε expression (**A,C**) and its quantitative evaluation normalized to GAPDH in the LV ((**B**), left panel) and RV ((**D**), right panel) of non-irradiated and post-irradiated Wistar rats with and without treatment. Abbreviations—PKCε: protein kinase C epsilon; GAPDH: housekeeping protein; C: control non-irradiated rats; C-A: control + aspirin; C-AT: control + atorvastatin; C-S: control + sildenafil; I: post-irradiated rats; I-A: post-irradiated + aspirin; I-AT: post-irradiated + atorvastatin; I-S: post-irradiated + sildenafil. Data are means ± SD of 6 hearts. [a] $p < 0.05$ vs. C (C vs. C-A, C-AT, C-S, I); [b] $p < 0.05$ vs. I (I vs. I-A, I-AT, I-S).

2.4. Expression of Protein Kinase C δ in Control and Irradiated Wistar Rats

Similar to PKCε, the myocardial expression of PKCδ (Figure 3) was increased in response to irradiation. The increase was significant in the LV (Figure 3A,B) while no change was observed in the RV (Figure 3C,D) when compared to the to non-irradiated controls. Treatment with drugs had no effect on the expression of PKCδ in the LV or RV in the irradiated and non-irradiated groups (Figure 3).

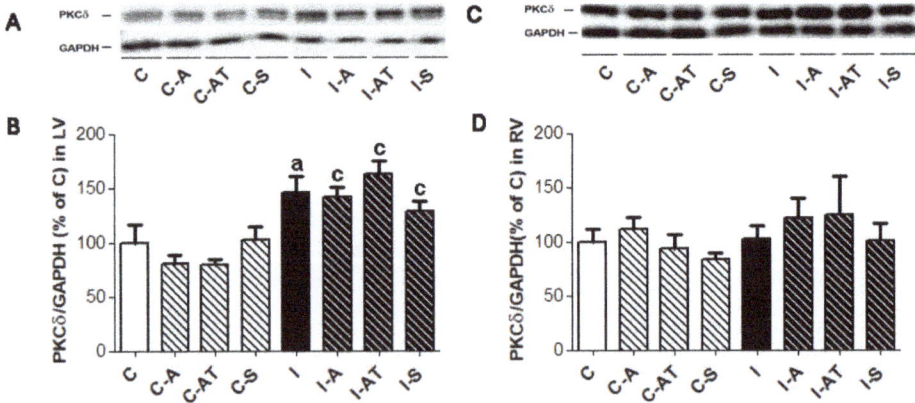

Figure 3. Representative immunoblots of PKCδ expression (**A,C**) and quantitative evaluation normalized to GAPDH in the LV ((**B**), left panel) and RV ((**D**), right panel) of non-irradiated and post-irradiated Wistar rats. Abbreviations—PKCδ: protein kinase C delta; GAPDH: housekeeping protein; C: control non-irradiated rats; C-A: control + aspirin; C-AT: control + atorvastatin; C-S: control + sildenafil; I: post-irradiated rats; I-A: post-irradiated + aspirin; I-AT: post-irradiated + atorvastatin; I-S: post-irradiated + sildenafil. Data are means ± SD of 6 hearts. [a] $p < 0.05$ vs. C (C vs. C-A, C-AT, C-S, I); [c] $p < 0.05$ treated C vs. treated I (C-A vs. I-A; C-AT vs. I-AT; C-S vs. I-S).

2.5. Myocardial Expression of miR-1 in Control and Irradiated Wistar Rats

miR-1 level decreased following six weeks after irradiation (Figure 4), which was signficant in the LV (Figure 4A) but not in the RV (Figure 4B). Treatment with the drugs had no significant effect on miR-1 level in post-irradiated as well as in non-irradiated groups (Figure 4).

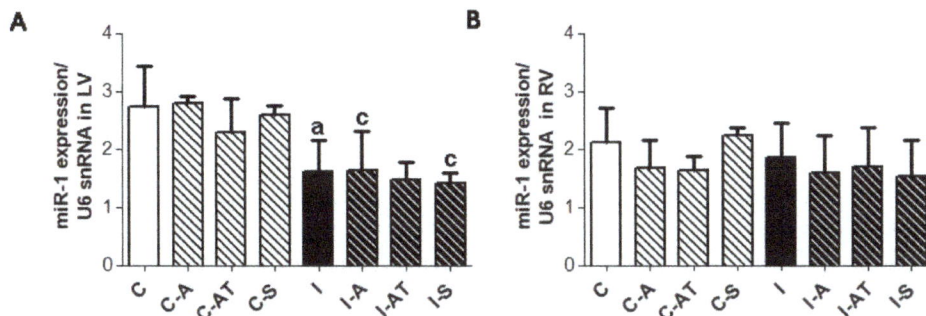

Figure 4. Myocardial expression of miR-1 in the LV ((**A**), left panel) and RV ((**B**), right panel) of non-irradiated and post-irradiated Wistar rats with and without treatment. Endogenous U6 small nuclear RNA (U6snRNA) was used to normalize miR-1. Abbreviations—C: control non-irradiated rats; C-A: control + aspirin; C-AT: control + atorvastatin; C-S: control + sildenafil; I: post-irradiated rats; I-A: post-irradiated + aspirin; I-AT: post-irradiated + atorvastatin; I-S: post-irradiated + sildenafil. Data are means ± SD of 5 hearts. [a] $p < 0.05$ vs. C (C vs. C-A, C-AT, C-S, I); [c] $p < 0.05$ treated C vs. treated I (C-A vs. I-A; C-AT vs. I-AT; C-S vs. I-S).

2.6. Expression of miR-21 in Control and Irradiated Wistar Rats

The expression of myocardial miR-21 was significantly increased in both the LV and RV of the post-irradiated rats compared to the non-irradiated controls (Figure 5). Six weeks of treatment with the selected drugs significantly suppressed miR-21 expression in the left (Figure 5A), although to a lesser extent in the RV (Figure 5B) following irradiation. There was no effect of drugs on miR-21 levels in the non-irradiated control groups (Figure 5).

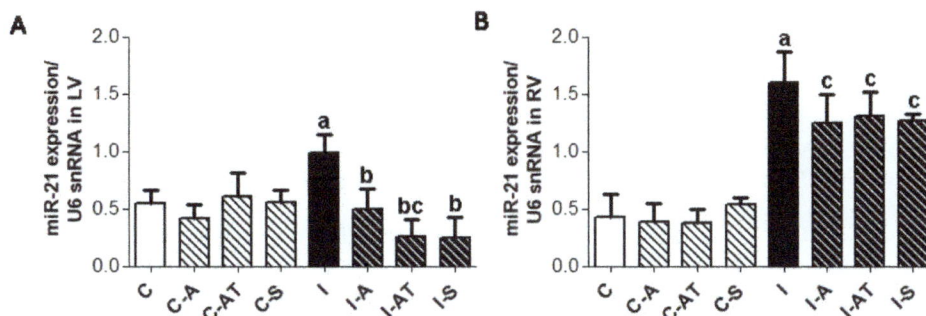

Figure 5. Myocardial expression of miR-21 in the LV (**A**) and RV (**B**) of non-irradiated and post-irradiated Wistar rats with and without treatment. Endogenous U6 small nuclear RNA (U6snRNA) was used to normalize miR-21. Abbreviations—C: control non-irradiated rats; C-A: control + aspirin; C-AT: control + atorvastatin; C-S: control + sildenafil; I: post-irradiated rats; I-A: post-irradiated + aspirin; I-AT: post-irradiated + atorvastatin; I-S: post-irradiated + sildenafil. Data are means ± SD of 5 hearts. [a] $p < 0.05$ vs. C (C vs. C-A, C-AT, C-S, I); [b] $p < 0.05$ vs. I (I vs. I-A, I-AT, I-S); [c] $p < 0.05$ treated C vs. treated I (C-A vs. I-A; C-AT vs. I-AT; C-S vs. I-S).

3. Discussion

In the present study, we showed that exposure of rats to a single dose of chest irradiation at 25 Gy caused up-regulation in the expression of Cx43, PKCε, and PKCδ in the LV following six weeks. In parallel, miR-1, which is known to repress GJA1 for Cx43 [3,6] was decreased. While miR-21, which is involved in myocardial remodeling and apoptosis [3,6], was increased following irradiation. These results are in accordance with our previous findings [6,10] and in line with the reported enhancement of myocardial Cx43 protein and mRNA expression found in the rabbit heart in response to heavy ion radiation [8,9]. Importantly, these alterations were associated with protection of the heart against arrhythmia and infarction [8,10].

In the present study, we have also demonstrated that irradiation did not induce significant changes in the expression of total Cx43 in the RV, although there was a trend towards an increase. This may be in part due to the miR-1, which was not changed in the RV in contrast to its significant suppression associated with the enhancement of Cx43 in the LV. Furthermore, neither PKCε nor PKCδ expression was altered in the RV unlike its significant elevation in LV following irradiation. Whether such distinct responses to irradiation in the LV and RV have any relationship with functional, metabolic, structural, and other differences of the heart chambers [17,18] needs to be investigated. Chamber related differences in Cx43 and PKC expression in response to altered thyroid status have been previously suggested [19]. A higher metabolic rate and mechanical load of the LV seems to be more prone to oxidative stress and injury. Nevertheless, our results suggest that cardiac response to chest irradiation is associated with the up-regulation of myocardial Cx43 connected with the suppression of miR-1 and enhanced PKCε and PKCδ signaling in the LV but not in the RV. On the other hand, the expression of miR-21 was significantly increased in both the LV and RV following irradiation. It is possible that the anti-apoptotic [20] and proliferation promoting effects of miR-21 [5] may be pro-survival to cardiac tissue during the early period in response to radiation-induced injury.

Our results also show that treatment of the irradiated rats with aspirin and atorvastatin prevented up-regulation of Cx43, i.e., its total and phosphorylated forms in the LV. Interestingly, none of the drugs reduced the level of miR-1 in the LV following irradiation. These data suggest that apart from the miR-1, other post-translational factors may modulate Cx43 levels in response to irradiation. Neither atorvastatin nor aspirin had any effect on the basal myocardial expression of Cx43 in the non-irradiated control rats.

The question arises, how aspirin or atorvastatin may affect changes in myocardial Cx43 protein levels induced by single chest irradiation? Considering that both compounds exert protection from oxidative stress and inflammation—the important culprits in irradiation-induced cardiac injury [1,2]—it is reasonable to speculate that these drugs may suppress these processes. If so, then cardiac stress would be attenuated and a subsequent compensatory response (most likely mediated by free radicals and pro-inflammatory molecules signaling) will be reduced. Indeed, the administration of antioxidants and/or scavengers of free radicals, such as hesperidin [21], melatonin [7] or selenium [22] prior to irradiation attenuated oxidative stress, inflammation, and cardiomyocyte necrosis. The potential use of statins as radioprotective agents has been recently reviewed, where signaling pathways targeting pro-inflammatory NF-κB might be implicated [1]. In support of this concept, it should be noted that unlike aspirin and atorvastatin, treatment with sildenafil (dominant vasculo-protective drug) did not affect the myocardial Cx43 levels in irradiated rats. Moreover, aspirin and atorvastatin but not sildenafil reduced expression of TNF-α in irradiated rats in our model, as reported previously [23].

In general, oxidative stress and/or inflammation contribute to the impairment of intercellular communication due to the acceleration of Cx43 and Cx43 interacting proteins that are in degradation and/or dysfunction [24]. Consequently, down-regulation of Cx43 due to chronic redox disorders and subclinical inflammation often accompanies cardiovascular disease [25,26]. On the other hand, acute heart injury (e.g., intermittent ischemia) triggers endogenous pro-survival molecular pathways, including up-regulation of Cx43, to protect heart function [27]. It appears that the heart may respond to stressors by compensatory/adaptive (such as up-regulation of Cx43) and by maladaptive/decompensatory changes (down-regulation of Cx43). Accordingly, enhanced Cx43 level observed in the compensatory hypertrophy was induced by pressure overload, while reduced Cx43 occurred in the decompensatory state [25].

Our results also show that the treatment of irradiated rats with aspirin and atorvastatin normalized the myocardial expression of PKCε, which is associated with the expression of phosphorylated forms of Cx43. These results suggest that oxidative stress and inflammation may modulate myocardial PKCε, signaling and its cardioprotective role as shown in various conditions [28]. Interestingly, treatment did not affect elevated myocardial PKCδ expression, i.e., its pro-hypertrophic signaling in irradiated rat heart. However, the expression of pro-fibrotic miR-21 was significantly decreased in the left heart ventricle by all tested drugs. Of note, miR-21 has been shown up-regulated in human fibroblast cells due to various stress-inducing conditions, including radiation [3,29]. Thus, more attention should be paid to elucidate the possible anti-fibrotic effects of atorvastatin, aspirin, and perhaps sildenafil. Importantly, microRNA-21 has been reported as a novel promising target in cancer radiation therapy [20] and atorvastatin was shown to suppress expansive remodeling by inhibition of macrophage infiltration [30].

Taken together, we assume that protection of the heart from oxidative stress and inflammation by appropriate drugs, including aspirin and atorvastatin or nonpharmacological compounds like melatonin [7], omega-PUFA [31] may counteract compensatory responses and attenuate adverse consequences of chest irradiation.

4. Materials and Methods

Animal experiments were approved by the Animal Research and Care Committee of the Institute for Heart Research, Slovak Academy of Sciences—Project 1873/11-221/3, approved on 30 September 2011 and in accordance with the rules issued by the State Veterinary Administration of the Slovak Republic, legislation No 289/2003. Rats were maintained on a 12:12 h's light/dark cycle with access to standard pellet and water *ad libitum*.

4.1. Experimental Model of Irradiation Induced Cardiac Injury

Three-month-old male Wistar rats were randomly divided into irradiated ($n = 24$) and non-irradiated ($n = 24$) groups. Rats were anesthetized with Narketan (115 mg/kg body weight) followed by myorelaxant Xylan (1 mg/kg body weight) and exposed to a single dose of 25 Gy of ionizing radiation given locally on mediastinum at the area of the heart using the electron linear accelerator UELR 5-1S (Producer NIIEFA St. Petersburg, RF, Russia), as described previously [6]. Control animals were shielded with lead plates. Non-irradiated controls (C) and irradiated (I) rats were treated one day before, the day of irradiation, and for the next six weeks with aspirin (A; 3 mg/day), atorvastatin (AT; 0.25 mg/day) or sildenafil (S; 0.3 mg/day) via a gastric tube. Doses of the drugs were calculated from the maximal therapeutic dose for humans in relation to the rat body weight.

At the end of the experiment, hearts were excised from the anesthetized animals (thiopental, 65 mg/kg body weight) followed by the registration of the whole heart as well as left (LV) and right ventricle (RV) weight. Frozen tissues from the heart ventricles were stored at $-80\,^{\circ}$C in freezer box and used for the analysis of Cx43, PKCε and PKCδ expression by the Western blot method and for miR-1 and miR-21 expression analysis by qRT-PCR.

4.2. Determination of Myocardial Cx43, PKCε and PKCδ Protein Expression

The western blot analysis was performed as previously described [6]. Briefly, the ventricular samples ($n = 6$ per group) were powdered and solubilized in SB20 (20% SDS, 10 mmol/L EDTA, 0.1 mol/L tris(hydroxymethyl) aminomethane (TRIS), pH 6.8 by sonicator UP 100H (Hielscher, Teltow, Germany). Total proteins (10–30 μg) from each sample were separated in 10% sodium dodecyl sulfate polyacrylamide gels, transferred onto a nitrocellulose membrane, and blocked with 5% non-fat dry milk in Tris-buffered saline. For the determination of Cx43, the membrane was incubated with a primary rabbit polyclonal antibody (diluted 1:4000; Anti-Connexin 43 C 6219; Sigma-Aldrich, St. Louis, MO, USA). For PKCε and PKCδ determination, the nitrocellulose membrane was incubated with primary rabbit polyclonal antibodies (PKCε Antibody, C-15: sc-214, Santa Cruz Biotechnology, Inc., PKCδ Antibody, C-17: sc-213, Santa Cruz Biotechnology, Inc., Santa Cruz, CA, USA) diluted 1:1000, overnight at 4 °C, followed by further incubation for 1 h at room temperature with a secondary donkey antibody (peroxidase-labeled anti-rabbit, 1:2000, Amersham Biosciences, Piscataway, NJ, USA). After ECL visualization, the densitometric analysis by Carestream Molecular Imaging Software (version 5.0, Carestream Health, New Haven, CT, USA.) was done using a KODAK In-Vivo Multispectral System FX. The measured values were normalized to the expression of GAPDH (glyceraldehyde-3-phosphate dehydrogenase) serving as a control [6].

4.3. Estimation of Myocardial miR-1 and miR-21 Levels

miR reverse transcription and TaqMan-based qRT-PCR analysis were performed as previously reported [6]. Total RNA including small RNA was isolated from the frozen ventricular tissue of the non-irradiated (n = 5 per group) and irradiated (n = 5 per group) Wistar rats using a miRNA mini kit according to the manufacturer's protocol (QIAGEN Sciences, Germantown, MD, USA). The concentration and purity of the isolated RNA was checked using a Nanodrop ND-1000 spectrophotometer (Agilent Technologies, Santa Clara, CA, USA). Briefly, 10 ng of total RNA were subjected for reverse transcription reaction with miRNA specific RT primers using microRNA reverse transcription kit (Applied Biosystems, Foster City, CA, USA) in accordance to the manufacturer's instructions. Real time PCR was performed using a Roche Light cycler 480 II (Roche Applied Science, Indianapolis, IN, USA). TaqMan miRNA assay probe (Applied Biosystems, Foster City, CA, USA) was applied to determine the expression level of miR-1 and miR-21. Endogenous U6 small nuclear RNA was used to normalize RNA content. Reverse transcription was performed using stem loop specific microRT primers under the following conditions: 16 °C for 30 min, 42 °C for 30 min, and 85 °C for 5 min. The obtained cDNA was diluted in 1:3 ratios and subjected to real-time PCR using a TaqMan amplicon specific assay probe under the following PCR cycle conditions: 95 °C for 10 min, 95 °C for 15 s, and 60 °C for 60 s [6].

4.4. Statistical Analysis

Data are expressed as means ± SD. One way two-tailed ANOVA and Tukey post hoc tests were used for statistical analysis. $p < 0.05$ was considered as statistically significant.

5. Conclusions

In conclusion, we have demonstrated that irradiation related changes are less pronounced in the RV as compared to the LV. Our results also suggest that treatment with aspirin and atorvastatin attenuate irradiation-induced up-regulation of myocardial Cx43 and PKCε signaling as well as miR-21 expression. Whether treatment related prevention or attenuation of irradiation-induced myocardial alterations at the early responsive period might help to prevent adverse late effects needs further investigations.

Acknowledgments: This study was supported by grants APVV-0241/11, APVV-15-0376 of Slovak Research and Development Agency and VEGA 2/0076/16, VEGA 2/0167/15 of Slovak Scientific Grant Agency and by EU Structural Fund ITMS 26230120006.

Author Contributions: Csilla Viczenczova: acquisition of data, data analysis/interpretation; Branislav Kura: acquisition of data, data analysis/interpretation; Tamara Egan Benova: acquisition of data, data analysis/interpretation; Chang Yin: acquisition of data, data analysis/interpretation; Rakesh C. Kukreja: formulation of the overall concept/design of experiments, editing of English language; Jan Slezak: formation of concept/design of experiment; Narcis Tribulova: formation of concept/design of experiment, critical revision of the manuscript, and approval of the article; Barbara Szeiffova Bacova: acquisition of data, data analysis/interpretation, approval of the article.

Conflicts of Interest: The authors declare no conflict of interest.

References

1. Cuomo, J.R.; Sharma, G.K.; Conger, P.D.; Weintraub, N.L. Novel concepts in radiation-induced cardiovascular disease. *World J. Cardiol.* **2016**, *8*, 504–519. [CrossRef] [PubMed]
2. Slezak, J.; Kura, B.; Ravingerova, T.; Tribulova, N.; Okruhlicova, L.; Barancik, M. Mechanisms of cardiac radiation injury and potential preventive approaches. *Can. J Physiol. Pharmacol.* **2015**, *27*, 1–17. [CrossRef] [PubMed]
3. Simone, N.L.; Soule, B.P.; Ly, D.; Saleh, A.D.; Savage, J.E.; Degraff, W.; Cook, J.; Harris, C.C.; Gius, D.; Mitchell, J.B. Ionizing radiation-induced oxidative stress alters miRNA expression. *PLoS ONE* **2009**, *4*, e6377. [CrossRef] [PubMed]
4. Kura, B.; Babal, P.; Slezak, J. Implication of microRNAs in the development and potential treatment of radiation—Induced heart disease. *Can. J. Physiol. Pharmacol.* **2017**, *95*, 1236–1244. [CrossRef] [PubMed]
5. Zhu, H.; Fan, G.C. Role of microRNAs in the reperfused myocardium towards post-infarct remodelling. *Cardiovasc. Res.* **2012**, *94*, 284–292. [CrossRef] [PubMed]
6. Viczenczova, C.; Szeiffova Bacova, B.; Egan Benova, T.; Kura, B.; Yin, C.; Weismann, P.; Kukreja, R.; Slezak, J.; Tribulova, N. Myocardial connexin-43 and PKC signalling are involved in adaptation of the heart to irradiation-induced injury: Implication of miR-1 and miR-21. *Gen. Physiol. Biophys.* **2016**, *35*, 215–222. [CrossRef] [PubMed]
7. Gürses, I.; Özeren, M.; Serin, M.; Yücel, N.; Erkal, H.S. Histopathological evaluation of melatonin as a protective agent in heart injury induced by radiation in a rat model. *Pathol. Res. Pract.* **2014**, *210*, 863–871. [CrossRef] [PubMed]
8. Amino, M.; Yoshioka, K.; Tanabe, T.; Tanaka, E.; Mori, H.; Furusawa, Y.; Zareba, W.; Yamazaki, M.; Nakagawa, H.; Honjo, H.; et al. Heavy ion radiation up-regulates connexin43 and ameliorates the arrhythmogenic substrates in rabbit hearts after myocardial infarction. *Cardiovasc. Res.* **2006**, *72*, 412–421. [CrossRef] [PubMed]
9. Amino, M.; Yoshioka, K.; Fujibayashi, D.; Hashida, T.; Furusawa, Y.; Zareba, W.; Ikari, Y.; Tanaka, E.; Mori, H.; Inokuchi, S.; et al. Year-long upregulation of connexin43 in rabbit hearts by heavy ion irradiation. *Am. J. Physiol. Heart Circ. Physiol.* **2010**, *298*, 1014–1021. [CrossRef] [PubMed]
10. Viczenczova, C.; Kura, B.; Chaudagar, K.K.; Szeiffova Bacova, B.; Egan Benova, T.; Barancik, M.; Knezl, V.; Ravingerova, T.; Tribulova, N.; Slezak, J. Myocardial connexin-43 is upregulated in response to acute cardiac injury in rats. *Can. J. Physiol. Pharmacol.* **2017**, *95*, 911–919. [CrossRef] [PubMed]
11. Song, J.; Yan, R.; Wu, Z.; Li, J.; Yan, M.; Hao, X.; Liu, J.; Li, S. ^{13}N-ammonia PET/CT detection of myocardial perfusion abnormalities in beagle dogs after local heart irradiation. *J. Nucl. Med.* **2017**, *58*, 605–610. [CrossRef] [PubMed]
12. Slezak, J.; Kura, B.; Babal, P.; Barancik, M.; Ferko, M.; Frimmel, K.; Kalocayova, B.; Kukreja, R.C.; Lazou, A.; Mezesova, L.; et al. Potential markers and metabolic processes involved in the mechanism of radiation-induced heart injury. *Can. J. Physiol. Pharmacol.* **2017**, *95*, 1190–1203. [CrossRef] [PubMed]
13. Shi, X.; Ding, M.; Dong, Z.; Chen, F.; Ye, J.; Wang, S.; Leonard, S.S.; Castranova, V.; Vallyathan, V. Antioxidant properties of aspirin: Characterization of the ability of aspirin to inhibit silica-induced lipid peroxidation, DNA damage, NF-kappaB activation, and TNF-alpha production. *Mol. Cell. Biochem.* **1999**, *199*, 93–102. [CrossRef] [PubMed]
14. Pignatelli, P.; Carnevale, R.; Pastori, D.; Cangemi, R.; Napoleone, L.; Bartimoccia, S.; Nocella, C.; Basili, S.; Violi, F. Immediate antioxidant and antiplatelet effect of atorvastatin via inhibition of Nox2. *Circulation* **2012**, *126*, 92–103. [CrossRef] [PubMed]
15. Lefer, A.M.; Scalia, R.; Lefer, D.J. Vascular effects of HMG CoA-reductase inhibitors (statins) unrelated to cholesterol lowering: New concepts for cardiovascular disease. *Cardiovasc. Res.* **2001**, *49*, 281–287. [CrossRef]
16. Webb, D.J.; Freestone, S.; Allen, M.J.; Muirhead, G.J. Sildenafil citrate and blood-pressure-lowering drugs: Results of drug interaction studies with an organic nitrate and a calcium antagonist. *Am. J. Cardiol.* **1999**, *83*, 21–28. [CrossRef]

17. Cihák, R.; Kolár, F.; Pelouch, V.; Procházka, J.; Ostádal, B.; Widimský, J. Functional changes in the right and left ventricle during development of cardiac hypertrophy and after its regression. *Cardiovasc. Res.* **1992**, *26*, 845–850. [CrossRef] [PubMed]

18. Noorman, M.; van Rijen, H.V.; van Veen, T.A.; de Bakker, J.M.; Stein, M. Differences in distribution of fibrosis in the ventricles underlie dominant arrhythmia vulnerability of the right ventricle in senescent mice. *Neth. Heart J.* **2008**, *16*, 356–358. [CrossRef] [PubMed]

19. Szeiffová Bačová, B.; Egan Beňová, T.; Viczenczová, C.; Soukup, T.; Rauchová, H.; Pavelka, S.; Knezl, V.; Barančík, M.; Tribulová, N. Cardiac connexin-43 and PKC signaling in rats with altered thyroid status without and with omega-3 fatty acids intake. *Physiol. Res.* **2016**, *65*, 77–90.

20. Liu, J.; Zhu, H.; Yang, X.; Ge, Y.; Zhang, C.; Qin, Q.; Lu, J.; Zhan, L.; Cheng, H.; Sun, X. MicroRNA-21 is a novel promising target in cancer radiation therapy. *Tumour Biol.* **2014**, *35*, 3975–3979. [CrossRef] [PubMed]

21. Rezaeyan, A.; Haddadi, G.H.; Hosseinzadeh, M.; Moradi, M.; Najafi, M. Radioprotective effects of hesperidin on oxidative damages and histopathological changes induced by X-irradiation in rats heart tissue. *J. Med. Phys.* **2016**, *41*, 182–191. [CrossRef] [PubMed]

22. Sieber, F.; Muir, S.A.; Cohen, E.P.; Fish, B.L.; Mäder, M.; Schock, A.M.; Althouse, B.J.; Moulder, J.E. Dietary selenium for the mitigation of radiation injury: Effects of selenium dose escalation and timing of supplementation. *Radiat. Res.* **2011**, *176*, 366–374. [CrossRef] [PubMed]

23. Kura, B.; Bagchi, A.S.; Akolkar, G.; Singal, P.K.; Slezak, J. Myocardial changes after mediastinal irradiation in rats: Molecular mechanisms and potential targets to minimize the adverse effects. In *Adaptation Biology and Medicine*; Kawai, Y., Hargens, A.R., Singal, P.K., Eds.; Narosa Publishing House Ltd.: New Delhi, India, 2017; pp. 93–122.

24. Smyth, J.W.; Hong, T.T.; Gao, D.; Vogan, J.M.; Jensen, B.C.; Fong, T.S.; Simpson, P.C.; Stainier, D.Y.; Chi, N.C.; Shaw, R.M. Limited forward trafficking of connexin 43 reduces cell-cell coupling in stressed human and mouse myocardium. *J. Clin. Investig.* **2010**, *120*, 266–279. [CrossRef] [PubMed]

25. Egan Benova, T.; Szeiffova Bacova, B.; Viczenczova, C.; Diez, E.; Barancik, M.; Tribulova, N. Protection of cardiac cell-to-cell coupling attenuate myocardial remodeling and proarrhythmia induced by hypertension. *Physiol. Res.* **2016**, *65*, 29–42.

26. Tribulova, N.; Szeiffova Bacova, B.; Benova, T.; Viczenczova, C. Can we protect from malignant arrhythmias by modulation of cardiac cell-to-cell coupling? *J. Electrocardiol.* **2015**, *48*, 434–440. [CrossRef] [PubMed]

27. Su, F.; Zhao, L.; Zhang, S.; Wang, J.; Chen, N.; Gong, Q.; Tang, J.; Wang, H.; Yao, J.; Wang, Q.; et al. Cardioprotection by PI3K-mediated signaling is required for anti-arrhythmia and myocardial repair in response to ischemic preconditioning in infarcted pig hearts. *Lab. Investig.* **2015**, *95*, 860–871. [CrossRef] [PubMed]

28. Jeyaraman, M.M.; Srisakuldee, W.; Nickel, B.E.; Kardami, E. Connexin43 phosphorylation and cytoprotection in the heart. *Biochim. Biophys. Acta* **2012**, *1818*, 2009–2013. [CrossRef] [PubMed]

29. Thum, T.; Gross, C.; Fiedler, J.; Fischer, T.; Kissler, S.; Bussen, M.; Galuppo, P.; Just, S.; Rottbauer, W.; Frantz, S.; et al. MicroRNA-21 contributes to myocardial disease by stimulating MAP kinase signalling in fibroblasts. *Nature* **2008**, *456*, 980–984. [CrossRef] [PubMed]

30. Qiang, B.; Toma, J.; Fujii, H.; Osherov, A.B.; Nili, N.; Sparkes, J.D.; Fefer, P.; Samuel, M.; Butany, J.; Leong-Poi, H.; et al. Statin therapy prevents expansive remodeling in venous bypass grafts. *Atherosclerosis* **2012**, *223*, 106–113. [CrossRef] [PubMed]

31. Fabian, C.J.; Kimler, B.F.; Hursting, S.D. Omega-3 fatty acids for breast cancer prevention and survivorship. *Breast Cancer Res.* **2015**, *17*, 62. [CrossRef] [PubMed]

International Journal of
Molecular Sciences

MDPI

Article

Pannexin-1 in Human Lymphatic Endothelial Cells Regulates Lymphangiogenesis

Jonathan Boucher [1], Claire Simonneau [1], Golthlay Denet [1], Jonathan Clarhaut [2,3],
Annie-Claire Balandre [1], Marc Mesnil [1], Laurent Cronier [1] and Arnaud Monvoisin [1,*]

[1] CNRS ERL 7003, Laboratoire "Signalisation & Transports Ioniques Membranaires", University of Poitiers,
    86073 Poitiers, France; jonathan.boucher@univ-poitiers.fr (J.B.); claire.simonneau86@gmail.com (C.S.);
    golthlay.denet@etu.univ-poitiers.fr (G.D.); annie-claire.balandre@univ-poitiers.fr (A.-C.B.);
    marc.mesnil@univ-poitiers.fr (M.M.); Laurent.Cronier@univ-poitiers.fr (L.C.)
[2] CNRS UMR 7285, Institut de Chimie des Milieux et des Matériaux de Poitiers (IC2MP),
    University of Poitiers, 86073 Poitiers, France; jonathan.clarhaut@chu-poitiers.fr
[3] CHU de Poitiers, 86021 Poitiers, France
[*] Correspondence: arnaud.monvoisin@univ-poitiers.fr; Tel.: +33-054-936-6385

Received: 27 April 2018; Accepted: 22 May 2018; Published: 24 May 2018

Abstract: The molecular mechanisms governing the formation of lymphatic vasculature are not yet well understood. Pannexins are transmembrane proteins that form channels which allow for diffusion of ions and small molecules (<1 kDa) between the extracellular space and the cytosol. The expression and function of pannexins in blood vessels have been studied in the last few decades. Meanwhile, no studies have been conducted to evaluate the role of pannexins during human lymphatic vessel formation. Here we show, using primary human dermal lymphatic endothelial cells (HDLECs), pharmacological tools (probenecid, Brilliant Blue FCF, mimetic peptides [^{10}Panx]) and siRNA-mediated knockdown that Pannexin-1 is necessary for capillary tube formation on Matrigel and for VEGF-C-induced invasion. These results newly identify Pannexin-1 as a protein highly expressed in HDLECs and its requirement during in vitro lymphangiogenesis.

Keywords: lymphatic endothelial cells; pannexins; Panx1; lymphangiogenesis; cell invasion; Vascular Endothelial Growth Factor-C (VEGF-C)

1. Introduction

Pannexin-1 (PANX1) is one of the three members of the Pannexin family with PANX2 and PANX3 discovered through homology to the invertebrate gap-junction forming proteins, innexins [1,2]. Pannexins (PANXs) and Connexins (CXs) share similar protein structure while they lack amino acid sequence homology [3]. By hexameric oligomerization PANX1 forms unopposed large-pore channels [4,5] which allow the release of molecules up to 1 kDa into the extracellular space such as ions, adenosine triphosphate (ATP) and other nucleotides [6,7]. PANX1 is ubiquitously expressed in several organs and tissues [2,8–12] and is the best characterized isoform of the PANX family. For a long time, PANX2 expression has been restricted to the central nervous system [13], but it is now well described that PANX2 is also ubiquitously distributed throughout the body [14]. Similarly, PANX3 has been mainly described in cartilage, bone and skin [2,11,15–21] but accumulating evidences show PANX3 expression in other tissues such as skeletal muscle, heart, cochlea, and arteries [2,15,18,22,23].

In line with the large PANX1 tissue distribution, this ATP release channel is directly or indirectly involved in numerous physiological functions or pathologies such as inflammatory diseases [24] or cancer [25,26]. However, little is currently known regarding the role of PANX1 in the vasculature. PANX1 has been found to be expressed in vivo within the vascular wall in arteries, arterioles, capillaries, veinules and smooth muscle cells (SMCs) but not in veins [12,27] and in vitro in isolated blood endothelial cells (ECs) [28–30]. Panx1-deficient mice showed significantly impaired endothelial function [31]. ATP release via PANX1 channels by either ECs or SMCs has been involved in the regulation of vascular tone, inflammation, and cerebral ischemic stroke [30,32,33]. While many data are available in blood vascular system, studies regarding roles of Panxs in regulating lymphatic vascular development are currently missing [34]. The main role of the lymphatic system is to transport in a unidirectional way extravasated fluids and macromolecules from tissues, through lymph nodes, back to the blood circulation to maintain homeostasis [35]. Lymphangiogenesis, the formation of new lymphatic vessels from preexisting ones [36,37], is associated with several diseases such as chronic inflammation, graft rejection and metastatic dissemination [38]. Since the identification of the Vascular Endothelial Growth Factor-C (VEGF-C) as the major lymphangiogenic factor [39], several other key genes and proteins involved in lymphatic development have been identified. Among them, it has recently been shown that at least three CX isoforms (CX37, CX43 and CX47) are expressed in developing and mature lymphatic vessels [40–42]. These studies showed that these CXs are necessary for the proper lymphatic valve development in collecting vessels and contribute to morphogenesis of the jugular lymph sac and thoracic duct. Moreover, CX mutations or deficiency in mouse and humans have been found to lead to lymphedema [40,43,44]. Regarding PANXs, no in vitro nor in vivo data were available onto their expression and/or function in human lymphatic vasculature and only one recent study has shown the expression of Panx1 in mouse LECs by qPCR [45]. This study aims to investigate the expression of PANXs in human lymphatic endothelial cells and more particularly the role of PANX1 during lymphangiogenesis.

2. Results

2.1. Human Lymphatic Endothelial Cells Express Pannexins

PANX1, -2 and -3 mRNA expression was examined by quantitative RT-PCR in human lymphatic endothelial cells. As shown in Figure 1A, the expression of *PANX1* was highest among the 3 PANX gene family while *PANX2* and *PANX3* were barely expressed.

Interestingly, Western blot analysis demonstrated that all three PANX isoforms were expressed in the HDLECs (Figure 1B). All PANXs were detected at the expected molecular weight. As previously described, a specific banding pattern of three bands was revealed for PANX1 indicating three different glycosylation states: Gly0, non-glycosylated core protein; Gly1, high-mannose species and Gly2, complex glycosylated species [4,46].

Confocal imaging of HDLECs showed a clear localization of PANX1 to the plasma membrane and in the perinuclear compartment (Figure 1C). Importantly, we found that VEGF-C, the main regulator of lymphangiogenesis, increased PANX1 expression in HDLECs after 6 and 24 h treatment by 78 ± 5% and 70 ± 5%, respectively (Figure 1D,E) whereas PANX2 and PANX3 expressions remained unaffected (Figure S1). Taken together, these results show that PANXs are expressed in HDLECs. Since PANX1 is the prevalent isoform and its expression is specifically modulated by VEGF-C, PANX1 is likely to be involved in lymphatic function.

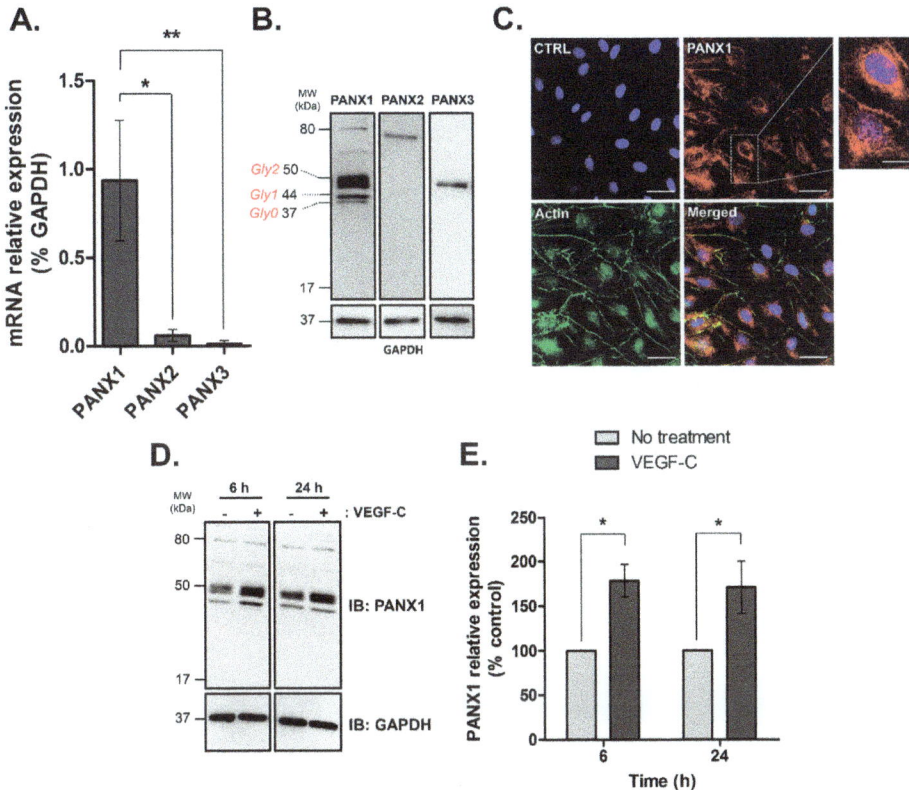

Figure 1. Pannexin isoforms expression in human dermal lymphatic endothelial cells (HDLECs). (**A**) PANXs mRNA expression in isolated HDLECs quantified by RT-PCR and normalized by GAPDH. The data represent mean ± SD from three independent experiments; (**B**) Western blot analysis of total protein extracts (20 µg/lane) from four independent HDLEC cultures demonstrating PANXs expression in HDLECs. Unglycosylated (Gly0) and glycosylated isoforms (Gly1 and Gly2) of PANX1 are indicated; (**C**) PANX1 immunofluorescence in HDLECs (red), F-actin was FITC-phalloidin stained (green) and nuclei were DAPI-stained (blue). CTRL: control immunofluorescence after omission of the primary antibody, Scale bar: 50 µm; Enlarged image marked by the white box shows higher magnification of PANX1 staining, scale bar 7 µm; (**D**) Representative Western blot analysis and (**E**) densitometric quantification of PANX1 expression normalized to GAPDH following 100 ng/mL VEGF-C treatment for the indicated times in HDLECs. Values are expressed as mean ± SD from three independent experiments. * $p < 0.05$ and ** $p < 0.01$.

2.2. Pharmacological Inhibitors of Pannexin-1 Modulate In Vitro Lymphangiogenesis

To investigate whether PANX1 might be involved in lymphangiogenesis we used the in vitro tube-formation assay which is a well-established test based on the ability of LECs to form three-dimensional capillary-like network when seeded on basement membrane extracts (Figure 2A, untreated). Treatment of HDLECs with Probenecid or Brilliant Blue FCF resulted in a disorganized tubular network (Figure 2A). Tube length complexes were significantly inhibited by 28 ± 4% and 20 ± 3% with probenecid at 0.1 and 1 mM respectively and inhibited by 21 ± 3% and 29 ± 4% with Brilliant Blue FCF at 1 and 5 µM respectively (Figure 2B). Number of junctions were also significantly inhibited by 28 ± 4% and 22 ± 3% with probenecid at 0.1 and 1 mM respectively and inhibited by 25 ± 4% and 33 ± 6% with Brilliant Blue FCF at 1 and 5 μM, respectively (Figure 2B).

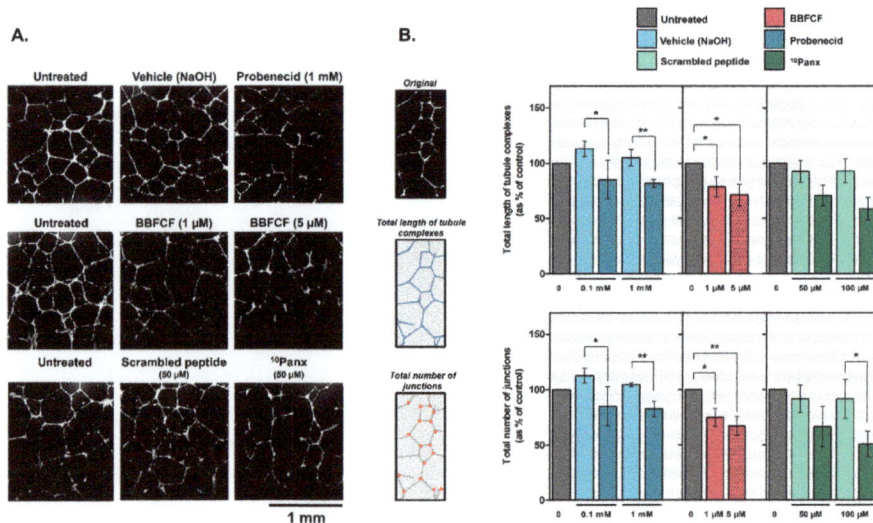

Figure 2. Inhibition of capillary-like formation in HDLECs by pharmacological inhibitors of Pannexin-1. (**A**) Representative images of capillary network formation by HDLECs seeded on Matrigel and treated with Probenecid, Brilliant Blue FCF or mimetic peptide [10]Panx; (**B**) Quantitative analysis for total length of tubule complexes and for total number of junctions in control and treated HDLECs. Data represent the mean ± SD from three independent experiments conducted in triplicate. * $p < 0.05$ and ** $p < 0.01$.

The implication of PANX1 in this process was confirmed by a second set of experiments using the mimetic inhibitory peptide [10]Panx instead. Results showed that the lengths of the capillary-like complexes and number of junctions in the HDLECs treated with [10]Panx at 50 or 100 μM were respectively 22 ± 2% or 34 ± 3% shorter and 25 ± 2% or 40 ± 4% lower than those observed in the control group (Figure 2B).

2.3. Pannexin-1 Silencing Inhibits In Vitro Lymphangiogenesis

To confirm the role of PANX1, we decided to inhibit its expression in HDLECs by siRNA silencing. Figure 3A shows that the siRNA significantly inhibited by 80 ± 11% the expression of PANX1 24 h after transfection compared to scramble. We observed no compensation by PANX2 nor PANX3 expression after PANX1 silencing (Figure 3B). Using the tube formation assay, PANX1 siRNA-transfected HDLECs showed less extensive capillary formation compared to control (Figure 3C). Once again, quantitative analyses showed that the total length of tubule complexes and the number of junctions formed by HDLECs were significantly inhibited by 24 ± 4% and 34 ± 3% respectively when PANX1 was silenced as compared with control (Figure 3D).

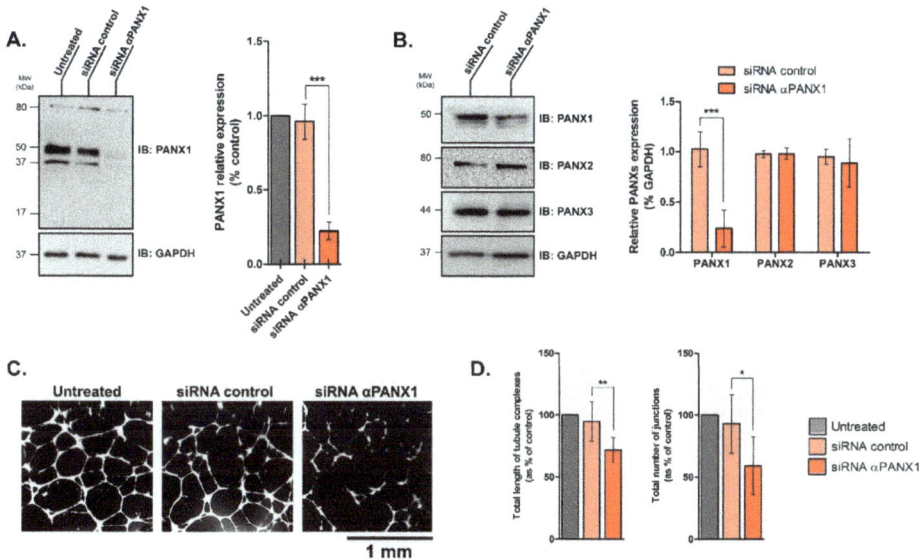

Figure 3. Silencing Pannexin-1 expression affects capillary-like formation by HDLECs. (**A**) Representative immunoblots of HDLECs extracts prepared 48 h after transfection with either the control or the PANX1-specific siRNAs. GAPDH blot served as the loading control; Bar graph shows the quantification of PANX1 expression loss 48 h after siRNA transfection. Data represent the mean ± SD from four independent experiments; (**B**) Representative immunoblots and densitometric quantification of PANX1, PANX2 and PANX3 expression from HDLECs extracts prepared 48 h after transfection with either the control or the PANX1-specific siRNAs. GAPDH blot served as the loading control; (**C**) Representative images of tube structure formation in HDLECs on Matrigel after transfection. Cells transfected with PANX1 siRNAs showed defects in capillary network formation; (**D**) Quantitative analysis for total length of tubule complexes and total number of junctions per field in control and PANX1 siRNA-transfected HDLECs. Data represent the mean ± SD from six independent experiments conducted in duplicate. * $p < 0.05$, ** $p < 0.01$ and *** $p < 0.001$.

2.4. Pannexin-1 Silencing Inhibits In Vitro HDLECs Invasion but Not Cell Proliferation

To investigate if PANX1 deficiency in HDLECs affected lymphangiogenesis by modulating cell proliferation, we used the BrdU incorporation assay. The results showed that knockdown of PANX1 had no effect on HDLECs proliferation when LECs were grown in EGM-V2 media (Figure 4A).

Finally, we tested the hypothesis that PANX1 is important for HDLECs invasion and we examined whether invasion of HDLECs induced by VEGF-C is affected by loss of PANX1 using a modified Boyden chamber assay. Figure 4B shows the results of a typical invasion experiment. The quantification revealed a significant inhibition of HDLECs invasion by 59 ± 6% after PANX1 expression silencing compared to scramble (Figure 4C) that can explain the disorganized capillary network we observed previously.

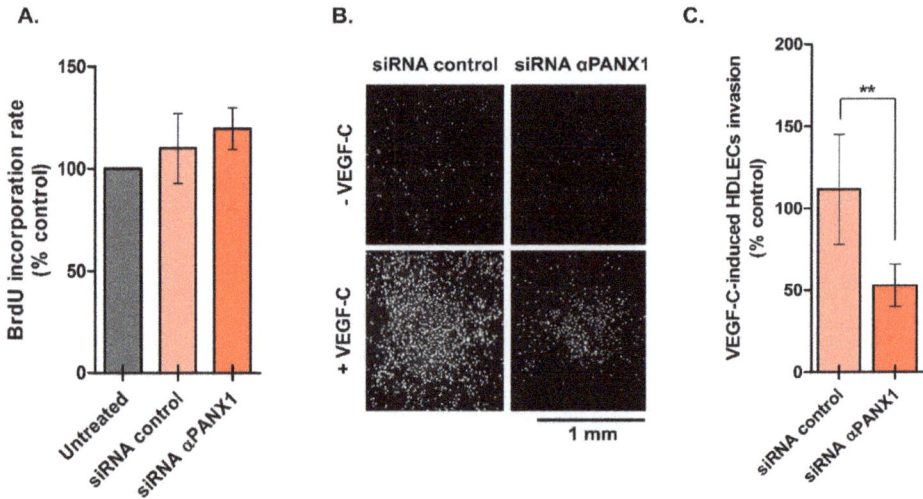

Figure 4. Loss of Pannexin-1 inhibits VEGF-C-mediated invasion of HDLECs. (**A**) HDLECs proliferation measurement 48 h after transfection with either control of PANX1 siRNAs in EGM-V2 media. Data represent the mean from three independent experiments conducted in triplicate (**B**) HDLECs were transfected with either control or PANX1 siRNAs and subjected to Boyden chamber assays in the presence or absence of VEGF-C (100 ng/mL). Representative images of HDLECs that invaded and migrated through the membrane pores after 18 h are shown; (**C**) Bar graph represents the mean number of invading cells. Results are expressed as the mean ± SD of three independent experiments conducted in triplicate. ** $p < 0.01$.

3. Discussion

In this present work, we tested the hypothesis that Pannexin-1 is important for human lymphatic endothelial cells to form capillary-like structures for extracellular matrix (ECM)-induced morphogenesis.

We expected PANX1 to be present in human LECs because it is ubiquitously expressed in vivo [2], in several cell lines [47], in human venous ECs [29,30,48] which share a common origin with LECs [36,37] and in murine LECs [45]. Indeed, by quantitative RT-PCR, we found that PANX1 was the most expressed member of the pannexin family in HDLECs. We also observed weak mRNA expression of PANX2 and PANX3 which could confirm their wider expressions in the human body as demonstrated by previous studies [14,15,18,22,23]. Additional work will be required to define the possible role of PANX2 and PANX3 in lymphatic development especially in pathological situations, since no major lymphatic phenotypic abnormalities are observed in adult mice deficient for these PANXs [19,20,32].

Our study reveals that PANX1 exhibited diverse localization patterns in HDLECs. As expected, based on its channel-forming ability, PANX1 localized at the plasma membrane [49,50]. PANX1 has also been found in the perinuclear compartment which has already been observed with endogenous PANX1 in other primary cell cultures or cell lines such as osteoblasts [49] or astrocytes and neurons [51–53]. Interestingly, when PANX1 was transfected into human bone marrow ECs, PANX1 localized in the endoplasmic reticulum (ER) and Golgi apparatus [8] and this pattern was observed in vivo in blood ECs of the lens [9]. It is well documented that this pattern reflects the intracellular PANX1 trafficking during which PANX1, as an unglycosylated core (Gly0), is glycosylated in the ER to a high-mannose form (Gly1), and then, in the Golgi to a complex glycosylated form (Gly2), before reaching the plasma membrane [4,46,54]. Our Western blot analysis obtained from HDLECs lysates revealed these multiple species of PANX1 which correlate with PANX1 distribution observed in HDLECs. Another possibility to explain this intracellular localization is that PANX1 may form Ca²⁺-permeable channels in the endoplasmic reticulum as observed in prostate cancer cells [55].

There is then no clear evidence that PANX1 might be essential for proper lymphatic vasculature development. Indeed, neither the Panx1-deficient mice [56,57] nor the only first patient with a PANX1 homozygous germline variant [58] display obvious phenotypes such as lymphedema that would suggest alteration in lymphatic vasculature function. Recently, Molica et al. showed, using double knock-out mice for Panx1 and Apoliprotein E (Apoe) to evaluate Panx1 role in atherosclerosis, that Panx1 in this context is necessary for lymphatic function by contributing to the drainage of interstitial fluid and to the uptake of dietary fat from the gut [47]. Nevertheless, in this study, there was no detail regarding density and morphology of the lymphatic vasculature in this Panx1-/-ApoE-/- mice as for the other Panx1-deficient models. In this present study, using three different strategies to inhibit PANX1, we find that this pannexin is required for in vitro lymphangiogenesis. In addition, we did not observe compensation by PANX2 or PANX3, in PANX1 siRNA-treated HDLECs that failed to form in vitro a well-organized capillaries network. Given the apparent normal lymphatic development in Panx1-deficient mice, it is possible that compensation by Panx2 or/and Panx3 arises in vivo in LECs. This has been observed in muscle [23] and arteries [59] where Panx1 deletion caused an increase in Panx3 expression which can also act as an ATP-release channel [60]. Additional work using Panx1-deficent mice to study the expression of all the members of the pannexin family in the lymphatic vasculature and in isolated LECs will be required to answer this question.

Lymphangiogenesis is a multistep process in which the proliferation and invasion abilities of LECs play a fundamental role. Depending on the cell types, PANX1 may promote [61], decrease [11,62] or have no effect on proliferation [23]. Our work shows that PANX1 knockdown did not change the proliferative rate of HDLECs, which suggests that the inhibition of the in vitro lymphangiogenesis that we observed is not due to an inhibitory effect on the cell cycle. Finally, we focused our work on the role of PANX1 in invasion which is a process that combines both ECM degradation and cell migration. We found out that PANX1, which was up-regulated in HDLECs by VEGF-C, is necessary for the VEGF-C-mediated invasion of HDLECs. Since PANX1 was found at the plasma membrane of HDLECs, it is reasonable to speculate that PANX1 acts as an ATP-release channel to explain its role during this step. We hypothesize that the release of ATP modulates LECs function such as migration through the activation of purinergic receptors as shown for other cell types [63,64]. Once bound to these receptors, ATP might induce Ca²⁺ release from intracellular Ca²⁺ stores, a signaling pathway that has been implicated in LECs migration and lymphangiogenesis [65,66]. LECs are known to express four P2 purinergic receptors (P2RX4, P2RX7, P2RY1 and P2RY11) with high expression of P2RX4 and P2RY1, and interestingly, inhibition of P2RY1 in presence of a specific antagonist impaired ATP-induced migration of HDLECs [67].

Meanwhile, we cannot exclude the hypothesis that PANX1 has channel-independent roles in regulating in vitro lymphangiogenesis. PANX1 directly interacts with the cytoskeleton through its association with actin and Arp2/3 and this association has been shown to modulate cell behavior [50,68,69]. Detailed molecular mechanism of PANX1-driven lymphangiogenesis is still unclear. Future studies using mouse model with lymphatic-specific deletion of PANX1, PANX2 or PANX3 will clarify their relative roles during developmental and pathological lymphangiogenesis.

4. Materials and Methods

4.1. Antibodies and Reagents

Rabbit polyclonal antibody against PANX1 was purchased from Sigma (HPA016930, St. Louis, MO, USA). Rabbit polyclonal anti-PANX2 was purchased from Santa Cruz Biotechnology (sc-133880, Santa Cruz, CA, USA). Mouse monoclonal anti-PANX3 was from R&D Systems (MAB8169, Minneapolis, MN, USA). Mouse monoclonal anti-GAPDH antibody (5G4) was supplied by HyTest

(Turku, Finland). Goat polyclonal HRP-conjugated secondary antibodies anti-Mouse and anti-Rabbit were from Agilent Dako (P044701-2, Santa Clara, CA, USA) and Sigma (A0545), respectively. Goat polyclonal Alexa Fluor 568-conjugated antibodies (anti-Mouse) were purchased from Molecular Probes, Thermo Fisher Scientific (A-11011, Waltham, MA, USA). FITC-Phalloidin (F432) and DAPI (D3571) were from Molecular Probes, Thermo Fisher Scientific. Calcein-AM was purchased from Sigma (C1359). Silencer Pre-designed siRNA against PANX1 (134470) and Silencer Select negative control siRNA (4390843) were obtained from Ambion, Thermo Fisher Scientific. Matrigel® matrix was obtained from Corning (354234, Corning, NY, USA). Recombinant Human Vascular Endothelial Growth Factor-C (VEGF-C) was obtained from Immunotools (11344692, Friesoythe, Germany) and reconstituted as a 100 µg/mL solution in sterile water. Probenecid (P8761) and Brilliant Blue FCF (80717) were purchased from Sigma and reconstituted as 50 mg/mL solution in 1M NaOH and as 30 mg/mL solution in sterile water, respectively. [10]Panx mimetic peptide (Trp-Arg-Gln-Ala-Ala-Phe-Val-Asp-Ser-Tyr) and Scrambled [10]Panx control peptide (Phe-Ser-Val-Tyr-Trp-Ala-Gln-Ala-Asp-Arg) were from Tocris Bioscience (Bristol, UK) and reconstituted as 0.5 mg/mL solution in PBS or water, respectively.

4.2. Cell Culture

Primary human dermal lymphatic endothelial cells (HDLECs) were obtained from Promocell (Heidelberg, Germany) and grown in EGM-V2 media which consists of EBM-2 basal media supplemented with 5% FCS and defined supplements such as epidermal growth factor (EGF), basic fibroblast growth factor (bFGF), insulin-like growth factor 1 (long R3 IGF-1), vascular endothelial growth factor A (VEGF-A), ascorbic acid and hydrocortisone. When HDLECs reached 80% confluency, cells were trypsinized following the procedure recommended by Promocell and then plated at 1×10^3 cells/cm^2. HDLECs were used up to passages 6–8 for all experiments.

4.3. Quantitative Real-Time PCR Analysis

1×10^5 HDLECs in 2 mL EGM-V2 medium were seeded in six-well plates and 24 h after, total RNA was extracted using the NucleoSpin RNA XS kit (Macherey-Nagel, Düren, Germany). Reverse transcription was performed with SuperScript II (Invitrogen, Thermo Fisher Scientific, Waltham, MA, USA) from 2 µg of total RNA according to the manufacturer's instructions. Gene expression was assessed relative to GAPDH by quantitative PCR with the GeneAmp 7000 Sequence Detection System and SYBR Green chemistry (Applied Biosystems, Thermo Fisher Scientific). Human GAPDH, PANX1, PANX2 and PANX3 primer sequences are listed in Table S1. Sensitivity and specificity of each primer couple were checked. For each primers, qPCR was also performed with plasmids containing the cDNA of PANX1, PANX2 and PANX3, serving as positive controls.

4.4. Western Blot

1×10^5 HDLECs in 2 mL EGM-V2 medium were seeded in six-well plates. 24 h after, cells were lysed with 30 µL extraction buffer (150 mM NaCl, 10 mM Tris-HCl, pH 8.0, 1 mM EDTA, 1 mM EGTA, 1% Triton X-100, 0.5% NP-40, 100 mM sodium orthovanadate) completed with 1× protease inhibitors cocktail (Roche Applied Science, Penzberg, Germany). Protein concentration was measured using DC protein assay kit (Bio-Rad Laboratories, Hercules, CA, USA). Afterwards 20 µg total proteins samples were mixed with an equal volume of 5X SDS gel-loading buffer (150 mM Tris-HCl, pH 6.8, 5% SDS, 12.5% 2-mercaptoethanol, 25% glycerol and 0.025% bromophenol blue). Proteins were resolved using 10% SDS-PAGE gels and transferred to PVDF membranes (Merck Millipore, Darmstadt, Germany). Membranes were blocked with 5% non-fat powdered milk in Tris-buffered saline-Tween (TBS-T; 25 mM Tris-HCl, pH 8.0, 15 mM NaCl, 0.01% Tween 20) for 3 h and incubated overnight at 4 °C with primary antibodies (anti-PANX1, 1:800; anti-PANX2, 1:350; anti-PANX3, 1:1000). Membranes were then washed with TBS-T and incubated with corresponding secondary horseradish peroxidase-conjugated antibodies (1:10,000) for 1 h. The anti-GAPDH antibodies are used as the loading control (1:10,000). Immunodetection was performed using chemiluminescent substrate Luminata Forte (Merck Millipore) and LAS-3000 imaging system (Fujifilm, Tokyo, Japan). Densitometric analysis of signals was carried out using ImageJ software (version 1.39o, National Institutes of Health, Bethesda, MD, USA). Preliminarily to expression experiments, positive controls samples were used to further validate the banding pattern of each antibody (Figure S2).

4.5. Immunofluorescence

HDLECs were plated on 0.1% gelatin-coated sterile glass coverslips in removable 12 well chambers (Ibidi GmbH, Martinsried, Germany) at 2×10^4 cells/well in EGM-V2. After 48 h, HDLECs were washed with 1X PBS and fixed in 4% PFA in 1X PBS at room temperature (RT) for 10 min. HDLECs were then washed with 1X PBS and coverslips were blocked with 1% BSA, 1% Triton X100 in 1X PBS at RT for 1 h. HDLECs were incubated overnight at 4 °C with anti-PANX1 primary antibodies prepared in blocking buffer at 1:100 dilution. A negative control was performed by omitting the primary antibody. The following day, HDLECs were incubated with Alexa Fluor 555 donkey anti-goat secondary antibodies (1:500) in blocking buffer for 2 h at RT. F-actin filaments and cell nuclei were stained with 1 nM FITC-phalloidin (Life Technologies, Thermo Fisher Scientific, Waltham, MA, USA) and 100 nM DAPI (Molecular Probes, Thermo Fisher Scientific) in 1X PBS for 30 min at RT. After extensive washing with 1X PBS, cover slips were mounted with Mowiol fluorescent mounting medium and Images were captured using a confocal microscope (Olympus FV1000, Tokyo, Japan). Control experiment omitting primary antibody was also performed.

4.6. siRNA Interference

The siRNA sequence used for small-interfering RNA-mediated inhibition of PANX1 was the following: PANX1 siRNA: 5′-AGGAUCCCUGAUUUGAUGCTG-3′. The siRNA sequence for the non-targeting control siRNA was undisclosed by the manufacturer. Transfections were performed using siPORT reagent according to the manufacturer's instructions. 2×10^5 HDLECs in 2.3 mL EGM-V2 medium were seeded in six-well plates. For each well, 5 μL siPORT Amine agent were diluted into 100 μL Opti-MEM medium and incubated for 10 min at RT. 7.5 μL and 12.5 μL of PANX1 or scramble siRNA at 10 and 1 μM respectively were diluted into Opti-MEM. Diluted siRNA and diluted siPORT Amine agent were mixed, incubated for 10 min at RT and finally dispensed onto HDLECs. Medium was finally removed 24 h after transfection prior experiments. The efficiency of the siRNA knockdown was determined by Western blot analysis 24 and 48 h post-transfection. XTT assay confirmed non-cytotoxic effects of both siPORT and oligos after 24 h treatment on HDLECs (Figure S3).

4.7. Cell Proliferation

BrdU incorporation assay was used to monitor the mitogenic effects of the siRNA silencing of PANX1. siRNA-transfected HDLECs were plated in 96-well plates at a density of 6×10^3 cells/well in 100 μL EGM-V2 and after 24 h bromodeoxyuridine (BrdU) was added into wells at a final concentration of 100 μM and incubated overnight at 37 °C. BrdU incorporation rate was measured using "Cell Proliferation ELISA, BrdU colorimetric kit" (Roche Applied Science, Penzberg, Germany) according to the manufacturer's instructions. Absorbance was measured at 450 nm with a 96-well microplate reader after adding sulfuric acid to stop colorimetric reaction.

4.8. Tube Formation

96-well plates were coated with 50 μL/well of cold Matrigel and allowed to solidify for 1 h at 37 °C. HDLECs were trypsinized and 15×10^3 cells in 200 μL of EGM-V2 medium were loaded on the solidified Matrigel in the presence or absence of PANX1 chemical inhibitors, Probenecid (at 1 or 2 mM final) or BBFCF (at 1 or 5 μM final). In a second set of experiments, HDLECs were mixed and loaded on Matrigel with the specific PANX1 mimetic peptide channel blocker, [10]Panx or the control peptide at 50 or 100 μM. Finally, PANX1 or control siRNA-transfected HDLECs were used in the same conditions than the native HDLECs in this assay. After 20 h, HDLECs were stained with 25 μM calcein-AM and then fixed with 2% PFA in 1X PBS. The 3-dimensional cell organization was photographed using Olympus MVX10 macroscope (objective $1 \times /1$; 22 °C; medium: PBS; Camera: Hamamatsu ORCA-03G; cellSens Dimension Version 1.4 software Olympus, Tokyo, Japan). Capillary-like structures (length of tubule complexes and number of junctions) were quantified by automatic counting in duplicate using the AngioQuant Version 1.33 software. Previously to these experiments we have determined that the concentrations of inhibitors used in this assay on the HDLECs were not cytotoxic (Figure S3).

4.9. Endothelial Cells Invasion

HDLECs migration was evaluated using 24-well cell culture inserts with 8-μm pores (BD Biosciences, Franklin Lakes, NJ, USA). 24 h after siRNA transfection, HDLECs were rinsed with EBM-2 media, detached with trypsin and seeded at 5×10^4 cells/100 μL 0.5% FCS EBM-2 onto Matrigel-coated (25 μL of 1 mg/mL) inserts. The inserts were then placed in the 24-well plates containing 500 μL of EBM-2 medium with or without 100 ng/mL recombinant human VEGF-C. The filters were removed following incubation for 17 h at 37 °C and 5% CO_2 and the Matrigel was wiped with a cotton swab. The invasive HDLECs were fixed in 4% PFA for 10 min prior to staining with DAPI. Filters were mounted with Mowiol fluorescent mounting medium and cells were photographed using an Olympus MVX10 macroscope (objective 2X, Tokyo, Japan). Cells were counted using ImageJ software after binarizing images.

4.10. Statistical Analysis

Statistical analyses were carried out on GraphPad Prism 5 software. All reported data are expressed as mean ± SD. ANOVA followed by Bonferroni post-tests was performed for analysis of 3 or more groups. Unpaired Student's *t*-test (Mann-Whitney) was used when only 2 experimental groups were analyzed.

5. Conclusions

In this study, we provide evidences that Pannexin-1 is expressed in human lymphatic endothelial cells and is involved in in vitro lymphangiogenesis.

Supplementary Materials: Supplementary materials can be found at http://www.mdpi.com/1422-0067/19/6/1558/s1.

Author Contributions: Conceptualization, A.M.; Data curation, J.B.; Formal analysis, J.B.; Funding acquisition, M.M., L.C. and A.M.; Investigation, J.B., C.S., G.D., A.-C.B., L.C. and A.M.; Methodology, J.B., C.S., G.D., J.C., A.-C.B. and A.M.; Project administration, A.M.; Resources, M.M., L.C. and A.M.; Supervision, A.M.; Validation, J.B. and J.C.; Visualization, J.B. and A.M.; Writing—original draft, A.M.; Writing—review & editing, M.M., L.C. and A.M.

Funding: This work was supported by funds from the "Ligue Nationale contre le cancer" (Comités de la Vienne, Deux-Sèvres and Charente-Maritime). J.B. was recipient of a doctoral fellowship from the "Ministère de l'enseignement supérieur et de la recherche".

Acknowledgments: Human primary keratinocytes were a kind gift from F. Morel (LITEC EA 4331, Université de Poitiers, France). We thank Anne Cantereau (ImageUP, University of Poitiers) for technical assistance in confocal microscopy.

Conflicts of Interest: The authors declare no conflict of interest.

Abbreviations

BrdU	Bromodeoxyuridine
CX	Connexin
DAPI	4′,6-diamidino-2-phenylindole
EBM	Endothelial cell basal medium
ECs	Endothelial cells
EGM	Endothelial cell growth medium
FCS	Fetal calf serum
GAPDH	Glyceraldehyde 3-phosphate dehydrogenase
HDLECs	Human dermal lymphatic endothelial cells
HUVECs	Human umbilical vein endothelial cells
PANX	Pannexin
PBS	Phosphate-buffered saline
PFA	Paraformaldehyde
RT	Room temperature
RT-PCR	Reverse transcription-polymerase chain reaction
siRNA	Small-interfering RNA
SMCs	Smooth muscle cells
TBS	Tris-buffered saline
VEGF-C	Vascular endothelial growth factor-C

References

1. Panchin, Y.; Kelmanson, I.; Matz, M.; Lukyanov, K.; Usman, N.; Lukyanov, S. A ubiquitous family of putative gap junction molecules. *Curr. Biol.* **2000**, *10*, R473–R474. [CrossRef]
2. Baranova, A.; Ivanov, D.; Petrash, N.; Pestova, A.; Skoblov, M.; Kelmanson, I.; Shagin, D.; Nazarenko, S.; Geraymovych, E.; Litvin, O.; et al. The mammalian pannexin family is homologous to the invertebrate innexin gap junction proteins. *Genomics* **2004**, *83*, 706–716. [CrossRef] [PubMed]
3. Ambrosi, C.; Gassmann, O.; Pranskevich, J.N.; Boassa, D.; Smock, A.; Wang, J.; Dahl, G.; Steinem, C.; Sosinsky, G.E. Pannexin1 and Pannexin2 channels show quaternary similarities to connexons and different oligomerization numbers from each other. *J. Biol. Chem.* **2010**, *285*, 24420–24431. [CrossRef] [PubMed]
4. Boassa, D.; Ambrosi, C.; Qiu, F.; Dahl, G.; Gaietta, G.; Sosinsky, G. Pannexin1 channels contain a glycosylation site that targets the hexamer to the plasma membrane. *J. Biol. Chem.* **2007**, *282*, 31733–31743. [CrossRef] [PubMed]
5. Sosinsky, G.E.; Boassa, D.; Dermietzel, R.; Duffy, H.S.; Laird, D.W.; MacVicar, B.; Naus, C.C.; Penuela, S.; Scemes, E.; Spray, D.C.; et al. Pannexin channels are not gap junction hemichannels. *Channels* **2011**, *5*, 193–197. [CrossRef] [PubMed]
6. Bao, L.; Locovei, S.; Dahl, G. Pannexin membrane channels are mechanosensitive conduits for ATP. *FEBS Lett.* **2004**, *572*, 65–68. [CrossRef] [PubMed]
7. Ma, W.; Compan, V.; Zheng, W.; Martin, E.; North, R.A.; Verkhratsky, A.; Surprenant, A. Pannexin 1 forms an anion-selective channel. *Pflugers Arch.* **2012**, *463*, 585–592. [CrossRef] [PubMed]
8. Dvoriantchikova, G.; Ivanov, D.; Panchin, Y.; Shestopalov, V.I. Expression of pannexin family of proteins in the retina. *FEBS Lett.* **2006**, *580*, 2178–2182. [CrossRef] [PubMed]
9. Dvoriantchikova, G.; Ivanov, D.; Pestova, A.; Shestopalov, V. Molecular characterization of pannexins in the lens. *Mol. Vis.* **2006**, *12*, 1417–1426. [PubMed]
10. Seminario-Vidal, L.; Kreda, S.; Jones, L.; O'Neal, W.; Trejo, J.; Boucher, R.C.; Lazarowski, E.R. Thrombin promotes release of ATP from lung epithelial cells through coordinated activation of rho- and Ca^{2+}-dependent signaling pathways. *J. Biol. Chem.* **2009**, *284*, 20638–20648. [CrossRef] [PubMed]
11. Celetti, S.J.; Cowan, K.N.; Penuela, S.; Shao, Q.; Churko, J.; Laird, D.W. Implications of pannexin 1 and pannexin 3 for keratinocyte differentiation. *J. Cell Sci.* **2010**, *123*, 1363–1372. [CrossRef] [PubMed]
12. Lohman, A.W.; Billaud, M.; Straub, A.C.; Johnstone, S.R.; Best, A.K.; Lee, M.; Barr, K.; Penuela, S.; Laird, D.W.; Isakson, B.E. Expression of pannexin isoforms in the systemic murine arterial network. *J. Vasc. Res.* **2012**, *49*, 405–416. [CrossRef] [PubMed]

13. Bruzzone, R.; Hormuzdi, S.G.; Barbe, M.T.; Herb, A.; Monyer, H. Pannexins, a family of gap junction proteins expressed in brain. *Proc. Natl. Acad. Sci. USA* **2003**, *100*, 13644–13649. [CrossRef] [PubMed]

14. Le Vasseur, M.; Lelowski, J.; Bechberger, J.F.; Sin, W.-C.; Naus, C.C. Pannexin 2 protein expression is not restricted to the CNS. *Front. Cell. Neurosci.* **2014**, *8*, 392. [CrossRef] [PubMed]

15. Penuela, S.; Bhalla, R.; Gong, X.-Q.; Cowan, K.N.; Celetti, S.J.; Cowan, B.J.; Bai, D.; Shao, Q.; Laird, D.W. Pannexin 1 and pannexin 3 are glycoproteins that exhibit many distinct characteristics from the connexin family of gap junction proteins. *J. Cell Sci.* **2007**, *120*, 3772–3783. [CrossRef] [PubMed]

16. Iwamoto, T.; Nakamura, T.; Doyle, A.; Ishikawa, M.; de Vega, S.; Fukumoto, S.; Yamada, Y. Pannexin 3 regulates intracellular ATP/cAMP levels and promotes chondrocyte differentiation. *J. Biol. Chem.* **2010**, *285*, 18948–18958. [CrossRef] [PubMed]

17. Bond, S.R.; Lau, A.; Penuela, S.; Sampaio, A.V.; Underhill, T.M.; Laird, D.W.; Naus, C.C. Pannexin 3 is a novel target for Runx2, expressed by osteoblasts and mature growth plate chondrocytes. *J. Bone Miner. Res.* **2011**, *26*, 2911–2922. [CrossRef] [PubMed]

18. Cowan, K.N.; Langlois, S.; Penuela, S.; Cowan, B.J.; Laird, D.W. Pannexin1 and Pannexin3 exhibit distinct localization patterns in human skin appendages and are regulated during keratinocyte differentiation and carcinogenesis. *Cell Commun. Adhes.* **2012**, *19*, 45–53. [CrossRef] [PubMed]

19. Oh, S.-K.; Shin, J.-O.; Baek, J.-I.; Lee, J.; Bae, J.W.; Ankamerddy, H.; Kim, M.-J.; Huh, T.-L.; Ryoo, Z.-Y.; Kim, U.-K.; et al. Pannexin 3 is required for normal progression of skeletal development in vertebrates. *FASEB J.* **2015**, *29*, 4473–4484. [CrossRef] [PubMed]

20. Caskenette, D.; Penuela, S.; Lee, V.; Barr, K.; Beier, F.; Laird, D.W.; Willmore, K.E. Global deletion of Panx3 produces multiple phenotypic effects in mouse humeri and femora. *J. Anat.* **2016**, *228*, 746–756. [CrossRef] [PubMed]

21. Ishikawa, M.; Yamada, Y. The Role of Pannexin 3 in Bone Biology. *J. Dent. Res.* **2017**, *96*, 372–379. [CrossRef] [PubMed]

22. Wang, X.-H.; Streeter, M.; Liu, Y.-P.; Zhao, H.-B. Identification and characterization of pannexin expression in the mammalian cochlea. *J. Comp. Neurol.* **2009**, *512*, 336–346. [CrossRef] [PubMed]

23. Langlois, S.; Xiang, X.; Young, K.; Cowan, B.J.; Penuela, S.; Cowan, K.N. Pannexin 1 and pannexin 3 channels regulate skeletal muscle myoblast proliferation and differentiation. *J. Biol. Chem.* **2014**, *289*, 30717–30731. [CrossRef] [PubMed]

24. Gulbransen, B.D.; Bashashati, M.; Hirota, S.A.; Gui, X.; Roberts, J.A.; MacDonald, J.A.; Muruve, D.A.; McKay, D.M.; Beck, P.L.; Mawe, G.M.; et al. Activation of neuronal P2X7 receptor-pannexin-1 mediates death of enteric neurons during colitis. *Nat. Med.* **2012**, *18*, 600–604. [CrossRef] [PubMed]

25. Penuela, S.; Gyenis, L.; Ablack, A.; Churko, J.M.; Berger, A.C.; Litchfield, D.W.; Lewis, J.D.; Laird, D.W. Loss of pannexin 1 attenuates melanoma progression by reversion to a melanocytic phenotype. *J. Biol. Chem.* **2012**, *287*, 29184–29193. [CrossRef] [PubMed]

26. Furlow, P.W.; Zhang, S.; Soong, T.D.; Halberg, N.; Goodarzi, H.; Mangrum, C.; Wu, Y.G.; Elemento, O.; Tavazoie, S.F. Mechanosensitive pannexin-1 channels mediate microvascular metastatic cell survival. *Nat. Cell Biol.* **2015**, *17*, 943–952. [CrossRef] [PubMed]

27. Billaud, M.; Lohman, A.W.; Straub, A.C.; Looft-Wilson, R.; Johnstone, S.R.; Araj, C.A.; Best, A.K.; Chekeni, F.B.; Ravichandran, K.S.; Penuela, S.; et al. Pannexin1 regulates α1-adrenergic receptor-mediated vasoconstriction. *Circ. Res.* **2011**, *109*, 80–85. [CrossRef] [PubMed]

28. Lohman, A.W.; Weaver, J.L.; Billaud, M.; Sandilos, J.K.; Griffiths, R.; Straub, A.C.; Penuela, S.; Leitinger, N.; Laird, D.W.; Bayliss, D.A.; et al. S-nitrosylation inhibits pannexin 1 channel function. *J. Biol. Chem.* **2012**, *287*, 39602–39612. [CrossRef] [PubMed]

29. Gödecke, S.; Roderigo, C.; Rose, C.R.; Rauch, B.H.; Gödecke, A.; Schrader, J. Thrombin-induced ATP release from human umbilical vein endothelial cells. *Am. J. Physiol.-Cell Physiol.* **2012**, *302*, C915–C923. [CrossRef] [PubMed]

30. Lohman, A.W.; Leskov, I.L.; Butcher, J.T.; Johnstone, S.R.; Stokes, T.A.; Begandt, D.; DeLalio, L.J.; Best, A.K.; Penuela, S.; Leitinger, N.; et al. Pannexin 1 channels regulate leukocyte emigration through the venous endothelium during acute inflammation. *Nat. Commun.* **2015**, *6*, 7965. [CrossRef] [PubMed]

31. Gaynullina, D.; Tarasova, O.S.; Kiryukhina, O.O.; Shestopalov, V.I.; Panchin, Y. Endothelial function is impaired in conduit arteries of pannexin1 knockout mice. *Biol. Direct* **2014**, *9*, 8. [CrossRef] [PubMed]

32. Bargiotas, P.; Krenz, A.; Hormuzdi, S.G.; Ridder, D.A.; Herb, A.; Barakat, W.; Penuela, S.; von Engelhardt, J.; Monyer, H.; Schwaninger, M. Pannexins in ischemia-induced neurodegeneration. *Proc. Natl. Acad. Sci. USA* **2011**, *108*, 20772–20777. [CrossRef] [PubMed]

33. Billaud, M.; Sandilos, J.K.; Isakson, B.E. Pannexin 1 in the regulation of vascular tone. *Trends Cardiovasc. Med.* **2012**, *22*, 68–72. [CrossRef] [PubMed]

34. Good, M.E.; Begandt, D.; DeLalio, L.J.; Keller, A.S.; Billaud, M.; Isakson, B.E. Emerging concepts regarding pannexin 1 in the vasculature. *Biochem. Soc. Trans.* **2015**, *43*, 495–501. [CrossRef] [PubMed]

35. Escobedo, N.; Oliver, G. Lymphangiogenesis: Origin, Specification, and Cell Fate Determination. *Annu. Rev. Cell Dev. Biol.* **2016**, *32*, 677–691. [CrossRef] [PubMed]

36. Sabin, F.R. On the origin of the lymphatic system from the veins and the development of the lymph hearts and thoracic duct in the pig. *Am. J. Anat.* **1902**, *1*, 367–389. [CrossRef]

37. Yang, Y.; Oliver, G. Development of the mammalian lymphatic vasculature. *J. Clin. Investig.* **2014**, *124*, 888–897. [CrossRef] [PubMed]

38. Venero Galanternik, M.; Stratman, A.N.; Jung, H.M.; Butler, M.G.; Weinstein, B.M. Building the drains: The lymphatic vasculature in health and disease. *Wiley Interdiscip. Rev. Dev. Biol.* **2016**, *5*, 689–710. [CrossRef] [PubMed]

39. Jeltsch, M.; Kaipainen, A.; Joukov, V.; Meng, X.; Lakso, M.; Rauvala, H.; Swartz, M.; Fukumura, D.; Jain, R.K.; Alitalo, K. Hyperplasia of lymphatic vessels in VEGF-C transgenic mice. *Science* **1997**, *276*, 1423–1425. [CrossRef] [PubMed]

40. Kanady, J.D.; Dellinger, M.T.; Munger, S.J.; Witte, M.H.; Simon, A.M. Connexin37 and Connexin43 deficiencies in mice disrupt lymphatic valve development and result in lymphatic disorders including lymphedema and chylothorax. *Dev. Biol.* **2011**, *354*, 253–266. [CrossRef] [PubMed]

41. Sabine, A.; Agalarov, Y.; Maby-El Hajjami, H.; Jaquet, M.; Hägerling, R.; Pollmann, C.; Bebber, D.; Pfenniger, A.; Miura, N.; Dormond, O.; et al. Mechanotransduction, PROX1, and FOXC2 cooperate to control connexin37 and calcineurin during lymphatic-valve formation. *Dev. Cell* **2012**, *22*, 430–445. [CrossRef] [PubMed]

42. Munger, S.J.; Kanady, J.D.; Simon, A.M. Absence of venous valves in mice lacking Connexin37. *Dev. Biol.* **2013**, *373*, 338–348. [CrossRef] [PubMed]

43. Ferrell, R.E.; Baty, C.J.; Kimak, M.A.; Karlsson, J.M.; Lawrence, E.C.; Franke-Snyder, M.; Meriney, S.D.; Feingold, E.; Finegold, D.N. GJC2 missense mutations cause human lymphedema. *Am. J. Hum. Genet.* **2010**, *86*, 943–948. [CrossRef] [PubMed]

44. Ostergaard, P.; Simpson, M.A.; Brice, G.; Mansour, S.; Connell, F.C.; Onoufriadis, A.; Child, A.H.; Hwang, J.; Kalidas, K.; Mortimer, P.S.; et al. Rapid identification of mutations in GJC2 in primary lymphoedema using whole exome sequencing combined with linkage analysis with delineation of the phenotype. *J. Med. Genet.* **2011**, *48*, 251–255. [CrossRef] [PubMed]

45. Molica, F.; Meens, M.J.; Dubrot, J.; Ehrlich, A.; Roth, C.L.; Morel, S.; Pelli, G.; Vinet, L.; Braunersreuther, V.; Ratib, O.; et al. Pannexin1 links lymphatic function to lipid metabolism and atherosclerosis. *Sci. Rep.* **2017**, *7*, 13706. [CrossRef] [PubMed]

46. Penuela, S.; Bhalla, R.; Nag, K.; Laird, D.W. Glycosylation regulates pannexin intermixing and cellular localization. *Mol. Biol. Cell* **2009**, *20*, 4313–4323. [CrossRef] [PubMed]

47. Penuela, S.; Gehi, R.; Laird, D.W. The biochemistry and function of pannexin channels. *Biochim. Biophys. Acta* **2013**, *1828*, 15–22. [CrossRef] [PubMed]

48. Shestopalov, V.I.; Panchin, Y. Pannexins and gap junction protein diversity. *Cell. Mol. Life Sci.* **2008**, *65*, 376–394. [CrossRef] [PubMed]

49. Penuela, S.; Celetti, S.J.; Bhalla, R.; Shao, Q.; Laird, D.W. Diverse Subcellular Distribution Profiles of Pannexin1 and Pannexin3. *Cell Commun. Adhes.* **2008**, *15*, 133–142. [CrossRef] [PubMed]

50. Bhalla-Gehi, R.; Penuela, S.; Churko, J.M.; Shao, Q.; Laird, D.W. Pannexin1 and pannexin3 delivery, cell surface dynamics, and cytoskeletal interactions. *J. Biol. Chem.* **2010**, *285*, 9147–9160. [CrossRef] [PubMed]

51. Zappalà, A.; Cicero, D.; Serapide, M.F.; Paz, C.; Catania, M.V.; Falchi, M.; Parenti, R.; Pantò, M.R.; La Delia, F.; Cicirata, F. Expression of pannexin1 in the CNS of adult mouse: Cellular localization and effect of 4-aminopyridine-induced seizures. *Neuroscience* **2006**, *141*, 167–178. [CrossRef] [PubMed]

52. Huang, Y.; Grinspan, J.B.; Abrams, C.K.; Scherer, S.S. Pannexin1 is expressed by neurons and glia but does not form functional gap junctions. *Glia* **2007**, *55*, 46–56. [CrossRef] [PubMed]

53. Boassa, D.; Nguyen, P.; Hu, J.; Ellisman, M.H.; Sosinsky, G.E. Pannexin2 oligomers localize in the membranes of endosomal vesicles in mammalian cells while Pannexin1 channels traffic to the plasma membrane. *Front. Cell. Neurosci.* **2014**, *8*, 468. [CrossRef] [PubMed]

54. Boyce, A.K.J.; Epp, A.L.; Nagarajan, A.; Swayne, L.A. Transcriptional and post-translational regulation of pannexins. *Biochim. Biophys. Acta* **2018**, *1860*, 72–82. [CrossRef] [PubMed]

55. Vanden Abeele, F.; Bidaux, G.; Gordienko, D.; Beck, B.; Panchin, Y.V.; Baranova, A.V.; Ivanov, D.V.; Skryma, R.; Prevarskaya, N. Functional implications of calcium permeability of the channel formed by pannexin 1. *J. Cell Biol.* **2006**, *174*, 535–546. [CrossRef] [PubMed]

56. Anselmi, F.; Hernandez, V.H.; Crispino, G.; Seydel, A.; Ortolano, S.; Roper, S.D.; Kessaris, N.; Richardson, W.; Rickheit, G.; Filippov, M.A.; et al. ATP release through connexin hemichannels and gap junction transfer of second messengers propagate Ca^{2+} signals across the inner ear. *Proc. Natl. Acad. Sci. USA* **2008**, *105*, 18770–18775. [CrossRef] [PubMed]

57. Qu, Y.; Misaghi, S.; Newton, K.; Gilmour, L.L.; Louie, S.; Cupp, J.E.; Dubyak, G.R.; Hackos, D.; Dixit, V.M. Pannexin-1 Is Required for ATP Release during Apoptosis but Not for Inflammasome Activation. *J. Immunol.* **2011**, *186*, 6553–6561. [CrossRef] [PubMed]

58. Shao, Q.; Lindstrom, K.; Shi, R.; Kelly, J.; Schroeder, A.; Juusola, J.; Levine, K.L.; Esseltine, J.L.; Penuela, S.; Jackson, M.F.; et al. A Germline Variant in the PANX1 Gene Has Reduced Channel Function and Is Associated with Multisystem Dysfunction. *J. Biol. Chem.* **2016**, *291*, 12432–12443. [CrossRef] [PubMed]

59. Lohman, A.W.; Isakson, B.E. Differentiating connexin hemichannels and pannexin channels in cellular ATP release. *FEBS Lett.* **2014**, *588*, 1379–1388. [CrossRef] [PubMed]

60. Ishikawa, M.; Iwamoto, T.; Nakamura, T.; Doyle, A.; Fukumoto, S.; Yamada, Y. Pannexin 3 functions as an ER Ca^{2+} channel, hemichannel, and gap junction to promote osteoblast differentiation. *J. Cell Biol.* **2011**, *193*, 1257–1274. [CrossRef] [PubMed]

61. Wicki-Stordeur, L.E.; Dzugalo, A.D.; Swansburg, R.M.; Suits, J.M.; Swayne, L.A. Pannexin 1 regulates postnatal neural stem and progenitor cell proliferation. *Neural Dev.* **2012**, *7*, 11. [CrossRef] [PubMed]

62. Wei, L.; Yang, X.; Shi, X.; Chen, Y. Pannexin-1 silencing inhibits the proliferation of U87-MG cells. *Mol. Med. Rep.* **2015**, *11*, 3487–3492. [CrossRef] [PubMed]

63. Alvarez, A.; Lagos-Cabré, R.; Kong, M.; Cárdenas, A.; Burgos-Bravo, F.; Schneider, P.; Quest, A.F.G.; Leyton, L. Integrin-mediated transactivation of P2X7R via hemichannel-dependent ATP release stimulates astrocyte migration. *Biochim. Biophys. Acta* **2016**, *1863*, 2175–2188. [CrossRef] [PubMed]

64. Sáez, P.J.; Vargas, P.; Shoji, K.F.; Harcha, P.A.; Lennon-Duménil, A.-M.; Sáez, J.C. ATP promotes the fast migration of dendritic cells through the activity of pannexin 1 channels and P2X7 receptors. *Sci. Signal.* **2017**, *10*, eaah7107. [CrossRef] [PubMed]

65. Yoon, C.M.; Hong, B.S.; Moon, H.G.; Lim, S.; Suh, P.-G.; Kim, Y.-K.; Chae, C.-B.; Gho, Y.S. Sphingosine-1-phosphate promotes lymphangiogenesis by stimulating $S1P1/G_i/PLC/Ca^{2+}$ signaling pathways. *Blood* **2008**, *112*, 1129–1138. [CrossRef] [PubMed]

66. Choi, D.; Park, E.; Jung, E.; Seong, Y.J.; Yoo, J.; Lee, E.; Hong, M.; Lee, S.; Ishida, H.; Burford, J.; et al. Laminar flow downregulates Notch activity to promote lymphatic sprouting. *J. Clin. Investig.* **2017**, *127*, 1225–1240. [CrossRef] [PubMed]

67. Niimi, K.; Ueda, M.; Fukumoto, M.; Kohara, M.; Sawano, T.; Tsuchihashi, R.; Shibata, S.; Inagaki, S.; Furuyama, T. Transcription factor FOXO1 promotes cell migration toward exogenous ATP via controlling P2Y1 receptor expression in lymphatic endothelial cells. *Biochem. Biophys. Res. Commun.* **2017**, *489*, 413–419. [CrossRef] [PubMed]

68. Bao, B.A.; Lai, C.P.; Naus, C.C.; Morgan, J.R. Pannexin1 drives multicellular aggregate compaction via a signaling cascade that remodels the actin cytoskeleton. *J. Biol. Chem.* **2012**, *287*, 8407–8416. [CrossRef] [PubMed]

69. Boyce, A.K.J.; Wicki-Stordeur, L.E.; Swayne, L.A. Powerful partnership: Crosstalk between pannexin 1 and the cytoskeleton. *Front. Physiol.* **2014**, *5*, 27. [CrossRef] [PubMed]

International Journal of
Molecular Sciences

MDPI

Article

Connexin 43 Plays a Role in Pulmonary Vascular Reactivity in Mice

Myo Htet, Jane E. Nally, Andrew Shaw, Bradley E. Foote, Patricia E. Martin and Yvonne Dempsie *

Department of Life Sciences, School of Health and Life Sciences, Glasgow Caledonian University,
Glasgow G4 0BA, UK; Myo.Htet@gcu.ac.uk (M.H.); J.E.Nally@gcu.ac.uk (J.E.N.); andrew1994@live.co.uk (A.S.);
BFOOTE200@caledonian.ac.uk (B.E.F.); patricia.martin@gcu.ac.uk (P.E.M.)
* Correspondence: yvonne.dempsie@gcu.ac.uk; Tel.: +44-141-273-1930

Received: 23 April 2018; Accepted: 20 June 2018; Published: 27 June 2018

Abstract: Pulmonary arterial hypertension (PAH) is a chronic condition characterized by vascular remodeling and increased vaso-reactivity. PAH is more common in females than in males (~3:1). Connexin (Cx)43 has been shown to be involved in cellular communication within the pulmonary vasculature. Therefore, we investigated the role of Cx43 in pulmonary vascular reactivity using *Cx43* heterozygous (*Cx43*$^{+/-}$) mice and ^{37,43}Gap27, which is a pharmacological inhibitor of Cx37 and Cx43. Contraction and relaxation responses were studied in intra-lobar pulmonary arteries (IPAs) derived from normoxic mice and hypoxic mice using wire myography. IPAs from male *Cx43*$^{+/-}$ mice displayed a small but significant increase in the contractile response to endothelin-1 (but not 5-hydroxytryptamine) under both normoxic and hypoxic conditions. There was no difference in the contractile response to endothelin-1 (ET-1) or 5-hydroxytryptamine (5-HT) in IPAs derived from female *Cx43*$^{+/-}$ mice compared to wildtype mice. Relaxation responses to methacholine (MCh) were attenuated in IPAs from male and female *Cx43*$^{+/-}$ mice or by pre-incubation of IPAs with ^{37,43}Gap27. N$_\omega$-Nitro-L-arginine methyl ester (L-NAME) fully inhibited MCh-induced relaxation. In conclusion, Cx43 is involved in nitric oxide (NO)-induced pulmonary vascular relaxation and plays a gender-specific and agonist-specific role in pulmonary vascular contractility. Therefore, reduced Cx43 signaling may contribute to pulmonary vascular dysfunction.

Keywords: pulmonary arterial hypertension (PAH); connexin43 (Cx43); gap junction; vascular reactivity; nitric oxide; serotonin; endothelin-1; isoprenaline

1. Introduction

Pulmonary arterial hypertension (PAH) is a progressive disease in which the mean resting pulmonary artery pressure rises above 25 mmHg with a mean resting capillary wedge pressure lower than 15 mmHg [1]. This increase in pressure is associated with both constriction and remodeling of the distal pulmonary vasculature and eventually leads to right-sided heart failure. Prognosis is poor and survival has been reported to only 68% after three years on therapy [2]. PAH is far more common in females than in males (~3:1) [3]. Dysregulation of cell-to-cell communication particularly between pulmonary artery endothelial cells (PAECs) and pulmonary artery smooth muscle cells (PASMCs) is thought to play an important role in both constriction and remodeling of the pulmonary vasculature in PAH [4]. For example, PAECs from patients with PAH have increased gene and protein expression of tryptophan hydroxylase 1 (Tph1), which is the rate-limiting enzyme in the synthesis of 5-hydroxytryptamine (5-HT) [5]. 5-HT causes contraction of pulmonary arteries and proliferation of PASMCs [6]. In addition, PAECs from PAH patients produce decreased amounts of nitric oxide, which is a potent vasodilator that suppresses proliferation of PASMCs [7].

Connexins are transmembrane proteins that can oligomerize to form a pore in the cell membrane known as a hemi-channel with small regulatory molecules such as 3′,5′-cyclic adenosine monophosphate (cAMP), adenosine 5′ triphosphate (ATP), calcium (Ca^{2+}), and inositol 1,4,5-triphosphate (IP_3), which can pass directly through the cell membrane. Hemi-channels on adjacent cells can align to form gap junctions that allow these mediators to pass directly from the cytoplasm of one cell to that of another. Connexins 37 (Cx37), 40 (Cx40), 43 (Cx43), and 45 (Cx45) are expressed in diverse networks throughout the vasculature [8–10]. In the systemic vasculature, connexins have established roles in regulating vascular tone, vascular cell growth, angiogenesis, cell differentiation, and development [11]. Pannexins belong to the same super-family as connexins and also form membrane-associated channels that can mediate the release of small signaling molecules [12]. Pannexin-1 (Panx1) is expressed throughout the vasculature in the endothelium and also in the medial layer of small resistance arteries [13] and has been shown to play an important role in the α-adrenoceptor mediated contractile response [14].

Recent evidence suggests that dysregulated connexin signaling is involved in the pathophysiology of PAH. For example, blood outgrowth endothelial cells from patients with PAH show abnormal gap junctional communication. In addition, the nitric oxide synthase (NOS) inhibitor asymmetric dimethyl arginine (ADMA), is upregulated in PAH patients and inhibits gap junctional communication in human PAECs. The effects of ADMA are prevented by over-expression of Cx43 or by treatment with rotigaptide, which enhances connexin coupling [15]. In line with this, Cx37 and Cx40 are down-regulated in PAECs from PAH patients. Restoration of the transcription factor myocyte enhancer factor 2 leads to increased expression of various transcriptional targets including Cx37 and Cx40. It also rescues pulmonary hypertension in experimental models [16]. Mice genetically deficient in Cx40 are protected against hypoxic-induced pulmonary hypertension. Both genetic knockdown and pharmacological inhibition of Cx40 attenuates hypoxic pulmonary vasoconstriction in the mouse isolated perfused lung [17]. Moreover, contraction to phenylephrine (an α-1 adrenoceptor agonist) in pulmonary arteries taken from chronic hypoxic and monocrotaline (MCT) treated rats is inhibited by both [37,43]Gap27 and [40]Gap27 (connexin mimetic peptide blockers of Cxs37/ 43 and Cx40, respectively) [18].

Furthermore, Cx43 has been shown to play a role in 5-HT mediated signaling in the pulmonary vasculature. 5-HT can be synthesized in human PAECs by Tph1 and then released to act on neighboring PASMCs to mediate both proliferation and contraction [6,19–21]. Studies have shown in experiments using co-cultures of rat PAECs and rat PASMCs that 5-HT can pass through myoendothelial gap junctions formed principally by Cx43 [22,23]. In line with this, 5-HT-induced contraction of isolated rat pulmonary arteries was attenuated by [37,43]Gap27 [18].

In the current study, we have assessed the role of Cx43 in pulmonary vascular reactivity using pulmonary arteries from *Cx43* heterozygous mice (*Cx43*[+/−] mice). Heterozygous mice were used as a complete genetic knockdown of Cx43, which is lethal in mice [24]. In addition to assessing vasoreactivity in *Cx43*[+/−] mice, we have also assessed vasoreactivity in the presence of [37,43]Gap27. Since PAH is more common in females than males [3], the role of Cx43 in vascular reactivity was assessed in both male and female mice under normoxic conditions. Paradoxically, male mice have previously been shown to develop more pronounced hypoxic-induced PAH than females [25] and, therefore, our hypoxic experiments were conducted in male mice. Gene expression of *Cx37* (encoded by *GJA4*), *Cx40* (encoded by *GJA5*), *Cx45* (encoded by *GJC1*), and *Panx1* was assessed in pulmonary arteries from *Cx43*[+/−] mice along with gene expression of various mediators known to play a role in the development of PAH: *Tph1*, endothelial nitric oxide synthase (*eNOS*, encoded by *NOS3*), and bone morphogenetic receptor type II (*BMPRII*, encoded by *BMPR2*). *BMPRII* mutations have been found in patients with various forms of PAH [26]. Dysregulated bone morphogenetic protein (BMP) signaling is thought to be pivotal to the pathophysiology of PAH [27].

2. Results

2.1. Gene Expression of Connexins in Pulmonary Arteries from Cx43 heterozygous Mice

First, quantitative real time PCR (qPCR) was performed to confirm reduced gene expression of Cx43 in pulmonary arteries of male and female $Cx43^{+/-}$ mice. $Cx43$ expression was higher in females than in males in both WT and $Cx43^{+/-}$ mice (Figure 1A). Afterward, pulmonary arterial gene expression levels of $Cx43$ with $Cx37$, $Cx40$, and $Cx45$ in wildtype mice were compared. $Cx43$ was the predominant vascular connexin in female mice. In male mice, there was a trend towards $Cx43$ being expressed at greater levels than $Cx37$ and $Cx40$, but this was not significant. $Cx45$ was expressed at lower levels in both male and female mice (Figure 1B).

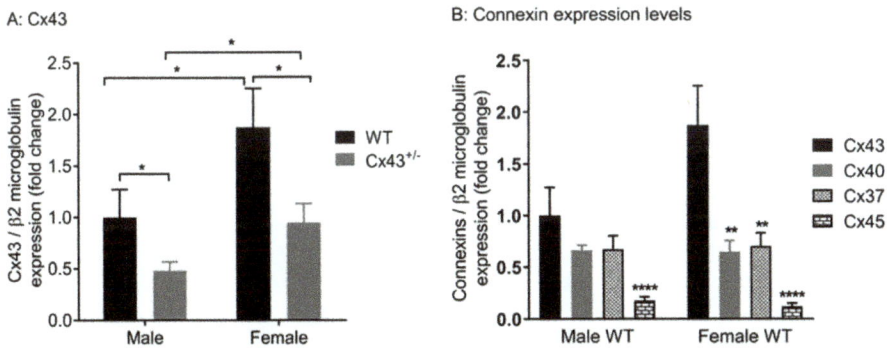

Figure 1. Connexin gene expression in pulmonary arteries from male and female mice. $Cx43$ gene expression is reduced in pulmonary arteries from $Cx43$ heterozygous ($Cx43^{+/-}$) mice (**A**). Female mice have increased levels of $Cx43$ compared to males (**A**). $Cx43$ is the predominant vascular connexin in female mice while male mice show similar levels of $Cx43$, $Cx40$, and $Cx37$. $Cx45$ is expressed in lower levels than $Cxs\ 43$, 40, and 37 in both male and female mice (**B**). Data are presented as mean $\pm$ S.E.M. and were analyzed by two-way ANOVA. $n = 6$ with each sample run in triplicate. **A**: * $p < 0.05$, **B**: ** $p < 0.01$, **** $p < 0.0001$ compared to $Cx43$.

We explored the effects of genetic knockdown of $Cx43$ on gene expression of $Cx37$, $Cx40$, $Cx45$, and $Panx1$, which are all expressed in the vasculature. Male $Cx43^{+/-}$ mice displayed reduced gene expression of $Cx37$, $Cx40$, $Cx45$, and $Panx1$ mRNA (Figure 2A–D) when compared to WT littermates. Female $Cx43^{+/-}$ mice, however, displayed no changes in gene expression of $Cx37$, $Cx40$, $Cx45$, or $Panx1$ compared to WT littermates (Figure 2A–D). The housekeeping gene $\beta2$-microglobulin was expressed at similar levels in all four groups of mice studied (Ct values: male WT 24.9 $\pm$ 0.2, male $Cx43^{+/-}$ 24.4 $\pm$ 0.2, female WT 25.0 $\pm$ 0.2, female $Cx43^{+/-}$ 24.6 $\pm$ 0.3, $p = 0.86$).

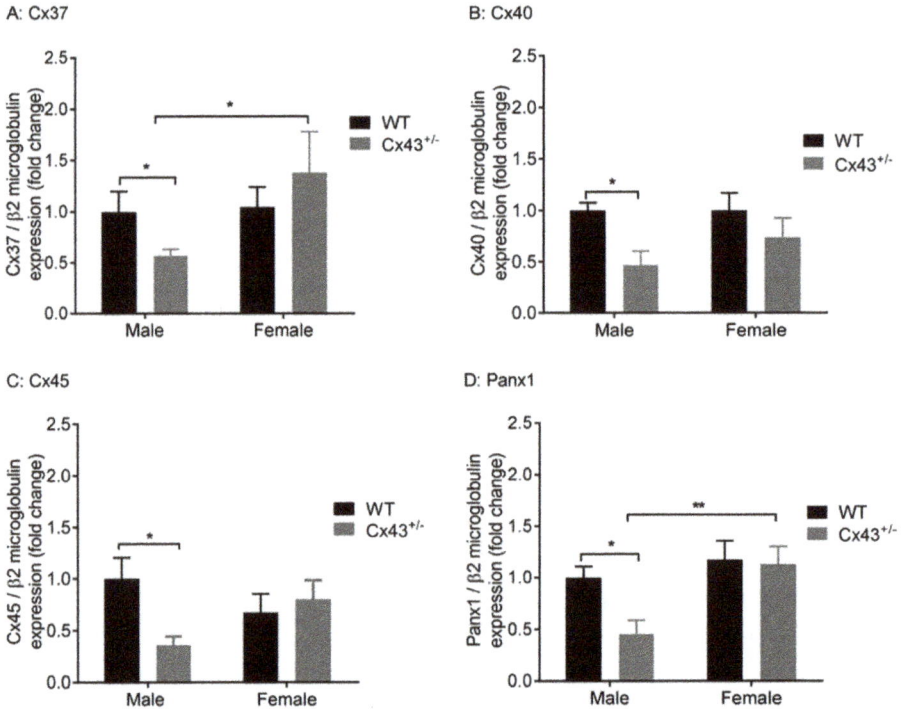

Figure 2. Gene expression of *Cx37* (**A**), *Cx40* (**B**), *Cx45* (**C**), and *Panx1* (**D**) in pulmonary arteries from male and female wildtype (WT) and *Cx43* heterozygous (*Cx43$^{+/-}$*) mice. Data are presented as mean ± S.E.M. and were analyzed by two-way ANOVA. * $p < 0.05$, ** $p < 0.01$, $n = 6$ per group with each sample analyzed in triplicate.

2.2. Pulmonary Arterial Contractile Responses

In the intra-lobar pulmonary arteries (IPAs) of male mice, endothelin-1 (ET-1) was more potent (had a lower median effective concentration or EC_{50} value) in *Cx43$^{+/-}$* mice compared to WT mice (Figure 3A; Table 1). Maximal response to ET-1 (E_{max}) was, however, unchanged between WT and *Cx43$^{+/-}$* mice. There was no global shift in the concentration response curve (Figure 3A; Table 1). IPAs from male *Cx43$^{+/-}$* mice showed similar contractile responses to 5-hydroxytryptamine (5-HT) as those from WT mice (Figure 3B; Table 1). Contractile responses to both ET-1 and 5-HT were similar in IPAs from female *Cx43$^{+/-}$* mice compared to female WT mice (Figure 3C,D; Table 1).

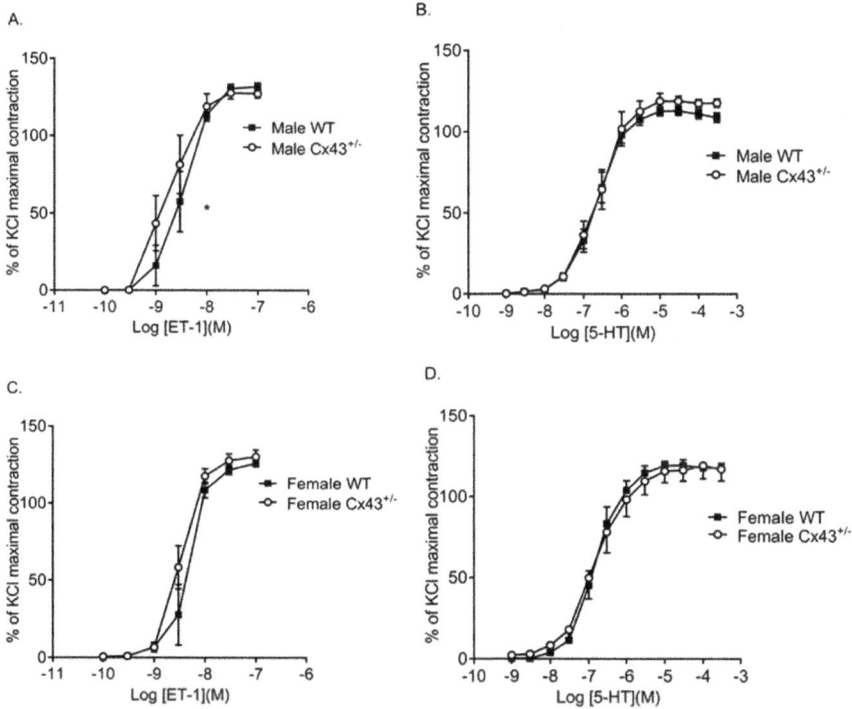

Figure 3. Pulmonary vascular contractility to ET-1 and 5-HT in intralobar pulmonary arteries (IPAs) from male and female wildtype (WT) and *Cx43* heterozygous (*Cx43*[+/−]) mice. ET-1 was more potent in IPAs from male *Cx43*[+/−] mice than WT mice (**A**). There was no difference in contractile response to 5-HT in IPAs from male WT and *Cx43*[+/−] mice (**B**). There was no difference in ET-1 (**C**) or 5-HT (**D**) induced contractile response in IPAs from female WT or *Cx43*[+/−] mice. Data are shown as mean ± S.E.M. Global differences in concentration response curves were compared by two-way ANOVA. Changes in logarithm of median effective concentration (Log EC_{50}) and maximal contractile responses (E_{max}) between two different groups were analyzed by using the Student's *t*-test. * EC_{50} is significantly ($p < 0.05$) reduced in male *Cx43*[+/−] mice, $n = 5$–7 per group.

Table 1. The contractile effects of ET-1 and 5-HT in IPAs of male and female wildtype (WT) and *Cx43* heterozygous (*Cx43*[+/−]) mice.

Agonists	Groups	Log EC_{50} (M)	E_{max} (%)	Global Shift in Concentration Response Curve	n
ET-1	Male WT	−8.46 ± 0.08	131.8 ± 2.5	NS	6
	Male *Cx43*[+/−]	−8.77 ± 0.12 *	127.4 ± 2.8		5
ET-1	Female WT	−8.31 ± 0.06	126 ± 2.2	NS	5
	Female *Cx43*[+/−]	−8.48 ± 0.04	130.3 ± 4.5		7
5-HT	Male WT	−6.67 ± 0.05	112.7 ± 2.8	NS	6
	Male *Cx43*[+/−]	−6.64 ± 0.07	118.9 ± 4.8		5
5-HT	Female WT	−6.82 ± 0.05	119.6 ± 3.5	NS	5
	Female *Cx43*[+/−]	−6.83 ± 0.1	116.6 ± 6.7		5

Log EC_{50} indicates logarithm of median effective concentration. E_{max} maximal contractile effect. Global differences in concentration response curves were compared by two-way ANOVA. Changes in logarithm of median effective concentration (Log EC_{50}) and maximal contractile responses (E_{max}) between two different groups were analyzed by the Student's *t*-test. NS: not significant. * $p < 0.05$ compared to male WT mice. Data are shown as mean ± S.E.M.

2.3. Pulmonary Arterial Relaxation Responses

The relaxation response produced by methacholine (MCh) was significantly reduced in IPAs of both male and female $Cx43^{+/-}$ mice when compared to WT mice (Figure 4A,B; Table 2). Pharmacological inhibition of Cx43 with ^{37,43}Gap27 also significantly attenuated the relaxation responses produced by MCh in IPAs of both male and female mice (Figure 4C,D; Table 2).

We then confirmed that MCh-induced relaxation responses were dependent upon nitric oxide (NO) since the L-NAME completely inhibited MCh-induced relaxation (Figure 4E; Table 2). Then we assessed the role of Cx43 in isoprenaline-induced relaxation. Isoprenaline is classically thought to induce relaxation via the cAMP pathway. In these experiments, the relaxation induced by isoprenaline was partially inhibited by ^{37,43}Gap27 (Figure 4F; Table 2). Furthermore, we showed nitric oxide plays a role in isoprenaline-induced relaxation since the L-NAME partially attenuated isoprenaline-induced relaxation (Figure 4G; Table 2).

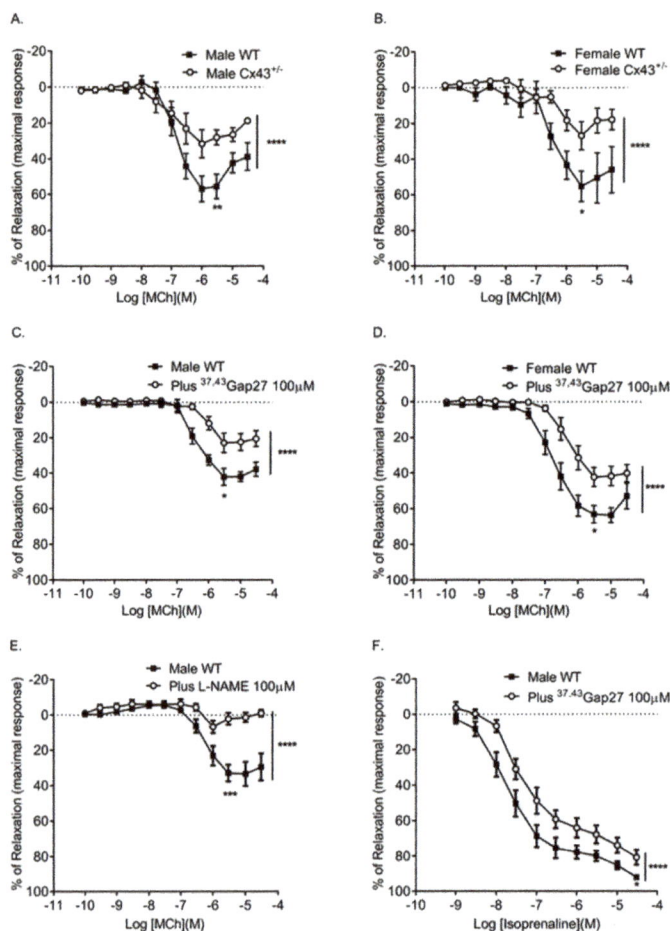

Figure 4. *Cont.*

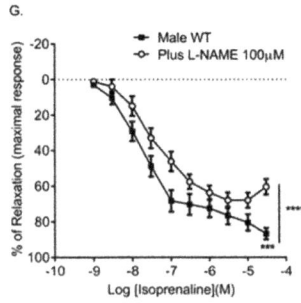

Figure 4. Effects of genetic reduction or pharmacological inhibition of Cx43 on pulmonary vascular relaxation responses. Male (**A**) and female (**B**) *Cx43* heterozygous (*Cx43*$^{+/-}$) mice show reduced relaxation in response to MCh. Pre-incubation with ^{37,43}Gap27 also reduced the relaxation response in male (**C**) and female (**D**) mice. MCh-induced relaxation was ablated in the presence of L-NAME (**E**). ^{37,43}Gap27 partially inhibited isoprenaline-induced relaxation in IPAs from male mice (**F**) as did L-NAME (**G**). Global differences in concentration response curves were compared by two-way ANOVA. Changes in logarithm of median effective concentration (Log EC$_{50}$) and maximal relaxation responses (R$_{max}$) between two different groups were analyzed by using the Student's *t*-test. Data are shown as mean ± S.E.M. * $p < 0.05$, ** $p < 0.01$, *** $p < 0.001$, **** $p < 0.0001$, $n = 5$–6 per group. Statistical symbols shown on the right hand side of graphs indicate global shifts in the concentration response curves. Statistical symbols underneath curves indicate changes in maximal relaxation (R$_{max}$) values.

Table 2. Log EC$_{50}$ (logarithm of median effective concentration) and R$_{max}$ (maximal relaxation) values for MCh-induced or isoprenaline-induced relaxation.

Agonists	Groups	Log EC$_{50}$ (M)	R$_{max}$ (%)	Global Shift in Concentration Response Curve	*n*
MCh	Male WT	−6.93 ± 0.08	55.6 ± 6.8	****	5
	Male *Cx43*$^{+/-}$	−7.14 ± 0.21 **	28 ± 4.1 **		5
MCh	Female WT	−6.52 ± 0.14	55.3 ± 8.5	****	6
	Female *Cx43*$^{+/-}$	−6.48 ± 0.24	26.8 ± 7.8 *		6
MCh	Male WT	−6.39 ± 0.2	42.1 ± 4.6	****	6
	Plus ^{37,43}Gap27	−6.08 ± 0.1 ****	22.9 ± 5.6 *		6
MCh	Female WT	−6.8 ± 0.08	63.1 ± 4.8	****	6
	Plus ^{37,43}Gap27	−6.36 ± 0.1 **	42.2 ± 5.2 *		6
MCh	Male WT	−6.27 ± 0.11	32.8 ± 4.6	****	6
	Plus L-NAME	not applicable	2.2 ± 3.4 ***		6
Isoprenaline	Male WT	−7.77 ± 0.17	92.1 ± 1.19	****	5
	Plus ^{37,43}Gap27	−7.41 ± 0.16	80.8 ± 4.2 *		5
Isoprenaline	Male WT	−7.84 ± 0.2	86.7 ± 3.2	****	6
	Plus L-NAME	−7.5 ± 0.13	60.4 ± 4.3 ***		6

Global differences in concentration response curves were compared by two-way ANOVA. Changes in logarithm of median effective concentration (Log EC$_{50}$) and maximal relaxation responses (R$_{max}$) between two different groups were analyzed by using the Student's *t*-test. Data are shown as mean ± S.E.M. * $p < 0.05$, ** $p < 0.01$, *** $p < 0.001$ and **** $p < 0.0001$.

2.4. Gene Expression of Bone Morphogenetic Protein Receptor Type II, Tryptophan Hydroxylase 1, and Endothelial Nitric Oxide Synthase in Pulmonary Arteries from Cx43 Heterozygous Mice

Since *Cx43*$^{+/-}$ mice displayed dysregulated pulmonary vascular reactivity, gene expression of *BMPRII* (encoded by *BMPR2*), *eNOS* (encoded by *NOS3*), and *Tph-1*, mediatorsimportant for regulating pulmonary vascular function, were assessed. Expression of *BMPR2*, *NOS3* , and *Tph-1* were not significantly altered in either male or female *Cx43*$^{+/-}$ mice compared to WT littermates (Figure 5A–C).

The expression of *NOS3* and *BMPR2* were significantly lower in female WT mice when compared to male WT mice (Figure 5A,B).

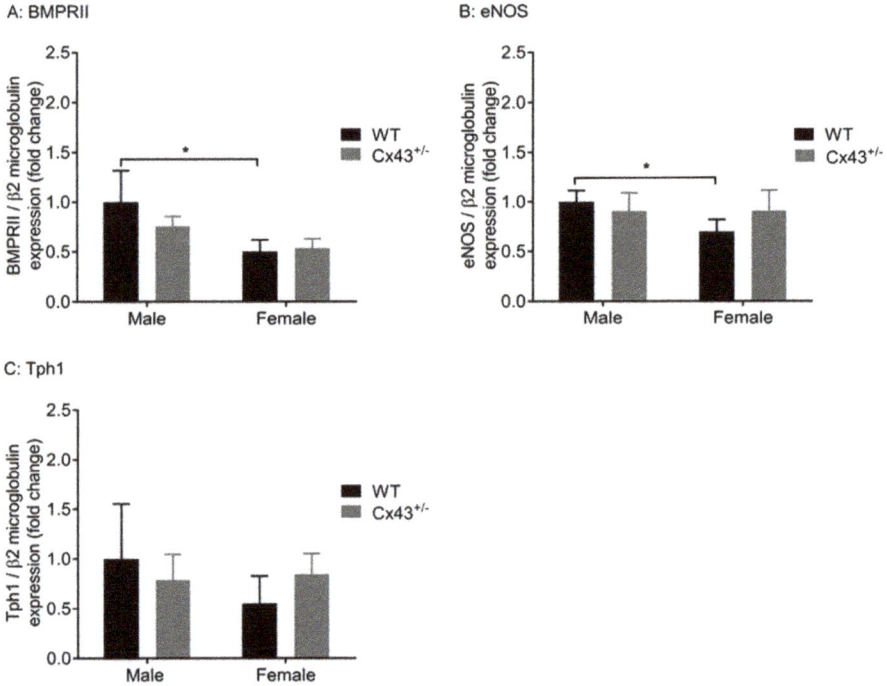

Figure 5. Gene expression of BMPRII (encoded by *BMPR2*) (**A**), eNOS (encoded by *NOS3*) (**B**), and *Tph1* (**C**) in pulmonary arteries of male and female WT and *Cx43*[+/−] mice. No differences were observed in expression of *BMPR2*, *NOS3*, or *Tph1* between WT and Cx43[+/−] mice (either male or female). *BMPR2* and *NOS3* were significantly downregulated in female WT mice compared to male WTs. Data are presented as mean ± S.E.M. and were analysed by using two-way ANOVA. * $p < 0.05$, $n = 6$ per group with each sample run in triplicate.

2.5. Effects of Hypoxia on Pulmonary Vascular Contractility in Male Cx43 Heterozugous Mice

The effects of chronic hypoxia on pulmonary vascular reactivity in *Cx43*[+/−] mice were investigated. The hypoxic experiments were carried out in male mice since it has been previously shown that male mice are more susceptible to hypoxic-induced PH than female mice [25]. Both WT and *Cx43*[+/−] mice developed the right ventricular hypertrophy after two weeks of chronic hypoxic exposure, which verifies the mouse hypoxic model (Figure 6A). IPAs derived from chronic hypoxic *Cx43*[+/−] mice showed an increased sensitivity to ET-1, which was assessed by a reduced EC_{50} value and a global leftward shift in the contractile response. In addition, the maximal contractile effect produced by ET-1 was significantly greater in IPAs from hypoxic *Cx43*[+/−] mice than from hypoxic WT mice (Figure 6B; Table 3). There was no difference in a contractile response to 5-HT in IPAs from hypoxic *Cx43*[+/−] mice (Figure 6C; Table 3).

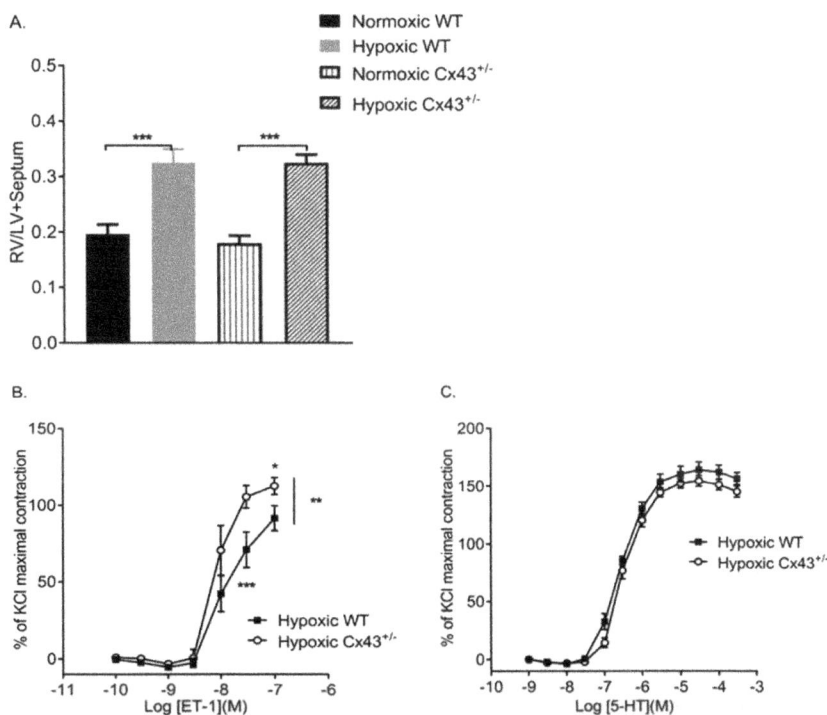

Figure 6. Assessment of the development of right ventricular hypertrophy by right ventricular weight ratio left ventricle plus septal weight (RV/LV+Septum) (**A**). Concentration response curves (CRCs) to ET-1 (**B**) and 5-HT (**C**) in intra-lobar pulmonary arteries derived from hypoxic wildtype (WT) and hypoxic *Cx43* heterozygous (*Cx43$^{+/-}$*) mice. Data are shown as mean ± S.E.M. Data in panel **A** were analyzed by two-way ANOVA. In panels **B** and **C**, global differences in CRCs were compared by two-way ANOVA. Changes in the logarithm of median effective concentration (Log EC$_{50}$) and maximal contractile responses (E$_{max}$) between two different groups were analyzed by using the Student's *t*-test * $p < 0.05$, ** $p < 0.01$, *** $p < 0.001$, n = 5–7 per group. The statistical symbol shown on the right hand side of graph B indicates a global shift in the CRC. The symbol underneath the curve indicates changes in the median effective concentration (EC$_{50}$) while the symbol above the curve indicates changes in the maximal response (E$_{max}$).

Table 3. The contractile effects of ET-1 and 5-HT in intra-lobar pulmonary arteries of hypoxic wildtype and *Cx43* heterozygous mice.

Agonists	Genotypes	Log EC$_{50}$ (M)	E$_{max}$ (%)	Global Shift in Concentration Response Curve	n
ET-1	Hypoxic WT	-7.95 ± 0.06	91.7 ± 8.1	**	6
	Hypoxic *Cx43$^{+/-}$*	-8.07 ± 0.09 ***	112.7 ± 5.6 *		7
5-HT	Hypoxic WT	-6.56 ± 0.04	126 ± 2.2	NS	7
	Hypoxic *Cx43$^{+/-}$*	-6.5 ± 0.03	130.3 ± 4.5		7

Log EC$_{50}$ indicates the logarithm of median effective concentration, E$_{max}$ maximal contractile effect. Global differences in concentration response curves were compared by using two-way ANOVA. Changes in logarithm of median effective concentration (Log EC$_{50}$) and maximal contractile responses (E$_{max}$) between two different groups were analyzed by using the Student's *t*-test. NS: not significant * $p < 0.05$, ** $p < 0.01$, *** $p < 0.001$. Data are shown as mean ± S.E.M.

2.6. Effects of Hypoxia on Expression of Cx43 in Mouse Lung and Pulmonary Artery

Afterward, the effects of chronic hypoxia on Cx43 gene and protein expression were assessed. Using qPCR, it was shown that hypoxia significantly down-regulated *Cx43* gene expression in both WT and $Cx43^{+/-}$ male mice (Figure 7A). We then used immunofluorescence to visualize the effects of hypoxia on Cx43 protein expression. Lung expression of Cx43 (green fluorescence) was reduced in hypoxic WT mice when compared to normoxic WT mice (Figure 7B). Furthermore, Cx43 immunoreactivity was further reduced in lungs of hypoxic $Cx43^{+/-}$ mice when compared to normoxic $Cx43^{+/-}$ mice (Figure 7B).

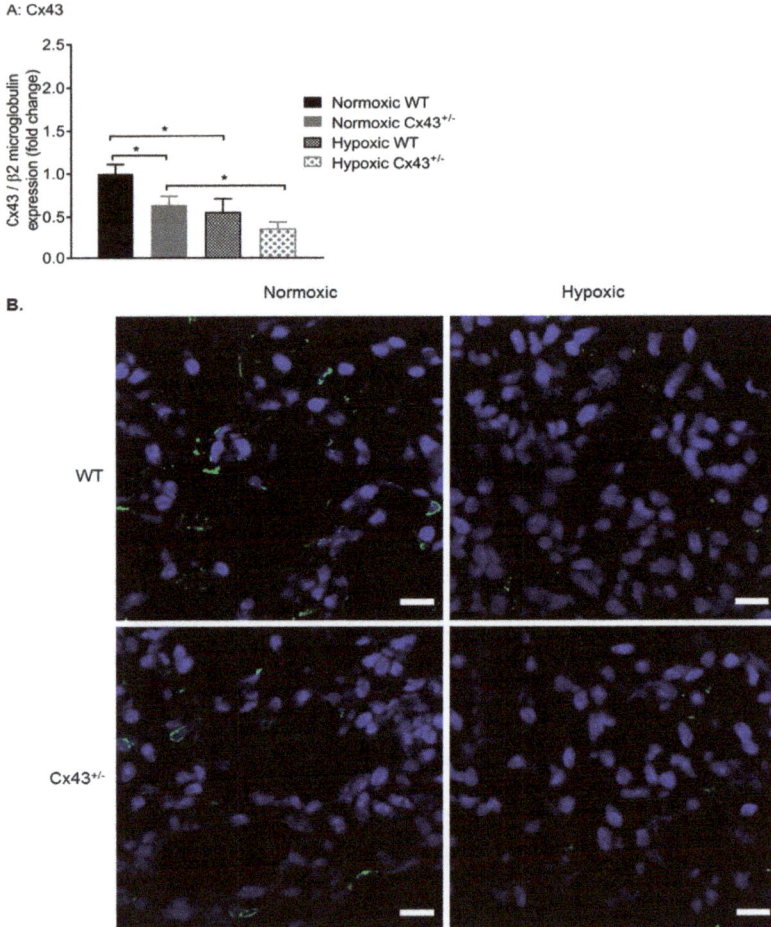

Figure 7. Cx43 expression in pulmonary arteries of wildtype (WT) and *Cx43* heterozygous ($Cx43^{+/-}$) mice under normoxic and chronic hypoxic conditions. *Cx43* gene expression is reduced by hypoxia in both WT and $Cx43^{+/-}$ mice (**A**). Confocal images of Cx43 immunofluorescence staining in normoxic and hypoxic WT and $Cx43^{+/-}$ mouse lung tissue sections are shown in (**B**). Green fluorescence punctate findings represent Cx43 immunoreactivity and blue staining represents nuclei. Scale bars represent 10 μm. For panel **A**, data is presented as mean ± S.E.M. and was analyzed by two-way ANOVA, * $p < 0.05$, $n = 6$ with each sample run in triplicate.

2.7. Effects of Hypoxia on Gene Expression in Pulmonary Arteries Derived from WT and Cx43$^{+/-}$ Mice

As chronic hypoxia mediated increased vascular reactivity to ET-1 and decreased Cx43 expression, the effects of chronic hypoxia on the expression of other vascular connexins and *Panx1* were assessed (Figure 8 A–D). Hypoxia mediated a downregulation of *Cx40* gene expression in WT mice (Figure 8B) while gene expression of *Cx45* was up-regulated by hypoxia in *Cx43*$^{+/-}$ mice (Figure 8C). Hypoxia had no effect on the expression of *Cx37* or *Panx1* in WT or *Cx43*$^{+/-}$ mice (Figure 8A,D).

Gene expression of *BMPR2*, *Tph1*, and *NOS3* in response to hypoxia in *Cx43*$^{+/-}$ mice was also assessed. *BMPR2* was downregulated by hypoxia in WT mice and *Tph1* was up-regulated by hypoxia in *Cx43*$^{+/-}$ mice. There were no differences in *BMPR2*, *NOS3*, or *Tph1* expression between hypoxic WT and hypoxic *Cx43*$^{+/-}$ mice (Figure 8E–G).

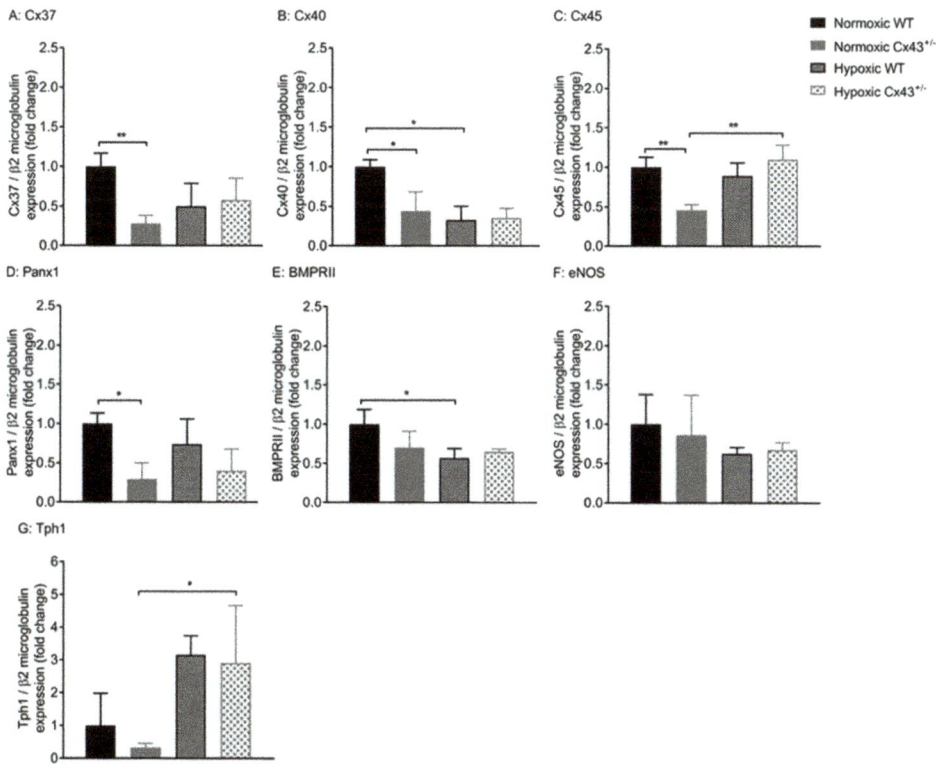

Figure 8. Gene expression of *Cx37* (**A**), *Cx40* (**B**), *Cx45* (**C**), *Panx1* (**D**), *BMPRII* (**E**), *eNOS* (**F**) and *Tph1* (**G**) in pulmonary arteries of wildtype (WT) and *Cx43* heterozygous (*Cx43*$^{+/-}$) mice under normoxic and chronic hypoxic conditions. All data are presented as mean ± S.E.M. and were analyzed by two-way ANOVA. * $p < 0.05$, ** $p < 0.01$, $n = 6$ per group with each sample run in triplicate.

3. Discussion

To our knowledge, this is the first study to show that mice genetically deficient in *Cx43* develop pulmonary vascular dysfunction. It is also the first study to show that Cx43 plays a role in NO-induced pulmonary vascular relaxation. These findings add to the growing body of evidence that suggests that Cx43 is involved in regulation of the pulmonary vasculature. Since PAH is associated with altered pulmonary vascular reactivity, these data suggest Cx43 is worthy of further investigation as a novel therapeutic target for PAH.

Relaxation responses to MCh were reduced in IPAs from both male and female $Cx43^{+/-}$ mice. Subsequently, we found that pharmacological inhibition of Cx43 using ^{37,43}Gap27 also attenuated MCh-induced relaxation. ^{37,43}Gap 27 inhibits both Cx37 and Cx43 and both Cx37 and Cx43 are down-regulated in male $Cx43^{+/-}$ mice. However, the vascular connexins of only Cx43 are downregulated in female $Cx43^{+/-}$ mice. Therefore, reduction of MCh-induced relaxation in female $Cx43^{+/-}$ mice suggests an important role for Cx43 in this effect. Our studies confirmed that MCh-induced relaxation was NO-dependent since the MCh mediated relaxation responses were completely abolished in the presence of L-NAME. It is widely considered that NO can diffuse through the endothelial and smooth muscle cell membranes to activate guanylate cyclase within the smooth muscle cell and mediate vasodilation. It has been shown, however, that diffusion of NO across the vascular cell membrane requires specific connexin channels [28]. NO opened and permeated hemichannels expressed in HeLa cells transfected and selected to express Cx43, Cx40, or Cx37. In addition, the blockade of connexin channels abolished myoendothelial NO transfer and NO-dependent vasodilation induced by acetylcholine in rat mesenteric arteries [28]. There is also mounting evidence that NO can interact with gap junctions in a complex and inter-dependent fashion [29]. Cx43 is constitutively S-nitrosylated at cysteine 271 by NO at the myoendothelial gap junctions. This nitrosylation keeps the myoendothelial gap junction open and denitrosylation closes the gap junction channel [30]. Additionally, NO has been shown to enhance gap junction coupling and increase trafficking of Cx40 to the membrane in endothelial cells via the protein kinase A activation [31]. Cx40 has been shown to co-localize with eNOS in the mouse aorta and is involved in conducting vasodilation [32]. *Cx40* knock out ($Cx40^{-/-}$) mice showed reduced basal and acetycholine induced NO release and reduced eNOS expression in aortas [32]. However, NO can act as a negative regulator of Cx37, which reduces Cx37-mediated dye transfer and electrical coupling in human umbilical vascular endothelial cells (HUVEC) and mouse microvascular endothelial cells [33,34]. Therefore, it is possible that Cx43 and NO interact in such a fashion that when Cx43 is genetically downregulated or pharmacologically inhibited, it leads to a reduction in NO-mediated vasorelaxation. In the present study, pulmonary arterial gene expression of *eNOS* was unchanged in male and female $Cx43^{+/-}$ mice when compared to WT controls. It would, however, be of interest to investigate eNOS protein levels and also MCh-induced NO release in pulmonary arteries from WT and $Cx43^{+/-}$ mice. It would also be of interest to investigate potential interactions between eNOS and Cx43 using co-immunoprecipitation.

^{37,43}Gap27 also mediated a small but significant inhibition of isoprenaline-induced relaxation. Isoprenaline is classically thought to act through the cAMP pathway. In the present study, isoprenaline-induced relaxation was, however, partially inhibited by L-NAME, which suggests that NO plays a role in isoprenaline-induced relaxation. A previous study has also shown β_2 adrenoceptor induced relaxation in mouse pulmonary arteries are attenuated by the L-NAME, endothelial denudation, or deletion of *eNOS* [35]. β_2 adrenoceptors have been detected in the mouse pulmonary endothelial layer by immunostaining [35]. The inhibitions of isoprenaline relaxation mediated by ^{37,43}Gap27 and L-NAME were of a similar magnitude. It is, therefore, possible that inhibition of connexin mediated communication via ^{37,43}Gap27 inhibits the NO component of isoprenaline relaxation. In future studies, it would be of interest to analyze whether smooth muscle responses to NO are affected by downregulation of Cx43. This could be achieved using an NO donor such as sodium nitroprusside.

Enhanced ET-1 mediated contraction was observed in IPAs derived from both normoxic and hypoxic male $Cx43^{+/-}$ mice compared to their WT counterparts. Normoxic female $Cx43^{+/-}$ mice did not show an increased contractile response to ET-1. However, male $Cx43^{+/-}$ mice showed downregulation of *Cx37, Cx40, Cx45*, and *Panx1*. These effects were not observed in female $Cx43^{+/-}$ mice. Downregulation of *Cx37, Cx40, Cx45*, or *Panx1* may, therefore, play a role in the increased contractile response to ET-1 observed in the male $Cx43^{+/-}$ mice. The reduced Cx43 expression observed in the pulmonary arteries of male mice compared to female mice observed in this study may also contribute to the increased effects of ET-1 in male $Cx43^{+/-}$ mice. Gender differences in the endothelin system may also contribute to contractile differences between male and female $Cx43^{+/-}$ mice [36]. Studies have shown that plasma endothelin levels were higher in men when compared to women [37,38]. At the receptor level, the ratio of ET_A and ET_B receptors in the endothelium of saphenous vein was greater in male subjects (3:1) than in female subjects (1:1) [39]. Haemodynamically, a rat model showed pressor responses induced by intravenous administration of ET-1 were greater in male rats than in female rats [40]. In vitro cell culture studies have also confirmed that 17-β oestradiol (E2) inhibited ET-1 gene expression through the extracellular signal regulated kinase (ERK) pathway [41].

5-HT mediated contractile responses in the IPAs were not affected by the partial loss of $Cx43$ in either male or female mice under either normoxic or hypoxic conditions. Therefore, the effects of reduced $Cx43$ gene expression varied according to the agonist used. These results agree with previously published data showing that the effects of pharmacological inhibition by ^{37,43}Gap27 on contractile responses in IPAs were dependent on the agonist used [18]. In line with the current results, Billaud et al. showed that ^{37,43}Gap27 had no effect on 5-HT induced contraction in IPAs derived from hypoxic rats. However, they did report that ^{37,43}Gap27 could inhibit 5-HT induced contraction in IPAs from normoxic rats. In contrast with the results reported here showing that downregulation of $Cx43$ enhanced ET-1 mediated contraction in IPAs form both normoxic and hypoxic $Cx43^{+/-}$ mice, Billaud et al. found that ^{37,43}Gap27 had no effect on ET-1 induced contraction in IPAs derived from normoxic or hypoxic rats [18]. 5-HT and ET-1 have different mechanisms of contraction. Since connexins are modulated by PKA, PKC, and calcium [42], the role of connexin mediated communication in the contractile response differs according to the agonist used. A possible reason for the disparity between the Billaud study and our own could be due to the ratiometric changes in the balance of Cx43:Cx40:Cx37 expression between rats and mice. A number of studies have reported tissue and species specific variation in connexin expression profiles and ^{37,43}Gap27 is defined as being specific to the SRPTEKTIFII sequence—conserved between Cx43 and Cx37 but different in Cx40 [43–46]. In addition, $Cx43^{+/-}$ mice will have compensatory changes in other genes, which may affect the contractile response while ^{37,43}Gap27 inhibits Cx37 as well as Cx43. This also potentially affects the contractile response.

In the gene expression studies, $Cx43$ expression was higher in female mice than in male mice. This is in keeping with a report which showed that Cx43 expression was higher in rat cardiomyocytes derived from females than males [47]. Furthermore, treatment with oestradiol increased Cx43 expression in human myometrium [48]. In addition to this, there is direct evidence that estrogen can regulate $Cx43$ gene expression since the promotor region of $Cx43$ gene contains an estrogen response element [49].

Chronic hypoxic rodents are a commonly used model for PAH. In the current study, the chronic hypoxia suppressed Cx43 gene and protein expression are located in the pulmonary arteries of both WT and $Cx43^{+/-}$ mice. Hypoxia can regulate Cx43 expression post-translationally by phosphorylation [50]. For example, one study showed exposure to hypoxia is associated with an increase in phosphorylation of Cx43- serine 368 (Ser368) in human microvascular endothelial cells and this phosphorylation was associated with downregulation Cx43 protein expression [51]. As hypoxia downregulates Cx43 expression, it would be of interest in future studies to assess the effects of hypoxia on MCh-induced relaxation in $Cx43^{+/-}$ mice. The current study showed $Cx40$ expression was also downregulated in hypoxic WT mice. Previous studies have shown that Cx40 expression was reduced during PAH in rats and treatment with sildenafil increased Cx40 expression via BMP signaling [52,53]. We found $Cx45$ expression to be upregulated in the $Cx43^{+/-}$ mice under chronic hypoxic conditions. The function of Cx45 in the vasculature remains unknown even though it has long been known to be expressed in vascular smooth muscle cells [54].

It is interesting to note that $Cx43^{+/-}$ mice did not develop increased right ventricular hypertrophy (RVH) in response to hypoxia compared to their wildtype counterparts. It has previously been shown by ourselves and others that changes in pulmonary vascular pressures and pulmonary vascular remodeling do not always lead to the development of RVH in mice. For example, increased pulmonary pressures and remodeling have been observed in the absence of RVH in mice that over-express the serotonin transporter, mice that over-express Mts 1, and mice that are dosed with dexfenfluramine [55–57]. In addition, Cx43 is highly expressed in cardiac myocytes and is thought to play an important role in hypertrophy of these cells. Expression and localization of Cx43 in cardiac myocytes has been shown to be dynamically regulated in various animal models of cardiac hypertrophy [58]. Therefore, it is possible that cardiac myocytes from Cx43$^{+/-}$ mice are functionally abnormal and, therefore, have an atypical hypertrophic response to hypoxia.

Female *Cx43*$^{+/-}$ mice did not exhibit the compensatory reduction in gene expression of *Cx37*, *Cx40*, *Cx45*, or *Panx1* that was observed in male *Cx43*$^{+/-}$ mice. Multiple lines of evidence show that *Cx43*, *Cx40*, and *Cx37* are interdependent on each other and compensatory changes occur upon connexin deletion. For instance, in *Cx40* knock out (*Cx40*$^{-/-}$) mice both total and smooth muscle Cx43 protein expression was reduced in the mouse aortas [59]. In addition, in *Cx40*$^{-/-}$ mice, the pericellular component of Cx43 staining was lost and there was increased redistribution of Cx43 in the perinuclear region [59]. This suggests deletion of *Cx40* not only leads to Cx43 downregulation but also affects its trafficking. Conversely, another group found that *Cx40*$^{-/-}$ mice showed upregulation of Cx37 and Cx43 in the aortic endothelium [60]. In *Cx40*$^{-/-}$ neonatal mice, Cx37 protein expression was downregulated in the endothelium and there was an increased Cx37 and Cx43 in the medial layer [61]. Genetic deletion of *Cx37* in mice showed reduction in endothelial Cx40 [61]. In the present study, we have assessed changes in gene expression in the whole pulmonary artery. It would be of interest to assess cell type specific changes in future studies.

Gene expression of *eNOS*, *BMPRII*, or *Tph1* were not affected by the loss of *Cx43* under normoxic or hypoxic conditions. Among the normoxic WT mice, *eNOS* and *BMPRII* expression was reduced in the females compared to male WT mice. The literature on eNOS expression in PAH patients is contradictory. eNOS was initially reported to be decreased in lungs of PAH patients [62], but evidence later found eNOS expression was unchanged or even increased in PAH patients [63,64]. Furthermore, NO levels have been shown to be reduced in PAH patients [65] and it has been reported that the activity of eNOS rather than its expression was altered in a murine model of PAH [63]. Our findings show that *BMPRII* expression is reduced in pulmonary arteries from female mice, which is in line with proof that activation of the estrogen response element in the promoter region of *BMPRII* can downregulate *BMPRII* expression [66].

In conclusion, this study has shown that Cx43 plays a role in NO-dependent vasodilation in the pulmonary vasculature. Cx43 is also involved in pulmonary vascular contractility. However, effects on contractility are gender-dependent and agonist-dependent. Hypoxia has been shown to decrease Cx43 expression in mouse pulmonary arteries, which is an effect that may contribute to the increased vasoreactivity observed under hypoxic conditions.

4. Materials and Methods

4.1. Ethical Statement

All experimental procedures were carried out in accordance with the United Kingdom Animal Procedures Act (1986) and with the "Guide for the Care and Use of Laboratory Animals" published by the US National Institutes of Health (NIH publication no.85, eighth edition). Ethical approval was granted by the Glasgow Caledonian University Animal Welfare and Ethics Committee (PPL70/7875, 16 September 2013).

4.2. Animals

Male and female wild-type (WT) and *Cx43* heterozygous (*Cx43*$^{+/-}$) mice (C57BL6, 5 to 9 months old) were used in this study. The generation of *Cx43*$^{+/-}$ mice was originally carried out by replacing exon 2 of the *Cx43* gene with the neomycin resistance gene, which was previously described [24].

Mice were grouped under standard laboratory conditions. All mice had access to a commercial diet and water ad libitum.

4.3. Genotyping

DNA was extracted from ear notch tissues derived from WT and $Cx43^{+/-}$ mice. Tissues were suspended in 300 µL TNES buffer [10 mM tris(hydroxymethyl)aminomethane (Tris), 0.4 M sodium chloride (NaCl), 100 mM ethylenediaminetetraacetic acid (EDTA), and 0.6% sodium dodecyl sulphate (SDS)] to which 1.5 µL proteinase K (Fisher Scientific, Loughborough, UK) was added. Samples were then incubated overnight at 55 °C. The next day, 84 µL of 5 M NaCl was added to each sample and samples were centrifuged for 10 min. The supernatant was collected and transferred to fresh tubes. DNA was precipitated by adding 200 µL ice cold 100% ethanol to each tube and vortexing. Samples were then centrifuged for 10 min, the supernatant was discarded, and the pellet was retained. Excess salt was removed by adding 200 µL ice cold 75% ethanol and samples were centrifuged again for 10 min. The supernatant was decanted gently and the pellet was allowed to air dry. The pellet was then re-suspended in 15 µL nuclease-free water and stored at −20 °C until use. A polymerase chain reaction (PCR) was then carried out to amplify the $Cx43$ (*GJA1*) and neomycin resistance (*neo^r*) genes. The primers used were as follows: *neo^r* forward 5'-GATCGGCCATTGAACAAGATG, melting temperature (Tm) = 56.4 °C, molecular weight (MW) = 6808.5, *neo^r* reverse: 5'-CCTGATGCTCTTCGTCCAGAT Tm = 57.2 °C, MW = 6637.3, *Cx43* forward: 5'-CAGTCTGCCTTTCGCTGT, Tm = 56 °C, MW = 5433; *Cx43* reverse: 5'-GTAGACCGCACTCAGGCT, Tm = 58 °C, MW = 5485. All primers were purchased from Integrated DNA technologies, Belgium. PCR reactions were performed in a thermal cycler (MJ Research PTC-100 Thermal Cycler, Watertown, MA, USA) and comprised an initial denaturation at 95 °C for 3 min followed by 40 cycles of: denaturation at 95 °C for 30 s, annealing at 55 °C for 30 s, and extension at 72 °C for 1 min. A final extension step was carried out at 72 °C for 15 min. Samples were then run in 2% agarose gel (v/v) for 45 min at 100 V.

4.4. Induction of Hypoxia

For induction of hypoxia, male WT and $Cx43^{+/-}$ mice were placed in a hypobaric chamber for 14 days, which is previously described [19]. The pressure was adjusted to 550 mbar (equivalent to 10% v/v O_2) slowly over two days to allow mice to acclimate. The temperature was maintained at 20–22 °C. Control littermates were kept in a normoxic environment.

4.5. Tissue Preparation

Mice were euthanized by injection of phenobarbitone (60 mg/kg i.p.). After death was confirmed, the chest walls were opened using the mediastinal approach and the hearts and lungs were dissected freely.

4.6. Wire Myography Studies

Pharmacological experiments were carried out in third generation intra-lobar pulmonary arteries (IPAs; ~300 µm internal diameter), which was previously described [67,68]. IPAs were mounted on a wire myograph (Danish Myo Technology, DMT) in freshly prepared Krebs-Henseleit Solution (composition (mmol/L) NaCl 119, KCl 4.7, $CaCl_2$ 2.5, $MgSO_4$ 1.2, $NaHCO_3$ 25, KH_2PO_4 1.2, and D-glucose 5.5) at (37 °C) and gassed with 95% O_2 / 5% CO_2. All chemicals required for Krebs-Henseleit solution were purchased from Fisher Scientific except $CaCl_2$ which was purchased from VWR International Ltd. (Lutterworth, Leicestershire, UK). Following equilibration for one hour, IPAs from normoxic mice were placed under pressures of 12–15 mmHg and IPAs from hypoxic mice that were placed under pressures of 30–35 mmHg to mimic the in vivo environment described previously [67]. Arteries were initially constricted with potassium chloride (KCl, 60 mM), which were then washed out. These processes were repeated two times before contractile or relaxation experiments were carried out. For contractile experiments, cumulative concentration response curves (CCRCs) to 5-HT (1 nM–300 µM) or ET-1 (0.1 nM–0.1 µM) were constructed. For relaxation experiments, vessels were pre-constricted with phenylephrine (PE; 3 µM) and CRCs to MCh (0.1 nM–30 µM) or isoprenaline (1 nM–30 µM) were constructed. In a subset of relaxation experiments, vessels were incubated with ^{37,43}Gap27 (100 µM) for 30 min prior to MCh or isoprenaline-induced relaxation responses. ^{37,43}Gap27 has previously been shown to have an IC_{50} of 31.5 ± 4.1 µM and to produce a maximum effect at 100 µM [69]. We previously used ^{37,43}Gap27 at 100 µM [70–72]. In another subset of relaxation experiments, the L-NAME (100 µM) was used to inhibit eNOS. In these experiments the L-NAME was applied for 30 min prior to pre-constriction with PE and the concentration of PE was reduced to 30 nM. It should be noted that 30 nM PE in the presence of L-NAME produced a similar contractile response to 3 µM PE alone. Changes in isometric tension were recorded on LabChart 7 software (AD Instruments Pty Ltd., Bella Vista, New South Wales, Australia). PE, L-NAME, MCh, ET and 5-HT were purchased from Sigma-Aldrich Company Ltd. (Gillingham, Dorset, UK).

4.7. Quantitative Real Time PCR (qPCR)

Main and branch pulmonary arteries (1st and 2nd order) were homogenized and RNA was extracted using the Nucleospin RNA kit (MACHEREY-NAGEL GmbH & Co. KG, Düren, Germany) as per the manufacturer instructions. RNA was then reverse transcribed to cDNA using the Precision nanoScript™ 2 Reverse Transcription Kit (Primerdesign Ltd., Chandler's Ford, UK), according to the manufacturer's protocols. qPCR was performed using Precision PLUS 2x qPCR master mix (Primerdesign Ltd.) II with taqman primers described in Table 4. β_2-microglobulin (B2M) (assay ID: HK-DD-mo-300, Primerdesign Ltd.) was used as the endogenous control and the sequences for the B2M were kept confidential by Primerdesign Ltd.

qPCR reactions were run in a real time PCR thermo-cycler machine (Viia™ 7 Real Time PCR System, ThermoFisher Scientific, Loughborough, UK) using the following conditions: 50 °C for 2 min and 95 °C for 10 min followed by 40 cycles of 95 °C for 15 s and 60 °C for 1 min. Gene expression was analyzed by the $2^{\Delta\Delta Ct}$ method. Samples from at least six mice were used for each group and reactions for each sample were run in triplicate.

Table 4. TaqMan probe primers and their sequences for qPCR reactions.

Gene	Sequence	Product Length (Base Pairs)	Tm (°C)
GJA1 (Cx43)	Sense-ACTGAGCCCATCCAAAGACT Antisense-CAGGAGGAGACATAGGTGAGAG	95	Sense-56.6 Antisense-57.3
GJA5 (Cx40)	Sense-ATGGTATACTCTCCTCAGCACTAC Antisense-CCAGTCATTGAGAAGACTCAGAAC	117	Sense-56.8 Antisense-57
GJA4 (Cx37)	Sense-ACACCCACCCTGATCTACCT Antisense-TCCCTCTTTCTGCCGCAAC	75	Sense-57.5 Antisense-58
GJC1 (Cx45)	Sense-CAGAGATGGAGTTAGAAAGCGAAA Antisense-AAGCCCACCTCAAACACAGT	148	Sense-57.8 Antisense-57.7
Panx1	Sense-TCAGCCTCATTAACCTCATTGTG Antisense-TGGGCAGGATTTCATACACTTTG	114	Sense-57.5 Antisense-58
NOS3 (eNOS)	Sense-GGAAGTAGCCAATGCAGTGAA Antisense-GCCAGTCTCAGAGCCATACA	97	Sense-56.8 Antisense-57.2
BMPR2 (BMPRII)	Sense-TGTTATCAGTGACTTTGGTTTATCC Antisense-CTTATAGCCCGCATTATCTTCTTCC	84	Sense-56.5 Antisense-56.3
Tph1	Sense-AATTCACGGAAGAAGAGATTAAGAC Antisense-CCAGTTGCGGGATGTTGTC	150	Sense-56.3 Antisense-56.9

4.8. Immunofluorescence Staining

Sagittal sections (7 µM) were cut from lung embedded at an optimal cutting temperature (OCT) compound using a cryostat (Cryostar™ NX70 Cryostat from, Thermofisher Scientific). Sections were

Int. J. Mol. Sci. **2018**, *19*, 1891

fixed in ice cold ($-20\,°C$) methanol for 20 min and rehydrated in phosphate buffer saline (PBS, pH = 7.4). Sections were permeablised in PBS containing 0.1% (v/v) Triton-X100 before being blocked in 5% (w/v) skimmed milk in PBS solution for 30 min at room temperature. Sections were stained with primary antibody (rabbit polyclonal anti-Cx43, 1:100; Sigma-Aldrich Company Ltd., Gillingham, Dorset, UK) and incubated at $4\,°C$ overnight. The next day, sections were washed in PBS for 30 min at room temperature and incubated with secondary antibody (goat anti-rabbit conjugated to Alexa flour 488, 1:500; Fisher Scientific) at $4\,°C$ for 2 h. Nuclei were counter-stained with DAPI (1:1000; Thermofisher. Loughborough, UK). Mounting medium was applied on tissue sections and cover slides were applied. Slides were then examined under the LSM 800 Carl ZEISS confocal microscope, (Königsallee, Germany) for immunoreactivity.

4.9. Statistical Analysis

All data were shown as mean $\pm$ S.E.M. Data for cumulative concentration response curves (CRCs) were analyzed using GraphPad Prism 6 software (La Jolla, CA, USA). Data were fitted to a logistic equation, CRCs were generated, and EC_{50} values were derived. Global differences in CRCs were compared by two-way ANOVA with the Bonferroni's post hoc test. Changes in the logarithm of median effective concentration (Log EC_{50}), maximal contractile responses (E_{max}), and maximal relaxation responses (R_{max}) between two different groups were analyzed by using the Student's *t*-test.

Author Contributions: Y.D., P.E.M., and J.E.N. conceived and designed the experiments. M.H. contributed to the design of experiments, performed the experiments, and analyzed the data. A.S. and B.E.F. contributed to myography experiments. M.H. and Y.D. wrote the manuscript with input from J.E.N. and P.E.M.

Acknowledgments: We would like to thank. Bjarne Due Larsen (Zealand Pharma) for supplying [37,43]Gap27 peptide as a kind gift. We would also like to thank Angus M Shaw for his advice on the myography experiments. This work was funded by grants from the British Pharmacological Society and Tenovus Scotland awarded to Yvonne Dempsie and Patricia Martin.

Conflicts of Interest: The authors declare no conflict of interest.

References

1. Hoeper, M.M.; Bogaard, H.J.; Condliffe, R.; Frantz, R.; Khanna, D.; Kurzyna, M.; Langleben, D.; Manes, A.; Satoh, T.; Torres, F. Definitions and Diagnosis of Pulmonary Hypertension. *J. Am. Coll. Cardiol.* **2013**, *62*, D42–D50. [CrossRef] [PubMed]
2. Benza, R.L.; Miller, D.P.; Barst, R.J.; Badesch, D.B.; Frost, A.E.; McGoon, M.D. An Evaluation of Long-Term Survival from Time of Diagnosis in Pulmonary Arterial Hypertension from the REVEAL Registry. *Chest* **2012**, *142*, 448–456. [CrossRef] [PubMed]
3. Dempsie, Y.; MacLean, M.R. The Influence of Gender on the Development of Pulmonary Arterial Hypertension. *Exp. Physiol.* **2013**, *98*, 1257–1261. [CrossRef] [PubMed]
4. Dempsie, Y.; Martin, P.; Upton, P.D. Connexin-Mediated Regulation of the Pulmonary Vasculature. *Biochem. Soc. Trans.* **2015**, *43*, 524–529. [CrossRef] [PubMed]
5. Eddahibi, S.; Guignabert, C.; Barlier-Mur, A.M.; Dewachter, L.; Fadel, E.; Dartevelle, P.; Humbert, M.; Simonneau, G.; Hanoun, N.; Saurini, F.; et al. Cross Talk between Endothelial and Smooth Muscle Cells in Pulmonary Hypertension: Critical Role for Serotonin-Induced Smooth Muscle Hyperplasia. *Circulation* **2006**, *113*, 1857–1864. [CrossRef] [PubMed]
6. MacLean, M.R.; Dempsie, Y. The serotonin hypothesis of pulmonary hypertension revisited. In *Membrane Receptors, Channels and Transporters in Pulmonary Circulation*; Humana Press: Totowa, NJ, USA, 2010; pp. 309–322.
7. Xu, W.; Kaneko, F.T.; Zheng, S.; Comhair, S.A.; Janocha, A.J.; Goggans, T.; Thunnissen, F.B.; Farver, C.; Hazen, S.L.; Jennings, C. Increased Arginase II and Decreased NO Synthesis in Endothelial Cells of Patients with Pulmonary Arterial Hypertension. *FASEB J.* **2004**, *18*, 1746–1748. [CrossRef] [PubMed]
8. Martin, P.E.; Evans, W.H. Incorporation of Connexins into Plasma Membranes and Gap Junctions. *Cardiovasc. Res.* **2004**, *62*, 378–387. [CrossRef] [PubMed]

9. Figueroa, X.F.; Duling, B.R. Gap Junctions in the Control of Vascular Function. Antioxid. *Redox Signal.* **2009**, *11*, 251–266. [CrossRef] [PubMed]
10. Brisset, A.C.; Isakson, B.E.; Kwak, B.R. Connexins in Vascular Physiology and Pathology. *Antioxid. Redox Signal.* **2009**, *11*, 267–282. [CrossRef] [PubMed]
11. Johnstone, S.; Isakson, B.; Locke, D. Biological and Biophysical Properties of Vascular Connexin Channels. *Int. Rev. Cell Mol. Biol.* **2009**, *278*, 69–118. [PubMed]
12. Sosinsky, G.E.; Boassa, D.; Dermietzel, R.; Duffy, H.S.; Laird, D.W.; MacVicar, B.; Naus, C.C.; Penuela, S.; Scemes, E.; Spray, D.C. Pannexin Channels are Not Gap Junction Hemichannels. *Channels* **2011**, *5*, 193–197. [CrossRef] [PubMed]
13. Lohman, A.W.; Billaud, M.; Straub, A.C.; Johnstone, S.R.; Best, A.K.; Lee, M.; Barr, K.; Penuela, S.; Laird, D.W.; Isakson, B.E. Expression of Pannexin Isoforms in the Systemic Murine Arterial Network. *J. Vasc. Res.* **2012**, *49*, 405–416. [CrossRef] [PubMed]
14. Billaud, M.; Chiu, Y.H.; Lohman, A.W.; Parpaite, T.; Butcher, J.T.; Mutchler, S.M.; DeLalio, L.J.; Artamonov, M.V.; Sandilos, J.K.; Best, A.K.; et al. A Molecular Signature in the pannexin1 Intracellular Loop Confers Channel Activation by the alpha1 Adrenoreceptor in Smooth Muscle Cells. *Sci. Signal.* **2015**, *8*, ra17. [CrossRef] [PubMed]
15. Tsang, H.; Leiper, J.; Lao, K.H.; Dowsett, L.; Delahaye, M.W.; Barnes, G.; Wharton, J.; Howard, L.; Iannone, L.; Lang, N.N. Role of Asymmetric Methylarginine and Connexin 43 in the Regulation of Pulmonary Endothelial Function. *Pulm. Circ.* **2013**, *3*, 675–691. [CrossRef] [PubMed]
16. Kim, J.; Hwangbo, C.; Hu, X.; Kang, Y.; Papangeli, I.; Mehrotra, D.; Park, H.; Ju, H.; McLean, D.L.; Comhair, S.A.; et al. Restoration of Impaired Endothelial Myocyte Enhancer Factor 2 Function Rescues Pulmonary Arterial Hypertension. *Circulation* **2015**, *131*, 190–199. [CrossRef] [PubMed]
17. Wang, L.; Yin, J.; Nickles, H.T.; Ranke, H.; Tabuchi, A.; Hoffmann, J.; Tabeling, C.; Barbosa-Sicard, E.; Chanson, M.; Kwak, B.R.; et al. Hypoxic Pulmonary Vasoconstriction Requires Connexin 40-Mediated Endothelial Signal Conduction. *J. Clin. Investig.* **2012**, *122*, 4218–4230. [CrossRef] [PubMed]
18. Billaud, M.; Dahan, D.; Marthan, R.; Savineau, J.; Guibert, C. Role of the Gap Junctions in the Contractile Response to Agonists in Pulmonary Artery from Two Rat Models of Pulmonary Hypertension. *Respir. Res.* **2011**, *12*, 30. [CrossRef] [PubMed]
19. Mair, K.; MacLean, M.; Morecroft, I.; Dempsie, Y.; Palmer, T. Novel Interactions between the 5-HT Transporter, 5-HT1B Receptors and Rho Kinase in Vivo and in Pulmonary Fibroblasts. *Br. J. Pharmacol.* **2008**, *155*, 606–616. [CrossRef] [PubMed]
20. Sullivan, C.C.; Du, L.; Chu, D.; Cho, A.J.; Kido, M.; Wolf, P.L.; Jamieson, S.W.; Thistlethwaite, P.A. Induction of Pulmonary Hypertension by an Angiopoietin 1/TIE2/serotonin Pathway. *Proc. Natl. Acad. Sci. USA* **2003**, *100*, 12331–12336. [CrossRef] [PubMed]
21. Morecroft, I.; Dempsie, Y.; Bader, M.; Walther, D.J.; Kotnik, K.; Loughlin, L.; Nilsen, M.; MacLean, M.R. Effect of Tryptophan Hydroxylase 1 Deficiency on the Development of Hypoxia-Induced Pulmonary Hypertension. *Hypertension* **2007**, *49*, 232–236. [CrossRef] [PubMed]
22. Gairhe, S.; Bauer, N.N.; Gebb, S.A.; McMurtry, I.F. Serotonin Passes through Myoendothelial Gap Junctions to Promote Pulmonary Arterial Smooth Muscle Cell Differentiation. *Am. J. Physiol. Lung Cell Mol. Physiol.* **2012**, *303*, L767–L777. [CrossRef] [PubMed]
23. Gairhe, S.; Bauer, N.N.; Gebb, S.A.; McMurtry, I.F. Myoendothelial Gap Junctional Signaling Induces Differentiation of Pulmonary Arterial Smooth Muscle Cells. *Am. J. Physiol. Lung Cell Mol. Physiol* **2011**, *301*, L527–L535. [CrossRef] [PubMed]
24. Reaume, A.G.; de Sousa, P.A.; Kulkarni, S.; Langille, B.L.; Zhu, D.; Davies, T.C.; Juneja, S.C.; Kidder, G.M.; Rossant, J. Cardiac Malformation in Neonatal Mice Lacking connexin43. *Science* **1995**, *267*, 1831–1834. [CrossRef] [PubMed]
25. Yang, Y.M.; Yuan, H.; Edwards, J.G.; Skayian, Y.; Ochani, K.; Miller, E.J.; Sehgal, P.B. Deletion of STAT5a/b in Vascular Smooth Muscle Abrogates the Male Bias in Hypoxic Pulmonary Hypertension in Mice: Implications in the Human Disease. *Mol. Med.* **2015**, *20*, 625–638. [PubMed]
26. Ma, L.; Chung, W.K. The Genetic Basis of Pulmonary Arterial Hypertension. *Hum. Genet.* **2014**, *133*, 471–479. [CrossRef] [PubMed]
27. Orriols, M.; Gomez-Puerto, M.C.; ten Dijke, P. BMP Type II Receptor as a Therapeutic Target in Pulmonary Arterial Hypertension. *Cell. Mol. Life Sci.* **2017**, *74*, 2979–2995. [CrossRef] [PubMed]

28. Figueroa, X.F.; Lillo, M.A.; Gaete, P.S.; Riquelme, M.A.; Sáez, J.C. Diffusion of Nitric Oxide Across Cell Membranes of the Vascular Wall Requires Specific Connexin-Based Channels. *Neuropharmacology* **2013**, *75*, 471–478. [CrossRef] [PubMed]

29. Looft-Wilson, R.; Billaud, M.; Johnstone, S.; Straub, A.; Isakson, B. Interaction between Nitric Oxide Signaling and Gap Junctions: Effects on Vascular Function. *Biochim. Biophys. Acta* **2012**, *1818*, 1895–1902. [CrossRef] [PubMed]

30. Straub, A.C.; Billaud, M.; Johnstone, S.R.; Best, A.K.; Yemen, S.; Dwyer, S.T.; Looft-Wilson, R.; Lysiak, J.J.; Gaston, B.; Palmer, L.; et al. Compartmentalized Connexin 43 S-nitrosylation/denitrosylation Regulates Heterocellular Communication in the Vessel Wall. *Arterioscler. Thromb. Vasc. Biol.* **2011**, *31*, 399–407. [CrossRef] [PubMed]

31. Hoffmann, A.; Gloe, T.; Pohl, U.; Zahler, S. Nitric Oxide Enhances De Novo Formation of Endothelial Gap Junctions. *Cardiovasc. Res.* **2003**, *60*, 421–430. [CrossRef] [PubMed]

32. Alonso, F.; Boittin, F.X.; Beny, J.L.; Haefliger, J.A. Loss of connexin40 is Associated with Decreased Endothelium-Dependent Relaxations and eNOS Levels in the Mouse Aorta. *Am. J. Physiol. Heart Circ. Physiol.* **2010**, *299*, H1365–H1373. [CrossRef] [PubMed]

33. Kameritsch, P.; Khandoga, N.; Nagel, W.; Hundhausen, C.; Lidington, D.; Pohl, U. Nitric Oxide Specifically Reduces the Permeability of Cx37-containing Gap Junctions to Small Molecules. *J. Cell. Physiol.* **2005**, *203*, 233–242. [CrossRef] [PubMed]

34. McKinnon, R.L.; Bolon, M.L.; Wang, H.X.; Swarbreck, S.; Kidder, G.M.; Simon, A.M.; Tyml, K. Reduction of Electrical Coupling between Microvascular Endothelial Cells by NO Depends on connexin37. *Am. J. Physiol. Heart Circ. Physiol.* **2009**, *297*, H93–H101. [CrossRef] [PubMed]

35. Leblais, V.; Delannoy, E.; Fresquet, F.; Bégueret, H.; Bellance, N.; Banquet, S.; Allieres, C.; Leroux, L.; Desgranges, C.; Gadeau, A. B-Adrenergic Relaxation in Pulmonary Arteries: Preservation of the Endothelial Nitric Oxide-Dependent β2 Component in Pulmonary Hypertension. *Cardiovasc. Res.* **2007**, *77*, 202–210. [CrossRef] [PubMed]

36. Mair, K.; Johansen, A.; Wright, A.; Wallace, E.; MacLean, M. Pulmonary Arterial Hypertension: Basis of Sex Differences in Incidence and Treatment Response. *Br. J. Pharmacol.* **2014**, *171*, 567–579. [CrossRef] [PubMed]

37. Polderman, K.H.; Stehouwer, C.D.; van Kamp, G.J.; Dekker, G.A.; Verheugt, F.W.; Gooren, L.J. Influence of Sex Hormones on Plasma Endothelin Levels. *Ann. Intern. Med.* **1993**, *118*, 429–432. [CrossRef] [PubMed]

38. Miyauchi, T.; Yanagisawa, M.; Iida, K.; Ajisaka, R.; Suzuki, N.; Fujino, M.; Goto, K.; Masaki, T.; Sugishita, Y. Age-and Sex-Related Variation of Plasma Endothelin-1 Concentration in Normal and Hypertensive Subjects. *Am. Heart J.* **1992**, *123*, 1092–1093. [CrossRef]

39. Ergul, A.; Shoemaker, K.; Puett, D.; Tackett, R.L. Gender Differences in the Expression of Endothelin Receptors in Human Saphenous Veins in Vitro. *J. Pharmacol. Exp. Ther.* **1998**, *285*, 511–517. [PubMed]

40. Tatchum-Talom, R.; Martel, C.; Labrie, C.; Labrie, F.; Marette, A. Gender Differences in Hemodynamic Responses to Endothelin-1. *J. Cardiovasc. Pharmacol.* **2000**, *36*, S102–S104. [CrossRef] [PubMed]

41. Juan, S.; Chen, J.; Chen, C.; Lin, H.; Cheng, C.; Liu, J.; Hsieh, M.; Chen, Y.; Chao, H.; Chen, T. 17β-Estradiol Inhibits Cyclic Strain-Induced Endothelin-1 Gene Expression within Vascular Endothelial Cells. *Am. J. Physiol.* **2004**, *287*, H1254–H1261. [CrossRef] [PubMed]

42. Axelsen, L.N.; Calloe, K.; Holstein-Rathlou, N.; Nielsen, M.S. Managing the Complexity of Communication: Regulation of Gap Junctions by Post-Translational Modification. *Front. Pharmacol.* **2013**, *4*, 130. [CrossRef] [PubMed]

43. Van Kempen, M.J.; Jongsma, H.J. Distribution of connexin37, connexin40 and connexin43 in the Aorta and Coronary Artery of several Mammals. *Histochem. Cell Biol.* **1999**, *112*, 479–486.

44. Hill, C.; Rummery, N.; Hickey, H.; Sandow, S.L. Heterogeneity in the Distribution of Vascular Gap Junctions and Connexins: Implications for Function. *Clin. Exp. Pharmacol. Physiol.* **2002**, *29*, 620–625. [CrossRef] [PubMed]

45. Yeh, H.I.; Rothery, S.; Dupont, E.; Coppen, S.R.; Severs, N.J. Individual Gap Junction Plaques Contain Multiple Connexins in Arterial Endothelium. *Circ. Res.* **1998**, *83*, 1248–1263. [CrossRef] [PubMed]

46. Chaytor, A.T.; Bakker, L.M.; Edwards, D.H.; Griffith, T.M. Connexin-mimetic Peptides Dissociate Electrotonic EDHF-type Signalling Via Myoendothelial and Smooth Muscle Gap Junctions in the Rabbit Iliac Artery. *Br. J. Pharmacol.* **2005**, *144*, 108–114. [CrossRef] [PubMed]

47. Stauffer, B.L.; Sobus, R.D.; Sucharov, C.C. Sex Differences in Cardiomyocyte connexin43 Expression. *J. Cardiovasc. Pharmacol.* **2011**, *58*, 32–39. [CrossRef] [PubMed]

48. Di, W.; Lachelin, G.C.; McGarrigle, H.; Thomas, N.; Becker, D. Oestriol and Oestradiol Increase Cell to Cell Communication and connexin43 Protein Expression in Human Myometrium. *Mol. Hum. Reprod.* **2001**, *7*, 671–679. [CrossRef] [PubMed]

49. Yu, W.; Dahl, G.; Werner, R. The connexin43 Gene is Responsive to Oestrogen. *Proc. Biol. Sci.* **1994**, *255*, 125–132. [CrossRef] [PubMed]

50. Matsushita, S.; Kurihara, H.; Watanabe, M.; Okada, T.; Sakai, T.; Amano, A. Alterations of Phosphorylation State of Connexin 43 during Hypoxia and Reoxygenation are Associated with Cardiac Function. *J. Histochem. Cytochem.* **2006**, *54*, 343–353. [CrossRef] [PubMed]

51. Faigle, M.; Seessle, J.; Zug, S.; El Kasmi, K.C.; Eltzschig, H.K. ATP Release from Vascular Endothelia Occurs Across Cx43 Hemichannels and is Attenuated during Hypoxia. *PLoS ONE* **2008**, *3*, e2801. [CrossRef] [PubMed]

52. Li, N.; Dai, D.; Dai, Y. CPU86017 and its Isomers Improve Hypoxic Pulmonary Hypertension by Attenuating Increased ETA Receptor Expression and Extracellular Matrix Accumulation. *Naunyn Schmiedebergs Arch. Pharmacol.* **2008**, *378*, 541. [CrossRef] [PubMed]

53. Yang, L.; Yin, N.; Hu, L.; Fan, H.; Yu, D.; Zhang, W.; Wang, S.; Feng, Y.; Fan, C.; Cao, F.; et al. Sildenefil Increases Connexin 40 in Smooth Muscle Cells through Activation of BMP Pathways in Pulmonary Arterial Hypertension. *Int. J. Clin. Exp. Pathol.* **2014**, *7*, 4674–4684. [PubMed]

54. Schmidt, V.J.; Jobs, A.; von Maltzahn, J.; Wörsdörfer, P.; Willecke, K.; de Wit, C. Connexin45 is Expressed in Vascular Smooth Muscle but its Function Remains Elusive. *PLoS ONE* **2012**, *7*, e42287. [CrossRef] [PubMed]

55. MacLean, M.R.; Deuchar, G.A.; Hicks, M.N.; Morecroft, I.; Shen, S.; Sheward, J.; Colston, J.; Loughlin, L.; Nilsen, M.; Dempsie, Y.; et al. Overexpression of the 5-Hydroxytryptamine Transporter Gene: Effect on Pulmonary Hemodynamics and Hypoxia-Induced Pulmonary Hypertension. *Circulation* **2004**, *109*, 2150–2155. [CrossRef] [PubMed]

56. Dempsie, Y.; Nilsen, M.; White, K.; Mair, K.M.; Loughlin, L.; Ambartsumian, N.; Rabinovitch, M.; MacLean, M.R. Development of Pulmonary Arterial Hypertension in Mice Over-Expressing S100A4/Mts1 is Specific to Females. *Respir. Res.* **2011**, *12*, 159. [CrossRef] [PubMed]

57. Dempsie, Y.; MacRitchie, N.A.; White, K.; Morecroft, I.; Wright, A.F.; Nilsen, M.; Loughlin, L.; Mair, K.M.; MacLean, M.R. Dexfenfluramine and the Oestrogen-Metabolizing Enzyme CYP1B1 in the Development of Pulmonary Arterial Hypertension. *Cardiovasc. Res.* **2013**, *99*, 24–34. [CrossRef] [PubMed]

58. Michela, P.; Velia, V.; Aldo, P.; Ada, P. Role of Connexin 43 in Cardiovascular Diseases. *Eur. J. Pharmacol.* **2015**, *768*, 71–76. [CrossRef] [PubMed]

59. Isakson, B.E.; Damon, D.N.; Day, K.H.; Liao, Y.; Duling, B.R. Connexin40 and connexin43 in Mouse Aortic Endothelium: Evidence for Coordinated Regulation. *Am J. Physiol.* **2006**, *290*, H1199–H1205. [CrossRef] [PubMed]

60. Kruger, O.; Beny, J.L.; Chabaud, F.; Traub, O.; Theis, M.; Brix, K.; Kirchhoff, S.; Willecke, K. Altered Dye Diffusion and Upregulation of connexin37 in Mouse Aortic Endothelium Deficient in connexin40. *J. Vasc. Res.* **2002**, *39*, 160–172. [CrossRef] [PubMed]

61. Simon, A.M.; McWhorter, A.R. Decreased Intercellular Dye-Transfer and Downregulation of Non-Ablated Connexins in Aortic Endothelium Deficient in connexin37 or connexin40. *J. Cell Sci.* **2003**, *116*, 2223–2236. [CrossRef] [PubMed]

62. Giaid, A.; Saleh, D. Reduced Expression of Endothelial Nitric Oxide Synthase in the Lungs of Patients with Pulmonary Hypertension. *N. Engl. J. Med.* **1995**, *333*, 214–221. [CrossRef] [PubMed]

63. Zhao, Y.Y.; Zhao, Y.D.; Mirza, M.K.; Huang, J.H.; Potula, H.H.; Vogel, S.M.; Brovkovych, V.; Yuan, J.X.; Wharton, J.; Malik, A.B. Persistent eNOS Activation Secondary to Caveolin-1 Deficiency Induces Pulmonary Hypertension in Mice and Humans through PKG Nitration. *J. Clin. Investig.* **2009**, *119*, 2009–2018. [CrossRef] [PubMed]

64. Mason, N.A.; Springall, D.R.; Burke, M.; Pollock, J.; Mikhail, G.; Yacoub, M.H.; Polak, J.M. High Expression of Endothelial Nitric Oxide Synthase in Plexiform Lesions of Pulmonary Hypertension. *J. Pathol.* **1998**, *185*, 313–318. [CrossRef]

65. Kaneko, F.T.; Arroliga, A.C.; Dweik, R.A.; Comhair, S.A.; Laskowski, D.; Oppedisano, R.; Thomassen, M.J.; Erzurum, S.C. Biochemical Reaction Products of Nitric Oxide as Quantitative Markers of Primary Pulmonary Hypertension. *Am. J. Respir. Crit. Care Med.* **1998**, *158*, 917–923. [CrossRef] [PubMed]

66. Austin, E.D.; Hamid, R.; Hemnes, A.R.; Loyd, J.E.; Blackwell, T.; Yu, C.; Phillips III, J.A.; Gaddipati, R.; Gladson, S.; Gu, E. BMPR2 Expression is Suppressed by Signaling through the Estrogen Receptor. *Biol. Sex Differ.* **2012**, *3*, 6. [CrossRef] [PubMed]

67. Keegan, A.; Morecroft, I.; Smillie, D.; Hicks, M.N.; MacLean, M.R. Contribution of the 5-HT(1B) Receptor to Hypoxia-Induced Pulmonary Hypertension: Converging Evidence using 5-HT(1B)-Receptor Knockout Mice and the 5-HT(1B/1D)-Receptor Antagonist GR127935. *Circ. Res.* **2001**, *89*, 1231–1239. [CrossRef] [PubMed]

68. Morecroft, I.; Pang, L.; Baranowska, M.; Nilsen, M.; Loughlin, L.; Dempsie, Y.; Millet, C.; MacLean, M.R. In Vivo Effects of a Combined 5-HT1B receptor/SERT Antagonist in Experimental Pulmonary Hypertension. *Cardiovasc. Res.* **2009**, *85*, 593–603. [CrossRef] [PubMed]

69. Chaytor, A.T.; Evans, W.H.; Griffith, T.; Thornbury, K. Peptides Homologous to Extracellular Loop Motifs of Connexin 43 Reversibly Abolish Rhythmic Contractile Activity in Rabbit Arteries. *J. Physiol.* **1997**, *503*, 99–110. [CrossRef] [PubMed]

70. Glass, B.J.; Hu, R.G.; Phillips, A.R.; Becker, D.L. The Action of Mimetic Peptides on Connexins Protects Fibroblasts from the Negative Effects of Ischemia Reperfusion. *Biol. Open* **2015**, *4*, 1473–1480. [CrossRef] [PubMed]

71. Faniku, C.; O'Shaughnessy, E.; Lorraine, C.; Johnstone, S.R.; Graham, A.; Greenhough, S.; Martin, P.E. The Connexin Mimetic Peptide Gap27 and Cx43-Knockdown Reveal Differential Roles for Connexin43 in Wound Closure Events in Skin Model Systems. *Int. J. Mol. Sci.* **2018**, *19*, 604. [CrossRef] [PubMed]

72. Wright, C.S.; Pollok, S.; Flint, D.J.; Brandner, J.M.; Martin, P.E. The Connexin Mimetic Peptide Gap27 Increases Human Dermal Fibroblast Migration in Hyperglycemic and Hyperinsulinemic Conditions in Vitro. *J. Cell. Physiol.* **2012**, *227*, 77–87. [CrossRef] [PubMed]

International Journal of
Molecular Sciences

MDPI

Review

Connexins and Pannexins: Important Players in Tumorigenesis, Metastasis and Potential Therapeutics

Sheila V. Graham [1,†], Jean X. Jiang [2,†] and Marc Mesnil [3,*]

[1] Centre for Virus Research, Institute of Infection Immunity and Inflammation, College of Medical Veterinary and Life Sciences, Room 254 Jarrett Building, Garscube Estate, University of Glasgow, Glasgow G61 1QH, UK; Sheila.Graham@gla.ac.uk

[2] Department of Biochemistry and Structural Biology, University of Texas, Health Science Center, 7703 Floyd Curl Drive, San Antonio, TX 78229, USA; jiangj@uthscsa.edu

[3] STIM Laboratory, ERL 7003CNRS/Building B36, University of Poitiers, 1 rue Georges BonnetTSA 51 106, 86073 Poitiers, CEDEX 09, France

*   Correspondence: marc.mesnil@univ-poitiers.fr; Tel.: +33-516-01-23-60

†   These authors contributed equally to this work.

Received: 28 April 2018; Accepted: 28 May 2018; Published: 1 June 2018

Abstract: Since their characterization more than five decades ago, gap junctions and their structural proteins—the connexins—have been associated with cancer cell growth. During that period, the accumulation of data and molecular knowledge about this association revealed an apparent contradictory relationship between them and cancer. It appeared that if gap junctions or connexins can down regulate cancer cell growth they can be also implied in the migration, invasion and metastatic dissemination of cancer cells. Interestingly, in all these situations, connexins seem to be involved through various mechanisms in which they can act either as gap-junctional intercellular communication mediators, modulators of signalling pathways through their interactome, or as hemichannels, which mediate autocrine/paracrine communication. This complex involvement of connexins in cancer progression is even more complicated by the fact that their hemichannel function may overlap with other gap junction-related proteins, the pannexins. Despite this complexity, the possible involvements of connexins and pannexins in cancer progression and the elucidation of the mechanisms they control may lead to use them as new targets to control cancer progression. In this review, the involvements of connexins and pannexins in these different topics (cancer cell growth, invasion/metastasis process, possible cancer therapeutic targets) are discussed.

Keywords: cancer; connexin; growth control; invasion; metastasis; pannexin; therapeutics

---

1. Introduction

The majority of cancers in adults are solid tumours [1]. Whatever their tissue origin, those tumours are characterized by two fundamental properties, which are, first, an uncontrolled cell proliferation forming the tumour itself and then an acquired invasion capacity leading to the dissemination of cancer cells in the organism. Fifty years of investigation have shown involvement of gap junctions (GJs) or their molecular components, the connexins (Cxs), in these two fundamental characteristics of cancer progression [2–4]. More recently, it appeared that the involvement of Cxs could be complicated by the fact that they can act independently from the establishment of gap-junctional intercellular communication (GJIC). For instance, Cxs may be involved in these mechanisms through their interactome to modulate signalling pathways [5] or by acting as hemichannels (Hcs) mediating autocrine/paracrine communication [6]. This last activity may overlap with pannexins (Panxs) which are Cx-related proteins (Figure 1) [7].

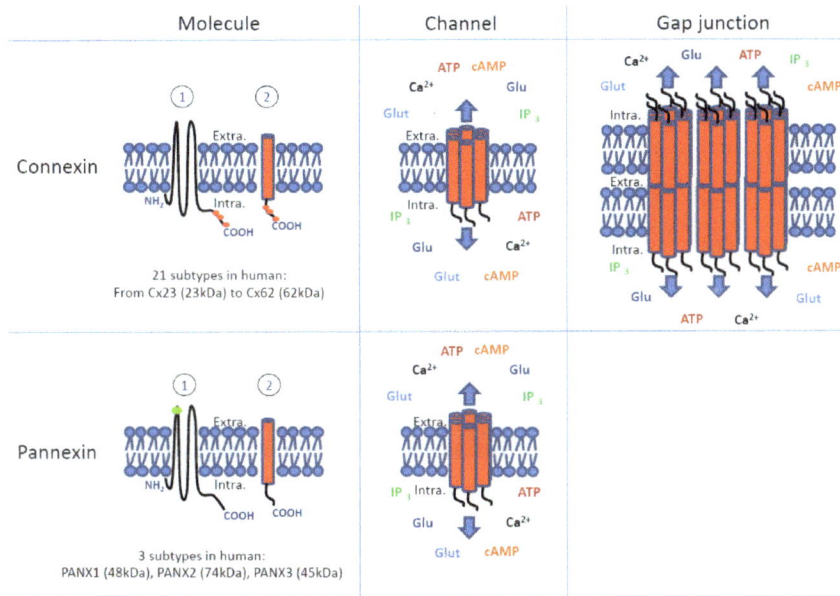

Figure 1. Connexin and pannexin molecules and channels formed by these molecules. As molecules, connexins (Cx) and pannexins (Panx) have similar topology with four transmembrane and intracellular (Intra.) NH_2 and COOH domains. In the left panels, both kinds of molecules are shown in a "spread" way to distinguish their topology (1) and in a "condensed" way (2) to better represent as transmembrane subunits of channels (centre panels) and gap junctions (right panel). In humans, 21 subtypes of connexins have been characterized, which are differentially expressed in tissues [8]. They are named according to their expected molecular weight (kDa) from the smallest connexin (Cx23: 23 kDa) to the largest one (Cx62: 62 kDa). The best-known member of the connexin family is the connexin43 (Cx43) which is the most common in the organism. Only 3 pannexin subtypes are known in human (PANX1, PANX2, PANX3) [9,10]. Except for Cx26, connexins can be phosphorylated mostly at their intracellular COOH tail (red spots) [11]. The level of phosphorylation potentially modifies channel gating, interaction with intracellular or other membrane proteins (connexin interactome) and thus their function and life cycle [11,12]. So far, pannexins do not appear to be regulated by phosphorylation as connexins are but they are more characterized as potentially N-glycosylated (green spots) molecules at their extracellular (Extra.) domain. Both connexins and pannexins can aggregate to form hexameric transmembrane channels permitting the passive passage of ions (e.g., Ca^{2+}) and small (<1–1.5 kDa) hydrophilic molecules such as nutrients (e.g., glucose: Glu), amino acids (e.g., glutamate: Glut), nucleotides (e.g., ATP) and second messengers (e.g., cAMP and IP_3). Theoretically, connexin-made channels (connexons also called hemichannels) and pannexin-made channels (pannexons) are permeable to the same type of ions and molecules even if pannexons permeability has been mostly studied for ATP, Ca^{2+} and glutamate (Glut). Moreover, connexons from one cell can dock with connexons of juxtaposed cells forming intercellular channels aggregated in gap junctions which permit the direct intercellular transfer from cytosol to cytosol (gap-junctional intercellular communication, GJIC) of same ions and molecules as isolated connexons. So far, no pannexon-made gap junctions have been described in physiological/pathological conditions. The term connexon is mostly used to define the transmembrane unit of gap junctions. When isolated in the plasma membrane, connexons are usually called hemichannels and can open with various stimuli such as, for example, hypoxia. For clarity in the figure, putative phosphorylation sites (red spots) and N-glycosylated sites (green spots) are not shown in channels and gap junctions.

Possible involvements of Cxs and Panxs in cancer progression and the elucidation of the mechanisms they control lead to their use as new possible targets to control cancer progression [13,14]. Here, we will review the involvement of Cxs and Panxs in these different topics, which are cancer cell proliferation, invasion/metastasis process and as possible targets for cancer control.

2. Connexins and Pannexins Involvement in Tumour Cell Growth

2.1. Connexins Involvement in Tumour Cell Growth

Shortly after their characterization, GJs were thought to be involved in growth regulation [15]. This assumption was the consequence of the possibility to estimate GJ functions through electrical coupling or diffusion of small hydrophilic fluorescent tracers [16,17]. By using such approaches, it rapidly appeared that cells derived from solid tumours (hepatoma, thyroid tumours, etc.) were not able to communicate through GJs [18,19]. These seminal studies introduced the notion that lack of GJ coupling could be a fundamental process in cancer leading to the formation of solid tumours by uncontrolled cell growth [15]. In other terms, growth regulation was the very first physiological role attributed to GJs and their mediation of a direct intercellular communication.

During the following decades, the involvement of GJIC in cancer cell growth regulation has been supported by a wide range of data. An early observation was about tumour promoter agents acting as inhibitors of GJIC [20,21]. This was observed in several models and reinforced the parallel between decreased GJIC and increased cell growth [22,23]. This parallel was extended to all kinds of phenomena able to inhibit GJIC such as cancer-causing viruses [24]. And such a phenomenon was so widely observed that it has been proposed that any GJIC inhibitor could be a potential tumour promoter [25]. If the tumour promoting effects of these chemicals were mostly known from in vitro studies, in some cases, GJIC inhibition effect could also be observed in vivo with transgenic mice exhibiting higher tumour susceptibility when defective for specific Cxs [26,27]. One of these best examples is liver since Cx32 gene knockout (KO) mice were shown to be more susceptible than wild-type mice to liver carcinogenesis after chemical treatments or even spontaneously [26]. This example was relevant to rodent and human situations for which liver tumours were correlated with lack of GJIC either by loss of expression or aberrant cytoplasmic localization of Cx32, respectively [28,29].

Conversely, strategies permitting the recovery of GJIC, by increasing Cx expression from non-communicating cancer cells, were expected to decrease cell growth. And indeed, globally this was the case as shown by approaches using chemical treatments or cDNA transfection. Chemicals known to be putative chemopreventing agents (flavonoids, carotenoids, retinoic acids, etc.) appeared to act on transformed cell lines by inducing GJIC and decreasing cell growth [30,31]. Cx cDNA transfection in GJIC-defective cancer cell lines brought similar conclusions that Cx expression is accompanied by decreased cell growth. This was observed in a variety of cancer cells (hepatoma, glioma, breast, etc.) in vitro and in vivo [32]. However, the type of transfected Cx was important since such an effect was mostly observed when the Cx of the normal tissue (before transformation or cancer progression) was re-expressed [33,34]. These results suggested that a recovery of GJIC is not sufficient by itself to have tumour suppressive effects but should be specifically controlled by the Cx subtype depending probably on the permeability capacity of the GJ it forms.

Thus, significant data accumulated over 50 years supported a similar conclusion. Whatever the models or the approaches (in situ detection of Cxs in tumours, cancer cell lines, chemical treatments, transgenic mice, cDNA transfection, etc.), the global conclusion is that Cx expression/GJIC is inversely correlated to cell growth. All these data have been analysed and synthesized in many reviews during past decades [2–4,32]. By considering all these observations, two kinds of molecular mechanisms can explain the involvement of Cxs in tumour cell growth regulation. The first one is to describe how Cxs, when present, can control cell growth. The second kind of molecular mechanisms, which are also needed for explaining the link between Cxs and cell growth, has to elucidate the origin of the lack of

Cx expression or function which is observed in tumour cells. These are the two kinds of mechanisms that will be reviewed below.

2.1.1. How Can the Presence of Connexins Regulate Cell Growth?

Most data attempting to elucidate how Cxs control cell growth came from Cx cDNA transfection in cancer cell lines. And from such approaches, whatever the cell types which were used (osteosarcoma, liver or lung carcinoma cells, etc.), a constant observation was that the increased expression of the original Cxs was followed by a longer G1 cell cycle phase slowing down the cell proliferation rate (Figure 2). A global analysis of these results suggests that this effect was the consequence of p27 accumulation [35,36]. From this common fact, diverse observations were made such as inhibition of enzymatic activity of Cyclin-dependent kinases (CDK) [36] and decreased amount of Cyclin D1 [37] and S-phase kinase-associated protein 2 (Skp2) [36,38,39]. To our knowledge, so far, no direct molecular link between Cx presence and the regulation of cell cycle has been demonstrated. Besides such an effect of Cxs on nuclear regulation of the cell cycle, it has been shown that Cxs can also act on the level of expression of growth factors. For instance, Cx43 re-expression but not Cx32, in C6 glioma cells is related to a decreased amount of milk fat globule-EGF factor 8 (MFG-E8) mRNA through an unknown mechanism (Figure 2) [40]. Therefore, Cx expression is mostly related to change of expression of growth factors or/and cell cycle regulators (p27, Cyclin D1, etc.). The most obvious scenarios for explaining how Cxs, when localized at the plasma membrane, can control gene expression might be through two major pathways. A first one would be through the Cx interactome by controlling growth transduction signalling and the second would be by diffusing growth regulators through GJs. Interestingly, as reported in the literature, both mechanisms have been observed and can explain the specificity of cell growth control induced by Cxs.

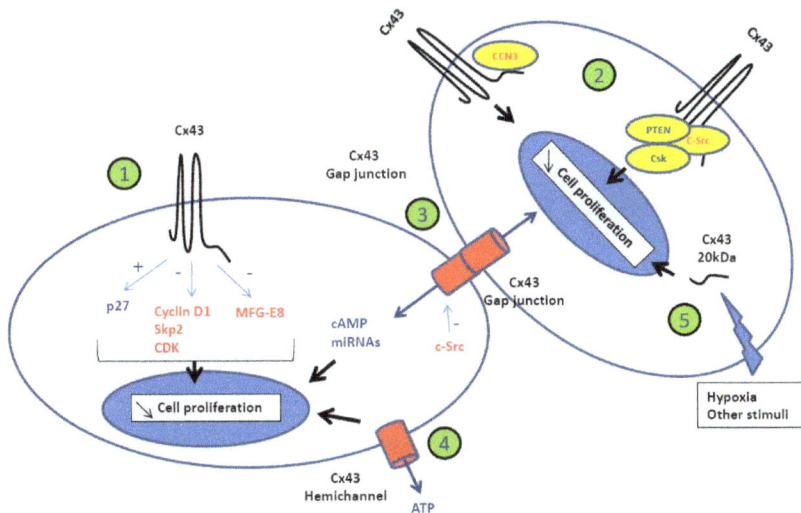

Figure 2. Connexin-mediated negative control of cell proliferation. Cx43 negatively regulates cell growth by acting differently on activators (red) and inhibitors (blue) of cell proliferation. This regulation is mediated through various mechanisms in which Cx43 acts by itself (1), as a sequestrator (2) of growth regulators (e.g., CCN3, PTEN, Csk, c-Src), as a mediator of GJIC (3), through hemichannel activity (4) or its 20 kDa carboxyl tail (CT)-domain (5). These various mechanisms act on the nucleus (thick black arrows) to decrease cell proliferation. Some of these mechanisms are mediated by hemichannel or gap-junction permeability (thick blue arrows). Positive (+)/negative (−) effects of Cx43 on cell cycle regulators (p27, Cyclin D1, etc.) and c-Src effect on Cx43 are also shown (thin blue arrows).

Gap-Junctional Intercellular Communication and Cell Growth Control

As mentioned above, the effect of Cx43 and Cx32 on cell growth has been extensively studied through various experimental models (cDNA transfection, transgenic mice, etc.) and appears to be specific. This specificity can be explained by their differential permeability which is illustrated by adenosine whose permeability is shifted from Cx32 to Cx43 channels by adding phosphate residues [41]. From such an observation, Cx channels appear as putative filters of intercellular signals that can be the consequence of the channel itself (diameter, amino-acid composition) or the configuration of the carboxyl tail (CT) which is sensitive to phosphorylation such as Cx43 channels closed by Src activation [42]. To our knowledge, a direct link between GJIC and growth regulation can be found in three situations. The first one is about the osteoblastic model in which extracellular growth stimulation induces the synthesis of second messengers that transit through GJs to activate extracellular signal-regulated kinase (ERK) and phosphatidylinositol-4,5-bisphosphate 3-kinase (PI3)/Akt serine/threonine kinase 1 (Akt) pathways. The translocation of ERK into the nucleus activates transcription factors that recognize a Cx-response element (CxRE) and induce osteocalcin and collagen I-1 expression [43]. Another example finally could explain the specific tumour suppressor effect of Cx26 on HeLa cells that was described two decades ago [33]. This effect seems to be the consequence of the maintenance of Cx26-mediated GJIC during the G2/M phase which permits intercellular cyclic 3',5'-adenosine monophosphate (cAMP) redistribution able to delay the cell cycle progression (Figure 2) [44]. And more recently, it was shown that not only metabolites like cAMP could act as growth regulators passing through GJIC but also microRNAs (miRNAs) (Figure 2). As an example, the transmission of anti-proliferative effects from miR-124-3p-transfected to non-transfected glioma cells was mediated by GJIC [45]. Similarly, GJIC was shown to inhibit cancer cell growth by transferring miRNAs from endothelial cells in vitro [46]. And interestingly, it was observed that miRNA transfer can occur also by delivering from exosomes in which Cx43 facilitates the release of content into target cells [47].

Cell Growth Control Independent from Gap-Junctional Intercellular Communication

The specific effect of Cxs in cell growth control can come also from their cytoplasmic domains (internal loop and CT domain) which are unique in length and amino-acid sequences [8]. It has been known for a long time that these parts and in particular the CT domain, can interact directly with cytosolic/membrane proteins. Such interactomes have been mostly described for Cx43 for which about 40 different proteins have been identified as interacting ones [12,48]. From such observations, it became clear that the interactome may participate both to cell growth regulation by controlling channel permeability (i.e., channel closure due to Src-induced tyrosine phosphorylation of the Cx43 CT domain) or by modulating signalling pathways from the plasma membrane to the nucleus. For this last case, it has been postulated that the CT domain of Cxs could control, through sequestration, the translocation of putative transcription factors from the cytosol to the nucleus (Figure 2). Such a behaviour has been described for Cx32 with Discs large homolog 1 (hDlg1) in hepatocytes [49] and for Cx43 with CCN3 in rat C6 glioma cells [50,51]. In this last case, down regulation of Cx43 permits the translocation of CCN3 to the nucleus which activates cell growth (Figure 2) [50]. Such a situation can explain why glioma cell growth is higher when Cx43 expression is repressed and vice-versa. A similar situation has been shown for the transcription factor ZO-1–associated nucleic acid–binding protein (ZONAB) [52]. More recently, the tumour suppressive effect of Cx43 expression could be explained by the region 266–283 in the CT domain of Cx43 which is able to recruit PTEN and C-Terminal src kinase (Csk) to inhibit the oncogenic activity of c-Src (Figure 2) [53]. It is also possible that such a phenomenon could still happen when Cx43 is localized in the cytoplasm. Even if it has not been described yet, it would explain the down regulation of growth which was observed in human glioblastoma cells after transfection of Cx43 which was mainly localized in the cytoplasm [54].

In this last example, Cx43 signal was also detected in the nucleus of the cells [54]. The anti-proliferative effect associated with a nuclear signal of Cx43 is more intriguing. This effect could be due to the Cx43 CT domain since the transfection of that part only was followed by decreased growth in several cell types (HeLa, Neuro2a and HEK293 cells) [55–57]. It has been suggested that the Cx43 CT domain would then act as a transcription factor but this hypothesis has not been proven yet (Figure 2). However, a 20 kDa isoform which corresponds to the Cx43 CT domain is known to be translated in some cell types under certain conditions activated in cancer cells and hypoxia [58]. Its function is not known yet even if it has been shown to act as a chaperone protein for trafficking of Cx43 to the cell membrane [59] and for microtubule dependent mitochondrial transport [60].

Finally, to be complete, Cxs are known to form Hcs in the plasma membrane (Figure 1) [61]. Study of those Hcs has been growing this last decade, especially for Cx43 but their link with cell proliferation is still not obvious even if adenosine triphosphate (ATP) release and modulation of Ca^{2+} concentrations were correlated with decreased cell proliferation in several cell types [62]. In osteocytes, they have been found to be involved in suppression of breast cancer cell growth and bone metastasis using transgenic mouse models expressing dominant-negative mutants inhibiting either GJIC and/or Hcs [63]. With recent development of new research tools, such as Cx-interacting peptides, antibodies and dominant-negative mutants, the distinctive mechanisms of GJs versus Hcs, although still limited start to be elucidated. However, the action of Cx Hcs can still be confounded with Panx channels (Figure 1).

2.1.2. What Does Prevent Connexin Expression or Function during Tumour Progression?

The expression of Cxs is often decreased in tumours whatever their origin [32]. Such a decreased expression may then participate to increase tumour growth by preventing the molecular mechanisms controlled by the presence of Cxs that were reviewed in the previous section. The molecular events leading to the disappearance of Cxs are not known precisely but could come from two mechanisms acting either at the transcriptional or at the post-transcriptional levels of Cx expression.

At the transcriptional level, similar to other genes which are shut down during tumour progression, Cx genes could be the target of epigenetic control. However, data about such a transcriptional control of Cx expression are not abundant in the cancer context even if it was suggested two decades ago [64]. In HeLa cells, silencing of the Cx43 gene was thought to be controlled by DNA methylation [65]. Loss of Cx32 function through hypermethylation is necessary for the development of renal cell carcinoma at the early carcinogenic process [66,67]. The CpG island hypermethylation level was associated with heavy smoking, poorly-differentiated tumour and low expression of Cx43 in non-small cell lung cancer [68]. More recently, hypermethylation of the Cx45 gene has been linked to its reduced expression in colon cancer [69]. This field of research is probably under investigated and would reveal if pursued that epigenetic phenomena are more involved than expected in the control of Cx expression.

At the post transcriptional level, Cx function can be regulated by ubiquitination, glycosylation, S-nitrosylation and in particular, phosphorylation of the CT domain. This has been mostly studied for Cx43 whose phosphorylation regulates GJIC through different mechanisms such as Cx trafficking, connexon assembly, channel gating and GJ degradation [11]. And indeed, in the cancer context, many oncogenes encode for kinases (i.e., c-Src) or proteins activating kinases (growth factor receptors) that are known to phosphorylate Cx43 and modulate its function [70]. As an example among others, epidermal growth factor (EGF) inhibits GJIC by inducing mitogen-activated protein kinase (MAPK)-mediated phosphorylation of Cx43 [71,72]. A similar effect has been observed for platelet-derived growth factor (PDGF) which activates MAPK and protein kinase C (PKC) pathways [73]. Interestingly, such a phosphorylation of the Cx43 CT domain establishes a direct link between growth stimulation and GJIC inhibition, which appears to be either the consequence of channel gating or Cx degradation [74].

Still at the post transcriptional level, an emerging field is about repression of Cx expression by miRNAs. For instance, mi-R-221/222 complex and miR-125b have been shown to downregulate Cx43 expression in glioma [75,76] or miR-20a in prostate cancer [77]. This field is still emerging and no doubt that it will be more involved in Cx gene regulation in future years.

Finally, the lack of expression or function of Cxs could be also theoretically the consequence of mutations affecting either the coding region of the Cx genes or their promoters. However, contrary to classical tumour suppressors (p53, Rb, etc.), such mutations have been rarely reported in the cancer context [32]. The most convincing result revealed a mutation affecting the Cx43 CT domain in human colon adenocarcinomas, which resulted in a restricted expression in invasive parts of the tumours [78]. To our knowledge, such an observation has not been confirmed. The fact that Cx mutations are not involved in human cancer is intriguing when considering their involvement in several human hereditary diseases [79]. So far, none of these diseases are known to be associated with a particular cancer susceptibility except for Cx26 mutations in the case of keratitis ichthyosis deafness (KID) syndrome which are associated with squamous cell carcinomas in 15% of patients [79]. The apparent general lack of association with cancer is probably the consequence of a lack of follow up of such patients.

2.2. Pannexins Involvement in Tumour Cell Growth

Originally, Panxs (3 members in mammals: Panxs1, 2 and 3) were identified as GJ proteins exhibiting homology with the invertebrate GJ proteins, the innexins [9]. Present in chordates, contrary to Cxs and despite a similar topology, they are not able to form functional GJs but form single membrane channels releasing autocrine and paracrine signals similar to Cx Hcs [10].

Data about a possible relationship between Panx expression and cancer progression or cancer cell growth are not so developed as they are for Cxs. In general, it seems that Panxs exhibit a so-called tumour suppressive effect similar to what is observed with Cxs. Such an analogy started during the last decade with the analysis of the brain cancer gene expression database REMBRANDT which revealed that the expression level of PANX2 and also PANX1 is positively correlated to post diagnosis survival of glioma patients [80]. To some extent, these observations were confirmed by the tumour suppressive effect induced both by Panx1 and Panx2 overexpression in rat C6 glioma cells in vitro and in vivo conditions [81,82]. In those cells, Panx1 expression had a wide range of anti-tumour activity by reducing in vitro cell proliferation, cell motility, anchorage-independent growth and tumour growth in nude mice. Interestingly, these effects, which are globally similar (except cell motility) with those observed after Cx43 transfection, were accompanied by an increased GJIC [81].

Similar observations have been obtained from skin where PANX1 and PANX3 levels are reduced both in human keratinocyte-derived basal cell carcinomas and squamous cell carcinomas [83]. This is in line with studies showing that those Panxs reduce growth of rat epidermal keratinocytes when overexpressed [84]. Such a growth inhibition was also observed for Panx3 in chondrocytes and osteoprogenitor cells by inhibiting the WNT pathway and via calcium-mediated regulation of p21 [85,86]. Recently, Panx3 was shown to inhibit the odontoblast proliferation through AMP-activated protein kinase (AMPK)/p21 signalling pathway and promote cell differentiation by bone morphogenetic protein (BMP)/Smad signalling pathway [87].

However, the situation is not so clear and probably depends on the cell type by considering melanocytes in which Panx1 expression is low whereas increased expression is correlated with melanoma aggressiveness [88]. More data are necessary before understanding the real involvement of Panxs in cancer cell growth control.

3. Connexins and Pannexins: Involvement in Tumour Metastasis and Microenvironment

3.1. The Process of Metastasis

In order to become metastatic, a clone of cancer cells must acquire aggressive growth properties and/or stem cell-like properties and the tumour microenvironment can drive acquisition of migratory and invasive properties through epithelial to mesenchymal transition (EMT). In the majority of tumours, which are epithelial in origin, cells must be able to breach the basement membrane, invade into the stroma and into blood vessels (intravasation) that infiltrate the tumour site. In the vasculature, they will adhere to blood vessel walls and be transported to distant sites where they emerge from the circulation (extravasation) to initiated new tumours. Finally, establishment of metastatic tumours requires survival and growth in the new tissue microenvironment. During all of these processes metastatic cells must evade the anti-tumoral immune response (Figure 3).

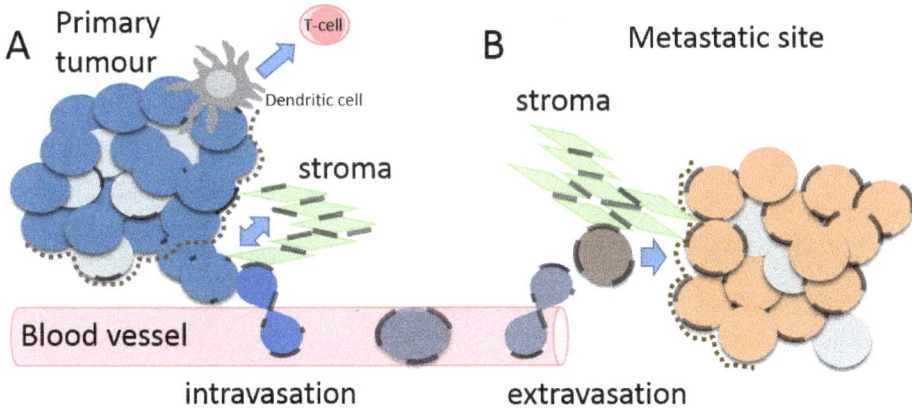

Figure 3. Gap-junctional intercellular communication in the tumour microenvironment and upon metastasis. (**A**) The tumour microenvironment consists of tumour cells (blue cells), non-tumour cells (light grey cells), immune cells including dendritic cells (dark grey cell) and CD8+ T-cells (pink cell), the basement membrane (brown dotted line) and the stroma (green cells). Tumour cells often display reduced gap junctions (transparent black lines) but can form heterotypic gap junctions with dendritic cells. Once they invade through to the stroma they can also form junctions with stromal cells. Upon intravasation into blood vessels tumour cells create gap junctions with endothelial cells lining the blood vessels; (**B**) Upon extravasation into a metastatic site, metastatic cells (orange cells) initiate gap-junctional intercellular communication with stromal cells and with other cells in the metastatic tumour microenvironment (light grey cells) and this may facilitate establishment of metastases. Depending on the site of metastasis, tumour cells may interact with cells of the immune system.

3.2. Connexin Involvement in Tumour Metastasis and Microenvironment

3.2.1. The Role of Connexins in Cancer Progression

Cxs can change expression levels, be re-localized [89–92] and/or exhibit altered phosphorylation upon progression to invasive tumour (Table 1) [93]. The resulting loss of functional GJs could alter tumour cell interaction with its microenvironment and promote EMT and migration from the primary tumour. Conversely, Cx expression can facilitate intravasation and adhesion to endothelial cells, enabling increased survival in the circulation. There is also evidence that Cx expression promotes exit from blood vessels into the metastatic site, where GJIC may be reinitiated [94] (Figure 3). However, Cxs may both promote tumour cell dormancy and cell survival, at metastatic sites. These effects may be reliant on tumour/stromal interactions and cooperation between invasive/metastatic cells and GJ formation in the tumour microenvironment and are likely to be Cx type, tumour type and cancer-stage-specific.

Int. J. Mol. Sci. **2018**, *19*, 1645

Table 1. Selected representative examples of changes in connexins during tumour progression and metastasis.

TISSUE	ORGANISM	CONNEXIN	REGULATION	REFERENCE
PRECANCERS AND PRIMARY TUMOURS				
PANCREATIC DUCTAL ADENOCARCINOMA	Mouse	Cx43	Increased levels Changes in phosphorylation	[89]
CERVICAL CANCER	Human	Cx26, Cx30, Cx43	Loss of connexin expression	[95–97]
BREAST CANCER	Human	Cx26, Cx43	Loss of Cx43 gap junctions	[98–101]
PROSTATE CANCER	Human	Cx32, Cx43	Decreased expression	[102]
COLON CANCER	Human	Cx32, Cx43	Gradual loss of expression	[92]
MELANOMA	Human	Cx26, Cx30	Increased expression	[103]
PRIMARY TUMOUR TO METASTASIS				
BREAST CANCER	Human	Cx26, Cx43		[101,104–108]
BRAIN	Human, rat	Cx30	Reduced expression	[109]
	Human	Cx43		[110,111]
PROSTATE	Human cell lines	Cx43	Increased Cx43 associated with increased invasion	[112]
LIVER	Rat cell lines	Cx43	Cx43 overexpression	[113]
	Human	Cx26	High expression	[114]
MELANOMA	Human Human cell lines	Cx26	Increased expression	[103,115] [116]

3.2.2. Invasion and the Local Microenvironment

E-cadherin is required for invasion in EMT and its loss is a marker of tumour progression. During invasion, cells display decreased GJIC, modification of cell-matrix interactions and acquisition of proteolytic properties to degrade the basal laminal proteins. Following this, altered stromal cells and microenvironment facilitate the motility of invasive cells through the extracellular matrix. All of these processes could be potentially altered by changes in Cx expression. For example, transfection of poorly coupled mouse epidermal cells with an E-cadherin expression construct increased GJIC [117]. Conversely, in prostate cancer cells, Cx43 levels correlated with levels of the transcription factor Snail-1 that inhibits expression of E-cadherin to promote EMT [118]. High levels of Cx43 and Snail-1 resulted in increased tumour cell invasion and Cx43 was downregulated upon Snail-1 silencing and vice versa. In keeping with these findings of a Cx43-Snail-1 axis controlling tumour cell behaviour, Cx43 expression could reverse A549 lung tumour cell resistance to the chemotherapeutic drug cisplatin by downregulating E-cadherin and EMT, while siRNA depletion of Cx43 initiated EMT [119]. Melanoma, breast, prostate and gastric cancers all display upregulated Cx43 and Cx26 in invasive lesions and metastases (Table 1) [101,103,104,112,115].

3.2.3. Promoting Metastasis: Connexins and Cell Motility

Early studies revealed that HeLa cervical cancer cells overexpressing Cx43 gained invasive properties in a chicken heart spheroid assay [120]. In a mouse melanoma model of metastasis following subcutaneous injection, clone F10 was less metastatic than the high Cx26-expressing clone BL6 but became as metastatic as BL6 upon Cx26 overexpression and BL6 cells expressing dominant negative Cx26 showed reduced metastatic potential [116]. γ-irradiation of C6 glioma cells induced Cx43 expression and increased ERK signalling and cell migration and a high Cx43 expressing clone displayed increased motility and invasion [121]. Conversely, knocking down Cx43 abrogated p38 MAPK activation and radiation-induced C6 cell migration [122]. Although GJIC was decreased upon Cx43 small interfering RNA (siRNA) depletion in the high Cx43 expressing C6 cells, GJ inhibitors did not alter motility indicating that Cx43 itself was responsible for the pro-metastatic effects [121]. Similarly, in a six-cell model of hepatocellular carcinoma, following injection into the tail vein of mice, only those lines with high metastatic potential formed foci in the lungs of the animals and this was reversible by depletion of Cx43 expression [113]. Another study found that blocking GJIC in GL15 glioblastoma cells increased motility in an in vitro 3D culture model [123]. However, blocking heterologous GJIC in ex vivo brain tissue by carbenoxolone reduced cell migration [123].

Cxs can facilitate adhesion of migrating cells to the endothelial layer of blood vessels and/or to specific distal sites (Figure 3). For instance, metastatic lung cancer cells could adhere to endothelial cells through GJs [124] as could metastasis-enabled melanoma cells ectopically expressing Cx26 in in vitro cultured vein segments [116]. In the case of colon cancer cells, conditioned medium from primary tumour cells enhanced phosphorylation of Cx43 and GJ formation between tumour and endothelial cells via the molecular chaperone heat shock protein 27 (HSP27), while metastatic colon cancer cells induced expression of Cx32 through action of the chemokine receptor CXCR2 [125]. Breast cancer cells that formed functional Cx43 GJs with endothelial cells facilitated migration out of the endothelial layer in in vitro culture [126] implicating Cx43 in the extravasation phase of metastasis. In zebrafish and chick embryo models, breast cancer and melanoma cell metastasis was dependent upon Cx43 and Cx26 to initiate brain metastatic lesions in association with the vasculature. Inhibition of Cx43-mediated GJIC inhibited extravasation, as did knock down of the EMT transcription factor twist [127].

3.2.4. Involvement of Gap Junctions and Hemichannels in Metastasis

Apparently contradictory effects of Cx43 in metastasis have been observed in different studies. When a functional null mutant Cx43 mouse line (G60S: that also has dominant negative effects on endogenous Cx43 activity) was crossed with erythroblastic leukemia viral oncogene homologue (ErbB) overexpressing mice [128], there was delayed onset and fewer and smaller primary breast tumours than in wild type mice but increased metastases to the lung [128]. In contrast, Cx43 overexpression in highly metastatic lung cancer cells reversed the metastatic tumour phenotype [129] but decreased Cx43 gene expression yielded breast cancer cells with increased metastatic potential [130,131]. In a two-cell model of prostate cancer, overexpressed Cx43 was present only in the cytoplasm and repressed proliferation, adhesion and invasion of normally invasive PC-3 cells. In contrast, overexpression of Cx43 in poorly metastatic LNCaP cells, re-established GJIC and increased bone metastasis in mice [132]. Stable overexpression of Cx43 in the MDA-MB-435 breast cancer cell line did not alter GJIC, invasion or migration in vitro. However, when injected into mice, the cells exhibited a reduced growth rate and fewer lung metastases [106]. This phenomenon was found to be GJIC-independent and it was suggested that it could be related to reduced N-cadherin expression, which would inhibit EMT. In another study, GJIC was restored in the same metastatic breast cancer cell line upon ectopic expression of the breast cancer metastasis suppressor gene BRMS1 [130]. The BRMS1-expressing cells showed increased levels of Cx43 but reduced Cx32, leading to loss of GJIC between breast cancer cells and between them and breast epithelial cells [130]. An in vivo murine study revealed that metastatic breast cancer cells in the bone formed more active GJs with osteoblasts than with themselves and BRMS1 expression increased homotypic GJIC. The breast cancer cells with increased heterotypic, relative to homotypic, GJ channels with osteoblasts were more metastatic than those that did not [105]. This suggests that the relative percentage of homo- and heterotypic GJ channels in tumour cells can influence metastasis. Moreover, it suggests that heterotypic GJs could be an important survival mechanism of tumour cells in the metastatic tumour microenvironment. It can be concluded that the precise timing of elevated or reduced Cx expression could be key to any effects during tumour progression.

Compared to GJs changes in Hc activity can produce different effects in metastasizing tumour cells. In a bone metastatic clone of MDA-MB-231 breast cancer cells, decreased Cx26 and Cx43 levels correlated with metastatic potential partly through alterations in Hc activity [107]. Similarly, a recent study reported suppression of breast cancer cell metastasis to the bone through osteocytic Cx43 Hcs [63]. Drug or mechanically-induced opening of Cx43 Hcs to release ATP from osteocytes led to inhibition of invasion and migration of the cancer cells. Analysis of a dominant negative Cx43 mutant that blocks GJs but not Hcs, revealed that Cx43 Hcs protected against tumour progression and metastasis [63]. The precise role of Cxs in tumour progression and metastasis might depend on the nature of the tumour, the properties of the cancer cell itself, the site of metastasis and the possibility of forming functional GJs at that site. It is clear that the tumour microenvironment drives cancer metastasis and Cx43 seems to stimulate growth of brain metastases after extravasation and tumour vasculature remodelling [133]. Protocadherin 7, a brain-specific cadherin, promoted Cx43-GJ assembly between breast and lung tumour cells and astrocytes. These GJs allowed cyclic guanosine monophosphate (cGAMP) to activate the stimulator of interferon genes (STING) pathway in astrocytes to induce an interferon response. The resulting changes in cell signalling could enhance growth of metastatic cells [133].

3.2.5. The Tumour Microenvironment

The tumour microenvironment, whether at the primary or secondary sites, is key to tumour cell survival and tumour progression [134]. In agreement with the hypothesis that Cxs control the microenvironment, Cx43-transfected glioma cells, which formed GJs with astrocytes in the striata of rats, were able to disseminate throughout the brain parenchyma. Cx43 itself, unlinked to GJIC, was shown to induce adhesive properties in the malignant glioma cells, which formed aggregates and were more invasive [135]. Also in rats, formation of GJIC with fibroblasts in co-culture stimulated prostate cancer cell migration [136,137]. However, Cx32 expression in metastatic renal cancer cells caused abrogation of invasive capacity via inactivation of c-Src signalling [138]. Tumour-associated immune cells are components of the tumour microenvironment. Heterotypic Cx43-GJs between tumour cells and dendritic cells can transmit melanoma antigenic peptides leading to activation of cytotoxic T-cells in vitro [139]. In vivo demonstration of Cx43-GJ transmission of antigenic peptides between antigen presenting cells has also been demonstrated [140]. GJ transmission of miRNAs between immune cells in the microenvironment and tumour cells is also expected to be a major regulator of metastasis because of the key role of many miRNAs in tumour suppression, while others can promote tumour progression [141].

3.3. Pannexins and Metastasis

The potential role of Panxs in metastasis is relatively unexplored. However, high levels of PANX1 mRNA were associated with metastatic spread in a two-cell model of hepatocellular carcinoma [142]. A key advance in understanding the role of Panxs in metastasis came from a study of the isogenic melanoma cell lines, F10 and BL6, mentioned previously. PANX1 levels were greatest in the most metastatic BL6 line [88]. PANX1 knock down reverted BL6 cells to a more normal melanocyte phenotype and these cells had reduced levels of vimentin and β-catenin, both markers of melanoma progression [88]. Importantly, in vivo data in a chick embryo xenograft model showed that reducing PANX1 expression reduced tumour growth and metastasis to the liver. A recent RNASeq analysis of breast cancer cells with different metastatic capacities revealed that cell lines with high metastatic potential had significantly enriched mutant mRNA encoding a N-terminal truncated PANX1 channel [143]. Truncated PANX-1, in association with wild type PANX1, seemed to confer a gain-of-function to channel activity and was found to promote metastatic cell survival. This appeared to be due to protection of tumour cells exiting the microvasculature via restrictive spaces between endothelial cells by enhancing ATP release from the Panx channels stimulated by mechanical deformation and abrogation of cell death [143]. In melanomas, P2X7/PANX1 channel activity has been linked to regulation of the NLRP3 inflammasome, which can result in release of pro-inflammatory, tumour promoting cytokines. Downstream effects on the tumour microenvironment could stimulate tumour growth and invasion. Of course, like Cxs, Panxs might also be found in future to repress tumour progression and metastasis.

4. Connexin and Pannexin Channels in Potential Cancer Therapeutics

4.1. Connexin Channels in Potential Cancer Therapeutics

The usefulness of Cxs and GJs as potential therapeutic targets for treating cancer has been studied for over four decades [4,144,145]. In recent years, several approaches have been developed in animal models to determine treatment modality by manipulating Cx channels. Although preclinical studies targeting connexin channels are still in their infancy, they hold great promise as de novo targets for cancer treatment.

4.1.1. Chemical Compounds in Modulating Connexins and Potential Cancer Therapy

Major attempts have focused on enhancement of GJIC function due to its impairment in primary cancer cells. Multiple chemical compounds have been used (e.g., retinoids, vitamin D, carotenoids, cAMP and lovastatin), which can fully or partially reverse the deficiency of GJIC in tumorigenic cells [146]. Lypopene, a carotenoid stimulates GJIC and Cx43 expression and inhibits the growth of the breast cancer MCF-7 cell line [147]. Extracts from the zooxanthellate jellyfish that show antioxidant activity exhibit higher levels of GJIC and cytotoxicity in MCF-7 cells than human epidermal keratinocytes [148].

An experimental approach was developed that killed tumorigenic cells based on GJIC selectively formed between them. In this study, tumorigenic BALB/c 3T3 and rat liver cells were loaded with Lucifer yellow (LY) and co-cultured with non-tumorigenic cells. By irradiation with blue light, only tumorigenic cells containing LY died but not the surrounding non-tumorigenic cells without LY [149]. This study further showed that when dibutyryl cAMP, retinoic acid, fluocinolone acetonide or dexamethasone were used during cell transformation, there was a reduction of transformed BALB/c 3T3 cell foci. These chemicals also increased and established GJIC between tumour cells and surrounding non-tumour cells, suggesting that the effects of chemicals on reversing the phenotypes of transformed cells rely on the establishment or enhancement of GJIC between tumour and normal cells.

Several cholesterol-lowering statin drugs (lovastatin, simvastatin, etc.) are suggested as anticancer reagents and high levels of mevalonate production are documented in various types of malignancies. Therefore, inhibition of the mevalonate producer, β-Hydroxy β-methylglutaryl-CoA (HMG-CoA) reductase, by statins offers a great potential for cancer treatment [150]. An earlier study shows that lovastatin increases GJIC in transformed E9 mouse lung carcinoma cells through the inhibition of PKC, although Cx43 expression and phosphorylation are not affected [151]. Moreover, apigenin, a flavonoid and lovastatin that is known to increase GJIC enhances bystander effect of the herpes simplex virus thymidine kinase/ganciclovir with reduction of cancer cell recovery on MCA38 adenocarcinoma cells, while neither chemical alone has such effect [152]. In vivo injection of both chemicals achieves 60–70% complete remission of tumour implanted in mice [152]. Simvastatin induced up-regulation of GJIC in Leydig tumour cells and this upregulation sensitized tumour cells to etoposide, a chemotherapeutic drug [153]. Simvastatin inhibited Cx43 phosphorylation by PKC and enhanced Cx43 membrane localization to promote formation of GJs (Ser368 phosphorylation promotes Cx43 internalization). However, a follow up study by the same group reported a protective function of simvastatin against toxicity by cisplatin on normal Sertoli cells [154]. This effect occurs at high cell density where GJIC forms and decreased GJIC by inhibitors or knocking down Cx43 by siRNA attenuates cell protective role of simvastatin. These two studies elucidate differential roles of GJIC by statins in chemotherapy by sensitizing drug effect on cancer cells and ameliorating toxicity in normal cells.

For Cx43 Hcs in cancer development, carbon monoxide (CO), a promising molecule to treat several diseases including cancer has been shown to inhibit their function [155]. CO donors inhibit Hc uptake in tumour cell lines (MCF-7 and HeLa cells) expressing exogenous Cx43 or Cx46 [156]. However, in general, scarce information is currently available describing the involvement of Cx Hcs in cancer cells.

Cxs can directly mediate the effect of chemotherapeutic drugs on cytotoxicity and apoptosis of cancer cells. Upregulation of Cx43 by cisplatin improves its resistance in a mesothelioma cell line (H28) [157]. GJIC inhibition fails to abrogate this effect but it is Cx43-dependent through the suppression of c-Src activation. Cx43 is increased in H28 cells by sunitinib treatment, which promotes apoptosis via the inhibition of receptor tyrosine kinase (RTK) signalling. This effect is likely to be mediated through direct interaction of Cx43 with an apoptotic related protein, Bax [158]. The Cx43 enhanced apoptotic effect of sunitinib was via enhancement of activation of Bax localized at the mitochondrial membrane and the phosphorylation of c-Jun N-terminal kinase (JNK) [159]. Several studies focus on the strategy of enhancing Cx expression in cancer cells. *Ganoderma lucidum*, an herbal mushroom known to inhibit tumour growth can increase Cx43 expression as well as vascular endothelial growth factor (VEGF) and inhibit growth of human ovarian cancer cells [160]. Such effect was abrogated by knocking down Cx43 expression. The bioactive substance sulforaphane inhibits cancer stem cells in aggressive pancreatic ductal adenocarcinoma through increased Cx43 and E-cadherin expression [161]. This treatment also inhibits the cancer stem cell markers c-Met and CD133, alters activation of several kinases and substrates, Glycogen synthase kinase 3 (GSK3), JNK and PKC and enhances GJ channels. Therefore, chemicals that can enhance GJs and Cx expression exhibit a high potency in suppressing cancer cell proliferation and tumour growth.

4.1.2. Connexin-Targeting Strategies in Potential Cancer Therapy

In recent years, several Cx mimetic peptides that reproduced portions of Cx sequences have been widely used in basic research as well as preclinical and therapeutic development [145]. Cx43-GJIC is decreased in breast cancer cells and efforts have been made to restore GJIC in these cells. αCT1, a mimetic peptide that targets CT domain of Cx43 can sustain and enhance GJIC function and has shown a great promise in promoting wound healing in skin by reducing scar formation [162]. A recent study shows that this peptide enhances Cx43 GJIC and reduces proliferation or survival of MCF7 and MDA-MB231 breast cancer cells but has no effect on MCF10A non-transformed cells [163]. A combination of αCT1 with tamoxifen or lapatinib augmented their effects on oestrogen receptor-positive MCF7 or Her2-positive BT474 breast cancer cells. Furthermore, treatment with αCT1 peptide sensitized human O-6-methylguanine-DNA methyltransferase (MGMT)-deficient and chemotherapeutic agent temozolomide (TMZ)-resistant glioblastoma (GBM) cells and combined treatment with the peptide and TMZ further incur autophagy and apoptosis of TMZ-resistance GMB cells [164]. A recent study shows that a cell-penetrating Cx mimetic peptide, TAT-Cx43(266-283) inhibits c-Src and focal adhesion kinase (FAK), upregulates phosphatase and tensin homology and reduces the growth, migration and survival of glioma stem cells (GSCs) from patients [165]. A Cx43 mimetic peptide juxtamembrane 2 (JM2) that is based on the Cx43 microtubule-binding domain inhibits Cx43 trafficking to the cell surface by promoting microtubule polymerization and reduces Hc numbers in the membrane for proinflammatory function. The authors imply that this peptide may have therapeutic value in treating proliferative diseases and cancer [166]. However, it is important to note that the recovery of GJIC does not consistently entail normalization of the tumour cells.

There are several reports concerning use of antibodies against Cxs. When a labelled monoclonal antibody against the second Cx43 extracellular loop domain was intravenously injected into rats with intracranial C6 glioma, antibody signals were detected in reactive, glial fibrillar acidic protein (GFAP)-positive astrocytes [167]. PEGylated immunoliposomes carrying monoclonal antibodies against GFAP and the above-described Cx43 monoclonal antibody were detected at the periphery of the glioma using either fluorescent or a paramagnetic probe [168]. These studies imply that these antibodies could potentially be used for targeted delivery of drugs to the zone of high-grade gliomas. Furthermore, magnetic resonance imaging data show that weekly administration of this Cx43 antibody at a dose of 5 mg/kg significantly reduces low-differentiated glioma volume and increases lifespan with a full recovery without delayed relapses in 19% animals [169]. Both Cx43 and brain-specific anion transporter (BSAT1) are preferably expressed in the brain tumour and peritumoral areas. Cisplatin-loaded nanogel conjugated with monoclonal Cx43 antibody [170] and BSAT1 was used to treat rats bearing tumours and the median survival was greater than control groups [171]. Vector nanogels seemed to reduce systemic toxicity of cisplatin [170]. Intriguingly, a combination of this Cx43 antibody with TMZ completely abolishes the antitumor effect of this antibody while combination treatment with γ-irradiation greatly inhibits tumour development and prolongs survival median to 60 days versus 38 days [172]. Recently, a magnetic resonance imaging (MRI) study further shows that uptake of Gd-based contrast agent with the same monoclonal Cx43 antibody is more than 4 times higher than nonspecific IgG-contrast agent and this Cx43 antibody conjugated agent markedly enhances visualization of glioma in vivo [173]. Although the specific molecular mechanism of this antibody is unknown, this Cx43-targeting monoclonal antibody could be developed as a potential drug and/or diagnostic agent for glioma therapies.

Finally, recombinant lentiviruses carrying siRNA were used to knockdown Cx37 expression in subcutaneous gastric tumours in mice [174]. Reduced levels of Cx37 are associated with higher apoptotic index of tumour cells in vivo. Cx46 is also detected in GBM cancer stem cells, while Cx43 is predominantly expressed in non-stem cells [175]. Besides Cx43, Cx46 is shown to express in GBM cancer stem cells (CSCs) that forms GJIC, while Cx43 is present in non-CSCs. During cancer differentiation, Cx46 is reduced associated with an increase of Cx43 and knocking down Cx46 by short hairpin RNA (shRNA) reduces stem cell maintenance.

Drug resistance is a major challenge for cancer treatment. Cisplatin is a commonly used chemotherapeutic agent for advanced non-small cell lung cancer but prolonged treatment leads to resistance due to development of EMT [119]. Overexpression of Cx43 reverses EMT and cisplatin resistance while Cx43 deletion initiates EMT and drug resistance in human lung cancer cell line A549. Patients with GBM, an aggressive adult primary brain tumour with poor prognosis, develop resistance to TMZ chemotherapy. In contrast to the situation in lung cancer, Cx43 is increased with the formation of GJIC in the resistant tumour cells and this increase is induced by epidermal growth factor receptor (EGFR) activated JNK-ERK1/2-AP-1 signalling [176]. Moreover, Cx43 expression in human glioma cells enhances resistance to TMZ via a mitochondrial apoptosis pathway by the reduction in Bax/Bcl-2 ratio and the release of cytochrome C [177]. Consistently, a recent study [178] showed that TMZ-resistant subline of U251 human GBM cells exhibited elevated Cx43 level compared to parental U251 cells, which was companied with increased EMT markers including vimentin, N-cadherin and β-catenin and decreased cell migration, monocyte adhesion and levels of vascular cell adhesion molecule (VCAM)-1. These studies suggest that depending upon cancer types, Cx43 expression and GJIC could be involved in either promoting or inhibiting sensitization of resistant cells to the chemotherapy. However, the underlying mechanisms remain elusive.

Recently, a new paradigm was proposed based on the data obtained in chronic inflammatory disorders and trauma in the eye that protecting cancer vasculature leads to reduced tumour hypoxia and promote survival of normal cells [179]. Given that Cx43 Hcs are involved in vascular leakage and endothelial cell death [180], modulation of these channels may provide an alternative for cancer treatment. Together, with advanced understanding of the mechanism of Cx channels in various types and stages of cancer development and metastasis, new lines of drugs that target them in cancer therapy are moving closer to reality.

4.2. Pannexin Channels in Potential Cancer Therapeutics

A great progress has been made in recent years for our understanding of Panx biology and physiology. However, compared to that of Cxs, the potential therapeutic application of Panxs in cancer is still limited.

4.2.1. Pannexin Channel Activation and Potential Cancer Therapy

Panx channels mediate ATP release and anti-tumour immune responses are associated with such a release from apoptotic cancer cells to engage P2 purinergic receptor signalling in leukocytes. A study shows that apoptotic reagents activate Panx1 channels via caspase-3 cleavage, which leads to ATP release in Jurkat T cell acute lymphocytes in chemotherapeutic drug-induced apoptosis [181,182]. Panx1 level is much higher in leukemic T lymphocytes than untransformed T lymphoblasts. Interestingly, chemotherapeutic drugs also cause ATP release with inhibition of caspase activation, which implies a Panx-independent mechanism. This study suggests that Panx1 channels and ATP release may mediate paracrine interaction between dying tumour cells and leukocytes in anti-tumour responses. A follow up study by the same group shows that activation of Panx1 channels by ATP is determined by expression level of particular ectonnucleotidases in tumour cell variants in Jurkat cell lines with and without the Fas-associated death domain (FADD) or receptor-interacting protein kinase 1 (RIP1) cell death regulatory proteins [183]. They noticed that robust levels of extracellular ATP/AMP were accumulated in apoptosis-deficient cells, not in apoptotic cells with the activation of Panx1 channels in response to chemotherapeutic drugs. Panx1 channel assists in accumulating immune-stimulatory ATP versus immunosuppressive adenosine within the tumour microenvironment. In support of the role of ATP and Panx1 channels in mediating immune response, a very recent study shows that ATP increases migration of dendritic cells through the activation of Panx1 channel and P2X7 receptor (P2X7R) [183]. In this study, they show that ATP actives P2X7R, which leads to opening of Panx1 channels and consequently results in more ATP release, re-organization of the actin cytoskeleton and faster migration of dendritic cells. Additionally, in vivo data show that Panx1 channels are required for the homing of dendritic cells to lymph nodes but not for maturation. Therefore, given that ATP acts as danger signal that recruits phagocytes including dendritic cells to cancer sites, activation of Panx channels through therapeutic drugs could hinder tumour growth and metastasis. Moreover, an US Food and Drug Administration (FDA)-approved anti-parasitic drug, Ivermectin allosterically regulates P2X4 receptors in breast cancer cells through opening of the P2X4/P2X7-gated Panx1 channels, which is associated with ATP release and consequently, cancer cell death [184]. Additionally, Ivermectin induces activation of autophagy and enrichment of inflammation mediators, ATP and high-mobility-group B (HMGB), suggesting that modulation of purinergic receptor signalling could be used as a platform for cancer immunotherapy [185].

4.2.2. Pannexin in Potential Cancer Diagnosis

A clinical report shows high relative expression of Panx3 in a patient with primary cutaneous sweat gland carcinomas with histologic features of a high-grade osteosarcoma [186]. By using quantitative trait loci (QTL) analysis, sequence comparison between strains and gene network analysis, this report links both body mass index (BMI) and tumorigenesis with Panx3 as a candidate gene in a genetically heterogeneous mouse model with carcinogen-induced cancer. A mutation encoding a truncated Panx1 (1–89) was identified which was enriched in highly metastatic breast cancer cells [143]. This truncated form of Panx1 further enhanced ATP release. In contrast to general belief of Panx channels in promoting cancer cell death, this paper suggests that ATP release by Panx1 suppresses deformation-induced apoptosis through P2Y receptor signalling and inhibition of Panx1 channels could reduce the efficiency of breast cancer metastasis. This could be partially explained by excess release of ATP by mutated Panx1 channels. Panx1 is present in skin melanocytes and is upregulated during melanoma tumour progression and tumorigenesis [88]. Knockdown of Panx1 in tumour cells decreases tumour cell growth, which indicates Panx1 as a potential target for treating melanoma. More studies are required to assess the expression levels of Panx subtypes in various types and stages of cancer.

4.2.3. Pannexin Channels in Pain Management Related to Cancer Treatment

Repeated treatment with the chemotherapeutic drug oxaliplatin is limited due to the development of a neuropathic pain in cancer patients. Functional recruitment of Panx1 mediates the increase of P2X7Rs in cerebrocortical nerve terminal in oxaliplatin-treated rats. Moreover, P2X7R antagonists and Panx1 inhibitors, Erioglaucine and [10]Panx peptide reverts neuropathic pain caused by oxaliplatin, while Panx1 inhibitors do not interfere the cytotoxic effect of oxaliplatin on human colon cancer cells HT-29 [187]. Consistently, a recent study shows that Panx1 expressed in immune cells plays a critical role for pain-like effects after nerve injury and this response is abrogated in Panx1 gene deficient mice [188]. These studies suggest that therapeutic modulation of Panx1 could be useful for treating neuropathic pain associated with cancer and cancer treatment.

5. Discussion and Conclusions

The involvement in cancer of GJs and their structural proteins, the Cxs, is a long story [4]. It rose just after the discovery of these particular intercellular junctions, which appeared to be absent in cancer cells. These very first observations suggested that the lack of GJIC could contribute to the lack of cell growth control which characterizes tumorigenesis [18]. Therefore, cell growth control was assumed to be one of the fundamental roles played by GJs. However, if this implication was assumed fifty years ago, the precise molecular mechanisms controlling cell growth came very late and are still unclear. There is a kind of paradox between the amount of observations accumulated for decades confirming a possible role of GJs as guardians of cellular homeostasis and replication and the lack of sufficient evidence explaining such a phenomenon.

Indeed, despite few exceptions, all kinds of observations were suggesting that the lack of GJs or Cxs is correlated to the lack of cell growth control and vice versa This consensus was supported by observations collected from a tremendous variety of models (cancer cell lines, primary cultures of tumour cells, in situ from biopsies, Cx-cDNA transfected cells, Cx-KO mice, chemical treatments decreasing or upregulating GJIC, etc.) whatever the species origins [32]. However, despite this consensus of observations, the assumption that GJs and Cxs were so-called tumour suppressors was not fully supported by several facts. First, GJs and Cxs did not behave as classical tumour suppressors since Cx gene mutations never appeared in tumours as commonly shown as for p53, Rb and so forth [32]. Second, no clear molecular mechanisms underlying the growth control that GJs and Cxs could exert has been established contrary to what was observed with classical tumour suppressors. These two aspects probably restricted GJs and Cxs to be considered as a real hallmark of cancer despite all the consensus studies we mentioned above [189]. The few molecular mechanisms that could explain the cell growth control exerted by GJs and Cxs seems to be "diffuse" and not so straightforward as growth signalling pathways described for oncogenes and tumour suppressors. Indeed, the involvement of Cxs in cell growth control is not clear at the molecular level and appears to be either GJIC-dependent or not. When this involvement was found to depend on GJIC, Cxs permit the intercellular diffusion of metabolites acting on cell growth control (i.e., Cx26 and the diffusion of cAMP, all along the cell cycle phases in HeLa cells) [44]. When this growth control is GJIC-independent, Cxs seem to act through their CT domain as a sequestrator preventing the nuclear translocation of cell growth regulators [53]. By comparing to our knowledge about cell growth control, Cxs seems to be a "helper" instead of a master regulator of cell growth control. Hopefully, future studies will bring more clear-cut information about the real involvement of GJs and Cxs in cancer cell growth [190].

In addition, we have also to consider that exceptions were observed in the consensus supporting the parallel between GJIC and cell growth control. These exceptions, supported by experimental observations, led to the hypothesis that Cxs could be protumoral actors when expressed at late stage of cancer progression. Indeed, from about twenty years ago, it appeared that Cxs could favour migration and invasion of cancer cells and participate to their dissemination [2,94]. A new wave of data then confirmed this new assumption that Cxs are actively involved in the late stages of carcinogenesis and participate to the aggressiveness of solid tumours. Very interestingly, from this more recent domain of investigation, Cxs were shown to play a role not only on migration and invasion of cancer cells but also on metastasis development by acting on intravasation, extravasation and dormancy of the metastatic cells. Within a few years, the Cx cancer statute has been changed then from tumour suppressor to tumour enhancer. Contrary to what appears at a first glance, this is not contradictory since an inverted correlation is often observed between cell proliferation and invasion capacity [191].

Int. J. Mol. Sci. **2018**, *19*, 1645

As for cell growth control, the molecular mechanisms underlying the involvement of Cxs in cell migration are not very clear. Once again, Cxs seem to control cell migration either through channel-dependent or –independent mechanisms. In the first case, the establishment of heterologous GJIC between cancer cells and cells of the tumour microenvironment may increase motility (such as glioma cells communicating with astrocytes) and further, helps to intravasation and extravasation [124–127,192]. When isolated in extracellular matrix, Cxs act on motility through GJIC-independent mechanisms by its CT domain. This has been particularly studied for Cx43 and even if the precise molecular events are not elucidated yet, it seems that the CT domain is involved by interacting with the actin cytoskeleton and helps to manage directional migration of the cancer cells [193]. Interestingly, such a phenomenon would not be pathological by itself since this process is present in migrations occurring in normal situations such as embryogenesis (neuron precursors migrating to the cortex) and leucocyte migration [194–196]. Other data also suggest that Cxs could be involved in formation of invadopodia and secretion of proteases during invasion process and also in metastasis targeting [132]. The molecular processes of all these phenomena are far from being to be elucidated. More data are needed to explain at the molecular level how Cxs can control cancer cell invasion and metastasis. These data are necessary for targeting Cxs to prevent eventually cancer invasion. This is of fundamental importance when considering that the majority of cancer deaths are the consequence of metastasis [1].

To conclude, there are globally sufficient data showing that Cxs are involved in carcinogenesis, especially in the progression of solid tumours. However, despite these data, the molecular mechanisms of the Cx involvement in carcinogenesis are not sufficiently elucidated yet. This lack of knowledge limits to use them as general therapeutical targets for cancer control. Moreover, the multifunctional sides of Cxs able to act as mediator of GJIC, through their interactome or even as Hcs make difficult to define their real implication in cancer. In addition, the similarity of Panxs with Cx Hcs adds another complexity to this area of research since this family of proteins seems to share functions with Cxs both in cell proliferation control and invasion. Facing this complexity, the only way to decipher the real impact of Cxs or Panxs in the cancer cell behaviour is to consider their involvement specifically in particular types of tumours but not globally [190]. One strategy could be by increasing in situ observations in order to localize precisely Cx/Panx expressions in the complex heterogeneity of specific human tumours and reveal the possible links of Cx/Panx localizations with the tumour behaviour. In particular, it could prove definitively the apparently opposed roles of Cxs in cell growth control and in cancer cell invasion through their differential expression either in the core of the tumour or in its invasive edges [111]. Therefore, due to uniqueness of the action of subtypes of Cxs and Cx channels on various types and stages of cancers, therapeutic approaches ought to be developed based on precise mechanism elucidated with more targeting approaches. This aligns with the current trend of drug development in treating cancer with precision medicine.

Acknowledgments: J.X.J. was supported by US National Institutes of Health grant CA196214, US Department of Defence (DoD) grant BC161273 and Welch Foundation grant AQ-1507; M.M. was supported by Ligue contre le Cancer (Comités de la Charente, Charente-Maritime, Deux-Sèvres, Morbihan et Vienne); S.V.G. was supported by Worldwide Cancer Research grant 08-0159 and acknowledges funding from the Medical Research Council as core funding for the MRC University of Glasgow Centre for Virus Research.

Conflicts of Interest: The authors declare no conflict of interest.

References

1. Forman, D.; Ferlay, J. The global and regional burden of cancer. In *World Cancer Report*; Stewart, B.W., Wild, C.P., Eds.; International Agency for Research on Cancer: Lyon, France, 2014; pp. 16–53, ISBN 978-92-832-0429-9.
2. Cronier, L.; Crespin, S.; Strale, P.O.; Defamie, N.; Mesnil, M. Gap junctions and cancer: New functions for an old story. *Antioxid. Redox Signal.* **2009**, *11*, 323–338. [CrossRef] [PubMed]

3. Naus, C.C.; Laird, D.W. Implications and challenges of connexin connections to cancer. *Nat. Rev. Cancer* **2010**, *10*, 435–441. [CrossRef] [PubMed]

4. Aasen, T.; Mesnil, M.; Naus, C.C.; Lampe, P.D.; Laird, D.W. Gap junctions and cancer: Communicating for 50 years. *Nat. Rev. Cancer* **2016**, *16*, 775–788. [CrossRef] [PubMed]

5. Hervé, J.C.; Bourmeyster, N.; Sarrouilhe, D.; Duffy, H.S. Gap junctional complexes: From partners to functions. *Prog. Biophys. Mol. Biol.* **2007**, *94*, 29–65. [CrossRef] [PubMed]

6. Aasen, T. Connexins: Junctional and non-junctional modulators of proliferation. *Cell Tissue Res.* **2015**, *360*, 685–699. [CrossRef] [PubMed]

7. Schalper, K.A.; Carvajal-Hausdorf, D.; Oyarzo, M.P. Possible role of hemichannels in cancer. *Front. Physiol.* **2014**, *5*, 237. [CrossRef] [PubMed]

8. Willecke, K.; Eiberger, J.; Degen, J.; Eckardt, D.; Romualdi, A.; Güldenagel, M.; Deutsch, U.; Söhl, G. Structural and functional diversity of connexin genes in the mouse and human genome. *Biol. Chem.* **2002**, *383*, 725–737. [CrossRef] [PubMed]

9. Panchin, Y.; Kelmanson, I.; Matz, M.; Lukyanov, K.; Usman, N.; Lukyanov, S. A ubiquitous family of putative gap junction molecules. *Curr. Biol.* **2000**, *10*, R473–474. [CrossRef]

10. Bond, S.R.; Naus, C.C. The pannexins: Past and present. *Front. Physiol.* **2014**, *5*, 58. [CrossRef] [PubMed]

11. Solan, J.L.; Lampe, P.D. Spatio-temporal regulation of connexin43 phosphorylation and gap junction dynamics. *Biochim. Biophys. Acta* **2018**, *1860*, 83–90. [CrossRef] [PubMed]

12. Hervé, J.C.; Derangeon, M.; Sarrouilhe, D.; Giepmans, B.N.; Bourmeyster, N. Gap junctional channels are parts of multiprotein complexes. *Biochim. Biophys. Acta* **2012**, *1818*, 1844–1865. [CrossRef] [PubMed]

13. Naus, C.C.; Giaume, C. Bridging the gap to therapeutic strategies based on connexin/pannexin biology. *J. Transl. Med.* **2016**, *14*, 330. [CrossRef] [PubMed]

14. Kandouz, M.; Batist, G. Gap junctions and connexins as therapeutic targets in cancer. *Expert Opin. Ther. Targets* **2010**, *14*, 681–692. [CrossRef] [PubMed]

15. Loewenstein, W.R. Junctional intercellular communication and the control of growth. *Biochim. Biophys. Acta* **1979**, *560*, 1–65. [CrossRef]

16. Loewenstein, W.R.; Socolar, S.J.; Higashino, S.; Kanno, Y.; Davidson, N. Intercellular Communication: Renal, Urinary Bladder, Sensory, and Salivary Gland Cells. *Science* **1965**, *149*, 295–298. [CrossRef] [PubMed]

17. Kanno, Y.; Loewenstein, W.R. Cell-to-cell passage of large molecules. *Nature* **1966**, *212*, 629–630. [CrossRef] [PubMed]

18. Loewenstein, W.R.; Kanno, Y. Intercellular communication and the control of tissue growth: Lack of communication between cancer cells. *Nature* **1966**, *209*, 1248–1249. [CrossRef] [PubMed]

19. Loewenstein, W.R.; Kanno, Y. Intercellular communication and tissue growth. I. Cancerous growth. *J. Cell Biol.* **1967**, *33*, 225–234. [CrossRef] [PubMed]

20. Yotti, L.P.; Chang, C.C.; Trosko, J.E. Elimination of metabolic cooperation in Chinese hamster cells by a tumor promoter. *Science* **1979**, *206*, 1089–1091. [CrossRef] [PubMed]

21. Murray, A.W.; Fitzgerald, D.J. Tumor promoters inhibit metabolic cooperation in cocultures of epidermal and 3T3 cells. *Biochem. Biophys. Res. Commun.* **1979**, *91*, 395–401. [CrossRef]

22. Yamasaki, H. Cell-cell interaction and carcinogenesis. *Toxicol. Pathol.* **1986**, *14*, 363–369. [CrossRef] [PubMed]

23. Trosko, J.E.; Jone, C.; Chang, C.C. Oncogenes, inhibited intercellular communication and tumor promotion. *Princess Takamatsu Symp.* **1983**, *14*, 101–113. [PubMed]

24. Atkinson, M.M.; Anderson, S.K.; Sheridan, J.D. Modification of gap junctions in cells transformed by a temperature-sensitive mutant of Rous sarcoma virus. *J. Membr. Biol.* **1986**, *91*, 53–64. [CrossRef] [PubMed]

25. Barrett, J.C.; Kakunaga, T.; Kuroki, T.; Neubert, D.; Trosko, J.E.; Vasiliev, J.M.; Williams, G.M.; Yamasaki, H. Short-term assays to predict carcinogenicity. In-vitro assays that may be predictive of tumour-promoting agents. *IARC Sci. Publ.* **1986**, 287–302.

26. Temme, A.; Buchmann, A.; Gabriel, H.D.; Nelles, E.; Schwarz, M.; Willecke, K. High incidence of spontaneous and chemically induced liver tumors in mice deficient for connexin32. *Curr. Biol.* **1997**, *7*, 713–716. [CrossRef]

27. Avanzo, J.L.; Mesnil, M.; Hernandez-Blazquez, F.J.; Mackowiak, I.I.; Mori, C.M.; da Silva, T.C.; Oloris, S.C.; Gárate, A.P.; Massironi, S.M.; Yamasaki, H.; et al. Increased susceptibility to urethane-induced lung tumors in mice with decreased expression of connexin43. *Carcinogenesis* **2004**, *25*, 1973–1982. [CrossRef] [PubMed]

28. Fitzgerald, D.J.; Mesnil, M.; Oyamada, M.; Tsuda, H.; Ito, N.; Yamasaki, H. Changes in gap junction protein (connexin 32) gene expression during rat liver carcinogenesis. *J. Cell Biochem.* **1989**, *41*, 97–102. [CrossRef] [PubMed]

29. Krutovskikh, V.; Mazzoleni, G.; Mironov, N.; Omori, Y.; Aguelon, A.M.; Mesnil, M.; Berger, F.; Partensky, C.; Yamasaki, H. Altered homologous and heterologous gap-junctional intercellular communication in primary human liver tumors associated with aberrant protein localization but not gene mutation of connexin 32. *Int. J. Cancer* **1994**, *56*, 87–94. [CrossRef] [PubMed]

30. Hossain, M.Z.; Wilkens, L.R.; Mehta, P.P.; Loewenstein, W.; Bertram, J.S. Enhancement of gap junctional communication by retinoids correlates with their ability to inhibit neoplastic transformation. *Carcinogenesis* **1989**, *10*, 1743–1748. [CrossRef] [PubMed]

31. Vine, A.L.; Bertram, J.S. Cancer chemoprevention by connexins. *Cancer Metastasis Rev.* **2002**, *21*, 199–216. [CrossRef] [PubMed]

32. Mesnil, M.; Crespin, S.; Avanzo, J.L.; Zaidan-Dagli, M.L. Defective gap junctional intercellular communication in the carcinogenic process. *Biochim. Biophys. Acta* **2005**, *1719*, 125–145. [CrossRef] [PubMed]

33. Mesnil, M.; Krutovskikh, V.; Piccoli, C.; Elfgang, C.; Traub, O.; Willecke, K.; Yamasaki, H. Negative growth control of HeLa cells by connexin genes: Connexin species specificity. *Cancer Res.* **1995**, *55*, 629–639. [PubMed]

34. Mesnil, M. Connexins and cancer. *Biol. Cell* **2002**, *94*, 493–500. [CrossRef]

35. Chen, S.C.; Pelletier, D.B.; Ao, P.; Boynton, A.L. Connexin43 reverses the phenotype of transformed cells and alters their expression of cyclin/cyclin-dependent kinases. *Cell Growth Differ.* **1995**, *6*, 681–690. [PubMed]

36. Zhang, Y.W.; Morita, I.; Ikeda, M.; Ma, K.W.; Murota, S. Connexin43 suppresses proliferation of osteosarcoma U2OS cells through post-transcriptional regulation of p27. *Oncogene* **2001**, *20*, 4138–4149. [CrossRef] [PubMed]

37. Koffler, L.; Roshong, S.; Kyu Park, I.; Cesen-Cummings, K.; Thompson, D.C.; Dwyer-Nield, L.D.; Rice, P.; Mamay, C.; Malkinson, A.M.; Ruch, R.J. Growth inhibition in G(1) and altered expression of cyclin D1 and p27(kip-1) after forced connexin expression in lung and liver carcinoma cells. *J. Cell Biochem.* **2000**, *79*, 347–354. [CrossRef]

38. Zhang, Y.W.; Kaneda, M.; Morita, I. The gap junction-independent tumor-suppressing effect of connexin 43. *J. Biol. Chem.* **2003**, *278*, 44852–44856. [CrossRef] [PubMed]

39. Zhang, Y.W.; Nakayama, K.; Nakayama, K.; Morita, I. A novel route for connexin 43 to inhibit cell proliferation: Negative regulation of S-phase kinase-associated protein (Skp2). *Cancer Res.* **2003**, *63*, 1623–1630. [PubMed]

40. Goldberg, G.S.; Bechberger, J.F.; Tajima, Y.; Merritt, M.; Omori, Y.; Gawinowicz, M.A.; Narayanan, R.; Tan, Y.; Sanai, Y.; Yamasaki, H.; et al. Connexin43 suppresses MFG-E8 while inducing contact growth inhibition of glioma cells. *Cancer Res.* **2000**, *60*, 6018–6026. [PubMed]

41. Goldberg, G.S.; Moreno, A.P.; Lampe, P.D. Gap junctions between cells expressing connexin 43 or 32 show inverse permselectivity to adenosine and ATP. *J. Biol. Chem.* **2002**, *277*, 36725–36730. [CrossRef] [PubMed]

42. Swenson, K.I.; Piwnica-Worms, H.; McNamee, H.; Paul, D.L. Tyrosine phosphorylation of the gap junction protein connexin43 is required for the pp60v-src-induced inhibition of communication. *Cell Regul.* **1990**, *1*, 989–1002. [CrossRef] [PubMed]

43. Stains, J.P.; Civitelli, R. Cell-to-cell interactions in bone. *Biochem. Biophys. Res. Commun.* **2005**, *328*, 721–727. [CrossRef] [PubMed]

44. Chandrasekhar, A.; Kalmykov, E.A.; Polusani, S.R.; Mathis, S.A.; Zucker, S.N.; Nicholson, B.J. Intercellular redistribution of cAMP underlies selective suppression of cancer cell growth by connexin26. *PLoS ONE* **2013**, *8*, e82335. [CrossRef] [PubMed]

45. Suzhi, Z.; Liang, T.; Yuexia, P.; Lucy, L.; Xiaoting, H.; Yuan, Z.; Qin, W. Gap Junctions Enhance the Antiproliferative Effect of MicroRNA-124–3p in Glioblastoma Cells. *J. Cell Physiol.* **2015**, *230*, 2476–2488. [CrossRef] [PubMed]

46. Thuringer, D.; Jego, G.; Berthenet, K.; Hammann, A.; Solary, E.; Garrido, C. Gap junction-mediated transfer of miR-145–5p from microvascular endothelial cells to colon cancer cells inhibits angiogenesis. *Oncotarget* **2016**, *7*, 28160–28168. [CrossRef] [PubMed]

47. Soares, A.R.; Martins-Marques, T.; Ribeiro-Rodrigues, T.; Ferreira, J.V.; Catarino, S.; Pinho, M.J.; Zuzarte, M.; Isabel Anjo, S.; Manadas, B.P.G.; Sluijter, J.; et al. Gap junctional protein Cx43 is involved in the communication between extracellular vesicles and mammalian cells. *Sci. Rep.* **2015**, *5*, 13243. [CrossRef] [PubMed]

48. Laird, D.W. The gap junction proteome and its relationship to disease. *Trends Cell Biol.* **2010**, *20*, 92–101. [CrossRef] [PubMed]

49. Duffy, H.S.; Iacobas, I.; Hotchkiss, K.; Hirst-Jensen, B.J.; Bosco, A.; Dandachi, N.; Dermietzel, R.; Sorgen, P.L.; Spray, D.C. The gap junction protein connexin32 interacts with the Src homology3/hook domain of discs large homolog 1. *J. Biol. Chem.* **2007**, *282*, 9789–9796. [CrossRef] [PubMed]

50. Fu, C.T.; Bechberger, J.F.; Ozog, M.A.; Perbal, B.; Naus, C.C. CCN3 (NOV) interacts with connexin43 in C6 glioma cells: Possible mechanism of connexin-mediated growth suppression. *J. Biol. Chem.* **2004**, *279*, 36943–36950. [CrossRef] [PubMed]

51. Gellhaus, A.; Dong, X.; Propson, S.; Maass, K.; Klein-Hitpass, L.; Kibschull, M.; Traub, O.; Willecke, K.; Perbal, B.; Lye, S.J.; et al. Connexin43 interacts with NOV: A possible mechanism for negative regulation of cell growth in choriocarcinoma cells. *J. Biol. Chem.* **2004**, *279*, 36931–36942. [CrossRef] [PubMed]

52. Penes, M.C.; Li, X.; Nagy, J.I. Expression of zonula occludens-1 (ZO-1) and the transcription factor ZO-1-associated nucleic acid-binding protein (ZONAB)-MsY3 in glial cells and colocalization at oligodendrocyte and astrocyte gap junctions in mouse brain. *Eur. J. Neurosci.* **2005**, *22*, 404–418. [CrossRef] [PubMed]

53. González-Sánchez, A.; Jaraíz-Rodríguez, M.; Domínguez-Prieto, M.; Herrero-González, S.; Medina, J.M.; Tabernero, A. Connexin43 recruits PTEN and Csk to inhibit c-Src activity in glioma cells and astrocytes. *Oncotarget* **2016**, *7*, 49819–49833. [CrossRef] [PubMed]

54. Huang, R.P.; Fan, Y.; Hossain, M.Z.; Peng, A.; Zeng, Z.L.; Boynton, A.L. Reversion of the neoplastic phenotype of human glioblastoma cells by connexin 43 (cx43). *Cancer Res.* **1998**, *58*, 5089–5096. [PubMed]

55. Dang, X.; Doble, B.W.; Kardami, E. The carboxy-tail of connexin-43 localizes to the nucleus and inhibits cell growth. *Mol. Cell Biochem.* **2003**, *242*, 35–38. [CrossRef] [PubMed]

56. Moorby, C.; Patel, M. Dual functions for connexins: Cx43 regulates growth independently of gap junction formation. *Exp. Cell Res.* **2001**, *271*, 238–248. [CrossRef] [PubMed]

57. Dang, X.; Jeyaraman, M.; Kardami, E. Regulation of connexin-43-mediated growth inhibition by a phosphorylatable amino-acid is independent of gap junction-forming ability. *Mol. Cell. Biochem.* **2006**, *289*, 201–217. [CrossRef] [PubMed]

58. Ul-Hussain, M.; Olk, S.; Schoenebeck, B.; Wasielewski, B.; Meier, C.; Prochnow, N.; May, C.; Galozzi, S.; Marcus, K.; Zoidl, G.; et al. Internal ribosomal entry site (IRES) activity generates endogenous carboxyl-terminal domains of Cx43 and is responsive to hypoxic conditions. *J. Biol. Chem.* **2014**, *289*, 20979–20990. [CrossRef] [PubMed]

59. Smyth, J.W.; Shaw, R.M. Autoregulation of connexin43 gap junction formation by internally translated isoforms. *Cell Rep.* **2013**, *5*, 611–618. [CrossRef] [PubMed]

60. Fu, Y.; Zhang, S.S.; Xiao, S.; Basheer, W.A.; Baum, R.; Epifantseva, I.; Hong, T.; Shaw, R.M. Cx43 Isoform GJA1–20k Promotes Microtubule Dependent Mitochondrial Transport. *Front. Physiol.* **2017**, *8*, 905. [CrossRef] [PubMed]

61. Bruzzone, S.; Guida, L.; Zocchi, E.; Franco, L.; De Flora, A. Connexin 43 hemi channels mediate Ca^{2+}-regulated transmembrane NAD+ fluxes in intact cells. *FASEB J.* **2001**, *15*, 10–12. [CrossRef] [PubMed]

62. Song, D.; Liu, X.; Liu, R.; Yang, L.; Zuo, J.; Liu, W. Connexin 43 hemichannel regulates H9c2 cell proliferation by modulating intracellular ATP and (Ca^{2+}). *Acta Biochim. Biophys. Sin.* **2010**, *42*, 472–482. [CrossRef] [PubMed]

63. Zhou, J.Z.; Riquelme, M.A.; Gu, S.; Kar, R.; Gao, X.; Sun, L.; Jiang, J.X. Osteocytic connexin hemichannels suppress breast cancer growth and bone metastasis. *Oncogene* **2016**, *35*, 5597–5607. [CrossRef] [PubMed]

64. Yamasaki, H.; Omori, Y.; Zaidan-Dagli, M.L.; Mironov, N.; Mesnil, M.; Krutovskikh, V. Genetic and epigenetic changes of intercellular communication genes during multistage carcinogenesis. *Cancer Detect Prev.* **1999**, *23*, 273–279. [CrossRef] [PubMed]

65. King, T.J.; Fukushima, L.H.; Donlon, T.A.; Hieber, A.D.; Shimabukuro, K.A.; Bertram, J.S. Correlation between growth control, neoplastic potential and endogenous connexin43 expression in HeLa cell lines: Implications for tumor progression. *Carcinogenesis* **2000**, *21*, 311–315. [CrossRef] [PubMed]

66. Yano, T.; Ito, F.; Yamasaki, H.; Hagiwara, K.; Ozasa, H.; Nakazawa, H.; Toma, H. Epigenetic inactivation of connexin 32 in renal cell carcinoma from hemodialytic patients. *Kidney Int.* **2004**, *65*, 1519. [CrossRef] [PubMed]

67. Sumiko, S.; Hiromi, H.; Hiromi, S.; Keiko, F.; Shigeto, K.; Taiichiro, S.; Toyohiko, A.; Kiyokazu, H.; Hiroshi, Y.; Tomohiro, Y. Prevention of renal cell carcinoma from hemodialysis patients by regulating epigenetic factors. *Kidney Int.* **2005**, *67*, 2506–2507. [CrossRef] [PubMed]

68. Jinn, Y.; Inase, N. Connexin 43, E-cadherin, beta-catenin and ZO-1 expression, and aberrant methylation of the connexin 43 gene in NSCLC. *Anticancer Res.* **2010**, *30*, 2271–2278. [PubMed]

69. Sirnes, S.; Honne, H.; Ahmed, D.; Danielsen, S.A.; Rognum, T.O.; Meling, G.I.; Leithe, E.; Rivedal, E.; Lothe, R.A.; Lind, G.E. DNA methylation analyses of the connexin gene family reveal silencing of GJC1 (Connexin45) by promoter hypermethylation in colorectal cancer. *Epigenetics* **2011**, *6*, 602–609. [CrossRef] [PubMed]

70. Warn-Cramer, B.J.; Lau, A.F. Regulation of gap junctions by tyrosine protein kinases. *Biochim. Biophys. Acta* **2004**, *1662*, 81–95. [CrossRef] [PubMed]

71. Warn-Cramer, B.J.; Cottrell, G.T.; Burt, J.M.; Lau, A.F. Regulation of connexin-43 gap junctional intercellular communication by mitogen-activated protein kinase. *J. Biol. Chem.* **1998**, *273*, 9188–9196. [CrossRef] [PubMed]

72. Lau, A.F.; Kanemitsu, M.Y.; Kurata, W.E.; Danesh, S.; Boynton, A.L. Epidermal growth factor disrupts gap-junctional communication and induces phosphorylation of connexin43 on serine. *Mol. Biol. Cell* **1992**, *3*, 865–874. [CrossRef] [PubMed]

73. Hossain, M.Z.; Ao, P.; Boynton, A.L. Platelet-derived growth factor-induced disruption of gap junctional communication and phosphorylation of connexin43 involves protein kinase C and mitogen-activated protein kinase. *J. Cell Physiol.* **1998**, *176*, 332–341. [CrossRef]

74. Leithe, E.; Mesnil, M.; Aasen, T. The connexin43 C-terminus: A tail of many tales. *Biochim. Biophys. Acta* **2018**, *1860*, 48–64. [CrossRef] [PubMed]

75. Hao, J.; Zhang, C.; Zhang, A.; Wang, K.; Jia, Z.; Wang, G.; Han, L.; Kang, C.; Pu, P. miR-221/222 is the regulator of Cx43 expression in human glioblastoma cells. *Oncol. Rep.* **2012**, *27*, 1504–1510. [PubMed]

76. Jin, Z.; Xu, S.; Yu, H.; Yang, B.; Zhao, H.; Zhao, G. miR-125b inhibits Connexin43 and promotes glioma growth. *Cell Mol. Neurobiol.* **2013**, *33*, 1143–1148. [CrossRef] [PubMed]

77. Li, X.; Pan, J.H.; Song, B.; Xiong, E.Q.; Chen, Z.W.; Zhou, Z.S.; Su, Y.P. Suppression of CX43 expression by miR-20a in the progression of human prostate cancer. *Cancer Biol. Ther.* **2012**, *13*, 890–898. [CrossRef] [PubMed]

78. Dubina, M.V.; Iatckii, N.A.; Popov, D.E.; Vasiliev, S.V.; Krutovskikh, V.A. Connexin 43, but not connexin 32, is mutated at advanced stages of human sporadic colon cancer. *Oncogene* **2002**, *21*, 4992–4996. [CrossRef] [PubMed]

79. Srinivas, M.; Verselis, V.K.; White, T.W. Human diseases associated with connexin mutations. *Biochim. Biophys. Acta* **2018**, *1860*, 192–201. [CrossRef] [PubMed]

80. Litvin, O.; Tiunova, A.; Connell-Alberts, Y.; Panchin, Y.; Baranova, A. What is hidden in the pannexin treasure trove: The sneak peek and the guesswork. *J. Cell Mol. Med.* **2006**, *10*, 613–634. [CrossRef] [PubMed]

81. Lai, C.P.; Bechberger, J.F.; Thompson, R.J.; MacVicar, B.A.; Bruzzone, R.; Naus, C.C. Tumor-suppressive effects of pannexin 1 in C6 glioma cells. *Cancer Res.* **2007**, *67*, 1545–1554. [CrossRef] [PubMed]

82. Lai, C.P.; Bechberger, J.F.; Naus, C.C. Pannexin2 as a novel growth regulator in C6 glioma cells. *Oncogene* **2009**, *28*, 4402–4408. [CrossRef] [PubMed]

83. Cowan, K.N.; Langlois, S.; Penuela, S.; Cowan, B.J.; Laird, D.W. Pannexin1 and Pannexin3 exhibit distinct localization patterns in human skin appendages and are regulated during keratinocyte differentiation and carcinogenesis. *Cell Commun. Adhes.* **2012**, *19*, 45–53. [CrossRef] [PubMed]

84. Celetti, S.J.; Cowan, K.N.; Penuela, S.; Shao, Q.; Churko, J.; Laird, D.W. Implications of pannexin 1 and pannexin 3 for keratinocyte differentiation. *J. Cell Sci.* **2010**, *123*, 1363–1372. [CrossRef] [PubMed]

85. Iwamoto, T.; Nakamura, T.; Doyle, A.; Ishikawa, M.; de Vega, S.; Fukumoto, S.; Yamada, Y. Pannexin 3 regulates intracellular ATP/cAMP levels and promotes chondrocyte differentiation. *J. Biol. Chem.* **2010**, *285*, 18948–18958. [CrossRef] [PubMed]

86. Ishikawa, M.; Iwamoto, T.; Fukumoto, S.; Yamada, Y. Pannexin 3 inhibits proliferation of osteoprogenitor cells by regulating Wnt and p21 signaling. *J. Biol. Chem.* **2014**, *289*, 2839–2851. [CrossRef] [PubMed]

87. Iwamoto, T.; Nakamura, T.; Ishikawa, M.; Yoshizaki, K.; Sugimoto, A.; Ida-Yonemochi, H.; Ohshima, H.; Saito, M.; Yamada, Y.; Fukumoto, S. Pannexin 3 regulates proliferation and differentiation of odontoblasts via its hemichannel activities. *PLoS ONE* **2017**, *12*, e0177557. [CrossRef] [PubMed]

88. Penuela, S.; Gyenis, L.; Ablack, A.; Churko, J.M.; Berger, A.C.; Litchfield, D.W.; Lewis, J.D.; Laird, D.W. Loss of pannexin 1 attenuates melanoma progression by reversion to a melanocytic phenotype. *J. Biol. Chem.* **2012**, *287*, 29184–29193. [CrossRef] [PubMed]

89. Solan, J.L.; Hingorani, S.R.; Lampe, P.D. Changes in Connexin43 Expression and Localization during Pancreatic Cancer Progression. *J. Membr. Biol.* **2012**, *245*, 255–262. [CrossRef] [PubMed]

90. Aasen, T.; Hodgins, M.B.; Edward, M.; Graham, S.V. The relationship between connexins, gap junctions, tissue architecture and tumour invasion, as studied in a novel in vitro model of HPV-16-associated cervical cancer progression. *Oncogene* **2003**, *22*, 7969–7980. [CrossRef] [PubMed]

91. Kanczuga-Koda, L.; Sulkowski, S.; Lenczewski, A.; Koda, M.; Wincewicz, A.; Baltaziak, M.; Sulkowska, M. Increased expression of connexins 26 and 43 in lymph node metastases of breast cancer. *J. Clin. Pathol.* **2006**, *59*, 429–433. [CrossRef] [PubMed]

92. Kanczuga-Koda, L.; Koda, M.; Sulkowski, S.; Wincewicz, A.; Zalewski, B.; Sulkowska, M. Gradual loss of functional gap junction within progression of colorectal cancer—A shift from membranous Cx32 and Cx43 expression to cytoplasmic pattern during colorectal carcinogenesis. *In Vivo* **2010**, *24*, 101–107. [PubMed]

93. Lowenstein, W.R. Junctional intercellular communication by phosphorylation. *Biochem. Soc. Symp.* **1985**, *50*, 43–58. [PubMed]

94. Defamie, N.; Chepied, A.; Mesnil, M. Connexins, gap junctions and tissue invasion. *FEBS Lett.* **2014**, *588*, 1331–1338. [CrossRef] [PubMed]

95. McNutt, N.; Hershberg, R.; Weinstein, R. Further observations on the occurrence of nexuses in benign and malignant human cervical epithelium. *J. Cell Biol.* **1971**, *51*, 805–825. [CrossRef] [PubMed]

96. King, T.; Fukushima, L.; Hieber, A.; Shimabukuro, A.; Sakr, W.; Bertram, J. Reduced levels of connexin 43 in cervical dysplasia: Inducible expression in a cervical carcinoma line decreases neoplastic potential with implications for tumour progression. *Carcinogenesis* **2000**, *21*, 1097–1109. [CrossRef] [PubMed]

97. Aasen, T.; Graham, S.V.; Edward, M.; Hodgins, M.B. Reduced expression of multiple gap junction proteins is a feature of cervical dysplasia. *Mol. Cancer* **2005**, *4*, 1–5. [CrossRef] [PubMed]

98. Jamieson, S.; Going, J.J.; D'Arcy, R.; George, W.D. Expression of gap junction proteins connexin 26 and connexin 43 in normal human breast and in breast tumours. *J. Pathol.* **1998**, *184*, 37–43. [CrossRef]

99. Laird, D.W.; Fistouris, P.; Batist, G.; Alpert, L.; Huynh, H.T.; Carystinos, G.; Alaoui-Jamali, M.A. Deficiency of connexin43 gap junctions is an independent marker for breast tumors. *Cancer Res.* **1999**, *59*, 4104–4110. [PubMed]

100. Singal, R.; Tu, Z.; Vanwert, J.; Ginder, G.; Kiang, D. Modulation of the connexin26 tumour suppressor gene expression through methylation in human mammary epithelial cell lines. *Anticancer Res.* **2000**, *20*, 59–64. [PubMed]

101. Naoi, Y.; Miyoshi, Y.; Taguchi, T.; Kim, S.J.; Arai, T.; Tamaki, Y.; Noguchi, S. Connexin26 expression is associated with lymphatic vessel invasion and poor prognosis in human breast cancer. *Breast Cancer Res. Treat.* **2007**, *106*, 11–17. [CrossRef] [PubMed]

102. Habermann, H.; Ray, V.; Habermann, W.; Prins, G.S. Alterations in gap junction protein expression in human benign prostatic hyperplasia and prostate cancer. *J. Urol.* **2002**, *167*, 655–660. [CrossRef]

103. Haass, N.K.; Ripperger, D.; Wladykowski, E.; Dawson, P.; Gimotty, P.A.; Blome, C.; Fischer, F.; Schmage, P.; Moll, I.; Brandner, J.M. Melanoma progression exhibits a significant impact on connexin expression patterns in the epidermal tumor microenvironment. *Histochem. Cell Biol.* **2009**, *133*, 113–124. [CrossRef] [PubMed]

104. Elzarrad, M.K.; Haroon, A.; Willecke, K.; Dobrowolski, R.; Gillespie, M.N.; Al-Mehdi, A.-B. Connexin-43 upregulation in micrometastases and tumor vasculature and its role in tumor cell attachment to pulmonary endothelium. *BMC Med.* **2008**, *6*, 20. [CrossRef] [PubMed]

105. Kapoor, P.; Saunders, M.M.; Li, Z.; Zhou, Z.; Sheaffer, N.; Kunze, E.L.; Samant, R.S.; Welch, D.R.; Donahue, H.J. Breast cancer metastatic potential: Correlation with increased heterotypic gap junctional intercellular communication between breast cancer cells and osteoblastic cells. *Int. J. Cancer* **2004**, *111*, 693–697. [CrossRef] [PubMed]

106. Li, Z.; Zhou, Z.; Donahue, H.J. Alterations in Cx43 and OB-cadherin affect breast cancer cell metastatic potential. *Clin. Exp. Metastasis* **2008**, *25*, 265–272. [CrossRef] [PubMed]

107. Li, Z.; Zhou, Z.; Welch, D.R.; Donahue, H.J. Expressing connexin 43 in breast cancer cells reduces their metastasis to lungs. *Clin. Exp. Metastasis* **2008**, *25*, 893–901. [CrossRef] [PubMed]

108. Chao, Y.; Wu, Q.; Acquafondata, M.; Dhir, R.; Wells, A. Partial Mesenchymal to Epithelial Reverting Transition in Breast and Prostate Cancer Metastases. *Cancer Microenviron.* **2012**, *5*, 19–28. [CrossRef] [PubMed]

109. Arun, S.; Ravisankar, S.; Vanisree, A.J. Implication of connexin30 on the stemness of glioma: Connexin30 reverses the malignant phenotype of glioma by modulating IGF-1R, CD133 and cMyc. *J. Neuro-Oncol.* **2017**, *135*, 473–485. [CrossRef] [PubMed]

110. Dong, H.; Zhou, X.; Wang, X.; Yang, Y.; Luo, J.; Liu, Y.; Mao, Q. Complex role of connexin 43 in astrocytic tumors and possible promotion of gliom-associated epileptic discharge. *Mol. Med. Rep.* **2017**, *16*, 7890–7900. [CrossRef] [PubMed]

111. Crespin, S.; Fromont, G.; Wager, M.; Levillain, P.; Cronier, L.; Monvoisin, A.; Defamie, N.; Mesnil, M. Expression of a gap junction protein, connexin43, in a large panel of human gliomas: New insights. *Cancer Med.* **2016**, *5*, 1742–1752. [CrossRef] [PubMed]

112. Zhang, A.; Hitomi, M.; Bar-Shain, N.; Dalimov, Z.; Ellis, L.; Velpula, K.K.; Fraizer, G.C.; Gourdie, R.G.; Lathia, J.D. Connexin 43 expression is associated with increased malignancy in prostate cancer cell lines and functions to promote migration. *Oncotarget* **2015**, *6*, 11640–11651. [CrossRef] [PubMed]

113. Ogawa, K.; Pitchakarn, P.; Suzuki, S.; Chewonarin, T.; Tang, M.; Takahashi, S.; Naiki-Ito, A.; Sato, S.; Takahashi, S.; Asamoto, M.; et al. Silencing of connexin 43 suppresses invasion, migration and lung metastasis of rat hepatocellular carcinoma cells. *Cancer Sci.* **2012**, *103*, 860–867. [CrossRef] [PubMed]

114. Ezumi, K.; Yamamoto, H.; Murata, K.; Higashiyama, M.; Damdinsuren, B.; Nakamura, Y.; Kyo, N.; Okami, J.; Ngan, C.Y.; Takemasa, I.; et al. Aberrant Expression of Connexin 26 Is Associated with Lung Metastasis of Colorectal Cancer. *Clin. Cancer Res.* **2008**, *14*, 677–684. [CrossRef] [PubMed]

115. Saito-Katsuragi, M.; Asada, H.; Niizeki, H.; Katoh, F.; Masuzawa, M.; Tsutsumi, M.; Kuniyasu, H.; Ito, A.; Nojima, H.; Miyagawa, S. Role for connexin 26 in metastasis of human malignant melanoma. *Cancer* **2007**, *110*, 1162–1172. [CrossRef] [PubMed]

116. Ito, A.; Katoh, F.; Kataoka, T.R.; Okada, M.; Tsubota, N.; Asada, H.; Yoshikawa, K.; Maeda, S.; Kitamura, Y.; Yamasaki, H.; et al. A role for heterologous gap junctions between melanoma and endothelial cells in metastasis. *J. Clin. Investig.* **2000**, *105*, 1189–1197. [CrossRef] [PubMed]

117. Jongen, W.M.; Fitzgerald, D.J.; Asamoto, M.; Piccoli, C.; Slaga, T.J.; Gros, D.; Takeichi, M.; Yamasaki, H. Regulation of connexin 43-mediated gap junctional intercellular communication by Ca^{2+} in mouse epidermal cells is controlled by E-cadherin. *J. Cell Biol.* **1991**, *114*, 545–555. [CrossRef] [PubMed]

118. Ryszawy, D.; Sarna, M.; Rak, M.; Szpak, K.; Kędracka-Krok, S.; Michalik, M.; Siedlar, M.; Zuba-Surma, E.; Burda, K.; Korohoda, W.; et al. Functional links between Snail-1 and Cx43 account for the recruitment of Cx43-positive cells into the invasive front of prostate cancer. *Carcinogenesis* **2014**, *35*, 1920–1930. [CrossRef] [PubMed]

119. Yu, M.; Zhang, C.; Li, L.; Dong, S.; Zhang, N.; Tong, X. Cx43 reverses the resistance of A549 lung adenocarcinoma cells to cisplatin by inhibiting EMT. *Oncol. Rep.* **2014**, *31*, 2751–2758. [CrossRef] [PubMed]

120. Graeber, S.H.M.; Hülser, D.F. Connexin Transfection Induces Invasive Properties in HeLa Cells. *Exp. Cell Res.* **1998**, *243*, 142–149. [CrossRef] [PubMed]

121. Bates, D.C.; Sin, W.C.; Aftab, Q.; Naus, C.C. Connexin43 enhances glioma invasion by a mechanism involving the carboxy terminus. *Glia* **2007**, *55*, 1554–1564. [CrossRef] [PubMed]

122. Ghosh, S.; Kumar, A.; Tripathi, R.P.; Chandna, S. Connexin-43 regulates p38-mediated cell migration and invasion induced selectively in tumour cells by low doses of γ-radiation in an ERK-1/2-independent manner. *Carcinogenesis* **2014**, *35*, 383–395. [CrossRef] [PubMed]

123. Oliveira, R.; Christov, C.; Guillamo, J.S.; de Boüard, S.; Palfi, S.; Venance, L.; Tardy, M.; Peschanski, M. Contribution of gap junctional communication between tumor cells and astroglia to the invasion of the brain parenchyma by human glioblastomas. *BMC Cell Biol.* **2005**, *6*, 7. [CrossRef] [PubMed]

124. El-Sabban, M.E.; Pauli, B.U. Adhesion-mediated gap junctional communication between lung metastatic cancer cells and endothelium. *Invasion Metastasis* **1994**, *14*, 164–176. [PubMed]

125. Thuringer, D.; Berthenet, K.; Cronier, L.; Solary, E.; Garrido, C. Primary tumor- and metastasis-derived colon cancer cells differently modulate connexin expression and function in human capillary endothelial cells. *Oncotarget* **2015**, *6*, 28800–28815. [CrossRef] [PubMed]

126. Pollmann, M.-A.; Shao, Q.; Laird, D.W.; Sandig, M. Connexin 43 mediated gap junctional communication enhances breast tumor cell diapedesis in culture. *Breast Cancer Res.* **2005**, *7*, R522–R534. [CrossRef] [PubMed]

127. Stoletov, K.; Strnadel, J.; Zardouzian, E.; Momiyama, M.; Park, F.D.; Kelber, J.A.; Pizzo, D.P.; Hoffman, R.; VandenBerg, S.R.; Klemke, R.L. Role of connexins in metastatic breast cancer and melanoma brain colonization. *J. Cell Sci.* **2013**, *126*, 904–913. [CrossRef] [PubMed]

128. Plante, I.; Stewart, M.K.G.; Barr, K.; Allan, A.L.; Laird, D.W. Cx43 suppresses mammary tumor metastasis to the lung in a Cx43 mutant mouse model of human disease. *Oncogene* **2011**, *30*, 1681–1692. [CrossRef] [PubMed]

129. Zhang, Z.Q.; Zhang, W.; Wang, N.Q.; Bani-Yaqhoub, M.; Lin, Z.X.; Naus, C.C. Suppression of tumorigenicity of human lung carcinoma cells after transfection with connexin 43. *Carcinogenesis* **1998**, *19*, 1889–1894. [CrossRef] [PubMed]

130. Saunders, M.M.; Seraj, M.J.; Li, Z.; Zhou, Z.; Winter, C.R.; Welch, D.R.; Donahue, H.J. Breast Cancer Metastatic Potential Correlates with a Breakdown in Homospecific and Heterospecific Gap Junctional Intercellular Communication. *Cancer Res.* **2001**, *61*, 1765–1767. [PubMed]

131. Shao, Q.; Wang, H.; McLachlan, E.; Veitch, G.I.L.; Laird, D.W. Down-regulation of Cx43 by Retroviral Delivery of Small Interfering RNA Promotes an Aggressive Breast Cancer Cell Phenotype. *Cancer Res.* **2005**, *65*, 2705–2711. [CrossRef] [PubMed]

132. Lamiche, C.; Clarhaut, J.; Strale, P.-O.; Crespin, S.; Pedretti, N.; Bernard, F.X.; Naus, C.C.; Chen, V.C.; Foster, L.J.; Defamie, N.; et al. The gap junction protein Cx43 is involved in the bone-targeted metastatic behaviour of human prostate cancer cells. *Clin. Exp. Metastasis* **2012**, *29*, 111–122. [CrossRef] [PubMed]

133. Chen, Q.; Boire, A.; Jin, X.; Valiente, M.; Er, E.E.; Lopez-Soto, A.; Jacob, L.S.; Patwa, R.; Shah, H.; Xu, K.; et al. Carcinoma–astrocyte gap junctions promote brain metastasis by cGAMP transfer. *Nature* **2016**, *533*, 493–498. [CrossRef] [PubMed]

134. Ramón y Cajal, S.; Capdevila, C.; Hernandez-Losa, J.; De Mattos-Arruda, L.; Ghosh, A.; Lorent, J.; Larsson, O.; Aasen, T.; Postovit, L.M.; Topisirovic, I. Cancer as an ecomolecular disease and a neoplastic consortium. *Biochim. Biophys. Acta (BBA) Rev. Cancer* **2017**, *1868*, 484–499. [CrossRef] [PubMed]

135. Lin, J.H.C.; Takano, T.; Cotrina, M.L.; Arcuino, G.; Kang, J.; Liu, S.; Gao, Q.; Jiang, L.; Li, F.; Lichtenberg-Frate, H.; et al. Connexin 43 Enhances the Adhesivity and Mediates the Invasion of Malignant Glioma Cells. *J. Neurosci.* **2002**, *22*, 4302–4311. [CrossRef] [PubMed]

136. Miekus, K.; Czernik, M.; Sroka, J.; Czyz, J.; Madeja, Z. Contact stimulation of prostate cancer cell migration: The role of gap junctional coupling and migration stimulated by heterotypic cell-to-cell contacts in determination of the metastatic phenotype of Dunning rat prostate cancer cells. *Biol. Cell* **2005**, *97*, 893–903. [CrossRef] [PubMed]

137. Stuhlmann, D.; Ale-Agha, N.; Reinehr, R.; Steinbrenner, H.; Ramos, M.C.; Sies, H.; Brenneisen, P. Modulation of homologous gap junctional intercellular communication of human dermal fibroblasts via a paracrine factor(s) generated by squamous tumor cells. *Carcinogenesis* **2003**, *24*, 1737–1748. [CrossRef] [PubMed]

138. Fujimoto, E.; Sato, H.; Nagashima, Y.; Negishi, E.; Shirai, S.; Fukumoto, K.; Hagiwara, H.; Hagiwara, K.; Ueno, K.; Yano, T. A Src family inhibitor (PP1) potentiates tumor-suppressive effect of connexin 32 gene in renal cancer cells. *Life Sci.* **2005**, *76*, 2711–2720. [CrossRef] [PubMed]

139. Mendoza-Naranjo, A.; Sáez, P.J.; Johansson, C.C.; Ramírez, M.; Mandaković, D.; Pereda, C.; López, M.N.; Kiessling, R.; Sáez, J.C.; Salazar-Onfray, F. Functional Gap Junctions Facilitate Melanoma Antigen Transfer and Cross-Presentation between Human Dendritic Cells. *J. Immunol.* **2007**, *178*, 6949–6957. [CrossRef] [PubMed]

140. Mazzini, E.; Massimiliano, L.; Penna, G.; Rescigno, M. Oral Tolerance Can Be Established via Gap Junction Transfer of Fed Antigens from CX3CR1+ Macrophages to CD103+ Dendritic Cells. *Immunity.* **2014**, *40*, 248–261. [CrossRef] [PubMed]

141. Aucher, A.; Rudnicka, D.; Davis, D.M. MicroRNAs Transfer from Human Macrophages to Hepato-Carcinoma Cells and Inhibit Proliferation. *J. Immunol.* **2013**, *191*, 6250–6260. [CrossRef] [PubMed]

142. Song, B.; Tang, J.-W.; Wang, B.; Cui, X.-N.; Hou, L.; Sun, L.; Mao, L.M.; Zhou, C.H.; Du, Y.; Wang, L.H.; et al. Identify lymphatic metastasis-associated genes in mouse hepatocarcinoma cell lines using gene chip. *World J. Gastroenterol. WJG* **2005**, *11*, 1463–1472. [CrossRef] [PubMed]

143. Furlow, P.W.; Zhang, S.; Soong, T.D.; Halberg, N.; Goodarzi, H.; Mangrum, C.; Wu, Y.G.; Elemento, O.; Tavazoie, S.F. Mechanosensitive pannexin-1 channels mediate microvascular metastatic cell survival. *Nat. Cell Biol.* **2015**, *17*, 943–952. [CrossRef] [PubMed]

144. Jiang, J.X.; Penuela, S. Connexin and pannexin channels in cancer. *BMC Cell Biol.* **2016**, *17*, 12. [CrossRef] [PubMed]

145. Willebrords, J.; Maes, M.; Crespo Yanguas, S.; Vinken, M. Inhibitors of connexin and pannexin channels as potential therapeutics. *Pharmacol. Ther.* **2017**, *180*, 144–160. [CrossRef] [PubMed]

146. Trosko, J.E.; Chang, C.C. Mechanism of up-regulated gap junctional intercellular communication during chemoprevention and chemotherapy of cancer. *Mutat. Res.* **2001**, *480–481*, 219–229. [CrossRef]

147. Fornelli, F.; Leone, A.; Verdesca, I.; Minervini, F.; Zacheo, G. The influence of lycopene on the proliferation of human breast cell line (MCF-7). *Toxicol. In Vitro* **2007**, *21*, 217–223. [CrossRef] [PubMed]

148. Leone, A.; Lecci, R.M.; Durante, M.; Piraino, S. Extract from the zooxanthellate jellyfish Cotylorhiza tuberculata modulates gap junction intercellular communication in human cell cultures. *Mar. Drugs* **2013**, *11*, 1728–1762. [CrossRef] [PubMed]

149. Yamasaki, H.; Katoh, F. Novel method for selective killing of transformed rodent cells through intercellular communication, with possible therapeutic applications. *Cancer Res.* **1988**, *48*, 3203–3207. [PubMed]

150. Safwat, S.; Ishak, R.A.; Hathout, R.M.; Mortada, N.D. Statins anticancer targeted delivery systems: Re-purposing an old molecule. *J. Pharm. Pharmacol.* **2017**, *69*, 613–624. [CrossRef] [PubMed]

151. Cesen-Cummings, K.; Warner, K.A.; Ruch, R.J. Role of protein kinase C in the deficient gap junctional intercellular communication of K-ras-transformed murine lung epithelial cells. *Anticancer Res.* **1998**, *18*, 4343–4346. [PubMed]

152. Touraine, R.L.; Vahanian, N.; Ramsey, W.J.; Blaese, R.M. Enhancement of the herpes simplex virus thymidine kinase/ganciclovir bystander effect and its antitumor efficacy in vivo by pharmacologic manipulation of gap junctions. *Hum. Gene Ther.* **1998**, *9*, 2385–2391. [CrossRef] [PubMed]

153. Wang, L.; Fu, Y.; Peng, J.; Wu, D.; Yu, M.; Xu, C.; Wang, Q.; Tao, L. Simvastatin-induced up-regulation of gap junctions composed of connexin 43 sensitize Leydig tumor cells to etoposide: An involvement of PKC pathway. *Toxicology* **2013**, *312*, 149–157. [CrossRef] [PubMed]

154. Wang, L.; Peng, J.; Huang, H.; Wang, Q.; Yu, M.; Tao, L. Simvastatin protects Sertoli cells against cisplatin cytotoxicity through enhanced gap junction intercellular communication. *Oncol. Rep.* **2015**, *34*, 2133–2141. [CrossRef] [PubMed]

155. Retamal, M.A.; Leon-Paravic, C.G.; Ezquer, M.; Ezquer, F.; Rio, R.D.; Pupo, A.; Martinez, A.D.; Gonzalez, C. Carbon monoxide: A new player in the redox regulation of connexin hemichannels. *IUBMB Life* **2015**, *67*, 428–437. [CrossRef] [PubMed]

156. Leon-Paravic, C.G.; Figueroa, V.A.; Guzman, D.J.; Valderrama, C.F.; Vallejos, A.A.; Fiori, M.C.; Altenberg, G.A.; Reuss, L.; Retamal, M.A. Carbon monoxide (CO) is a novel inhibitor of connexin hemichannels. *J. Biol. Chem.* **2014**, *289*, 36150–36157. [CrossRef] [PubMed]

157. Sato, H.; Iwata, H.; Takano, Y.; Yamada, R.; Okuzawa, H.; Nagashima, Y.; Yamaura, K.; Ueno, K.; Yano, T. Enhanced effect of connexin 43 on cisplatin-induced cytotoxicity in mesothelioma cells. *J. Pharmacol. Sci.* **2009**, *110*, 466–475. [CrossRef] [PubMed]

158. Uzu, M.; Sato, H.; Yamada, R.; Kashiba, T.; Shibata, Y.; Yamaura, K.; Ueno, K. Effect of enhanced expression of connexin 43 on sunitinib-induced cytotoxicity in mesothelioma cells. *J. Pharmacol. Sci.* **2015**, *128*, 17–26. [CrossRef] [PubMed]

159. Uzu, M.; Sato, H.; Shimizu, A.; Shibata, Y.; Ueno, K.; Hisaka, A. Connexin 43 enhances Bax activation via JNK activation in sunitinib-induced apoptosis in mesothelioma cells. *J. Pharmacol. Sci.* **2017**, *134*, 101–107. [CrossRef] [PubMed]

160. Dai, S.; Liu, J.; Sun, X.; Wang, N. Ganoderma lucidum inhibits proliferation of human ovarian cancer cells by suppressing VEGF expression and up-regulating the expression of connexin 43. *BMC Complement. Altern. Med.* **2014**, *14*, 434. [CrossRef] [PubMed]

161. Forster, T.; Rausch, V.; Zhang, Y.; Isayev, O.; Heilmann, K.; Schoensiegel, F.; Liu, L.; Nessling, M.; Richter, K.; Labsch, S.; et al. Sulforaphane counteracts aggressiveness of pancreatic cancer driven by dysregulated Cx43-mediated gap junctional intercellular communication. *Oncotarget* **2014**, *5*, 1621–1634. [CrossRef] [PubMed]

162. Ghatnekar, G.S.; O'Quinn, M.P.; Jourdan, L.J.; Gurjarpadhye, A.A.; Draughn, R.L.; Gourdie, R.G. Connexin43 carboxyl-terminal peptides reduce scar progenitor and promote regenerative healing following skin wounding. *Regen. Med.* **2009**, *4*, 205–223. [CrossRef] [PubMed]

163. Grek, C.L.; Rhett, J.M.; Bruce, J.S.; Abt, M.A.; Ghatnekar, G.S.; Yeh, E.S. Targeting connexin 43 with alpha-connexin carboxyl-terminal (ACT1) peptide enhances the activity of the targeted inhibitors, tamoxifen and lapatinib, in breast cancer: Clinical implication for ACT1. *BMC Cancer* **2015**, *15*, 296. [CrossRef] [PubMed]

164. Murphy, S.F.; Varghese, R.T.; Lamouille, S.; Guo, S.; Pridham, K.J.; Kanabur, P.; Osimani, A.M.; Sharma, S.; Jourdan, J.; Rodgers, C.M.; et al. Connexin 43 Inhibition Sensitizes Chemoresistant Glioblastoma Cells to Temozolomide. *Cancer Res.* **2016**, *76*, 139–149. [CrossRef] [PubMed]

165. Jaraiz-Rodriguez, M.; Tabernero, M.D.; Gonzalez-Tablas, M.; Otero, A.; Orfao, A.; Medina, J.M.; Tabernero, A. A Short Region of Connexin43 Reduces Human Glioma Stem Cell Migration, Invasion, and Survival through Src, PTEN, and FAK. *Stem Cell Rep.* **2017**, *9*, 451–463. [CrossRef] [PubMed]

166. Rhett, J.M.; Calder, B.W.; Fann, S.A.; Bainbridge, H.; Gourdie, R.G.; Yost, M.J. Mechanism of action of the anti-inflammatory connexin43 mimetic peptide JM2. *Am. J. Physiol. Cell Physiol.* **2017**, *313*, C314–C326. [CrossRef] [PubMed]

167. Baklaushev, V.P.; Yusubalieva, G.M.; Tsitrin, E.B.; Gurina, O.I.; Grinenko, N.P.; Victorov, I.V.; Chekhonin, V.P. Visualization of Connexin 43-positive cells of glioma and the periglioma zone by means of intercenously injected monoclonal antibodies. *Drug Deliv.* **2011**, *18*, 331–337. [CrossRef] [PubMed]

168. Chekhonin, V.P.; Baklaushev, V.P.; Yusubalieva, G.M.; Belorusova, A.E.; Gulyaev, M.V.; Tsitrin, E.B.; Grinenko, N.F.; Gurina, O.I.; Pirogov, Y.A. Targeted delivery of liposomal nanocontainers to the peritumoral zone of glioma by means of monoclonal antibodies against GFAP and the extracellular loop of Cx43. *Nanomedicine* **2012**, *8*, 63–70. [CrossRef] [PubMed]

169. Yusubalieva, G.M.; Baklaushev, V.P.; Gurina, O.I.; Gulyaev, M.V.; Pirogov, Y.A.; Chekhonin, V.P. Antitumor effects of monoclonal antibodies to connexin 43 extracellular fragment in induced low-differentiated glioma. *Bull. Exp. Biol. Med.* **2012**, *153*, 163–169. [CrossRef] [PubMed]

170. Nukolova, N.V.; Baklaushev, V.P.; Abakumova, T.O.; Mel'nikov, P.A.; Abakumov, M.A.; Yusubalieva, G.M.; Bychkov, D.A.; Kabanov, A.V.; Chekhonin, V.P. Targeted delivery of cisplatin by small es, Cyrilliconnexin 43 vector nanogels to the focus of experimental glioma C6. *Bull. Exp. Biol. Med.* **2014**, *157*, 524–529. [CrossRef] [PubMed]

171. Baklaushev, V.P.; Nukolova, N.N.; Khalansky, A.S.; Gurina, O.I.; Yusubalieva, G.M.; Grinenko, N.P.; Gubskiy, I.L.; Melnikov, P.A.; Kardashova, K.; Kabanov, A.V.; et al. Treatment of glioma by cisplatin-loaded nanogels conjugated with monoclonal antibodies against Cx43 and BSAT1. *Drug Deliv.* **2015**, *22*, 276–285. [CrossRef] [PubMed]

172. Yusubalieva, G.M.; Baklaushev, V.P.; Gurina, O.I.; Zorkina, Y.A.; Gubskii, I.L.; Kobyakov, G.L.; Golanov, A.V.; Goryainov, S.A.; Gorlachev, G.E.; Konovalov, A.N.; et al. Treatment of poorly differentiated glioma using a combination of monoclonal antibodies to extracellular connexin-43 fragment, temozolomide, and radiotherapy. *Bull. Exp. Biol. Med.* **2014**, *157*, 510–515. [CrossRef] [PubMed]

173. Abakumova, T.; Abakumov, M.; Shein, S.; Chelushkin, P.; Bychkov, D.; Mukhin, V.; Yusubalieva, G.; Grinenko, N.; Kabanov, A.; Nukolova, N.; et al. Connexin 43-targeted T1 contrast agent for MRI diagnosis of glioma. *Contrast Media Mol. Imaging* **2016**, *11*, 15–23. [CrossRef] [PubMed]

174. Jing, Y.; Guo, S.; Zhang, X.; Sun, A.; Tao, F.; Ju, H.; Qian, H. Effects of small interfering RNA interference of connexin 37 on subcutaneous gastric tumours in mice. *Mol. Med. Rep.* **2014**, *10*, 2955–2960. [CrossRef] [PubMed]

175. Hitomi, M.; Deleyrolle, L.P.; Mulkearns-Hubert, E.E.; Jarrar, A.; Li, M.; Sinyuk, M.; Otvos, B.; Brunet, S.; Flavahan, W.A.; Hubert, C.G.; et al. Differential connexin function enhances self-renewal in glioblastoma. *Cell Rep.* **2015**, *11*, 1031–1042. [CrossRef] [PubMed]

176. Munoz, J.L.; Rodriguez-Cruz, V.; Greco, S.J.; Ramkissoon, S.H.; Ligon, K.L.; Rameshwar, P. Temozolomide resistance in glioblastoma cells occurs partly through epidermal growth factor receptor-mediated induction of connexin 43. *Cell Death Dis.* **2014**, *5*, e1145. [CrossRef] [PubMed]

177. Gielen, P.R.; Aftab, Q.; Ma, N.; Chen, V.C.; Hong, X.; Lozinsky, S.; Naus, C.C.; Sin, W.C. Connexin43 confers Temozolomide resistance in human glioma cells by modulating the mitochondrial apoptosis pathway. *Neuropharmacology* **2013**, *75*, 539–548. [CrossRef] [PubMed]

178. Lai, S.W.; Huang, B.R.; Liu, Y.S.; Lin, H.Y.; Chen, C.C.; Tsai, C.F.; Lu, D.Y.; Lin, C. Differential Characterization of Temozolomide-Resistant Human Glioma Cells. *Int. J. Mol. Sci.* **2018**, *19*, E127. [CrossRef] [PubMed]

179. Zhang, J.; O'Carroll, S.J.; Henare, K.; Ching, L.M.; Ormonde, S.; Nicholson, L.F.; Danesh-Meyer, H.V.; Green, C.R. Connexin hemichannel induced vascular leak suggests a new paradigm for cancer therapy. *FEBS Lett.* **2014**, *588*, 1365–1371. [CrossRef] [PubMed]

180. Danesh-Meyer, H.V.; Kerr, N.M.; Zhang, J.; Eady, E.K.; O'Carroll, S.J.; Nicholson, L.F.; Johnson, C.S.; Green, C.R. Connexin43 mimetic peptide reduces vascular leak and retinal ganglion cell death following retinal ischaemia. *Brain* **2012**, *135*, 506–520. [CrossRef] [PubMed]

181. Boyd-Tressler, A.; Penuela, S.; Laird, D.W.; Dubyak, G.R. Chemotherapeutic drugs induce ATP release via caspase-gated pannexin-1 channels and a caspase/pannexin-1- independent mechanism. *J. Biol. Chem.* **2014**, *289*, 27246–27263. [CrossRef] [PubMed]

182. Boyd-Tressler, A.M.; Lane, G.S.; Dubyak, G.R. Up-regulated Ectonucleotidases in Fas-Associated Death Domain Protein- and Receptor-Interacting Protein Kinase 1-Deficient Jurkat Leukemia Cells Counteract Extracellular ATP/AMP Accumulation via Pannexin-1 Channels during Chemotherapeutic Drug-Induced Apoptosis. *Mol. Pharmacol.* **2017**, *92*, 30–47. [PubMed]

183. Saez, P.J.; Vargas, P.; Shoji, K.F.; Harcha, P.A.; Lennon-Dumenil, A.M.; Saez, J.C. ATP promotes the fast migration of dendritic cells through the activity of pannexin 1 channels and P2X7 receptors. *Sci. Signal* **2017**, *10*. [CrossRef] [PubMed]

184. Draganov, D.; Gopalakrishna-Pillai, S.; Chen, Y.R.; Zuckerman, N.; Moeller, S.; Wang, C.; Ann, D.; Lee, P.P. Modulation of P2X4/P2X7/Pannexin-1 sensitivity to extracellular ATP via Ivermectin induces a non-apoptotic and inflammatory form of cancer cell death. *Sci. Rep.* **2015**, *5*, 16222. [CrossRef] [PubMed]

185. Halliwill, K.D.; Quigley, D.A.; Kang, H.C.; Del Rosario, R.; Ginzinger, D.; Balmain, A. Panx3 links body mass index and tumorigenesis in a genetically heterogeneous mouse model of carcinogen-induced cancer. *Genome Med.* **2016**, *8*, 83. [CrossRef] [PubMed]

186. Romano, R.C.; Gardner, J.M.; Shalin, S.C.; Ram, R.; Govindarajan, R.; Montgomery, C.O.; Gilley, J.H.; Nicholas, R.W. High Relative Expression of Pannexin 3 (PANX3) in an Axillary Sweat Gland Carcinoma With Osteosarcomatous Transformation. *Am. J. Dermatopathol.* **2016**, *38*, 846–851. [CrossRef] [PubMed]

187. Di Cesare Mannelli, L.; Marcoli, M.; Micheli, L.; Zanardelli, M.; Maura, G.; Ghelardini, C.; Cervetto, C. Oxaliplatin evokes P2X7-dependent glutamate release in the cerebral cortex: A pain mechanism mediated by Pannexin 1. *Neuropharmacology* **2015**, *97*, 133–141. [CrossRef] [PubMed]

188. Weaver, J.L.; Arandjelovic, S.; Brown Mendu, K.; Schappe, S.; Buckley, M.W.; Chiu, Y.H.; Shu, S.; Kim, J.K.; Chung, J.; Chung, J.; et al. Hematopoietic pannexin 1 function is critical for neuropathic pain. *Sci. Rep.* **2017**, *7*, 42550. [CrossRef] [PubMed]

189. Hanahan, D.; Weinberg, R.A. Hallmarks of cancer: The next generation. *Cell* **2011**, *144*, 646–674. [CrossRef] [PubMed]

190. Mesnil, M.; Aasen, T.; Boucher, J.; Chépied, A.; Cronier, L.; Defamie, N.; Kameritsch, P.; Laird, D.W.; Lampe, P.D.; Lathia, J.D.; et al. An update on minding the gap in cancer. *Biochim. Biophys. Acta* **2018**, *1860*, 237–243. [CrossRef] [PubMed]

191. Sin, W.C.; Crespin, S.; Mesnil, M. Opposing roles of connexin43 in glioma progression. *Biochim. Biophys. Acta* **2012**, *1818*, 2058–2067. [CrossRef] [PubMed]

192. Sin, W.C.; Aftab, Q.; Bechberger, J.F.; Leung, J.H.; Chen, H.; Naus, C.C. Astrocytes promote glioma invasion via the gap junction protein connexin43. *Oncogene* **2016**, *35*, 1504–1516. [CrossRef] [PubMed]

193. Matsuuchi, L.; Naus, C.C. Gap junction proteins on the move: Connexins, the cytoskeleton and migration. *Biochim. Biophys. Acta* **2013**, *1828*, 94–108. [CrossRef] [PubMed]

194. Machtaler, S.; Choi, K.; Dang-Lawson, M.; Falk, L.; Pournia, F.; Naus, C.C.; Matsuuchi, L. The role of the gap junction protein connexin43 in B lymphocyte motility and migration. *FEBS Lett.* **2014**, *588*, 1249–1258. [CrossRef] [PubMed]

195. Elias, L.A.; Wang, D.D.; Kriegstein, A.R. Gap junction adhesion is necessary for radial migration in the neocortex. *Nature* **2007**, *448*, 901–907. [CrossRef] [PubMed]

196. Naus, C.C.; Aftab, Q.; Sin, W.C. Common mechanisms linking connexin43 to neural progenitor cell migration and glioma invasion. *Semin. Cell Dev. Biol.* **2016**, *50*, 59–66. [CrossRef] [PubMed]

International Journal of
Molecular Sciences

MDPI

Article

The Complex Subtype-Dependent Role of Connexin 43 (*GJA1*) in Breast Cancer

Mélanie Busby [1], Michael T. Hallett [2] and Isabelle Plante [1,*]

[1] INRS-Institut Armand-Frappier, Laval, QC H7V 1B7, Canada; melanie.busby@inrs.iaf.ca
[2] Centre for Structural and Functional Genomics, Department of Biology, Concordia University, Montreal, QC H4B 1R6, Canada; michael.hallett@concordia.ca
*   Correspondence: Isabelle.Plante@iaf.inrs.ca

Received: 31 January 2018; Accepted: 26 February 2018; Published: 28 February 2018

Abstract: Gap junction transmembrane channels allow the transfer of small molecules between the cytoplasm of adjacent cells. They are formed by proteins named connexins (Cxs) that have long been considered as a tumor suppressor. This widespread view has been challenged by recent studies suggesting that the role of Connexin 43 (Cx43) in cancer is tissue- and stage-specific and can even promote tumor progression. High throughput profiling of invasive breast cancer has allowed for the construction of subtyping schemes that partition patients into at least four distinct intrinsic subtypes. This study characterizes Cx43 expression during cancer progression with each of the tumor subtypes using a compendium of publicly available gene expression data. In particular, we show that Cx43 expression depends greatly on intrinsic subtype. Tumor grade also co-varies with patient subtype, resulting in Cx43 co-expression with grade in a subtype-dependent manner. Better survival was associated with a high expression of Cx43 in unstratified and luminal tumors but with a low expression in Her2e subtype. A better understanding of Cx43 regulation in a subtype-dependent manner is needed to clarify the context in which Cx43 is associated with tumor suppression or cancer progression.

Keywords: gap junctions; Connexin 43 (Cx43); breast cancer; survival analysis; intrinsic subtype

1. Introduction

Connexin 43 (Cx43), a protein encoded by the Gap Junction protein alpha 1 gene (*GJA1*), forms gap junction transmembrane channels facilitating communication between the cytoplasm of two adjacent cells. Small molecules, including metabolites, second messengers and electrical signals pass through these channels in a process called Gap Junction Intercellular Communication (GJIC). Cx43 transcription is thought to be regulated both by transcription factors and by epigenetic mechanisms [1], but is also regulated at the protein level by post-transcriptional modifications, trafficking to and from the plasma membrane and gating of the channels [2].

The breast epithelium is composed of two layers of cells: an inner layer of luminal cells surrounded by an outer layer of basal cells, composed mainly of myoepithelial cells but also comprising stem and progenitor cell populations [3]. It is well established that Cx43 is expressed mainly in the basal layer; however, a few studies showed Cx43 expression in luminal cells [4–6]. A study using transmission electron microscopy reported gap junctions to be present between the basal and the luminal layers in normal breast tissues, although the exact connexin involved was not determined [7]. A few studies have also demonstrated the expression of Cx43 in fibroblasts surrounding the breast epithelium and in endothelial cells [8–10].

The role of Cx43 in breast cancer is controversial. On the one hand, Cx43 has long been considered a tumor suppressor [11] with studies demonstrating it was under-expressed at the mRNA and the protein level in cancer cell lines [12,13] or aberrant localization and phosphorylation in tumors [12–16].

Cx43 has also been linked to the control of processes associated with breast cancer progression and metastasis such as proliferation, invasion, migration and apoptosis [17]. Moreover, it was shown in vivo and in vitro that metastatic capacity was increased in tumors cells showing a weak GJIC capacity and a lower number of gap junction plaques [18,19]. Re-expression of Cx43 in tumor cells led to reduced growth of tumors in nude mice and fewer metastases to the lungs [20,21]. Mice expressing a mutant form of Cx43 (G60S) also showed increased breast tumor metastasis to the lung [3].

On the other hand, much evidence suggested that Cx43 is involved in later stages of breast cancer progression. For instance, it has been suggested that Cx43 mediates the interaction between tumor and endothelial cells to facilitate adhesion and extravasation at secondary sites [22–24]. Cx43 has also been found to be expressed at higher levels in lymph node metastasis than in the corresponding primary tumor [25]. The context of expression that allows Cx43 to act as a tumor suppressor or promoter has not been elucidated and therefore precludes its targeting in breast cancer therapies [11].

Breast cancer is highly heterogeneous, with both intra- and inter-tumoral molecular variability. During the last decade, high throughput techniques have generated a body of new data in many diseases including breast cancer. Genome-wide gene expression profiling has produced classification schemes including the intrinsic subtypes consisting of luminal A (LumA), luminal B (LumB), basal-like and HER2-enriched (Her2e) tumors. Luminal tumors are generally characterized by the expression of the estrogen receptor alpha (ERα) and the progesterone receptors (PR). Most Her2e tumors harbor a genomic amplification of chromosome 17q12 that contains the erb-b2 receptor tyrosine kinase 2 gene (ERBB2/HER2). Approximately half of Her2e tumors express ERα. Basal-like tumors are often negative for ERα or PR receptors as well as for HER2 and also express basal cytokeratins [26,27].

This study aims to investigate Cx43's ambiguous role during cancer progression with each of the breast tumor intrinsic subtypes using a compendium of publicly available gene expression data with large samples. Here, we report that Cx43 expression depends greatly on intrinsic subtype. Tumor grade also co-varies with patient subtype, resulting in Cx43 co-expression with grade in a subtype-dependent manner. Better survival was associated with a high expression of Cx43 in unstratified and luminal tumors but with a low expression in the Her2e subtype.

2. Results

2.1. GJA1 Expression and Localization in the Breast

We first investigated the tissue localization and level of expression of Cx43 protein in human samples of both morphologically normal breast tissue and tumors using the Human Protein Atlas. This is a public database containing a large collection of normal and cancer tissue slides which have been probed with various antibodies followed by a hematoxylin counterstain [28]. Cx43 is a membrane channel and is usually considered to be expressed in the myoepithelial cell. A typical punctate staining of junctional plaques formed by Cx43 channels was observed for normal tissues. The staining could be observed in the myoepithelial layer, as expected, but also in some luminal cells (Figure 1a). Although an under-expression of Cx43 protein is observed in some of the 21 cancer samples available (Figure 1a), others show a clear over-expression, mostly in well differentiated luminal-like neoplastic cells, which did not appeared to be associated with a basal layer (Figure 1c). In other samples, Cx43 could also be seen in a layer of cells separating neoplastic tissue from stroma, although this layer sometimes adhered poorly to both adjacent compartments (Figure 1d,e). Cx43 was also observed in samples with poorly differentiated cell and tissue morphology (Figure 1f,g). Interestingly, Cx43 protein could also be found in spindle-shaped cells in the stroma (Figure 1h). Overall, some normal punctate patterns could be observed in some tumors (Figure 1d) while the majority of the samples showed either a downregulation or an aberrant cytoplasmic localization of Cx43 in tumor cells.

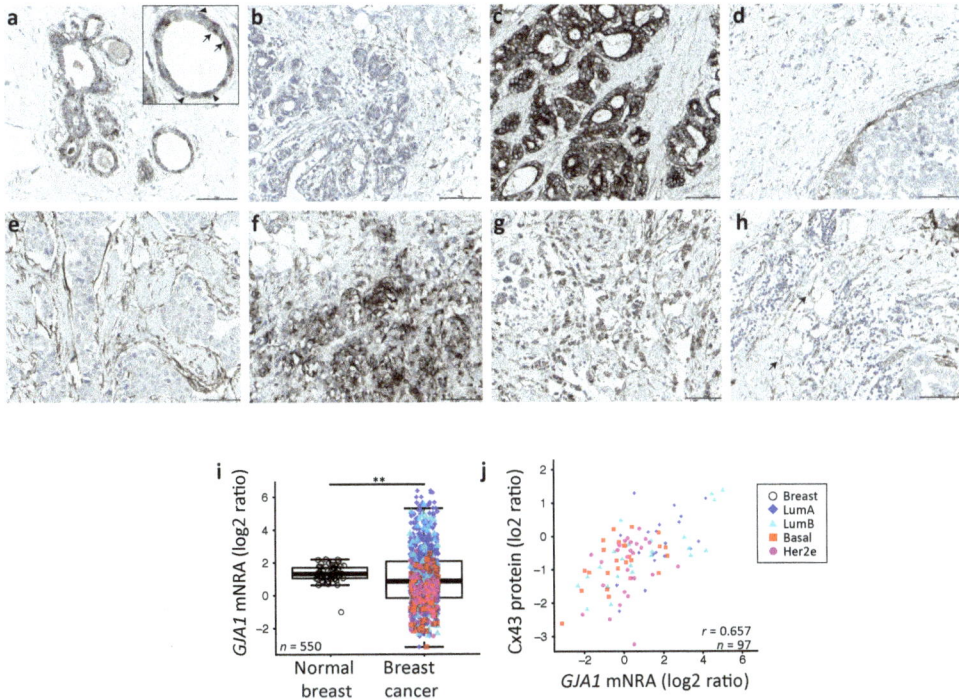

Figure 1. (**a–h**) The Human Protein Atlas normal and breast cancer tissue staining by immunohistochemistry for Connexin 43 (Cx43) (CAB010753 antibody). (**a**) Normal breast. Insert: Arrow head: myoepithelial cell's staining; arrow: luminal cell's staining. (**b–h**) Breast cancer tissue, (**h**) arrow: staining of spindle shaped stromal cells. Scale bar = 100 μm. (**i**) *GJA1* mRNA expression in breast tumor vs. adjacent normal breast tissue in the The Cancer Genome Atlas (TCGA) dataset. *p* value: * <0.05; ** <0.01; *** <0.001. (**j**) Scatter plot showing Cx43 protein and *GJA1* mRNA level in tumors. In the legend, "Breast" indicates adjacent normal breast tissue. Pearson's correlation coefficient is given (r).

We next compared transcript levels of *GJA1* in breast tumor samples and in the non-cancerous adjacent tissues using microarray-based data of The Cancer Genome Atlas project (TCGA) breast invasive carcinoma cohort (BRCA) of clinical samples. We observed a far greater variance in mRNA expression in tumor samples compared to tumor adjacent morphologically normal breast tissue (Figure 1i, the Fligner–Killeen test of homogeneity of variances, *p* value < 10^{-12}). We also used whole sample Cx43 protein levels obtained for 105 TCGA samples by mass spectrometry. *GJA1* mRNA and protein level are significantly correlated (Figure 1j, (Pearson correlation rho = 0.6515, *p* value < e^{-13}). Our results confirm that, in breast cancer, *GJA1* is concurrently dysregulated at both the protein and the mRNA level.

2.2. GJA1 Expression Varies with Breast Cancer Subtype

We then speculated that *GJA1* variability could be linked to the molecular heterogeneity of breast cancer. When we compared *GJA1* mRNA expression after stratifying patient samples by their intrinsic subtype (Pam50 by genefu [29]) (Figure 2a), the increase in variance in gene expression of tumor samples relative to normal tissue was observed in every subtype. The LumA had a mean expression level statistically indistinguishable from morphologically normal samples, but a small significant progressive decrease in expression is observed from LumB to basal and Her2e subtypes (Figure 2a).

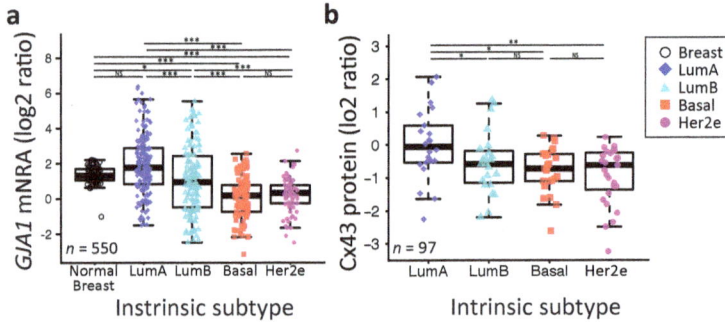

Figure 2. The *GJA1* expression level is more variable in breast tumor than in normal tissue and varies with subtype. (**a**) The *GJA1* mRNA level in normal breast tissue and in each tumor intrinsic subtype. (**b**) The Cx43 protein level in each intrinsic subtype. In the legend, "Breast" indicates adjacent normal breast tissue. All data are from the TCGA dataset. *p* value: * <0.05; ** <0.01; *** <0.001; NS Not statistically significant.

A similar pattern was observed in the four other datasets, although normal breast tissues were only used in TCGA and Curtis datasets (Figure A1). A similar pattern was also observed at the protein level in the TCGA dataset (Figure 2b). Together, these results suggest that the expression of *GJA1* is strongly associated with tumor subtype and is more variable in each subtype in comparison to morphologically normal tissue.

2.3. Somatic DNA-Level Events of GJA1 Do Not Drive Expression Changes of GJA1 in Breast Cancer

We next asked if underlying DNA-level somatic copy number changes in the genomic loci harboring *GJA1* influence gene expression levels. For the TCGA dataset, a few tumors had amplification or deletion of *GJA1*, compared to genes known to be amplified in breast cancer (Figure 3a).

Moreover, tumors with *GJA1* amplification did not show an increase in expression while only deep deletions reduced expression (−2 in called copy number, as shown in Figure 3c). Most luminal tumors with the highest expression of *GJA1* were found to have either a normal copy number or single deletion (Figure 3b–d). Moreover, in tumors with a *GJA1* gain or amplification, a slight but significant decrease in expression, rather than an increase, could be observed compared to normal tissues (Figure 3c). *GJA1* mRNA also weakly negatively correlated with DNA copy number (Figure 3c), suggesting that Cx43 over-expression in breast cancer is not driven by DNA amplification. To validate our procedure, we used the *HSF2* gene, a close neighbor of *GJA1* on chr6q22 which shares similar copy number in 99% of TCGA's breast cancer cases. In contrast to what was observed for *GJA1*, *HSF2* mRNA was positively correlated with the copy number of its own gene (Figure 3c). Moreover, somatic point mutation data showed that, in the TCGA cohort, only three breast cancer patients out of 977 harbored at least one *GJA1* mutation, accounting for 0.31% of the tumors (Figure 3e). Only one tumor with an extremely high number of total mutations (TCGA-AN-A046) was found to have both a *GJA1* mutation and a slightly higher expression of the gene (2.76) compared to normal range (−1.01 to 2.20) (Figure 3c,e). Together, these results argue that loss or amplification of the *GJA1* gene likely does not dictate mRNA and protein dysregulation in breast cancer.

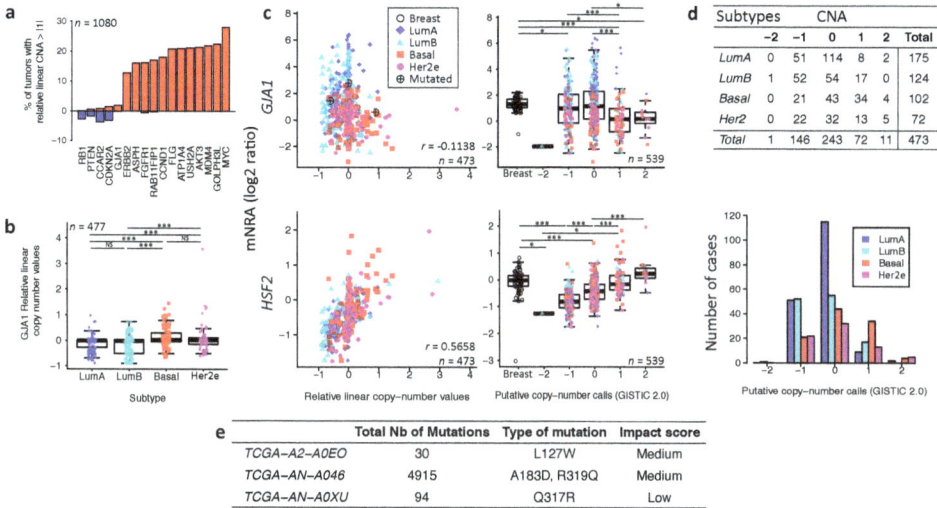

Figure 3. Somatic DNA-level events of *GJA1* do not drive expression changes of *GJA1* in breast cancer. (**a**) Percentage of tumors with a relative linear copy number >1 (amplification, in red) and <−1 (hemi- or homozygous deletion, in blue) for *GJA1* compared to other genes known to be altered in breast cancer. (**b**) Relative linear copy number value for each breast cancer subtype. (**c**) Somatic copy number alteration and putative copy number calls against mRNA expression for *GJA1* and *HSF2*. Copy number calls were computed by TCGA using GISTIC 2.0 (−2, Homozygous deletion; −1, Hemizygous deletion; 0, Neutral/no change; 1, Gain; 2, High level amplification). Pearson's correlation coefficient between relative linear copy number value and mRNA expression is given *GJA1* and *HSF2*. In the legend, "Breast" indicates adjacent normal breast tissue. (**d**) Contingency table and barplot showing the distribution of copy number alteration (CNA) by subtype. Due to the small number of samples in −2 and 2 CNA, Fisher's exact test was applied on −1, 0 and 1 CNA. (**e**) Total number and *GJA1* mutations observed in the 3 cases out of 988 patients. All data from the TCGA dataset. *p* value: * <0.05; ** <0.01; *** <0.001; NS Not statistically significant.

2.4. GJA1 Level Is Dependent on Hormonal Receptor Status

Because Cx43 level varies through the mammary gland development and the reproductive cycle, it has been suggested that it could be regulated by hormones, similar to what has been observed in other tissues [4,30–32]. We thus next investigated whether the *GJA1* mRNA level was directly linked with hormonal receptors status.

Consistent with the subtype-specific expression of Cx43, ERα- or PR-positive breast tumors had a significantly higher expression of *GJA1* mRNA compared to ERα- or PR-negative tumors (Figures 4a and A1–A3). Results were similar for all five datasets, except for PR in the NKI dataset where the low number of samples did not allow statistical significance to be reached (Figures 4a, A2a and A3a). However, there were no strong correlations between *GJA1* expression and *ESR1* mRNA, with total protein level (ERα), or with the activated form of ERα phosphorylated on serine 118 (ERα_pS118) (Figures 4b–e and A1b,c). While only weak correlations were observed in most individual subtypes, a stronger correlation between *GJA1* and *ESR1* mRNA and protein was observed when the tumors were pooled (Figure 4b–d). As expected, *ESR1* mRNA was better correlated with total ERα (Pearson's rho = 0.9011, Spearman's rho = 0.8969) than with ERα_pS118 proteins (Pearson's rho = 0.5459, Spearman's rho = 0.6407).

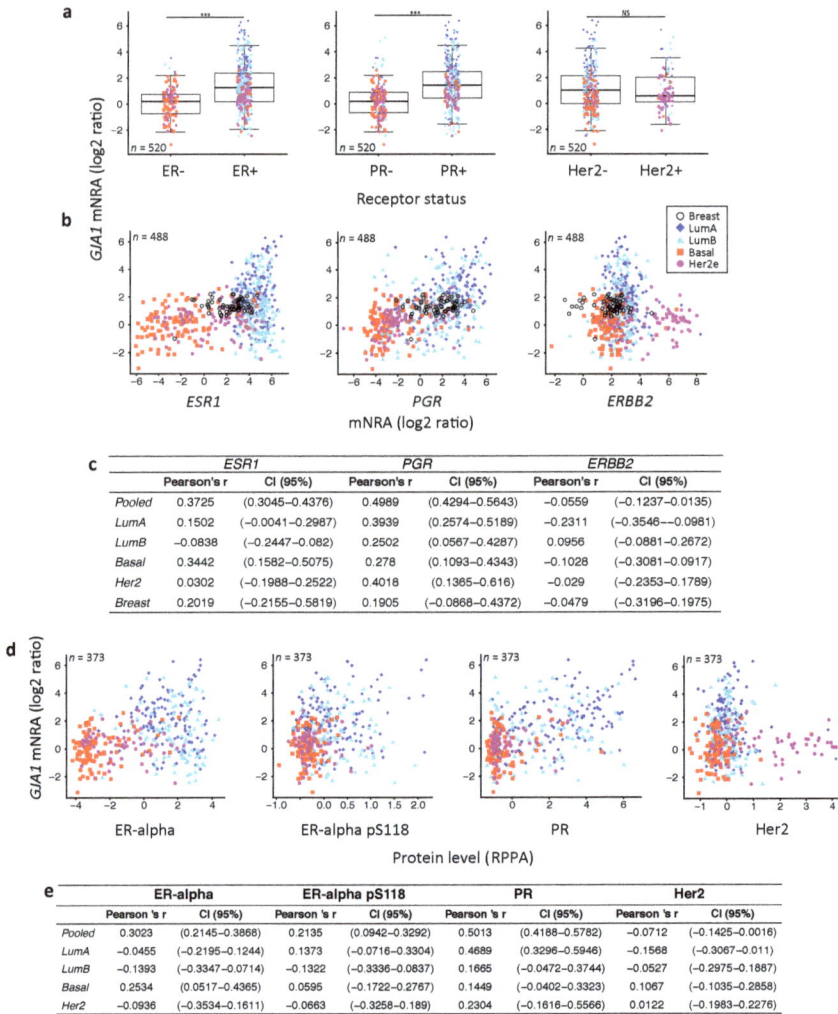

Figure 4. Depending on the receptor status, *GJA1* is associated with different mRNA levels in clinical samples. (**a**) Expression of *GJA1* mRNA stratified by estrogen receptor alpha (ERα), progesterone receptor (PR) and erb-b2 receptor tyrosine kinase 2 (ERBB2/Her2) status in the TCGA dataset. *p* value: * <0.05; ** <0.01; *** <0.001; NS Not statistically significant. (**b**) Plot of *GJA1* vs. *ESR1*, *PGR* and *ERBB2* mRNA (microarray) level in each subtype and in normal breast tissue. In the legend, "Breast" indicates adjacent normal breast tissue. (**c**) Bootstrapped correlations between *ESR1*, *PGR* or *ERBB2* and *GJA1* mRNA level either in pooled breast cancer tumors or in individual breast cancer intrinsic subtypes and in normal breast tissue. (**d**) Plot of *GJA1* mRNA vs. ER-alpha, ER-alpha pS118, PR or HER2 protein level assessed by reverse phase protein assay (RPPA). (**e**) Bootstrapped correlations between *GJA1* mRNA and ER-alpha, ER-alpha pS118, PR or HER2 protein level (RPPA). All data are from TCGA's BRCA dataset.

Stronger correlations between *PGR* mRNA and protein levels and *GJA1* mRNA levels were observed, not only in unstratified (pooled) analysis, but also in individual subtypes within most datasets (Figures 4b–e and A1b,c). This association was stronger in cancer samples than in normal breast tissues in all datasets for which normal tissues were available. Similar to ERα, total PR protein was well correlated with *PGR* mRNA (Pearson' rho = 0.8593 and Spearman's rho = 0.8723).

Tumors positive for the HER2 receptor by histochemistry (TCGA dataset) did not express significantly different levels of Cx43 mRNA. However, when HER2 status was given by HER2 amplicon probes or HER2 mRNA expression (Vanvliet, NKI and Curtis datasets), HER2+ tumors had a significantly lower level of *GJA1*. No direct correlation was observed between *GJA1* and the HER2 (ERBB2) mRNA (Figures 4a–e and A4a–d). A good correlation was observed between HER2 mRNA and total HER2 protein level (Pearson's rho = 0.8344, Spearman's rho 0.68634). The correlation between HER2 protein (HER2 and HER2_pY1248 activated form) and *GJA1* mRNA was not stronger than that observed for HER2 mRNA (Figures 4b–e and A3d).

Together, the significant differences observed in *GJA1* mRNA level in individual subtypes and with receptor status suggest that *GJA1* level is dependent on the molecular context provided by such subtypes. In addition, *GJA1* does not vary directly with *ESR1* and HER2 mRNA and protein levels but shows a stronger correlation with *PGR* mRNA and PR protein in tumor samples.

2.5. GJA1 mRNA Is Dysregulated at the Early Stages of Breast Cancer and Is Reduced with Grade When Tumors Are Pooled

To reconcile evidence supporting both tumor-suppressive and -promoting roles, it has been suggested that Cx43 function could depend on tissue type or evolve with tumor stage [11]. We therefore investigated whether *GJA1* expression in primary breast tumor changed with stage and grade at the mRNA level in breast cancer. Since grade/stage are strongly associated with subtype, we first stratified our cohorts by intrinsic subtype. We used the Curtis dataset, as *GJA1* expression was available for invasive tumors (stages 0 to IV) and "normal" adjacent tissue for numerous samples. A significant dysregulation of *GJA1* expression occurred at the early stages in all breast cancer subtypes, although both over-expression and downregulation could be observed (Figures 5 and A5). Most of the *GJA1* over-expressing luminal tumors were found to be of low stage (0–II). However, a reduction was observed in early stage basal-like and Her2e tumors (Figures 5 and A5). A significant increase in *GJA1* was also observed in the invasive stage I compared with stage 0 in all subtypes (Figure 5).

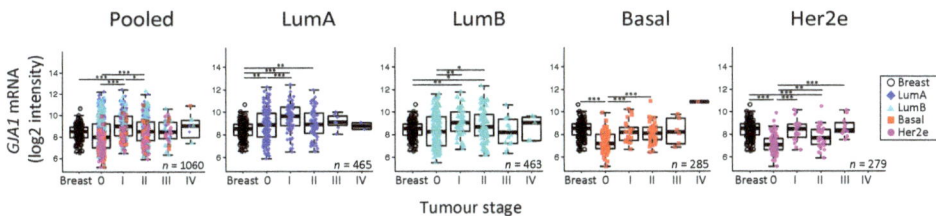

Figure 5. *GJA1* mRNA level is dysregulated in clinical samples at the early stages of breast cancer. *GJA1* mRNA level for each tumor stage and for normal breast in the Curtis Discovery dataset, either in pooled breast tumors or stratified by intrinsic subtype. In the legend, "Breast" indicates adjacent normal breast tissue. *p* value: * <0.05; ** <0.01; *** <0.001.

We then investigated whether or not the expression of *GJA1* could be linked to tumors' grade. Our analysis revealed that *GJA1* mRNA expression was significantly decreased with grade when all tumors were pooled, but not when they were stratified by intrinsic subtype (Figures 6 and A6). A significant decrease in *GJA1* with grades in LumB tumors could be observed only in the Vanvliet's dataset but not in other datasets analyzed (Figures 6 and A6).

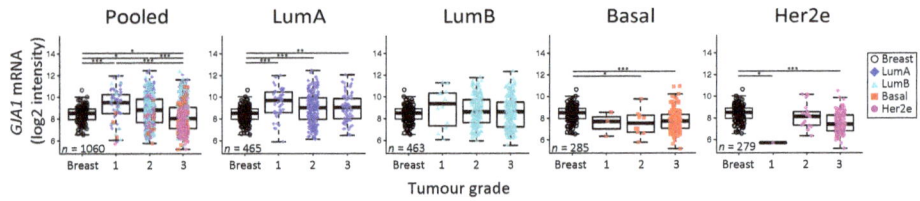

Figure 6. *GJA1* mRNA level is downregulated with grade in clinical samples only in pooled tumors but not in individual subtypes. *GJA1* mRNA level for each tumor grade in the Curtis Discovery dataset, either in pooled breast tumors or stratified by intrinsic subtype. In the legend, "Breast" indicates adjacent normal breast tissue. *p* value: * <0.05; ** <0.01; *** <0.001.

Interestingly, basal and Her2e tumors, which express a low level of *GJA1* (Figure 2), account for an important proportion of grade 3 tumors, thus reducing the mean *GJA1* expression for this grade (Figure 6). Moreover, grade 1 tumors are mostly luminal A and B, with a subset of *GJA1* over-expressing tumors, introducing an upward bias in this grade. Grade 2 tumors consist of a more balanced mix of all the subtypes (Figures 2 and 6). These results suggest that an observed reduction in *GJA1* with grade in pooled tumors is likely a bias induced by the pooling of the tumors' subtypes.

2.6. In Her2e Breast Tumors, a Low Expression of GJA1 Is Associated with a Better Prognosis

To gain further insight into the role of Cx43 in breast cancer, we analyzed how the level of *GJA1* mRNA expression in each subtype was associated with outcome. Observations that Cx43 expression was associated with a worse prognostic in ER-negative [33] and Her2e [34] tumors have been previously reported using the web-based platform KMPlotter [35] while ER-positive tumors had a better prognosis [33]. Investigating further the results of BreastMark and KMPlotter Web platforms, survival analysis showed that pooled and luminal tumors with high levels of *GJA1* mRNA were associated with a better prognostic (hazard ratio < 1), although results were not always statistically significant (Figures 7 and A7a,b). Conversely, basal-like and Her2e tumors followed an opposite trend (hazard ratio > 1), with high expression of *GJA1* strongly associated with a worse prognosis in the Her2e subtype (Figures 7 and A7a,b).

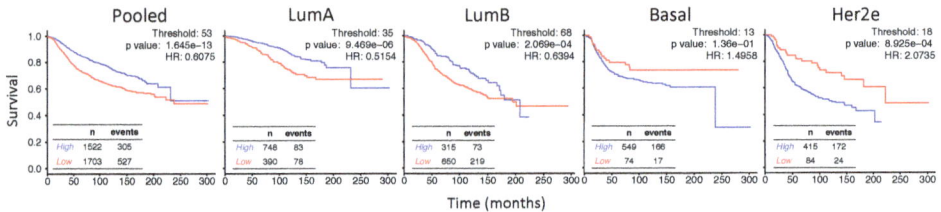

Figure 7. *GJA1* is associated with a diverging outcome depending on breast cancer subtype. The Kaplan–Meyer plots show survival curves for patients with breast tumors expressing either high (blue) or low (red) levels of *GJA1* mRNA in pooled tumors or in individual intrinsic subtypes. TCGA, NKI, Vanvliet and both Curtis datasets were aggregated for the analysis. The best cutoff was determined as the percentile lending the lowest log rank test *p* value (Figure A12) and was 53 in Pooled tumors, 35 in luminal A (LumA) tumors, 68 in luminal B (LumB), 13 in Basal and 18 in Her2e tumors.

Since the aggregation of several datasets in BreastMark and KMPlotter platforms could lead to artifacts in survival analysis, we went further by performing our own survival analysis for each subtypes for either aggregated (Figure 7) or individual datasets (Figures A9 and A12) following the determination of the best cutoff either by the receiver operating characteristic (ROC) curve (Figure A8a–c) or by the smallest *p* value of the log rank test for different thresholds (10–90) (Figure A10).

ROC curves have shown that the highest area Under the curve (AUC) for *GJA1* was obtained when tumors were pooled (Figure A8a) and *GJA1* was then ranked, at worst, in the eleven first percentiles when compared to all the probes present in the five datasets (Figure A8c). The log rank test was highly significant for all the analyses (Figures 7, A9 and A12) and for a vast range of cutoffs (Figure A10c), suggesting that *GJA1* has the greatest discriminating power when cohorts are unstratified. This is in line with a differential expression of *GJA1* in luminal vs. basal and Her2e tumors that also have diverging prognostics (Figure 2a).

When analyzing individual subtypes, a high expression was also significantly associated with a better prognosis in all analyses for LumA and for most analyses for LumB tumors (Figures 7, A9 and A12). However, survival curves in most analyses as well as the hazard ratio for a wide range of cutoffs (Figure A11) showed that this tendency is reversed in Basal and Her2e tumors where *GJA1* is mostly associated with a worse prognosis. This result was most significant in Her2e tumors, especially with smaller cutoffs while significance was rarely reached for Basal tumors.

However, *GJA1* ROC curves showed that *GJA1* did not consistently identify bad prognosis tumors with a high specificity and sensitivity (Figure A8a). These results suggest that although stratifying tumors revealed that the role of *GJA1* possibly differs in different breast cancer subtypes, *GJA1* should not be used as a clinical marker. These results also highlight once again how analyses using pooled tumor subtypes might induce biases and hide diverging results that are subtype-specific.

3. Discussion

Traditionally, Cx43 was considered as a tumor suppressor in the breast, with many studies reporting decreased Cx43 expression in tumor compared to normal breast tissue via both in vivo and in vitro studies [3,12,13,18–21]. However, other studies contradict these findings [8,22–25]. This recent evidence has cast doubt on Cxs tumor's suppressive role, suggesting that the Cxs function in cancer was tissue- and tumor stage-dependent [11,17]. At least four different subtypes of breast cancer have been identified, each having unique molecular profiles, responses to treatment and prognostics. Our evidence suggests that the role of Cx43 is dependent on subtype.

3.1. Cx43 Expression Is Dysregulated in Breast Cancer

Early studies first showed a dramatic downregulation of *GJA1* at the mRNA and the protein level in breast cancer cell lines as well as in rat and human breast tumors [12,13]. Conversely, other studies showed an increase in a subset of tumors [15]. Most of these studies analyzed a limited number of samples and were conducted either prior to the intrinsic subtype classification of breast cancer or did not use such classification. Our results, with several large cohorts of breast cancer clinical samples, reconcile these contradictory data by demonstrating that the observed dysregulation can involve both he increased and decreased expression of the Cx43 protein and mRNA. These observations are consistent with more recent reports at the protein level [15,16,25,36].

3.2. Dysregulation of Cx43 Is Linked to Hormonal Receptor Status and Tumor Subtype

Our results showed that the expression of Cx43 in breast tumors was lower in basal and Her2e than in normal tissues and that Cx43 levels vary greatly within luminal subtypes. This subtype-dependent expression was also shown by more recent studies, the result of which also support a higher expression of Cx43 mRNA and protein in luminal tumors than in basal-like and Her2e subtypes [36,37]. Because the intrinsic subtypes are characterized by, among others, hormonal receptor status, we wanted to evaluate whether a functional link could be captured in whole-tumor expression profiles between Cx43 and ERα, PR or HER2. Whole-tumor expression has been used by others, both to assess the content of specific cell types in samples and to decipher functional links between genes [38,39]. Using this method, we showed that *GJA1* mRNA increases in a subset of ERα- and PR-positive tumors and in the luminal subtypes, which are largely ERα- and PR-positive. These results were not surprising as much evidence supports a link between Cx43 and hormones in breast tissue [32] and in other tissues [40–43]. *GJA1* is also expressed at lower levels when HER2 status is positive and within the Her2e breast cancer subtype, except in the TCGA dataset. In an early study, it was reported that Cx43 gap junctions were dramatically reduced in breast tumors, and that this reduction was considered to occur regardless of ERα, PR or HER2 status [12]. More recent studies have reported that Cx43 protein expression correlated positively with PR and ERα status [44,45] and negatively with HER2 protein expression [45]. However, Conklin et al. reported that no correlation was observed between Cx43 and HER2 protein in tissue microarrays [44].

Our results suggest a direct relationship between *GJA1* and PR expression in breast cancer samples. Our analysis shows that *GJA1* level correlates with PR mRNA and protein in several subtypes. These results suggest that either PR or *GJA1* levels are dependent on the relative amount of some cell types co-expressing both genes, or that a functional link exists in the regulation of these genes in the same cell type or via paracrine signaling. Accumulating evidence has shown that ERα and PR are expressed in cell populations that do not totally overlap. *GJA1* is usually associated with basal cells while PR is thought to be expressed mainly in hormone-responsive luminal cell [1]. However, PR has been detected in some human breast basal cells, especially within immature lobules [1], suggesting an expression in primitive basal progenitor cells. PR has been suggested to coordinate basal cell proliferation, either via paracrine or autocrine stimulation [1]. It was also reported that the unliganded progesterone receptor isoform A (PRA) could activate Cx43 transcription by interacting with AP-1 heterodimers composed of FRA2 and JUND [42]. More studies are needed to better understand Cx43 localization and regulation, as well as its potential link with hormones. This knowledge is essential to further understand mammary gland morphogenesis and how Cx43 and hormones are involved in breast cancer.

Several other questions remain unanswered regarding the link between *GJA1* and ERα, PR and HER2. While the receptor's protein and mRNA levels were well correlated in our study, their functional status in the samples is unknown. Protein expression data for some phosphorylated forms of ER (ERα_pS118) and HER2 (HER2_pY1248) receptor were available. Beyond single phosphorylation, the activation of these receptors is mostly dependent on complex post-transcriptional processing which affects receptors' specific functions and gene transcription. As a result, prognostic significance of ERα has been shown to be phosphorylation site-specific [46]. Therefore, it cannot be excluded that *GJA1* mRNA expression can be regulated by ERα or HER2 and that these links could not be captured by expression profiles from breast cancer samples. Regardless of the precise nature of the link between *GJA1* and hormone receptors, our results suggest that *GJA1* level is dependent on the overall molecular context provided by each breast cancer subtype and that this might relate to *PGR* level, at least in some subtypes.

3.3. Upregulation of Cx43 mRNA Is Not Driven by DNA Amplification in Breast Cancer

Somatic DNA-level chromosomal aberrations are a defining characteristic of cancer and are common in breast carcinoma. Genomic loss and amplification cause decreases and increases in the transcription of genes in the region and often with concomitant effects of protein expression. We found that *GJA1* is rarely the target of such somatic events, and when it occurred it was often in Her2e and basal subtypes, consistent with the observation that these two subtypes generally have an increased amount of genomic instability in comparison to the luminal subtypes. Our results are also in accordance with previous studies that have shown that the region of human chromosome 6 where *GJA1* is located (6q22.31) has a relatively low level of amplification and deletions [47]. Cx43 was rarely mutated in breast cancer samples. Together, these results suggest that *GJA1* dysregulation at the mRNA and protein levels involves a dysregulation of other factors impacting the transcription (epigenetics, transcription factors) or mRNA stability.

3.4. Cx43 mRNA Level Is Dysregulated at the Early Stages of Breast Cancer

It was previously reported that, in primary tumors, Cx43 protein expression correlated with clinical stages [45]. Our analysis of microarray data from large cohorts of primary tumors suggested that Cx43 is decreased in a subset of tumors during early carcinogenesis (stage 0) and is re-expressed at higher levels in stage I tumors. While stage 0 of the luminal subtypes showed an increased variance of expression, those of basal-like and Her2e tumors had a significantly reduced expression compared to normal tissues. However, we could not observe a robust mRNA reduction, or increase, in later stages compared either to early tumor stages or to normal tissues.

A previous study investigated immunohistochemistry for Cx43 protein expression in ductal carcinoma in situ (DCIS), DCIS with microinvasion, DCIS with invasive ductal carcinoma (IDC) and IDC alone. In pooled tumors as well as in most subtypes, the lowest expression of Cx43 protein occurred neither in DCIS nor IDC alone but precisely in DCIS with microinvasion where only three out of thirty-seven cases (8%) were positive [36]. On the other side, out of 193 invasive lesions, sixty-three (33%) expressed Cx43. When looking specifically at the Her2e subtype, Cx43 was not expressed in a lower number of DCIS with microinvasion as in other subtypes since Cx43 was rarely expressed. Cx43 was present in only one of twenty IDC samples while the remaining twenty-six samples of other groups (DCIS, DCIS with microinvasion and DCIS with IDC) were all negatives. Whether or not the stromal compartment was included in the analysis was not specified. These results are consistent with our observation that, in all subtypes, DCIS (typically stage 0) had a lower expression than invasive stage I tumors. Together, these result point to Cx43 dysregulation as an early event in tumorigenesis, similar to what has been observed in the early stages of cervix, endometrial and thyroid cancers [48].

Moreover, it should be noted that while breast cancer stages are based on the size and the spreading of the disease in the tissue or to distant sites, the mRNA expression profiles we used only account for gene expression in whole primary tumors. Important morphologic information is therefore lost. During cancer progression, localized neoplastic cells acquire the capacity to invade surrounding tissues, and eventually reach the blood or lymphatic vasculature, allowing them to spread to other organs [49]. Depending on the stage and their location within the tumor or the tissue, these tumor cell populations face different challenges depending on the processes accomplished and on the microenvironment surrounding them [49]. Microarray data do not make it possible to either finely assess the expression of specific cells according to their specific localization in the tumor or to distinguish tumor gene expression from the stroma. A comprehensive study of events occurring early in carcinogenesis and accounting for the geographical localization within the tumors at primary or distant sites, and for the different cell populations in a subtype-dependent manner, is therefore the next logical step in further understanding Cx43's role in tumor progression.

3.5. The Apparent Grade-Dependent Decrease in Cx43 Is Linked to Its Low Expression in More Aggressive Subtypes

Tumor grade is a measure of the degree of abnormality of tumor cells and of dedifferentiation of cancer tissues compared to normal breast tissue. Our results showed that, when all tumors are pooled, the *GJA1* mRNA level seems to increase in grade 1 tumors compared to normal tissues and gradually decrease with increasing grade. However, stratifying the tumors by subtype showed that within an intrinsic subtype, the distribution of the tumors within each grade varies considerably. Indeed, luminal A and B tumors are more frequently of grade 1 or 2 and some of them over-express *GJA1* (Figure 2), while most basal and Her2e tumors, that express a low level of Cx43, are mostly of grade 3. As a result, *GJA1* mRNA is not lost with grade in individual subtypes. These results suggest that the observed correlation in pooled tumors is, in fact, a bias attributable to the pooling of the tumors, and reflects the high grade of basal and Her2e tumors. These results also highlight how pooling the different intrinsic subtypes, expressing varying degrees of *GJA1*, can introduce important biases in cohort analysis and will likely yield different results depending on the composition, in terms of the subtypes, of the cohorts studied.

3.6. High Expression of Cx43 Is Associated with a Good Prognostic in Luminal Subtypes, but with a Worse Prognostic in Her2e Tumors

Our results showed that, consistent with its ascribed role as a tumor suppressor in breast cancer, Cx43 was expressed at lower levels in more aggressive basal and Her2e subtypes than in luminal subtypes. Survival analysis of pooled tumors therefore showed a better survival of tumors highly expressing Cx43. However, as with grade, pooling breast cancer subtypes to analyze the effect of *GJA1* on the outcome introduces biases. Tumors expressing low levels of *GJA1* are overrepresented in aggressive basal and Her2e tumors, likely dragging down the survival of the group expressing a low level of *GJA1*. Therefore, performing survival analysis on pooled tumors, a good prognosis patient is automatically segregated into the curve of tumors highly expressing *GJA1*, and vice versa.

Paradoxically, the prognostic associated with Cx43 expression diverged depending on the intrinsic subtype, with a good prognosis in luminal tumors and an opposite trend in Her2e tumors. A previous study using immunohistochemistry found no correlation between Cx43 protein level and patient outcome [44]. However, similar to our results, more recent studies using expression array-based survival curves found that a high *GJA1* expression was associated with a better prognosis in ERα-positive breast cancer tumors, while an opposite trend was observed in ERα-negative tumors [33] and Her2e tumors [34]. The worse prognosis associated with *GJA1* in Her2e tumors suggests that *GJA1* function in breast cancer might not just be tissue- and stage-dependent, as suggested by others [11,17], but might also be subtype-dependent.

Cx43 has been reported to be expressed both in epithelial and stromal cells types. The molecular landscape provided by different cell types and/or by different breast cancer subtypes might provide different context, possibly allowing Cx43 to assume different functions and leading to different outcomes. It could be hypothesized that such context may provide different sets of interacting partners for *GJA1*, and its expression might even be driven by a different set of transcriptional or epigenetic regulators. In addition, an important determinant of the capacity of Cx43 to assume its channel function is unarguably its proper membrane localization. From array based mRNA expression data, it is until now impossible to assess neither the cellular localization nor the functional status of Cx43. It is very likely that these important and relevant information would contribute to a more complete understanding of the functions of Cx43 in breast cancer. For instance, a recent study demonstrated that over-expressing Cx43 in two different HER2-positive breast cancer cell lines lead to a diverging ability to proliferate, migrate, form mammospheres and form tumors in mice. Tumorigenic characteristics of the cancer cells were enhanced when functional gap junction channels could not be formed upon Cx43 over-expression, but were reduced when membrane gap junctions plaques allowed cells to communicate [34]. These aspects of Cx43 biology might explain its different roles according to subtypes but also possibly within subtypes and should therefore be addressed. Additional researches are required to better understand the context that allows Cx43 to suppress or promote carcinogenesis in different intrinsic subtypes.

4. Materials and Methods

4.1. Gene Expression

We used 4K samples over different expression platforms. Vanvliet used Affymetrix Human Genome U133A (data processed with Robust Multi-Array Average (RMA)) [50]. Curtis discovery and Curtis validation used the Illumina HT-12 v3 platform (expression given as a Log2 intensity level) [47]. The Cancer Genome Atlas (TCGA) used a custom Agilent G4502A 244K array (expression given as Log2 Lowess normalized ratio) [51]. NKI used a Hu25K Agilent platform (samples were hybridized against a pool of equal amount of RNA from each patient and gene expression is given as a log10 of intensity ratio) [52]. Normalized signal per probe or probe set mRNA expression was downloaded for tumor samples for all five datasets. Breast cancer intrinsic subtype was assigned to each sample with the Pam50 molecular subtyping algorithm using the R genefu package [29]. Survival for each case was determined as in [27].

ERα, PR and HER2 status provided in the original publication of the dataset was used. For TCGA, ER, PR and HER2 status was obtained by immunohistochemistry (IHC). For NKI, ER and PR status was determined by IHC and the sample was considered positive if at least 10% of the cells were positive. For Vanvliet, ER and PR status was determined with the Bioconductor package ROCR based on the expression of the probe 205225_at and validated with IHC when available. NKI and Vanvliet HER2 status was determined using the probes of the HER2 amplicon genes. For Curtis datasets, ER, PR and HER2 status was based on mRNA expression. In the TCGA dataset, the level of some proteins has been investigated with reverse phase protein assay (RPPA). Data were available for total ERα, PR and HER2 as well as for the phosphorylated forms of ERα (pS118) and HER2 (pY1248) that are at least partially indicative of the activation status [46,53]. *GJA1* protein level obtained by mass spectrometry for 105 TCGA samples was retrieved from the protein report found at the Clinical Proteomic Tumor Analysis Consortium (CPTAC) data portal [54]. Levels are given as the log2 of the ratio of each sample with respect to a pooled reporter sample.

4.2. DNA Alteration

Copy number alterations (CNAs) were measured in the TCGA dataset with Affymetrix 6.0 single nucleotide polymorphism (SNP) arrays and segmented using Circular Binary Segmentation (labeled here as Relative linear copy number values) [51]. Data were further processed by TCGA using GISTIC 2.0 to assign the Putative copy number calls per gene (-2: Homozygous deletion, -1: Hemizygous deletion, 0: Neutral/no change, 1: Gain, 2: High level amplification) [51]. Mutations were detected using whole-exome sequencing after controlling for germline and normal adjacent tissue mutations [51]. Linear and called CNA data as well as mutation data for the TCGA dataset were retrieved using R via cBioportal [55,56]. A total of 977 patients had data for mutations [51].

4.3. Survival Analysis

Survival data was available for all five datasets. The log-rank test was used to estimate significance and hazard ratios (95% CI) were computed via Cox regression using survival package [57]. ROC curves were computed using the pROC package [58]. Kaplan–Meier plots were used to visualize the data. The best cutoff to determine tumors expressing high or low levels of Cx43 was selected using either the ROC curves or based on the smallest *p* value of the log rank test computed for each threshold between 10 and 90. For aggregated datasets analysis, each cohort was first split into groups based on the selected threshold and datasets were pooled only after splitting.

In addition, survival analyses were computed using the BreastMark and KMPlotter web platforms that use several well-known dataset [35,59]. BreastMark allows thresholds of 25, 50 and 75 percentile to be selected to split the different cohorts used before aggregating them. For each analysis, we selected the threshold giving the best results. Since less samples were available in KMPlotter for statistical computation, we only included analyses for which there were at least 100 samples to draw both high and low expression curves for each subtype.

4.4. Statistical Analysis

All statistical analyses were carried out with R version 3.4.3 [60]. For two class comparisons, (cancer vs. normal tissues; positive vs. negative hormonal status) the Wilcoxon-Mann-Witney test was used. When more than two classes were compared, the Kruskall-Wallis test was used followed by the Dunn post-hoc test to assess the statistical significance for each pair of samples (to compare subtypes). For stages and grades, differential gene expression was assessed using Limma package [61]. A Benjamini-Hochberg correction was applied to adjust the *p* values for multiple testing. Because each subtype had a different number of patients, when correlation tests were performed between the expression of *GJA1* and the expression of other genes, a non-parametric bootstrap procedure was used for each subtype to derive the mean correlation coefficient and a percent confidence interval.

5. Conclusions

Our study has clarified the expression pattern of *GJA1* mRNA in breast cancer and showed that *GJA1* expression, as well as its prognostic significance, is dependent on breast cancer subtype. We also highlighted important biases that are introduced in analyzing pooled tumors. These biases need to be taken into consideration when studying *GJA1*, but also numerous other genes that are known to be linked, for instance, to ERα expression. Breast cancers are heterogeneous and genetically diverse and the lack of recognition of this molecular heterogeneity might explain the conflicting results from the literature, not only for *GJA1*, but potentially for other tumor suppressors or oncogenes. Overall, these results clearly showed that the molecular context where Cx43 is expressed in general, and the tumor subtypes of breast cancer in particular, should be taken into account when investigating Cx43's role in carcinogenesis.

Acknowledgments: We gratefully acknowledge financial support: MB was supported by Alexander Graham Bell Canada Graduate Scholarships-Doctoral and Master's Programs from the Natural Sciences and Engineering Research Council of Canada (NSERC), by the Fonds de Recherche du Québec—Nature et Technologies (FRQNT) and by the Fonds de Recherche du Québec-Santé (FRQS). IP is funded by the Quebec Breast Cancer Foundation, by the FRQS and the NSERC. MTH is funded by NSERC and the Canadian Institutes of Health Research (CIHR).

Author Contributions: Mélanie Busby, Isabelle Plante and Michael T. Hallett conceived and designed the experiments. Michael T. Hallett provided the access to breast cancer datasets. Mélanie Busby performed the experiments, analyzed the data and wrote the paper. Isabelle Plante and Michael T. Hallett helped draft the manuscript. All authors read and approved the final manuscript.

Conflicts of Interest: The authors declare no conflict of interest.

Abbreviations

CNA	Copy number alterations
Cx43	Connexin 43
DCIS	Ductal carcinoma in situ
ERα	Estrogen receptor alpha
ERBB2	erb-b2 receptor tyrosine kinase 2
GJIC	Gap Junction Intercellular Communication
GJA1	Gap Junction protein alpha 1, the gene encoding Connexin 43
Her2e	HER2-enriched
IDC	Invasive ductal carcinoma
LumA	Luminal A
LumB	Luminal B
PR	Progesterone Receptor
RPPA	Reverse phase protein assay
TCGA	The Cancer Genome Atlas

Appendix A

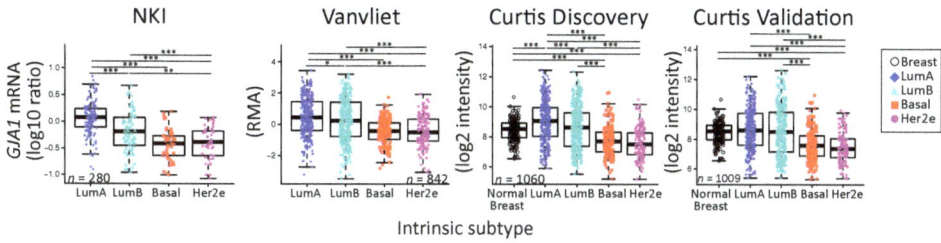

Figure A1. *GJA1* mRNA level in breast cancer clinical samples is dependent on breast cancer subtype in four datasets. *GJA1* mRNA level in tumors stratified by subtype in Vanvliet, NKI, Curtis Discovery and Curtis Validation datasets. In the legend, "Breast" indicates adjacent normal breast tissue. *p* value: * <0.05; ** <0.01; *** <0.001.

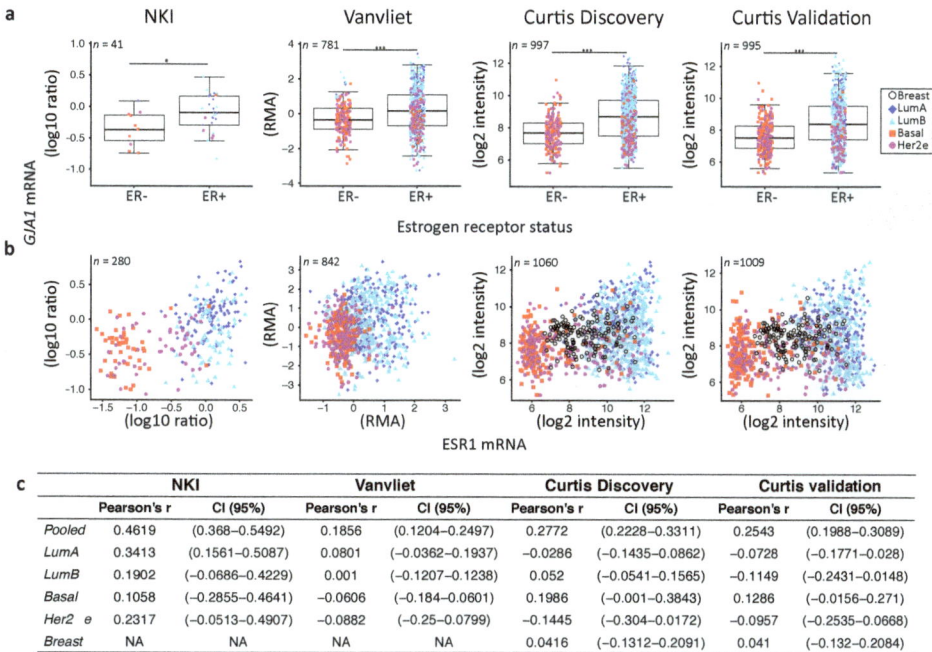

	NKI		Vanvliet		Curtis Discovery		Curtis validation	
	Pearson's r	CI (95%)	Pearson's r	CI (95%)	Pearson's r	CI (95%)	Pearson's r	CI (95%)
Pooled	0.4619	(0.368–0.5492)	0.1856	(0.1204–0.2497)	0.2772	(0.2228–0.3311)	0.2543	(0.1988–0.3089)
LumA	0.3413	(0.1561–0.5087)	0.0801	(−0.0362–0.1937)	−0.0286	(−0.1435–0.0862)	−0.0728	(−0.1771–0.028)
LumB	0.1902	(−0.0686–0.4229)	0.001	(−0.1207–0.1238)	0.052	(−0.0541–0.1565)	−0.1149	(−0.2431–0.0148)
Basal	0.1058	(−0.2855–0.4641)	−0.0606	(−0.184–0.0601)	0.1986	(−0.001–0.3843)	0.1286	(−0.0156–0.271)
Her2 e	0.2317	(−0.0513–0.4907)	−0.0882	(−0.25–0.0799)	−0.1445	(−0.304–0.0172)	−0.0957	(−0.2535–0.0668)
Breast	NA	NA	NA	NA	0.0416	(−0.1312–0.2091)	0.041	(−0.132–0.2084)

Figure A2. *GJA1* mRNA level in breast cancer clinical samples is higher in ER-positive tumors in four datasets. (**a**) *GJA1* mRNA level in breast tumors stratified by estrogen receptor status determined as described in Material and Methods. *p* value: * <0.05; ** <0.01; *** <0.001. (**b**) Expression of *GJA1* vs. estrogen receptor alpha (*ESR1*) mRNA in each subtype and in normal breast tissue. (**c**) Bootstrapped correlations between *ESR1* and *GJA1* mRNA level either in pooled breast cancer tumors or in individual intrinsic subtypes and in normal breast tissue. Data from Vanvliet, NKI, Curtis Discovery and Curtis Validation datasets.

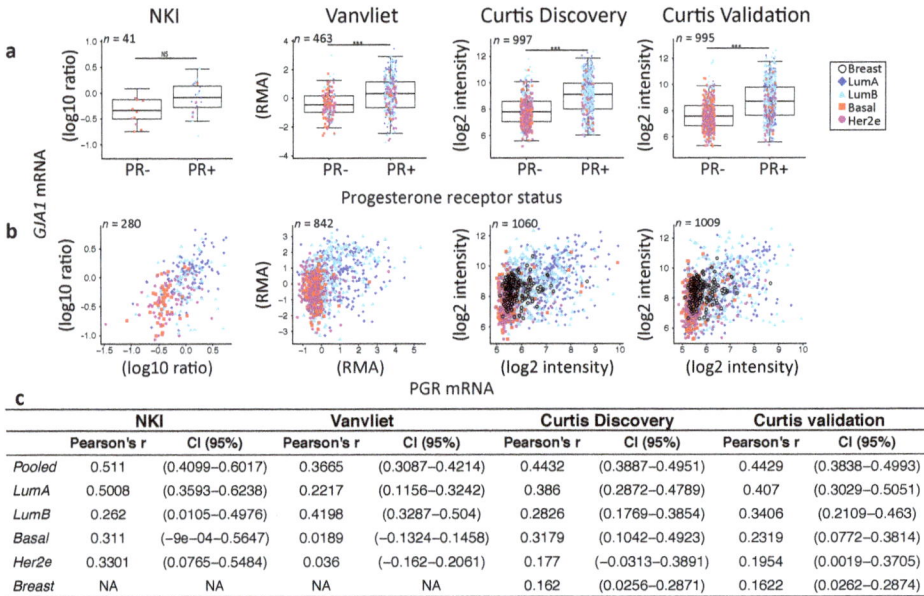

a

b

PGR mRNA

c

	NKI		Vanvliet		Curtis Discovery		Curtis validation	
	Pearson's r	CI (95%)	Pearson's r	CI (95%)	Pearson's r	CI (95%)	Pearson's r	CI (95%)
Pooled	0.511	(0.4099–0.6017)	0.3665	(0.3087–0.4214)	0.4432	(0.3887–0.4951)	0.4429	(0.3838–0.4993)
LumA	0.5008	(0.3593–0.6238)	0.2217	(0.1156–0.3242)	0.386	(0.2872–0.4789)	0.407	(0.3029–0.5051)
LumB	0.262	(0.0105–0.4976)	0.4198	(0.3287–0.504)	0.2826	(0.1769–0.3854)	0.3406	(0.2109–0.463)
Basal	0.311	(−9e−04–0.5647)	0.0189	(−0.1324–0.1458)	0.3179	(0.1042–0.4923)	0.2319	(0.0772–0.3814)
Her2e	0.3301	(0.0765–0.5484)	0.036	(−0.162–0.2061)	0.177	(−0.0313–0.3891)	0.1954	(0.0019–0.3705)
Breast	NA	NA	NA	NA	0.162	(0.0256–0.2871)	0.1622	(0.0262–0.2874)

Figure A3. *GJA1* mRNA level in breast cancer clinical samples increases with *PGR* expression in four datasets. (**a**) *GJA1* mRNA level in breast tumors stratified by PR status determined as described in Material and Methods. In the legend, ""Breast" indicates adjacent normal breast tissue. *p* value: * <0.05; ** <0.01; *** <0.001; NS Not statistically significant. (**b**) Expression of *GJA1* vs. progesterone receptor (PGR) mRNA in each subtype and in normal breast tissue. (**c**) Bootstrapped correlations between *PGR* and *GJA1* mRNA level either in pooled breast tumors or in individual breast cancer intrinsic subtypes. Data from Vanvliet, NKI, Curtis Discovery and Curtis Validation datasets.

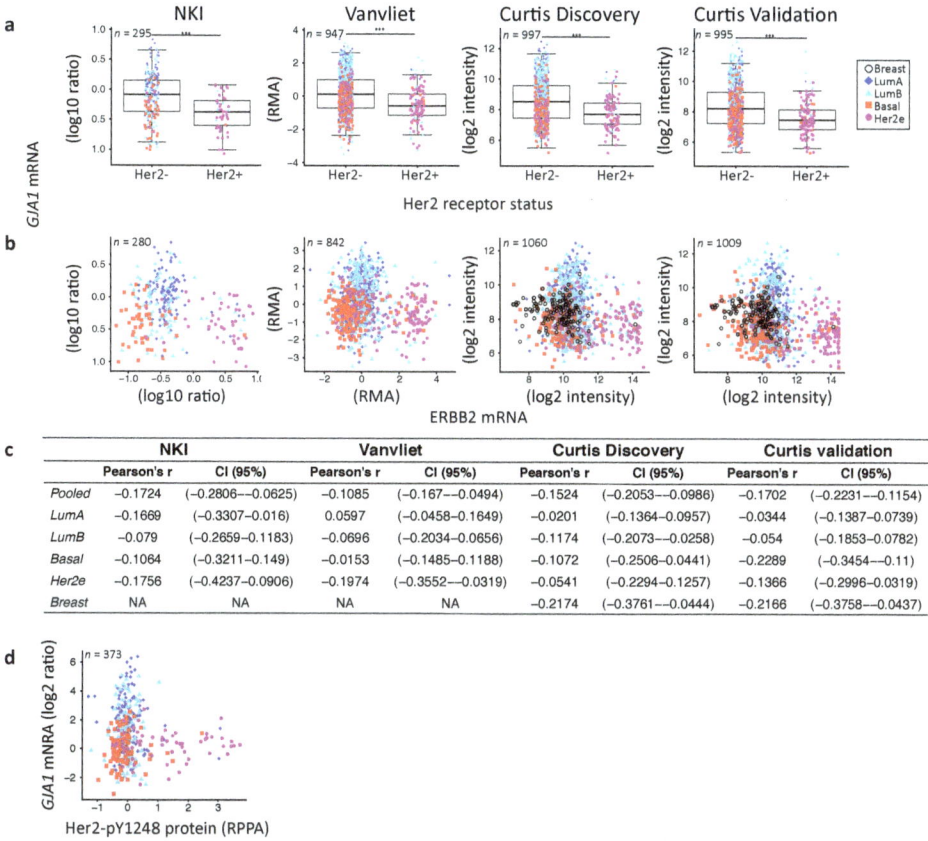

Figure A4. *GJA1* mRNA level in breast cancer clinical samples is lower in HER2-positive breast tumors in four datasets. (**a**) *GJA1* mRNA level in breast tumors stratified by HER2 status determined as described in Material and Methods. In the legend, "Breast" indicates adjacent normal breast tissue. *p* value: * <0.05; ** <0.01; *** <0.001. (**b**) Expression of *GJA1* vs. *ESR1* mRNA in each subtype and in normal breast tissue. (**c**) Bootstrapped correlations between *ERBB2* and *GJA1* mRNA level either in pooled breast cancer tumors or in individual intrinsic subtypes and in normal breast tissue. Data from Vanvliet, NKI, Curtis Discovery and Curtis validation datasets. (**d**) *GJA1* mRNA vs. pY1248 phosphorylated form of HER2 protein level determined by RPPA in breast tumors from the TCGA dataset.

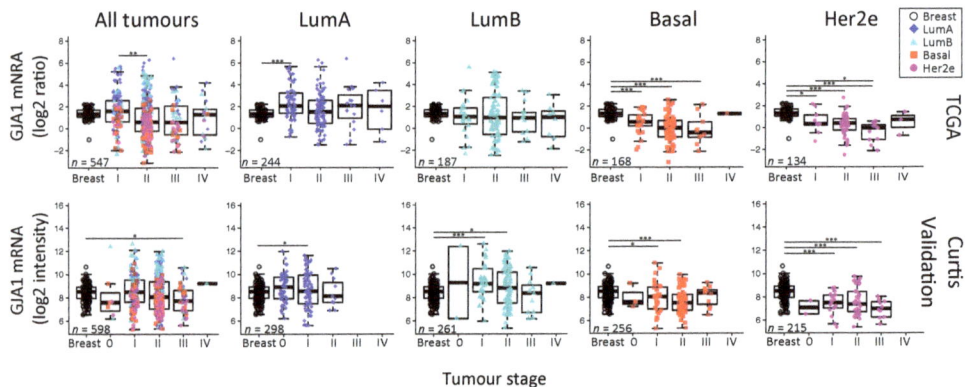

Figure A5. *GJA1* mRNA level is dysregulated in the early stages of breast cancer in clinical samples. *GJA1* mRNA level for each tumor stage either in pooled breast tumors or stratified by intrinsic subtype. Data from TCGA and Curtis Validation datasets. In the legend, "Breast" indicates adjacent normal breast tissue. *p* value: * <0.05; ** <0.01; *** <0.001.

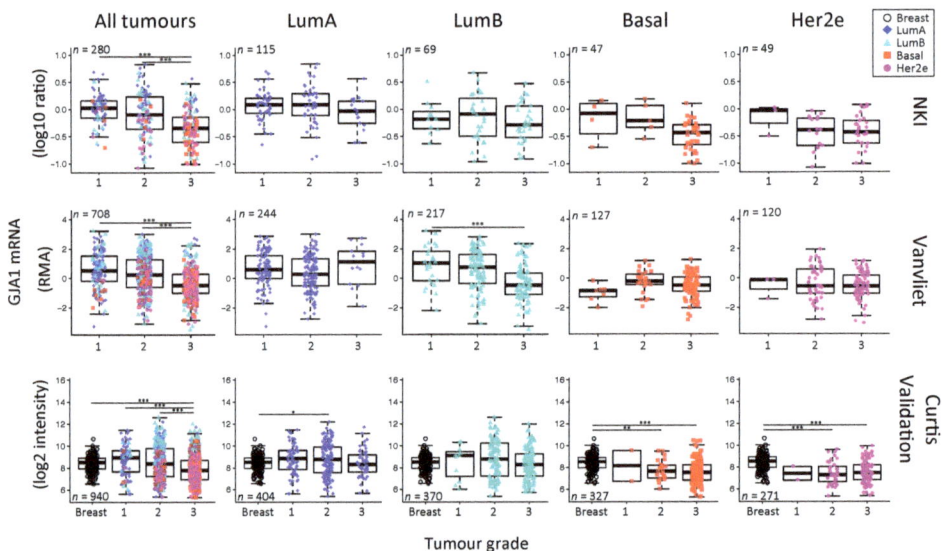

Figure A6. *GJA1* mRNA level is reduced with grade in breast cancer clinical samples but only in pooled tumors. *GJA1* mRNA level for each tumor grade, in Vanvliet, NKI and Curtis Validation datasets, either in pooled breast tumors or stratified by intrinsic subtype. In the legend, "Breast" indicates adjacent normal breast tissue. *p* value: * <0.05; ** <0.01; *** <0.001.

a

		Pooled	LumA	LumB	Basal	Her2
DFS	n	2592	807	987	419	275
	n events	1093	243	473	213	135
	Hazard ratio	0.828 (0.7351 – 0.9325)	0.8154 (0.6313 – 1.053)	0.7394 (0.6014 – 0.9091)	1.411 (0.9417 – 2.113)	1.478 (1.041 – 2.1)
	Score (logrank) test	9.71	2.45	0.7394	1.411	4.82
	p value	0.001832	0.1173	0.004042	0.09353	0.02813
	Cutoff	0.5	0.5	0.75	0.75	0.5
DDFS	n	2223	821	698	393	189
	n events	593	150	215	132	78
	Hazard ratio	0.6972 (0.5695 – 0.8535)	0.7004 (0.5073 – 0.967)	0.7667 (0.5622 – 1.046)	1.348 (0.9345 – 1.945)	3.146 (1.81 – 5.47)
	Score (logrank) test	12.34	4.73	2.83	2.57	18.31
	p value	0.0004422	0.02962	0.09237	0.1088	0.00001875
	Cutoff	0.75	0.5	0.75	0.25	0.25
OS	n	2091	800	678	293	237
	n events	539	106	217	98	98
	Hazard ratio	0.5214 (0.4149 – 0.6552)	0.4926 (0.3362 – 0.7217)	0.5532 (0.3972 – 0.7705)	1.264 (0.8412 – 1.899)	1.992 (1.308 – 3.032)
	Score (logrank) test	32.33	13.76	12.63	1.28	10.73
	p value	0.00000001302	0.0002082	0.0003797	0.2585	0.001053
	Cutoff	0.75	0.5	0.75	0.25	0.25
Combined	n	4580	1644	1474	798	459
	n events	1607	394	620	337	204
	Hazard ratio	0.761 (0.6896 – 0.8396)	0.7071 (0.5891 – 0.848)	0.7071 (0.5891 – 0.8488)	1.204 (0.9607 – 1.509)	1.63 (1.212 – 2.191)
	Score (logrank) test	29.79	13.97	13.97	2.61	10.68
	p value	0.00000004807	0.0001857	0.0001857	0.1064	0.001083
	Cutoff	0.5	0.75	0.75	0.25	0.25

b

		Pooled	LumA	LumB	Basal	Her2
RFS	n	3554	1764	1002	580	208
	Hazard ratio	0.78 (0.69 – 0.89)	0.77 (0.64 – 0.92)	0.91 (0.73 – 1.13)	1.59 (1.23 – 2.07)	1.96 (1.18 – 3.25)
	p value	0.0002	0.0032	0.387	0.00046	0.0085
DMFS	n	1609	918	361	219	111
	Hazard ratio	0.6 (0.48 – 0.74)	0.62 (0.46 – 0.84)	0.53 (0.34 – 0.83)	1.65 (0.98 – 2.76)	2.1 (1.08 – 4.1)
	p value	0.0000032	0.0018	0.0051	0.0559	0.0259

Hazard Ratio	< 1	> 1
p < 0.01		
p < 0.05		
N.S.		

Figure A7. *GJA1* is associated with different outcomes depending on the breast cancer subtype. (**a**) BreastMark's; and (**b**) KMPlotter's web interface were used to complete the survival analysis for *GJA1* expression. Disease-Free Survival (DFS), Distant Disease-Free survival (DDFS), Overall Survival (OS) and Combined Survival (Combined) are given either for pooled tumors or for individual breast cancer subtypes in Breast Mark (**a**). Relapse-Free Survival (RFS) and Distant Metastasis-Free survival are given for *GJA1* probe 201667_at in KMPlotter (**b**). The best cutoff was determined manually for BreastMark (25, 50 or 75 percentile) based on *p* value and logrank score and automatically for KMPlotter.

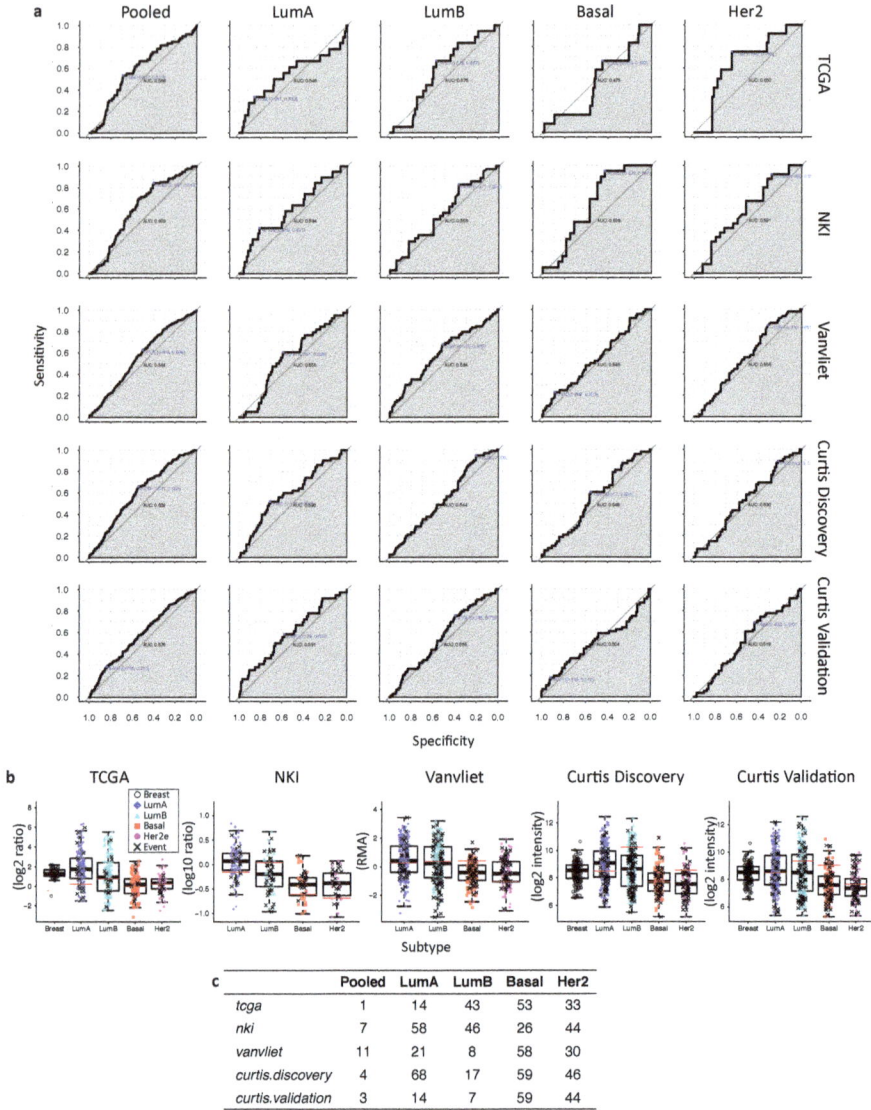

Figure A8. (**a**) Determination of the best cutoff for *GJA1* using ROC curves in individual subtypes of breast cancer. ROC curves evaluating the sensitivity vs. the specificity of *GJA1* as a marker of prognostic (occurrence of events in five years after tumor removal) in pooled tumors and each individual subtype of breast cancer of five datasets (TCGA, NKI, Vanvliet and Curtis discovery and Curtis validation). AUC, Area under the curve. The best cutoff is given in blue and is used in a subsequent survival analysis in Figure A9. (**b**) Expression of *GJA1* in pooled tumors and individual subtypes in five datasets. The red lines indicate the cutoff as determined by the ROC curves. In the legend, "Breast" indicates adjacent normal breast tissue. X indicates tumors with subsequent recurrence events. (**c**) Rank of the *GJA1* probe in pooled tumors and each individual subtype for five datasets. Rank given in percentile of the area under the ROC curve compared to all the other probes.

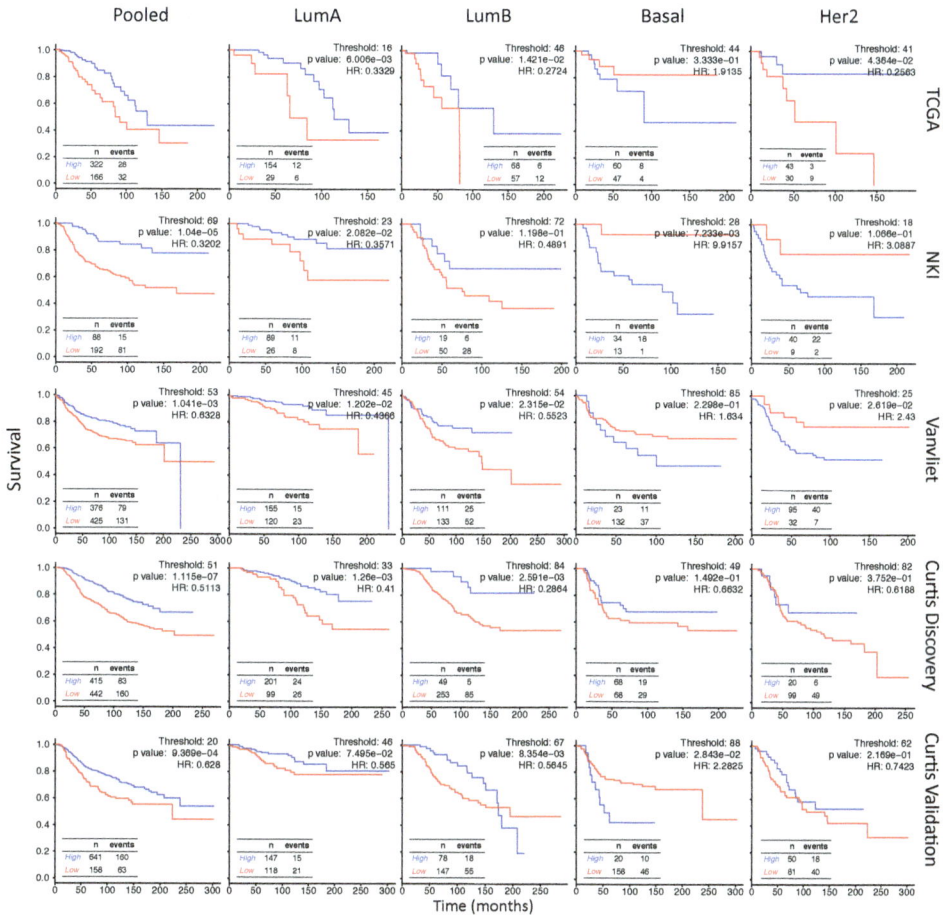

Figure A9. Survival analysis of *GJA1* in individual subtypes of breast cancer with cutoff selected according to the ROC curve. Kaplan–Meyer plots show survival curves for patients with breast tumors expressing either high (blue) or low (red) levels of *GJA1* mRNA in pooled tumors or in individual intrinsic subtypes. TCGA, NKI, Vanvliet and both Curtis datasets were used for the analysis. The best cutoff was determined from ROC curves as in Figure A8.

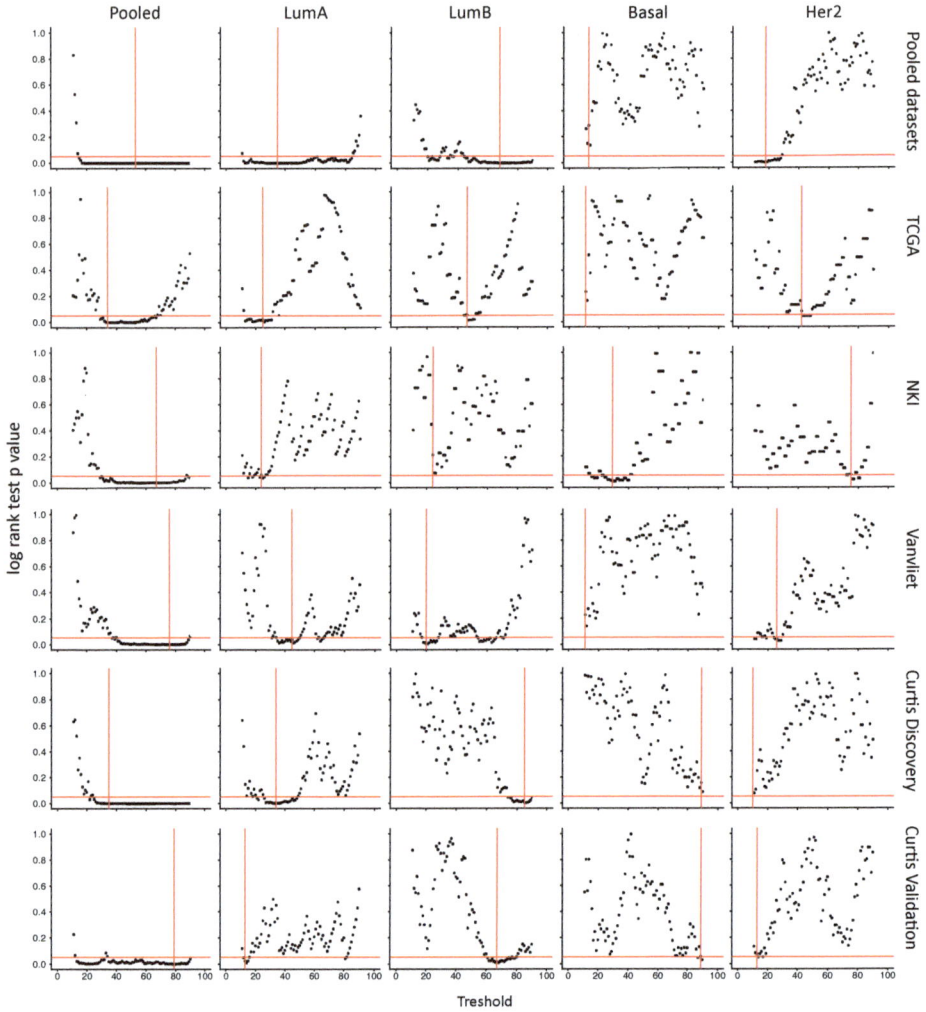

Figure A10. Determination of the best cutoff for *GJA1* using the log rank test *p* value in individual subtypes of breast cancer. Graph of the *p* value of the log rank test to compare the survival of patients with Cx43 expressed at high or low levels according to varying thresholds (10–90 percentile). Horizontal red lines indicate 0.05 *p* value. Vertical red lines indicate the threshold with the lowest *p* value for the log rank test, used for subsequent survival analysis in Figures 7 and A12. Results given for pooled tumors or individual breast cancer subtypes in our five datasets (TCGA, NKI, Vanvliet and Curtis discovery and Curtis validation) as well as in aggregated datasets (Pooled datasets).

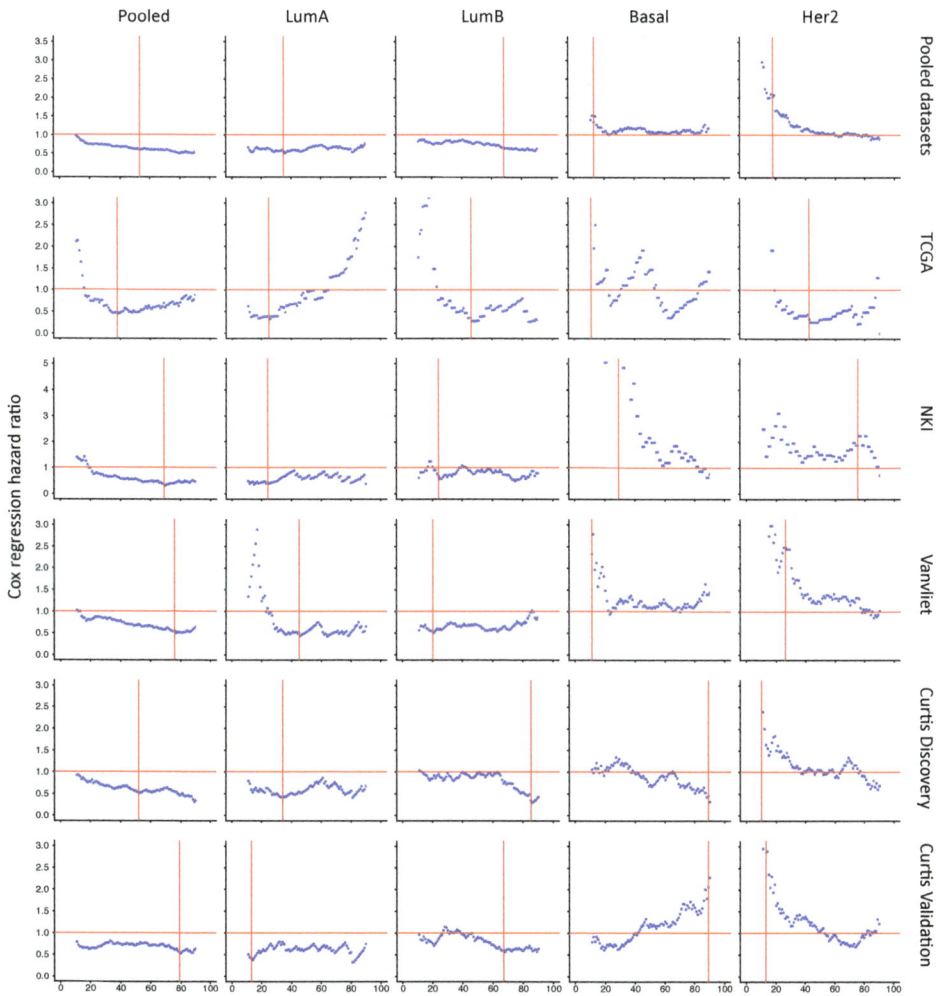

Figure A11. Cox regression hazard ratio associated with *GJA1* for different thresholds. Graph of the Cox regression hazard ratio to compare the survival of patients with Cx43 expressed at high or low levels according to varying thresholds (10–90 percentile). Horizontal red lines indicate a neutral hazard ratio of 1. Vertical red lines indicate the threshold with the lowest log rank test *p* value as determined in Figure A10, used for subsequent survival analysis. Results given for pooled tumors or individual breast cancer subtypes in our five datasets (TCGA, NKI, Vanvliet and Curtis discovery and Curtis validation) as well as in aggregated datasets (Pooled datasets).

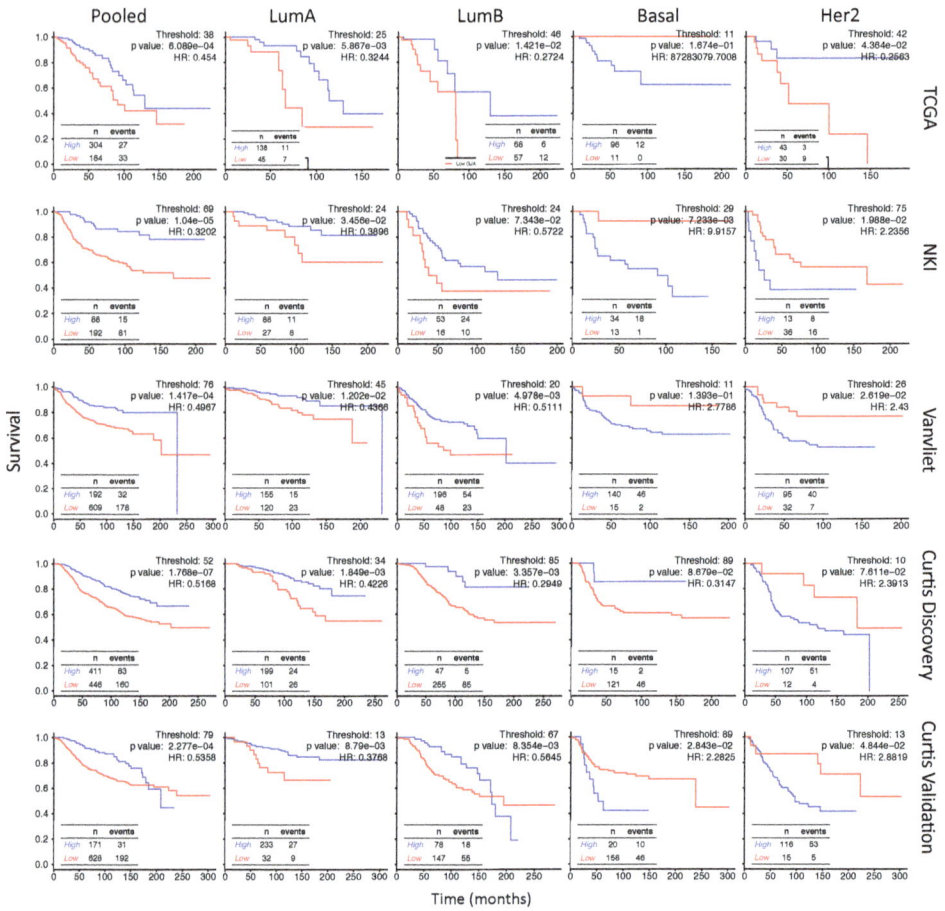

Figure A12. Survival analysis of *GJA1* in individual subtypes of breast cancer with the cutoff selected according to the log rank test *p* value. The Kaplan–Meyer plot shows survival curves for patients with breast tumors expressing either high (blue) or low (red) levels of *GJA1* mRNA in pooled tumors or in individual intrinsic subtypes. TCGA, NKI, Vanvliet and both Curtis datasets were used separately for the analysis. The best cutoff was determined as the percentile lending the lowest log rank test *p* value as in Figure A10.

References

1. Arendt, L.M.; Kuperwasser, C. Form and function: how estrogen and progesterone regulate the mammary epithelial hierarchy. *J. Mammary Gland Biol. Neoplasia* **2015**, *20*, 9–25.
2. Su, V.; Lau, A.F. Connexins: Mechanisms regulating protein levels and intercellular communication. *FEBS Lett.* **2014**, *588*, 1212–1220.
3. Plante, I.; Stewart, M.K.; Barr, K.; Allan, A.L.; Laird, D.W. Cx43 suppresses mammary tumor metastasis to the lung in a Cx43 mutant mouse model of human disease. *Oncogene* **2011**, *30*, 1681–1692.
4. Dianati, E.; Poiraud, J.; Weber-Ouellette, A.; Plante, I. Connexins, E-cadherin, Claudin-7 and beta-catenin transiently form junctional nexuses during the post-natal mammary gland development. *Dev. Biol.* **2016**, *416*, 52–68.

5. Monaghan, P.; Clarke, C.; Perusinghe, N.P.; Moss, D.W.; Chen, X.Y.; Evans, W.H. Gap junction distribution and connexin expression in human breast. *Exp. Cell Res.* **1996**, *223*, 29–38.

6. Talhouk, R.S.; Elble, R.C.; Bassam, R.; Daher, M.; Sfeir, A.; Mosleh, L.A.; El-Khoury, H.; Hamoui, S.; Pauli, B.U.; El-Sabban, M.E. Developmental expression patterns and regulation of connexins in the mouse mammary gland: Expression of connexin30 in lactogenesis. *Cell Tissue Res.* **2005**, *319*, 49–59.

7. Pitelka, D.R.; Hamamoto, S.T.; Duafala, J.G.; Nemanic, M.K. Cell contacts in the mouse mammary gland. I. Normal gland in postnatal development and the secretory cycle. *J. Cell Biol.* **1973**, *56*, 797–818.

8. Pollmann, M.A.; Shao, Q.; Laird, D.W.; Sandig, M. Connexin 43 mediated gap junctional communication enhances breast tumor cell diapedesis in culture. *Breast Cancer Res.* **2005**, *7*, R522–R534.

9. Tomasetto, C.; Neveu, M.J.; Daley, J.; Horan, P.K.; Sager, R. Specificity of gap junction communication among human mammary cells and connexin transfectants in culture. *J. Cell Biol.* **1993**, *122*, 157–167.

10. Woodward, T.L.; Sia, M.A.; Blaschuk, O.W.; Turner, J.D.; Laird, D.W. Deficient epithelial-fibroblast heterocellular gap junction communication can be overcome by co-culture with an intermediate cell type but not by E-cadherin transgene expression. *J. Cell Sci.* **1998**, *111*, 3529–3539.

11. Naus, C.C.; Laird, D.W. Implications and challenges of connexin connections to cancer. *Nat. Rev. Cancer* **2010**, *10*, 435–441.

12. Laird, D.W.; Fistouris, P.; Batist, G.; Alpert, L.; Huynh, H.T.; Carystinos, G.D.; Alaoui-Jamali, M.A. Deficiency of connexin43 gap junctions is an independent marker for breast tumors. *Cancer Res.* **1999**, *59*, 4104–4110.

13. Lee, S.W.; Tomasetto, C.; Paul, D.; Keyomarsi, K.; Sager, R. Transcriptional downregulation of gap-junction proteins blocks junctional communication in human mammary tumor cell lines. *J. Cell Biol.* **1992**, *118*, 1213–1221.

14. Gould, V.E.; Mosquera, J.M.; Leykauf, K.; Gattuso, P.; Durst, M.; Alonso, A. The phosphorylated form of connexin43 is up-regulated in breast hyperplasias and carcinomas and in their neoformed capillaries. *Hum. Pathol.* **2005**, *36*, 536–545.

15. Jamieson, S.; Going, J.J.; D'Arcy, R.; George, W.D. Expression of gap junction proteins connexin 26 and connexin 43 in normal human breast and in breast tumours. *J. Pathol.* **1998**, *184*, 37–43.

16. Kanczuga-Koda, L.; Sulkowska, M.; Koda, M.; Reszec, J.; Famulski, W.; Baltaziak, M.; Sulkowski, S. Expression of connexin 43 in breast cancer in comparison with mammary dysplasia and the normal mammary gland. *Folia Morphol. (Warsz)* **2003**, *62*, 439–442.

17. El-Saghir, J.A.; El-Habre, E.T.; El-Sabban, M.E.; Talhouk, R.S. Connexins: A junctional crossroad to breast cancer. *Int. J. Dev. Biol.* **2011**, *55*, 773–780.

18. Nicolson, G.L.; Dulski, K.M.; Trosko, J.E. Loss of intercellular junctional communication correlates with metastatic potential in mammary adenocarcinoma cells. *Proc. Natl. Acad. Sci. USA* **1988**, *85*, 473–476.

19. Ren, J.; Hamada, J.; Takeichi, N.; Fujikawa, S.; Kobayashi, H. Ultrastructural differences in junctional intercellular communication between highly and weakly metastatic clones derived from rat mammary carcinoma. *Cancer Res.* **1990**, *50*, 358–362.

20. Li, Z.; Zhou, Z.; Welch, D.R.; Donahue, H.J. Expressing connexin 43 in breast cancer cells reduces their metastasis to lungs. *Clin. Exp. Metastasis* **2008**, *25*, 893–901.

21. Qin, H.; Shao, Q.; Curtis, H.; Galipeau, J.; Belliveau, D.J.; Wang, T.; Alaoui-Jamali, M.A.; Laird, D.W. Retroviral delivery of connexin genes to human breast tumor cells inhibits in vivo tumor growth by a mechanism that is independent of significant gap junctional intercellular communication. *J. Biol. Chem.* **2002**, *277*, 29132–29138.

22. El Sabban, M.E.; Pauli, B.U. Adhesion-mediated gap junctional communication between lung-metastatic cancer cells and endothelium. *Invasion Metastasis* **1994**, *14*, 164–176.

23. Elzarrad, M.K.; Haroon, A.; Willecke, K.; Dobrowolski, R.; Gillespie, M.N.; Al-Mehdi, A.B. Connexin-43 upregulation in micrometastases and tumor vasculature and its role in tumor cell attachment to pulmonary endothelium. *BMC Med.* **2008**, *6*, 20–36.

24. Stoletov, K.; Strnadel, J.; Zardouzian, E.; Momiyama, M.; Park, F.D.; Kelber, J.A.; Pizzo, D.P.; Hoffman, R.; Vandenberg, S.R.; Klemke, R.L. Role of connexins in metastatic breast cancer and melanoma brain colonization. *J. Cell Sci.* **2013**, *126*, 904–913.

25. Kanczuga-Koda, L.; Sulkowski, S.; Lenczewski, A.; Koda, M.; Wincewicz, A.; Baltaziak, M.; Sulkowska, M. Increased expression of connexins 26 and 43 in lymph node metastases of breast cancer. *J. Clin. Pathol.* **2006**, *59*, 429–433.

26. Sorlie, T.; Perou, C.M.; Tibshirani, R.; Aas, T.; Geisler, S.; Johnsen, H.; Hastie, T.; Eisen, M.B.; van de Rijn, M.; Jeffrey, S.S.; et al. Gene expression patterns of breast carcinomas distinguish tumor subclasses with clinical implications. *Proc. Natl. Acad. Sci. USA* **2001**, *98*, 10869–10874.

27. Tofigh, A.; Suderman, M.; Paquet, E.R.; Livingstone, J.; Bertos, N.; Saleh, S.M.; Zhao, H.; Souleimanova, M.; Cory, S.; Lesurf, R.; et al. The prognostic ease and difficulty of invasive breast carcinoma. *Cell Rep.* **2014**, *9*, 129–142.

28. Uhlen, M.; Bjorling, E.; Agaton, C.; Szigyarto, C.A.; Amini, B.; Andersen, E.; Andersson, A.C.; Angelidou, P.; Asplund, A.; Asplund, C.; et al. A human protein atlas for normal and cancer tissues based on antibody proteomics. *Mol. Cell Proteom.* **2005**, *4*, 1920–1932.

29. Gendoo, D.M.; Ratanasirigulchai, N.; Schroder, M.S.; Pare, L.; Parker, J.S.; Prat, A.; Haibe-Kains, B. Genefu: An R/Bioconductor package for computation of gene expression-based signatures in breast cancer. *Bioinformatics* **2016**, *32*, 1097–1099.

30. Mitra, S.; Annamalai, L.; Chakraborty, S.; Johnson, K.; Song, X.H.; Batra, S.K.; Mehta, P.P. Androgen-regulated formation and degradation of gap junctions in androgen-responsive human prostate cancer cells. *Mol. Biol. Cell* **2006**, *17*, 5400–5416.

31. Ren, J.; Wang, X.H.; Wang, G.C.; Wu, J.H. 17beta estradiol regulation of connexin 43-based gap junction and mechanosensitivity through classical estrogen receptor pathway in osteocyte-like MLO-Y4 cells. *Bone* **2013**, *53*, 587–596.

32. Stewart, M.; Simek, J.; Laird, D. Insights into the role of Connexins in Mammary Gland Morphogenesis and Function. *Reproduction* **2015**, *149*, R279–R290.

33. Teleki, I.; Szasz, A.M.; Maros, M.E.; Gyorffy, B.; Kulka, J.; Meggyeshazi, N.; Kiszner, G.; Balla, P.; Samu, A.; Krenacs, T. Correlations of differentially expressed gap junction connexins cx26, cx30, cx32, cx43 and cx46 with breast cancer progression and prognosis. *PLoS ONE* **2014**, *9*, e112541.

34. Yeh, E.S.; Williams, C.J.; Williams, C.B.; Bonilla, I.V.; Klauber-DeMore, N.; Phillips, S.L. Dysregulated connexin 43 in HER2-positive drug resistant breast cancer cells enhances proliferation and migration. *Oncotarget* **2017**, *8*, 109358–109369.

35. Gyorffy, B.; Lanczky, A.; Eklund, A.C.; Denkert, C.; Budczies, J.; Li, Q.; Szallasi, Z. An online survival analysis tool to rapidly assess the effect of 22,277 genes on breast cancer prognosis using microarray data of 1809 patients. *Breast Cancer Res. Treat.* **2010**, *123*, 725–731.

36. Park, S.Y.; Lee, H.E.; Li, H.; Shipitsin, M.; Gelman, R.; Polyak, K. Heterogeneity for Stem Cell-Related Markers According to Tumor Subtype and Histologic Stage in Breast Cancer. *Clin. Cancer Res.* **2010**, *16*, 876–887.

37. Fu, Y.; Shao, Z.M.; He, Q.Z.; Jiang, B.Q.; Wu, Y.; Zhuang, Z.G. Hsa-miR-206 represses the proliferation and invasion of breast cancer cells by targeting Cx43. *Eur. Rev. Med. Pharmacol. Sci.* **2015**, *19*, 2091–2104.

38. Clarke, C.; Madden, S.F.; Doolan, P.; Aherne, S.T.; Joyce, H.; O'Driscoll, L.; Gallagher, W.M.; Hennessy, B.T.; Moriarty, M.; Crown, J.; et al. Correlating transcriptional networks to breast cancer survival: A large-scale coexpression analysis. *Carcinogenesis* **2013**, *34*, 2300–2308.

39. Yoshihara, K.; Shahmoradgoli, M.; Martinez, E.; Vegesna, R.; Kim, H.; Torres-Garcia, W.; Trevino, V.; Shen, H.; Laird, P.W.; Levine, D.A.; et al. Inferring tumour purity and stromal and immune cell admixture from expression data. *Nat. Commun.* **2013**, *4*, 2612–2622.

40. Grummer, R.; Traub, O.; Winterhager, E. Gap junction connexin genes cx26 and cx43 are differentially regulated by ovarian steroid hormones in rat endometrium. *Endocrinology* **1999**, *140*, 2509–2516.

41. Gulinello, M.; Etgen, A.M. Sexually dimorphic hormonal regulation of the gap junction protein, CX43, in rats and altered female reproductive function in CX43+/− mice. *Brain Res.* **2005**, *1045*, 107–115.

42. Nadeem, L.; Shynlova, O.; Matysiak-Zablocki, E.; Mesiano, S.; Dong, X.; Lye, S. Molecular evidence of functional progesterone withdrawal in human myometrium. *Nat. Commun.* **2016**, *7*, 11565.

43. Yu, J.; Berga, S.L.; Johnston-MacAnanny, E.B.; Sidell, N.; Bagchi, I.C.; Bagchi, M.K.; Taylor, R.N. Endometrial Stromal Decidualization Responds Reversibly to Hormone Stimulation and Withdrawal. *Endocrinology* **2016**, *157*, 2432–2446.

44. Conklin, C.; Huntsman, D.; Yorida, E.; Makretsov, N.; Turbin, D.; Bechberger, J.F.; Sin, W.C.; Naus, C.C. Tissue microarray analysis of connexin expression and its prognostic significance in human breast cancer. *Cancer Lett.* **2007**, *255*, 284–294.

45. Teleki, I.; Krenacs, T.; Szasz, M.A.; Kulka, J.; Wichmann, B.; Leo, C.; Papassotiropoulos, B.; Riemenschnitter, C.; Moch, H.; Varga, Z. The potential prognostic value of connexin 26 and 46 expression in neoadjuvant-treated breast cancer. *BMC Cancer* **2013**, *13*, 50.

46. Murphy, L.C.; Seekallu, S.V.; Watson, P.H. Clinical significance of estrogen receptor phosphorylation. *Endocr. Relat. Cancer* **2011**, *18*, R1–R14.

47. Curtis, C.; Shah, S.P.; Chin, S.F.; Turashvili, G.; Rueda, O.M.; Dunning, M.J.; Speed, D.; Lynch, A.G.; Samarajiwa, S.; Yuan, Y.; et al. The genomic and transcriptomic architecture of 2000 breast tumours reveals novel subgroups. *Nature* **2012**, *486*, 346–352.

48. Mesnil, M.; Crespin, S.; Avanzo, J.L.; Zaidan-Dagli, M.L. Defective gap junctional intercellular communication in the carcinogenic process. *Biochim. Biophys. Acta* **2005**, *1719*, 125–145.

49. Valastyan, S.; Weinberg, R.A. Tumor metastasis: Molecular insights and evolving paradigms. *Cell* **2011**, *147*, 275–292.

50. Van Vliet, M.H.; Reyal, F.; Horlings, H.M.; van de Vijver, M.J.; Reinders, M.J.; Wessels, L.F. Pooling breast cancer datasets has a synergetic effect on classification performance and improves signature stability. *BMC Genom.* **2008**, *9*, 375–397.

51. TCGA. Comprehensive molecular portraits of human breast tumours. *Nature* **2012**, *490*, 61–70.

52. Van de Vijver, M.J.; He, Y.D.; van't Veer, L.J.; Dai, H.; Hart, A.A.; Voskuil, D.W.; Schreiber, G.J.; Peterse, J.L.; Roberts, C.; Marton, M.J.; et al. A gene-expression signature as a predictor of survival in breast cancer. *N. Engl. J. Med.* **2002**, *347*, 1999–2009.

53. Hazan, R.; Margolis, B.; Dombalagian, M.; Ullrich, A.; Zilberstein, A.; Schlessinger, J. Identification of autophosphorylation sites of HER2/neu. *Cell Growth Differ.* **1990**, *1*, 3–7.

54. Mertins, P.; Mani, D.R.; Ruggles, K.V.; Gillette, M.A.; Clauser, K.R.; Wang, P.; Wang, X.; Qiao, J.W.; Cao, S.; Petralia, F.; et al. Proteogenomics connects somatic mutations to signalling in breast cancer. *Nature* **2016**, *534*, 55–62.

55. Ciriello, G.; Gatza, M.L.; Beck, A.H.; Wilkerson, M.D.; Rhie, S.K.; Pastore, A.; Zhang, H.; McLellan, M.; Yau, C.; Kandoth, C.; et al. Comprehensive Molecular Portraits of Invasive Lobular Breast Cancer. *Cell* **2015**, *163*, 506–519.

56. Gao, J.; Aksoy, B.A.; Dogrusoz, U.; Dresdner, G.; Gross, B.; Sumer, S.O.; Sun, Y.; Jacobsen, A.; Sinha, R.; Larsson, E.; et al. Integrative analysis of complex cancer genomics and clinical profiles using the cBioPortal. *Sci. Signal.* **2013**, *6*, 11–30.

57. Therneau, T.M.; Grambsch, P.M. *Modeling Survival Data: Extending the Cox Model*; Springer: New York, NY, USA, 2000.

58. Robin, X.; Turck, N.; Hainard, A.; Tiberti, N.; Lisacek, F.; Sanchez, J.C.; Muller, M. pROC: An open-source package for R and S+ to analyze and compare ROC curves. *BMC Bioinform.* **2011**, *12*, 77.

59. Madden, S.F.; Clarke, C.; Gaule, P.; Aherne, S.T.; O'Donovan, N.; Clynes, M.; Crown, J.; Gallagher, W.M. BreastMark: An integrated approach to mining publicly available transcriptomic datasets relating to breast cancer outcome. *Breast Cancer Res.* **2013**, *15*, R52–R66.

60. R Core Team. *R: A Language and Environment for Statistical Computing*; R Foundation for Statistical Computing: Vienna, Austria, 2016.

61. Ritchie, M.E.; Phipson, B.; Wu, D.; Hu, Y.; Law, C.W.; Shi, W.; Smyth, G.K. limma powers differential expression analyses for RNA-sequencing and microarray studies. *Nucl. Acids Res.* **2015**, *43*, e47.

International Journal of
Molecular Sciences

MDPI

Communication

Intrinsic Oncogenic Function of Intracellular Connexin26 Protein in Head and Neck Squamous Cell Carcinoma Cells

Nobuko Iikawa [1,2], **Yohei Yamamoto** [1], **Yohei Kawasaki** [2], **Aki Nishijima-Matsunobu** [1], **Maya Suzuki** [1], **Takechiyo Yamada** [2] **and Yasufumi Omori** [1,*]

[1] Department of Molecular and Tumour Pathology, Akita University Graduate School of Medicine, Akita 010-8543, Japan; iikawan@med.akita-u.ac.jp (N.I.); cificap@med.akita-u.ac.jp (Y.Y.); akinishijima@med.akita-u.ac.jp (A.N.-M.); maya@med.akita-u.ac.jp (M.S.)
[2] Department of Otorhinolaryngology and Head-and-Neck Surgery, Akita University Graduate School of Medicine, Akita 010-8543, Japan; kawa0807@med.akita-u.ac.jp (Y.K.); ymdtkcy@med.akita-u.ac.jp (T.Y.)
* Correspondence: yasu@med.akita-u.ac.jp; Tel.: +81-18-884-6059

Received: 6 May 2018; Accepted: 21 July 2018; Published: 23 July 2018

Abstract: It has long been known that the gap junction is down-regulated in many tumours. One of the downregulation mechanisms is the translocation of connexin, a gap junction protein, from cell membrane into cytoplasm, nucleus, or Golgi apparatus. Interestingly, as tumours progress and reinforce their malignant phenotype, the amount of aberrantly-localised connexin increases in different malignant tumours including oesophageal squamous cell carcinoma, thus suggesting that such an aberrantly-localised connexin should be oncogenic, although gap junctional connexins are often tumour-suppressive. To define the dual roles of connexin in head and neck squamous cell carcinoma (HNSCC), we introduced the wild-type connexin26 (wtCx26) or the mutant Cx26 (icCx26) gene, the product of which carries the amino acid sequence AKKFF, an endoplasmic reticulum-Golgi retention signal, at the C-terminus and is not sorted to cell membrane, into the human FaDu hypopharyngeal cancer cell line that had severely impaired the expression of connexin during carcinogenesis. wtCx26 protein was trafficked to the cell membrane and formed gap junction, which successfully exerted cell-cell communication. On the other hand, the icCx26 protein was co-localised with a Golgi marker, as revealed by immunofluorescence, and thus was retained on the way to the cell membrane. While the forced expression of wtCx26 suppressed both cell proliferation in vitro and tumorigenicity in mice in vivo, icCx26 significantly enhanced both cell proliferation and tumorigenicity compared with the mock control clones, indicating that an excessive accumulation of connexin protein in intracellular domains should be involved in cancer progression and that restoration of proper subcellular sorting of connexin might be a therapeutic strategy to control HNSCC.

Keywords: gap junction; cell-cell communication; connexin26; head and neck squamous cell carcinoma; ER-Golgi retention signal; cancer progression

---

1. Introduction

Gap junction (GJ) is an intercellular channel directly connecting the cytoplasms of two adjacent cells, which then exchange water-soluble small molecules (<1 kDa) through GJ [1]. Serving as a tool for cell-cell communication, GJ plays essential roles in the homeostasis of cellular society. To the contrary, it has been well known that failure in gap junctional intercellular communication (GJIC) is profoundly involved in serious disorders and diseases of various organs, including arrhythmia [2], sensorineural hearing loss [3], Charcot-Marie-Tooth disease [4], and even cancers [5]. A gap junctional channel comprises two membrane-integrated hemichannels provided by each of two adjacent cells, and each hemichannel is composed of hexameric connexin molecules to become functional. In mammals, more than 20 molecular species have so far been identified in a connexin protein family [6,7]. Usually, several different connexin proteins are co-expressed in a single cell and the combination of the expressed connexin proteins varies tissue to tissue, i.e., while the keratinocytes of the epidermis express many connexin proteins such as connexin26, 30, 30.3, 31.1, 32, 43, and 45 [8], the cardiomyocytes express connexin40, 43, and 45 [9].

A large number of reports from us and others have established that GJIC suppresses tumour promotion in carcinogenesis by restraining the cancer-initiated cells in cellular society and that downregulation of GJIC leads to cancer development, as proven by a considerable number of experiments both in vitro and in vivo [10]. In this context, since the connexin protein localised in cytoplasm cannot participate in GJ formation, translocation of connexin protein from cell membrane to cytoplasm is considered to be loss-of-function of GJ in terms of tumour suppression. However, there is a growing body of evidence indicating that an excessive accumulation of connexin protein in cytoplasm and/or organelles enhances cancer progression such as invasion and metastasis [11–14]. Notably in squamous cell carcinoma of the head and neck and the oesophagus, the expression level of connexin26 (Cx26) in various intracellular domains correlates to the grade of malignancy or the extent of lymph node metastasis [15,16]. Furthermore, we have previously reported that accumulation of connexin32 (Cx32) in Golgi apparatus increases cancer stem cells in number and enhances the metastatic ability of the cell lines derived from human hepatocellular carcinoma [17].

In the present study, to define the roles of connexin protein localised in a Golgi area in the malignant phenotype of human head and neck squamous cell carcinoma (HNSCC), we transduced human FaDu hypopharyngeal cancer cells with the retrovirus vector carrying the mutant Cx26 cDNA which encoded intracellular Cx26 (icCx26) protein and compared their cancerous behaviours with those of the clones transduced with the wild-type Cx26 (wtCx26) or the empty vector. Our different analyses finally indicated that while wtCx26 protein integrated into GJ at cell membrane functioned tumour-suppressively, icCx26 protein rather reinforced malignant phenotype in FaDu cells.

2. Results

2.1. AKKFF Amino Acid Sequence Successfully Retains Cx26 Protein in a Golgi Area in FaDu Cells

As previously reported, an excessive accumulation of Cx32 protein in Golgi apparatus reinforces the malignant phenotype of human hepatoma HuH7 cells, suggesting that connexin protein in Golgi apparatus might have a distinct function from GJ [18]. Thus, to define the roles of intracellular connexin protein in the malignant phenotype in the context of HNSCC, the mutant Cx26 protein in which the amino acid sequence AKKFF, an endoplasmic reticulum (ER)-Golgi retention signal [19], was added to the C-terminus was overexpressed in human FaDu hypopharyngeal HNSCC cells by retroviral transduction. As shown in Figure 1a, the control mock-transduced FaDu cells express a negligible amount of endogenous Cx26 protein. On the other hand, the clones transduced by either wtCx26 or icCx26 construct overexpress respective corresponding proteins. Consistently with the immunoblotting, wtCx26 protein successfully provides punctuate strong fluorescent signals at a cell-cell contact area, indicating an efficient formation of GJ plaques (Figure 1b). In contrast, the icCx26 protein covalently-conjugated with the ER-Golgi retention signal fails to locate in cell membrane and is co-localised with a Golgi marker GM130, indicating a dense accumulation of icCx26 protein in a Golgi area. The amino acid sequence AKKFF we used as an ER-Golgi retention signal has been reported to target connexin43 (Cx43) protein to the ER-Golgi intermediate compartment (ERGIC) [19]. As shown in Figure 1b, immunofluorescent signals given by icCx26 protein appear much larger than ERGIC [20]. We thus interpret that icCx26 protein is localised not only in ERGIC but also inside or on Golgi apparatus. It has been known that Cx26, unlike Cx32 and Cx43, takes an alternative pathway instead of the secretory pathway as a membrane trafficking route [21]. However, as shown in Figure A1,

GJ plaques composed of wtCx26 are disrupted in the presence of Brefeldin A, suggesting that wtCx26 could take the secretory pathway in our FaDu cells as observed in mouse keratinocytes [22].

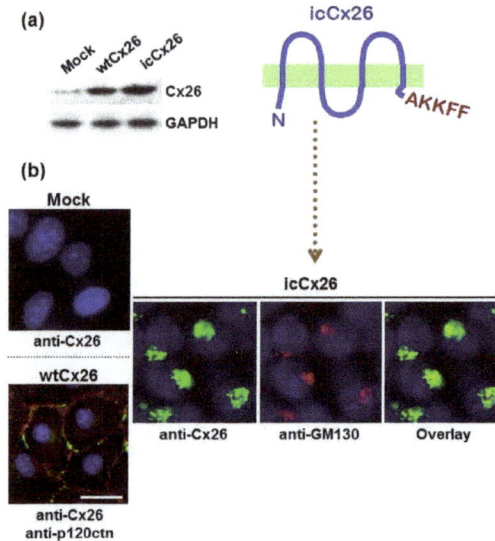

Figure 1. Expression and subcellular localisation of wtCx26 or icCx26 protein in the FaDu clones retrovirally-transduced by each construct examined. (**a**) Immunoblotting of Cx26 protein expressed in the wtCx26, icCx26, and mock clones. The expression of glyceraldehyde-3-phosphate dehydrogenase (GAPDH) was examined as a loading control. (**b**) Indirect immunofluorescence of Cx26, p120catenin (p120ctn), and GM130 proteins in the FaDu clones. The fluorescent signals of p120ctn protein were visualised by Alexa-568 (orange to red) and indicate a juxtamembrane area in the wtCx26 clone. Nuclei were stained with diamidine phenylindole dihydrochloride (DAPI). Note that signals of both Cx26 and GM130 proteins are co-localised in the icCx26 clone (overlay). Scale bar, 20 μm.

2.2. wtCx26 but Not icCx26 Protein Has the Ability to Exert GJIC

As mentioned above, wtCx26 protein is sorted to cell membrane and is capable of forming GJ plaques in a cell-cell contact area of FaDu cells. To examine whether the wild type-mediated GJs are indeed functional in FaDu cells, we performed a scrape loading dye-coupling assay. As shown in Figure 2a, while the primarily-scraped cells are co-stained by rhodamine B isothiocyanate (RITC)-dextran and Lucifer yellow in all the clones examined, the clone transduced with wtCx26 but neither icCx26 nor the mock construct is positive only for Lucifer yellow in a zone adjacent to RITC-positive cells. Since, unlike RITC-dextran, Lucifer yellow can pass through GJ, the cells in the Lucifer yellow-positive zone are considered to have received the dye through functional GJs. As predicted, these results clearly indicate that icCx26 protein cannot contribute to GJ formation in cell membrane and is thus non-functional as a GJ protein (Figure 2b).

2.3. wtCx26 and icCx26 Proteins Regulate Cell Proliferation and Invasion in a Reciprocal Manner

To examine the effects of intracellular accumulation of Cx26 protein on cell proliferation, each clone transduced with wtCx26, icCx26, or the mock construct was plated in 60-mm dishes in triplicate and the cell number was counted with hemocytometer. As shown by growth curve (Figure 3a), the proliferation rate of FaDu cell clone overexpressing wtCx26 protein is significantly lower than that of the mock clone. More interestingly, overexpression of icCx26 protein retained in a Golgi area has remarkably elevated the proliferation rate compared with the mock clone. It has been known that

GJ-mediated modulation of cell proliferation is often most obvious in the alteration of saturation density [23,24]. It is also the case with our experiments, i.e., the saturation densities of wtCx26- and icCx26-transduced clones are approximately 60% and 180% of that of the mock clone, respectively (Figure 3a). Taken together, while cell proliferation is suppressed by Cx26 protein integrated into GJ, GJ-independent Cx26 protein localised in a Golgi area enhances cell proliferation.

We further investigated whether overexpression of icCx26 protein could affect invasiveness of FaDu cells by evaluating the ability of each clone to invade the basement membrane matrix. Similarly to other malignant phenotypes, Figure 3b demonstrates that overexpression of icCx26 and wtCx26 proteins enhances and declines the invasiveness of FaDu cells, respectively.

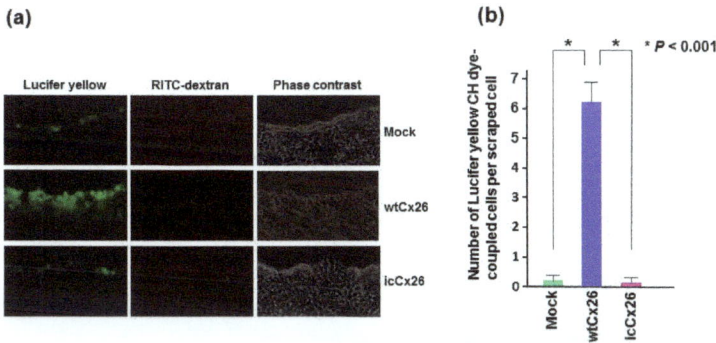

Figure 2. Scrape-loading dye-transfer assay to measure GJIC ability. The wtCx26, icCx26, and mock clones were soaked in a cocktail of Lucifer yellow CH and RITC-dextran, scraped by a micropipette tip, and observed under a fluorescence microscope after 5 min of incubation. (**a**) Representative micrographs of 3 different clones. The same fields of each clone were captured. Note that dye-coupled cells with Lucifer yellow CH were observed only in the wtCx26 clone. (**b**) Histogram showing the mean GJIC capacity of each clone. Error bars represent the SD ($n = 6$).

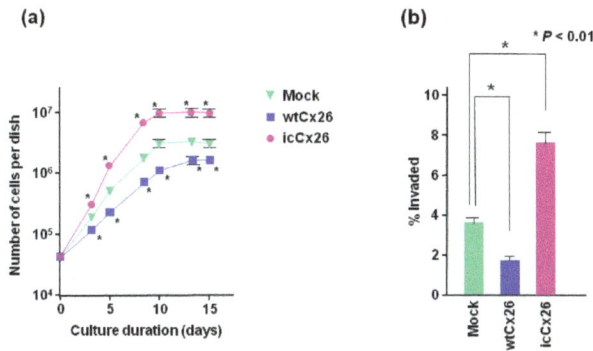

Figure 3. Effects of wtCx26 and icCx26 proteins on cell proliferation and invasion capacity in vitro of FaDu cells. (**a**) Growth curve of each clone of FaDu cells. The wtCx26, icCx26, and mock clones were cultured for the indicated periods. The cells were counted every 2 or 3 days in triplicate dishes. Error bars represent the SD ($n = 3$). No error bar is indicated when the SD is too small to show. * $p < 0.001$ (significantly different from the mock clone at the corresponding time point). (**b**) Invasion capacity of each clone into the matrix basement membrane. The cells were seeded onto Matrigel, which had been settled on cell culture inserts in advance. The cells that infiltrated into the Matrigel layer were counted and their proportion to the total cell number is indicated. Error bars represent the SD ($n = 6$).

2.4. icCx26 Protein Reinforces Tumorigenicity of FaDu Cells in Nude Mice

To assess the effect of intracellular accumulation of Cx26 protein on tumorigenicity in vivo, 1×10^6 cells each of the three clones transduced with wtCx26, icCx26, or the mock construct were implanted subcutaneously into the backs of 6 male nude mice per clone. All of the 18 mice examined developed xenograft-derived subcutaneous tumours (Figure 4a). The growth curves of tumours show that the clone overexpressing icCx26 protein manifests a greatly higher growth rate of tumours compared with the mock clone (Figure 4a,b). Consistent with many other papers, the growth rate of tumours was significantly declined by overexpression of wtCx26 protein (Figure 4b), which forms GJ plaques at a cell-cell contact area (Figure 1b).

Furthermore, the tumours derived from each clone were subjected to immunohistochemistry to determine subcellular localisation of Cx26 protein in the tumours. Figure 4c shows that wtCx26 and icCx26 proteins are localised in cell membrane and cytoplasm, respectively, in the corresponding tumours. Regardless of expressed types of Cx26 protein, the behaviours of the Cx26 proteins examined are not different between in vitro and in vivo. These results clearly indicate that oncogenic roles of icCx26 protein have been confirmed both in vitro and in vivo.

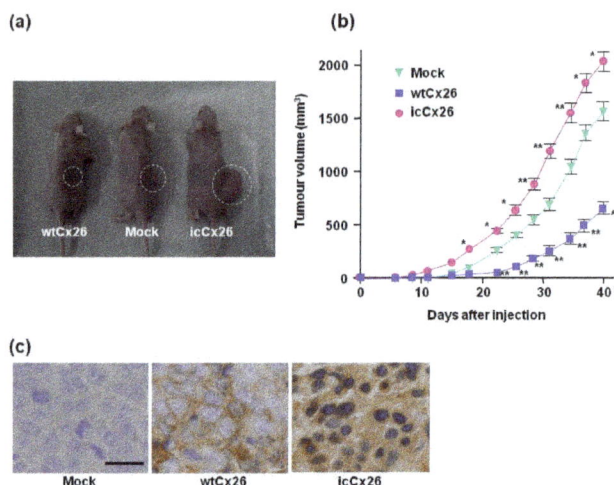

Figure 4. Xenografts of the wtCx26, icCx26, and mock clones into nude mice and tumorigenicity assay in vivo. (**a**) Representative mice bearing tumours raised from 1×10^6 cells of each clone. (**b**) Tumorigenicity in vivo of each clone. The size of each tumour was measured every 2 or 3 days. Error bars represent the SD ($n = 6$). No error bar is indicated when the SD is too small to show. * $p < 0.03$, ** $p < 0.001$ (significantly different from the mock clone at the corresponding time point). (**c**) Expression and subcellular localisation of Cx26 protein in the tumours raised from the xenografts. As revealed by immnohistochemistry, wtCx26 was localised in a cell-cell boundary area. Scale bar, 20 µm.

3. Discussion

It has been established by many convincing evidences that GJ is, in general, a tumour-suppressive cellular apparatus. As such, are connexin proteins, an exclusive component of GJ, considered to be a tumour suppressor? When connexin proteins serve as GJ components, they are usually tumour-suppressive. However, connexin proteins often translocate from cell membrane into an intracellular site in different histological types of malignant tumours [14]. Although such an aberrantly-localised connexin protein cannot function as GJs, their intracellular translocation might generate an unexpected intrinsic function in connexin molecules and make some contribution to tumour progression. To address such a question, we have previously demonstrated that Cx32 protein is not localised in plasma membrane but in the Golgi-apparatus in human HuH7 hepatoma cells and that accumulation of Cx32 protein in the Golgi-apparatus reinforces different malignant phenotypes of HuH7 cells, resulting in the induction of metastasis in the mice xenografted with HuH7 cells overexpressing Cx32 protein in the Golgi-apparatus [17,18].

HuH7 cells are, by nature, incapable of forming GJs due to retention of Cx32 protein in the Golgi-apparatus. In the present study, we employed FaDu HNSCC cells, which express almost no connexin protein, but which can support membrane sorting of a normal connexin protein to generate functional GJs (Figure 1). In other words, FaDu cells are quite normal in terms of the GJ system. HNSCCs are raised from the basal cells, which express mainly Cx26 among different connexin proteins, in the stratified squamous epithelium. Therefore, using FaDu cells and the mutant Cx26 construct coding Cx26 protein conjugated with an ER-Golgi retention signal, we could successfully compare functions between icCx26 and wtCx26 proteins in terms of GJIC, cell proliferation, invasion, and tumorigenicity and find out that icCx26 protein had a GJ-independent intrinsic oncogenic function.

We have been unable to provide the mechanism of how our icCx26 protein behaves in a pro-oncogenic manner. Since icCx26 protein is localised in a Golgi area, ER-stress response may be involved in the mechanism. ER-stress induces two contradictory responses called "adaptive response" and "destructive response" [25]. The adaptive response can be pro-oncogenic. While many proteins related to ER-stress response function in a Golgi-independent manner, ER-resident ATF6 protein is translocated to Golgi apparatus, activated there, then imported into nucleus, and finally induces ER-stress response [26]. icCx26 protein might be involved in such a pathway. More directly, icCx26 protein might play a role in recently-unravelled Golgi stress [27].

It has long been proposed that connexin in tumour has dual or even multiple functions [28,29]. Although intracellular connexin proteins including intra-Golgi, cytoplasmic, and nuclear connexins are rather common in tumours [30–33], there has been little examination thus far of their existence and roles in a physiological condition [34]. Thus, this study has contributed to proving a pathological significance of intracellular connexin proteins. Furthermore, membrane-sorting mechanism of connexin proteins still remains controversial. From aspects of cancer control, mechanism of intrinsic function of connexin proteins in a pathological condition and improvement of membrane sorting of connexin proteins should become targets to be elucidated.

4. Materials and Methods

4.1. Vector Construct

To add the amino acid sequence AKKFF, an ER-Golgi retention signal [19], to C-terminus of Cx26, the fragment containing the coding sequence of human Cx26 (*GJB2*) cDNA [35] was amplified by polymerase chain reaction with the following set of primers: Forward, 5′-ACACAAGCATCTTCTTC-3′; Reverse, 5′-GCGAATTCTTAGAAGAACTTCTTGGCAACTGGCTTTTTTGACTTCCCAGA-3′. It was then digested by the restriction enzymes Bsp119I and EcoRI. The Bsp119I-EcoRI fragment was exchanged with the corresponding fragment of the previously prepared human Cx26/pQCXIN construct, resulting in the mutant Cx26 cDNA, coding icCx26 protein, cloned into pQCXIN retrovirus vector (Clontech Laboratories, Mountain View, CA, USA).

4.2. Cell Culture and Retroviral Transduction

Human FaDu hypopharyngeal squamous cell carcinoma cell line was supplied by American Type Culture Collection (ATCC, Manassas, VA, USA). It has been confirmed that our FaDu cells express no detectable level of Cx43 protein as revealed by immunoblotting and immunofluorescence (Figure A2). The cells and their established subclones were cultured in RPMI1640 medium (Nissui Pharmaceutical, Tokyo, Japan) containing 10% foetal calf serum (FCS), 100 units/mL penicillin and 100 μg/mL streptomycin. PT-67 packaging cells were grown in Dulbecco modified Eagle medium (Thermo Fisher Scientific, Rockford, IL, USA), 10% FCS, 100 U/mL penicillin and 100 μg/mL streptomycin.

All the cells were incubated at 37 °C in a humidified atmosphere containing 5% CO_2 in air. To determine cell proliferation, 5×10^4 cells were seeded into 60-mm dishes in triplicate in 4 mL of medium with 10% FCS. The cells were grown under the aforementioned conditions and counted every 2 or 3 days with a haemocytometer. Dead cells, as determined by trypan blue staining, were left out of the count. FaDu cells expressing wtCx26 or icCx26 protein were established as follows. wtCx26/pQCXIN, icCx26/pQCXIN construct, or pQCXIN empty vector was transfected with FuGENE HD Transfection Reagent (Promega, Madison, WI, USA) into the packaging cell PT-67, and stable transformants were selected with 400 μg/mL G418. FaDu cells were then infected with virus-containing supernatant, supplemented with 4 μg/mL of polybrene, from PT-67 cells transfected with each of the 3 constructs. After 3 weeks of selection with 400 μg/mL G418, G418-resistant FaDu transductants were subcloned by limiting the dilution method. Randomly-selected 8 clones each from wtCx26 and icCx26 stable transductants as well as 5 clones from mock transductants were subjected to a preliminary cell proliferation assay to measure population doubling time. Since all the mock clones showed similar population doubling times without variation, native clonal variation of FaDu cells is considered to be small (Figure A3). 2 and 1 clones of wtCx26 and icCx26 transductants, respectively, showed a population doubling time indistinguishable from that of the mock clones (Figure A3). Omitting these 3 clones, we used a clone indicating a median value from each of three groups for later experiments. It has been confirmed by immunoblotting that the omitted 3 clones express no exogenous Cx26 protein.

4.3. Immunoblotting

Immunoblotting analysis was performed mostly as previously described [23]. As primary antibodies, anti-Cx26 polyclonal antibody (pAb) (Thermo Fisher Scientific) and anti-GAPDH monoclonal antibody (mAb) clone 6C5 (HyTest, Turku, Finland) were applied after diluted at 1:500 and 1:10,000, respectively. Then as second antibodies, horseradish peroxidase (HRP)-conjugated anti-rabbit and anti-mouse IgG antibodies (GE Healthcare Bio-Sciences, Piscataway, NJ, USA) were applied at dilution ratios 1:2000 and 1:5000, respectively. Finally, the protein-antibody complex was chemiluminated with a WEST-one Western Blot Detection System (iNtRON Biotechnology, Seoul, Korea) following the manufacturer's protocol.

4.4. Indirect Immunofluorescence

Indirect immunofluorescence was performed as described previously. Anti-Cx26 mAb clone CX-12H10 (Thermo Fisher Scientific) and anti-p120 Catenin pAb (Sigma-Aldrich, St. Louis, MO, USA), anti-GM130 pAb (Sigma-Aldrich) were diluted at 1:150, 1:200, and 1:3500, respectively. After fixation with acetone, cells are incubated with the diluted primary antibodies. Specific signals were revealed by anti-mouse IgG-Alexa 488 (Thermo Fisher Scientific) and anti-rabbit IgG-Alexa 568 (Thermo Fisher Scientific). Nuclei were stained with DAPI (KPL, Gaithersburg, MD, USA) at a concentration of 0.5 μg/mL.

4.5. Scrape-Loading Dye-Transfer Assay

The assay was performed as described in el-Fouly et al. [36] with modification. The confluent cells on 60-mm dishes were washed with PBS containing 1 mM $CaCl_2$ and immersed in 3 mL of dye cocktail composed of 0.1% Lucifer yellow CH (Sigma-Aldrich) and 0.1% RITC-dextran (Sigma-Aldrich). Several parallel scrape lines were then made with a micropipette tip, and the cells were incubated for 5 min at 37 °C. After rinsed with PBS, the cells dye-coupled with Lucifer yellow were detected under a fluorescence microscope. The cells positive for RITC-dextran were considered to be primarily-scraped but not dye-coupled cells.

4.6. Invasion Assay

Invasion capacity was evaluated quantitatively with Falcon Permeable Support with 8-μm-pore filter (Corning Inc. Life Sciences, Tewksbury, MA, USA). The filters of cell culture inserts were precoated with 500 µg Matrigel (Corning Inc. Life Sciences), dried for 24 h in an incubator and rehydrated by RPMI1640 for 1 h before inoculation of cells. The cell culture inserts were set on 4 mL of FCS-supplemented RPMI1640 poured for the lower compartments of 6-well plates. 5.0×10^5 cells resuspended in 2 mL of serum-free RPMI1640 containing 0.01% bovine serum albumin were seeded to each upper well. After 72 h of incubation, cells were trypsinised and collected separately from the top of the membrane, the underside of the membrane, and the lower compartment. Invasion was quantified as the percentage of cells recovered from the underside of the membrane and the lower compartment over the total cell number.

4.7. Xenograft into Nude Mice

1×10^6 cells suspended in 200 µL of PBS were injected subcutaneously into the backs of 6 male BALB/c-nu/nu mice of 6 weeks of age per clone. Two perpendicular diameters (d_1 and d_2) of each tumour were measured every 2 or 3 days and converted to tumour volume (mm^3) according to the formula: $V = (\pi/6)(d_1 \times d_2)^{3/2}$ [37]. 40 days after injection, the mice were euthanized. A portion of each tumour was frozen, and the rest was fixed in 10% buffered formalin for further analysis. The protocol of the animal work was approved (No. 14016, 31/Jan/2014) by the Committee for Ethics of Animal Experimentation and in accordance with the Guidelines of Animal Experiments of Akita University.

4.8. Immunohistochemistry

Formalin fixed paraffin-embedded sections on slide glass were deparaffinised, then immersed in 3% hydrogen peroxide/methanol at room temperature for 15 min. The slides were incubated with anti-Cx26 mAb clone CX-12H10 (Thermo Fisher Scientific) in a humidified chamber at 4 °C overnight. Specific signals were visualised by employing HRP-labelled polymer method as follows, the slides were reacted with EnVision+ system-HRP for mouse (Agilent, Santa Clara, CA, USA) at room temperature for 30 min and finally 3,3′-diaminobenzidine was oxidized for signal detection.

4.9. Statistical Analysis

The student's *t*-test was performed for the estimation of statistical significance. *p* values are two-tailed. All experiments were independently repeated at least 3 times except for the tumorigenicity assay of xenografts in mice, which was performed only once.

Author Contributions: N.I., Y.Y., Y.K., A.N.-M., T.Y., and Y.O. conceived and designed the experiments; N.I., M.S., and Y.O. performed the experiments; N.I., Y.Y., Y.K., and Y.O. analyzed the data; N.I. and Y.O. wrote the paper.

Acknowledgments: This work was supported in part by JSPS KAKENHI Grant Numbers 21590427 and 24590472 to Y.O. and 25861525 to Y.K. The authors are very grateful to Reiko Ito, Yuko Doi, and Yusuke Ono for their technical assistance, and to Eriko Kumagai for her secretarial work.

Conflicts of Interest: The authors declare no conflict of interest.

Abbreviations

GJ	gap junction
GJIC	gap junctional intercellular communication
HNSCC	head and neck squamous cell carcinoma
wtCx26	wild-type connexin26
icCx26	intracellular conexin26
ERGIC	ER-Golgi intermediate compartment
PBS	phosphate-buffered saline
GAPDH	glyceraldehyde-3-phosphate dehydrogenase
HRP	horseradish peroxidase
DAPI	diamidine phenylindole dihydrochloride
SD	standard deviation

Appendix A

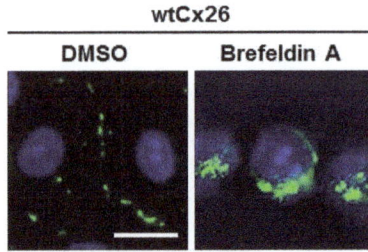

Figure A1. Effect of Brefeldin A on wtCx26-mdiated GJ plaques. Indirect immunofluorescence of Cx26 protein in wtCx26 clone in the presence or absence of 5 µg/mL Brefeldin A/DMSO. The fluorescent signals of Cx26 protein were visualised by Alexa-488. Nuclei were stained with DAPI. Scale bar, 20 µm.

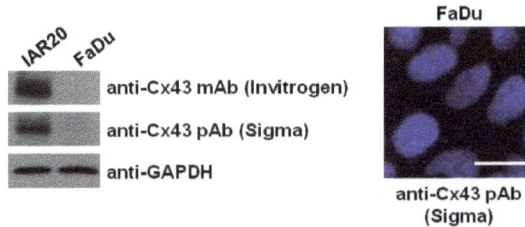

Figure A2. Expression of Cx43 protein in FaDu cells. Immunoblotting was performed with two different anti-Cx43 antibodies, mAb (Invitrogen; clone CX-1B1) and pAb (Sigma-Aldrich). As a positive control, IAR20 cells were used. For indirect immunofluorescence, the fluorescent signals of Cx26 protein were visualised by Alexa-488. Nuclei were stained with DAPI. Note that no positive signals for Cx26 protein are detected. Scale bar, 20 µm.

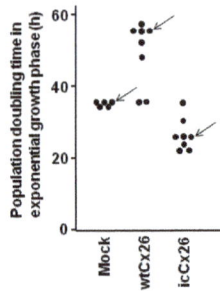

Figure A3. Population doubling time of randomly-selected clones of stable transductants. Randomly-selected 8 clones each from wtCx26 and icCx26 stable transductants as well as 5 clones from mock transductants were subjected to a preliminary cell proliferation assay to measure population doubling time. Each plot corresponds to a single clone. Note that none of the mock clones show native clonal variation. 2 and 1 clones of wtCx26 and icCx26 transductants, respectively, show a population doubling time indistinguishable from that of the mock clones. It has been confirmed that these 3 clones express Cx26 protein at a similar level to that of parental FaDu cells as revealed by immunoblotting. Therefore, omitting the 3 clones, a clone (arrow) indicating a median value was chosen from each of three groups for later experiments. Clonal variation between the clones is due to difference in expression levels of the transgene products.

References

1. Graham, S.V.; Jiang, J.X.; Mesnil, M. Connexins and Pannexins: Important Players in Tumorigenesis, Metastasis and Potential Therapeutics. *Int. J. Mol. Sci.* **2018**, *19*, 1645. [CrossRef] [PubMed]
2. Severs, N.J. Gap junction remodeling and cardiac arrhythmogenesis: Cause or coincidence? *J. Cell. Mol. Med.* **2001**, *5*, 355–366. [CrossRef] [PubMed]
3. Kelsell, D.P.; Dunlop, J.; Stevens, H.P.; Lench, N.J.; Liang, J.N.; Parry, G.; Mueller, R.F.; Leigh, I.M. Connexin 26 mutations in hereditary non-syndromic sensorineural deafness. *Nature* **1997**, *387*, 80–83. [CrossRef] [PubMed]
4. Bergoffen, J.; Scherer, S.S.; Wang, S.; Scott, M.O.; Bone, L.J.; Paul, D.L.; Chen, K.; Lensch, M.W.; Chance, P.F.; Fischbeck, K.H. Connexin mutations in X-linked Charcot-Marie-Tooth disease. *Science* **1993**, *262*, 2039–2042. [CrossRef] [PubMed]
5. Aasen, T.; Mesnil, M.; Naus, C.C.; Lampe, P.D.; Laird, D.W. Gap junctions and cancer: Communicating for 50 years. *Nat. Rev. Cancer* **2016**, *16*, 775–788. [CrossRef] [PubMed]
6. Beyer, E.C.; Berthoud, V.M. The family of connexin genes. In *Connexins*; Harris, A.L., Locke, D., Eds.; Humana Press: New York, NY, USA, 2009; pp. 3–26.
7. Sohl, G.; Willecke, K. An update on connexin genes and their nomenclature in mouse and man. *Cell Commun. Adhes.* **2003**, *10*, 173–180. [CrossRef] [PubMed]
8. Aasen, T.; Kelsell, D.P. Connexins in skin biology. In *Connexins*; Harris, A.L., Locke, D., Eds.; Humana Press: New York, NY, USA, 2009; pp. 307–321.
9. Severs, N.J. Connexins in the heart. In *Connexins*; Harris, A.L., Locke, D., Eds.; Humana Press: New York, NY, USA, 2009; pp. 435–456.
10. Mesnil, M.; Aasen, T.; Boucher, J.; Chepied, A.; Cronier, L.; Defamie, N.; Kameritsch, P.; Laird, D.W.; Lampe, P.D.; Lathia, J.D.; et al. An update on minding the gap in cancer. *Biochim. Biophys. Acta* **2018**, *1860*, 237–243. [CrossRef] [PubMed]
11. Kanczuga-Koda, L.; Wincewicz, A.; Fudala, A.; Abrycki, T.; Famulski, W.; Baltaziak, M.; Sulkowski, S.; Koda, M. E-cadherin and beta-catenin adhesion proteins correlate positively with connexins in colorectal cancer. *Oncol. Lett.* **2014**, *7*, 1863–1870. [CrossRef] [PubMed]
12. Naoi, Y.; Miyoshi, Y.; Taguchi, T.; Kim, S.J.; Arai, T.; Maruyama, N.; Tamaki, Y.; Noguchi, S. Connexin26 expression is associated with aggressive phenotype in human papillary and follicular thyroid cancers. *Cancer Lett.* **2008**, *262*, 248–256. [CrossRef] [PubMed]
13. Naoi, Y.; Miyoshi, Y.; Taguchi, T.; Kim, S.J.; Arai, T.; Tamaki, Y.; Noguchi, S. Connexin26 expression is associated with lymphatic vessel invasion and poor prognosis in human breast cancer. *Breast Cancer Res. Treat.* **2007**, *106*, 11–17. [CrossRef] [PubMed]
14. Omori, Y.; Li, Q.; Nishikawa, Y.; Yoshioka, T.; Yoshida, M.; Nishimura, T.; Enomoto, K. Pathological significance of intracytoplasmic connexin proteins: Implication in tumor progression. *J. Membr. Biol.* **2007**, *218*, 73–77. [CrossRef] [PubMed]
15. Inose, T.; Kato, H.; Kimura, H.; Faried, A.; Tanaka, N.; Sakai, M.; Sano, A.; Sohda, M.; Nakajima, M.; Fukai, Y.; et al. Correlation between connexin 26 expression and poor prognosis of esophageal squamous cell carcinoma. *Ann. Surg. Oncol.* **2009**, *16*, 1704–1710. [CrossRef] [PubMed]
16. Ozawa, H.; Matsunaga, T.; Kamiya, K.; Tokumaru, Y.; Fujii, M.; Tomita, T.; Ogawa, K. Decreased expression of connexin-30 and aberrant expression of connexin-26 in human head and neck cancer. *Anticancer Res.* **2007**, *27*, 2189–2195. [PubMed]
17. Kawasaki, Y.; Omori, Y.; Li, Q.; Nishikawa, Y.; Yoshioka, T.; Yoshida, M.; Ishikawa, K.; Enomoto, K. Cytoplasmic accumulation of connexin32 expands cancer stem cell population in human HuH7 hepatoma cells by enhancing its self-renewal. *Int. J. Cancer* **2011**, *128*, 51–62. [CrossRef] [PubMed]
18. Li, Q.; Omori, Y.; Nishikawa, Y.; Yoshioka, T.; Yamamoto, Y.; Enomoto, K. Cytoplasmic accumulation of connexin32 protein enhances motility and metastatic ability of human hepatoma cells in vitro and in vivo. *Int. J. Cancer* **2007**, *121*, 536–546. [CrossRef] [PubMed]
19. Sarma, J.D.; Wang, F.; Koval, M. Targeted gap junction protein constructs reveal connexin-specific differences in oligomerization. *J. Biol. Chem.* **2002**, *277*, 20911–20918. [CrossRef] [PubMed]
20. Brandizzi, F.; Barlowe, C. Organization of the ER-Golgi interface for membrane traffic control. *Nat. Rev. Mol. Cell Biol.* **2013**, *14*, 382–392. [CrossRef] [PubMed]

21. Martin, P.E.; Blundell, G.; Ahmad, S.; Errington, R.J.; Evans, W.H. Multiple pathways in the trafficking and assembly of connexin 26, 32 and 43 into gap junction intercellular communication channels. *J. Cell Sci.* **2001**, *114*, 3845–3855. [PubMed]

22. Hernandez-Blazquez, F.J.; Joazeiro, P.P.; Omori, Y.; Yamasaki, H. Control of intracellular movement of connexins by E-cadherin in murine skin papilloma cells. *Exp. Cell Res.* **2001**, *270*, 235–247. [CrossRef] [PubMed]

23. Momiyama, M.; Omori, Y.; Ishizaki, Y.; Nishikawa, Y.; Tokairin, T.; Ogawa, J.; Enomoto, K. Connexin26-mediated gap junctional communication reverses the malignant phenotype of MCF-7 breast cancer cells. *Cancer Sci.* **2003**, *94*, 501–507. [CrossRef] [PubMed]

24. Omori, Y.; Yamasaki, H. Gap junction proteins connexin32 and connexin43 partially acquire growth-suppressive function in HeLa cells by deletion of their C-terminal tails. *Carcinogenesis* **1999**, *20*, 1913–1918. [CrossRef] [PubMed]

25. Joshi, S.; Wang, T.; Araujo, T.L.S.; Sharma, S.; Brodsky, J.L.; Chiosis, G. Adapting to stress—Chaperome networks in cancer. *Nat. Rev. Cancer* **2018**. [CrossRef] [PubMed]

26. Ghemrawi, R.; Battaglia-Hsu, S.F.; Arnold, C. Endoplasmic Reticulum Stress in Metabolic Disorders. *Cells* **2018**, *7*. [CrossRef] [PubMed]

27. Taniguchi, M.; Yoshida, H. TFE3, HSP47, and CREB3 Pathways of the Mammalian Golgi Stress Response. *Cell Struct. Funct.* **2017**, *42*, 27–36. [CrossRef] [PubMed]

28. Naus, C.C.; Laird, D.W. Implications and challenges of connexin connections to cancer. *Nat. Rev. Cancer* **2010**, *10*, 435–441. [CrossRef] [PubMed]

29. Omori, Y.; Kawasaki, Y.; Li, Q.; Yoshioka, T.; Yamamoto, Y.; Enomoto, K. Cytoplasmic connexin32 and self-renewal of cancer stem cells: Implication in metastasis. In *Hepatocellular Carcinoma—Basic Research*; Lau, W.Y., Ed.; InTech: Rijeka, Croatia, 2012; pp. 235–252.

30. Jamieson, S.; Going, J.J.; D'Arcy, R.; George, W.D. Expression of gap junction proteins connexin 26 and connexin 43 in normal human breast and in breast tumours. *J. Pathol.* **1998**, *184*, 37–43. [CrossRef]

31. Krutovskikh, V.; Mazzoleni, G.; Mironov, N.; Omori, Y.; Aguelon, A.M.; Mesnil, M.; Berger, F.; Partensky, C.; Yamasaki, H. Altered homologous and heterologous gap-junctional intercellular communication in primary human liver tumors associated with aberrant protein localization but not gene mutation of connexin 32. *Int. J. Cancer* **1994**, *56*, 87–94. [CrossRef] [PubMed]

32. Mehta, P.P.; Perez-Stable, C.; Nadji, M.; Mian, M.; Asotra, K.; Roos, B.A. Suppression of human prostate cancer cell growth by forced expression of connexin genes. *Dev. Genet.* **1999**, *24*, 91–110. [CrossRef]

33. Omori, Y.; Krutovskikh, V.; Mironov, N.; Tsuda, H.; Yamasaki, H. Cx32 gene mutation in a chemically induced rat liver tumour. *Carcinogenesis* **1996**, *17*, 2077–2080. [CrossRef] [PubMed]

34. Goodenough, D.A.; Paul, D.L. Beyond the gap: Functions of unpaired connexon channels. *Nat. Rev. Mol. Cell Biol.* **2003**, *4*, 285–294. [CrossRef] [PubMed]

35. Lee, S.W.; Tomasetto, C.; Paul, D.; Keyomarsi, K.; Sager, R. Transcriptional downregulation of gap-junction proteins blocks junctional communication in human mammary tumor cell lines. *J. Cell Biol.* **1992**, *118*, 1213–1221. [CrossRef] [PubMed]

36. El-Fouly, M.H.; Trosko, J.E.; Chang, C.C. Scrape-loading and dye transfer. A rapid and simple technique to study gap junctional intercellular communication. *Exp. Cell Res.* **1987**, *168*, 422–430. [CrossRef]

37. Warri, A.M.; Huovinen, R.L.; Laine, A.M.; Martikainen, P.M.; Harkonen, P.L. Apoptosis in toremifene-induced growth inhibition of human breast cancer cells in vivo and in vitro. *J. Natl. Cancer Inst.* **1993**, *85*, 1412–1418. [CrossRef] [PubMed]

International Journal of
Molecular Sciences

MDPI

Review

Connexin Communication Compartments and Wound Repair in Epithelial Tissue

Marc Chanson [1,†], Masakatsu Watanabe [2,†], Erin M. O'Shaughnessy [3], Alice Zoso [1]
and Patricia E. Martin [3,*,†]

1   Department of Pediatrics and Cell Physiology & Metabolism, Geneva University Hospitals and University
    of Geneva, 1211 Geneva, Switzerland; marc.chanson@unige.ch (M.C.); alice.zoso@unige.ch (A.Z.)
2   Graduate School of Frontier Biosciences, Osaka University, Osaka 565-0871, Japan;
    watanabe-m@fbs.osaka-u.ac.jp
3   Department of Life Sciences, School of Health and Life Sciences, Glasgow Caledonian University,
    Glasgow G4 0BA, UK; Erin.OShaughnessy@gcu.ac.uk
*   Correspondence: patricia.martin@gcu.ac.uk; Tel.: +44-141-331-3726
†   These authors contributed equally to this work.

Received: 27 March 2018; Accepted: 26 April 2018; Published: 3 May 2018

Abstract: Epithelial tissues line the lumen of tracts and ducts connecting to the external environment. They are critical in forming an interface between the internal and external environment and, following assault from environmental factors and pathogens, they must rapidly repair to maintain cellular homeostasis. These tissue networks, that range from a single cell layer, such as in airway epithelium, to highly stratified and differentiated epithelial surfaces, such as the epidermis, are held together by a junctional nexus of proteins including adherens, tight and gap junctions, often forming unique and localised communication compartments activated for localised tissue repair. This review focuses on the dynamic changes that occur in connexins, the constituent proteins of the intercellular gap junction channel, during wound-healing processes and in localised inflammation, with an emphasis on the lung and skin. Current developments in targeting connexins as corrective therapies to improve wound closure and resolve localised inflammation are also discussed. Finally, we consider the emergence of the zebrafish as a concerted whole-animal model to study, visualise and track the events of wound repair and regeneration in real-time living model systems.

Keywords: epithelial tissue; connexin; pannexin; wound healing; inflammation; zebrafish models

1. Introduction

Connexins are a highly conserved group of transmembrane proteins, with 21 subtypes expressed in a human, which form gap junctions with neighbouring cells to enable intercellular communication and metabolite exchange. Connexin hemichannels in the plasma membrane are typically in a default-closed position but can be induced to open under conditions of cell stress to release adenosine triphosphate (ATP) and other small signalling molecules [1,2]. Pannexins, a family of three sister proteins to the connexins, also form membrane channels and are increasingly associated with inflammatory mediated events [3]. Connexins and pannexins share a common topology consisting of four transmembrane domains that span the plasma membrane; two highly conserved extracellular loops, and intracellular C- and N-terminal domains. Six connexin subunits oligomerise to form a connexon that is trafficked to and inserted into the plasma membrane. Connexins laterally accrete to dock with other hemichannels from neighbouring cells to form an intercellular gap junction [1]. Pannexins also oligomerise to form channels and trafficking to the plasma membrane is dependent on interaction with the actin cytoskeleton [4,5]. Both connexins and pannexins play a key role in coordinating processes that mediate the development and maintenance of tissues in multicellular organisms. Distinct connexin and pannexin expression profiles are observed in different tissues with multiple connexins often expressed within the same cell. For example, in the cardiovascular system connexins (Cxs) 37, 40, 43 and 45 are expressed. In epithelial tissue, such as the airway epithelium and stratified epidermis, Cx43 is still the predominant connexin present with Cxs 26, 30 and 31 in particular emerging to play important roles as discussed below. Each connexin forms transmembrane channels that have unique permeability properties in terms of size, ionic permeability and selectivity [6,7]. Thus, the pattern in which these proteins are expressed is very important to cellular function, with each combination of heteromeric channel conferring unique signalling properties. Heterotypic channels are limited in function by connexin compatibility, predicted to prevent cells following different differential pathways to 'communicate' [8]. Connexins have a short half-life, resulting in rapid turnover and, together, these properties enable specialised tissue-specific, spatial and temporal communication compartments. It is well established that connexins participate in liver regeneration after injury [9]. However, less is known about the mechanisms of airway epithelium repair in the adult respiratory system. This review will focus on tissue-specific communication compartments during the dynamic events that occur during the regeneration and repair of epithelial tissue networks including the lung airway epithelium and the epidermis. Wound repair in the whole zebrafish model will also be addressed.

2. Connexins in Normal and Repairing Airway Epithelium

2.1. Connexins in the Airway Epithelium

The airway epithelium is a fundamental component of the innate immune system by protecting the lung against invading pathogens. The concerted action of ciliated and mucin-secreting cells maintain efficient mucociliary clearance, and regulates the production of anti-inflammatory and antimicrobial molecules [10]. Connexin- and pannexin (Panx)-channels provide a complex communication network that maintains lung homeostasis and modulates host defences in both conductive and respiratory (alveoli) airways [11,12]. The upper airway epithelium and submucosal glands express (identified at the mRNA level and/or protein level) about 10 connexins (Cx26, Cx30, Cx30.3, Cx31, Cx31.1, Cx32, Cx37, Cx43, Cx46). Cx32, Cx43 and Cx46 are found in the alveolar epithelium while Cx37, Cx40 and Cx43 are expressed in the alveolar endothelium. Panx1 is ubiquitously expressed, while little information is yet available for Panx2 and Panx3.

In the conducting airways, the communication network plays key roles within the mucosa barrier. For example, cell-to-cell transfer of ions and second messengers is thought to contribute to mucociliary clearance by hydrating the luminal surface and controlling ciliary beat frequency [13,14]. The latter mechanisms are fine-tuned by Panx1-mediated release of ATP which, in an autocrine manner, regulates mucin and water secretion as well as cilia activity [15–17]. Gap junctions are also components of the innate immunity defence system by mediating the cell-to-cell spread of pro-inflammatory and pro-apoptotic signals according to the pathogen recognition receptors (PRRs) activated [18,19]. Hence, connexins and Panx1 are deregulated in terms of activity and/or expression in several pathologies, including chronic rhinosinusitis [20] and cystic fibrosis [21,22].

In the respiratory alveoli, the intercellular network participates in the production of surfactant in response to stretch on the epithelial side of the alveoli [23]. On the endothelial side, connexin and Panx1 channels mediate hypoxia-induced vasoconstriction, leukocyte adhesion and transmigration across the endothelial-alveolar wall [24–26]. Finally, it has been proposed that alveolar macrophages attached to the alveolar surface may communicate immunosuppressive signals to alveolar epithelial cells, since leukocyte-specific knockdown of Cx43 enhanced endotoxin-induced lung inflammation [27].

Epithelial repair is a multi-step process initiated by migration of basal cells (BCs) at the leading edge of the wound along with the induction of proliferation to repopulate the injured area [28]. This phase of newly proliferating cells, or blastema, is followed by cell differentiation and patterning (Figure 1). Stem cells and BCs contribute to the regeneration of intact airway epithelium. The BC population is highly heterogeneous comprising morphologically indistinguishable multipotent stem cells and committed precursors [28,29]. The mechanisms that underlie cell-fate specification in repairing airway epithelia are poorly known but several transcription factors, including Yap/Sox and Fox family members as well as Notch family receptors and specific microRNAs, has been reported [28–34]. For example, Notch signalling is required for the transition of mature BCs to early progenitor cells (EPs), and at later phases for differentiation into goblet/secretory cells (GCs) and ciliated cells (CCs) (Figure 1).

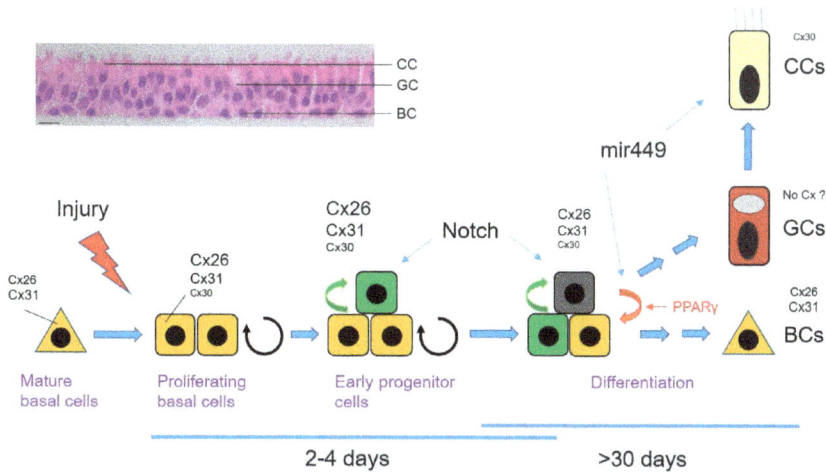

Figure 1. The different phases of airway epithelium regeneration after wounding. The top left images show the histology of the pseudostratified airway epithelium after culturing human airway epithelial cells for 1.5 months on Transwell filter, and the capacity of the epithelium to repair after wounding. BC: basal cell; CC: ciliated cell; GC: goblet cell. The scheme illustrates the steps involved in airway epithelium repair after injury (blue arrows); wound closure is reached within 3–4 days. Orange triangle: CK5-expressing quiescent basal cells; orange square: CK5 and CK14-activated basal cells; light green square: early progenitor cells; dark green square: late progenitor cells. Early differentiation (passage from proliferating cells to early progenitors) is dictated in part by Notch activation (green arrows). Cell division arrest and later differentiation requires increased expression of miR-449. The relative changes in connexin expression (Cx26, Cx30, Cx31) is illustrated for the different stages of the repair process by the size of the fonts. PPARγ signalling also contributes to BC differentiation and decreases connexin expression (red arrow). Bar: 50 μm.

2.2. Repair Research in the Adult Airway Epithelium

Re-establishment of the integrity of the airway epithelium is a prerequisite to restoration of the tissue homeostasis and host-defence mechanisms [20]. The tracheobronchial airway epithelium is composed of BCs, GCs and CCs.

2.3. Connexins in Wound Repair of the Airway Epithelium

The pattern of connexin expression by the human airway epithelium depends on cell phenotype and the stage of differentiation. Several connexin isoforms are expressed in the undifferentiated airway epithelium of which Cx26 and Cx43 rapidly disappear upon differentiation [35,36]. In the well-polarized epithelium, cell-specific localization of connexin expression is found. Thus, Cx31 connects BCs while Cx30 connects CCs and perhaps BCs to CCs, suggesting distinct compartments of intercellular communication within the airway epithelium. Remnants of Cx26 could be detected, however, in a subset of BCs [36]. At least two populations of BCs can be distinguished by their expression of cytokeratins (CK5 and CK14) whereas CK8 is a marker of EPs and of fully differentiated BCs and CCs. Interestingly, Cx26 expression is strongly induced in CK14-expressing BCs undergoing proliferation in response to wounding [36]. The sustained increase of Cx26 is maintained until wound closure, a point at which the protein progressively returns to its basal level of expression with epithelium differentiation (Figure 1). It is proposed that induction of Cx26-mediated intercellular communication by proliferative signals in activated BCs may represent a means to contain their over-proliferation, which would make them permissive for further differentiation signals. Not only Cx26 but also Cx30 and Cx31 are subjected to modulation during wound repair (Figure 1). At the mRNA level, both Cx30 and Cx31 are upregulated during the proliferation phase to reach a plateau at the time of wound closure. Cx30 detection coincided with the apparition of EPs. The expression of both connexins decreased back to basal levels during the differentiation period (Figure 1). Although an apparent relationship between connexin expression and proliferation can be proposed, the mechanisms involved have not yet been elucidated.

Among the multitude intracellular signalling pathways that control cell-fate specification in lung development and lung-related diseases, a role for the lipid-responsive transcription factor peroxisome proliferator-activated receptor gamma (PPARγ) has been proposed [37,38]. PPARγ is endogenously activated by 15-keto prostaglandin E2 (15kPGE2), which is generated from PGE2 by the activity of hydroxyprostaglandin dehydrogenase (HPGD). In a recent report, Bou Saab and collaborators found that HPGD is involved in the PPARγ-dependent control of the BC population during the repair process [39]. Interestingly, Cx26 expressed in activated BCs was found to be highly sensitive to the differentiating signals mediated by PPARγ [40]. One hypothesis is that PPARγ-induced downregulation of Cx26 may promote in a subpopulation of activated BCs their exit of the cell cycle for further differentiation and/or trigger their return to their original state. Clearly, additional studies are required to understand the regulatory mechanisms and the role fulfilled by the dynamic changes in connexin expression during airway epithelium repair. Gene-silencing approaches using clustered regularly interspaced short palindromic repeats-Cas9 (CRISPR-Cas9) in primary cultures of human airway epithelial cells are anticipated to bring important information on the role of gap junctional intercellular communication in cell fate during airway mucosal regeneration.

3. Connexins in the Epidermis

The epidermis is a highly specialised, stratified epithelial layer forming a tough barrier to the external environment. It plays a vital protective role from environmental insult and is highly subject to localised trauma and injury resulting in a need for rapid repair of the tissue [41]. The key components of the epidermis are keratinocytes, a specialised subset of cells that attach firmly to the basement membrane of the skin that provides an interface between the vascular dermal layers and the avascular epidermis. A small subset of epidermal stem cells, resident in the hair follicle and interfollicular regions are responsible for the continual renewal of the epidermis [42]. Cells on the basal membrane are, under normal conditions, the only proliferative cells within the epidermis, although only about 15% of cells actively participate in this process, this increases when enhanced proliferation is needed such as in wound healing. Stratification occurs as subsets of the basal layer keratinocytes undergo asymmetrical cell division and enter a differentiation programme resulting in the stratified and cornified epidermis and watertight epidermal barrier (Figure 2). These differentiated layers are characterised by a complex differentiation process typified by changes in the expression profile of keratins, key intermediate filaments of the epidermis. Keratins form pairs with basic and acidic partner complexes; the CK5/CK14 complex is associated with basal keratinocytes and CK1/CK10 expression with cells committed to terminal differentiation pathways [43]. Keratinocyte differentiation results in the transformation of cellular morphology and protein expression, ultimately resulting in a loss of nuclei and the production of keratinohyalin granules in the stratum corneum layer. Keratins interact with a range of proteins including desmosomes, fillagrin, loricrin and keratolinin to give a highly ordered and structured epidermis and the stratum corneum its flattened shape (Figure 2). There is a balance between the renewal and desquamation of the epidermis, a process that takes about 28 days in a man with a normal epidermal profile of about 10 cell layers [43].

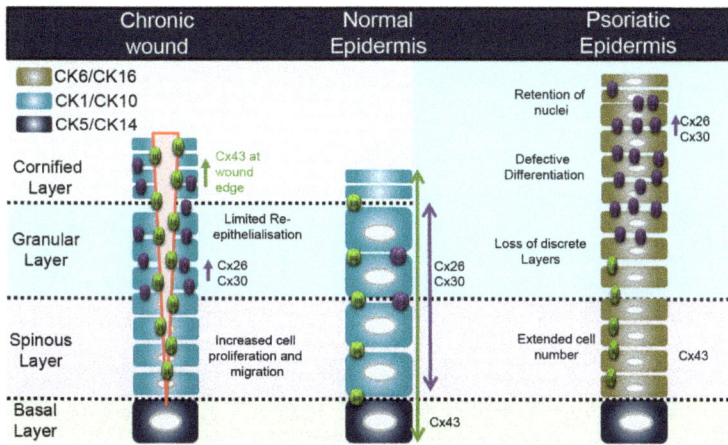

Figure 2. Connexin expression profile in the normal epidermis, at a chronic wound edge, and representative state in a psoriatic epidermis.

Up to 10 different connexin isoforms are expressed with the profile characterised by the differentiation status (Figure 2). Cx43 is the predominant connexin in the basal layers, associated with proliferation, while Cx26 and Cx30 are associated with upper differentiated cell layers. Cx31 tends to be expressed in similar layers to Cx43 while Cx31.1 is only found in the upper differentiated layers and linked with apoptosis and cell shedding [44,45]. Due to the avascular nature of the epidermis, it is generally accepted that gap junctional intercellular communication is central to ensuring correct signalling between the inner and outer layers. Specialised communication compartments have been suggested due to the predicted nature of the α and β connexin interactions and selective permeability properties [6,46]. Recent evidence suggests that Cx43 (a member of the α connexin subgroup) and Cx31 (a member of the β connexin subgroup) can form heteromeric channels, indicating that such interactions could act as a bridging link between epidermal areas where the incompatible Cx43 and β-connexins Cx26 and Cx30 are the predominant connexins expressed [47,48]. In the human epidermis, the two main connexins are Cx26 and Cx43. Cx26 expression is also found in hair follicles and eccrine sweat glands, whereas Cx43 is located in the interfollicular epidermis [44,49,50]. The importance of connexins in the epidermis is further highlighted by the plethora of mutations that are associated with both inflammatory and non-inflammatory hyperproliferative epidermal disorders (recently reviewed by [51,52]). Pannexins 1 and 3 are also expressed in keratinocytes where knockout mouse models revealed an important role for Panx1 in epidermal formation [53–55].

3.1. Connexins in Epidermal Wound Healing

Skin integrity is dependent on interactions between keratinocytes and the extracellular matrix (ECM), and these are the first cell type to sense injury. The introduction of a wound to the epidermis triggers an acute inflammatory response, keratinocyte migration and proliferation, and induces changes in the cytoskeleton and keratinocyte adhesion [56,57]. In the context of connexins, a key event in normal wound healing is the downregulation of Cx43 at the wound edge within 6 h of injury [58]. This is associated with the activation of migration of keratinocytes into the gap. There is also evidence of specialised spatial communication compartments behind the wound edge where Cx43 is subject to phosphorylation, particularly at P-ser368 on the carboxyl terminal domain [59,60].

The localised switch in phosphorylation probably changes the functional signalling parameters of Cx43 channels, further enhancing migration of wound edge keratinocytes. This may include the induction of signalling pathways, including Transforming Growth Factor-β TGF-β and ECM deposition, all of which are key events in wound closure [61–63]. Upon re-stratification, levels of Cx43 are re-established [45,58,64,65]. In other studies, Panx1 was also shown to be elevated at the wound edge during the early stages of wound healing and overall rates of wound closure were reduced in Panx1 Knockout (KO) mice models [66].

3.2. Connexins and Inflammation in the Epidermis: Chronic Wounds and Psoriasis

Dysregulation of connexin expression in the epidermis is associated with a variety of conditions, re-enforcing the importance of these proteins in maintaining epidermal integrity. This is most evident for Cx43 and Cx26 where alteration in the fine balance of control of expression is associated with pathological conditions including chronic non-healing wounds, psoriasis and a range of connexin-channelopathies linked with mutations in β-connexins causing skin disease [8,51].

Chronic non-healing wounds are maintained in a high inflammatory state and are subject to infection. These are increasingly associated with situations such as diabetic ulcers and lower limb disease, but also with conditions such as pressure sores, prevalent in ageing societies, increased obesity and diabetes rates [57,67]. Such events place enormous burdens on healthcare resources and management. Although many novel therapies are under development, including hyperbaric oxygen therapy, growth factors and stem-cell implantation, current treatments only manage the condition by frequent debridement therapy and pressure off-loading [68]. In terms of connexin expression, Cx43 is significantly upregulated at the wound edge of chronic non-healing wounds in both diabetic and non-diabetic patients [58,69,70]. Such changes in expression significantly alter the dynamic and subtle crosstalk between cells with the local area; it is proposed that it plays a vital role in the sustainment of the non-healing wound state. Thus, Cx43 in particular has become a prime therapeutic target to improve wound healing with both antisense Cx43 and Cx43 peptidomimetic strategies exhibiting exciting opportunities [71] (see Section 5). In chronic non-healing wound margins, reports have also determined that Cx26/Cx30 expression is significantly enhanced and associated with the hyper-proliferative and inflammatory skin phenotype [70].

Psoriasis is another chronic inflammatory skin condition affecting 2–3% of the population. The classic phenotype of psoriasis can be likened to that of a chronic wound with its persistent inflammation and changes to the healing profile of the epidermis [72]. Persistent plaques that vary in size and depth form the prototypic form of psoriasis, Psoriasis vulgaris, often described as the Koebner phenomenon with well defined, silvery-white scaly skin areas of skin lesions. The leading hypothesis on disease initiation in psoriasis is T-cell driven via the immune system (inside-out hypothesis); however, recent data suggest that the environment may play a role (outside-in), including shifts in the skin microbiome [73,74], epidemiological factors such as geographical location [75], and epidermal barrier disruption [76,77]. Genetic factors also play a role, with several loci identified as psoriasis susceptibility risk factors. The most prominent is *PSOR1*, a Major Histocompatibility Complex (MHC) Class 1 region on the chromosome 6p21 [78], and a number of reports suggest that polymorphisms on *GJB2* represent suitable markers of susceptibility [79–82].

Enhanced Cx26 and Cx30 protein expression is highly evident in psoriatic plaques and a transcriptome analysis revealed *GJB2*, to be among the top 100 genes upregulated in psoriasis, with its expression levels increased up to 18-fold in psoriatic lesions compared to normal tissue [50,51,83–85]. Epidermal hyperplasia in the psoriatic epidermis drives an accelerated growth and altered differentiation of the keratinocytes that results in a loss of the discrete epidermal layers. The highly proliferative basal layer leads to an extended cell number in the spinous layer and a merging of the granular and cornified layers. Nuclei are retained in the outer layers and there is an overall defective terminal differentiation, which drives a change in keratin expression from the normal CK1 and CK10 to CK6 and CK16, characteristic markers for psoriasis [72,86]. In support of a role for the over-expression of Cx26 in driving some of these events, transgenic mice studies over-expressing Cx26 in suprabasal keratinocytes exhibited pathological features similar to those seen in psoriasis, with the suggestion that enhanced Cx26 hemichannel activity plays a central role [87]. Several further lines of evidence suggest that susceptibility to psoriasis is linked to the epidermal differentiation complex, including downregulation of E-cadherin and Cx43 and enhanced expression of Claudin13, a key component of epidermal tight junctions [76,88–92].

4. Disruption of Cx43:Cx26 Balance in Epithelial Tissue: Connexins and the Environment

Maintaining homeostasis with the microbiome is a critical feature of all epithelial tissues [93,94]. Dysbiosis of the skin microflora has been associated with many dermatological conditions including but not limited to psoriasis and chronic non-healing diabetic wounds. Bacterial colonisation of psoriatic skin shows a shift from commensal organisms, such as *Staphylococcus epidermidis*, to more opportunistic pathogens such as *S. aureus* [73,95] and shifts in skin flora of diabetic patients and non-healing wounds are highly evident. The microbiome of the respiratory system is less well characterized but recent reports showed its alteration in diseases such as Cystic Fibrosis CF. The lungs of CF patients are normal *in utero* and in the newborn period, and are usually colonized by a variety of opportunistic bacterial, viral and fungal pathogens in an age-dependent sequence [94]. The most frequently found organism during colonization of CF airways is *S. aureus*, followed later by *Pseudomonas aeruginosa*, which remains the critical determinant of pulmonary pathology in the late stages of the disease.

Accumulating evidence suggests that exposure of epithelial cells to components of opportunistic pathogens, such as *S. aureus*, *P. aeruginosa* and *Shigella flexneri*, can alter connexin hemichannel activity and connexin expression levels [77,96,97]. In the context of the epidermis, we previously determined that exposure of keratinocytes to peptidoglycan isolated from the cell wall of the opportunistic skin pathogen *S. aureus* induces Cx26 expression with associated links to inflammation, while that isolated from the commensal *S. epidermidis* is without effect [77]. In parallel studies in other tissue networks, bacterial colonisation has been shown to trigger connexin signalling including the glial, intestine and lung epithelia [98]. Hence, controlling the level of Cx expression and/or function in these specialised tissue niches and regulating the balance of Cx43 and Cx26 is a critical focus for maintaining epithelial integrity. Further studies suggest that acute exposure to pro-inflammatory mediators, such as peptidoglycan or lipopolysaccharide, is sufficient to trigger connexin hemichannel activity and release of secondary messengers such as ATP, nicotinamide adenine dinucleotide (NAD$^+$), glutamate and prostaglandins [99,100]. Following the release of ATP, the activation of purinergic signalling cascades occurs that plays a central role in differentiation and proliferation of the epidermis as well as regulation of innate immune responses [101]. In conditions such as psoriasis or chronic non-healing wounds where connexin expression levels are excessive, the localised release of ATP could be in part responsible for exacerbated pro-inflammatory responses, altering intracellular calcium dynamics leading to changes in the terminal differentiation programme and a hyperproliferative state [102–105].

Similarly, the release of extracellular ATP is thought to amplify the inflammatory response evoked by *P. aeruginosa*-dependent infection of the airway epithelium. Interestingly, both connexins and Panx1 were found to be involved in ATP release at the airway mucosa [106,107]. However, protective or deleterious outcomes of ATP on inflammation may depend on the activated purinergic receptors. For example, the release of ATP from airway epithelial cell via Panx1 channel and subsequent activation of P2Y11 contributes to the resolution of inflammation and triggers wound repair [21].

5. Connexins as Therapeutic Targets in Epithelial Tissues

De-regulation of connexin expression is thus a key event in epithelial pathology. To enable the dissection of the molecular mechanisms underpinning these events, a range of studies utilising knockout and antisense technologies and connexin peptidomimetics have provided extensive information and revealed exiting therapeutic strategies [108].

In targeted epidermal ablation of Cx43 in mouse models, wound-healing rates were significantly faster than in normal mice. The reduction of Cx43 showed that the keratinocyte layer was much thinner with a wound closure rate that more than doubled, reaching or surpassing the rates of the untreated controls [65,109]. Early studies by Becker and Green developed a topical application of an antisense oligonucleotide targeted to Cx43, reducing Cx43 expression and improving wound closure rates in both "normal" and diabetic rat wound-healing models. Day one observations of epidermal regrowth determined that, after injury, diabetic skin showed no re-growth compared to the controls, but when treated with the Cx43 antisense oligonucleotide the regrowth matched that of the controls [110–113]. This strategy was taken forward to clinical trials providing powerful evidence for the development of Cx-therapeutic strategies. Significantly, these studies provided evidence that reduction of Cx43 reduced the inflammatory status, altered ECM deposition and reduced scarring [114–116].

Although an antisense approach is applicable, there is much controversy over channel versus non-channel function. A peptide targeted to the carboxyl terminal domain of Cx43, thereby interfering with ZO-1 interactions, was identified as a potent regulator of Cx43 channel function. Application of this peptide, ACT-1, to wounds in animal models determined that it enhanced wound-closure rates and reduced inflammation, ECM deposition and scarring, without altering *Cx43* gene expression. It is the first in its class of connexin peptidomimetics to be successfully applied in clinical trials showing profound improvement in healing rates of venous leg ulcers [117–120]. Further mimetic peptides that mimic the extracellular loops of connexins (Gap26 and Gap27) have been reported to enhance wound-closure rates in in vitro 2D and 3D human and mouse epidermal models. Gap27 blocks hemichannel and gap junctional communication without influence on *Cx43* gene expression in keratinocytes [64,121–123]. It enhances keratinocyte migration rates, without altering cell proliferation supporting concepts that hemichannel activity is involved in keratinocyte galvanotaixis [61,124].

Less is known regarding the targeting of connexins and Panx1 in the airway epithelium. In a murine bleomycin model of acute lung injury, intravenous administration of a Panx1 mimetic peptide reduced the presence of leukocytes in the bronchoalveolar lavage fluid [125]. Similarly, neutrophils' recruitment to the airspace in response to lung lipopolysaccharide (LPS) administration was reduced by intratracheal instillation of Gap26 during the course of inflammation [26]. Whether this inhibition occurs at the epithelium, endothelium or leukocyte level is not clear. Finally, the effects of peptidomimetics on airway epithelium repair have not yet been reported.

6. Zebrafish Connexins in Wound Repair and Regeneration

Zebrafish are an emergent model organism used to study development and regeneration because of their unique advantages, including their small body length (2–3 cm), easy breeding, large number of fertilised eggs (over 100) obtained from a single pair of zebrafish at one time, and short generation time (3 months). The embryos usually hatch and start swimming at 3 days after fertilisation and begin feeding about 5 days after fertilisation, indicating that the development of the nervous system, motor function, and digestive organs is almost complete in this short period. Furthermore, zebrafish have a transparent embryo body, facilitating the observation of development by live imaging. In addition, genome information and genetic-modification methods, including Tol2-mediated gene transfer and CRISPR gene knockout, have been utilised in zebrafish [126,127]. These tools have enabled confirmation of these phenomena, as shown in other organisms, and have accelerated biological research.

6.1. Wound-Repair Research in Zebrafish

In addition to skin and bone, which are common targets in wound-repair and regeneration studies, nerves, heart, retina, tendon and muscles are actively evaluated in studies utilising the high regeneration capacity of zebrafish [128,129]. Furthermore, by exploiting the transparency characteristics of zebrafish embryos, live cell imaging of wound repair has been performed. In a recent report, for example, the H_2O_2 concentration gradient occurring in skin wounds was successfully visualised, and macrophages accumulated at the wound point on the skin in an H_2O_2 concentration-dependent manner [130].

6.2. Zebrafish Connexins

With regard to gap junction genes, 38 connexin and four pannexin genes are predicted in the zebrafish genome. The higher number of connexin genes in zebrafish compared with that in humans is due to gene duplication events, which occurred in the teleost lineage during fish evolution [131]. Because expression analysis and functional analysis have not been performed for all of these genes, it is unclear how many genes are actually functional. Cx43 is the most abundantly expressed connexin in many organs and is being actively studied. Zebrafish (Zf) Cx43 and human Cx43 have a very high identity of 80% at the amino acid level, and functional conservation between ZfCx43 and mammalian Cx43 was analysed and confirmed [132,133]. As a connexin specific to the teleost lineage, ZfCx39.4 is not present in the genome of amphibians or higher organisms. Interestingly, ZfCx39.4 is involved in skin pattern formation in zebrafish, together with ZfCx41.8, which is an orthologue of mammalian Cx40 [134,135].

6.3. Zebrafish Connexins in the Heart

Unlike epithelial tissues discussed above, mammalian hearts do not regenerate because cell division ceases after birth; however, recent studies have confirmed that partial cell division ability is retained [136]. On the other hand, amphibians and fish have higher regeneration ability than mammals, and wounded hearts regenerate well in these organisms because differentiated cardiomyocytes can undergo dedifferentiation and proliferation throughout the organisms' lifetime [137].

The expression of ZfCx36.7 (an orthologue of human Cx31.9), ZfCx43 (an orthologue of Cx43), ZfCx41.8 (an orthologue of Cx40), ZfCx45.6 (an orthologue of Cx45), and ZfCx48.5 (an orthologue of Cx 46) have been reported in zebrafish hearts [138–141]. Among these proteins, mutations in ZfCx36.7 and ZfCx48.5 have been isolated from mutant zebrafish with heart defects, indicating that these connexins are important for maintaining heart function. In contrast, no defects have been detected in the hearts of ZfCx41.8 and ZfCx45.6 mutants, although knockout of *Cx40* in mammals causes severe defects in heart function. With regard to Cx43 in zebrafish, expression was detected in normal and regenerating hearts; however, the role of Cx43 in regeneration is still unclear [142,143].

6.4. Zebrafish Connexins in the Fin

The fin of the zebrafish is composed of segmented hemirays of bone matrix that surround mesenchymal cells. The amputation of the fin follows subsequent steps, wound healing, blastema formation and regenerative outgrowth. During this process, osteoblasts function both in fin bone elongation and joint formation, and the fin returns to its original form in about 10–14 days [137].

As a connexin involved in the development and regeneration of the fin, Cx43 has been isolated from a zebrafish mutant, *short-of-fin* (*sof*), which exhibits shorter fin segment length compared with that of wild-type fish [133]. Four alleles of the *sof* mutant have been isolated. One is a mutant having low expression of *Cx43* mRNA but no amino acid substitution in the protein, and the other three have amino acid substitutions in Cx43. Electrophysiological analysis has been performed for each of the three mutants, indicating that the electrical properties of gap junctions are well linked to caudal fin length in adult fish [144].

In a model of Cx43 function in fin regeneration, it was proposed that Cx43 controls the switch for elongation of fin length or for initiation of joint formation [145]. Cx43 is expressed in regenerating fin mesenchyme and cells surrounding the bone joint, and expression of *cx43* was correlated with cell proliferation during fin regeneration. In the *cx43* mutant, decreased expression levels were detected and the cell proliferation rate was also decreased, resulting in shortening of fin length [146]. Loss of Cx43 function also causes premature joint formation, leading to shorter fin ray segments [145–147]. Several genes acting downstream of *cx43* were identified. The *evx1* gene is required for joint formation in the zebrafish fin [148]. Indeed, *evx1* mutant fish do not form joints, although normal development and regeneration in fin formation are observed [149]. Recent studies determined that Cx43 can regulate *evx1* expression [150]. Independent of *evx1* function, Cx43 controls the switching of fin growth and joint formation through Semapholin3d (Sema3d) function [148,151]. This process involves Cx43 activating Sema3d, which suppresses joint formation through the receptor, PlexinA3. At the same time, Sema3d also induces cell proliferation by suppressing the Neuropilin2a receptor, which has negative effects on cell proliferation. In contrast, when Cx43 function is suppressed, cell proliferation is also suppressed because Neurpilin2a is activated, and joint formation is activated by negative regulation of PlexinA3. Furthermore, with regard to cell proliferation, Hela1n1a protein is involved in the stabilisation of Sema3d function [152,153].

As a zebrafish-specific phenomenon, Cx43 function may be influenced by ZfCx40.8, which is a paralogue of Cx43. As described above, ZfCx43 has 80% identity at the amino acid level with human Cx43, whereas ZfCx40.8 has only 63% identity with human Cx43. This lower identity is due to the presence of a low homology region in the C-terminal domain of ZfCx40.8, suggesting that factors causing functional diversification of ZfCx40.8 exist in this region. Moreover, ZfCx40.8 is thought to be localised to and function in the cell membrane at the time of ontogeny, but remains in the Golgi at the time of fin regeneration. In previous in vitro experiments, the C-terminal domain of ZfCx40.8, which shows low homology to Cx43, was found to control the difference in membrane localisation of ZfCx40.8 between generation and regeneration of fins. Differences in membrane localisation of ZfCx40.8 may be involved in controlling the promoting function of ZfCx43 for cell division or differentiation [154,155].

7. Concluding Remarks

Connexin expression profiles within epithelial tissue networks are dynamically regulated following wounding and assault from external pathogens. These events are localised and result in changes in spatial, tissue-specific, communication compartments that allow tissues to respond to repair processes. We have reviewed recent understanding of the connexin communication compartments within the airway epithelia and epidermal networks. Another stratified epithelium surface, continually exposed to the external environment and renewed via a subset of specialised stem cells, is the cornea [42]. Corneal wound healing is a complex process that has many similarities to epithelial, stromal and endothelial cells. Wound healing of the corneal epithelium involves cell migration, cellular proliferation, adhesion, differentiation and cell-layer stratification much like is observed in the epidermis, and injury of the cornea again results in dysregulation of Cx43 expression. In addition to the studies in the skin, accumulating evidence highlights connexins as advanced therapeutic targets for corneal wound repair with both antisense Cx43 and peptidomimetic strategies proving advantageous, accelerating wound healing and reducing inflammation [156–158]. Furthermore, Gap27 was recently reported to promote migration during corneal repair, but had a limited effect on deeper vascular tissue and inflammation, again illustrating the importance of these complementary strategies in determining channel versus non-channel functions in specific subcellular compartments [159].

In conclusion, the molecular mechanisms that are driven by altered signalling caused by shifts in connexin expression profiles are now beginning to be resolved. The use of 3D ex vivo model systems, CRISPR-Cas9 technologies, and the emergence of the zebrafish and its genetic modification capabilities, together with the panel of tools to modify connexin function, provide pivotal tools for future studies. Although transfection of primary epithelial cells is challenging, lentiviral vectors can be used to introduce large transgenes into dividing or quiescent cells; thus, knockdown of target genes by delivery of CRIPR-Cas9 has already been achieved [160].

Author Contributions: The first drafts of the manuscripts were compiled by Ph.D. students A.Z. and E.M.O. who also contributed the figures. The main body of the text was written by M.C. (lung epithelia); P.E.M. (epidermis) and M.W. (Zebra fish). The final draft was collated by P.E.M. and proof read by all authors. M.C. and M.W. share equal first authorship.

Acknowledgments: We thank our funding bodies that supported underpinning work leading to this review. This includes current projects funded by the Psoriasis Association (Grant No: ST3 15) to P.E.M.; the Swiss National Science Foundation (Grant No: 310030-172909/1, the Swiss Cystic Fibrosis Foundation, and Vaincre la Mucoviscidose to M.C.; the Ministry of Education, Culture, Sports, Science, and Technology in Japan KAKENHI Grant 26291049 & 17H03683 to M.W. No funding costs were used for publication as this is an invited review for the Special Issue—The interplay of connexins and pannexins in health and disease.

Conflicts of Interest: The authors declare no conflicts of interest

Abbreviations

15kPGE2	15-keto prostaglandin E2
ATP	adenosine triphosphate
BC	basal cell
CC	ciliated cell
CF	cystic fibrosis
CK	cytokeratin
CRISPR-cas9	clustered regularly interspaced short palindromic repeats-Cas9
Cx	connexin
ECM	extracellular matrix
EP	early progenitor cell
GC	goblet /secretory cell
HPGD	hydroxyprostaglandin dehydrogenase
KO	knockout
LPS	lipopolysaccharide
Panx	pannexin
PGN	peptidoglycan
PPARγ	peroxisome proliferator-activated receptor gamma
TGF-β	transforming growth factor β
Zf	zebrafish

References

1. Laird, D.W. Life cycle of connexins in health and disease. *Biochem. J.* **2006**, *394*, 527–543. [CrossRef] [PubMed]
2. Evans, W.H.; Martin, P.E. Gap junctions: Structure and function (Review). *Mol. Membr. Biol.* **2002**, *19*, 121–136. [CrossRef] [PubMed]
3. Koval, M.; Isakson, B.E.; Gourdie, R.G. Connexins, Pannexins and innexins: Protein cousins with overlapping functions. *FEBS Lett.* **2014**, *588*, 1185. [CrossRef] [PubMed]
4. Chiu, Y.H.; Jin, X.; Medina, C.B.; Leonhardt, S.A.; Kiessling, V.; Bennett, B.C.; Shu, S.; Tamm, L.K.; Yeager, M.; Ravichandran, K.S.; et al. A quantized mechanism for activation of pannexin channels. *Nat. Commun.* **2017**, *8*, 14324. [CrossRef] [PubMed]
5. Bhalla-Gehi, R.; Penuela, S.; Churko, J.M.; Shao, Q.; Laird, D.W. Pannexin1 and pannexin3 delivery, cell surface dynamics, and cytoskeletal interactions. *J. Biol. Chem.* **2010**, *285*, 9147–9160. [CrossRef] [PubMed]
6. Nicholson, B.J.; Weber, P.A.; Cao, F.; Chang, H.; Lampe, P.; Goldberg, G. The molecular basis of selective permeability of connexins is complex and includes both size and charge. *Braz. J. Med. Biol. Res.* **2000**, *33*, 369–378. [CrossRef] [PubMed]
7. Ek-Vitorin, J.F.; Burt, J.M. Structural basis for the selective permeability of channels made of communicating junction proteins. *Biochim. Biophys. Acta* **2013**, *1828*, 51–68. [CrossRef] [PubMed]
8. Laird, D.W.; Naus, C.C.; Lampe, P.D. SnapShot: Connexins and disease. *Cell* **2017**, *170*, 1260. [CrossRef] [PubMed]

9. Crespo Yanguas, S.; Willebrords, J.; Maes, M.; da Silva, T.C.; Veloso Alves Pereira, I.; Cogliati, B.; Zaidan Dagli, M.L.; Vinken, M. Connexins and pannexins in liver damage. *EXCLI J.* **2016**, *15*, 177–186. [PubMed]

10. Whitsett, J.A.; Alenghat, T. Respiratory epithelial cells orchestrate pulmonary innate immunity. *Nat. Immunol.* **2015**, *16*, 27–35. [CrossRef] [PubMed]

11. Johnson, L.N.; Koval, M. Cross-talk between pulmonary injury, oxidant stress, and gap junctional communication. *Antioxid. Redox Signal.* **2009**, *11*, 355–367. [CrossRef] [PubMed]

12. Losa, D.; Chanson, M. The lung communication network. *Cell. Mol. Life Sci.* **2015**, *72*, 2793–2808. [CrossRef] [PubMed]

13. Scheckenbach, K.E.; Crespin, S.; Kwak, B.R.; Chanson, M. Connexin channel-dependent signaling pathways in inflammation. *J. Vasc. Res.* **2011**, *48*, 91–103. [CrossRef] [PubMed]

14. Boitano, S.; Dirksen, E.R.; Sanderson, M.J. Intercellular propagation of calcium waves mediated by inositol trisphosphate. *Science* **1992**, *258*, 292–295. [CrossRef] [PubMed]

15. Droguett, K.; Rios, M.; Carreno, D.V.; Navarrete, C.; Fuentes, C.; Villalon, M.; Barrera, N.P. An autocrine ATP release mechanism regulates basal ciliary activity in airway epithelium. *J. Physiol.* **2017**, *595*, 4755–4767. [CrossRef] [PubMed]

16. Shishikura, Y.; Koarai, A.; Aizawa, H.; Yamaya, M.; Sugiura, H.; Watanabe, M.; Hashimoto, Y.; Numakura, T.; Makiguti, T.; Abe, K.; et al. Extracellular ATP is involved in dsRNA-induced MUC5AC production via P2Y2R in human airway epithelium. *Respir. Res.* **2016**, *17*, 121. [CrossRef] [PubMed]

17. Okada, S.F.; Zhang, L.; Kreda, S.M.; Abdullah, L.H.; Davis, C.W.; Pickles, R.J.; Lazarowski, E.R.; Boucher, R.C. Coupled nucleotide and mucin hypersecretion from goblet-cell metaplastic human airway epithelium. *Am. J. Respir. Cell Mol. Biol.* **2011**, *45*, 253–260. [CrossRef] [PubMed]

18. Losa, D.; Kohler, T.; Bellec, J.; Dudez, T.; Crespin, S.; Bacchetta, M.; Boulanger, P.; Hong, S.S.; Morel, S.; Nguyen, T.H.; et al. Pseudomonas aeruginosa-induced apoptosis in airway epithelial cells is mediated by gap junctional communication in a JNK-dependent manner. *J. Immunol.* **2014**, *192*, 4804–4812. [CrossRef] [PubMed]

19. Martin, F.J.; Prince, A.S. TLR2 regulates gap junction intercellular communication in airway cells. *J. Immunol.* **2008**, *180*, 4986–4993. [CrossRef] [PubMed]

20. Kim, R.; Chang, G.; Hu, R.; Phillips, A.; Douglas, R. Connexin gap junction channels and chronic rhinosinusitis. *Int. Forum Allergy Rhinol.* **2016**, *6*, 611–617. [CrossRef] [PubMed]

21. Higgins, G.; Ringholz, F.; Buchanan, P.; McNally, P.; Urbach, V. Physiological impact of abnormal lipoxin A(4) production on cystic fibrosis airway epithelium and therapeutic potential. *BioMed Res. Int.* **2015**, *2015*, 781087. [CrossRef] [PubMed]

22. Molina, S.A.; Stauffer, B.; Moriarty, H.K.; Kim, A.H.; McCarty, N.A.; Koval, M. Junctional abnormalities in human airway epithelial cells expressing F508del CFTR. *Am. J. Physiol. Lung Cell. Mol. Physiol.* **2015**, *309*, L475–L487. [CrossRef] [PubMed]

23. Ashino, Y.; Ying, X.; Dobbs, L.G.; Bhattacharya, J. [Ca(2+)](i) oscillations regulate type II cell exocytosis in the pulmonary alveolus. *Am. J. Physiol. Lung Cell. Mol. Physiol.* **2000**, *279*, L5–L13. [CrossRef] [PubMed]

24. Wang, L.; Yin, J.; Nickles, H.T.; Ranke, H.; Tabuchi, A.; Hoffmann, J.; Tabeling, C.; Barbosa-Sicard, E.; Chanson, M.; Kwak, B.R.; et al. Hypoxic pulmonary vasoconstriction requires connexin 40-mediated endothelial signal conduction. *J. Clin. Investig.* **2012**, *122*, 4218–4230. [CrossRef] [PubMed]

25. Chadjichristos, C.E.; Scheckenbach, K.E.; van Veen, T.A.; Richani Sarieddine, M.Z.; de Wit, C.; Yang, Z.; Roth, I.; Bacchetta, M.; Viswambharan, H.; Foglia, B.; et al. Endothelial-specific deletion of connexin40 promotes atherosclerosis by increasing CD73-dependent leukocyte adhesion. *Circulation* **2010**, *121*, 123–131. [CrossRef] [PubMed]

26. Sarieddine, M.Z.; Scheckenbach, K.E.; Foglia, B.; Maass, K.; Garcia, I.; Kwak, B.R.; Chanson, M. Connexin43 modulates neutrophil recruitment to the lung. *J. Cell. Mol. Med.* **2009**, *13*, 4560–4570. [CrossRef] [PubMed]

27. Westphalen, K.; Gusarova, G.A.; Islam, M.N.; Subramanian, M.; Cohen, T.S.; Prince, A.S.; Bhattacharya, J. Sessile alveolar macrophages communicate with alveolar epithelium to modulate immunity. *Nature* **2014**, *506*, 503–506. [CrossRef] [PubMed]

28. Hogan, B.L.; Barkauskas, C.E.; Chapman, H.A.; Epstein, J.A.; Jain, R.; Hsia, C.C.; Niklason, L.; Calle, E.; Le, A.; Randell, S.H.; et al. Repair and regeneration of the respiratory system: Complexity, plasticity, and mechanisms of lung stem cell function. *Cell Stem Cell* **2014**, *15*, 123–138. [CrossRef] [PubMed]

29. Watson, J.K.; Rulands, S.; Wilkinson, A.C.; Wuidart, A.; Ousset, M.; van Keymeulen, A.; Gottgens, B.; Blanpain, C.; Simons, B.D.; Rawlins, E.L. Clonal dynamics reveal two distinct populations of basal cells in slow-turnover airway epithelium. *Cell Rep.* **2015**, *12*, 90–101. [CrossRef] [PubMed]

30. Morrisey, E.E.; Cardoso, W.V.; Lane, R.H.; Rabinovitch, M.; Abman, S.H.; Ai, X.; Albertine, K.H.; Bland, R.D.; Chapman, H.A.; Checkley, W.; et al. Molecular determinants of lung development. *Ann. Am. Thorac. Soc.* **2013**, *10*, S12–S16. [CrossRef] [PubMed]

31. Marcet, B.; Chevalier, B.; Coraux, C.; Kodjabachian, L.; Barbry, P. MicroRNA-based silencing of Delta/Notch signaling promotes multiple cilia formation. *Cell Cycle* **2011**, *10*, 2858–2864. [CrossRef] [PubMed]

32. Zhao, R.; Fallon, T.R.; Saladi, S.V.; Pardo-Saganta, A.; Villoria, J.; Mou, H.; Vinarsky, V.; Gonzalez-Celeiro, M.; Nunna, N.; Hariri, L.P.; et al. Yap tunes airway epithelial size and architecture by regulating the identity, maintenance, and self-renewal of stem cells. *Dev. Cell* **2014**, *30*, 151–165. [CrossRef] [PubMed]

33. Mahoney, J.E.; Mori, M.; Szymaniak, A.D.; Varelas, X.; Cardoso, W.V. The hippo pathway effector Yap controls patterning and differentiation of airway epithelial progenitors. *Dev. Cell* **2014**, *30*, 137–150. [CrossRef] [PubMed]

34. Carson, J.L.; Reed, W.; Moats-Staats, B.M.; Brighton, L.E.; Gambling, T.M.; Hu, S.C.; Collier, A.M. Connexin 26 expression in human and ferret airways and lung during development. *Am. J. Respir. Cell Mol. Biol.* **1998**, *18*, 111–119. [CrossRef] [PubMed]

35. Wiszniewski, L.; Sanz, J.; Scerri, I.; Gasparotto, E.; Dudez, T.; Lacroix, J.S.; Suter, S.; Gallati, S.; Chanson, M. Functional expression of connexin30 and connexin31 in the polarized human airway epithelium. *Differentiation* **2007**, *75*, 382–392. [CrossRef] [PubMed]

36. Crespin, S.; Bacchetta, M.; Bou Saab, J.; Tantilipikorn, P.; Bellec, J.; Dudez, T.; Nguyen, T.H.; Kwak, B.R.; Lacroix, J.S.; Huang, S.; et al. Cx26 regulates proliferation of repairing basal airway epithelial cells. *Int. J. Biochem. Cell Biol.* **2014**, *52*, 152–160. [CrossRef] [PubMed]

37. Huang, T.H.; Razmovski-Naumovski, V.; Kota, B.P.; Lin, D.S.; Roufogalis, B.D. The pathophysiological function of peroxisome proliferator-activated receptor-γ in lung-related diseases. *Respir. Res.* **2005**, *6*, 102. [CrossRef] [PubMed]

38. Dekkers, J.F.; van der Ent, C.K.; Kalkhoven, E.; Beekman, J.M. PPARγ as a therapeutic target in cystic fibrosis. *Trends Mol. Med.* **2012**, *18*, 283–291. [CrossRef] [PubMed]

39. Bou Saab, J.; Bacchetta, M.; Chanson, M. Ineffective correction of PPARγ signaling in cystic fibrosis airway epithelial cells undergoing repair. *Int. J. Biochem. Cell Biol.* **2016**, *78*, 361–369. [CrossRef] [PubMed]

40. Bou Saab, J. Connexin26 and PPAR-Gamma Signaling Pathway in Human Airway Epithelial Cells. Ph.D. Thesis, University of Geneva, Geneva, Switzerland, 18 December 2015.

41. Proksch, E.; Brandner, J.M.; Jensen, J.M. The skin: An indispensable barrier. *Exp. Dermatol.* **2008**, *17*, 1063–1072. [CrossRef] [PubMed]

42. Blanpain, C.; Horsley, V.; Fuchs, E. Epithelial stem cells: Turning over new leaves. *Cell* **2007**, *128*, 445–458. [CrossRef] [PubMed]

43. Simpson, C.L.; Patel, D.M.; Green, K.J. Deconstructing the skin: Cytoarchitectural determinants of epidermal morphogenesis. *Nat. Rev. Mol. Cell Biol.* **2011**, *12*, 565–580. [CrossRef] [PubMed]

44. Di, W.L.; Rugg, E.L.; Leigh, I.M.; Kelsell, D.P. Multiple epidermal connexins are expressed in different keratinocyte subpopulations including connexin 31. *J. Investig. Dermatol.* **2001**, *117*, 958–964. [CrossRef] [PubMed]

45. Martin, P.E.; Easton, J.A.; Hodgins, M.B.; Wright, C.S. Connexins: Sensors of epidermal integrity that are therapeutic targets. *FEBS Lett.* **2014**, *588*, 1304–1314. [CrossRef] [PubMed]

46. Mesnil, M.; Krutovskikh, V.; Piccoli, C.; Elfgang, C.; Traub, O.; Willecke, K.; Yamasaki, H. Negative growth control of HeLa cells by connexin genes: Connexin species specificity. *Cancer Res.* **1995**, *55*, 629–639. [PubMed]

47. Garcia, I.E.; Maripillan, J.; Jara, O.; Ceriani, R.; Palacios-Munoz, A.; Ramachandran, J.; Olivero, P.; Perez-Acle, T.; Gonzalez, C.; Saez, J.C.; et al. Keratitis-ichthyosis-deafness syndrome-associated cx26 mutants produce nonfunctional gap junctions but hyperactive hemichannels when co-expressed with wild type cx43. *J. Investig. Dermatol.* **2015**, *135*, 1338–1347. [CrossRef] [PubMed]

48. Easton, J.A.; Alboulshi, A.K.; Kamps, M.A.F.; Broers, G.H.M.R.; Coull, B.J.; Oji, V.; van Geel, M.; van Steensel, M.A.M.; Martin, P.E. A rare missense mutation in GJB3 (Cx31G45E) is associated with a unique cellular phenotype resulting in necrotic cell death. *Exp. Dermatol.* **2018**. [CrossRef] [PubMed]

49. Kam, E.; Hodgins, M.B. Communication compartments in hair follicles and their implication in differentiative control. *Development* **1992**, *114*, 389–393. [PubMed]

50. Lucke, T.; Choudhry, R.; Thom, R.; Selmer, I.S.; Burden, A.D.; Hodgins, M.B. Upregulation of connexin 26 is a feature of keratinocyte differentiation in hyperproliferative epidermis, vaginal epithelium, and buccal epithelium. *J. Investig. Dermatol.* **1999**, *112*, 354–361. [CrossRef] [PubMed]

51. Martin, P.E.; van Steensel, M. Connexins and skin disease: Insights into the role of β connexins in skin homeostasis. *Cell Tissue Res.* **2015**, *360*, 645–658. [CrossRef] [PubMed]

52. Faniku, C.; Wright, C.S.; Martin, P.E. Connexins and pannexins in the integumentary system: The skin and appendages. *Cell. Mol. Life Sci.* **2015**, *72*, 2937–2947. [CrossRef] [PubMed]

53. Celetti, S.J.; Cowan, K.N.; Penuela, S.; Shao, Q.; Churko, J.; Laird, D.W. Implications of pannexin 1 and pannexin 3 for keratinocyte differentiation. *J. Cell Sci.* **2010**, *123*, 1363–1372. [CrossRef] [PubMed]

54. Penuela, S.; Celetti, S.J.; Bhalla, R.; Shao, Q.; Laird, D.W. Diverse subcellular distribution profiles of pannexin 1 and pannexin 3. *Cell. Commun Adhes.* **2008**, *15*, 133–142. [CrossRef] [PubMed]

55. Cowan, K.N.; Langlois, S.; Penuela, S.; Cowan, B.J.; Laird, D.W. Pannexin1 and Pannexin3 exhibit distinct localisation patterns in human skin appendages and are regulated during keratinocyte differentiation and carcinogenesis. *Cell Commun. Adhes.* **2012**, *19*, 45–53. [CrossRef] [PubMed]

56. Martin, P. Wound healing—Aiming for perfect skin regeneration. *Science* **1997**, *276*, 75–81. [CrossRef] [PubMed]

57. Stojadinovic, O.; Brem, H.; Vouthounis, C.; Lee, B.; Fallon, J.; Stallcup, M.; Merchant, A.; Galiano, R.D.; Tomic-Canic, M. Molecular pathogenesis of chronic wounds: The role of β-catenin and c-myc in the inhibition of epithelialization and wound healing. *Am. J. Pathol.* **2005**, *167*, 59–69. [CrossRef]

58. Brandner, J.M.; Houdek, P.; Husing, B.; Kaiser, C.; Moll, I. Connexins 26, 30, and 43: Differences among spontaneous, chronic, and accelerated human wound healing. *J. Investig. Dermatol.* **2004**, *122*, 1310–1320. [CrossRef] [PubMed]

59. Solan, J.L.; Lampe, P.D. Kinase programs spatiotemporally regulate gap junction assembly and disassembly: Effects on wound repair. *Semin. Cell Dev. Biol.* **2016**, *50*, 40–48. [CrossRef] [PubMed]

60. Marquez-Rosado, L.; Singh, D.; Rincon-Arano, H.; Solan, J.L.; Lampe, P.D. CASK (LIN2) interacts with Cx43 in wounded skin and their coexpression affects cell migration. *J. Cell Sci.* **2012**, *125*, 695–702. [CrossRef] [PubMed]

61. Faniku, C.; O'Shaughnessy, E.; Lorraine, C.; Johnstone, S.R.; Graham, A.; Greenhough, S.; Martin, P.E.M. The connexin mimetic peptide Gap27 and Cx43-knockdown reveal differential roles for Connexin43 in wound closure events in skin model systems. *Int. J. Mol. Sci.* **2018**, *19*, e604. [CrossRef] [PubMed]

62. Hills, C.E.; Siamantouras, E.; Smith, S.W.; Cockwell, P.; Liu, K.K.; Squires, P.E. TGF-β modulates cell-to-cell communication in early epithelial-to-mesenchymal transition. *Diabetologia* **2012**, *55*, 812–824. [CrossRef] [PubMed]

63. Dai, P.; Nakagami, T.; Tanaka, H.; Hitomi, T.; Takamatsu, T. Cx43 mediates TGF-β signaling through competitive Smads binding to microtubules. *Mol. Biol Cell* **2007**, *18*, 2264–2273. [CrossRef] [PubMed]

64. Kandyba, E.E.; Hodgins, M.B.; Martin, P.E. A murine living skin equivalent amenable to live-cell imaging: Analysis of the roles of connexins in the epidermis. *J. Investig. Dermatol.* **2008**, *128*, 1039–1049. [CrossRef] [PubMed]

65. Kretz, M.; Euwens, C.; Hombach, S.; Eckardt, D.; Teubner, B.; Traub, O.; Willecke, K.; Ott, T. Altered connexin expression and wound healing in the epidermis of connexin-deficient mice. *J. Cell Sci.* **2003**, *116*, 3443–3452. [CrossRef] [PubMed]

66. Penuela, S.; Kelly, J.J.; Churko, J.M.; Barr, K.J.; Berger, A.C.; Laird, D.W. Panx1 regulates cellular properties of keratinocytes and dermal fibroblasts in skin development and wound healing. *J. Investig. Dermatol.* **2014**, *134*, 2026–2035. [CrossRef] [PubMed]

67. Falanga, V. Wound healing and its impairment in the diabetic foot. *Lancet* **2005**, *366*, 1736–1743. [CrossRef]

68. Game, F.L.; Apelqvist, J.; Attinger, C.; Hartemann, A.; Hinchliffe, R.J.; Londahl, M.; Price, P.E.; Jeffcoate, W.J. Effectiveness of interventions to enhance healing of chronic ulcers of the foot in diabetes: A systematic review. *Diabetes Metab. Res. Rev.* **2016**, *32*, 154–168. [CrossRef] [PubMed]

69. Becker, D.L.; Thrasivoulou, C.; Phillips, A.R. Connexins in wound healing; perspectives in diabetic patients. *Biochim. Biophys. Acta* **2012**, *1818*, 2068–2075. [CrossRef] [PubMed]

70. Sutcliffe, J.E.; Chin, K.Y.; Thrasivoulou, C.; Serena, T.E.; O'Neil, S.; Hu, R.; White, A.M.; Madden, L.; Richards, T.; Phillips, A.R.; et al. Abnormal connexin expression in human chronic wounds. *Br. J. Dermatol.* **2015**, *173*, 1205–1215. [CrossRef] [PubMed]

71. Martin, P.E. Connexins help fill the Gap: Markers and therapeutic targets for chronic nonhealing wounds. *Br. J. Dermatol.* **2015**, *173*, 1123–1124. [CrossRef] [PubMed]

72. Bowcock, A.M.; Krueger, J.G. Getting under the skin: The immunogenetics of psoriasis. *Nat. Rev. Immunol.* **2005**, *5*, 699–711. [CrossRef] [PubMed]

73. Buchau, A.S.; Gallo, R.L. Innate immunity and antimicrobial defense systems in psoriasis. *Clin. Dermatol.* **2007**, *25*, 616–624. [CrossRef] [PubMed]

74. Gallo, R.L.; Nakatsuji, T. Microbial symbiosis with the innate immune defense system of the skin. *J. Investig. Dermatol.* **2011**, *131*, 1974–1980. [CrossRef] [PubMed]

75. Enamandram, M.; Kimball, A.B. Psoriasis epidemiology: The interplay of genes and the environment. *J. Investig. Dermatol.* **2013**, *133*, 287–289. [CrossRef] [PubMed]

76. Kirschner, N.; Poetzl, C.; von den Driesch, P.; Wladykowski, E.; Moll, I.; Behne, M.J.; Brandner, J.M. Alteration of tight junction proteins is an early event in psoriasis: Putative involvement of proinflammatory cytokines. *Am. J. Pathol.* **2009**, *175*, 1095–1106. [CrossRef] [PubMed]

77. Donnelly, S.; English, G.; de Zwart-Storm, E.A.; Lang, S.; van Steensel, M.A.; Martin, P.E. Differential susceptibility of Cx26 mutations associated with epidermal dysplasias to peptidoglycan derived from Staphylococcus aureus and Staphylococcus epidermidis. *Exp. Dermatol.* **2012**, *21*, 592–598. [CrossRef] [PubMed]

78. Nair, R.P.; Stuart, P.E.; Nistor, I.; Hiremagalore, R.; Chia, N.V.; Jenisch, S.; Weichenthal, M.; Abecasis, G.R.; Lim, H.W.; Christophers, E.; et al. Sequence and haplotype analysis supports HLA-C as the psoriasis susceptibility 1 gene. *Am. J. Hum. Genet.* **2006**, *78*, 827–851. [CrossRef] [PubMed]

79. Sun, L.D.; Cheng, H.; Wang, Z.X.; Zhang, A.P.; Wang, P.G.; Xu, J.H.; Zhu, Q.X.; Zhou, H.S.; Ellinghaus, E.; Zhang, F.R.; et al. Association analyses identify six new psoriasis susceptibility loci in the Chinese population. *Nat. Genet.* **2010**, *42*, 1005–1009. [CrossRef] [PubMed]

80. Liu, Q.P.; Wu, L.S.; Li, F.F.; Liu, S.; Su, J.; Kuang, Y.H.; Chen, C.; Xie, X.Y.; Jiang, M.H.; Zhao, S.; et al. The association between GJB2 gene polymorphism and psoriasis: A verification study. *Arch. Dermatol. Res.* **2012**, *304*, 769–772. [CrossRef] [PubMed]

81. Wang, X.; Ramirez, A.; Budunova, I. Overexpression of connexin26 in the basal keratinocytes reduces sensitivity to tumor promoter TPA. *Exp. Dermatol.* **2010**, *19*, 633–640. [CrossRef] [PubMed]

82. Yao, F.; Yue, M.; Zhang, C.; Zuo, X.; Zheng, X.; Zhang, A.; Wang, Z.; Liu, S.; Li, H.; Meng, L.; et al. A genetic coding variant rs72474224 in GJB2 is associated with clinical features of psoriasis vulgaris in a Chinese Han population. *Tissue Antigens* **2015**, *86*, 134–138. [CrossRef] [PubMed]

83. Li, B.; Tsoi, L.C.; Swindell, W.R.; Gudjonsson, J.E.; Tejasvi, T.; Johnston, A.; Ding, J.; Stuart, P.E.; Xing, X.; Kochkodan, J.J.; et al. Transcriptome analysis of psoriasis in a large case-control sample: RNA-seq provides insights into disease mechanisms. *J. Investig. Dermatol.* **2014**, *134*, 1828–1838. [CrossRef] [PubMed]

84. Labarthe, M.P.; Bosco, D.; Saurat, J.H.; Meda, P.; Salomon, D. Upregulation of connexin 26 between keratinocytes of psoriatic lesions. *J. Investig. Dermatol.* **1998**, *111*, 72–76. [CrossRef] [PubMed]

85. Shaker, O.; Abdel-Halim, M. Connexin 26 in psoriatic skin before and after two conventional therapeutic modalities: Methotrexate and PUVA. *Eur. J. Dermatol.* **2012**, *22*, 218–224. [PubMed]

86. Moravcova, M.; Libra, A.; Dvorakova, J.; Viskova, A.; Muthny, T.; Velebny, V.; Kubala, L. Modulation of keratin 1, 10 and involucrin expression as part of the complex response of the human keratinocyte cell line HaCaT to ultraviolet radiation. *Interdiscip. Toxicol.* **2013**, *6*, 203–208. [CrossRef] [PubMed]

87. Djalilian, A.R.; McGaughey, D.; Patel, S.; Seo, E.Y.; Yang, C.; Cheng, J.; Tomic, M.; Sinha, S.; Ishida-Yamamoto, A.; Segre, J.A. Connexin 26 regulates epidermal barrier and wound remodeling and promotes psoriasiform response. *J. Clin. Investig.* **2006**, *116*, 1243–1253. [CrossRef] [PubMed]

88. Chung, E.; Cook, P.W.; Parkos, C.A.; Park, Y.K.; Pittelkow, M.R.; Coffey, R.J. Amphiregulin causes functional downregulation of adherens junctions in psoriasis. *J. Investig. Dermatol.* **2005**, *124*, 1134–1140. [CrossRef] [PubMed]

89. Hampton, P.J.; Ross, O.K.; Reynolds, N.J. Increased nuclear β-catenin in suprabasal involved psoriatic epidermis. *Br. J. Dermatol.* **2007**, *157*, 1168–1177. [CrossRef] [PubMed]

90. Li, Z.; Peng, Z.; Wang, Y.; Geng, S.; Ji, F. Decreased expression of E-cadherin and β-catenin in the lesional skin of patients with active psoriasis. *Int. J. Dermatol.* **2008**, *47*, 207–209. [CrossRef] [PubMed]

91. Tschachler, E. ; Psoriasis: The epidermal component. *Clin. Dermatol.* **2007**, *25*, 589–595. [CrossRef] [PubMed]

92. Hivnor, C.; Williams, N.; Singh, F.; VanVoorhees, A.; Dzubow, L.; Baldwin, D.; Seykora, J. Gene expression profiling of porokeratosis demonstrates similarities with psoriasis. *J. Cutan. Pathol.* **2004**, *31*, 657–664. [CrossRef] [PubMed]

93. Grice, E.A.; Kong, H.H.; Conlan, S.; Deming, C.B.; Davis, J.; Young, A.C.; Bouffard, G.G.; Blakesley, R.W.; Murray, P.R.; Green, E.D.; et al. Topographical and temporal diversity of the human skin microbiome. *Science* **2009**, *324*, 1190–1192. [CrossRef] [PubMed]

94. Acosta, N.; Whelan, F.J.; Somayaji, R.; Poonja, A.; Surette, M.G.; Rabin, H.R.; Parkins, M.D. The evolving cystic fibrosis microbiome: A comparative cohort study spanning 16 years. *Ann. Am. Thorac. Soc.* **2017**, *14*, 1288–1297. [CrossRef] [PubMed]

95. Castelino, M.; Eyre, S.; Upton, M.; Ho, P.; Barton, A. The bacterial skin microbiome in psoriatic arthritis, an unexplored link in pathogenesis: Challenges and opportunities offered by recent technological advances. *Rheumatology* **2014**, *53*, 777–784. [CrossRef] [PubMed]

96. Man, Y.K.; Trolove, C.; Tattersall, D.; Thomas, A.C.; Papakonstantinopoulou, A.; Patel, D.; Scott, C.; Chong, J.; Jagger, D.J.; O'Toole, E.A.; et al. A deafness-associated mutant human connexin 26 improves the epithelial barrier in vitro. *J. Membr. Biol.* **2007**, *218*, 29–37. [CrossRef] [PubMed]

97. Losa, D.; Kohler, T.; Bacchetta, M.; Saab, J.B.; Frieden, M.; van Delden, C.; Chanson, M. Airway Epithelial cell integrity protects from cytotoxicity of pseudomonas aeruginosa quorum-sensing signals. *Am. J. Respir. Cell Mol. Biol.* **2015**, *53*, 265–275. [CrossRef] [PubMed]

98. Bou Saab, J.; Losa, D.; Chanson, M.; Ruez, R. Connexins in respiratory and gastrointestinal mucosal immunity. *FEBS Lett.* **2014**, *588*, 1288–1296. [CrossRef] [PubMed]

99. Braet, K.; Vandamme, W.; Martin, P.E.; Evans, W.H.; Leybaert, L. Photoliberating inositol-1,4,5-trisphosphate triggers ATP release that is blocked by the connexin mimetic peptide gap 26. *Cell. Calcium* **2003**, *33*, 37–48. [CrossRef]

100. De Vuyst, E.; Decrock, E.; de Bock, M.; Yamasaki, H.; Naus, C.C.; Evans, W.H.; Leybaert, L. Connexin hemichannels and gap junction channels are differentially influenced by lipopolysaccharide and basic fibroblast growth factor. *Mol. Biol. Cell* **2007**, *18*, 34–46. [CrossRef] [PubMed]

101. Burnstock, G. Purinergic signalling. *Br. J. Pharmacol.* **2006**, *147*, S172–S181. [CrossRef] [PubMed]

102. Dixon, C.J.; Bowler, W.B.; Littlewood-Evans, A.; Dillon, J.P.; Bilbe, G.; Sharpe, G.R.; Gallagher, J.A. Regulation of epidermal homeostasis through P2Y2 receptors. *Br. J. Pharmacol.* **1999**, *127*, 1680–1686. [CrossRef] [PubMed]

103. Pillai, S.; Bikle, D.D. Adenosine triphosphate stimulates phosphoinositide metabolism, mobilizes intracellular calcium, and inhibits terminal differentiation of human epidermal keratinocytes. *J. Clin. Investig.* **1992**, *90*, 42–51. [CrossRef] [PubMed]

104. Denda, M.; Inoue, K.; Fuziwara, S.; Denda, S. P2X purinergic receptor antagonist accelerates skin barrier repair and prevents epidermal hyperplasia induced by skin barrier disruption. *J. Investig. Dermatol.* **2002**, *119*, 1034–1040. [CrossRef] [PubMed]

105. Denda, M.; Denda, S. Air-exposed keratinocytes exhibited intracellular calcium oscillation. *Skin Res. Technol.* **2007**, *13*, 195–201. [CrossRef] [PubMed]

106. Ransford, G.A.; Fregien, N.; Qiu, F.; Dahl, G.; Conner, G.E.; Salathe, M. Pannexin 1 contributes to ATP release in airway epithelia. *Am. J. Respir. Cell Mol. Biol.* **2009**, *41*, 525–534. [CrossRef] [PubMed]

107. Seminario-Vidal, L.; Okada, S.F.; Sesma, J.I.; Kreda, S.M.; van Heusden, C.A.; Zhu, Y.; Jones, L.C.; O'Neal, W.K.; Penuela, S.; Laird, D.W.; et al. Rho signaling regulates pannexin 1-mediated ATP release from airway epithelia. *J. Biol. Chem.* **2011**, *286*, 26277–26286. [CrossRef] [PubMed]

108. Evans, W.H.; Bultynck, G.; Leybaert, L. Manipulating connexin communication channels: Use of peptidomimetics and the translational outputs. *J. Membr. Biol.* **2012**, *245*, 437–449. [CrossRef] [PubMed]

109. Kretz, M.; Maass, K.; Willecke, K. Expression and function of connexins in the epidermis, analyzed with transgenic mouse mutants. *Eur. J. Cell Biol.* **2004**, *83*, 647–654. [CrossRef] [PubMed]

110. Qiu, C.; Coutinho, P.; Frank, S.; Franke, S.; Law, L.Y.; Martin, P.; Green, C.R.; Becker, D.L. Targeting connexin43 expression accelerates the rate of wound repair. *Curr. Biol.* **2003**, *13*, 1697–1703. [CrossRef] [PubMed]

111. Wang, C.M.; Lincoln, J.; Cook, J.E.; Becker, D.L. Abnormal connexin expression underlies delayed wound healing in diabetic skin. *Diabetes* **2007**, *56*, 2809–2817. [CrossRef] [PubMed]

112. Law, L.Y.; Zhang, W.V.; Stott, N.S.; Becker, D.L.; Green, C.R. In vitro optimization of antisense oligodeoxynucleotide design: An example using the connexin gene family. *J. Biomol. Tech.* **2006**, *17*, 270–282. [PubMed]

113. Mori, R.; Power, K.T.; Wang, C.M.; Martin, P.; Becker, D.L. Acute downregulation of connexin43 at wound sites leads to a reduced inflammatory response, enhanced keratinocyte proliferation and wound fibroblast migration. *J. Cell Sci.* **2006**, *119*, 5193–5203. [CrossRef] [PubMed]

114. Cronin, M.; Anderson, P.N.; Cook, J.E.; Green, C.R.; Becker, D.L. Blocking connexin43 expression reduces inflammation and improves functional recovery after spinal cord injury. *Mol. Cell. Neurosci.* **2008**, *39*, 152–160. [CrossRef] [PubMed]

115. Danesh-Meyer, H.V.; Huang, R.; Nicholson, L.F.; Green, C.R. Connexin43 antisense oligodeoxynucleotide treatment down-regulates the inflammatory response in an in vitro interphase organotypic culture model of optic nerve ischaemia. *J. Clin. Neurosci.* **2008**, *15*, 1253–1263. [CrossRef] [PubMed]

116. Green, C.R.; Nicholson, L.F. Interrupting the inflammatory cycle in chronic diseases- do gap junctions provide the answer? *Cell Biol. Int.* **2008**, *32*, 1578–1583. [CrossRef] [PubMed]

117. Ghatnekar, G.S.; Grek, C.L.; Armstrong, D.G.; Desai, S.C.; Gourdie, R.G. The effect of a connexin43-based Peptide on the healing of chronic venous leg ulcers: A multicenter, randomized trial. *J. Investig. Dermatol.* **2015**, *135*, 289–298. [CrossRef] [PubMed]

118. Ghatnekar, G.S.; O'Quinn, M.P.; Jourdan, L.J.; Gurjarpadhye, A.A.; Draughn, R.L.; Gourdie, R.G. Connexin43 carboxyl-terminal peptides reduce scar progenitor and promote regenerative healing following skin wounding. *RegenMed* **2009**, *4*, 205–223. [CrossRef] [PubMed]

119. Ongstad, E.L.; O'Quinn, M.P.; Ghatnekar, G.S.; Yost, M.J.; Gourdie, R.G. A Connexin43 Mimetic Peptide Promotes Regenerative Healing and Improves Mechanical Properties in Skin and Heart. *Adv. Wound Care* **2013**, *2*, 55–62. [CrossRef] [PubMed]

120. Rhett, J.M.; Ghatnekar, G.S.; Palatinus, J.A.; O'Quinn, M.; Yost, M.J.; Gourdie, R.G. Novel therapies for scar reduction and regenerative healing of skin wounds. *Trends Biotechnol.* **2008**, *26*, 173–180. [CrossRef] [PubMed]

121. Wright, C.S.; Berends, R.F.; Flint, D.J.; Martin, P.E. Cell motility in models of wounded human skin is improved by Gap27 despite raised glucose, insulin and IGFBP-5. *Exp. Cell Res.* **2013**, *319*, 390–401. [CrossRef] [PubMed]

122. Wright, C.S.; Pollok, S.; Flint, D.J.; Brandner, J.M.; Martin, P.E. The connexin mimetic peptide Gap27 increases human dermal fibroblast migration in hyperglycemic and hyperinsulinemic conditions in vitro. *J. Cell. Physiol.* **2012**, *227*, 77–87. [CrossRef] [PubMed]

123. Wright, C.S.; van Steensel, M.A.; Hodgins, M.B.; Martin, P.E. Connexin mimetic peptides improve cell migration rates of human epidermal keratinocytes and dermal fibroblasts in vitro. *Wound Repair Regen.* **2009**, *17*, 240–249. [CrossRef] [PubMed]

124. Riding, A.; Pullar, C.E. ATP Release and P2 Y Receptor signaling are essential for keratinocyte galvanotaxis. *J. Cell. Physiol.* **2016**, *231*, 181–191. [CrossRef] [PubMed]

125. Riteau, N.; Gasse, P.; Fauconnier, L.; Gombault, A.; Couegnat, M.; Fick, L.; Kanellopoulos, J.; Quesniaux, V.F.; Marchand-Adam, S.; Crestani, B.; et al. Extracellular ATP is a danger signal activating P2X7 receptor in lung inflammation and fibrosis. *Am. J. Respir. Crit. Care Med.* **2011**, *182*, 774–783. [CrossRef] [PubMed]

126. Hwang, W.Y.; Fu, Y.; Reyon, D.; Maeder, M.L.; Tsai, S.Q.; Sander, J.D.; Peterson, R.T.; Yeh, J.R.; Joung, J.K. Efficient genome editing in zebrafish using a CRISPR-Cas system. *Nat. Biotechnol.* **2013**, *31*, 227–229. [CrossRef] [PubMed]

127. Urasaki, A.; Morvan, G.; Kawakami, K. Functional dissection of the Tol2 transposable element identified the minimal cis-sequence and a highly repetitive sequence in the subterminal region essential for transposition. *Genetics* **2006**, *174*, 639–649. [CrossRef] [PubMed]

128. Li, Q.; Frank, M.; Thisse, C.I.; Thisse, B.V.; Uitto, J. Zebrafish: A model system to study heritable skin diseases. *J. Investig. Dermatol.* **2011**, *131*, 565–571. [CrossRef] [PubMed]

129. Li, Q.; Uitto, J. Zebrafish as a model system to study skin biology and pathology. *J. Investig. Dermatol.* **2014**, *134*, 1–6. [CrossRef] [PubMed]

130. Niethammer, P.; Grabher, C.; Look, A.T.; Mitchison, T.J. A tissue-scale gradient of hydrogen peroxide mediates rapid wound detection in zebrafish. *Nature* **2009**, *459*, 996–999. [CrossRef] [PubMed]

131. Watanabe, M. Gap junction in the teleost fish lineage: duplicated connexins may contribute to skin pattern formation and body shape determination. *Front. Cell Dev. Biol.* **2017**, *5*, 13. [CrossRef] [PubMed]

132. Chatterjee, B.; Chin, A.J.; Valdimarsson, G.; Finis, C.; Sonntag, J.M.; Choi, B.Y.; Tao, L.; Balasubramanian, K.; Bell, C.; Krufka, A.; et al. Developmental regulation and expression of the zebrafish connexin43 gene. *Dev. Dyn.* **2005**, *233*, 890–906. [CrossRef] [PubMed]

133. Iovine, M.K.; Higgins, E.P.; Hindes, A.; Coblitz, B.; Johnson, S.L. Mutations in connexin43 (GJA1) perturb bone growth in zebrafish fins. *Dev. Biol.* **2005**, *278*, 208–219. [CrossRef] [PubMed]

134. Irion, U.; Frohnhofer, H.G.; Krauss, J.; Colak Champollion, T.; Maischein, H.M.; Geiger-Rudolph, S.; Weiler, C.; Nusslein-Volhard, C. Gap junctions composed of connexins 41.8 and 39.4 are essential for colour pattern formation in zebrafish. *eLife* **2014**, *3*, e05125. [CrossRef] [PubMed]

135. Watanabe, M.; Sawada, R.; Aramaki, T.; Skerrett, I.M.; Kondo, S. The physiological characterization of connexin41.8 and connexin39.4, which are involved in the striped pattern formation of zebrafish. *J. Biol. Chem.* **2016**, *291*, 1053–1063. [CrossRef] [PubMed]

136. Nakada, Y.; Canseco, D.C.; Thet, S.; Abdisalaam, S.; Asaithamby, A.; Santos, C.X.; Shah, A.M.; Zhang, H.; Faber, J.E.; Kinter, M.T.; et al. Hypoxia induces heart regeneration in adult mice. *Nature* **2017**, *541*, 222–227. [CrossRef] [PubMed]

137. Sehring, I.M.; Jahn, C.; Weidinger, G. Zebrafish fin and heart: What's special about regeneration? *Curr. Opin. Genet. Dev.* **2016**, *40*, 48–56. [CrossRef] [PubMed]

138. Cheng, S.; Shakespeare, T.; Mui, R.; White, T.W.; Valdimarsson, G. Connexin 48.5 is required for normal cardiovascular function and lens development in zebrafish embryos. *J. Biol. Chem.* **2004**, *279*, 36993–37003. [CrossRef] [PubMed]

139. Christie, T.L.; Mui, R.; White, T.W.; Valdimarsson, G. Molecular cloning, functional analysis, and RNA expression analysis of connexin45.6: A zebrafish cardiovascular connexin. *Am. J. Physiol. Heart Circ. Physiol.* **2004**, *286*, H1623–H1632. [CrossRef] [PubMed]

140. Sultana, N.; Nag, K.; Hoshijima, K.; Laird, D.W.; Kawakami, A.; Hirose, S. Zebrafish early cardiac connexin, Cx36.7/Ecx, regulates myofibril orientation and heart morphogenesis by establishing Nkx2.5 expression. *Proc. Natl. Acad. Sci. USA* **2008**, *105*, 4763–4768. [CrossRef] [PubMed]

141. Watanabe, M.; Iwashita, M.; Ishii, M.; Kurachi, Y.; Kawakami, A.; Kondo, S.; Okada, N. Spot pattern of leopard Danio is caused by mutation in the zebrafish connexin41.8 gene. *EMBO Rep.* **2006**, *7*, 893–897. [CrossRef] [PubMed]

142. Sallin, P.; de Preux Charles, A.S.; Duruz, V.; Pfefferli, C.; Jazwinska, A. A dual epimorphic and compensatory mode of heart regeneration in zebrafish. *Dev. Biol.* **2015**, *399*, 27–40. [CrossRef] [PubMed]

143. Yu, F.; Li, R.; Parks, E.; Takabe, W.; Hsiai, T.K. Electrocardiogram signals to assess zebrafish heart regeneration: Implication of long QT intervals. *Ann. Biomed. Eng.* **2010**, *38*, 2346–2357. [CrossRef] [PubMed]

144. Hoptak-Solga, A.D.; Klein, K.A.; DeRosa, A.M.; White, T.W.; Iovine, M.K. Zebrafish short fin mutations in connexin43 lead to aberrant gap junctional intercellular communication. *FEBS Lett.* **2007**, *581*, 3297–3302. [CrossRef] [PubMed]

145. Sims, K., Jr.; Eble, D.M.; Iovine, M.K. Connexin43 regulates joint location in zebrafish fins. *Dev. Biol.* **2009**, *327*, 410–418. [CrossRef] [PubMed]

146. Brown, A.M.; Fisher, S.; Iovine, M.K. Osteoblast maturation occurs in overlapping proximal-distal compartments during fin regeneration in zebrafish. *Dev. Dyn.* **2009**, *238*, 2922–2928. [CrossRef] [PubMed]

147. Hoptak-Solga, A.D.; Nielsen, S.; Jain, I.; Thummel, R.; Hyde, D.R.; Iovine, M.K. Connexin43 (GJA1) is required in the population of dividing cells during fin regeneration. *Dev. Biol.* **2008**, *317*, 541–548. [CrossRef] [PubMed]

148. Ton, Q.V.; Iovine, M.K. Identification of an evx1-dependent joint-formation pathway during FIN regeneration. *PLoS ONE* **2013**, *8*, e81240. [CrossRef] [PubMed]

149. Schulte, C.J.; Allen, C.; England, S. J.; Juarez-Morales, J.L.; Lewis, K.E. Evx1 is required for joint formation in zebrafish fin dermoskeleton. *Dev. Dyn.* **2011**, *240*, 1240–1248. [CrossRef] [PubMed]

150. Dardis, G.; Tryon, R.; Ton, Q.; Johnson, S.L.; Iovine, M.K. Cx43 suppresses evx1 expression to regulate joint initiation in the regenerating fin. *Dev. Dyn.* **2017**, *246*, 691–699. [CrossRef] [PubMed]

151. Ton, Q.V.; Kathryn Iovine, M. Semaphorin3d mediates Cx43-dependent phenotypes during fin regeneration. *Dev. Biol.* **2012**, *366*, 195–203. [CrossRef] [PubMed]

152. Govindan, J.; Iovine, M.K. Hapln1a is required for connexin43-dependent growth and patterning in the regenerating fin skeleton. *PLoS ONE* **2014**, *9*, e88574. [CrossRef] [PubMed]

153. Govindan, J.; Tun, K.M.; Iovine, M.K. Cx43-dependent skeletal phenotypes are mediated by interactions between the Hapln1a-ECM and Sema3d during fin regeneration. *PLoS ONE* **2016**, *11*, e0148202. [CrossRef] [PubMed]

154. Gerhart, S.V.; Eble, D.M.; Burger, R.M.; Oline, S.N.; Vacaru, A.; Sadler, K.C.; Jefferis, R.; Iovine, M.K. The Cx43-like connexin protein Cx40.8 is differentially localized during fin ontogeny and fin regeneration. *PLoS ONE* **2012**, *7*, e31364. [CrossRef] [PubMed]

155. Gerhart, S.V.; Jefferis, R.; Iovine, M.K. Cx40.8, a Cx43-like protein, forms gap junction channels inefficiently and may require Cx43 for its association at the plasma membrane. *FEBS Lett.* **2009**, *583*, 3419–3424. [CrossRef] [PubMed]

156. Grupcheva, C.N.; Laux, W.T.; Rupenthal, I.D.; McGhee, J.; McGhee, C.N.; Green, C.R. Improved corneal wound healing through modulation of gap junction communication using connexin43-specific antisense oligodeoxynucleotides. *Investig. Ophthalmol. Vis. Sci.* **2012**, *53*, 1130–1138. [CrossRef] [PubMed]

157. Moore, K.; Bryant, Z.J.; Ghatnekar, G.; Singh, U.P.; Gourdie, R.G.; Potts, J.D. A synthetic connexin 43 mimetic peptide augments corneal wound healing. *Exp. Eye Res.* **2013**, *115*, 178–188. [CrossRef] [PubMed]

158. Ormonde, S.; Chou, C.Y.; Goold, L.; Petsoglou, C.; Al-Taie, R.; Sherwin, T.; McGhee, C.N.; Green, C.R. Regulation of connexin43 gap junction protein triggers vascular recovery and healing in human ocular persistent epithelial defect wounds. *J. Membr. Biol.* **2012**, *245*, 381–388. [CrossRef] [PubMed]

159. Elbadawy, H.M.; Mirabelli, P.; Xeroudaki, M.; Parekh, M.; Bertolin, M.; Breda, C.; Cagini, C.; Ponzin, D.; Lagali, N.; Ferrari, S. Effect of connexin 43 inhibition by the mimetic peptide Gap27 on corneal wound healing, inflammation and neovascularization. *Br. J. Pharmacol.* **2016**, *173*, 2880–2893. [CrossRef] [PubMed]

160. Bellec, J.; Bacchetta, M.; Losa, D.; Anegon, I.; Chanson, M.; Nguyen, T.H. CFTR inactivation by lentiviral vector-mediated RNA interference and CRISPR-Cas9 genome editing in human airway epithelial cells. *Curr. Gene Ther.* **2015**, *15*, 447–459. [CrossRef] [PubMed]

International Journal of
Molecular Sciences

MDPI

Article

The Connexin Mimetic Peptide Gap27 and Cx43-Knockdown Reveal Differential Roles for Connexin43 in Wound Closure Events in Skin Model Systems

Chrysovalantou Faniku [1], Erin O'Shaughnessy [1], Claire Lorraine [1], Scott R. Johnstone [1,2,3], Annette Graham [1], Sebastian Greenhough [1,†] and Patricia E. M. Martin [1,*]

[1]   Department of Life Sciences, School of Health and Life Sciences,
      Glasgow Caledonian University, Glasgow G4 0BA, UK; Chrysovalantou.Faniku@gcu.ac.uk (C.F.);
      Erin.OShaughnessy@gcu.ac.uk (E.O.); clairelorraine@hotmail.co.uk (C.L.); srj6n@eservices.virginia.edu (S.R.J.);
      ann.graham@gcu.ac.uk (A.G.); S.Greenhough@beatson.gla.ac.uk (S.G.)
[2]   Robert M. Berne Cardiovascular Research Center, University of Virginia School of Medicine,
      P.O. Box 801394, Charlottesville, VA 22908, USA
[3]   Institute of Cardiovascular and Medical Sciences, College of Medical, Veterinary and Life Sciences,
      University of Glasgow, Glasgow G12 8TT, UK
*   Correspondence: patricia.martin@gcu.ac.uk; Tel.: +44-141-331-3726
†   Current Address: Cancer Research UK Beatson Institute, Garscube Estate, Switchback Road, Bearsden,
      Glasgow G61 1BD, UK

Received: 19 January 2018; Accepted: 9 February 2018; Published: 18 February 2018

Abstract: In the epidermis, remodelling of Connexin43 is a key event in wound closure. However, controversy between the role of connexin channel and non-channel functions exist. We compared the impact of SiRNA targeted to Connexin43 and the connexin mimetic peptide Gap27 on scrape wound closure rates and hemichannel signalling in adult keratinocytes (AK) and fibroblasts sourced from juvenile foreskin (JFF), human neonatal fibroblasts (HNDF) and adult dermal tissue (ADF). The impact of these agents, following 24 h exposure, on *GJA1* (encoding Connexin43), *Ki67* and *TGF-β1* gene expression, and Connexin43 and pSmad3 protein expression levels, were examined by qPCR and Western Blot respectively. In all cell types Gap27 (100 nM–100 µM) attenuated hemichannel activity. In AK and JFF cells, Gap27 (100 nM–100 µM) enhanced scrape wound closure rates by ~50% but did not influence movement in HNDF or ADF cells. In both JF and AK cells, exposure to Gap27 for 24 h reduced the level of Cx43 protein expression but did not affect the level in ADF and HNDF cells. Connexin43-SiRNA enhanced scrape wound closure in all the cell types under investigation. In HDNF and ADF, Connexin43-SiRNA enhanced cell proliferation rates, with enhanced proliferation also observed following exposure of HDNF to Gap27. By contrast, in JFF and AK cells no changes in proliferation occurred. In JFF cells, Connexin43-SiRNA enhanced *TGF-β1* levels and in JFF and ADF cells both Connexin43-SiRNA and Gap27 enhanced pSmad3 protein expression levels. We conclude that Connexin43 signalling plays an important role in cell migration in keratinocytes and foreskin derived fibroblasts, however, different pathways are evoked and in dermal derived adult and neonatal fibroblasts, inhibition of Connexin43 signalling plays a more significant role in regulating cell proliferation than cell migration.

Keywords: wound healing; connexin mimetic peptide; connexin hemichannel; cell migration; SiRNA

1. Introduction

Connexin43 (Cx43) is expressed in nearly every tissue in the body where it forms hemichannels and intercellular gap junctions and plays diverse roles in coordinating cellular activities [1].

The connexin mimetic peptide (CMP) Gap27, targeted to the SRPTEKTIFFI sequence (amino acids 204–214) on the second extracellular loop of Cx43 is a versatile inhibitor of connexin-mediated communication (CMC) in tissue networks [2–4]. Early investigations with these peptides by Evans, Griffiths and colleagues [5–8] led to advancement of understanding of the role of connexins in the vasculature and identification of heterocellular communication at the myoendothelial gap junction [9,10]. Other studies employing Gap27 in excitable tissue networks have identified the role of connexins in the coordination of cardiomyocyte activities and calcium wave propagation [11,12], at neuronal synapses and more recently in pathological processes such as epilepsy [13]. In non-excitable tissues, Gap27 blocks the passive exchange of small gap junction permeable dyes such as calcein AM and determined a role for intercellular communication during transendothelial migration [14,15].

Many of the "acute" studies using CMPs have provided evidence for connexin channel signalling in coordinating cellular activities [2]. However, connexins also have reported "non-channel" functions and controversy exists in longer term studies as to whether channel or non-channel activities play key roles in events such as cell adhesion and migration [16]. This is no less evident in the skin where dynamic changes in connexin expression occur during wound healing [17–19]. Antisense oligonucleotides targeted to Cx43 provided the first evidence that connexin based therapies could improve wound healing and resolve inflammation [18,20,21]. Gap27 and other connexin mimetic peptides targeted to the carboxyl terminal domain of Cx43 (such as αCT-1) also improved cell migration rates in 2D and 3D organotypic skin wound model systems [22–25]. In previous studies [22,23], we determined that while Gap27 enhanced migration rates in keratinocytes and fibroblasts isolated from juvenile foreskin discards it was less effective in fibroblasts isolated from adult dermal explants.

In the present work we compared the effect of Gap27 and SiRNA targeted to Cx43, on cell migration in adult keratinocytes (AK) and adult dermal (ADF), juvenile foreskin (JFF) and neonatal foreskin (HDNF) derived fibroblasts. Our findings provide new insights into the effects of Cx43 channel inhibition versus Cx43 gene expression on cell migration, and our results also show that the response of such behaviour varies between cell types (keratinocyte versus fibroblast) and between cells of the same type (skin fibroblasts) but of different tissue origins.

2. Results

2.1. The Impact of Gap27 on Cell Migration Rates in Juvenile Foreskin Fibroblasts

Previously we determined that Gap27 inhibits CMC and enhanced scrape wound closure in fibroblasts and keratinocytes derived from juvenile foreskin explants [22,24]. To further explore the effect of 100 μM Gap27 on cell motility, JFF cells were subject to time-lapse microscopy and images captured every 15 min over a 48 h migration period. The speed of cell movement in non-treated and Gap27 (100 μM) treated cells was analysed: Gap27 treated cells reached 50% scrape closure in approximately half the time taken by non-treated cells (Figure 1A).

Image trajectory analysis was performed to elucidate if the differences observed in JFF scrape closure ± Gap27 were due to variation in cell directionality or speed of movement. Graphical representation of the XY co-ordinate data obtained from tracking the movement of 18 individual cells for each set of JFF images (±100 μM Gap27) illustrated differences in cell migration between the control and peptide treated cells, most noticeably an increase in distance travelled by the Gap27 treated cells, compared to controls, into the scraped area. The data also suggest that the majority of peptide treated cells migrate in straighter lines towards the scraped area compared to controls which had a more lateral movement. Cell tracking data determined that Gap27 treatment in JFF cells significantly increased the average cell velocity over 48 h by 2.5 μm/h compared to controls; the average velocity was 0.23 ± 0.003 μm/min (13.8 μm/h) in control cells and 0.27 ± 0.004 μm/min (16.3 μm/h) in peptide treated cells (Figure 1B). To further explore these differences, data sets of the rate of cell movement at the leading edge (Figure 1C), 0–50 μm (Figure 1D) and 50–100 μm (Figure 1E) behind the wound edge were analysed. At the leading wound edge, the migration of control and peptide treated cells were comparable (Figure 1C). However, in cells located 0–50 μm behind the wound edge cell migration rates in Gap27 was faster, with the greatest difference in velocity occurring 50–100 μm behind the wound edge, where the Gap27 treated cells migrated at a rate of 0.273 ± 0.006 μm/min compared to the non-treated cells that migrated at rates of 0.214 ± 0.004 μm/min (Figure 1D,E). These data indicate that during scrape wound closure in JFF cell monolayers, cell velocity is greater in wound edge cells compared to cells behind the wound edge. However, Gap27 treatment enhances the scrape wound closure in JFF cell monolayers in vitro by increasing cell velocity in cells behind the wound edge.

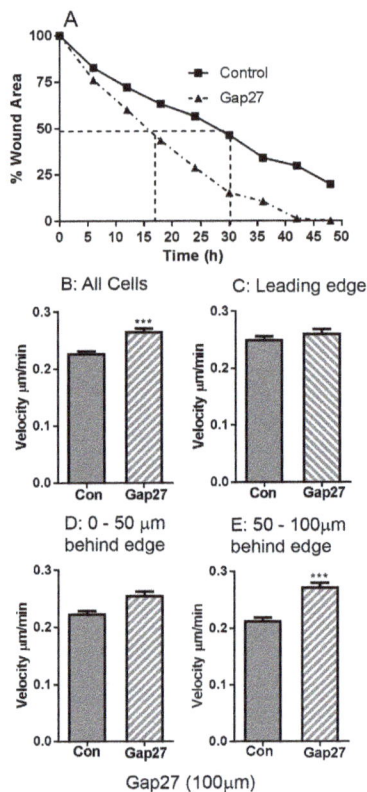

Figure 1. Gap27 (100 μM) influences the speed of Juvenile Foreskin Fibroblasts (JFF) cell migration. (**A**) time-lapse migration data of JFF cells; (**B**) Average cell velocity of JFF cells; (**C**) Cell velocity of JFF cells at the leading edge of the scrape wound; (**D**) Cell velocity of JFF cells 0–50 μm behind the wound edge; (**E**) Cell velocity of JFF cells 50–100 μm behind the wound edge. *n* = 18 cells were tracked in total with 6 cells from each specific area. *** $p < 0.005$.

2.2. Impact of Gap27 and SiRNA Targeted to Cx43 on Cell Migration in Skin Model Systems

While both CMPs and antisense Cx43 knockdown strategies are widely accepted to enhance wound closure rates [20,22,26], a direct comparison of their effects on cell migration events has not been reported. We thus explored the impact of Gap27 and SiRNA targeted to Cx43 on scrape wound closure rates in keratinocytes and fibroblasts isolated from adult skin biopsies and compared this to cells derived from juvenile foreskin and neonatal human fibroblasts. Initially a dose response of Gap27 determined that the peptide effectively enhanced scrape wound closure rates in primary adult keratinocytes at 100 nM–100 μM, but was without effect at lower doses (Figure 2A). In these AK cells, SiRNA targeted to Cx43 significantly enhanced the rate of scrape wound closure (Figure 2B). In JFF cells, 100 nM Gap27 and SiRNA targeted to Cx43 significantly enhanced the rate of scrape wound closure with 50% closure rates more than two times faster than non-treated samples (Figure 2C). Multiple studies performed in adult fibroblasts demonstrated that 100 nM–100 µM Gap27 had limited impact on cell migration responses (Figure 2D); by contrast, significant increase in 50% closure rates occurred in adult fibroblasts transfected with SiRNA targeted to Cx43, compared with the SiRNA control (Figure 2D).

To further compare the efficacy of Gap27 on fibroblast cell migration events, scrape wound assays were also performed using commercially sourced human neonatal dermal fibroblasts. Treatment with Gap27 was ineffective in enhancing migration responses at concentrations of 100 nM or 100 µM (Figure 2E,F). By contrast, inhibition of Cx43 expression by transfecting the cells with SiRNA targeted to Cx43 significantly improved cell migration rates suggesting a "non-channel" role for Cx43 in migration of these neonatal fibroblasts (Figure 2E). We also explored the migration responses of HNDF cells on various extracellular matrix components in the presence and absence of 100 µM Gap27. Although the cells migrated slightly faster on both fibronectin and collagen matrices, Gap27 still did not enhance cell migration rates [27].

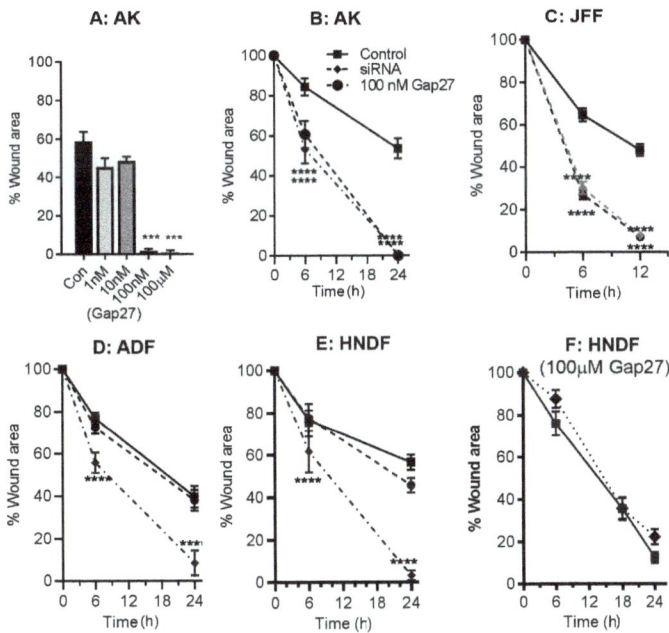

Figure 2. Gap27 and SiRNA targeted to Cx43 have differential effects on scrape wound closure rates in skin cells. Dose response of Gap27 in AK cells (**A**); SiRNA targeted to Cx43 and 100 nM Gap27 enhance scrape wound closure in AK (**B**) and JFF cells (**C**); Gap27 does not enhance cell migration rates in ADF or HNDF cells (**D–F**). $n = 3$, ** $p < 0.01$; *** $p < 0.005$.

2.3. Gap27 Attenuates Hemichannel Signalling at Lower Doses than Gap Junction Coupling

In previous studies, microinjection analysis with Alexa 488 determined that Gap27 effectively inhibits gap junction coupling at concentrations of 50 µM in keratinocytes and HeLa43 cells [23,24]. We also previously reported that Gap27 inhibits ATP release in both keratinocytes and fibroblasts isolated from juvenile foreskin tissue discards at concentrations of 100 µM [22,23,28]. In view of the stark contrast in the impact of Gap27 on cell migration rates between the different cell types, we further explored the ability of Gap27 to attenuate hemichannel activity. In all the cell types, Gap27 effectively inhibited ATP release in a dose responsive manner, effective at 10–100 µM concentrations (Figure 3A–D). This data further suggests that in the ADF and HNDF, hemichannel signalling events are unlikely to be involved in controlling cell migration.

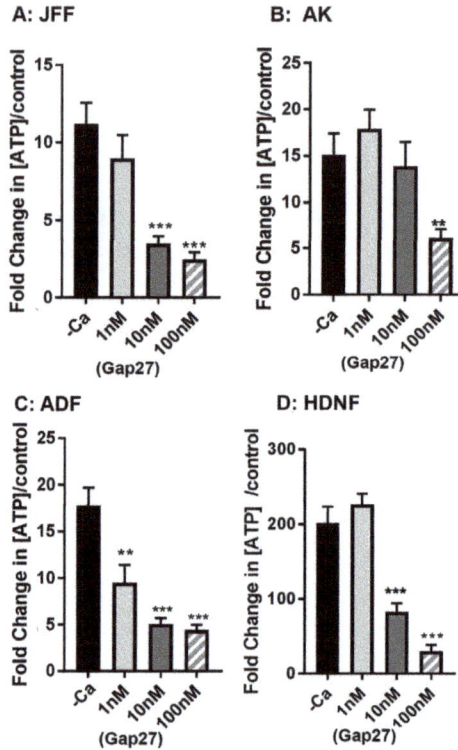

Figure 3. Gap27 inhibits hemichannel signalling. JFF cells (**A**); Keratinocytes (**B**); ADF (**C**) and HNDF (**D**) cells were exposed to a dose response of Gap27 and ATP release assays performed following calcium deprivation. Data are presented at the Fold change in ATP release over control cells that were not subject to calcium challenge. $n = 3$, *** $p < 0.005$, ** $p < 0.01$.

2.4. The Impact of Gap27 and SiRNA Targeted to Cx43 on Gene and Protein Expression Profiles in Skin Model Systems

At the end point of the cell migration assays, RNA and protein were extracted and subject to qPCR and Western blot analysis to determine any significant changes in gene and or protein expression profiles.

SiRNA targeted to Cx43 reduced the level of *Cx43* gene and protein expression by >50% in JFF, AK, ADF and HNDF cells (Figure 4A–D). Exposure to 100 nM Gap27 for up to 24 h reduced *Cx43* gene expression levels in JFF cells but had limited impact on *Cx43* gene expression levels in the other cell types (Figure 4A–D (panel 1)). In ADF and HNDF cells, 100 nM Gap27 did not influence the level of Cx43 protein expression (Figure 4C,D (panels 2 and 3)). However, in JFF cells exposure to 100 nM Gap27 for 24 h caused a >2-fold reduction in the level of Cx43 protein expression and a similar trend, although not as pronounced, was observed in AK cells (Figure 4A,B (panels 2 and 3)).

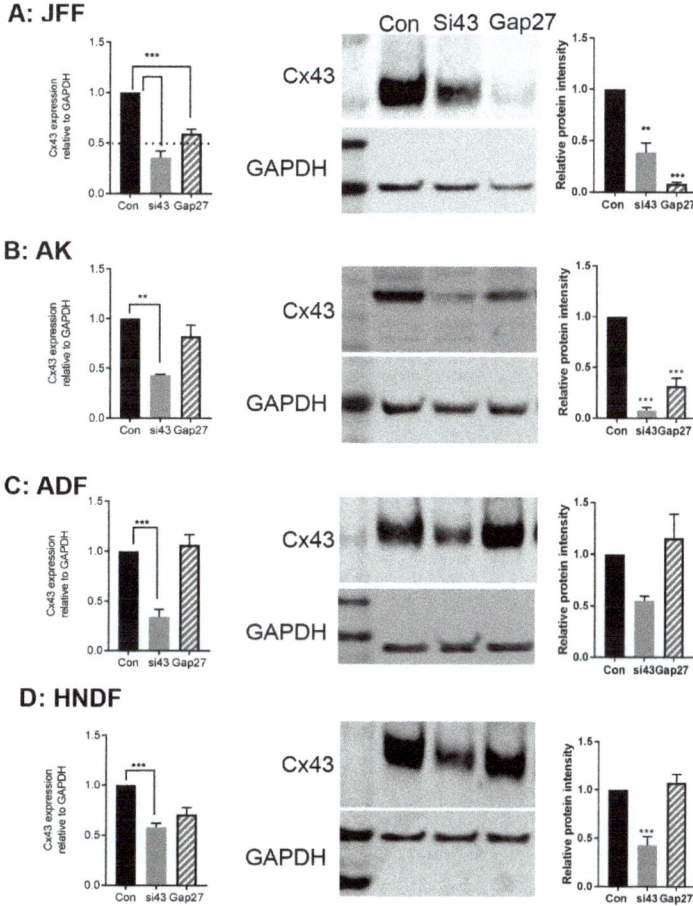

Figure 4. The impact of Gap27 and SiRNA targeted to Cx43 on gene expression. At the end of scrape wound closure assays, RNA and protein was harvested from cells and subject to real time PCR or Western blot analysis to determine levels of Cx43 expression. JFF (**A**); AK (**B**); AF (**C**) and HNDF (**D**). Panel 1 represents changes in gene expression, Panel 2 represents a typical Western blot, Panel 3 represents densitometric analysis of three Western blots. $n = 3$, *** $p < 0.005$, ** $p < 0.01$ a threshold of two fold increase or decrease in expression was considered significant.

2.5. The Impact of Gap27 and SiRNA Targeted to Cx43 on Cell Proliferation, TGF-β1 and SMAD3 Signalling Pathways

To determine if the differences observed in cell migration responses between the various cell groups were related to changes in cell proliferation, the level of *Ki67* gene expression in each of the cell types and treatment groups was determined. In JFF and AK cells, none of the treatments evoked a greater than two-fold increase in the level of *Ki67* gene expression (Figure 5A, AK and JFF panels). By contrast, in ADF cells a 10-fold increase in *Ki67* gene expression was observed in cells exposed to SiRNA targeted to Cx43 but not in those exposed to Gap27 (Figure 5A, ADF panel). In HNDF cells, proliferation was dramatically enhanced in all treatment groups (5-10 fold) compared to non-treated cells (Figure 5A, HNDF panel).

Transforming growth factor β1 (TGF-β1) is a major transcription factor regulating cell signalling events involved in migration. The level of *TGF-β1* gene expression was enhanced ~3–4 fold in JFF cells following knockdown of *Cx43* gene expression and following treatment with 100 nM Gap27 for 24 h (Figure 5B, JFF panel). By contrast, in AK, AF and HNDF cells, Gap27 and SiRNA targeted to Cx43 had limited impact on the level of *TGF-β1* gene expression levels (Figure 5B) at the 24 h time point.

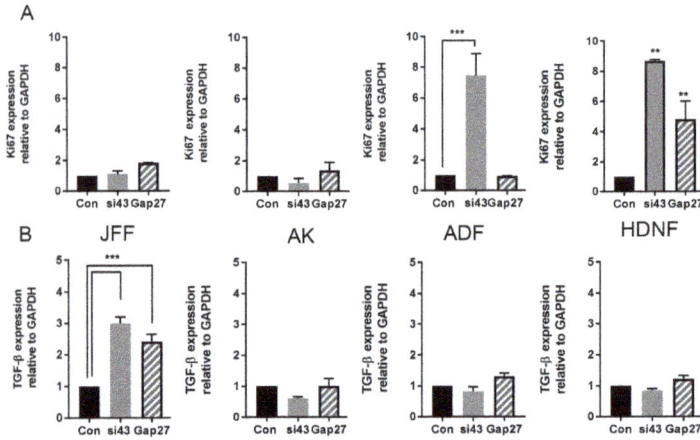

Figure 5. The impact of Gap27 and SiRNA targeted to Cx43 on *Ki67* and *TGF-β1* gene expression. At the end of scrape wound closure assays RNA was harvested from cells and subject to real time PCR analysis to determine changes in gene expression of (**A**) *Ki67* and (**B**) *TGF-β1* $n = 3$, *** $p < 0.005$, ** $p < 0.01$ a threshold of two fold increase or decrease in gene expression was considered significant.

Finally, previous reports identified that phosphorylation of Smad3 is associated with exposure of mucosal derived fibroblasts to Gap27 [29]. Probing the Western blots with an antibody targeting pSmad3 identified that in JFF and ADF cells exposure to SiRNA targeted to Cx43 and 100 nM Gap27 for 24 h enhanced the level of pSmad3 expression (Figure 6A,C). By contrast, in HNDF and AK cells the level remained constant (Figure 6B,D), further suggesting differential signalling pathways are triggered in different compartments of the skin following remodelling of Cx43.

A summary of the combined data is presented in Table 1.

Table 1. Summary of the effect of 100 nM Gap27 (Gap27) and Cx43-SiRNA (SiRNA) on cellular events related to wound closure in JFF, AK, ADF and HNDF cells. Hemichannel activity was monitored following 90 min exposure to peptide and 15 min challenge with Calcium free media. All other assays were recorded 24 h post scrape wounding in the presence or absence of 100 nM Gap27 or Cx43-SiRNA. $n = 3$ in all cases. ND: not determined; NE: no effect; ↑ enhanced effect; ↓ inhibitory effect. For details of experimental design and statistics see text and figures.

Cell Type	Treatment	HC	Migration	Cx43 Protein	Proliferation	TGF-β1	pSmad3
JFF	SiRNA	ND	↑	↓	NE	↑	↑
	Gap27	↓	↑	↓	NE	↑	↑
AK	SiRNA	ND	↑	↓	NE	NE	NE
	Gap27	↓	↑	↓	NE	NE	NE
ADF	SiRNA	ND	↑	↓	↑	↑	↑
	Gap27	↓	NE	NE	NE	NE	↑
HNDF	SiRNA	ND	↑	↓	↑	NE	NE
	Gap27	↓	NE	NE	↑	NE	NE

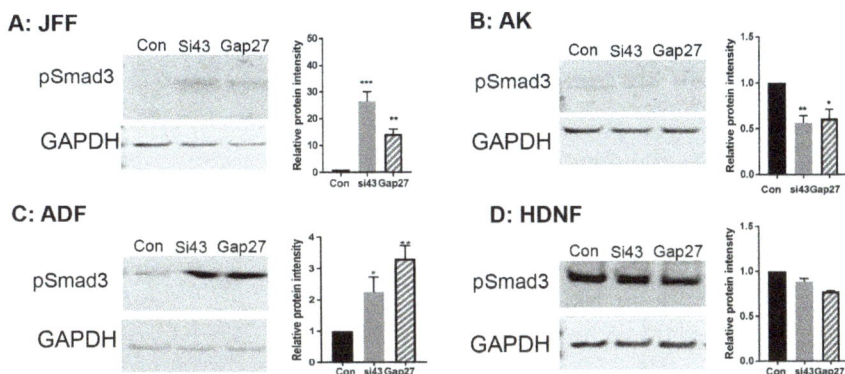

Figure 6. The impact of Gap27 and SiRNA targeted to Cx43 on pSmad3 expression. At the end of scrape, wound closure assays protein was harvested from cells and subject to Western blot analysis. Blots were probed with an antibody targeted to pSmad3 and GAPDH. Representative blots are presented for each cell type JFF (**A**); AK (**B**); AF (**C**) and HNDF (**D**) (panel 1). Panel 2 represents densitometric analysis of triplicate blots and relative pSmad3 protein levels compared to GAPDH. $n = 3$, *** $p < 0.005$, ** $p < 0.01$; * $p < 0.5$.

3. Discussion

In the present study we compared the effect of Gap27 and SiRNA targeted to Cx43 on scrape wound closure rates in fibroblasts derived from neonatal, juvenile foreskin and adult dermal explants and matched adult keratinocytes. Both of these reagents enhanced scrape wound closure in the JFF cells and adult keratinocytes with Gap27 enhancing cell migration rates at concentrations of 100 nM, reflecting the dose of peptide that inhibited hemichannel signalling and supporting the concept that ATP release via hemichannels is required for keratinocyte galvanotaxis [30]. By contrast, in the AF and HNDF cells, SiRNA targeted to Cx43 enhanced scrape wound closure, but treatment with Gap27 was without effect, extending studies where we reported limited effects of Gap27 on adult fibroblast migration rates [22,23]. Profound differences in cell migration, proliferation and the TGF-β1/pSmad3 signalling axis occurred between the cells isolated from different skin compartments. The data provides new insights into the controversy surrounding Cx43 channel versus non-channel functions in cell migration and wound repair responses [16] and suggests that inhibition of hemichannel signalling alone is insufficient to modify cell migration events in adult dermal fibroblasts.

Within epithelial tissues, including the skin and cornea, it is widely accepted that downregulation of Cx43 is favourable to wound closure, as reported in the skin of connexin-deficient mice [19] and by the development of Cx43 anti-sense oligonucleotides, that have proven effective in rat models and in human clinical trials [18,26,31,32]. Other studies by Gourdie and colleagues used a peptide targeting the carboxyl tail of Cx43 and its binding site with the PDZ domain to improve wound healing, resolve inflammation and reduce scarring in rat models and recently in human clinical trials [33–35]. Studies using Pep5, based on the Gap27 sequence, have shown remarkable effects on tissue repair and inflammation in the retina, cornea and spinal cord [36–38]. Recent studies using Gap27 have also shown that this peptide is effective in improving rabbit corneal wound healing [39] and in primary human gingival fibroblasts, isolated from donors aged 26–48 years of age [29]. A further peptide TAT-Gap19, has also recently been reported to enhance scrape wound closure of human gingival fibroblasts [40]. TAT-Gap19, targeted to the intracellular loop of Cx43, was designed as a cell permeant peptide and has been extensively used to characterise interactions between the intracellular loop of Cx43 and the carboxy terminal tail [41,42]. This peptide effectively blocks hemichannel activity but has no effect on gap junction coupling, in contrast to Gap27 which blocks all forms of connexin mediated communication. Thus, inhibiting *Cx43* gene expression and blocking channel function both have a positive influence on wound closure events; however, comparisons of the mechanisms underlying the modes of action of these different means of remodelling Cx43 remains unresolved.

In the present study we have identified profound differences in wound healing events in skin cells isolated from different sources, and between adult epidermal and stromal derived fibroblasts, which relate to whether Cx43 channel function or gene expression is regulated.

In the case of JFF cells, derived from juvenile foreskin discards (a thin layer of tissue) and in adult keratinocytes, both Gap27 and SiRNA targeted to Cx43 enhanced scrape wound closure. Further, in both of these cell types at the end point of the migration time course, neither knockdown of Cx43 expression or inhibition of channel function by Gap27 had any effect on cell proliferation as monitored by *Ki67* gene expression. This re-enforces our previous findings where irradiated juvenile fibroblasts were used and Gap27 effectively enhanced wound closure [22]. Hence in JFF cells and keratinocytes proliferation factors are not involved in the enhanced cell migration response, suggesting hemichannel signalling plays an important role in co-ordinating cellular responses [30,40].

By contrast, in ADF and HNDF while SiRNA targeted to Cx43 accelerated the rates of scrape wound closure, exposure to Gap27 had a limited effect on cell migration rates. In AF cells, Gap27 did not influence the gene expression of *Ki67* but decreasing *Cx43* gene expression enhanced cell proliferation. In the HNDF cells, exposure to Gap27 and to SiRNA targeted to Cx43 enhanced cell proliferation, but only SiRNA targeted to Cx43 enhanced cell migration rates. It is also noteworthy that at the end point of the migration assays, while SiRNA targeted to Cx43 reduced Cx43 gene and protein expression in all cell types by >50% exposure to Gap27 had no effect on Cx43 protein levels in AF cells. By contrast, in the JFF and AK cells Gap27 reduced the level of Cx43 protein expression by ~50%, however it had a limited effect on *Cx43* gene expression, suggesting Cx43 changes were post-transcriptional. A similar effect was also observed in adult human gingival fibroblasts where Gap27 also enhanced wound closure rates [29]. Studies using mouse NIH3T3 fibroblasts also determined that Gap27 reduced Cx43 protein expression in line with our observations in JFF cells [43]. Studies in our lab also determined that Gap27 improved cell migration rates in primary neonatal mouse fibroblasts [27] and keratinocytes [44]. Taken together, this data suggests that in human adult dermal fibroblasts and neonatal fibroblasts, non-channel functions of Cx43 may be more important than acute hemichannel signalling in regulating cell migratory behaviour.

Previously, we identified changes in expression of a panel of genes associated with extracellular matrix (ECM) deposition in JFF cells following exposure to Gap27, including metalloproteinase 9 (MMP-9) and connective tissue growth factor (CTGF) [23]. Several other reports have indicated a link between altered Cx43 expression and ECM regulation, including in fibroblasts isolated from a patient harbouring a non-functional Cx43 mutation associated with the Cx-channelopathy oculodentodigital dysplasia [45]. Modifying Cx43 expression or function by antisense oligonucleotides and peptide αCTI have also been associated with alterations in ECM deposition [20,33]. Further studies by Tarzemany et al. systematically reviewed the impact of Gap27 and more recently TAT-Gap19 on gene expression and cell signalling pathways involved in wound closure events in human adult gingival fibroblasts [29,40]. Both peptides effectively enhanced wound closure rates and gene array analysis after 24 h indicated changes in expression of a panel of genes related to ECM deposition including a number of MMP proteins and CTGF, in agreement with our previous studies on JFF cells [23].

The TGF-β signalling pathway is important in controlling cell migration events, and a number of studies have suggested links between TGF-β1 and Cx43 expression in wound healing scenarios [20,46]. Treatment with either Gap27 or SiRNA targeted to Cx43 enhanced the gene expression of *TGF-β1* in JFF cells, but had little impact following 24 h exposure to these reagents in the adult keratinocytes and fibroblasts. However, it remains possible that expression of *TGF-β1* is transiently induced at earlier time points, since TGF-β1 has been reported to stimulate chemotactic migration of human fibroblasts [47].

The TGF-β/Smad3 signalling axis plays an important role in cell migration events, and has been linked with Cx43 expression. In line with this in both JFF and ADF cells, levels of pSmad3 were increased following 24 h exposure to Gap27 and SiRNA targeted to Cx43. A profound increase in pSmad3 was also reported in gingival fibroblasts exposed to Gap27 for 24 h [29]. Smad3 and the carboxyl terminal tail of Cx43 both compete for a similar 'microtubule binding domain' [48]. Thus, alteration in Cx43 expression levels or function may modify the interaction. Upon translocation to the nucleus, Smad3 is phosphorylated to pSmad3 and exerts transcriptional control on a range of genes involved in regulation of inflammation, cell proliferation and re-epithelisation in both positive and negative tissue-specific ways. Given the diverse pathways that pSmad3 can regulate, it is highly likely that in the ADF cells, where Gap27 did not influence cell migration, other key cellular events may be affected. Although no evidence of induction of pSmad3 expression was observed in the keratinocytes, a host of other signalling pathways including the ERK1/2 and JNK pathways may be influenced [49] and this is subject to further investigation. In the present studies, the influence of modulation of Cx43 was assessed in monocultures. In the future it will be important to exploit our 3D organotypic models as it is well established that keratinocytes and fibroblasts can influence responses of adjacent cells as part of a coordinated tissue event [50].

4. Materials and Methods

4.1. Cell Culture

Primary juvenile human dermal fibroblasts (JFF cells) were derived from paediatric foreskins discarded at surgery following informed consent with ethical approval by Yorkhill Hospital Trust Research Ethics Committee, Glasgow, or were kindly gifted by Prof J Brandner, University Hamburg, with their use approved by the ethics committee of the Aerztekammer Hamburg (060900) as previously described [22]. Human neonatal dermal fibroblasts (HNDF) were sourced from Invitrogen (Cat No.: C0045C, Paisley, UK) and human adult dermal fibroblasts (AF) and keratinocytes (AK) were obtained from the GCU Skin Research Tissue Bank, which has NHS and GCU research ethical approval (NHS REC Ref 16/ES/0069). Fibroblasts and keratinocytes were isolated from tissue explants and cultured as previously described [22,24]. All cells were maintained at 37 °C, 5% CO_2. Monolayers of all fibroblasts were maintained in DMEM (Lonza, Wokingham, UK) supplemented with 10% (v/v) foetal calf serum, 2 mM glutamine, 50 Units/mL penicillin/streptomycin (Lonza, Wokingham, UK), hereafter termed "complete" DMEM (cDMEM). Keratinocytes were maintained in EPILIFE medium as previously described [51] (Thermo Fisher, Paisley, UK). Cells were seeded at appropriate densities on 24-well plates for ATP assays (~0.5×10^5 per well) or 6-well plates (~1×10^6 cells per well) for all other assays.

4.2. Inhibition of Connexin Mediated Communication

For the purpose of this study, Gap27 (MW 1305) (Zealand Pharma, Glostrop, Denmark) was used in aqueous solution at doses ranging from 1–100 μM for 15 min–48 h depending on experimental requirements. The peptide and media were replaced at 8 h intervals.

4.3. Knockdown of Cx43 Expression by siRNA

SiRNA duplex sequences targeted to Cx43 (TriFECTa®RNAi Kit from Integrated DNA Technologies (Tyne & Wear, UK)) along with a fluorescently-labelled scrambled transfection control duplex: TYE 563™ was used to knockdown Cx43 expression. SiRNA transfection was carried out using Lipofectamine 3000 transfection reagent (Invitrogen, Paisley, UK). Transfection reagents and siRNAs were combined and incubated for 20 min at room temperature for complex formation. Cells were transfected with a final concentration of 5 nM siRNA diluted in 1.5 mL of EPILIFE medium for primary keratinocytes and serum free DMEM (SFM) for fibroblasts. Twenty hours post transfection scrape wound assays were performed and cells subsequently harvested for endpoint analysis as described below. Transfection efficiency of control SiRNA, determined by fluorescent microscopy analysis of the scrambled SiRNA control was ~90% for all cell types.

4.4. Hemichannel Functionality Assays

Hemichannel activity was assessed by ATP release assays with minor modifications to that previously described [51]. Briefly, cells were seeded on 24-well plates and grown to ~80% confluency overnight. Cells were then washed three times in SFM and incubated in SFM for 1 h prior to exposure to Gap27 for 90 min. Following this, one half of the plate was challenged with Ca^{2+} and Mg^{2+} free PBS in the presence or absence of Gap27 for 15 min. Supernatants were collected and microcentrifuged at 10,000 rpm for 5 min prior to addition of 25 μL of supernatant to a well of an opaque-walled Nunc 96 well plate containing 25 μL ATP assay mix diluted 1:10 with ATP dilution buffer (Sigma-Aldrich, Gillingham, UK). ATP standards (25 μL) 0–10 nM diluted in SFM, and in PBS, were also added in duplicate. Luminescence was measured in relative luminescence units (RLU) using the Fluostar Optima plate reader (BMG Labtech, Aylesbury, UK). Experiments were performed in triplicate per treatment group, and each experiment was carried out a minimum of three times ($n = 3$). Data is represented as the fold change in ATP released between control and peptide treated wells.

4.5. Scrape Wound Assays and Time-Lapse Microscopic Analysis of Cell Migration

Cells were pre-exposed to peptide or SFM for 90 min, or were transfected with SiRNA targeting Cx43 for 20 h, prior to introducing a scrape wound to confluent cell monolayers using a sterile 100 μL pipette tip. Cell migration was monitored by taking triplicate images of wound area 0, 6, 12 and 24 h post scraping on a CMEX-3200 camera [51]. The scrape wound area was measured at each time point using Image J software. Values were normalised by comparing with the corresponding initial wound size. For time-lapse microscopy analysis, images were recorded on a Zeiss Axiovert 100 microscope (Cambridge, UK) linked up to a CCD camera (Nikon Eclipse TS10, Kingston Upon Thames, UK). Image capture was controlled by AQMsoftware (Kinetic Imaging Ltd., Nottingham, UK). Images were captured every 15 min for up to 48 h [52]. The movement of 18 individual cells for each set of JFF time-lapse images in the presence or absence of 100 μM Gap27, were tracked using Image J software tracking plug-in. The size of each image was 512 × 512 pixels with a diameter of ~500 μm, therefore, at ×100 magnification each pixel represented approximately 1 μm. This value was used as the x/y calibration value with a time interval value of 15 min. Cells were tracked by clicking on the leading edge of a cell on sequential images representing every 15 min over the 48 h period. Six cells were randomly chosen from each of (1) wound edge; (2) 0–50 μm from the wound edge and (3) 50–100 μm from the wound edge. The data output produced by Image J software included the XY co-ordinates together with distance and velocity values. The XY co-ordinates were plotted on a graph using Excel software, providing an individual track for each cell, enabling visualisation of cell movement over 48 h.

4.6. RNA Extraction and Real Time PCR

The Bioline ISOLATE RNA Kit (Bioline, London, UK) was used according to manufacturer's instructions to extract RNA from cell monolayers (usually 2 wells for a 6 well plate). RNA concentrations were determined using a Nanodrop ND-100 at 260/280 nm. cDNA was prepared from the RNA samples using cDNA synthesis kit from Primerdesign and real-time PCR was performed using Primerdesign Master Mix kit (Primer Design, Chandlers Ford, UK). Primers amplifying human *Cx43*, human *Ki67*, human *TGF-β1* and the house keeping gene *GAPDH* were purchased from IDT (Tyne & Wear, UK) (Table S1). All reactions were performed in an ABI 7500 FA Real-Time PCR system (Applied Biosystems, Warrington, UK). The mRNA expression level for each gene was determined using the ΔC_t method and each sample was run in triplicate.

The CT value obtained for the target gene in all samples was first normalised with the CT value obtained for the housekeeping gene. The resulting change in CT (ΔC_t) calculated for test samples was then normalised with the ΔC_t calculated for the control sample, giving a $\Delta\Delta C_t$ value. The gene expression ratio was calculated using $2^{-\Delta\Delta Ct}$, providing a fold increase or decrease in gene expression compared to the control sample. Gene fold changes $\geq \pm 2$ were considered significant.

4.7. Western Blot Analysis

Protein was harvested from cells in 100 μL lysis buffer (1% (v/v) SDS, 30 mM Na_3VO_4, 1 μM DTT, protease inhibitor cocktail (Sigma-Aldrich) and phenylmethanosulfonylfluoride (PMSF)) prepared in 1xPBS as previously described [52].

Equivalent amounts of protein (30–80 μg) were mixed with 5 μL loading buffer (NuPAGE® LDS Sample Buffer (4X)) and the volume adjusted to 20 μL with lysis buffer. Samples were mixed for 15 min at 20 rpm followed by brief centrifugation and separated by 4–12% sodium dodecyl sulphate polyacrylamide gel electrophoresis (SDS-PAGE) (NuPAGE® Novex® Bis-Tris Mini Gels; Thermo Fisher) followed by electrophoretic transfer to a nitrocellulose membrane using an I-Blot transfer system (Invitrogen) following manufacturer's instructions. Transfer efficiency was determined by staining the blots with Ponceau S (0.1% (w/v) in 5% acetic acid) (Sigma-Aldrich) for 15 s, prior to rinsing in distilled water and probing for relevant protein expression using appropriate primary antibodies as previously described [52]. Membranes were probed with primary antibodies to detect Cx43 (Rivedal polylclonal antibody 1:2000 dilution, kindly gifted by Edward Leithe [53]), GAPDH (mouse monoclonal antibody, Santa Cruz (LOCATION) (1:5000 dilution)) and pSmad3 (rabbit polyclonal antibody Abcam (Cambridge, UK) (1:2000 dilution)) expression as appropriate. Secondary antibodies were IRDye® 800CW goat anti-rabbit IgG or IRDye® 680CW goat anti-mouse IgG (Licor 1:15,000 dilution) as appropriate. Blots were developed by exposing the image for a period of 15 s to 5 min according to the intensity of the signal using an Odyssey FC Dual Mode Licor imaging system (LI-COR Biosceinces UK Ltd, Lincoln, UK). Densitometric values were quantified using the Odyssey software. To enable normalisation of the blots and comparison of the effect of different treatments on protein expression, the intensity of the protein bands were compared to the house keeping protein.

4.8. Statistical Analysis

Experiments were performed in triplicate per setting and on three separate occasions with at least 2 different patient samples. Results were compiled in GraphPad Prism software (La Jolla, San Diego, CA, USA) and all data is expressed as mean ± SEM unless otherwise stated. Statistical tests were performed on the data using Student's unpaired t-test or one-way ANOVA and Dunnett's post-test as appropriate, with statistical significance inferred at $p < 0.05$.

5. Conclusions

In conclusion, we provide an in-depth study on the comparative effects of Gap27 with a Cx43-SiRNA knockdown approach to improve wound healing and identify significant differences in the cell signalling pathways that are controlled by Cx43 in fibroblasts and keratinocytes. Further work is now warranted to define the molecular pathways by which Cx43 exerts its effects in the skin which will aide in identifying new therapeutic strategies and applications for specific types of wounds.

Supplementary Materials: Supplementary materials can be found at www.mdpi.com/1422-0067/19/2/604/s1.

Acknowledgments: We thank Professor Joanna Brandner, University of Hamburg for the supply of the Juvenile Foreskin Fibroblasts used in the latter part of this project. Chrysovalantou Faniku and Claire Lorraine were funded by GCU studentships and Erin O'Shaughnessy by a Ph.D. studentship from the Psoriasis Association (Grant No.: ST3 15). We thank, Professor Rachel Errington (University of Cardiff) for access to time lapse microscopy, Dr. Bjarne Due Larsen (Zealand Pharma) for supply of Gap27 and a grant from Animal Free Research (AG) for support of the GCU Skin Research Tissue Bank. We are also indebted to Professor Malcolm Hodgins for helpful discussions and inputs. No funds for covering the costs to publish in open access were awarded.

Author Contributions: Patricia E.M. Martin, Claire Lorraine, Chrysovalantou Faniku conceived and designed the experiments; Claire Lorraine, Chrysovalantou Faniku and Erin O'Shaughnessy performed the experiments and Scott R. Johnstone performed the initial timelapse experiments; Claire Lorraine, Chrysovalantou Faniku, Erin O'Shaughnessy and Patricia E.M. Martin analysed the data; Sebastian Greenhough, Annette Graham and Scott R. Johnstone contributed reagents/materials/analysis tools and Sebastian Greenhough, funded by Animal Free Research, did not participate in any experiments involving animals or animal tissue; Claire Lorraine and Chrysovalantou Faniku contributed initial drafts and Patricia E.M. Martin wrote the final manuscript that was critically proof read by Annette Graham.

Conflicts of Interest: The authors declare no conflicts of interest.

Abbreviations

Cx	connexin
CMP	connexin mimetic peptide
CMC	connexin mediated communication
JFF	human juvenile foreskin derived fibroblasts
HNDF	human neonatal dermal fibroblasts
AF	adult dermal derived fibroblasts
AK	adult keratinocytes
SFM	serum free media

References

1. Evans, W.H.; Martin, P.E. Gap junctions: Structure and function (Review). *Mol. Membr. Biol.* **2002**, *19*, 121–136. [CrossRef] [PubMed]
2. Evans, W.H.; Bultynck, G.; Leybaert, L. Manipulating connexin communication channels: Use of peptidomimetics and the translational outputs. *J. Mem. Biol.* **2012**, *245*, 437–449. [CrossRef] [PubMed]
3. Evans, W.H.; Leybaert, L. Mimetic peptides as blockers of connexin channel-facilitated intercellular communication. *Cell. Commun. Adhes.* **2007**, *14*, 265–273. [CrossRef] [PubMed]
4. Willebrords, J.; Maes, M.; Crespo Yanguas, S.; Vinken, M. Inhibitors of connexin and pannexin channels as potential therapeutics. *Pharmacol. Ther.* **2017**, *180*, 144–160. [CrossRef] [PubMed]
5. Chaytor, A.T.; Evans, W.H.; Griffith, T.M. Peptides homologous to extracellular loop motifs of connexin 43 reversibly abolish rhythmic contractile activity in rabbit arteries. *J. Physiol.* **1997**, *503 Pt 1*, 99–110. [CrossRef] [PubMed]
6. Chaytor, A.T.; Martin, P.E.; Evans, W.H.; Randall, M.D.; Griffith, T.M. The endothelial component of cannabinoid-induced relaxation in rabbit mesenteric artery depends on gap junctional communication. *J. Physiol.* **1999**, *520*, 539–550. [CrossRef] [PubMed]
7. Dora, K.A.; Martin, P.E.; Chaytor, A.T.; Evans, W.H.; Garland, C.J.; Griffith, T.M. Role of heterocellular Gap junctional communication in endothelium-dependent smooth muscle hyperpolarization: inhibition by a connexin-mimetic peptide. *Biochem. Biophys. Res. Commun.* **1999**, *254*, 27–31. [CrossRef] [PubMed]
8. Hutcheson, I.R.; Chaytor, A.T.; Evans, W.H.; Griffith, T.M. Nitric oxide-independent relaxations to acetylcholine and A23187 involve different routes of heterocellular communication. Role of Gap junctions and phospholipase A2. *Circ. Res.* **1999**, *84*, 53–63. [CrossRef] [PubMed]
9. Griffith, T.M.; Chaytor, A.T.; Edwards, D.H. The obligatory link: Role of gap junctional communication in endothelium-dependent smooth muscle hyperpolarization. *Pharmacol. Res.* **2004**, *49*, 551–564. [CrossRef] [PubMed]
10. Straub, A.C.; Zeigler, A.C.; Isakson, B.E. The myoendothelial junction: connections that deliver the message. *Physiology* **2014**, *29*, 242–249. [CrossRef] [PubMed]
11. Verma, V.; Hallett, M.B.; Leybaert, L.; Martin, P.E.; Evans, W.H. Perturbing plasma membrane hemichannels attenuates calcium signalling in cardiac cells and HeLa cells expressing connexins. *Eur. J. Cell. Biol.* **2009**, *88*, 79–90. [CrossRef] [PubMed]
12. Boitano, S.; Evans, W.H. Connexin mimetic peptides reversibly inhibit Ca^{2+} signaling through gap junctions in airway cells. *Am. J. Physiol. Lung Cell. Mol. Physiol.* **2000**, *279*, L623–L630. [CrossRef] [PubMed]
13. Yoon, J.J.; Nicholson, L.F.; Feng, S.X.; Vis, J.C.; Green, C.R. A novel method of organotypic brain slice culture using connexin-specific antisense oligodeoxynucleotides to improve neuronal survival. *Brain Res.* **2010**, *1353*, 194–203. [CrossRef] [PubMed]
14. Oviedo-Orta, E.; Errington, R.J.; Evans, W.H. Gap junction intercellular communication during lymphocyte transendothelial migration. *Cell. Biol. Int.* **2002**, *26*, 253–263. [CrossRef] [PubMed]
15. Oviedo-Orta, E.; Evans, W.H. Gap junctions and connexins: potential contributors to the immunological synapse. *J. Leukoc. Biol.* **2002**, *72*, 636–642. [PubMed]
16. Kameritsch, P.; Pogoda, K.; Pohl, U. Channel-independent influence of connexin 43 on cell migration. *BBA-Bioenergetics* **2012**, *1818*, 1993–2001. [CrossRef] [PubMed]

17. Brandner, J.M.; Houdek, P.; Husing, B.; Kaiser, C.; Moll, I. Connexins 26, 30, and 43: Differences among spontaneous, chronic, and accelerated human wound healing. *J. Investig. Dermatol.* **2004**, *122*, 1310–1320. [CrossRef] [PubMed]

18. Becker, D.L.; Thrasivoulou, C.; Phillips, A.R. Connexins in wound healing; perspectives in diabetic patients. *BBA-Bioenergetics* **2012**, *1818*, 2068–2075. [CrossRef] [PubMed]

19. Kretz, M.; Euwens, C.; Hombach, S.; Eckardt, D.; Teubner, B.; Traub, O.; Willecke, K.; Ott, T. Altered connexin expression and wound healing in the epidermis of connexin-deficient mice. *J. Cell. Sci.* **2003**, *116 Pt 16*, 3443–3452. [CrossRef] [PubMed]

20. Mori, R.; Power, K.T.; Wang, C.M.; Martin, P.; Becker, D.L. Acute downregulation of connexin43 at wound sites leads to a reduced inflammatory response, enhanced keratinocyte proliferation and wound fibroblast migration. *J. Cell. Sci.* **2006**, *119 Pt 24*, 5193–5203. [CrossRef] [PubMed]

21. Mendoza-Naranjo, A.; Cormie, P.; Serrano, A.E.; Hu, R.; O'Neill, S.; Wang, C.M.; Thrasivoulou, C.; Power, K.T.; White, A.; Serena, T.; et al. Targeting Cx43 and N-cadherin, which are abnormally upregulated in venous leg ulcers, influences migration, adhesion and activation of Rho GTPases. *PLoS ONE* **2012**, *7*, e37374. [CrossRef] [PubMed]

22. Pollok, S.; Pfeiffer, A.C.; Lobmann, R.; Wright, C.S.; Moll, I.; Martin, P.E.; Brandner, J.M. Connexin 43 mimetic peptide Gap27 reveals potential differences in the role of Cx43 in wound repair between diabetic and non-diabetic cells. *J. Cell. Mol. Med.* **2011**, *15*, 861–873. [CrossRef] [PubMed]

23. Wright, C.S.; Pollok, S.; Flint, D.J.; Brandner, J.M.; Martin, P.E. The connexin mimetic peptide Gap27 increases human dermal fibroblast migration in hyperglycemic and hyperinsulinemic conditions in vitro. *J. Cell. Physiol.* **2012**, *227*, 77–87. [CrossRef] [PubMed]

24. Wright, C.S.; van Steensel, M.A.; Hodgins, M.B.; Martin, P.E. Connexin mimetic peptides improve cell migration rates of human epidermal keratinocytes and dermal fibroblasts in vitro. *Wound Repair Regen.* **2009**, *17*, 240–249. [CrossRef] [PubMed]

25. Ghatnekar, G.S.; Grek, C.L.; Armstrong, D.G.; Desai, S.C.; Gourdie, R.G. The effect of a connexin43-based Peptide on the healing of chronic venous leg ulcers: a multicenter, randomized trial. *J. Investig. Dermatol.* **2015**, *135*, 289–298. [CrossRef] [PubMed]

26. Qiu, C.; Coutinho, P.; Frank, S.; Franke, S.; Law, L.Y.; Martin, P.; Green, C.R.; Becker, D.L. Targeting connexin43 expression accelerates the rate of wound repair. *Curr. Biol.* **2003**, *13*, 1697–1703. [CrossRef] [PubMed]

27. Lorraine, C. The role of connexins in skin wound healing events. Ph.D. Thesis, Glasgow Caledonian University, Glasgow, UK, available through British Library Electronic Theses Online System. 2015.

28. Wright, J.A.; Richards, T.; Becker, D.L. Connexins and diabetes. *Cardiol. Res. Pract.* **2012**, *2012*, 496904. [CrossRef] [PubMed]

29. Tarzemany, R.; Jiang, G.; Larjava, H.; Hakkinen, L. Expression and function of connexin 43 in human gingival wound healing and fibroblasts. *PLoS ONE* **2015**, *10*, e0115524. [CrossRef] [PubMed]

30. Riding, A.; Pullar, C.E. ATP Release and P2 Y Receptor Signaling are Essential for Keratinocyte Galvanotaxis. *J. Cell. Physiol.* **2016**, *231*, 181–191. [CrossRef] [PubMed]

31. Grupcheva, C.N.; Laux, W.T.; Rupenthal, I.D.; McGhee, J.; McGhee, C.N.; Green, C.R. Improved corneal wound healing through modulation of gap junction communication using connexin43-specific antisense oligodeoxynucleotides. *Investig. Ophthalmol. Vis. Sci.* **2012**, *53*, 1130–1138. [CrossRef] [PubMed]

32. Ormonde, S.; Chou, C.Y.; Goold, L.; Petsoglou, C.; Al-Taie, R.; Sherwin, T.; McGhee, C.N.; Green, C.R. Regulation of connexin43 gap junction protein triggers vascular recovery and healing in human ocular persistent epithelial defect wounds. *J. Memb. Biol.* **2012**, *245*, 381–388. [CrossRef] [PubMed]

33. Ghatnekar, G.S.; O'Quinn, M.P.; Jourdan, L.J.; Gurjarpadhye, A.A.; Draughn, R.L.; Gourdie, R.G. Connexin43 carboxyl-terminal peptides reduce scar progenitor and promote regenerative healing following skin wounding. *Regen. Med.* **2009**, *4*, 205–223. [CrossRef] [PubMed]

34. Moore, K.; Bryant, Z.J.; Ghatnekar, G.; Singh, U.P.; Gourdie, R.G.; Potts, J.D. A synthetic connexin 43 mimetic peptide augments corneal wound healing. *Exp. Eye Res.* **2013**, *115*, 178–188. [CrossRef] [PubMed]

35. Soder, B.L.; Propst, J.T.; Brooks, T.M.; Goodwin, R.L.; Friedman, H.I.; Yost, M.J.; Gourdie, R.G. The connexin43 carboxyl-terminal peptide ACT1 modulates the biological response to silicone implants. *Plast. Reconstr. Surg.* **2009**, *123*, 1440–1451. [CrossRef] [PubMed]

36. Danesh-Meyer, H.V.; Kerr, N.M.; Zhang, J.; Eady, E.K.; O'Carroll, S.J.; Nicholson, L.F.; Johnson, C.S.; Green, C.R. Connexin43 mimetic peptide reduces vascular leak and retinal ganglion cell death following retinal ischaemia. *Brain* **2012**, *135 Pt 2*, 506–520. [CrossRef] [PubMed]

37. Guo, C.X.; Mat Nor, M.N.; Danesh-Meyer, H.V.; Vessey, K.A.; Fletcher, E.L.; O'Carroll, S.J.; Acosta, M.L.; Green, C.R. Connexin43 Mimetic Peptide Improves Retinal Function and Reduces Inflammation in a Light-Damaged Albino Rat Model. *Investig. Ophthalmol. Vis.Sci.* **2016**, *57*, 3961–3973. [CrossRef] [PubMed]

38. Mao, Y.; Nguyen, T.; Tonkin, R.S.; Lees, J.G.; Warren, C.; O'Carroll, S.J.; Nicholson, L.F.B.; Green, C.R.; Moalem-Taylor, G.; Gorrie, C.A. Characterisation of Peptide5 systemic administration for treating traumatic spinal cord injured rats. *Exp. Brain Res.* **2017**, *235*, 3033–3048. [CrossRef] [PubMed]

39. Elbadawy, H.M.; Mirabelli, P.; Xeroudaki, M.; Parekh, M.; Bertolin, M.; Breda, C.; Cagini, C.; Ponzin, D.; Lagali, N.; Ferrari, S. Effect of connexin43 inhibition by the mimetic peptide Gap27 on corneal wound healing, inflammation and neovascularization. *Br. J. Pharmacol.* **2016**, *173*, 2880–2893. [CrossRef] [PubMed]

40. Tarzemany, R.; Jiang, G.; Jiang, J.X.; Larjava, H.; Hakkinen, L. Connexin43 Hemichannels Regulate the Expression of Wound Healing-Associated Genes in Human Gingival Fibroblasts. *Sci. Rep.* **2017**, *7*, 14157. [CrossRef] [PubMed]

41. Abudara, V.; Bechberger, J.; Freitas-Andrade, M.; De Bock, M.; Wang, N.; Bultynck, G.; Naus, C.C.; Leybaert, L.; Giaume, C. The connexin43 mimetic peptide Gap19 inhibits hemichannels without altering gap junctional communication in astrocytes. *Front. Cell. Neurosci.* **2014**, *8*, 306. [CrossRef] [PubMed]

42. Iyyathurai, J.; Wang, N.; D'Hondt, C.; Jiang, J.X.; Leybaert, L.; Bultynck, G. The SH3—Binding domain of Cx43 participates in loop/tail interactions critical for Cx43—hemichannel activity. *CMLS* **2017**. [CrossRef] [PubMed]

43. Glass, B.J.; Hu, R.G.; Phillips, A.R.; Becker, D.L. The action of mimetic peptides on connexins protects fibroblasts from the negative effects of ischemia reperfusion. *Biol. Open* **2015**, *4*, 1473–1480. [CrossRef] [PubMed]

44. Kandyba, E.E.; Hodgins, M.B.; Martin, P.E. A murine living skin equivalent amenable to live-cell imaging: Analysis of the roles of connexins in the epidermis. *J. Investig. Dermatol.* **2008**, *128*, 1039–1049. [CrossRef] [PubMed]

45. Kelly, J.J.; Esseltine, J.L.; Shao, Q.; Jabs, E.W.; Sampson, J.; Auranen, M.; Bai, D.; Laird, D.W. Specific functional pathologies of Cx43 mutations associated with oculodentodigital dysplasia. *Mol. Biol. Cell.* **2016**, *27*, 2172–2185. [CrossRef] [PubMed]

46. Hills, C.E.; Siamantouras, E.; Smith, S.W.; Cockwell, P.; Liu, K.K.; Squires, P.E. TGFbeta modulates cell-to-cell communication in early epithelial-to-mesenchymal transition. *Diabetologia* **2012**, *55*, 812–824. [CrossRef] [PubMed]

47. Postlethwaite, A.E.; Keski-Oja, J.; Moses, H.L.; Kang, A.H. Stimulation of the chemotactic migration of human fibroblasts by transforming growth factor beta. *J. Exp. Med.* **1987**, *165*, 251–256. [CrossRef] [PubMed]

48. Dai, P.; Nakagami, T.; Tanaka, H.; Hitomi, T.; Takamatsu, T. Cx43 mediates TGF-beta signaling through competitive Smads binding to microtubules. *Mol. Biol. Cell.* **2007**, *18*, 2264–2273. [CrossRef] [PubMed]

49. Leivonen, S.K.; Lazaridis, K.; Decock, J.; Chantry, A.; Edwards, D.R.; Kahari, V.M. TGF-β-elicited induction of tissue inhibitor of metalloproteinases (TIMP)-3 expression in fibroblasts involves complex interplay between Smad3, p38α, and ERK1/2. *PLoS ONE* **2013**, *8*, e57474. [CrossRef] [PubMed]

50. Huang, P.; Bi, J.; Owen, G.R.; Chen, W.; Rokka, A.; Koivisto, L.; Heino, J.; Hakkinen, L.; Larjava, H. Keratinocyte Microvesicles Regulate the Expression of Multiple Genes in Dermal Fibroblasts. *J. Investig. Dermatol.* **2015**, *135*, 3051–3059. [CrossRef] [PubMed]

51. Wright, C.S.; Berends, R.F.; Flint, D.J.; Martin, P.E. Cell motility in models of wounded human skin is improved by Gap27 despite raised glucose, insulin and IGFBP-5. *Exp. Cell. Res.* **2013**, *319*, 390–401. [CrossRef] [PubMed]

52. Johnstone, S.R.; Best, A.K.; Wright, C.S.; Isakson, B.E.; Errington, R.J.; Martin, P.E. Enhanced connexin 43 expression delays intra-mitotic duration and cell cycle traverse independently of gap junction channel function. *J. Cell. Biochem.* **2010**, *110*, 772–782. [CrossRef] [PubMed]

53. Leithe, E.; Rivedal, E. Ubiquitination and down-regulation of gap junction protein connexin-43 in response to 12-*O*-tetradecanoylphorbol 13-acetate treatment. *J. Biol. Chem.* **2004**, *279*, 50089–50096. [CrossRef] [PubMed]

International Journal of
Molecular Sciences

MDPI

Article

Knockout of Pannexin-1 Induces Hearing Loss

Jin Chen, Chun Liang, Liang Zong, Yan Zhu and Hong-Bo Zhao *

Department of Otolaryngology, University of Kentucky Medical Center, 800 Rose Street, Lexington,
KY 40536, USA; catkin19832002@163.com (J.C.); chunliang13@yeah.net (C.L.); cell-099@163.com (L.Z.);
yan.zhu@uky.edu (Y.Z.)
* Correspondence: hzhao2@uky.edu; Tel.: +1-859-257-5097

Received: 12 March 2018; Accepted: 23 April 2018; Published: 30 April 2018

Abstract: Mutations of gap junction connexin genes induce a high incidence of nonsyndromic hearing loss. Pannexin genes also encode gap junctional proteins in vertebrates. Recent studies demonstrated that Pannexin-1 (Panx1) deficiency in mice and mutation in humans are also associated with hearing loss. So far, several Panx1 knockout (KO) mouse lines were established. In general, these Panx1 KO mouse lines demonstrate consistent phenotypes in most aspects, including hearing loss. However, a recent study reported that a Panx1 KO mouse line, which was created by Genentech Inc., had no hearing loss as measured by the auditory brainstem response (ABR) threshold at low-frequency range (<24 kHz). Here, we used multiple auditory function tests and re-examined hearing function in the Genentech Panx1 (Gen-Panx1) KO mouse. We found that ABR thresholds in the Gen-Panx1 KO mouse were significantly increased, in particular, in the high-frequency region. Moreover, consistent with the increase in ABR threshold, distortion product otoacoustic emission (DPOAE) and cochlear microphonics (CM), which reflect active cochlear amplification and auditory receptor current, respectively, were significantly reduced. These data demonstrated that the Gen-Panx1 KO mouse has hearing loss and further confirmed that Panx1 deficiency can cause deafness.

Keywords: Panx1; deafness; hearing; gap junction; inner ear; ABR; DPOAE; CM

1. Introduction

Gap junctions play a critical role in hearing. It has been found that mutations in the Connexin26 (Cx26, *GJB2*) gap junctional gene induce a high incidence of hearing loss, accounting for more than 50% of the cases of nonsyndromic hearing loss [1–5]. The pannexin gene family also encodes gap junctional proteins in vertebrates. So far, three pannexin isoforms (Panx1, 2, and 3) have been cloned from the human and mouse genomes [6,7]. Like connexins, pannexins have ubiquitous expression in almost all tissues. In the mammalian inner ear, pannexins also have extensive expression; all three pannexin isoforms have expression in the inner ear [8]. Panx1 predominantly expresses at the supporting cells in the organ of Corti, the interdental cells in the spiral limbus, inner and outer sulcus cells, and fibrocytes in the cochlear lateral wall. Panx2 predominantly expresses at the basal cells in the stria vascularis, and Panx3 mainly expresses at the bony structure of the cochlea. However, auditory sensory hair cells have no expression of pannexins [8]. These distinctive cellular distributions strongly suggest that pannexins may have important functions in hearing.

Unlike connexins that form intercellular gap junctional channels, pannexins usually function as undocked gap junctional channels on the plasma membrane to provide an intracellular-extracellular conduit. Due to the relatively large pore size, Panx1 channels in many organs and tissues act as conduits that allow small molecules, such as ATP [9–11], to pass through in order to participate in many physiological functions and pathological processes [12–20]. Our previous study also demonstrated that Panx1 channels dominate ATP release in the cochlea [20,21]. ATP in the cochlea can mediate hair cells' sound transduction and neurotransmission [22], outer hair cell (OHC) electromotility [23,24], hearing dynamic range [25], synchronization of auditory nerve activity during development [26,27], gap junctional coupling [28], K^+-sinking [29], and endocochlear potential (EP) generation [21].

It is well-known that mutations in connexin genes can induce hearing loss [1–5]. However, it was for a long-time undetermined whether Panx deficiency can induce hearing loss, until recent studies showing that Panx1-deficient mice [21,30] and mutation in humans [31] are associated with deafness. So far, several Panx1 knockout (KO) mouse lines have been established [15,21,30,32,33]. In general, these Panx1 KO mice demonstrated consistent phenotypes. However, one Panx1 KO mouse line, which was created by Genentech Inc. (South San Francisco, CA, USA) [15], was reported to have no hearing loss, as measured by auditory brainstem response (ABR) recording at low-frequency range (<24 kHz) [34]. In this study, we re-examined hearing function of this Genentech Panx1 (Gen-Panx1) KO mouse line using multiple auditory functional tests. We found that the Gen-Panx1 KO mice have hearing loss, particularly in the high-frequency region, which is consistent with hearing loss observed in Foxg1-Panx1 conditional knockout (cKO) mice (a line which was created by crossing with a Foxg1-Cre line) [21], and in Pax2-Panx1 cKO mice (a line which was created by crossing with a Pax2-Cre mouse line) [30]. These new data further confirmed that Panx1 deficiency can induce hearing loss.

2. Results

2.1. Panx1 Deletion in the Cochlea in Gen-Panx1 KO Mice

As previously reported [8], Panx1 had predominant expression in the cochlea, including the organ of Corti and the lateral wall (Figure 1a). In the Gen-Panx1 KO mouse, Panx1 expression at the cochlear lateral wall was deleted (Figure 1b,d); Panx1 labeling was completely absent at the spiral ligament (SPL). However, Panx1 labeling at the organ of Corti (OC), the spiral limbus (SLM), and outer sulcus cells (OSCs) remained intense (Figure 1b–d). This deletion pattern in the Gen-Panx1 KO mice is similar to Panx1 deletion in the cochlea in the Foxg1-Panx1 cKO mice [21]. This location- and cell-specific knockout pattern also provided direct evidence for the specificity of the anti-Panx1 antibody used in this study.

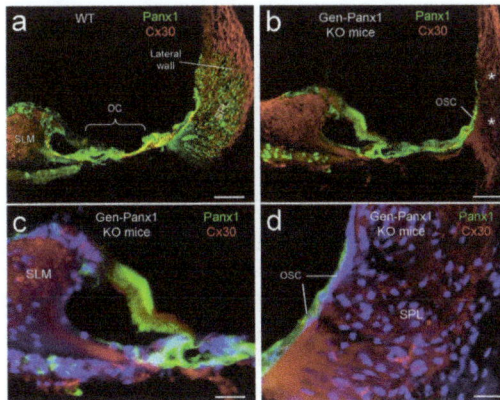

Figure 1. Deletion of Panx1 in the cochlea of Gen-Panx1 knockout (KO) mice. (**a**): Double immunofluorescent staining for Panx1 (green) and Cx30 (red) in the wild-type (WT) mouse cochlea. (**b–d**): Panx1 expression and deletion in the cochlea of Gen-Panx1 KO mice. White asterisks in panel (**b**) indicate the absence of Panx1 labeling at the spiral ligament (SPL) in the lateral wall of the Gen-Panx1 KO mice. OC: organ of Corti; OSC: outer sulcus cell; SLM: spiral limbus. Scale bars: (**a,b**): 50 μm, (**c,d**): 25 μm.

Co-labeling for Cx30 showed that there was no apparent difference between Gen-Panx1 KO mice and wild-type (WT) mice (Figure 1). Cx26 expression in the cochlea also appeared normal in the Gen-Panx1 KO mice (data not shown).

2.2. Hearing Loss in Gen-Panx1 KO Mice

Figure 2a-c shows ABR recording in the Gen-Panx1 KO mice. ABR thresholds at 8, 16, 24, 32, and 40 kHz were 60.0 ± 2.28, 50.0 ± 2.28, 57.1 ± 3.30, 86.4 ± 3.21, and 93.9 ± 0.77 dB SPL, respectively (Figure 2b). In comparison with WT mice, ABR thresholds at 8, 16, 24, 32, and 40 kHz in the Gen-Panx1 KO mice were significantly increased by 14.0 ± 2.88, 15.5 ± 2.88, 12.1 ± 3.30, 34.4 ± 3.21, and 38.4 ± 1.77 dB SPL, respectively ($p < 0.001$, one-way ANOVA with a Bonferroni correction). In the high-frequency region, ABR thresholds had larger increases in comparison with those in the low-frequency region (Figure 2b). However, there was no significant difference in ABR thresholds between male and female Gen-Panx1 KO mice (Figure 2c). Thus, we did not separate different genders in data analyses in the following experiments.

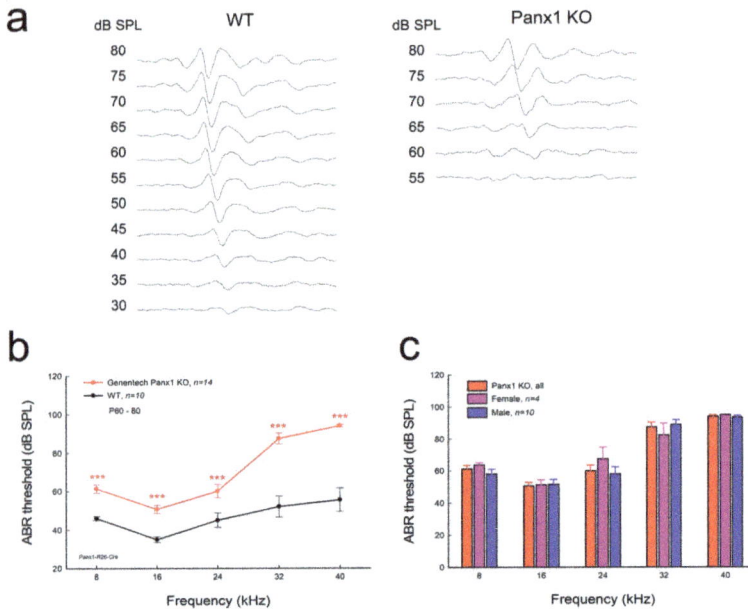

Figure 2. Hearing loss in Gen-Panx1 KO mice. (**a**): auditory brainstem responses (ABRs) in Gen-Panx1 KO and WT mice. The ABR was evoked by a 16 kHz tone burst. (**b**): ABR thresholds measured in Gen-Panx1 KO and WT mice. ABR thresholds in Gen-Panx1 KO mice were increased. *** $p < 0.001$ as determined by one-way ANOVA with a Bonferroni correction. (**c**): There is no significant difference in ABR thresholds between different genders in the Gen-Panx1 KO mice.

2.3. Reduction of Distortion Product Otoacoustic Emission in Gen-Panx1 KO Mice

Consistent with increases in ABR threshold (Figure 2), distortion product otoacoustic emission (DPOAE) in the Gen-Panx1 KO mouse was significantly reduced (Figure 3a-d). DPOAEs at f_0 of 20 kHz in the Gen-Panx1 KO mice and WT mice were 25.6 ± 2.76 and 37.1 ± 1.83 dB SPL, respectively (Figure 3c). In comparison with WT mice, DPOAEs in the Gen-Panx1 KO mice at $f_0 = 4$, 8, 16, and 20 kHz were reduced by -1.32 ± 0.63, -4.43 ± 2.18, -8.71 ± 2.76, and -11.5 ± 2.73 dB, respectively (Figure 3b). As shown in our previous publication [30], the reduction was increased as sound intensity was increased (Figure 3c,d). In I-O plot (Figure 3c), DPOAEs in the Gen-Panx1 KO mice at the stimulus intensity range from 40 to 60 dB SPL were significantly reduced. In comparison with WT mice, the reduction was $\geq$10 dB SPL ($p < 0.001$, one-way ANOVA with a Bonferroni correction) (Figure 3d). However, at the stimulus intensity lower than 40 dB SPL, DPOAEs in the Gen-Panx1 KO mice and WT mice were not apparently different (Figure 3c,d).

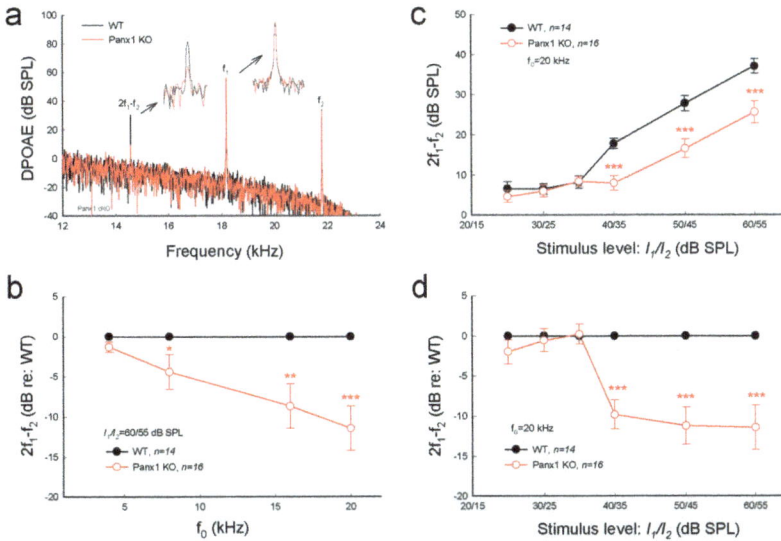

Figure 3. Distortion product otoacoustic emission (DPOAE) reduction in Gen-Panx1 KO mice. (**a**): Evoked spectra of acoustic emission in Panx1 KO mice and in WT mice. Inset: High-magnification plot of $2f_1 - f_2$ and f_1 peaks. (**b**): Reduction of DPOAE in frequency responses in the Panx1 KO mice. Magnitudes of $2f_1 - f_2$ in the Panx1 KO mice ware normalized to those in WT mice. (**c,d**): I-O function of DPOAE in Panx1 KO and WT mice. DPOAEs in panel (**d**) were normalized to those in WT mice. * $p < 0.05$, ** $p < 0.01$, and *** $p < 0.001$ as determined by one-way ANOVA with a Bonferroni correction.

2.4. Reduction of Auditory Receptor Potential in Gen-Panx1 KO Mice

Cochlear microphonics (CM) is the auditory receptor current/potential. Figure 4 shows that the CM in the Gen-Panx1 KO mice was reduced. CM in WT mice was 40–60 µV, while CM in the Gen-Panx1 KO mice was 10–30 µV, thus showing a reduction by more than 50% ($p < 0.001$, one-way ANOVA with a Bonferroni correction).

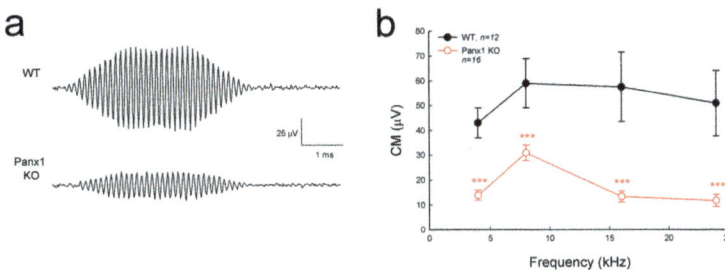

Figure 4. Cochlear microphonics (CM) reduction in the Gen-Panx1 KO mouse. (**a**): CM waveforms recorded from Gen-Panx1 KO and WT mice. (**b**): Reduction of CM in the Panx1 KO mouse. *** $p < 0.001$ as determined by one-way ANOVA with a Bonferroni correction.

2.5. No Apparent Hair Cell Loss in Gen-Panx1 KO Mice

Figure 5 shows that there was no substantial hair cell loss in Gen-Panx1 KO mice. The loss of hair cells in the Gen-Panx1 KO mice was less than 5%. There was no significant difference of hair cell loss between WT and Gen-Panx1 KO mice. This is consistent with the previous report that there is no significant hair cell loss in Foxg1-Panx1 cKO mice [21].

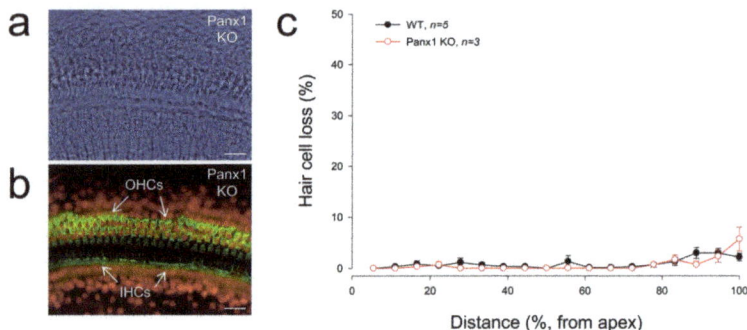

Figure 5. No apparent hair cells in Gen-Panx1 KO mice. (**a,b**): The cochlear sensory epithelia of the Gen-Panx1 KO mice in whole-mounting preparation with staining of phalloidin-Alexa Fluor-488 (green) and propidium iodide (red). OHCs: outer hair cells; IHCs: inner hair cells. Scale bars: 20 μm. (**c**): There is no apparent hair cell loss in the Gen-Panx1 KO mice.

3. Discussion

In this study, we found that the Gen-Panx1 KO mouse had hearing loss. ABR thresholds were significantly increased, and DPOAE and CM were significantly reduced (Figures 2–4). These data are consistent with previous findings that Panx1-deficient mice have hearing loss [21,30] and are also consistent with the fact that Panx1 mutation in humans can cause deafness [31].

In this experiment, in addition to the increases in ABR threshold, we found that DPOAE in the Gen-Panx1 KO mice was reduced (Figure 3). This finding is consistent with our previous findings of DPOAE reduction and hearing loss in Pax2-Panx1 cKO mice [30]. DPOAE reflects the activity of active cochlear amplification, which can increase hearing sensitivity and is required for normal hearing. DPOAE reduction suggests that active cochlear amplification is impaired, which can induce hearing loss and increase ABR threshold (Figure 2). Moreover, in line with the observed reduction of CM in the Foxg1-Panx1 cKO mice [21], CM in this Gen-Panx1 KO mouse line was also reduced (Figure 4). CM is the auditory receptor current/potential. CM reduction can consequently decrease active cochlear amplification, thereby eventually leading to hearing loss. Thus, consistent with ABR recordings, these auditory function tests also demonstrated hearing loss in the Gen-Panx1 KO mice.

However, these findings are inconsistent with a recent report that the Gen-Panx1 KO mice had no hearing loss in ABR recordings [34]. Several factors could contribute to this discrepancy. First, ABR in that report [34] was recorded at frequency range <24 kHz. However, as shown in Figure 2 and in our previous reports [21,30], the Panx1 deficiency-induced hearing loss was most severe at high-frequency (>24 kHz). By recording at these low frequencies, hearing loss at higher frequencies may have been missed by the authors. Second, the study used C57BL/6 mice rather than WT littermates as control. Although the Gen-Panx1 KO mice have similar genetic backgrounds with C57BL/6 mice, this still could produce significant differences in hearing functional tests. In particular, hearing loss in the Gen-Panx1 KO mice in the middle- and low-frequency range is not as large as that in the high-frequency range (Figure 2b). Third, unlike in the previous report [34], which only considered ABR recordings, in the present study, we also recorded CM and DPOAE (Figures 3 and 4). All of these auditory functional tests demonstrated hearing loss in the Gen-Panx1 KO mouse.

In the previous report [34], the susceptibility of the Gen-Panx1 KO mouse to noise was also examined. Mice were exposed to a high intensity tone (12 kHz, 115 dB SPL, 1 h), and there was no significant difference in ABR thresholds between the Panx1-deficient mice and WT mice measured at the post-exposure day 7 [34]. However, since both WT and KO mice had large ABR threshold shifts for this high-intensity exposure, the susceptibility of Panx1 KO mice to noise could not be assessed. In order to assess susceptibility, low- or moderate-intensity noise exposure could be used to test whether the Panx1-deficient mice have larger threshold shifts than WT mice. In addition, only recording ABR thresholds at the seventh day after noise exposure [34] may have been too short a time to assess a permanent threshold shift (PTS).

Recently, another Panx1 KO mouse line (B6;129-*Panx1*[tm1.Fam/Cnrm], European Mouse Mutant Archive (EMMA): E11476) [32] has also been reported to have no hearing loss [35]. However, it has been reported that the brain tissues of this EMMA Panx1 KO mouse line showed no negative reaction to Panx1 in Western blotting using several anti-Panx1 antibodies [36], including a chicken anti-human Panx1 antibody used in this study, whose specificity was widely validated in previous experiments by multiple assays in different Panx1 KO mouse lines [8,9,21,30,37,38]. This suggests that this EMMA Panx1 KO mouse line may have a hypomorphic phenotype. As shown in our immunofluorescent staining in this study (Figure 1) and also in previous studies in other Panx1 KO mouse lines [21,30], the location- and cell-specific deletion of Panx1 is clearly visible in the cochlea. This cell-specific deletion pattern also provides direct and unequivocal evidence that the used chicken anti-human antibody is specific to Panx1. Indeed, a hypomorphic phenotype has been found in the KOMP (Knockout Mouse Project, University of California, Davis, CA, USA) Panx1 KO mouse line, in which Panx1 is not completely deleted [33]. Finally, Panx1 deficiency-induced hearing loss is progressive [20,30]; the increase in ABR threshold is small at postnatal day 30 (P30) and becomes large and apparent after P60 [20,30]. The ABR threshold in the previous report [35] was averaged from P30 to P90. This could also attenuate the potential difference. Thus, as suggested by these inconsistent results and reports, Panx1 function including characterization of Panx1 KO mice and the specificity of anti-Panx1 antibodies still needs to receive further study in the future.

Our previous study demonstrated that the Panx1 deficiency reduced ATP release and EP generation in the cochlea [21]. Positive EP in the cochlear endolymph is generated in the cochlear lateral wall [21,39] and a driving force for K^+-ions passing through the mechano-transduction channels in hair cells to produce CM [39], although hair cells have neither connexin [40,41] nor pannexin expression [8]. EP reduction can reduce CM, thereby leading to the reduction of active cochlear amplification and eventually hearing loss. In this study, we did not measure EP and ATP release in the Gen-Panx1 KO mice. However, CM and DPOAE were reduced in the Gen-Panx1 KO mice (Figures 3 and 4). Thus, the Gen-Panx1 KO mice may share the same mechanism to reduce CM and active cochlear amplification, thereby leading to hearing loss.

4. Materials and Methods

4.1. Panx1 KO Mice and Genotyping

Gen-Panx1 KO mice were created by Vishva Dixit at Genentech Inc. with the loxP-Cre technique [15]. In this transgenic mouse strain, Exon2 of Panx1 was floxed with FLP recombinase and was deleted by crossing to the C57BL/6-Gt(ROSA)26Sortm16(Cre) Arte Cre deleter strain (Taconic Artemis, Rensselaer, NY, USA). Mice were genotyped by tail genomic DNA with PCR primers: 5′-TGA CCA CAG ACA GCA CTTAAG-3′ and 5′-CGT CTG AGA GCT CCC TGG CG-3′, which yield a 651-bp WT band and a 335-bp knockout band [15]. WT littermates served as control. All recordings were performed under anesthesia with Ketamine and Xylazine (8.5 mL saline + 1 mL (100 mg/mL) Ketamine + 0.55 mL (20 mg/mL) Xylazine, 0.1 mL/10 g body weight, ip) to minimize suffering. The experimental procedures were approved by the University of Kentucky's Animal Care and Use Committee (Protocol

No. 00902M2005, May 9, 2010) and conducted according to the standards of the NIH Guidelines for the Care and Use of Laboratory Animals.

4.2. Data Processing and Statistical Analysis

A total of 16 Gen-Panx1 KO mice (4 females and 12 males) were used in this study. Littermate WT (6 females and 8 males) mice were used as control. Mouse ages were between P60 to P90. For ABR, DPOAE, and CM recordings, 14 (4 females and 10 males), 16 (4 females and 12 males), and 16 (4 females and 12 males) of Gen-Panx1 KO mice were used, respectively. Exact numbers of mice used in each experiment were also indicated in the corresponding figures. Since there was no significant difference in ABR threshold between different genders in Gen-Panx1 KO mice (Figure 2c), we did not do gender analyses in other experiments.

The statistical analyses were performed by SPSS v18.0 (SPSS Inc., Chicago, IL, USA) using one-way ANOVA with a Bonferroni correction. The level of statistical significance was set at $p < 0.05$. Data were presented as mean $\pm$ SEM and plotted by SigmaPlot v10 (SPSS Inc., Chicago, IL, USA). All sample size and P values were reported in the figures or the figure legends.

4.3. ABR, CM, and DPOAE Recordings

As described in our previous publications [21,42–46], ABR was recorded by a Tucker-Davis ABR & DPOAE workstation with ES-1 high frequency speakers (Tucker-Davis Tech., Alachua, FL, USA). Mouse body temperature was maintained at 37–38 °C. ABR was measured by clicks (rate: 20/s) in alternative polarity and series tone pips (5 ms duration with 1 ms up and down ramp, rate: 20/s) from 80 to 10 dB SPL in a 5-dB step at 8, 16, 24, 32, and 40 kHz. The signal was amplified (50,000×), filtered (300–3000 Hz), and averaged by 500 times. The lowest level at which the ABR could be recognized was defined as the threshold. The high levels of acoustic stimuli (100 to 70 dB SPL) were used if the threshold was >75 dB SPL.

CM was recorded with the same electrode configuration as the ABR recording, i.e., one electrode was inserted at the vertex, one electrode was ventrolaterally inserted to the right or left ear, and the ground needle electrode was inserted in the right leg [42,46]. However, the signal evoked by tone bursts was amplified (50,000×), filtered (3–50 kHz), and averaged by 250 times, as described in our previous publications [42,46].

For DPOAE recording, an ear-plug that contained a small microphone and two small tubes was plugged into the outer ear cannel. Two-testing sounds were separately induced into the ear through two-small tubes [30,44–46]. The frequencies of two-testing sounds were determined by a geometric mean of f_1 and f_2 ($f_0 = (f_1 \times f_2)^{1/2}$) at $f_0 = 4, 8, 16$, and 20 kHz with $f_2/f_1 = 1.22$. The intensity of f_1 (I_1) was set at 5 dB SPL higher than that of f_2 (I_2). The distortion product was recorded with an average of 150 times. A cubic distortion product of $2f_1 - f_2$ was measured as DPOAE.

4.4. Cochlear Preparation and Immunofluorescent Staining

The detailed methods and procedures of immunofluorescent staining can be found in our previous reports [8,41]. After decapitation, the cochlea was isolated and fixed with 4% paraformaldehyde for 0.5–1 h. After decalcification with 10% EDTA for 2 days, the cochlea was embedded with OCT (Cat # 4583, Sakura Finetek USA Inc., Torrance, CA, USA), and cut into 10-μm thick sections at −22∼ 24 °C by a cryostat (Thermo Electron Corp., Waltham, MA, USA). For immunofluorescent staining, the sections were incubated in a blocking solution, which was composed of 10% goat serum and 1% bovine serum albumin, with 0.1% Triton X-100 for 30 min, and then reacted with chicken anti-human Panx1 antibody (1:500; #4515, a gift from Gerhard Dahl at the University of Miami Medical School) and polyclonal rabbit anti-Cx30 antibody (1:400, #71-2200, Invitrogen, Carlsbad, CA, USA) or monoclonal mouse anti-Cx26 (1:400, # 33-5800, Invitrogen, CA) at 4 °C overnight. The specificity of this Panx1 antibody was verified in previous publications by Western blotting and Panx1 KO mouse tissues in our laboratory and other laboratories [8,9,21,30,37,38]. After washout with PBS for three times, the sections were incubated with corresponding secondary antibodies (Alexa Fluor 488- and 568, 1:500, Molecular Probes, Eugene, OR, USA) at room temperature (23 °C) for 2 h to visualize the staining. The staining was observed under a fluorescence microscope (Nikon 2000, Nickon, Melville, NY, USA).

4.5. Cochlear Epithelium Whole-Mounting and Hair Cell Loss Accounting

As reported in our previous publications [21,44,45], the cochlear epithelia were isolated and stained with phalloidin-Alexa Fluor-488 and propidium iodide in the whole-mounting preparation. Hair cells were accounted under a 20× lens.

5. Conclusions

Our present study demonstrated that Gen-Panx1 KO mice have hearing loss. These data are consistent with previous observations from other Panx1 KO mouse lines [21,30] and the known occurrence of Panx1 mutation-induced hearing loss in humans [31]. These new data further confirm that Panx1 deficiency can induce hearing loss. However, Panx1 function in hearing still remains largely undetermined and needs to be further studied in the future. Also, as demonstrated by this study, multiple auditory function tests, careful characterization of Panx1-deficient mice, and specificity of anti-Panx1 antibody are important and required for assessing the role of Panx1 in hearing and other Panx1 functions.

Author Contributions: H.-B.Z. conceived the general framework of this study. J.C., C.L., L.Z., Y.Z., and H.-B.Z performed the experiments and analyzed data. H.-B.Z wrote the paper.

Funding: This work was supported by NIH R01 DC005989, DC017025, and R56 DC015019 to H.-B.Z, the National Natural Science Foundation of China (No. 81600795) to L.Z. and (No. 81500791) to J.C.

Acknowledgments: We are grateful to Vishva Dixit at Genentech Inc. for kindly providing Panx1 KO mice and Gerhard Dahl at Miami University for kindly providing anti-Panx1 antibody.

Conflicts of Interest: The authors declare no conflict of interest.

Abbreviations

ABR	auditory brainstem response
CM	cochlear microphonics
cKO	conditional knockout
DPOAE	distortion product otoacoustic emission
EMMA	European Mouse Mutant Archive
EP	endocochlear potential
KO	knockout
OC	organ of Corti
OHC	outer hair cell
OSC	outer sulcus cell
PTS	permanent threshold shift
SLM	spiral limbus
WT	wild-type

References

1. Castillo, F.J.; Castillo, I. The DFNB1 subtype of autosomal recessive non-syndromic hearing impairment. *Front. Biosci.* **2011**, *17*, 3252–3274. [CrossRef]
2. Castillo, F.J.; Castillo, I. DFNB1 non-syndromic hearing impairment: Diversity of mutations and associated phenotypes. *Front. Mol. Neurosci.* **2017**, *10*, 428. [CrossRef] [PubMed]
3. Chan, D.K.; Chang, K.W. GJB2-associated hearing loss: Systematic review of worldwide prevalence, genotype, and auditory phenotype. *Laryngoscope* **2014**, *124*, E34–E53. [CrossRef] [PubMed]

4. Wingard, J.C.; Zhao, H.B. Cellular and Deafness Mechanisms Underlying Connexin Mutation-Induced Hearing Loss—A Common Hereditary Deafness. *Front. Cell. Neurosci.* **2015**, *9*, 202. [CrossRef] [PubMed]

5. Zhao, H.B. Hypothesis of K^+-recycling defect is not a primary deafness mechanism for Cx26 (*GJB2*) deficiency. *Front. Mol. Neurosci.* **2017**, *10*, 162. [CrossRef] [PubMed]

6. Bruzzone, R.; Hormuzdi, S.G.; Barbe, M.T.; Herb, A.; Monyer, H. Pannexins, a family of gap junction proteins expressed in brain. *Proc. Natl. Acad. Sci. USA* **2003**, *100*, 13644–13649. [CrossRef] [PubMed]

7. Baranova, A.; Ivanov, D.; Petrash, N.; Pestova, A.; Skoblov, M.; Kelmanson, I.; Shagin, D.; Nazarenko, S.; Geraymovych, E.; Litvin, O.; et al. The mammalian pannexin family is homologous to the invertebrate innexin gap junction proteins. *Genomics* **2004**, *83*, 706–716. [CrossRef] [PubMed]

8. Wang, X.H.; Streeter, M.; Liu, Y.P.; Zhao, H.B. Identification and characterization of pannexin expression in the mammalian cochlea. *J. Comp. Neurol.* **2009**, *512*, 336–346. [CrossRef] [PubMed]

9. Locovei, S.; Bao, L.; Dahl, G. Pannexin 1 in erythrocytes: Function without a gap. *Proc. Natl. Acad. Sci. USA* **2006**, *103*, 7655–7659. [CrossRef] [PubMed]

10. Sosinsky, G.E.; Boassa, D.; Dermietzel, R.; Duffy, H.S.; Laird, D.W.; MacVicar, B.; Naus, C.C.; Penuela, S.; Scemes, E.; Spray, D.C.; et al. Pannexin channels are not gap junction hemichannels. *Channels (Austin)* **2011**, *5*, 193–197. [CrossRef] [PubMed]

11. Dahl, G. ATP release through pannexon channels. *Philos. Trans. R. Soc. Lond. B Biol. Sci.* **2015**, *370*, 1672. [CrossRef] [PubMed]

12. Thompson, R.J.; Zhou, N.; MacVicar, B.A. Ischemia opens neuronal gap junction hemichannels. *Science* **2006**, *312*, 924–927. [CrossRef] [PubMed]

13. Thompson, R.J.; Jackson, M.F.; Olah, M.E.; Rungta, R.L.; Hines, D.J.; Beazely, M.A.; MacDonald, J.F.; MacVicar, B.A. Activation of pannexin-1 hemichannels augments aberrant bursting in the hippocampus. *Science* **2008**, *322*, 1555–1559. [CrossRef] [PubMed]

14. Chekeni, F.B.; Elliott, M.R.; Sandilos, J.K.; Walk, S.F.; Kinchen, J.M.; Lazarowski, E.R.; Armstrong, A.J.; Penuela, S.; Laird, D.W.; Salvesen, G.S.; et al. Pannexin 1 channels mediate 'find-me' signal release and membrane permeability during apoptosis. *Nature* **2010**, *467*, 863–867. [CrossRef] [PubMed]

15. Qu, Y.; Misaghi, S.; Newton, K.; Gilmour, L.L.; Louie, S.; Cupp, J.E.; Dubyak, G.R.; Hackos, D.; Dixit, V.M. Pannexin-1 is required for ATP release during apoptosis but not for inflammasome activation. *J. Immunol.* **2011**, *186*, 6553–6561. [CrossRef] [PubMed]

16. Penuela, S.; Gehi, R.; Laird, D.W. The biochemistry and function of pannexin channels. *Biochim. Biophys. Acta* **2013**, *1828*, 15–22. [CrossRef] [PubMed]

17. Penuela, S.; Kelly, J.J.; Churko, J.M.; Barr, K.J.; Berger, A.C.; Laird, D.W. Panx1 regulates cellular properties of keratinocytes and dermal fibroblasts in skin development and wound healing. *J. Investig. Dermatol.* **2014**, *134*, 2026–2035. [CrossRef] [PubMed]

18. Poon, I.K.; Chiu, Y.H.; Armstrong, A.J.; Kinchen, J.M.; Juncadella, I.J.; Bayliss, D.A.; Ravichandran, K.S. Unexpected link between an antibiotic, pannexin channels and apoptosis. *Nature* **2014**, *507*, 329–334. [CrossRef] [PubMed]

19. Weilinger, N.L.; Lohman, A.W.; Rakai, B.D.; Ma, E.M.; Bialecki, J.; Maslieieva, V.; Rilea, T.; Bandet, M.V.; Ikuta, N.T.; Scott, L.; et al. Metabotropic NMDA receptor signaling couples Src family kinases to pannexin-1 during excitotoxicity. *Nat. Neurosci.* **2016**, *19*, 432–442. [CrossRef] [PubMed]

20. Zhao, H.B. Expression and function of pannexins in the inner ear and hearing. *BMC Cell Biol.* **2016**, *17*, 16. [CrossRef] [PubMed]

21. Chen, J.; Zhu, Y.; Liang, C.; Chen, J.; Zhao, H.B. Pannexin1 channels dominate ATP release in the cochlea ensuring endocochlear potential and auditory receptor potential generation and hearing. *Sci. Rep.* **2015**, *5*, 10762. [CrossRef] [PubMed]

22. Housley, G.D.; Bringmann, A.; Reichenbach, A. Purinergic signaling in special senses. *Trends Neurosci.* **2009**, *32*, 128–141. [CrossRef] [PubMed]

23. Zhao, H.B.; Yu, N.; Fleming, C.R. Gap junctional hemichannel-mediated ATP release and hearing controls in the inner ear. *Proc. Natl. Acad. Sci. USA* **2005**, *102*, 18724–18729. [CrossRef] [PubMed]

24. Yu, N.; Zhao, H.B. ATP activates P2x receptors and requires extracellular Ca^{++} participation to modify outer hair cell nonlinear capacitance. *Pflugers Arch.* **2008**, *457*, 453–461. [CrossRef] [PubMed]

25. Housley, G.D.; Morton-Jones, R.; Vlajkovic, S.M.; Telang, R.S.; Paramananthasivam, V.; Tadros, S.F.; Wong, A.C.; Froud, K.E.; Cederholm, J.M.; Sivakumaran, Y.; et al. ATP-gated ion channels mediate adaptation to elevated sound levels. *Proc. Natl. Acad. Sci. USA* **2013**, *110*, 7494–7499. [CrossRef] [PubMed]

26. Tritsch, N.X.; Yi, E.; Gale, J.E.; Glowatzki, E.; Bergles, D.E. The origin of spontaneous activity in the developing auditory system. *Nature* **2007**, *450*, 50–55. [CrossRef] [PubMed]

27. Tritsch, N.X.; Bergles, D.E. Developmental regulation of spontaneous activity in the mammalian cochlea. *J. Neurosci.* **2010**, *30*, 1539–1550. [CrossRef] [PubMed]

28. Zhu, Y.; Zhao, H.B. ATP activates P2X receptors to mediate gap junctional coupling in the cochlea. *Biochem. Biophys. Res. Commun.* **2012**, *426*, 528–532. [CrossRef] [PubMed]

29. Zhu, Y.; Zhao, H.B. ATP-mediated potassium recycling in the cochlear supporting cells. *Purinergic Signal.* **2010**, *6*, 221–229. [CrossRef] [PubMed]

30. Zhao, H.B.; Zhu, Y.; Liang, C.; Chen, J. Pannexin 1 deficiency can induce hearing loss. *Biochem. Biophys. Res. Commun.* **2015**, *463*, 143–147. [CrossRef] [PubMed]

31. Shao, Q.; Lindstrom, K.; Shi, R.; Kelly, J.; Schroeder, A.; Juusola, J.; Levine, K.L.; Esseltine, J.L.; Penuela, S.; Jackson, M.F.; et al. A germline variant in the PANX1 gene has reduced channel function and is associated with multisystem dysfunction. *J. Biol. Chem.* **2016**, *291*, 12432–12443. [CrossRef] [PubMed]

32. Anselmi, F.; Hernandez, V.H.; Crispino, G.; Seydel, A.; Ortolano, S.; Roper, S.D.; Kessaris, N.; Richardson, W.; Rickheit, G.; Filippov, M.A.; et al. ATP release through connexin hemichannels and gap junction transfer of second messengers propagate Ca^{2+} signals across the inner ear. *Proc. Natl. Acad. Sci. USA* **2008**, *105*, 18770–18775. [CrossRef] [PubMed]

33. Hanstein, R.; Negoro, H.; Patel, N.K.; Charollais, A.; Meda, P.; Spray, D.C.; Suadicani, S.O.; Scemes, E. Promises and pitfalls of a Pannexin1 transgenic mouse line. *Front. Pharmacol.* **2013**, *4*, 61. [CrossRef] [PubMed]

34. Abitbol, J.M.; Kelly, J.J.; Barr, K.; Schormans, A.L.; Laird, D.W.; Allman, B.L. Differential effects of pannexins on noise-induced hearing loss. *Biochem. J.* **2016**, *473*, 4665–4680. [CrossRef] [PubMed]

35. Zorzi, V.; Paciello, F.; Ziraldo, G.; Peres, C.; Mazzarda, F.; Nardin, C.; Pasquini, M.; Chiani, F.; Raspa, M.; Scavizzi, F.; et al. Mouse Panx1 is dispensable for hearing acquisition and auditory function. *Front. Mol. Neurosci.* **2017**, *10*, 379. [CrossRef] [PubMed]

36. Bargiotas, P.; Krenz, A.; Hormuzdi, S.G.; Ridder, D.A.; Herb, A.; Barakat, W.; Penuela, S.; von Engelhardt, J.; Monyer, H.; Schwaninger, M. Pannexins in ischemia-induced neurodegeneration. *Proc. Natl. Acad. Sci. USA* **2011**, *108*, 20772–20777. [CrossRef] [PubMed]

37. Dvoriantchikova, G.; Ivanov, D.; Panchin, Y.; Shestopalov, V.I. Expression of pannexin family of proteins in the retina. *FEBS Lett.* **2006**, *580*, 2178–2182. [CrossRef] [PubMed]

38. Zoidl, G.; Petrasch-Parwez, E.; Ray, A.; Meier, C.; Bunse, S.; Habbes, H.W.; Dahl, G.; Dermietzel, R. Localization of the pannexin1 protein at postsynaptic sites in the cerebral cortex and hippocampus. *Neuroscience* **2007**, *146*, 9–16. [CrossRef] [PubMed]

39. Chen, J.; Zhao, H.B. The role of an inwardly rectifying K^+ channel (Kir4.1) in the inner ear and hearing loss. *Neuroscience* **2014**, *265*, 137–146. [CrossRef] [PubMed]

40. Kikuchi, T.; Kimura, R.S.; Paul, D.L.; Adams, J.C. Gap junctions in the rat cochlea: Immunohistochemical and ultrastructural analysis. *Anat. Embryol.* **1995**, *191*, 101–118. [CrossRef] [PubMed]

41. Zhao, H.B.; Yu, N. Distinct and gradient distributions of connexin26 and connexin30 in the cochlear sensory epithelium of guinea pigs. *J. Comp. Neurol.* **2006**, *499*, 506–518. [CrossRef] [PubMed]

42. Liang, C.; Zhu, Y.; Zong, L.; Lu, G.J.; Zhao, H.B. Cell degeneration is not a primary causer for Connexin26 (*GJB2*) deficiency associated hearing loss. *Neurosci. Lett.* **2012**, *528*, 36–41. [CrossRef] [PubMed]

43. Chen, J.; Chen, J.; Zhu, Y.; Liang, C.; Zhao, H.B. Deafness induced by Connexin26 (GJB2) deficiency is not determined by endocochlear potential (EP) reduction but is associated with cochlear developmental disorders. *Biochem. Biophys. Res. Commun.* **2014**, *448*, 28–32. [CrossRef] [PubMed]

44. Zhu, Y.; Liang, C.; Chen, J.; Zong, L.; Chen, G.D.; Zhao, H.B. Active cochlear amplification is dependent on supporting cell gap junctions. *Nat. Commun.* **2013**, *4*, 1786. [CrossRef] [PubMed]

45. Zhu, Y.; Chen, J.; Liang, C.; Zong, L.; Chen, J.; Jones, R.O.; Zhao, H.B. Connexin26 (*GJB2*) deficiency reduces active cochlear amplification leading to late-onset hearing loss. *Neuroscience* **2015**, *284*, 719–729. [CrossRef] [PubMed]

46. Mei, L.; Chen, J.; Zong, L.; Zhu, Y.; Liang, C.; Jones, R.O.; Zhao, H.B. A deafness mechanism of digenic Cx26 (*GJB2*) and Cx30 (*GJB6*) mutations: Reduction of endocochlear potential by impairment of heterogeneous gap junctional function in the cochlear lateral wall. *Neurobiol. Dis.* **2017**, *108*, 195–203. [CrossRef] [PubMed]

International Journal of
Molecular Sciences

MDPI

Review

Connexin 43-Based Therapeutics for Dermal Wound Healing

Jade Montgomery [1,2], Gautam S. Ghatnekar [3], Christina L. Grek [3], Kurtis E. Moyer [4,5] and Robert G. Gourdie [1,2,6,*]

1   Virginia Tech Carilion Research Institute, Roanoke, VA 24016, USA; jmont@vt.edu
2   School of Biomedical Engineering and Sciences, Virginia Tech-Wake Forest University, Blacksburg, VA 24061, USA
3   FirstString Research, Inc., Mount Pleasant, SC 29464, USA; ghatnekar@firststringresearch.com (G.S.G.); grek@firststringresearch.com (C.L.G.)
4   Department of Surgery, Virginia Tech Carilion School of Medicine, Roanoke, VA 24016, USA; kemoyer@carilionclinic.org
5   Department of Surgery, Carilion Clinic, Roanoke, VA 24016, USA
6   Department of Emergency Medicine, Virginia Tech Carilion School of Medicine, Roanoke, VA 24016, USA
*   Correspondence: gourdier@vtc.vt.edu; Tel.: +1-843-860-8971

Received: 4 May 2018; Accepted: 12 June 2018; Published: 15 June 2018

Abstract: The most ubiquitous gap junction protein within the body, connexin 43 (Cx43), is a target of interest for modulating the dermal wound healing response. Observational studies found associations between Cx43 at the wound edge and poor healing response, and subsequent studies utilizing local knockdown of Cx43 found improvements in wound closure rate and final scar appearance. Further preclinical work conducted using Cx43-based peptide therapeutics, including alpha connexin carboxyl terminus 1 (αCT1), a peptide mimetic of the Cx43 carboxyl terminus, reported similar improvements in wound healing and scar formation. Clinical trials and further study into the mode of action have since been conducted on αCT1, and Phase III testing for treatment of diabetic foot ulcers is currently underway. Therapeutics targeting connexin activity show promise in beneficially modulating the human body's natural healing response for improved patient outcomes across a variety of injuries.

Keywords: gap junctions; hemichannels; connexins; skin; wound healing; scar formation; peptide

1. Gap Junctions, Connexins, and Skin Wound Healing

Gap junctions (GJs) are complexes of intercellular channels composed of proteins encoded by the connexin multigene family [1]. Gap junctions enable direct cytoplasmic coupling between cells, permitting the intercellular exchange of small molecules (<1000 Da). Additionally, undocked gap junctional connexons or hemichannels are also increasingly recognized as having roles in homeostasis and disease [2–5]. A number of connexin gene family members have been reported to be expressed in skin, including connexin 26 (Cx26), Cx30, Cx31.1, Cx30.3, Cx37, Cx40, Cx45 and Cx43 [6–10]. Of these, Cx43 (Figure 1) is the most abundant and ubiquitous, being present in both the epidermal and dermal cutaneous layers.

Wound healing typically progresses in four stages: Hemostasis, inflammation, proliferation, and maturation [11]. Hemostasis begins immediately after injury, with platelets and various clotting factors invading the wound space to create a fibrin clot that prevents further bleeding. After hemostasis is achieved, inflammation begins within an hour of the original injury. Blood vessels dilate and become more porous, enabling inflammatory leukocytes to invade the wound and phagocytize bacteria and dead/damaged cells. It is this inflammatory stage that is the key target of many wound healing experiments. Fetuses, which do not mount a mature inflammatory response, are able to heal wounds without leaving a lasting scar, particularly during mid-gestational stages [12–14]. Research has also found that eliminating the inflammatory response in an adult mouse resulted in wounds that healed efficiently with reduced scarring [15]. On the other hand, failure of the inflammatory phase to resolve appears to be an important aspect of why pathological skin wounds, such as diabetic foot ulcers, are slow to heal [16].

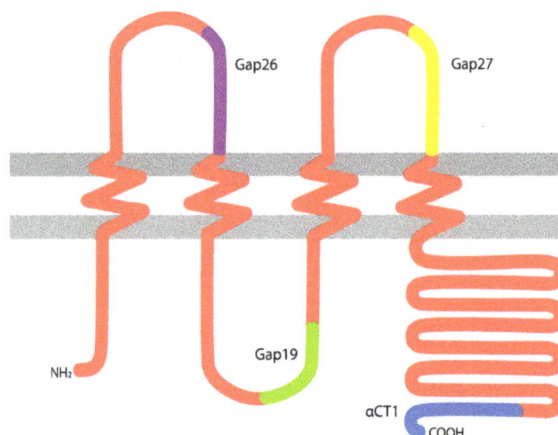

Figure 1. A diagram of connexin 43 (Cx43) spanning the cell membrane, with approximate locations highlighted from which several memetic peptides were derived. αCT1: alpha connexin carboxyl terminus 1.

Gap junction channel function and connexin activity have long been recognized as having important assignments in nearly all phases of skin wound healing, including in the coordination of the inflammatory response, propagation of injury signals between cells, wound closure, granulation-tissue formation, and scar remodeling after injury [6–10,17–28]. In this short review, we will summarize key findings on basic research into GJs and connexins in skin wound healing as well as recount recent progress on translating this fundamental knowledge to the clinic.

2. Early Work on the Role of Cx43 in Cutaneous Wound Healing

A key initial set of findings on the role of connexins in cutaneous wound healing was made in the laboratory of Paul and co-workers [26]. In studies in rodent models, it has been determined that Cx26, Cx31 and Cx43 undergo characteristic cell-specific changes during the wound healing progression [26, 27]. Of particular note, Cx43 expression, as well as GJ-mediated intercellular communication, decreases transiently in epidermal cells at the wound edge over the first 24 h following injury [26,27]. Cx43 was downregulated not only in the epidermis of the wound immediately following injury but in the epidermis surrounding the wound as well. A contrasting observation was made in the deep dermis. Here, Cx43 was found to be transiently upregulated in fibroblasts and other tissues in the hours immediately following wounding. After one week, and at later time-points associated with granulation tissue formation and remodeling, increased Cx43 was associated with increases in granulation tissue formation and maturity. Taken together, these results indicated the possibility that localized modulation of Cx43 levels, or certain aspects of Cx43 function, could be a potential method for beneficially altering the cutaneous healing response. This prospect is also suggested by studies of the buccal mucosa, the tissue that lines the inside of the cheeks and floor of the mouth [29], and gingival tissues lining the gum [24,30,31]. In these mucosal tissues, both of which heal more quickly and with significantly less scarring than skin, this relatively fetal-like healing response occurs in association with a rapid and strong downregulation of Cx43.

The first in vivo study of the potential benefits of manipulating wound-localized Cx43 levels was conducted by Green, Becker, and colleagues, who topically applied Cx43 antisense directly to healing wounds [20]. Application of a Cx43 antisense gel to adult rat wounds immediately after wounding increased the rate of Cx43 downregulation in the epidermis and prevented the upregulation of Cx43 in the dermis. This had the remarkable macroscopic effect of reducing inflammation at the wound site, increasing the rate of wound closure, and reducing the appearance of scars at 12 days post-wounding. These effects were particularly noticeable for incisional wounds, although excisional wounds were also improved. The transient upregulation of Cx43 in the smooth muscle cells and endothelial cells of the blood vessels after wounding has been suggested to increase vasodilation and allow the infiltration of inflammatory cells. The Cx43 antisense prevented this upregulation, resulting in a decrease in inflammatory neutrophil numbers and a reduction in the overall inflammatory response in the antisense-treated tissue. The granulation tissue area was also significantly decreased in the treated wounds. In a follow-up study, the group reported decreases in leukocytes and macrophages in Cx43 antisense-treated wounds, concomitant with reduced expression of CC chemokine ligand-2 and Tumor Necrosis Factor alpha, suggesting that the observed enhanced regeneration may have been mediated in part by attenuating inflammation at the wound site [32].

More recently, Martin and co-workers have used an alternate gene knockdown approach, short interfering RNAs (siRNAs) targeted to Cx43, together with the Cx43 mimetic peptide Gap27, to study channel-dependent and independent effects of Cx43 in the response of dermal cells to injury [33]. Their data indicates that the response to targeting Cx43 varies between cell types (keratinocyte versus fibroblast) and between cells of the same type (skin fibroblasts), but of different tissue origins. The knockdown of Cx43 via siRNA enhanced both scrape wound closure and cell proliferation in dermal fibroblasts of human adults and neonates, indicating roles for Cx43 in cell proliferation and migration in these cell types. By contrast, in adult keratinocytes and juvenile foreskin fibroblasts, only scrape wound closure was enhanced, indicating that in these tissue types Cx43 still has a significant effect on cell migration, but its knockdown does not enhance proliferation.

Similar to the Cx43 knockdown/anti-sense experiments, excisional wound studies of a Cx43 heterozygous knockout (Cx43$^{+/-}$) mouse showed decreased inflammation and increased wound closure in the Cx43$^{+/-}$ mouse model compared to wildtype littermates [23]. While no difference was reported in the collagen deposition of the granulation tissue, increased numbers of active dermal fibroblasts were found in the wound space of the Cx43$^{+/-}$ mice, indicating increased fibroblast infiltration/proliferation and activation. Gene expression of extracellular remodeling proteins, including collagen I and III, was also significantly increased in the Cx43$^{+/-}$ mice seven days post-injury; a time-point consistent with the initiation of the remodeling phase of scar tissue.

Research on chronic wounds, such as venous leg ulcers and diabetic foot ulcers, have shown that Cx43 may be a key participating protein in these conditions [8,17,21,34–42]. Studies of diabetic wounds found Cx43 persistence at the wound edge in these chronic wounds [34], while another study by the Becker group on biopsies taken from human venous leg ulcers determined that Cx43 was significantly overexpressed, not just at the edge of these chronic wounds, but throughout the entire dermis of the biopsies [35]. Cx43 over-expression has also been reported to show strong correlations to varicose vein severity in patients—a precursor to leg ulceration from venous insufficiency [36].

3. Preclinical Studies of αCT1 Peptide in Skin Wound Healing

During the last decade, a number of peptides targeting specific activities and functions of Cx43 have been developed and studied in the context of wound healing. These include peptides such as alpha connexin carboxyl terminus 1 (αCT1), Gap19, Gap26, Gap27, and more, each targeting different binding sites with varying specificity and size in attempts to narrow down the mode of action and assess therapeutic opportunity [7–9,20,30,37–42]. Although many of these peptides have been studied in depth for years, only one peptide thus far, αCT1, has moved forward to pivotal Phase III clinical testing (NCT02667327).

αCT1, also referred to as aCT1 or ACT1 in publications, incorporates the last nine amino acids of the Cx43 carboxyl terminus (CT) with an amino terminal antennapedia internalization vector to enable the peptide to penetrate the cell cytoplasm [43]. Originally developed to study the effects of binding between Cx43 and the actin-binding protein zonula occludens-1 (ZO-1), αCT1 peptide competitively inhibits the interaction between the Post synaptic density Drosophila disc large tumor suppressor Zonula occludens-1 (PDZ)-binding domain at the CT of Cx43 and the second PDZ domain of ZO-1 [43,44]. Ongoing work has indicated that the peptide also influences interactions within and/or between Cx43 molecules, with effects on the Cx43 phosphostatus [45,46]. Unlike Cx43 antisense and knockout models, αCT1 does not appear to affect Cx43 protein levels [7,43]. This suggests that its mode of action in wound healing is unlikely to be mediated by the direct effects on the abundance of Cx43, although the influence of αCT1 may well involve downstream alterations in the network of protein-protein interactions that flow from a decrease of Cx43 protein levels at the wound edge. The prevention of Cx43 recycling and transport facilitated by ZO-1 caused by αCT1, as well as non-PDZ based interactions involving αCT1, including binding the Cx43 molecule itself [45], are two likely candidates for causing this alteration.

Early observations in scratch wound assays of 3T3 fibroblasts treated with αCT1 noted that treated fibroblasts appeared more active, migrating with greater speed across the scratch [6]. When mouse excisional wounds were treated with αCT1, inflammation was reduced, and the wound closure rate was increased [7,10,47], mirroring the results of the Cx43 antisense and knockout experiments. Additionally, strength testing analyses of scar tissue 90 days post-wounding revealed that αCT1 treated scars had significantly improved mechanical properties compared to control [7]. Interestingly, the improvement in scar mechanical properties seen at 90 days was more marked than that at 30 days post-wounding, suggesting that the acute treatment by αCT1 had effects that continued long into the remodeling phase of scar formation [7,47]. Additional studies of αCT1 skin wound healing were conducted in a pig model [7]. Porcine models are considered the gold standard for wound healing studies as pig skin is thought to be the closest analogue to human skin [48]. Porcine dermal wounds treated with αCT1 showed decreased granulation tissue area size and increased sub-epidermal vascularity, somewhat regenerating the patterns of blood vessel distribution found in unwounded skin [7].

Additional therapeutic opportunities beyond undiseased dermal wound healing have been identified for αCT1. Topical ophthalmic delivery of αCT1 resulted in decreased inflammation and accelerated healing in both a standard rat model of corneal injury [49] and a diabetic rat model of corneal injury [50]. The Cx43-based peptide has also been found to modulate the biological response to silicone implants, attenuating neutrophil infiltration and increasing vascularity of the specialized internal scar tissue that forms around implants, also reducing the density of activated fibroblasts (myofibroblasts) and type I collagen deposition in this tissue [51]. Application of αCT1 directly to an infarcted heart improved cardiac contractility, reduced the propensity for arrhythmia, and maintained action potential conduction velocity at normal speeds [46,47]. As we will discuss in detail in the following section, treatment of chronic diabetic foot ulcers and venous leg ulcers with αCT1 significantly reduced ulcer size, increased the likelihood of complete ulcer closure, and decreased the time to ulcer closure [37,41].

4. Clinical Trials

Given the efficacy of αCT1 in animal studies, multiple Phase I and Phase II clinical trials have since been conducted on the wound healing and scar reduction capabilities of αCT1 [37–39,41,42]—see also Table 1. These clinical trials have studied the safety and efficacy of αCT1 in the treatment of venous leg ulcers, diabetic foot ulcers, and surgical wounds, with the peptide being tested as the active ingredient of a gel formulation branded as Granexin®.

Table 1. Summary of completed alpha connexin carboxyl terminus 1 (αCT1) clinical trials.

Clinical Trial Phase	Phase I	Phase II		
Wound Type	Healthy Human Dermal Wounds	Venous Leg Ulcers	Diabetic Foot Ulcers	Cutaneous Scarring/Laparoscopic Incisions
Treatment Regimen	Immediately after wounding and 24 h later	Twice during the 1st week and once a week thereafter	Twice during the 1st week and once a week thereafter	Immediately after wounding and 24 h later
Patients	49	92	92	91
No Adverse Effects	✓	✓	✓	✓
Mean Percent Ulcer Area Reduction at 12 Weeks	-	79% αCT1 vs. 36% control	94% αCT1 vs. 52% control	-
Incidence of Complete Ulcer Closure at 12 Weeks	-	57% αCT1 vs. 28% control	81% αCT1 vs. 50% control	-
Comparative Vancouver Scar Scale Scores at 9 Months	-	-	-	47% better for αCT1 compared to within-patient controls

The initial clinical trial on the effect of αCT1 on human dermal wound healing was performed on 49 healthy human volunteers in Switzerland in a randomized, double-blind Phase 1 study. As shown in Figure 2, on Day 1 a biopsy punch was used to create a wound in the unblemished skin underneath both arms. One underarm wound from the patient was treated with an αCT1 gel formulation, while the wound on the patient's other underarm was treated with a vehicle gel, enabling within-patient comparisons. Treatments were applied immediately after injury and again 24 h later. The αCT1 dosage in the gel depended on the cohort. Cohort 1 received a gel with 20 μM of αCT1; Cohort 2, 50 μM; Cohort 3, 100 μM, and Cohort 4, 200 μM of αCT1. Cohort 3, which received a gel with 100 μM of αCT1, was given the same dosage that was used in Phase II clinical trials and is considered the therapeutic dosage. Wound healing was followed for 29 days and recorded photographically. On Day 29, following final photograph collection, the healed scars were biopsied to permit examination of the histological features of the scar tissue. The biopsies were washed, placed in paraformaldehyde for 24 h, and embedded in paraffin for sectioning. Data collected from these biopsy sections revealed improved healing outcomes, in terms of collagen order, density and maturity, with αCT1 treatment and has formed the basis of ongoing studies into the αCT1 mode of action. Importantly, from the perspective of the safety focus of Phase I clinical testing, αCT1 usage showed no local or systemic adverse effects associated with treatment.

Continuing onto Phase II, two of the Phase II clinical trials involved studies of chronic skin wounds characterized by chronic inflammation and retarded re-epithelialization: Venous leg ulcers and diabetic foot wounds [37,41]. For treatment of chronic wounds, αCT1 was topically applied to the wound area twice during the first week, and then on a weekly basis thereafter. In the venous leg ulcer trial, $n = 92$ patients were randomized for study [37]. αCT1 treatment was associated with a significantly greater reduction in the mean percent ulcer area by 12 weeks (79% wound closure in the treatment group compared to 36% in the control), and a doubling in incidence of complete wound closure by 12 weeks (57% of the treatment group had completely healed wounds by the study end point of 12 weeks compared to only 28% of the control group). Venous leg wounds treated with αCT1 also showed a shorter time to 50% ($p = 0.014$), and 100% ($p = 0.041$) wound closure than the control group. The median time to 50% and 100% wound closure in the treatment group was 2.9 and 6.0 weeks, respectively, while the control group took an average of 6.9 and 12.1 weeks to reach the same milestones.

Similar results were found for diabetic foot ulcers in a separate clinical trial, in which $n = 92$ patients were randomized for study [41]. The mean percent ulcer area at 12 weeks was 94% for the treatment group versus 52% for the control, and the incidence of 100% wound closure at the study end point of 12 weeks was 81% for treatment versus 50% for control. Median time to complete wound closure was 6.0 weeks for the treatment group while the control group's estimated median time to closure was 14.6 weeks. Time to 50% ulcer closure for this study was not significantly different. To summarize the results of these two clinical trials, chronic wounds treated with αCT1 heal more quickly and are more likely to completely resolve within 12 weeks than the control group of chronic wounds treated with the current standard of care only.

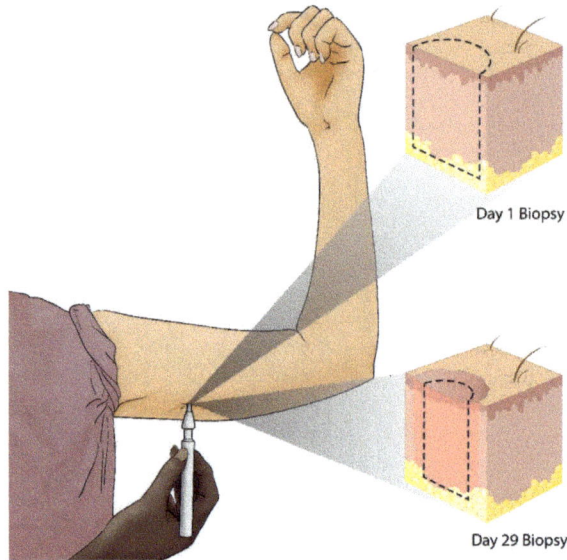

Day 1 Biopsy

Day 29 Biopsy

Figure 2. Alpha connexin carboxyl terminus 1 (αCT1) Phase I clinical trial sampling scheme, performed on healthy human volunteers.

In a third Phase II clinical trial, the potential of αCT1 in reducing post-surgical scarring was assessed [39]. Unlike the weekly αCT1 application protocol applied in the treatment of chronic wounds, αCT1 was applied to surgical wounds immediately after injury and then again 24 h later. This treatment regime was followed by scarring assessment over a nine-month study period. This acute treatment regime was similar to that used in the earlier animal studies involving therapeutic evaluation of the Cx43 antisense as well as αCT1 [7,20]. The clinical trial involved 91 patients who had received laparoscopic abdominal surgery involving two or more incisions, allowing within-patient controls with the surgical wound on one side of the abdomen treated with αCT1 and the control wound on the opposite side. Treatment versus control was randomized to patient sides, and both wounds were treated with identical conventional standard of care protocols. [39]. Scar appearance, as judged by the Vancouver Scar Scale clinical standard, was equivalent between treatment groups after one month, but after nine months αCT1 treated scars showed a highly significant, 47%, improvement ($p < 0.005$) in scar appearance (Figure 3). Since αCT1 was only applied in the first two days, the results raise interesting questions about the mechanism by which brief, transitory targeting of Cx43 and/or its activities after injury is able to induce a long-term modification in scarring outcome.

Preclinical results found improvements in the tensile strength of αCT1 treated wounds—a property directly linked to extracellular matrix composition and structure [7,47]. In line with this, we have identified structural changes in the extracellular matrix of αCT1 treated wounds from biopsies collected in Phase I clinical trial that suggest the peptide is prompting the deposition of an initial collagen matrix more similar to unwounded skin (Figure 4). Work exploring this hypothesis is currently ongoing.

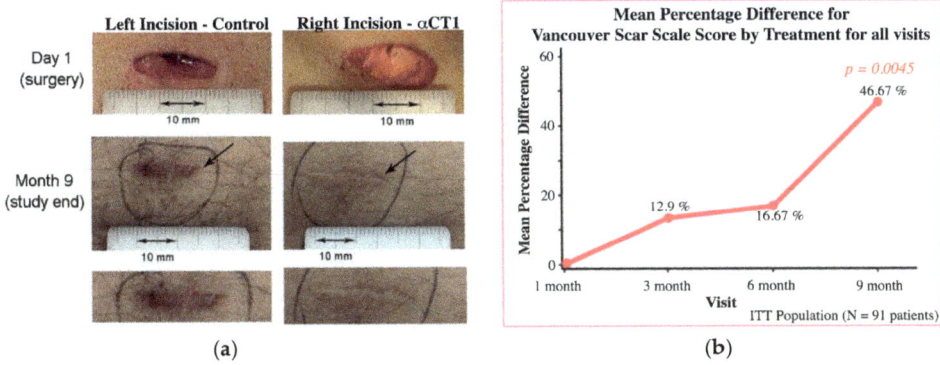

(a)

(b)

Figure 3. (**a**) Photographs of a single patient's αCT1 and control treated wounds immediately after surgery and at the study end point of nine months; (**b**) Mean percentage difference between treatment and control scar scores for the Phase II scar appearance clinical trials [35].

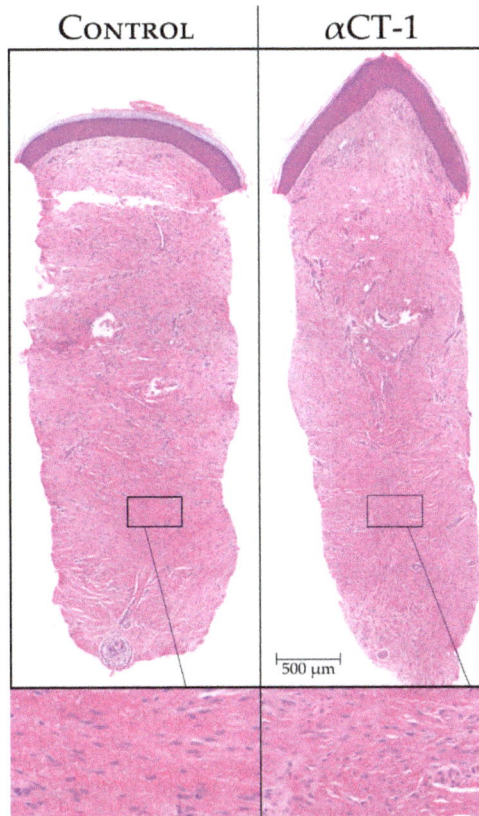

Figure 4. H&E stained whole sections of Phase I biopsies from a single patient at 29 days post-wounding. The left arm was treated with a vehicle (**left**), while the right arm was treated with 100 μM αCT1 (**right**). The boxed regions highlight areas of subtle variance in tissue organization deep within the dermis between the control and treated scars—magnified 4.67× from upper panels.

5. Conclusions

Preclinical and clinical studies of αCT1 have indicated that this peptide, based on the CT-most sequence of Cx43, beneficially modulates the healing of both undiseased and diseased, chronic skin wounds without detrimental side-effects. However, important work remains to be undertaken to characterize the details of the molecular and cellular mechanisms by which therapies targeting Cx43 function, such as αCT1, improve wound healing and mitigate scar formation. Identifying key parts of the cascade may allow the development of improved, targeted therapeutics, reducing the potential for off-target effects. Connexin-based therapeutics are also being explored in preclinical studies of injury to other organ systems and tissues, including the heart, eye, brain, and lungs [52–59]. Moving from topical delivery of drugs like αCT1 to internal administration in tissues such as the heart or brain will pose significant challenges. The regulatory bar for safety will be necessarily higher. There will also be questions on optimal route and mode of delivery, treatment regime, the stability of the drug in body fluids, the negotiation of the immune system, and obstacles such as the blood-brain barrier that will need to be addressed. This being said, ongoing research provides encouragement that connexin-based therapeutics could be a promising path towards future medical interventions in the healing of the human body.

Author Contributions: Writing-Original Draft Preparation, J.M., R.G.G, Writing-Review & Editing J.M., C.L.G., G.S.G., K.E.M., and R.G.G.

Funding: This research was funded by NIH grant number HL56728.

Conflicts of Interest: FirstString Research holds the exclusive patent for the composition and methods for the use of the patents for αCT1. G.S.G. and R.G.G. are inventors on this patent. G.S.G. and C.L.G. are employees of FirstString Research Inc. R.G.G. is a non-paid member of the scientific advisory board of FirstString Research Inc. G.S.G., C.L.G., and R.G.G. hold stock in FirstString Research Inc. J.M. and K.E.M. have no disclosures and have no competing interests.

Abbreviations

Cx43	Connexin 43
αCT1	Alpha connexin carboxyl terminus 1
GJ	Gap Junction
ZO-1	Zonula occludens-1
CT	Carboxyl terminus

References

1. Sohl, G.; Willecke, K. Gap junctions and the connexin protein family. *Cardiovasc. Res.* **2004**, *62*, 228–232. [CrossRef] [PubMed]
2. Schalper, K.A.; Orellana, J.A.; Berthoud, V.M.; Sáez, J.C. Dysfunctions of the diffusional membrane pathways mediated by hemichannels in inherited and acquired human diseases. *Curr. Vasc. Pharmacol.* **2009**, *7*, 486–505. [CrossRef] [PubMed]
3. Bruzzone, S.; Guida, L.; Zocchi, E.; Franco, L.; De Flora, A. Connexin 43 hemi channels mediate Ca^{2+}-regulated transmembrane NAD^+ fluxes in intact cells. *FASEB J.* **2001**, *15*, 10–12. [CrossRef] [PubMed]
4. Schalper, K.A.; Sánchez, H.A.; Lee, S.C.; Altenberg, G.A.; Nathanson, M.H.; Sáez, J.C. Connexin 43 hemichannels mediate the Ca^{2+} influx induced by extracellular alkalinization. *Am. J. Physiol.-Cell Physiol.* **2010**, *299*, C1504–C1515. [CrossRef] [PubMed]
5. De Bock, M.; Wang, N.; Bol, M.; Decrock, E.; Ponsaerts, R.; Bultynck, G.; Dupont, G.; Leybaert, L. Connexin 43 hemichannels contribute to cytoplasmic Ca^{2+} oscillations by providing a bimodal Ca^{2+}-dependent Ca^{2+} entry pathway. *J. Biol. Chem.* **2012**, *287*, 12250–12266. [CrossRef] [PubMed]
6. Rhett, J.M.; Ghatnekar, G.S.; Palatinus, J.A.; O'Quinn, M.; Yost, M.J.; Gourdie, R.G. Novel therapies for scar reduction and regenerative healing of skin wounds. *Trends Biotechnol.* **2008**, *26*, 173–180. [CrossRef] [PubMed]

7. Ghatnekar, G.S.; Quinn, M.P.O.; Jourdan, L.J.; Gurjarpadhye, A.A.; Draughn, R.L.; Gourdie, R.G. Connexin43 carboxyl-terminal peptides reduce scar progenitor and promote regenerative healing following skin wounding. *Future Med.* **2009**, *4*, 205–223. [CrossRef] [PubMed]
8. Martin, P.E.M.; Easton, J.A.; Hodgins, M.B.; Wright, C.S. Connexins: Sensors of epidermal integrity that are therapeutic targets. *FEBS Lett.* **2014**, *588*, 1304–1314. [CrossRef] [PubMed]
9. Churko, J.M.; Laird, D.W. Gap junction remodeling in skin repair following wounding and disease. *Physiology* **2013**, *28*, 190–198. [CrossRef] [PubMed]
10. Gourdie, R.G.; Ghatnekar, G.S.; O'Quinn, M.; Rhett, M.J.; Barker, R.J.; Zhu, C.; Jourdan, J.; Hunter, A.W. The unstoppable connexin43 carboxyl-terminus: New roles in gap junction organization and wound healing. *Ann. N. Y. Acad. Sci.* **2006**, *1080*, 49–62. [CrossRef] [PubMed]
11. Nguyen, D.T.; Orgill, D.P.; Murphy, G.F. Chapter 4: The Pathophysiologic Basis for Wound Healing and Cutaneous Regeneration. In *Biomaterials For Treating Skin Loss*; Woodhead Publishing: Sawston, UK; Cambridge, UK; CRC Press: Boca Raton, FL, USA, 2009; pp. 25–57. ISBN 978-1-84569-363-3.
12. Martin, P.; Lewis, J. Actin cables and epidermal movement in embryonic wound healing. *Nature* **1992**, *360*, 179–183. [CrossRef] [PubMed]
13. Martin, P.; D'Souza, D.; Martin, J.; Grose, R.; Cooper, L.; Maki, R.; McKercher, S.R. Wound healing in the PU. 1 null mouse—tissue repair is not dependent on inflammatory cells. *Curr. Biol.* **2003**, *13*, 1122–1128. [CrossRef]
14. Martin, P. Mechanisms of wound healing in the embryo and fetus. *Curr. Top. Dev. Biol.* **1996**, *32*, 175–203. [PubMed]
15. Ferguson, M.W.; O'Kane, S. Scar-free healing: From embryonic mechanisms to adult therapeutic intervention. *Philos. Trans. R. Soc. Lond. B. Biol. Sci.* **2004**, *359*, 839–850. [CrossRef] [PubMed]
16. Davis, F.M.; Kimball, A.; Boniakowski, A.; Gallagher, K. Dysfunctional Wound Healing in Diabetic Foot Ulcers: New Crossroads. *Curr. Diabetes Rep.* **2018**, *18*, 2. [CrossRef] [PubMed]
17. Wright, C.S.; Berends, R.F.; Flint, D.J.; Martin, P.E.M. Cell motility in models of wounded human skin is improved by Gap27 despite raised glucose, insulin and IGFBP-5. *Exp. Cell Res.* **2013**, *319*, 390–401. [CrossRef] [PubMed]
18. Churko, J.M.; Kelly, J.J.; Macdonald, A.; Lee, J.; Sampson, J.; Bai, D.; Laird, D.W. The G60S Cx43 mutant enhances keratinocyte proliferation and differentiation. *Exp. Dermatol.* **2012**, *21*, 612–618. [CrossRef] [PubMed]
19. Márquez-Rosado, L.; Singh, D.; Rincón-Arano, H.; Solan, J.L.; Lampe, P.D. CASK (LIN2) interacts with Cx43 in wounded skin and their coexpression affects cell migration. *J. Cell Sci.* **2012**, *125*, 695–702. [CrossRef] [PubMed]
20. Qiu, C.; Coutinho, P.; Frank, S.; Franke, S.; Law, L.; Martin, P.; Green, C.R.; Becker, D.L. Targeting connexin43 expression accelerates the rate of wound repair. *Curr. Biol.* **2003**, *13*, 1697–1703. [CrossRef] [PubMed]
21. Becker, D.L.; Thrasivoulou, C.; Phillips, A.R.J. Connexins in wound healing; perspectives in diabetic patients. *Biochim. Biophys. Acta Biomembr.* **2012**, *1818*, 2068–2075. [CrossRef] [PubMed]
22. Lorraine, C.; Wright, C.S.; Martin, P.E.M. Connexin43 plays diverse roles in co-ordinating cell migration and wound closure events. *Biochem. Soc. Trans.* **2015**, *43*, 482–488. [CrossRef] [PubMed]
23. Cogliati, B.; Vinken, M.; Silva, T.C.; Araújo, C.M.M.; Aloia, T.P.A.; Chaible, L.M.; Mori, C.M.C.; Dagli, M.L.Z. Connexin 43 deficiency accelerates skin wound healing and extracellular matrix remodeling in mice. *J. Dermatol. Sci.* **2015**, *79*, 50–56. [CrossRef] [PubMed]
24. Tarzemany, R.; Jiang, G.; Larjava, H.; Häkkinen, L. Expression and function of connexin 43 in human gingival wound healing and fibroblasts. *PLoS ONE* **2015**, *10*, e0115524. [CrossRef] [PubMed]
25. Scott, C.A.; Tattersall, D.; O'Toole, E.A.; Kelsell, D.P. Connexins in epidermal homeostasis and skin disease. *Biochim. Biophys. Acta* **2012**, *1818*, 1952–1961. [CrossRef] [PubMed]
26. Goliger, J.A.; Paul, D.L. Wounding alters epidermal connexin expression and gap junction-mediated intercellular communication. *Mol. Biol. Cell* **1995**, *6*, 1491–1501. [CrossRef] [PubMed]
27. Coutinho, P.; Qiu, C.; Frank, S.; Tamber, K.; Becker, D. Dynamic changes in connexin expression correlate with key events in the wound healing process. *Cell Biol. Int.* **2003**, *27*, 525–541. [CrossRef]
28. Moyer, K.E.; Davis, A.; Saggers, G.C.; Mackay, D.R.; Ehrlich, H.P. Wound healing: The role of gap junctional communication in rat granulation tissue maturation. *Exp. Mol. Pathol.* **2002**, *72*, 10–16. [CrossRef] [PubMed]

29. Davis, N.G.; Phillips, A.; Becker, D.L. Connexin dynamics in the privileged wound healing of the buccal mucosa. *Wound Repair Regen.* **2013**, *21*, 571–578. [CrossRef] [PubMed]

30. Tarzemany, R.; Jiang, G.; Jiang, J.X.; Larjava, H.; Häkkinen, L. Connexin 43 Hemichannels Regulate the Expression of Wound Healing-Associated Genes in Human Gingival Fibroblasts. *Sci. Rep.* **2017**, *7*, 14157. [CrossRef] [PubMed]

31. Tarzemany, R.; Jiang, G.; Jiang, J.X.; Gallant-Behm, C.; Wiebe, C.; Hart, D.A.; Larjava, H.; Häkkinen, L. Connexin 43 regulates the expression of wound healing-related genes in human gingival and skin fibroblasts. *Exp. Cell Res.* **2018**, *367*, 150–161. [CrossRef] [PubMed]

32. Mori, R.; Power, K.T.; Wang, C.M.; Martin, P.; Becker, D.L. Acute downregulation of connexin43 at wound sites leads to a reduced inflammatory response, enhanced keratinocyte proliferation and wound fibroblast migration. *J. Cell Sci.* **2006**, *119*, 5193–5203. [CrossRef] [PubMed]

33. Faniku, C.; O'Shaughnessy, E.; Lorraine, C.; Johnstone, S.R.; Graham, A.; Greenhough, S.; Martin, P.E.M. The Connexin Mimetic Peptide Gap27 and Cx43-Knockdown Reveal Differential Roles for Connexin43 in Wound Closure Events in Skin Model Systems. *Int. J. Mol. Sci.* **2018**, *19*, 604. [CrossRef] [PubMed]

34. Brandner, J.M.; Houdek, P.; Hüsing, B.; Kaiser, C.; Moll, I. Connexins 26, 30, and 43: Differences among spontaneous, chronic, and accelerated human wound healing. *J. Investig. Dermatol.* **2004**, *122*, 1310–1320. [CrossRef] [PubMed]

35. Mendoza-Naranjo, A.; Cormie, P.; Serrano, A.E.; Hu, R.; O'Neill, S.; Wang, C.M.; Thrasivoulou, C.; Power, K.T.; White, A.; Serena, T.; et al. Targeting Cx43 and N.-Cadherin, Which Are Abnormally Upregulated in Venous Leg Ulcers, Influences Migration, Adhesion and Activation of Rho GTPases. *PLoS ONE* **2012**, *7*, e37374. [CrossRef] [PubMed]

36. Kanapathy, M.; Simpson, R.; Madden, L.; Thrasivoulou, C.; Mosahebi, A.; Becker, D.L.; Richards, T. Upregulation of epidermal gap junctional proteins in patients with venous disease. *Br. J. Surg.* **2018**, *105*, 59–67. [CrossRef] [PubMed]

37. Ghatnekar, G.S.; Grek, C.L.; Armstrong, D.G.; Desai, S.C.; Gourdie, R.G. The effect of a connexin43-based Peptide on the healing of chronic venous leg ulcers: A multicenter, randomized trial. *J. Investig. Dermatol.* **2015**, *135*, 289–298. [CrossRef] [PubMed]

38. Kirsner, R.S.; Baquerizo Nole, K.L.; Fox, J.D.; Liu, S.N. Healing refractory venous ulcers: New treatments offer hope. *J. Investig. Dermatol.* **2015**, *135*, 19–23. [CrossRef] [PubMed]

39. Grek, C.L.; Montgomery, J.; Sharma, M.; Ravi, A.; Rajkumar, J.S.; Moyer, K.E.; Gourdie, R.G.; Ghatnekar, G.S. A Multicenter Randomized Controlled Trial Evaluating a Cx43-Mimetic Peptide in Cutaneous Scarring. *J. Investig. Dermatol.* **2017**, *137*, 620–630. [CrossRef] [PubMed]

40. Pollok, S.; Pfeiffer, A.-C.; Lobmann, R.; Wright, C.S.; Moll, I.; Martin, P.E.M.; Brandner, J.M. Connexin 43 mimetic peptide Gap27 reveals potential differences in the role of Cx43 in wound repair between diabetic and non-diabetic cells. *J. Cell. Mol. Med.* **2011**, *15*, 861–873. [CrossRef] [PubMed]

41. Grek, C.L.; Prasad, G.M.; Viswanathan, V.; Armstrong, D.G.; Gourdie, R.G.; Ghatnekar, G.S. Topical administration of a connexin43-based peptide augments healing of chronic neuropathic diabetic foot ulcers: A multicenter, randomized trial. *Wound Repair Regen.* **2015**, *23*, 203–212. [CrossRef] [PubMed]

42. Grek, C.L.; Rhett, J.M.; Ghatnekar, G.S. Cardiac to cancer: Connecting connexins to clinical opportunity. *FEBS Lett.* **2014**, *588*, 1349–1364. [CrossRef] [PubMed]

43. Hunter, A.W.; Barker, R.J.; Zhu, C.; Gourdie, R.G. Zonula Occludens-1 Alters Connexin43 Gap Junction Size and Organization by Influencing Channel Accretion. *Mol. Biol. Cell* **2005**, *16*, 5686–5698. [CrossRef] [PubMed]

44. Rhett, J.M.; Jourdan, J.; Gourdie, R.G. Connexin 43 connexon to gap junction transition is regulated by zonula occludens-1. *Mol. Biol. Cell* **2011**, *22*, 1516–1528. [CrossRef] [PubMed]

45. Jiang, J.; Palatinus, J.A.; He, H.; Iyyathuraia, J.; Jordan, J.; McGowan, F.X.; Schey, K.; Bultynck, G.; Zhang, Z.; Gourdie, R.G. Phosphorylation of Connexin43 at Serine368 is Necessary for Induction of Cardioprotection by a Connexin43 Carboxyl-Terminal Mimetic Peptide. *Circulation* **2016**, *134*, A16380.

46. O'Quinn, M.P.; Palatinus, J.A.; Harris, B.S.; Hewett, K.W.; Gourdie, R.G. A peptide mimetic of the connexin43 carboxyl terminus reduces gap junction remodeling and induced arrhythmia following ventricular injury. *Circ. Res.* **2011**, *108*, 704–715. [CrossRef] [PubMed]

47. Ongstad, E.L.; O'Quinn, M.P.; Ghatnekar, G.S.; Yost, M.J.; Gourdie, R.G. A Connexin43 Mimetic Peptide Promotes Regenerative Healing and Improves Mechanical Properties in Skin and Heart. *Adv. Wound Care* **2013**, *2*, 55–62. [CrossRef] [PubMed]

48. Schanz, J.; Pusch, J.; Hansmann, J.; Walles, H. Vascularised human tissue models: A new approach for the refinement of biomedical research. *J. Biotechnol.* **2010**, *148*, 56–63. [CrossRef] [PubMed]

49. Moore, K.; Bryant, Z.J.; Ghatnekar, G.; Singh, U.P.; Gourdie, R.G.; Potts, J.D. A synthetic connexin 43 mimetic peptide augments corneal wound healing. *Exp. Eye Res.* **2013**, *115*, 178–188. [CrossRef] [PubMed]

50. Moore, K.; Ghatnekar, G.; Gourdie, R.G.; Potts, J.D. Impact of the Controlled Release of a Connexin 43 Peptide on Corneal Wound Closure in an STZ Model of Type I Diabetes. *PLoS ONE* **2014**, *9*, e86570. [CrossRef] [PubMed]

51. Soder, B.L.; Propst, J.T.; Brooks, T.M.; Goodwin, R.L.; Friedman, H.I.; Yost, M.J.; Gourdie, R.G. The connexin43 carboxyl-terminal peptide ACT1 modulates the biological response to silicone implants. *Plast. Reconstr. Surg.* **2009**, *123*, 1440–1451. [CrossRef] [PubMed]

52. Yi, C.; Koulakoff, A.; Giaume, C. Astroglial Connexins as a Therapeutic Target for Alzheimer's Disease. *Curr. Pharm. Des.* **2017**, *23*, 4958–4968. [CrossRef] [PubMed]

53. Willebrords, J.; Maes, M.; Crespo Yanguas, S.; Vinken, M. Inhibitors of connexin and pannexin channels as potential therapeutics. *Pharmacol. Ther.* **2017**, *180*, 144–160. [CrossRef] [PubMed]

54. Leybaert, L.; Lampe, P.D.; Dhein, S.; Kwak, B.R.; Ferdinandy, P.; Beyer, E.C.; Laird, D.W.; Naus, C.C.; Green, C.R.; Schulz, R. Connexins in Cardiovascular and Neurovascular Health and Disease: Pharmacological Implications. *Pharmacol. Rev.* **2017**, *69*, 396–478. [CrossRef] [PubMed]

55. Becker, D.L.; Phillips, A.R.; Duft, B.J.; Kim, Y.; Green, C.R. Translating connexin biology into therapeutics. *Semin. Cell Dev. Biol.* **2016**, *50*, 49–58. [CrossRef] [PubMed]

56. Iyyathurai, J.; D'hondt, C.; Wang, N.; De Bock, M.; Himpens, B.; Retamal, M.A.; Stehberg, J.; Leybaert, L.; Bultynck, G. Peptides and peptide-derived molecules targeting the intracellular domains of Cx43: Gap junctions versus hemichannels. *Neuropharmacology* **2013**, *75*, 491–505. [CrossRef] [PubMed]

57. Obert, E.; Strauss, R.; Brandon, C.; Grek, C.L.; Ghatnekar, G.S.; Gourdie, R.G.; Rohrer, B. Targeting the tight junction protein, zonula occludens-1, with the connexin43 mimetic peptide, αCT1, reduces VEGF-dependent RPE pathophysiology. *J. Mol. Med.* **2017**, *95*, 535–552. [CrossRef] [PubMed]

58. Naus, C.C.; Giaume, C. Bridging the gap to therapeutic strategies based on connexin/pannexin biology. *J. Transl. Med.* **2016**, *14*, 330. [CrossRef] [PubMed]

59. Hsieh, Y.C.; Lin, J.C.; Hung, C.Y.; Li, C.H.; Lin, S.F.; Yeh, H.I.; Huang, J.L.; Lo, C.P.; Haugan, K.; Larsen, B.D.; et al. Gap junction modifier rotigaptide decreases the susceptibility to ventricular arrhythmia by enhancing conduction velocity and suppressing discordant alternans during therapeutic hypothermia in isolated rabbit hearts. *Heart Rhythm* **2016**, *13*, 251–261. [CrossRef] [PubMed]

MDPI

St. Alban-Anlage 66

4052 Basel

Switzerland

Tel. +41 61 683 77 34

Fax +41 61 302 89 18

www.mdpi.com

International Journal of Molecular Sciences Editorial Office

E-mail: ijms@mdpi.com

www.mdpi.com/journal/ijms

www.ingramcontent.com/pod-product-compliance
Lightning Source LLC
Chambersburg PA
CBHW051709210326
41597CB00032B/5416